Samuel Almond Miller

The American Palaeozoic Fossils

A Catalogue of the Genera and Species

Samuel Almond Miller

The American Palaeozoic Fossils
A Catalogue of the Genera and Species

ISBN/EAN: 9783337279561

Printed in Europe, USA, Canada, Australia, Japan

Cover: Foto ©berggeist007 / pixelio.de

More available books at **www.hansebooks.com**

THE

AMERICAN

PALAEOZOIC FOSSILS:

A CATALOGUE

OF THE

GENERA AND SPECIES,

WITH

NAMES OF AUTHORS, DATES, PLACES OF PUBLICATION, GROUPS OF ROCKS IN WHICH
FOUND, AND THE ETYMOLOGY AND SIGNIFICATION OF THE WORDS,

AND

AN INTRODUCTION

DEVOTED TO THE STRATIGRAPHICAL GEOLOGY OF THE PALEOZOIC ROCKS.

BY S. A. MILLER.

CINCINNATI, OHIO:
PUBLISHED BY THE AUTHOR, No. 8 W. THIRD ST.
1877.

PREFACE.

THE SCIENCE OF GEOLOGY rests upon the fossil contents of the rocks, and, therefore, there can be but little knowledge of the science without a knowledge of the fossils. Every one professes to take an interest in Geology, although the complaint against the technical names is quite general. This work may be regarded as a compilation of these technical names in alphabetical order, with the signification of each, and in this respect, it is a dictionary adapted to the use of every one, and designed to remove the objections against the hard words used in the science. The references to authors and publications are for the special benefit of the describers of fossils and the students of palæontology, while the Groups of rocks are given for the convenience of collectors and students generally. The general plan of the work will be, at once, apparent to the reader; a few remarks, however, in regard thereto, may not be inappropriate.

The state of the science, at this time, does not admit of a very great degree of certainty in the arrangement of Palæozoic Fossils into families; such arrangement must, therefore, be regarded as more or less provisional. And, in this work, where a class could not be arranged into families already limited and defined, with approximate correctness, the attempt has been omitted. No new families have been proposed, with the exception of three names, followed by the provisional interrogation point in the Class Echinodermata. The new family names in the Class Lamellibranchiata are used by Prof. James Hall in the fifth volume of the Palæontology of New York.

Each generic name is followed by the name of the author, the date of the first publication, the title of the book in which it first appeared, which is generally abbreviated, and the etymology of the word, which is included within brackets. Valid names are printed in Roman letters. Generic names, preoccupied, never defined, and where the fossils are unknown in American palæozoic rocks, or the names invalid for any other reason, are italicized. The generic name always begins with a capital letter, whether it is a valid or invalid name.

Each specific name, or as it was formerly, called the *trivial* name, is followed with the name of the author, the date of the first publication, the title of the book in which it appeared, which is generally abbreviated, the Group of rocks in which the fossil is found when the Group is known; but otherwise the formation alone is given, and the signification of the name included within brackets. The name of the Group of rocks is generally abbreviated. Where the author, in the first instance, referred his species to the wrong genus, such generic and specific name is included within parenthesis immediately following the date of the publication. For instance, *Delthyris* is a synonym for *Spirifera*, and many species have been erroneously referred to it; such are written in this manner:

Spirifera arenosa, Conrad, 1838, (Delthyris arenosa), Ann. Rep. N. Y., Oriskany sandstone. [Sig. sandy.]

Specific names never begin with capital letters.

The author has endeavored to quote from the original publications, and for kindness and assistance in obtaining access to them, he expresses his obligations to Prof. James Hall, of Albany, New York; Mr. C. D. Walcott, of Trenton Falls, New York; Prof. Leo. Lesquereux, of Columbus, Ohio, and to Mr. Thomas Vickers, the able and efficient librarian of the public library in the city of Cincinnati. A few of the works cited, however, were not within reach, and the references to these are therefore second hand. In the attempt to make the catalogue of names complete, within the scope of the work, the author visited several libraries and otherwise used reasonable efforts for that purpose, but he has no doubt that a few names have been omitted. The number of names in the book is as follows:

Generic names in Roman letters, - - - - 1,000
Generic names in Italic letters, - - - - - 200
Specific names in Roman letters, - - - - 8,000
Specific names in Italic letters, - - - - - 2,000

Total number of genera and species, including synonyms, etc., - - - - - - 11,200

Some of the authors of generic and specific names have not been familiar with the Latin and Greek languages, others have been careless in the exercise of their knowledge when forming the new words, and many names have been misprinted for want of careful proof-reading. For these and probably other reasons, the specific names under a single genus have been found with masculine, feminine, and neuter terminations; no uniformity has existed in the terminations added to proper names, and words have been otherwise incorrectly formed. So universally have these errors prevailed, that the author found in some Classes, twenty-five per cent. of the names defective. In a conversation with Prof. E. W. Claypole, of Antioch College, Yellow Springs, Ohio, (a gentleman thoroughly learned in the Latin and Greek languages) shortly before commencing the publication, with regard to the etymology of words, and the importance of following the laws of language in making the genders of the adjective specific names correspond with those of the generic terms, he very kindly proffered his assistance, for the purpose of making all proper corrections in this regard. Publication was commenced in February, and the proofs were sent to Prof. Claypole as rapidly as the matter was put in type for his inspection and correction, but the matter was published so rapidly that it frequently went to press before the return by mail, of his proof. The greater part of the work, therefore, did not have the benefit of his revision.

As an illustration, the genus *Macrocheilus* has been considered masculine by all the authors, and consequently the specific names have been made masculine. As Prof. Claypole did not read the proof, the specific names are published here in the masculine, as the authors made them; but on his authority the species under the genera *Temnocheilus*, *Solenocheilus*, etc., have been changed from the masculine to the neuter gender. It may be a question, however, whether usage has not made such genera masculine. No such doubt can arise in the case of the words ending in *nema* as, *Cyclonema*, *Loxonema*, *Dictyonema*, etc., which have been treated almost or quite uniformly by the authors as feminine; for *nema* is neuter, in both the Latin and the Greek languages, and there is no reason whatever, for using it as feminine. Where specific names have been formed in more genders than one under a single genus, there seems to be no responsibility in correcting the errors and making the specific names correspond in gender with the genus, and this has been done in several hundred instances in this work. Perfection, however, has not been reached in this publication, and for the purpose of rendering greater assistance in that regard, the "Index of Genera" has been made to indicate the gender of each genus, so that any one with a little knowledge of Latin can make the species conform in gender with the genus. The reader will know that generic names are usually coined from Greek words, and specific names from Latin words or proper names with Latin terminations.

In Latinizing specific proper names, no uniformity has ever existed, nor can the author claim to have accomplished it in this work. Some authors have used the terminations *ianus*, *iana* and *ianum*, while others have used *anus*, *ana* and *anum*. The first impression was that the former endings were proper, and consequently they were used in part of this work; but the best authors use only the latter, and upon reflection we were of the opinion that the latter endings are the proper ones, and they were thereafter used in this work exclusively. The reader is directed, therefore, to correct the proper names ending in *ianus*, *iana* and *ianum*, which occur, in some instances, from the Polypi to the Brachiopoda, by striking out the letter *i* and thus perfect this class of words.

Again, no uniformity exists in specific proper names put in the genitive case by adding the Latin ending *i*, where the proper names end in *e*, *y*, etc. There seems to be no difficulty where the proper name ends in a consonant, in accomplishing the desired purpose by adding a single *i*, and generally where words of more than one syllable end in *e*, the same purpose may be accomplished by changing the *e* to *i*; but there are words where positive difficulty exists, for instance ; Case, Casey, White, Whity, Moor, Moore, Hoy, etc. Some would change the *e* into *i* in Moore and add *i* to Moor, so that in each case we would have the same specific name *moori*, and so with White and Whity; but there are very strong reasons against such action. In this work, the words of one syllable ending in *e* have sometimes been written with an *i* added, and at other times made to end in *ii*, according to the rules of the British Association, as casei and casii. But the author is of the opinion that the single letter *i* in these words is to be preferred; thus casei, caseyi, whitei, whityi, moori, moorei, hoyi, etc., for the following reason. The use of such words is justified because the men have been in some way useful to the science, and the preservation of their names is part of the history of its growth, and can be defended with stronger reason than the use of mythological names, even within the limits warranted by the rules of the British Association, where a fanciful resemblance is made a test. A man may devote ten, twenty, or thirty years of his life to the collection of fossils, or he may devote as many years to the study and description of them. In either case his services may constitute the entire history of the fossils of a given locality, and in what way can the science be better subserved than by perpetuating his name among the fossils he discovered or defined? If the reason for the use of his name is a good one, it is equally strong to use it so that it will not be misunderstood, and can not be made to represent an individual of a different name. The fact that the name does not readily assume a Latin form is of slight weight, in the opinion of the author, compared with the preservation of the history of the science in perpetuating the names of those who have devoted themselves to discover and systematize it.

At the request of the author, Prof. Claypole has written an essay upon the rules of nomenclature, for this work, to which the reader is referred for further light in the construction of words.

CONSTRUCTION OF SYSTEMATIC NAMES IN PALÆONTOLOGY.*

By PROF. E. W. CLAYPOLE, B. A., B. Sc., (London), Antioch College, Yellow Springs, Ohio.

The Latin language is universally adopted in the scientific world for naming species, fossil and recent, in both the Animal and Vegetable Kingdoms. Linnaeus introduced the plan of limiting the name of a species to two words, the former of which was generic and the latter specific, and of making these terms in their formation comply with the rules of the Latin tongue. Since his time, this plan has become the rule. At first, terms were chosen expressing the character or characters which define the genus or species, and such names constitute the perfection of scientific nomenclature. But even Linnaeus soon found it impossible to maintain this standard, and himself resorted to Homer's Iliad to find names with which to christen the butterflies that thronged in upon him. Hence we have Troilus, Danaus, Philenor, &c. Such names are mere arbitrary, meaningless counters attached to the objects and only tenable by an act of memory. They recall no character and are recalled by none. Hence they are inferior to the former class of names. But if in Linnaeus' day their necessity was evident, much more so is it to-day, when almost every language and lexicon have been ransacked to find unpreoccupied names for the hosts of natural objects for which they are required. In some large genera, such for instance as Orthoceras, Orthis and Rhynchonella, it has become exceedingly difficult to coin descriptive names which have not been already thrown into circulation by some other author. This will be evident to any reader in the number of specific names in the catalogue marked "preoccupied," and one object is to provide a remedy for the difficulty by enabling any one to see at a glance whether or not any proposed name has been previously published by any American, and to some extent, by any European author.

The conventional Latin of the Sciences differs somewhat from the classical Latin of the schools. As a dead language, Latin is free from the changes to which all living languages are subject, and as the most widely studied of the dead languages, it forms an excellent means of communication between men of science in different countries. Words and combinations of letters have, however, been introduced alien to the genius of the language, and some of them most barbarous and unmusical.

---

*NOTE.—This paper has been prepared at the request of Mr. S. A. MILLER, the author of the AMERICAN PALÆOZOIC FOSSILS, and with the view of giving some assistance to palæontological students and workers, in avoiding errors and improving the nomenclature of science. E. W. C.

Indeed their introduction cannot well be opposed, as they often offer useful characters. But the laws of Latin should nevertheless control their formation. They should be made, as nearly as possible, what the Roman would have made them had they been introduced from outside into the language while yet living. In this way no violence is done to it, and in all countries words will be formed in a similar manner.

In the year 1862, a committee of the British Association for the Advancement of Science, published a set of rules and recommendations on the subject of Scientific nomenclature, the former of which may be said to be a summary of the practice of the best scientific writers in all lands, and therefore binding, and the latter have been, with a few exceptions, almost universally adopted. Thus, alluding to the point above mentioned, the committee in its " Recommendations " says :

b. Barbarous names.—Some authors protest strongly against the introduction of exotic words into our Latin nomenclature; others defend the practice with equal warmth. We may remark, first, that the practice is not contrary to classical usage, for the Greeks and Romans did occasionally, though with reluctance, introduce barbarous words in a modified form into their respective languages. Secondly, the preservation of the trivial names which animals bear in their native countries is often of great use to the traveler, in aiding him to discover and identify species. We do not, therefore, consider, if such words have a Latin termination given to them, that the occasional and judicious use of them as scientific terms can be justly objected to.

In addition to drawing up a mere list of the North American Palæozoic Fossils, an attempt has been made to give some help to the many workers in the palæontological field towards rightly naming their fossils at the outset. A mistake once in print is very difficult to eradicate. It reappears, it crops out in some unexpected place and with every reappearance has a new lease of life. Few who have not made some study of the subject have any conception of the confusion now existing in the nomenclature of science. The errors are of two kinds—scientific and grammatical. To such a length has synonymy or the giving of different names to the same species now arrived, that it not unfrequently happens that the common English name is more definite and vivid, and therefore more useful than the Scientific. Read, for instance, the twelve different names which have been given to the common Chain Coral, from Fougt, in 1765, to Bronn, in 1835. This, in many cases, arises from mistaken identification, which with all care must sometimes happen, but it arises in a still greater degree from the impossibility of learning what species have been found and what names have been given, when they were given and by whom.

Another source of confusion is the erroneous nomenclature often adopted. All whose duty calls them to the study of palæontology are not, and cannot be, classical scholars, and consequently not a few names are formed in violation of the recognised rules of language. Some errors of this kind have been corrected and a few rules are appended to aid in preventing others.

o. Corrupted words.—In the construction of compound Latin words, there are certain grammatical rules which have been known, and acted on for two thousand years, and which a naturalist is bound to acquaint himself with before he tries his skill in coining zoological terms. One of the chief of these rules is, that in compounding words, all the radical or essential parts of the constituent members must be retained, and no change made except in the variable terminations. But several generic names have been lately introduced which run counter to this rule, and form most unsightly objects to all who are conversant with the spirit of the Latin language. A name made up of the first half of one word, and the last half of another, is as deformed a monster in nomenclature as a mermaid or a centaur would be in zoology; yet we find examples in the names *Corcorax*, (from *Corvus* and *Pyrrhocorax*), *Cypsnagra* (from *Cypselus* and *Tanagra*), *Merulaxis* (*Merula* and *Lynallaxis*), *Loxigilla* (from *Loxia* and *Fringilla*), etc. In other cases, when the *commencement* of both the simple words is retained in the compound, a fault is still committed by cutting off too much of the radical and vital portions, as is the case in *Bucorvus* (from *Buceros* and *Corvus*), *Ninox* (*Nisus* and *Noctua*), etc.

l. Adjective generic names.—The names of genera are, in all cases, essentially substantive, and hence adjective terms can not be employed for them without doing violence to grammar. The generic names of *Hians*, *Criniger*, *Cursorius*, *Nitidula*, etc., are examples of this incorrect usage.—*Recommendations.*

If a writer is unable to coin such a generic term without falling into these or other errors, it would be wise to abandon the attempt and form one in another manner.

RULES FOR THE FORMATION OF NEW NAMES IN PALÆONTOLOGY.

a. Gender of generic terms.—A generic name should always be of that gender which the laws of the language from which it is taken demand. Definite rules cannot well be laid down, because the practice of various languages differs on this point. In general, where no change is made in the termination of the word which forms the end of a new name, the gender of that word will determine that of the name. Thus, the words, *stoma*, a mouth, and *ceras*, a horn, are both, in Greek, of the neuter gender; and, consequently, all the many compounds ending with these words, such as Orthoceras, Platystoma, etc., must be of that gender also, and have corresponding specific adjective terms. If the first founder of a genus were to take the pains to ascertain the right gender of his new name, and make his adjective terms accordingly, many errors would be avoided, for subsequent authors, apparently without a thought, in establishing a new species, coin a specific name of the same gender as that originally published, and thus perpetuate and increase error. For example, the word *desma*, a bond, like almost all Greek words ending in *ma*, is of the neuter gender; and yet, in a work standing so high as Woodward's Mollusca, we find *Lyrodesma plana*, followed shortly afterwards by *Cochlodesma praetenue.*

b. Gender of specific terms.—Every specific name must agree in gender with that of the genus to which it belongs; and yet, owing to the above and several other causes, this rule is incessantly overlooked or violated, even by writers whose classical attainments are beyond doubt. In few, if any, large genera do we fail to find two, if not all, the three genders among the specific names. For instance, in the genus Receptaculites we have *Receptaculites fungosum*, *R. globulare*, *R. formosus*, *R.*

reticulata. Such terms, being mere errors, pure and simple, and involving no disputed or debatable point, may be reduced to accord without difficulty.

c. *Latin and Greek terminations.*—The addition of these terminations to old words often requires a previous change in those words the following rules are offered as some assistance to those whose time does not allow the consultation of authorities, scientific and linguistic. It is not always easy for even a classical scholar to determine off-hand how the Greeks or Romans would have made a word, had it been coined by them while their languages were still living. Much examination and collection of examples is necessary before certainty can be reached, and in a few cases some doubt may then remain. These cases are, however, very few, especially as the terminations employed in palæontology are not numerous. The following rules will probably suffice to enable any one with an elementary knowledge of the classic tongues to form correctly a new specific term, but nothing can give full immunity from risk of error where some such knowledge is not possessed.

—*alis.* This Latin termination, implying resemblance, is seldom used, except in words already compounded in Greek and Latin, and when otherwise, it must be annexed to the stem of the word, as *rectilateralis, quadrilateralis.*

—*anus.* This Latin termination implies resemblance or association, and may be added to proper names, personal or local ; though in science its use is almost confined to the former. If the word be capable of taking a classic form, the termination should be simply annexed to the stem as *Linnaeus, linnaeanus; Lesquereux, (lescuria) lescurianus;* in conformity with classic usage; *pagus, paganus; Claudius, claudianus; Neapolis, neapolitanus.* In other cases, the addition of this termination must follow the same rule as those for *ensis,* as *America, americanus; Geinitz, geinitzanus ; Meek, meekanus ; Erie, erianus ; Italy, (ia) italianus.*

—*atus.* This Latin termination strictly implies the possession of the thing to the name of which it is added. It is, therefore, affixed to the stem of common names only, as *costa, costatus; galea, galeatus; fornix, fornicatus; sinus, sinuatus; stria, striatus; lobus, lobatus; rostrum, rostratus.* It is worthy of remark here that this termination sometimes loses its *at,* to shorten the word. The practice is not commendable from a linguistic standpoint, but some of the terms so made have become fixed in the nomenclature, as *Orthis biloba.*

—*formis.* This Latin termination implies resemblance of shape, and should be confined to Latin words, to the stem of which it should be joined by the connecting vowel *i,* as *laterna, laterniformis; pistillum, pistilliformis.* In forming terms, such as the first given above from Latin words ending in *a,* the error of using *ae* as the connecting vowel should be avoided; being inconsistent with classic usage, as well as more awkward and lengthy, thus we have from *terra, terricola; gemma, gemmifer; squama, squamiger; tuba, tubiformis;* etc.

—ensis. This is a Latin termination, expressive of locality, and cannot, therefore, be correctly employed except as an affix to the name of a place. This rule has been traversed in few real, but in many apparent instances. *Lingula morsensis* is an illustration of the former. In accordance with law this has been changed to *L. morsi*, being given in honor of Mr. Morse. *Zygospira cincinnatiensis, Pupa vermilionensis, Cardium napoleonense, Athyris hannibalensis* are apparent exceptions; but these terms are formed from words, which, though originally personal or trivial, have now become local names, and consequently no valid objection can be raised against them. In using this termination the following rules have been generally followed, and are therefore adopted here. The exceptions are very few, and have been reduced into conformity: 1st. If the name of the place ends in a consonant the termination is annexed to the word, as *Clinton, clintonensis*. 2d. If the name ends in *a* or *e*, these letters are dropped and the termination then annexed, as *Canada, canadensis; Nebraska, nebrascensis; Minnesota, minnesotensis; Iowa, iowensis; Indiana, indianensis; Lasalle, lasallensis; Erie, eriensis*. 3rd. If the name ends in *i*, *o* or *u*, that vowel is retained, as *Mississippi, mississippiensis; Missouri, missouriensis; Chicago, chicagoensis; Colorado, coloradoensis; Chouteau, chouteauensis*. 4th. If the name ends in *y*, that letter becomes *i* upon the addition of the termination, as *Kentucky, kentuckiensis; Alleghany, alleghaniensis;* in accordance with classic usage, as *Sicily, siciliensis*.

—i. The termination *i* is to be considered a mere indication of the Latin genitive case, and custom, rather than correctness, has, in some sense, legalized its addition to any name. In practice, however, it is almost restricted to proper names. Thus we have *Knighti, Littoni, Flemingi, Ivesi*. The Rule of the British Association on this matter runs thus: "In Latinizing proper names, the simplest rule seems to be to use the termination *us*, genitive *i*, when the name ends with a consonant, and *ius*, gen. *ii*, when it ends with a vowel."

—icus. This Greek termination implies resemblance and may be added to common names, under the same rules as those given for—*ensis*, except that in forming the word, a vowel is suppressed if it would precede the termination; thus, *Macedon, macedonicus; Italy (ia), italicus.* It is little used, except in words already existing, as *ellipticus*, and, therefore, needs no further notice.

—eus. This Latin termination has been occasionally employed, but as it implies "made of," it is evidently seldom, if ever, admissible in palæontology. The term *eboraceus*, from *eboracum*, the Latin name for York, is a misnomer and should have been *eboracensis*.

—idae. This Greek patronymic has come into general use as a convenient termination to express the resemblance running through a family.

B. It is recommended that the assemblages of genera, termed *families*, should be uniformly named, by adding the termination *idæ* to the name of the earliest known, or most typically characterized genus in them; and that their subdivisions, termed *subfamilies*, should be similarly constructed, with the termination *inæ*. These words are formed by changing the last syllable of the genitive case into *idæ* or *inæ*, as *strix, Strigis, Strigidæ, Buceros, Bucerotis, Bucerotidæ*, not *Strixidæ, Bucridæ.—[Recommendations.*

It should be borne in mind that in the termination *—idae*, the *i* is short, but in *—inae* it is long.

—inus. This Latin termination is of wider application and classic usage sanctions its application to both common and proper names, though less commonly to the names of persons. Latin usage, however, restricted its application more than modern scientific practice has done, and applied it mainly to proper names, local terms and living beings; as *caninus, alpinus*. Hence, such words as *rugatinus, sulcatinus, secalinus, taxinus* and *velutinus* are at best suspicious if not illegitimate. This termination, however employed, is subject to the laws given under*—ensis*.

—ites. This termination was early adopted by naturalists to express the fossil nature of the specimen and so prevent confusion, while at the same time expressing resemblance to some existing genus or species. It is a contraction of the Greek word *lithos*, a stone. In most instances, however, it coalesces with the last vowel of the root and becomes long in compensation. This and long usage in many words, such as Ammonites, Belemnites, Pyrites, have completely established the long *i*, while the gender is determined by that of the Greek word to be masculine. All specific terms in the genus must, therefore, be of this gender.

—oides. This Greek termination, signifying " like," should be added only to the stems of words of Greek origin. No connecting vowel is necessary, as one already exists in it. Thus we have *dactylos, dactyloides; discos, discoides*. A Latin form is often used, *—oideus*, which, however, obeys the same laws; except that the Greek termination is alike in all genders, while the Latin is inflected as Latin adjectives of similar termination.

d. Compound terms.—In forming compound terms as generic or specific names, care should be taken to connect them rightly. If an adjective of three terminations or a noun of the second Latin declension composes the former part of the word, either *i* or *o* may be employed as a connecting vowel, the choice being largely determined by the ear. Thus *sulcomarginatus* is better than *sulci-marginatus*, and *crassicaulis* than *crassocaulis*. If, however, the adjective has but one or two terminations, or the noun be of the first, third or fourth Latin declension, the connecting vowel *i* should always be employed as *tenuistriatus, pinniformis, ilicifolius, retiformis, cornifer*. The connecting vowel *o* is admissible by Greek usage in all declensions, as *Ulodendron, Cycloconcha, Syringodendron, Alethopteris, Dictyonema, Dictyopteris*, except that where the first part of the word is an adjective ending in*—ys*, it is shorter and at the same time consonant with classic usage to employ no connecting vowel at all; thus, *pachyderma, euryteines, oxynotus, Platystoma*, etc., are better than *pachyoderma, euryoteines, oxyonotus, Platyostoma*, etc.

e. Spelling of new names.—

In writing zoological names, the rules of Latin orthography must be adhered to.

In Latinizing Greek words, there are certain rules of orthography known to classical scholars which must never be departed from. For instance, the names which modern authors have written *Aipuknemia, Xenophasia, poiocephala*, must, according to the laws of etymology, be spelt *Æpycnæmia, Xenophasia*, and *pœcephala*. In Latinizing modern words, the rules of classic usage do not apply, and all that we can do is to give to such terms as classical an appearance as we can, consistently with the preservation of their etymology. In the case of European words, whose orthography is fixed, it is best to retain the original form, even though it may include letters and combinations unknown in Latin. Such words for instance as *Woodwardi, Knighti, Bullocki, Eschscholtzi*, would be quite unintelligible, if they were Latinized into *Vudvardi, Cnichti, Bullocci, Essolzi*, etc. But words of barbarous origin, having no fixed orthography, are more pliable, and hence, when adopted into the Latin, they should be rendered as classical in appearance as is consistent with the preservation of their original sound. Thus, the words *Tockus, awmree, argoondat, kundoo*, etc., should, when Latinized, have been written *Toccus, ansure, argunda, cundu*, etc. Such words ought, in all practicable cases, to have a Latin termination given them, especially if they are used generically.

This rule, with its limitations and exceptions has been not seldom departed from in naming fossils. Many names have gained currency, which are needlessly unclassical. At the same time, we meet with marked examples of compliance. Such names as the following stand out prominently—*sancti ludovici, cestriensis*, are far more correct and pleasing than *louisi* and *chesterensis, lesouriae* or *lesouri* than *lesquereuxi*, while *Lepidophloios*, of Sternberg, should be spelled *Lepidophlocus*, in accordance with Latin rather than Greek custom.

It is almost unnecessary to add that when a new term is derived from sources purely classical, care should be taken to spell it accurately, and not to disguise or conceal its origin by any attempts to eliminate a letter or two. This, however, has been sometimes done, as for instance, in the name *Stenocisma*, Conrad, (in which the etymology is masked by the misspelling) and which will be found written *Stenoschisma*.

f. Mythological names.—In regard to the numerous mythological names, especially those given by the late palaeontologist, to the Canadian Survey, Mr. E. Billings, the following recommendations of the British Association deserve notice:

d. Mythological or historical names.—When these have no perceptible reference or allusion to the characters of the object on which they are conferred, they may be properly regarded as unmeaning and in bad taste. Thus, the generic names *Lesbia, Leilus, Remus, Corydon, Pasiphæ*, have been applied to a humming-bird, a butterfly, a beetle, a parrot, and a crab, respectively, without any perceptible association of ideas. But mythological names may sometimes be used as generic, with the same propriety as technical ones, in cases where a direct allusion can be traced between the narrated actions of a personage and the observed habits or structure of an animal. Thus, when the name *Progne* is given to a swallow, *Clotho* to a spider, *Hydra* to a polyp, *Athene* to an owl, *Nestor* to a greyheaded parrot, etc., a pleasing and beneficial connection is established between classical literature and physical science.—*Recommendations.*

Mr. Billings was, probably, led to adopt the practice by the increasing difficulty of finding unpreoccupied names and an unwillingness to encumber the science with more synonyms.

g. Personal names.—

*g. Specific names derived from persons.—*So long as these complimentary designations are used with moderation, and are restricted to persons of eminence as scientific zoologists, they may be employed with propriety in cases where expressive or characteristic words are not to be found. But we fully concur with those who censure the practice of naming species after persons of no scientific reputation, as curiosity dealers (e. g. *Caniveti Boissoneauti*), Peruvian priestesses (*Cora amazilia*), or Hottentots (*Klassi*).

*h. Generic names derived from persons.—*Words of this class have been very extensively used in botany, and therefore it would have been well to exclude them wholly from zoology, for the sake of obtaining a *memoria technica* by which the name of a genus would at once tell us to which of the kingdoms of nature it belonged. Some few personal generic names have, however, crept into zoology, as *Cuvieria, Mulleria, Rossia, Lessonia*, etc., but they are very rare in comparison with those of botany, and it is, perhaps, desirable not to add to their number.—*Recommendations.*

Another objection to this kind of name, so many of which have crept into palæontological nomenclature of late years, is that they yield with difficulty, in many instances, to the plastic hand of the classical linguist. The collector or student who stands godfather to a newly discovered fossil, and seeks immortality for some friend or acquaintance by making it his namesake, naturally wishes to keep the name as little changed as possible, lest his purpose should in part fail. On the other hand the classical scholar would rather see the name reduced in some degree to Latin form so that it may offend the eye and ear as little as possible. Between these two a contest arises, and hence we have various forms for words which should be alike, as *Barrandi, barrandei; Moori, moorei*. Perhaps in this respect it would be wisest to adopt the recommendations of the British Association Committee, quoted on page xi, with a slight modification, and add *i* for a genitive case where the name ends in a consonant, in *a, o* or *u*, and *ii* when it ends in *e, i* or *y*, omitting in the latter case the final vowel of the name.

*h. Nouns as specific terms.—*In those cases, and they are numerous, where a noun is used as a specific name, no change can be made in its termination. Hence, however, inconsistent it may appear to some to see *Orthis biloba* and *Orthis circulus, Productus cora* and *Productus costatus*, side by side, such combinations are accurate. It may, however, be added that adjective terms are preferable, whenever possible.

*i. Inelegant names.—*With the following extract we leave this part of our subject:

*Names of harsh and inelegant pronunciation.—*These words are grating to the ear, either from inelegance of form, as *Huhua, Yuhina, Craxirex, Eschscholtzia*; or, from too great length, as *chirostrongylostinus, Opetiorhynchus, brachypodioides, thecodontosaurus*, not to mention the *Eualiolimnosaurus crocodilocephaloides* of a German naturalist. It is needless to enlarge on the advantage of consulting euphony in the construction of our language. As a general rule it may be recommended to avoid introducing words of more than five syllables.

We have now pointed out the principal rocks and shoals which lie in the path of the nomenclator, and it will be seen that the navigation through them is by no means easy. The task of constructing a language which shall supply the demands of scientific accuracy on the one hand, and of literary elegance on the other, is not to be inconsiderately undertaken by unqualified persons. Our nomenclature presents but too many flaws and inelegancies already, and as the stern law of priority forbids their removal, it follows that they must remain as monuments of the bad taste or bad scholarship of their authors to the latest ages in which zoology shall be studied.

Etymologies and types of new genera to be stated.—It is obvious that the names of genera would, in general, be far more carefully constructed, and their definitions would be rendered more exact, if authors would adopt the following suggestion:

It is recommended that in defining new genera, the etymology of the name should be always stated, and that one species should be invariably selected as a type or standard of reference.

The following extract, from the same code, also bears on the point in question:

Nonsense names.—Some authors having found difficulty in selecting generic names which have not been used before, have adopted the plan of coining words at random, without any derivation or meaning whatever. The following are examples: *Viralva, Xema, Azeca, Assiminia, Quedius, Spisula.* To the same class we may refer *anagrams* of other generic names, *Dacleo* and *Cedola* of *Alcedo, Zapornia* of *Dorzana,* etc. Such verbal trifling as this is in very bad taste, and is especially calculated to bring the science into contempt. It finds no precedent in the Augustan age of Latin, but can be compared only to the puerile quibblings of the Middle Ages. It is contrary to the genius of all languages, which appear never to produce new words by spontaneous generation, but always to derive them from some other source, however distant or obscure. And it is peculiarly annoying to the etymologist, who, after seeking in vain through the vast storehouse of human language for the parentage of such words, discovers at last that he has been pursuing an *ignis fatuus.*

Besides having often to follow such will-o'-the-wisps, any student who undertakes the task, will soon find it no easy one to ferret out the origin and meaning of veritable scientific names, often made up without distinct allusion to any conspicuous character of the fossil in question, and, occasionally, by men whose acquaintance with the languages is of the slightest kind. After all, care and pains, not a few malformed words and barbarous terms too long current to be now withdrawn from circulation, must ever remain, as corresponding words survive in our common English tongue; marking some age of folly or ignorance in by-gone days of the language. Such terms will serve as memorials, fossil relics, to show to future times the freaks of linguistic development in the early days of Palæontology.

INTRODUCTION.

The most ancient rocks known to man had their immediate origin in sedimentary deposition at the bottom of an ocean. At this point commences Geology; for previous to this period nothing has been ascertained as to the condition of the Earth. It commences at the base of the lowest rocks discovered, and from thence it investigates the overlying strata, the changes that have taken place, the lapse of time, and the development of animal life, down to the present moment.

Were it possible to obtain a transverse section of these rocks at their maximum thickness, at any particular locality, there would be presented to the view an exposure of more than thirty miles, representing the geological column, which is regarded as one continuous whole, from the base to the top.

Geologists have adopted various divisions and subdivisions of this column, for the purpose of illustrating the different aspects of the science. But that which subdivides it into groups seems to be the most natural and systematic, and is the classification now generally preferred.

In the infancy of geological science, the known strata were divided into formations, each of which was supposed to represent a geological period, during which no marked changes occurred in the condition of the Earth or in its organic life. The commencement of each formation was supposed to mark a new epoch of creation, and its close to represent a great cataclysm that destroyed all vegetable and animal organisms. Modern geologists, however, have determined that the evidence of special creations and cataclysms is entirely wanting, and that the formations do not mark geological periods of repose, nor are they any more distinct in their character, from each other, than are the groups into which they have been subdivided.

The palæozoic rocks, as understood by geologists, embrace the Permian group, and all that is below it. The subdivision into Archæan, Silurian, Devonian and Carboniferous formations is retained as a matter of convenience, for the purpose of directing attention, in a general way, to a particular quarter of these rocks. The Archæan (most ancient) are the metamorphic rocks. The Silurian is next in order. The Devonian, from Devonshire, England, is a much smaller formation, and is characterized by its fish remains. The Carboniferous formation, so-called from the fact that it contains the great coal deposits of Europe, concludes the series of palæozoic rocks. Other formations contain coal deposits, but they are otherwise so distinct, that they are never in danger of being classed with this formation as originally defined.

The names used in American Geology, for the purpose of the classification of the strata into groups, have been derived, generally, from the localities at which the rocks have been studied and described. This method of nomenclature is preferred to any other, because the name itself can never mislead as to the mineral structure or relative position of the rocks, and the geologist may visit and study the locality with the certainty that he is investigating the typical group. Prof. Rogers conceived the idea of improving the nomenclature of the palæozoic rocks by dividing them into fifteen parts and giving them names significant of their relative ages. This he did by using words suggesting metaphorically different parts of a day, as follows: Primal, Auroral, Matinal, Levant, Surgent, Scalent, Premeridian, Meridian, Post Meridian, Cadent, Vergent, Ponent, Vespertine, Umbral and Seral, meaning respectively the formations of the Dawn, Daybreak, Morning, Sunrise, Mounting Day, Climbing Day, Forenoon, Noon, Afternoon, Declining Day, Descending Day, Sunset, Evening, Dusk and Nightfall. Unfortunately for his attempt to substitute another, for the geographical nomenclature then quite well established and susceptible of indefinite expansion, without the use of conflicting terms or words that could mislead the student, there were several extensive groups of rocks full of the remains of animal life, as yet unexplored, and consequently quite unknown to his system. For obvious reasons the nomenclature suggested by Mr. Rogers has not been adopted, and in all probability never will be.

The Archæan formation is divided into the Laurentian and Huronian groups. The Laurentian series of metamorphic rocks forms the base of the geological column, and has an estimated thickness in Canada of 32,750 feet. It takes its name from the Laurentian mountains of Canada. The *Eozoon canadense*, a fossil Rhizopod, is found at the base of the Grenville band of limestone, which is near the middle of this series. The estimated depth of the Laurentian series to the lowest place at which this fossil has been found is 16,500 feet. Sir W. E. Logan describes the Grenville band as follows:

"The general character of the rock connected with the fossil produces the impression that it is a great foraminiferal reef, in which the pyroxene masses represent a more ancient portion, which having died, and become much broken up, and worn into cavities and deep recesses, afforded a seat for a new growth of foraminifera, represented by the calcareo serpentinous part. This in its turn became broken up, leaving, however, in some places, uninjured portions of the organic structure. The main difference between this foraminiferal reef, and more recent coral reefs, seems to be, that while with the latter are usually associated many shells and other organic remains, in the more ancient one the only remains yet found are those of the animal which built the reef."

The next series of rocks overlying the Laurentian is called the Huronian, which, on the north shore of Lake Huron and to the eastward, consists of quartzites, chloritic slates, bands of limestone chert, jasper and slate conglomerates, not less than 18,000 feet in thickness.

On Lake Superior, Sault Ste. Marie, Mamainse, and other places, it is exposed from 10,000 to 16,208 feet in thickness. An approximate estimate of the thickness of this series on Michipicoten Island, says Mr. McFarlane, is 18,500 feet. And if we compare the rocks of Michipicoten Island with those of

Mamainse, it would appear that the lower rocks of the latter series do not come to the surface at Michipicoten Island, and that the higher rocks of the Michipicoten series have not been developed at Mamainse, or lie beneath the waters of the lake to the southwest of the promontory. It would, therefore, appear just in estimating the thickness of the upper copper-bearing rocks of the eastern part of Lake Superior (which are Huronian), to add to the Mamainse series the above-mentioned 4,000 feet of resinous traps, or porphyrites, which would make the whole thickness at least 20,000 feet. (Geo. of Can., 1863, pp. 55, 67, 86; Geo. of Can., 1866, pp. 132, 141.)

In addition to the *Eozoon canadense*, only a few fossil species have been described from the Huronian rocks, and these have been placed in genera of uncertain affinity, to wit: *Aspidella*, *Stenotheca*, and *Scenella*.

The metamorphic strata, equivalent to the Laurentian and Huronian series of Canada are described in Safford's Geological Survey of Tennessee, as many thousand feet thick. Fossil remains of foraminifera have been found in them. They have also been found in the metamorphic rocks of Europe; so that these rocks are no longer called azoic.

Next above the Archæan subdivision lies the great Silurian formation, first determined by Sir R. I. Murchison, and named in memory of the ancient Silures, who inhabited Wales, where he first studied the exposure of the rocks. He subdivided it into the Lower Silurian and Upper Silurian formations, and these names have been adopted in this country. The Lower Silurian is much the most extensive, and is divided into groups or subdivisions, in ascending order as follows:

1. The St. John's Group, consisting of black shales and sandstones, resting conformably upon older schistose rocks, at St. John's, New Brunswick, 3,000 feet thick. (Geo. of Can., 1866, p. 235.)

This Group of rocks contains the remains of Paradoxides, Conocephalites, Arionellus, Microdiscus, Agnostus, Orthis, and other fossil genera. The Paradoxides beds near Boston are supposed to belong to it.

2. The Potsdam Group, which takes its name from Potsdam in Northern New York, where it is well developed, and consists of a fine-grained, even-bedded sandstone, traversed by parallel joints. The Potsdam Group is sometimes divided into Upper .Potsdam and Lower Potsdam Groups. The sandstones and limestones on the north shore of Belle Isle, and the rocks which, in the State of Vermont are called the Georgia slates, and the red sand-rock, belong to the Lower Potsdam Group. The Upper Potsdam Group is displayed in Minnesota, Wisconsin, New York, Pennsylvania, and other States. In Tennessee the Potsdam Group is divided into three great sub-groups, in ascending order as follows: 1. The Ococe conglomerate and slates; 2. Chilhowee sandstone; 3. Knox Group. The Ococe conglomerate and slates takes its name from Ococe river, and is 10,000 feet thick. The Chilhowee sandstone takes its name from the Chilhowee Mountain, and is 2,000 feet thick. The Knox Group takes its name from Knox county, and is 7,000 feet thick; making the total thickness of the Potsdam in Tennessee 19,000 feet. It is from 500 to 2,500 feet thick in Canada, 2,020 at Bonne Bay, Newfoundland, 1,147 feet at the Straits of Belle Isle, about 2,000 feet in Pennsylvania and West Virginia, and from 600 to 800 feet thick in Wisconsin and Minnesota.

The fossils which chiefly characterize the Potsdam Group belong to the genera Palæophycus, Scolithus, Archæocyathus; Obolella, Camarella, Dinobolus, Lingulella, Salterella, Bathyurus, Conocephalites, Olenellus and Dicellocephalus. Gasteropoda and Cephalopoda first make their appearance in the Upper Potsdam. Poor specimens of Cystidea and Crinoidea have been found in this Group.

3. The Calciferous Group, named from the calcareo-silicious character of the rocks. It was called the Calciferous sandrock and Transition rock by Eaton, and the Barnegat limestone, Newburgh limestone, Warwick limestone, Oolitic limestone, Fucoidal layers, and slaty limestone, in the early Geological Reports of New York. It is divided into the Lower and Upper Calciferous Groups. The Upper Calciferous is found in Newfoundland about 1,000 feet in thickness. The Lower Calciferous is the original Calciferous Group of New York. It is exposed in Minnesota, Wisconsin, Tennessee, Missouri, Pennsylvania, Virginia, Canada, and other places, usually less than 500 feet in thickness. It is found, however, in Missouri 1,315 feet thick, where it is subdivided into 1st, 2d, 3d and 4th Magnesian limestone, and in Newfoundland 1,839 feet thick. The St. Peter's sandstone of the Northwest, named from its great development on the St. Peter's river, belongs to this group. There is in Newfoundland an important series of strata, having a thickness of 2,061 feet, known as division N. of the Newfoundland rocks, which should probably be regarded as part of the Upper Calciferous Group. If so, it would give us a maximum thickness of about 4,000 feet for the Calciferous Group.

The Lamellibranchiata first make their appearance in this Group, while the Gasteropoda and Cephalopoda become numerous. *Pleurotomaria canadensis* is known to pass from the Potsdam into this Group.

4. The Quebec Group, which takes its name from the city of Quebec, in Canada, where it was first studied, and is subdivided into the Levis, Lauzon, and Sillery Groups. The Levis division, which takes its name from Point Levis in Canada, comprehends the Philipsburgh rocks which are 4,860 feet thick, in addition to 1,385 feet of the Orleans section, making the whole division 6,145 feet thick. The Lauzon division, which takes its name from Lauzon, in Canada, is about 4,000 feet in thickness. And the Sillery division, which takes its name from Sillery Cove in Canada, is 2,000 feet in thickness, making the maximum depth of the three divisions of this Group 12,145 feet. The Levis division is highly fossiliferous, while only a few fossils are known from other divisions. Some of the fossils of this Group are found both in the Chazy and Calciferous Groups, and the Canadian geologists for a time supposed it to be the equivalent, in some way or other, of these rocks; but later investigations have shown that it has a fauna of its own, and that it offers beds of passage from the Potsdam to the Trenton fauna, in addition to those of the Chazy and Calciferous.

5. The Chazy Group, which takes its name from Chazy, Clinton county, New York, and has an extensive geographical range over New York and Canada, is seldom found over 300 feet in thickness. The upper part of the 1st Magnesian limestone in Missouri may be the equivalent of this Group, or rather the dark bluish gray semi-crystalline limestones, interstratified with the grayish drab earthy magnesian varieties, destitute of chert, which crop out in some

places on top of the 1st Magnesian, and are usually classed with it. The Group has likewise been doubtfully identified in Wisconsin and other Northwestern localities.

6. The Birdseye Limestone, which is a well marked Group of rocks in New York. The rocks break with a conchoidal fracture, and the surface presents numerous crystalline spots due to calcspar in the tubes of *Tetradium fibratum*, which it contains in abundance, and from which the Group takes its name. It is not, however, so distinct in its fossil remains as to be characterized and determined elsewhere, and therefore some geologists treat the name as a synonym for the Black River Group.

7. The Black River Group (including the Birdseye limestone or its equivalent) which is found in New York, Pennsylvania, Canada and the Island of Anticosti, and is supposed to underlie Ohio, Indiana and Illinois, for it is again found cropping out under the Trenton Group in Missouri and other places west of the Mississippi. It has a very wide geographical range, and in Pennsylvania attains the important thickness of 5,500 feet.

The *Orthoceratites*, which commenced an existence in the Upper Potsdam, reach their greatest development in this Group of rocks. Some shells are found more than ten feet in length and exceeding a foot in diameter. Other Cephalopoda are found in abundance, and several new genera commence an existence.

The synonyms for the Black River Group in the early geological reports are *Mohawk limestone, Base of the Trenton limestone, Blue limestone, Black marble of Isle La Motte, Bald Mountain limestone, sparry limerocks, Transition chequered limestone, Seven foot Tier* and *Metalliferous limerock.*

8. The Trenton Group, which takes its name from Trenton, Oneida county, New York, is found almost everywhere on the Continent, where the Lower Silurian rocks are exposed. In New York, Pennsylvania, Kentucky, Illinois, Missouri and Canada, its greatest thickness does not exceed 1,000 feet; while on the Island of Anticosti it reaches 1,700, and in Tennessee 2,500 feet. The Galena limestone of Illinois and Wisconsin, and the Cape Girardeau limestone of Missouri, belong to it.

The limestones of this Group are literally a mass of fossils, and withal so well preserved, that several hundred species and many genera have been defined. No Group of palæozoic strata has been studied with more interest, or has yielded more facts beneficial to science. It is a magnificent museum of well preserved shells, representing almost every character of the ancient population of the sea.

9. The Utica Slate Group, which takes its name from Utica, New York, and seems to be confined in its Geographical range to Pennsylvania, New York and Canada, and reaches its greatest thickness in Pennsylvania at 400 feet. Its synonyms are *Black slate, Pulaski shales, Frankfort slate* and *Lorraine shales.*

10. The Hudson River Group, which takes its name from the exposure on Hudson River, New York, has an extensive geographical range, and reaches its greatest thickness in Canada, at 2,000 feet. The fossils which characterize it are nearly the same as those which characterize the Trenton Group. The intervention of the Utica slate in New York and Canada furnishes the only excuse for separating the two groups. In the Western and Southern States, where the Utica slate is absent from the strata, the upper part of the Lower Silurian is

generally called the Cincinnati Group, though in Tennessee it is known as the Nashville Group, or more generally it is known as the Blue Limestone. The Cincinnati Group, as exposed in Ohio, Indiana and Kentucky, with which I am better acquainted than with any other Group of rocks, will receive more than a passing notice. The total thickness of the exposure will scarcely exceed 1,000 feet, the lower part of which is probably the equivalent of the upper part of the Trenton Group, and the remainder belongs to the Hudson River Group. The Cincinnati Group, therefore, is of value as a technical name only so far as it expresses the absence of the Utica slate, and points to the locality of its exposure.

Some fossils, as *Bellerophon bilobatus*, *Strophomena alternata*, *Zygospira modesta*, *Leptena sericea*, *Buthotrephis gracilis*, *Beyrichia chambersi*, *Calymene senaria*, *Isotelus gigas* and *Isotelus megistos* pass entirely through the Group. *Trinucleus concentricus*, *Triarthrus becki*, *Orthis multisecta*, *O. emacerata*, *Streptorhynchus hallia*, *Ambonychia bellistriata*, *Modiolopsis cincinnatiensis*, *Cycloconcha mediocardinalis*, *Lichenocrinus crateriformis* and *Chetetes* (?) *jamesi*, are confined to the lower half of the group. *Glyptocrinus decadactylus*, *G. dyeri*, *G. nealli*, *G. fornshelli*, *Lichenocrinus tuberculatus*, *Streptorhynchus filitexta*, *S. subtenta*, *S. sulcata*, *S. sinuata*, *S. nutans*, *Orthis insculpta*, *O. subquadrata*, *Rhynchonella capax*, *R. dentata*, *Cypricardites haynesi*, *Anomalodonta gigantea*, *A. alata*, *Anodontopsis milleri*, *Favistella stellata*, *Tetradium fibratum* and *Streptelasma corniculum*, are found only in the upper part of the Group. Some fossils occupy only a few feet in vertical range, as *Orthis insculpta*, *Orthis retrorsa*, *O. emacerata*, *Glyptocrinus nealli* and *Streptorhynchus sulcata*. These facts teach us that during the deposition of the rocks, the fauna of the ocean was constantly changing. One form ceasing to exist at a given place at one time, and another at another time; a new species appearing at one period, and another at another period. Not, however, indicating either the extinction or creation of a new species, because though the *Orthis insculpta* has a vertical range of only about three feet, it is found in the Trenton Group, in New York, in much earlier strata; and substantially the same may be said of other forms.

The Group of rocks, throughout its entire thickness, is composed of alternate layers of blue marl and limestone, of varying thickness. In some places the marl is 6 or 8 feet thick, without a layer of stone. At other places, one layer of stone, 4, 6 or 8 inches in thickness, follows another, with intervening layers of marl, of much less thickness, for 40 or 50 feet. It is rare to find a layer of limestone more than a foot in thickness. All the layers are broken into small, irregular pieces, sufficiently large, however, for cellar and other light stone work for which they are used. When the blue marl is exposed for a few years to the action of the weather, it gradually loses its color, and finally presents a dull gray appearance. Where the marl in the bed, only a few feet from the surface, has been changed to the dull gray color, crystals of the sulphate of lime are found on the layers of stone and in the marl. The silicious matter prevails over the carbonate of lime in the layers of marl, while the carbonate of lime is much in excess of the silicious matter in the layers of stone, due in part, at least, to the fact that the stones are almost literally a mass of more or less comminuted shells, corals and crinoids. There is nothing, therefore, in the general character and appearance of the stones and marl to indicate the changing character of the fossil contents, to which I shall now call more particular attention.

The *Zygospira modesta* is found throughout the group, varying in size from a small pin-head to a pea. The largest size has been called *Zygospira cincinnatiensis*. The smaller specimens differ in their proportional length and breadth and in the proportional elevation of the middle of the ventral valve, and corresponding depression of the mesial sinus of the dorsal valve. The larger specimens differ in the same respect; and as the number of plications is but little increased, they become larger and coarser. The same species from the Trenton Group, at Ottawa, Canada, is more elongated, and more finely plicated than the Cincinnati forms; while specimens from the Trenton Group of Southern Minnesota are scarcely distinguishable from Cincinnati specimens of medium size. This species passes through the Trenton, Utica and Hudson River Groups, and is found in the Clinton Group.

Strophomena alternata is found throughout the Group. Specimens secured within 200 feet of low water mark at Cincinnati, are large, thin, frail, and somewhat flat, but in their markings resemble the more profound specimens from the Trenton Group of New York and Ottawa, Canada. Many specimens found from 350 to 450 feet above low water mark are peculiarly thick, firm and heavy. From 450 feet above low water mark to the top of the Group the specimens are, generally, proportionately longer on the hinge line and more distinctly eared than they are below, and frequently much larger. One form of these long eared specimens has been called variety *loxorhytis*. The variety *nasuta* is most distinctly marked at an elevation from 400 to 450 feet above low water mark, where it is thicker and deeper than the same variety from the Trenton Group of New York and Canada. The variety *alternistriata* is most common in the middle and upper part of the Group. The variety *fracta* is found only in a vertical range of a few feet about the middle of the Group. This species is widely distributed and ranges from the Chazy to the Clinton Groups, passing through a great many forms, which, if constant or characteristic of particular geological horizons, would be regarded as good species.

Leptena sericea is found throughout the Group, changing at times in size, length of the hinge line and comparative thickness. It is a common form in the Trenton, Hudson River and Clinton Groups.

Strophomena tenuistriata is frail and rare in the lower part of the Group, but quite common and well preserved in the upper part. This, including its nearly related forms, under the names of *rhomboidalis* and *depressa*, is almost world wide in its distribution, and ranges from the Trenton Group to the Lower Carboniferous. One could not hesitate, however, in separating the Lower Silurian from the Upper Silurian forms, and these again from the Devonian and Lower Carboniferous forms, while remarking the somewhat general resemblance between them.

Streptorhynchus hallia is found in the lower 200 feet of the Group, and is not known to occur elsewhere. *S. planoconvexa* occupies only a few feet in vertical range about the middle of the Group. *S. nutans*, *S. planumbona*, *S. subtenta* and *S. filitexta* are confined in their range to the upper part of the Group; though *S. subtenta* is found in the Hudson River Group at English Head, Anticosti, and *S. filitexta* in the Trenton Group of New York.

Streptorhynchus sinuata has a vertical range of only a few feet below the middle of the Group, and *S. sulcata* has a vertical range of only a few feet near the upper part of the Group.

Orthis bellula, *O. plicatella*, *O. fissicosta*, *O. ella* and *Cythere cincinnatiensis* are confined within a vertical range of about 100 feet, near the middle of the Group. *O. plicatella*, *O. triplicatella*, *O. fissicosta* and *O. jamesi* vary much in size and proportional length and breadth and general appearance, and sometimes run so close together that it is only by close observation that the species are separated. *Orthis ella* varies so much in size and number of plications that it could be separated into three forms sufficiently distinct to have specific names, if the forms were found only in distinct Groups of rocks. But, probably, no shell indicates the unceasing change and development of animal life during the deposition of the Group as much as the *Orthis lynx*. It is found of all dimensions, from 1-16 of an inch to 2 inches in length, breadth and thickness. The mesial sinus is usually occupied with three plications, and the mesial fold with four; but sometimes the mesial sinus has only two plications, and sometimes it has four and even five, while the mesial fold always contains one more than the sinus, if the specimen is regularly developed. The more profound the sinus, the fewer plications in it. Some specimens are much longer than they are wide; others much wider than long. Some specimens, with hinge-line shorter than the width of the shell, become globose and nearly as round as an apple; others have the hinge-line prolonged to double the width of the shell, and have nearly the form of *Spirifera mucronata*. Small specimens of the globose form are marked with about sixteen plications, while the long-eared forms have as many as forty on each valve. Some specimens have thin shells; while others, no larger, have very thick ones. These extreme varieties do not occupy the same layers of rock, but different strata. Considerable variation exists, however, in specimens occupying the same layer; and so many intermediate forms are found in different layers, that the extremes in the Group are linked together.

The maximum thickness of the lower Silurian, as shown by the Groups mentioned, is 48,745 feet; and the fossiliferous part of the metamorphic rocks, 36,500 feet; making a total of 85,245 feet, or a little over sixteen miles from the top of the Hudson River Group to the base of the fossiliferous rocks. In other words, if all the Metamorphic and Lower Silurian Groups were fully represented, at their greatest thickness, on the Hudson River or at Cincinnati, we would expect to find fossils by digging or boring at these places for sixteen miles. The fact is probable, however, that part of the earth was dry land, while another part was covered with an ocean; and that the dry land was worn away by the action of rain, and other causes, while the ocean bed, gradually filled up, as the Atlantic fills to-day, by sedimentary deposition. There is no evidence of dry land, during all this period; but the negative evidence to the contrary, in the total absence of land plants and animals. Dry land may have existed, however, in the shape of barren rocks and disintegrated matter, for mechanical deposition; but if it did not exist, it is presumed that deposition took place more rapidly at the bed of the ocean at one place than at another, and that the ocean currents removed what had been deposited at one place, and carried it to another; so that, in either case, the maximum thickness of each Group is the measure of the lapse of time

STRATIGRAPHICAL GEOLOGY.

that transpired during its deposition on the ocean-bed; and these, when placed together, in their order of deposition, constitute the true geological column.

The Upper Silurian rocks are subdivided in ascending order, as follows: 1. Oneida Conglomerate, which takes its name from Oneida county, New York. 2. Medina sandstone, which takes its name from Medina, New York. 3. Clinton Group, which takes its name from Clinton, New York. 4. Niagara Group, which takes its name from its development at the Falls of Niagara. 5. Onondaga Salt Group, which takes its name from Onondaga, New York, where the salt springs have been extensively wrought; and 6. Lower Helderberg Group, which takes its name from the Helderberg Mountains of New York.

The Oneida Conglomerate has been called the Shawangunk Grit, the Shawangunk Conglomerate and the Millstone Grit. The word Shawangunk, signifying, in the Indian language of the aborigines, White rock, has been regarded as quite appropriate, because it is expressive of the character of the rocks. The greatest thickness of this Group in New York and Pennsylvania is about 500 feet. The rocks are of such a character that they have not preserved the fossil remains with the exception of imperfect fucoidal impressions.

The Medina sandstone is usually of a red color, with the exception of a gray band near the top. Between the mouth of the Niagara river and Lewiston, it is 350 feet thick, though at Barton, Canada, it is 618 feet thick. Its dimensions in Pennsylvania are much increased. In the latter State it is subdivided into three Groups on lithological grounds; the lower, a compact greenish gray sandstone about 400 feet thick, the next a soft argillaceous red and brown sandstone and shale 700 feet thick, and the higher a white or light gray sandstone and shales, 450 thick; making a total thickness of 1,150 feet. Like other sandstones, it usually contains but few fossils, but in some localities it is highly fossiliferous, especially in the upper part.

The Clinton Group is only estimated, in Ohio and other Western localities, at 50 feet or less in thickness. In New York and Canada from 50 to 400 feet, and on the Island of Anticosti at 610 feet; but in Pennsylvania it reaches the great thickness of 1,620 feet. (Geo. of Penn., vol. i., p. 106.)

Prof. Hall says of this group: "In the Western portion of the State (N. Y.) the limit between the Medina sandstone and Clinton Group is well defined, and the materials very distinct; but, in the central part of the State, we find the same conditions which operated during the deposition of the Medina sandstone to have been continued into the Clinton Group. The latter commences by a shaly deposit, which is soon succeeded by alternations of sandstone, in character precisely like the Medina sandstone. The general character of the marine vegetation of the two periods is similar; and a peculiar type of plants commences its existence in the Medina sandstone, and terminates in the Clinton Group. When we examine the Clinton Group in the central part of the State, its analogies are chiefly with the Medina sandstone; and it is there a powerful and important formation, presenting, however, great variation in its successive beds and characters, in every respect truly protean. In its Western extension, the Clinton Group assimilates in character to the Niagara Group, and in the Western district has nearly lost the character which it presents in Oneida county. At the same time that the Group assumes a more calcareous character in its Western extension, it loses the fossils which were typical of it, and becomes charged with fossils peculiar to calcareous strata. Thus, while we find its lower beds, from Wayne county Westward to the Niagara river, characterized by peculiar fossils, we find the upper beds containing many species which pass upward into the Niagara Group. Indeed, there is no line which can be designated between these two Groups, which shall mark the limits of the organic products. It is true, nevertheless, that by far the greater part of the fossils of the two Groups are distinct; and the small number in the lower Group, of those which we regard as proper to the Niagara Group, are for the most part inconspicuous, and not so well developed as they are in the Niagara." Again, he says:

"In tracing the Clinton Group Westerly, we find its affinities more with the rocks below, or that the material and fossils recognized on the one side as the Clinton formation are not strongly separated from the upper beds of the Hudson River Group; and studied in these localities alone, they might be regarded as constituting part of the same. On the other hand, the Niagara becomes defined as a calcareous Group, and the line between it and the strata below is strongly drawn. The base of this limestone would everywhere be recognized as the base of the Upper Silurian Rocks, while the strata below are marked by fossils which belong to the Lower Silurian fauna."

The Niagara Group consists of shales and limestones, and may, for the purposes of this introduction, include the lenticular mass of dolomitic limestones found in Canada, and bearing the name Guelph Group. The Guelph Group takes its name from the town of Guelph in Canada, where it is about 160 feet in thickness. At Lockport and at Niagara Falls, the Niagara Group consists of about 80 feet of shales, and 164 feet of limestones. The Group is found exposed in Ohio, Indiana, Illinois and other Western States, rarely exceeding 400 feet in thickness; but in Tennessee it reaches 1,700 feet, and is subdivided as follows: 1st, Clinch Mountain sandstone, consisting of shales and sandstones, 700 feet; 2d, White Oak Mountain sandstone, 500 feet; 3d, Dyestone Group of shales and sandstones, which takes its name from an iron ore, which is sometimes used as a dyestone, 300 feet; and 4th, Meniscus Limestone, which takes its name from a lens or a meniscus-shaped fossil sponge, named by Roemer, *Astræosponyia meniscus*, 200 feet.

Prof. Hall says: "The rocks of this Group, where best developed in Western New York, consist of a mass of shale, succeeded by one of limestone, the passage from the former to the latter taking place by the gradual increase of calcareous matter. The upper or terminating limestone of the Clinton Group is succeeded by a soft argillo-calcareous shale, which maintains its character unchanged for a thickness of 80 to 100 feet. Throughout the greater part of this it abounds in fossils, nearly all of which are distinct from those in the beds of the Clinton Group. In the Western part of New York, the lithological characters of the Clinton and Niagara Groups are so similar, that they could well be united. The fossils also of the two Groups, though generally distinct, are nevertheless generically similar, and several species pass from the lower to the higher Group. Still farther West, the assimilation becomes more perfect, and there appears to be no line of separation between the two Groups. At the same time the fossils appear to be commingled."

STRATIGRAPHICAL GEOLOGY.

The Onondaga Salt Group is sometimes called the Onondaga limestone or Gypsiferous series. Its outcrop in New York is traced, says Prof. Hall, from Montgomery county, where the formation is represented by a thin band, Westward into Wayne county, where it attains a thickness of 1,000 feet, and thinning out towards Canada, it crosses the Niagara river, at 300 feet in thickness, whence it is traced Northwestwardly to Lake Huron, and thence to Mackinac. It is also exposed in Pennsylvania, where it was called the Surgent red marl, and in Western and Southern localities. In its lower part, it is made up chiefly of marls and thin shaly limestones, which include the gypsum and salt. Its upper portion consists of magnesian limestones, often yielding hydraulic or water lime, and is hence sometimes distinguished as the Water Lime Group, where it really forms part of the Onondaga Group.

The Lower Helderberg Group has a wide geographical range, but is not susceptible of subdivision into many Groups, at any great distance from the Helderberg Mountains, where Vanuxem separated it into ; 1st, Water Lime Group or Tentaculite limestone ; 2d, Pentamerus limestone ; 3d, Delthyris shaly limestone ; 4th, Encrinal limestone ; and 5th, Upper Pentamerus limestone. The Water Lime Group was so called from its yielding hydraulic cement, and is about 200 feet thick in New York, and thins out in Canada in a Northwesterly direction. Pentamerus limestone took its name from the *Pentamerus galeatus* found in it, in Cherry Valley, New York, where it is about 30 feet thick. The Delthyris shaly limestone was so named from the abundance of *Spirifera macropleura*, and *S. pachoptera*, formerly called *Delthyris*, found in it. It is about 70 feet thick. The Encrinal limestone is about 25 feet thick, and was so named from the quantity of broken encrinites it contains. It has also been called the *Scutella* limestone, from a shield-like pelvis of a crinoid found in it. The Upper Pentamerus limestone is about 75 feet thick, and was so named from the abundance of *Pentamerus pseudogaleatus* with which it is characterized.

The Lower Helderberg attains the greatest thickness of 2,000 feet, at Gaspe, Canada. It is 1,720 feet in Pennsylvania, 400 to 500 in New York, and from 100 to 200 feet thick in the Western States.

The fossils of this Group are quite similar to those of the Niagara, and mark a very gradual development from the former to the latter. The Water Lime Group is especially characterized by large crustaceans of the genera *Eurypterus* and *Pterygotus*, the highest forms of organized life, which, so far as we know, had, up to this period, appeared upon the earth.

The rocks of the Upper Silurian formation, as shown by the preceding estimates, are about 8,000 feet in thickness. They contain the fossil remains of no vertebrate animal so far as yet known. They show the uninterrupted course of oceanic life from one Group of rocks to the next, and the gradual appearance of higher organisms, and yet they are without land plants, save perhaps a species of Psilophyton, and vertebrate animals, even of the lowest oceanic types. The student of biology and the laws of evolution may pause here and reflect upon the fact, that from the geological horizon of the *Eozoon canadense*, we have passed upwards through nearly eighteen miles in thickness of oceanic deposits, which represent many millions of years as we understand the laws of deposition, and while the changes in the forms of life have been numerous and wonderful, and the development into higher forms constant throughout the whole period, yet that life had found its most complete development in a lowly organized articulate animal—the awkwardly constructed, loosely thrown together—flimsy *Eurypterus*.

The Devonian formation was so named by Murchison from Devonshire, England. It is subdivided in ascending order into: 1st, Oriskany Sandstone; 2d, Upper Helderberg Group; 3d, Hamilton Group; 4th, Portage Group; 5th, Chemung Group; and 6th, the Catskill Group.

The Oriskany Sandstone takes its name from Oriskany, in Oneida county, New York. It has a wide geographical range, being found in Canada, New York, Pennsylvania, Maryland, Illinois, and other States. Its maximum thickness is placed at 300 feet.

Prof. Hall says: "The line of demarkation between subordinate Groups, and the line of separation between systems, are equally strong, and that the whole series may be regarded as a succession of minor Groups; that the strong lines of division are almost always due to the absence of some formation, which if present, would show a gradation to the next; and these subdivisions into systems have been made dependent on the imperfection rather than the perfection of the sequence. Thus the strong line of demarkation between the Silurian and Devonian which exists where the lower Helderberg Group is absent, is softened to a gentle gradation through the intervention of these strata and the Oriskany sandstone. Where these are present in all their members, the line of separation becomes less sharply defined, and we have some evidence that there may exist other intermediate members, or a more full development of those now known between the two formations." (Pal. of N. Y., vol. iii., p. 35.)

In Southern Illinois, the Oriskany sandstone of the Devonian system is underlaid by a Group of silicious limestones, that in their upper beds contain well marked Devonian fossils, and below those that seem to be characteristic, Upper Silurian forms; thus forming beds of passage from the Upper Silurian to the Devonian systems. This Group seems to hold about the same relation to these two systems that the Anticosti Group of Canada holds between the Upper and Lower Silurian of that country. This Group is called the "Clear creek limestone," and is limited in its outcrop to the counties of Jackson, Union, and Alexander, first making its appearance in the bluffs of Mississippi, at the lower end of the ridge known as the "Devil's backbone," in Jackson county, and continuing along the river bluffs to Clear creek, in Union county, where they are fully developed, and where they probably attain their maximum thickness of from 250 to 350 feet. (Geo. of Ill., vol. i., p. 125.) Subsequent investigations, and a more complete collection of the fossils which belonged to the upper and lower divisions of the mass, led to the conclusion that the upper division represented, at least in part, the Oriskany period, and the lower, the Delthyris shaly beds of the Lower Helderberg series. And in accordance with this view, without any well marked line of separation on lithological grounds, but supported by an examination of the same beds in Perry county, Missouri, the upper 200 feet, at the maximum thickness, is placed in the lower division of the Oriskany period, and the lower 200 feet, at the maximum thickness, is placed in the Lower Helderberg period. (Geo. of Ill., vol. ii., p. 8; vol. iii., p. 24.)

The Upper Helderberg Group, in its fullest development, consists of four members, the Caudagalli grit, the Schoharie grit, the Onondaga and Corniferous limestones. The first, when characteristic, is a dark, gritty slate, which has a cleavage vertical to the line of deposition, and is generally destitute of fossils; but with surfaces, covered with curved, fucoid-like markings, which have given it its name. This rock constitutes beds of passage from the Oriskany sandstone, and graduates above into the Schoharie grit, which is an arenaceous limestone, weathering to a brownish color, and succeeded by the gray, subcrystalline, coralline formation, which is known in New York as the Onondaga limestone, while the Corniferous limestone consists of the higher dark-colored chert beds of the Group. (Hall's Pal., vol. iii., p. 43.)

The Caudagalli grit was named from a fucoid having some resemblance in form to the tail of a chicken cock. It has a small geographical range, and its maximum thickness in New York is placed at 70 feet.

The Schoharie grit, named from Schoharie, New York, has a small geographical range and no considerable thickness. In Pennsylvania and New Jersey, where the Caudagalli and Schoharie grits have been called the Post Meridian grits, they have a thickness of 300 feet. This Group in New York consists of a fine grained calcareous sandstone, somewhat resembling the Oriskany, but bearing quite different fossils.

No vertebrate animal has yet made its appearance.

The Onondaga limestone is only from 20 to 50 feet thick in New York, and, though traced over a great extent of country, rarely exceeds that thickness. In Missouri it is said to vary from 10 inches to 75 feet in thickness, but there are few if any Western localities where this Group can be separated from the Corniferous.

The Corniferous limestone was named from the chert found in it, which breaks with a corny fracture. It varies from 100 to 200 feet in thickness, in Ohio, Indiana, Illinois, Pennsylvania and New York. It is from 300 to 400 feet thick in Michigan, and reaches its maximum of 850 feet at Tilsonbury, Canada. At Louisville, Kentucky, this Group consists of a mass of fossil corals, in a bed of hard limestone. It has the appearance of having been a coral reef, and has been so designated, but the limestone is so firm that perfect specimens of the corals are not easily procured.

This is the Group of rocks in which the first remains of vertebrate animals are found. These remains consist generally of the teeth of fish, but other hard parts are also found. Some strata are known which are literally a mass of fish teeth cemented together in a compact limestone. Land plants become more common in this Group.

The maximum of these subdivisions of the Upper Helderberg Group is therefore 1,225 feet. Each subdivision in New York is characterized by distinct fossils, but in Canada several of the most characteristic species of the Oriskany sandstone ascend through each of the overlying Groups into the Corniferous.

From this time forward, the five sub-kingdoms in animal life are represented in every Group of rocks capable of their preservation, viz. : *Protista, Radiates, Mollusks, Articulates,* and *Vertebrates.* They all continue to change and develop, but the great field of evolution is well nigh surrendered to the *Vertebrates,* which have commenced an existence in the lowest forms of marine fish, soon to appear in higher states of perfection, and to be followed by a number of Batrachian or Reptilian forms before the close of palæozoic time.

The Hamilton Group was named by Vanuxem, from Hamilton, Madison county, New York. In its fullest development, it consists of the Marcellus shale, Ludlowville shale, Encrinal limestone, Moscow shale, Tully limestone, and Genessee slate. It is about 1,200 feet thick in Eastern New York, and 1,150 feet in Eastern Pennsylvania. It thins out Westerly and Southerly, but maintains a thickness of from 300 to 600 feet in Canada.

Prof. Hall says: "The Hamilton Group consists, in Eastern New York, at base of the black Marcellus shale, including some bands of Goniatite limestone. Next succeeds a hard, compact, calcareo-arenaceous shale, which, under atmospheric influences, crumbles into angular fragments. This is followed by more arenaceous bands, and by bands of soft slaty shale, with arenaceous shale or argillaceous sandstone, and with some thin bands of limestone, which are almost entirely composed of organic remains. Toward the Western part of New York the coarser materials gradually diminish, and we find an increasing proportion of soft shales, with a more general diffusion of the calcareous matter, and the mass is terminated by a limestone. Finally, from the Genessee river to the Western limits of the State, the entire Group, above the Marcellus shale, which is persistent, consists of dark, soft shales and bands of limestone. Thus the lithological characters are at the East, an olive shale and sandstone; at the West, a grayish-blue, calcareous shale, with bands of limestone. (Pal., vol. iii., p. 46.)

The Portage Group was named from Portage, New York. It is 1,400 feet thick in the Eastern part of the State, and 1,150 feet thick in Eastern Pennsylvania. It thins out Westerly and Northerly. The Black Slate or Huron Shale of Michigan and Ohio belongs to this Group, and is from 300 to 400 feet thick. It is from 100 to 200 feet thick at Louisville, Kentucky, and New Albany, Indiana, and 50 feet thick in Missouri. The Group ought not to be called the Huron Shale either in Ohio or elsewhere, because that name was appropriated by the early Canadian geologists, and applied to a Group of Metamorphic rocks. If one were to speak of the slate in the Huronian Group, it would be called the Huron Slate, and so would shale if found there, be properly designated as Huron Shale. The words Huron and Huronian have the same signification, and are too near alike to be used to designate widely separated Groups of rocks. The "Huron Shale" is a synonym for the "Portage Group," with nothing to commend its use, because it has neither geographical nor local significance.

The Chemung Group was named from Chemung, New York, and is about 2,000 feet thick in the Eastern part of that State, but at Huntington, Pennsylvania, it is 3,200 feet in thickness. It thins out to the West, and is estimated at only 400 feet in Ohio, where it is called the Erie Shale, and 200 feet in Missouri. While the Hamilton, Portage and Chemung Groups in New York are, combined, only about 4,000 feet thick, and in Pennsylvania do not much exceed 6,000, at Gaspe, Canada, they are 7,036 feet, though this estimate may include the Catskill Group.

The passage from the Silurian formation to the Devonian at Gaspe, Canada, where the rocks are exposed 9,000 feet in thickness, is not evidenced by any change in lithological character, and is hardly determinable from an examination of the fossils. The lower 2,000 feet is classed with the Helderberg Group in the Silurian, but it may include the Oriskany sandstone of the Devonian series. The upper 7,036 feet are supposed to represent all the other Groups in the Devonian formation of New York, but the divisions are not clearly defined as in New York, nor readily separable from an examination of either the fossils or the rocks. (Geo. of Canada, 1863, p. 396; do. 1866, p. 260; Hall's Pal., vol. iii., p. 45.)

A very interesting Group of rocks, because of the highly fossiliferous character and abundance of Goniatites, is exposed at Rockford, near Seymour, Indiana, which probably represents part of the Chemung Group; though Prof. Meek and the Illinois geologists have regarded it as the representative of the Kinderhook Group.

The shales and sandstones of the Catskill Group form in their greatest expansion at the Catskill Mountains, from which the Group takes its name, a mass of at least 3,000 feet in thickness. The Group is composed of red and greenish or olive shales and shaly sandstones, with some gray and mottled sandstones and conglomerates.

In Pennsylvania this Group is divided into: 1st, Ponent Red Sandstone, which is 5,000 feet thick in its Southeastern outcrops; 2d, Vespertine, Conglomerate and sandstone, 2,660 feet in thickness, near the Susquehanna, making a total thickness of 7,660 feet.

The rocks of the Devonian age are therefore 15,235 feet, or nearly three miles in thickness, and are connected together by their interlocking fossil contents, and united with those of Silurian age, precisely as the Lower Silurian Groups are related to each other.

The Devonian rocks are followed by the Carboniferous, which are divided into: 1st, Lower Carboniferous; 2d, Carboniferous Conglomerate; 3d, Coal Measures; and 4th, Permian.

The Lower Carboniferous Group, in Nova Scotia, consists of reddish and gray sandstones and shales, conglomerates and thick beds of limestone, with marine shells and gypsum, and is 7,636 feet in thickness. On the island of Bonaventure, it is about 2,000 feet in thickness, or with the Carboniferous conglomerate 2,766, and contains the *Eatonia peculiaris*, which is found in the Oriskany sandstone of New York. In Illinois, the Lower Carboniferous is subdivided into Groups in ascending order as follows: 1st, Kinderhook Group, from 100 to 150 feet; 2d, Burlington Group, from 25 to 200 feet; 3d, Keokuk Group, from 100 to 150 feet; 4th, St. Louis Group, from 50 to 200 feet; and 5th, Chester Group, from 500 to 800 feet. In Missouri, it is subdivided in ascending order, into: 1st, Encrinital limestone; 2d, Archimedes limestone; 3d, St. Louis limestone; and 4th, Ferruginous sandstone; the maximum thickness of which is only about 1,200 feet. The Burlington Group has been called the Encrinital limestone, and in Missouri it is 500 feet thick. The Keokuk limestone is the Archimedes limestone of Owen. The Warsaw limestone is sometimes called the second Archimedes limestone. The St. Louis limestone was called the Concretionary limestone by Owen. The Chester Group has been called Kaskaskia limestone, Upper Archimedes limestone, and Pentremital limestone. In Tennessee, the Lower Carboniferous is subdivided into the Mountain limestone and Siliceous Group.

The total thickness is about 1,200 feet. In Ohio the Group is subdivided into the Cleveland Shale, Bedford Shale, Berea Grit and Cuyahoga Shale, these together constitute what is also known as the Waverly Group, named from the quarries at Waverly, Ohio. In Michigan, Prof. Winchell subdivided the Group into the Marshall Group, Napoleon Group, Michigan Salt Group, and Carboniferous limestone, the total thickness of which is 550 feet. In the *Anthracite Coal* region of Pennsylvania this Group has a maximum thickness of 3,000 feet, and consists mainly of red shale; it thins out rapidly towards the Northwest, but maintains a great thickness Southwardly through Virginia and into Alabama, gradually changing its character, however, to a calcareous limestone.

The Lower Carboniferous rocks present us with a greater number of species of Crinoids, and these in greater profusion than all the other subdivisions of the Palæozoic rocks. A single locality at Burlington, Iowa, in the Burlington Group, has furnished about 350 species. Another locality at Crawfordsville, Indiana, has almost a world-wide reputation for the great beauty, perfection and abundancy of its crinoids. It is in the Keokuk Group.

The genus *Nautilus* among the Cephalopoda is clearly recognized and is quite abundant, while the genus *Orthoceras*, whose perfection represented an organization akin to the embryonic form of the *Nautilus* has become correspondingly rare.

The Carboniferous conglomerate is 1,400 feet in thickness in Pennsylvania, and entirely thins out before reaching the Mississippi river. It is only from 100 to 200 feet thick in Ohio.

Prof. Hall says: "It was evidently formed from the fragments of older formations, drifted, water-worn, rounded and deposited with the larger pieces at the base, and the whole cemented together with smaller pebbles and sand. The depth of the formation in Pennsylvania, and its thinning out to the North and West, shows the current to have been from Southeast to Northwest, and probably indicates the close proximity of the source in a Southeasterly direction. In Michigan the thinning out is toward the South, or in a contrary direction. In Illinois the formation thins out from the West toward the East. The character of this formation, its manner of deposition, the currents which must have existed to distribute it, all indicate that this continent was an archipelago at the era of the Carboniferous conglomerate."

In some places the conglomerate is a quartzose grit used for millstones, and it is hence called the Millstone grit.

The Coal Measures are 14,570 feet thick in Nova Scotia, 8,000 feet in Pennsylvania, 2,500 feet in Tennessee, 2,000 feet in Ohio, 1,200 feet in Illinois, 640 feet in Missouri, 2,000 feet in Kansas, and a greater thickness in Nebraska.

This Group is sometimes divided into Upper and Lower Coal Measures, a separation that seems to be founded upon the fossil contents in many places.

Land Plants, which began their existence in the Devonian era, if we except *Psilophyton princeps*, became abundant in the Coal Measures. They are distributed through the rocks, the shales, and the coal. Marine Vegetation, the growth of the Marsh, and the Flora of dry land, existed in immense quantities, and was widely distributed, but the higher orders of plants and forest trees were yet unknown on the face of the earth.

Among the Cephalopoda, numereus genera and species had disappeared. The *Nautilus* the highest form then developed, was common, and furnishes us with several subgeneric types, thus manifesting its prosperity, and pointing to its continuance in succeeding strata; while the *Orthoceras* had become rare and diminutive, preparatory to interring the last of the family in this Group of rocks.

Among the Gasteropoda there was a decided advancement from the marine forms, to the land snails of the genus Pupa.

Among the *Articulates* the progress in animal life was still more clearly manifested, not only in the appearance of the marine forms of the genera *Euproops*, *Acanthotelson*, and *Anthrapalæmon*, but in the appearance of terrestrial insects of the family *Neuropteridæ*; *Myriapods* of the genera *Eoscorpius*, *Mazonia* and *Architarbus;* while the *Trilobites* that swarmed in frail hulks in earlier days gradually became extinct.

The wheel of evolution rolled yet more rapidly among the Vertebrata. The fish became more diversified and more highly organized. Amphibian animals made their appearance in several families, some of them were protected by scales, others were not; some had long vertebral columns, others had short ones; some had limbs well developed, and in form were lizard-like, while others were destitute of limbs or possessed them weak and half developed.

The Permian Group was so named by Murchison from Perm, a government in Russia. This Group is known only in the country West of the Mississippi, and is so intimately connected with the Coal Measures as to be hardly separable. Its maximum thickness does not exceed 500 feet. The maximum thickness, therefore, of the Carboniferous Groups may be placed at 24,100 feet.

The Carboniferous rocks, as found in the Uinta Mountain Region, have been subdivided by Powell and others, into four Groups, viz.: Lodore Group, Red Wall Group, Lower Aubrey Group and Upper Aubrey Group.

The maximum thickness of the Groups forming the Palæozoic rocks of North America, as here shown, is as follows:

Laurentian	32,750 feet.
Huronian	20,000 "
Lower Silurian	48,745 "
Upper Silurian	8,000 "
Devonian	15,235 "
Carboniferous	24,100 "
Total	148,830 feet.

This is a little over 28 miles, all of which is known to be fossiliferous, except the three miles at the base. It may be, that the thickness of some of the Groups is overestimated, and it may be, that two Groups, which are estimated, were deposited at the same time, and that only one of them should be counted; but on the other hand, it may be, that some Groups are entirely omitted, and that others have not been measured at the place of their greatest thickness. The probability is, therefore, that the maximum thickness of the Groups, when more certainly ascertained, as it will be by future explorations and measurements, will not fall much, if at all, below the present estimate.

The limestones of the Palæozoic rocks were formed in clear ocean waters, from the remains of calcareous shells. Their formation must have been extremely slow, so slow, that a foot may represent a thousand years, or even more. The shales, clays and marls may have been deposited with greater rapidity; but when we consider, that the change from one kind of rock-forming material to another indicates a break in the continuity of time, and that great lapse of time was necessary for the growth of the marine and land vegetation, which formed the coal found between beds of clay and shale, we are led to the conclusion, that the time which elapsed between two separate beds of clay and shale, or marl, added to the time necessary for the deposition of their materials, will, on the whole, make their formation as slow as that of the limestones. The sandstones and conglomerates, particularly those of the Coal Measures, seem to have been made up of transported materials, and were therefore deposited much faster than the limestones, though but few of them appear to have been made with any rapidity. The evenness of the strata, over a great extent of country, indicates slowness in transportation and deposit. The fact, that the materials must have been taken from pre-existing rocks, by the water, before transportation, tends again to convince us of the slowness of their formation. From these considerations, it would not be extravagant to say, that palæozoic time represents more than one hundred millions of years, and we would close our eyes against the testimony of the rocks, were we to conclude that palæozoic time could be estimated by years less than many millions.

The vegetable kingdom began with the lowest of its kind, the algæ or sea weeds, and with the lowest forms of these. The development was as gradual as the deposition of the strata. It was not until the Devonian age, that land plants appeared of sufficient firmness for preservation, if we except Dawson's *Psilophyton*, which probably grew in a marsh. These were of the lowest classes. They became more diffuse and diversified with the lapse of time; but the palæozoic era closed without the appearance of any of the higher orders or classes.

The animal kingdom likewise began with the lowest of its kind, the *Eozoon canadense*. The learned Dr. Haeckel has established the fifth sub-kingdom in animal life to include forms below the *Radiata*, and therefore very nearly related to inorganic matter. This sub-kingdom he has called *Protista*. The *Eozoon canadense*, under this classification, belongs to the order *Polythalamia*, sub-class *Radiolaria*, class *Rhizopoda*, sub-kingdom *Protista*. Ages passed, about which we know very little, before the period of the St. John's Group, which ushered in the lower Silurian. At this time we find the lowest forms of the Radiates, Mollusks and Articulates. The Articulates are represented by the lowest forms of the Trilobites, which, in their perfect state, represented the embryonic condition of the existing *Limulus*. Millions of years pass by again, before the appearance of Gasteropoda, and Cephalopoda, in the Upper Potsdam Group; meantime the system of life, which commenced with the lowest forms, as if by spontaneous generation, by evolution, increases species and genera and reaches a higher and still higher grade of development. Later still, in the Calciferous Group, the *Lamellibranchiata* commenced its existence; a class that has fought its way through all succeeding time, and is even now in the height of its prosperity and advancement. All classes of life, which existed in the ocean, up to the first appearance of the *Lamellibranchiata*, continued to live, develop, increase their species and genera, and improve, through millions of years, before the Vertebrates first made their appearance in the Devonian seas. During all this time, the earth was barren and lifeless. Mighty changes had taken place, mud had been deposited on the bed of the ocean many miles in thickness, and life had grown from its mineral origin in the ocean, until it had nearly the strength to maintain itself on land; but all this had been accomplished as silently as the earth moves in her orbit. The same gradual development continued throughout each sub-kingdom to the close of the Carboniferous period. The evolution in animal forms was as slow as time, and quite as monotonous, except in the constant progress to a higher and more complicated existence. The Coal Measures furnish us with fish-like remains, having the limbs of a frog or the breathing capacity of a tailed batrachian. Several genera have been made from the fossil remains of this period, which bridge the chasm, from the Ganoid fish to the batrachian and the lacertian. The highest Palæozoic type of animal life, yet known, Prof. Dawson has called *Hylonomus lyelli*. The distinguished principal of McGill College says, that it presents characters partly allying it to the newts and other batrachians and partly to the true lizards. The structure of the skull and vertebræ resembling a batrachian, and the well developed ribs, broad pelvis, and cutaneous covering assimilating it to the true lizards.

The following inferences are therefore to be drawn from the testimony afforded by the Palæozoic rocks:

First. That the maximum thickness of the Groups of strata is from twenty-five to thirty miles.

Second. That it required many millions of years for the formation of these Groups.

Third. That both vegetable and animal life commenced an existence, in the lowest forms, such as might have been produced by a concentration of chemical forces, or, by what has been called spontaneous generation.

Fourth. That, by processes of evolution, vegetable life developed from marine forms to land plants.

Fifth. That animal life began in the sub-kingdom Protista; from this sub-kingdom, by processes of evolution and the survival of the fittest, there arose Radiates, Mollusks, Articulates and Vertebrates. Each of these sub-kingdoms is now in the highest state of its development, though many families and some orders in each sub-kingdom have had their day and become extinct, or have been on the decline for untold ages.

VEGETABLE KINGDOM.

PLANTÆ.

The genus *Prototaxites* was founded by Dawson upon fossil wood, supposed to resemble the genus *Taxites*. The same author founded the genus *Nematoxylon*, upon what he supposed to be nearly related fossil wood. Mr. Carruthers examined the same specimens and pronounced them both Algæ, and founded the new genus *Nematophycus*. Seeds, stems, roots and other organs of uncertain affinity, have received generic and specific names. Some fossils are classed by some authors with the Algæ, as *Cruziana*, *Scolithus*, etc., while others regard them as tracks of marine animals. Authors are not in accord, even on the arrangement of the fronds and branches of ferns into families. For these reasons, I have not undertaken to arrange the fossils of the vegetable kingdom, represented in the palæozoic rocks, into families.

ACANTHOPHYTON, Dawson, 1862, Quar. Jour. Geo. Soc., vol. 18. [Ety. *akantha*, a thorn; *phyton*, a plant.]
spinosum, Dawson, 1862, Quar. Jour. Geo. Soc., vol. 18, Chemung Gr. [Sig. thorny.]
ALETHOPTERIS, Sternberg, 1825, Vers. Darst. Flora der Vorwelt. [Ety. *alethos*, true; *pteris*, a fern.]
acuta, Brongniart, 1828, (Pecopteris acuta) Prodrome d'un Histoire des Vegetaux Fossiles, Coal Meas. [Sig. acute.]
aquilina, Brongniart, 1828, (Pecopteris aquilina) Prodr. Hist. Veg. Foss., Coal Meas. [Ety. *aquilinus*, like an eagle; from *Pteris aquilina*, the eagle brake.]
bunburyi, Andrews, 1875, Ohio Pal., vol. 2, Coal Meas. [Ety. proper name.]
coxana, Lesquereux, 1861, Geo. Sur. Ky., vol. 4, Coal Meas. [Ety. proper name.]
crenulata, Brongniart, 1828, (Pecopteris crenulata) Prodr. Hist. Veg. Foss. Coal Meas. [Sig. slightly crenulated or zigzagged.]
cristata, Gutbier, 1843, (Pecopteris cristata) in Gæa von Sachsen, Coal Meas. [Sig. crested.]
discrepans, Dawson, 1868, Acad. Geol., Devonian. [Sig. different.]
distans, Lesquereux, 1858, Geo. Sur. Penn., vol. 2, Coal Meas. [Ety. from the distant pinnules.]
emarginata, Gœppert, 1836, (Pecopteris emarginata) Syst. Filic. Foss., Coal Meas. [Sig. without a border.]
erosa, Gutbier, 1843, (Pecopteris erosa) in Gæa von Sachsen, Coal Meas. [Sig. eroded.]
falcata, Lesquereux, 1870, Geo. Sur. Ill., vol. 4, Coal Meas. [Sig. sickle-shaped.]

grandifolia, Newberry, 1873, Ohio Pal., vol. 1, Coal Meas. [Sig. large-leaved.]
grandis, Dawson, 1863, Can. Nat. & Geol., vol. 8, Coal Meas. [Sig. grand, large.]
halli, Lesquereux, 1870, Geo. Sur. Ill., vol. 4, Coal Meas. [Ety. proper name.]
heterophylla, Lindley & Hutton, 1833, (Pecopteris heterophylla) Foss. Flora, Coal Meas. [Sig. irregularly-leaved.]
holdeni, Andrews, 1875, Ohio Pal., vol. 2, Coal Meas. [Ety. proper name.]
hymenophylloides, Lesquereux, 1870, Geo. Sur. Ill., vol. 4, Coal Meas. [Ety. from resemblance to *Hymenophyllites*.]
inflata, Lesquereux, 1870, Geo. Sur. Ill., vol. 4, Coal Meas. [Sig. inflated.]
ingens, Dawson, 1868, Acad. Geol., Devonian. [Sig. huge.]
lævis, Lesquereux, 1858, Geo. Sur. Penn., vol. 2, Coal Meas. [Sig. smooth.]
lanceolata, Lesquereux, 1870, Geo. Sur. Ill., vol. 4, Coal Meas. [Sig. lance-shaped.]
lonchitidis, Sternberg, 1824, (Filicites lonchiticus) Vers. Darst. Flora der Vorwelt, Coal Meas. [Ety. from *Lonchitis*; the fern, "adder's tongue."]
longifolia, Presl, 1838, (Pecopteris longifolia) in Sternb. Flora der Vorwelt, Coal Meas. [Sig. long-leaved.]
macrophylla, Newberry, 1873, Ohio Pal. vol. 1, Coal Meas. [Sig. long-leaved.]
marginata, Brongniart, 1828, (Neuropteris marginata) Hist. Veg. Foss., Coal Meas. [Sig. bordered.]
massillonis, Lesquereux, 1866, Geo. Sur. Ill., vol. 2, Coal Meas. [Ety. proper name.]
maxima, Andrews, 1875, Ohio. Pal., vol. 2, Coal Meas. [Sig. the largest.]

mazonana, Lesquereux, 1870, Geo. Sur. Ill., vol. 4, Coal Meas. [Ety. proper name.]
muricata, Brongniart, 1828, (Pecopteris muricata) Hist. Veg. Foss., Coal Meas. [Sig. armed with thorns.]
nervosa, Brongniart, 1828, (Pecopteris nervosa) Hist. Veg. Foss., Coal Meas. [Sig. full of nerves.]
obscura, Lesquereux, 1858, Geo. Sur. Penn., vol. 2, Coal Meas. [Ety. from the obscure nerves.]
oweni, Lesquereux, 1860, Geo. Rep. Ark., vol. 2, Coal Meas. [Ety. proper name.]
pectinata, Lesquereux, 1866, Geo. Sur. Ill., vol. 2, Coal Meas. [Sig. sloping two ways, like a comb.]
pennsylvanica, Lesquereux, 1858, Geo. Sur. Penn., vol. 2, Coal Meas. [Ety. proper name.]
perleyi, Hart, 1868, Acad. Geol., Devonian. [Ety. proper name.]
pluckeneti, Schlotheim, 1820, (Filicites pluckeneti) Petref., Coal Meas. [Ety. proper name.]
preciosa, see Pecopteris preciosa.
pteroides, Brongniart, 1828, (Pecopteris pteroides) Hist. Veg. Foss., Coal Meas. [Sig. wing-formed.]
rugosa, Lesquereux, 1858, Catal. Pottsville Foss., Coal Meas. [Sig. wrinkled.]
serlii, Brongniart, 1828, (Pecopteris serlii) Hist. Veg. Foss., Coal Meas. [Ety. proper name.]
serrula, Lesquereux, 1858, Geo. Sur. Penn., vol. 2, Coal Meas. [Ety. diminutive of *serra*, a saw.]
serrulata, see Pecopteris serrulata.
solida, Lesquereux, 1870, Geo. Sur. Ill., vol. 4, Coal Meas. [Sig. solid.]
spinulosa, Lesquereux, 1870, Geo. Sur. Ill., vol. 4. [Sig. full of little spines.]
stellata, Lesquereux, 1866, Geo. Sur. Ill., vol. 2. [Sig. starred.]
taeniopteroidea, Bunbury, 1847, Quar. Jour. Geo. Soc., vol. 3, Coal Meas. [Sig. like *Tæniopteris*.]
urophylla, Brongniart, 1828, (Pecopteris urophylla) Hist. Veg. Foss., Coal Meas. [Sig. sharp-leaved.]
ANARTHROCANNA, Gœppert, 1845, in Tchih. Voy. [Ety. *an*, without; *arthros*, joint; *canna*, a plant.]
perryana, Dawson, 1863, Quar. Jour. Geo. Soc., vol. 19, Devonian. [Ety. proper name.]
ANEIMITES, Dawson, 1861, Quar. Jour. Geol. Soc., vol. 17—A subgenus of *Cyclopteris acadica*, see Cyclopteris acadica.
ANNULARIA, Sternberg, 1822, Vers. Darst. Flora der Vorwelt. [Ety. *annulus*, a ring.] Wood, in 1860, proposed the name Trochophyllum instead of Annularia, because the latter was preoccupied as a generic name in the sub-kingdom Mollusca.
acuminata, Dawson, 1861, Can. Nat., vol. 6, Devonian. [Sig. sharp-pointed.]

calamitoidea, Schimper, 1869, Pal. Veget., Coal Meas. [Sig. like a *Calamite*.]
dawsoni, Schimper, 1869, Palæontologie Vegetale, Devonian. [Ety. proper name.] Proposed for Asterophyllites latifolius, of Dawson, because that name was preoccupied.
equisetiformis, Lindley & Hutton, 1835, Foss. Flora, vol. 2, Coal Meas. [Ety. from the resemblance to *Equisetum*.]
fertilis, Sternb., 1824, Vers. Darst. Flora der Vorwelt, Coal Meas. [Sig. fertile.]
inflata, Lesquereux, 1870, Geo. Sur. Ill., vol. 4, Coal Meas. [Sig. inflated.]
laxa, Dawson, 1871, Foss. Plants Canada, Devonian. [Sig. loose, open.]
longifolia, Brongniart, 1828, Prodrome Hist. Veg. Foss., Coal Meas. [Ety. *longus*, long; *folium*, a leaf.]
minuta, Brongniart, 1828, Prodr. Hist. Veg. Foss., Coal Meas. [Sig. minute.]
sphenophylloides, Zenker, 1833, (Galium sphenophylloides) in Leonh. v. Bronn's Jahrb., Coal Meas. [Ety. *sphen*, a wedge; *phyllon*, a leaf; *eidos*, form.]
ANTHOLITHES, Brongniart, 1822, Mem. du Mus. d'Hist. Nat., vol. 8. [Ety. *anthos*, flower; *lithos*, stone.]
devonicus, Dawson, 1868, Acad. Geol., Devonian. [Ety. proper name.]
floridus, Dawson, 1871, Foss. Plants Can., Devonian. [Sig. full of flowers.]
priscus, Newberry, 1854, Ann. of Sci., vol. 2, Coal Meas. [Sig. ancient.]
pygmeus, Dawson, 1863, Can. Nat., vol. 8, Coal Meas. [Sig. dwarfish.]
rhabdocarpus, Dawson, 1863, Can. Nat., vol. 8, Coal Meas. [Ety. *rhabdos*, a rod; *karpos*, fruit.]
squamosus, Dawson, 1863, Can. Nat., vol. 8, Coal Meas. [Sig. scaly.]
spinosus, Dawson, 1868, Acad. Geol., Coal Meas. [Sig. full of spines.]
ARAUCARITES, Presl, 1838, in Sternberg, Vers. Darst. Flora der Vorwelt. [Ety. *araucarites*, from *Araucaria*, a genus of large trees growing in the Southern hemisphere, especially in Australia.]
gracilis, Dawson, 1863, Can. Nat., vol. 8, Coal Meas. [Sig. slender.]
ARCHÆOPTERIS, Dawson, 1863, Can. Nat., vol. 8. [Ety. *archaios*, ancient; *pteris*, fern.]
acadica, Dawson, 1863, Can. Nat., vol. 8, Coal Meas. [Ety. proper name.]
harti, Dawson, 1863, Can. Nat., vol. 8, Coal Meas. [Ety. proper name.]
obtusa, Lesquereux, 1858, (Nœggerathia obtusa) Geo. Sur. Penn., vol. 2, Devonian. [Sig. obtuse.]
stricta, Andrews, 1875, Ohio Pal., vol. 2, Coal Meas. [Sig. pressed together.]
ARTHRARIA, Billings, 1874, Pal. Foss., vol. 2. [Ety. *arthron*, a joint.]
antiquata, Billings, 1874, Pal. Foss., vol. 2, Potsdam Gr. [Sig. ancient.]
biclavata, S. A. Miller, 1875, Cin. Quar. Jour. Sci., vol. 2, Cin'ti Gr. [Sig. double-clubbed.]

ARTHROPHYCUS, Hall, 1852, Pal. N. Y., vol. 2. [Ety. *arthron*, a joint; *phykos*, a sea plant.]
  harlani, Conrad, 1838, (Fucoides harlani) Ann. Rep. N. Y., Medina sandstone. [Ety. proper name.]
ARTHROSTIGMA, Dawson, 1871, Foss. Plants Canada. [Ety. *arthron*, a joint; *stigma*, a dot or puncture.]
  gracile, Dawson, 1871, Foss. Plants Can., Devonian. [Sig. slender.]
ARTISIA, Sternberg, 1825, Vers. Darst. Flora der Vorwelt. [Ety. proper name.]
  transversa, Steinhaur, 1818, (Phytolithus transversus) Trans. Am. Phil. Assoc., Coal Meas. [Sig. transverse.]
ASPLENITES, Gœppert, 1836, Systema Filicum Fossilium. [Ety. *Asplenium*, a genus of ferns.]
  ruber, Lesquereux, 1858, Geo. Sur. Penn., vol. 2, Coal Meas. [Ety. *ruber*, red.]
ASTEROCARPUS, Gœppert, 1836, Syst. Fil. Foss. [Ety. *aster*, a star; *karpos*, fruit.]
  grandis, Lesquereux, 1870, Geo. Sur. Ill., vol. 4, Coal Meas. [Sig. great.]
  sternbergi, Gœppert, 1836, Syst. Filic. Foss., Coal Mens. [Ety. proper name.]
ASTEROPHYCUS, Lesquereux, 1876, 7th Ann. Rep. Geol. Sur. Ind. [Ety. *aster*, a star; *phykos*, a sea weed.]
  coxi, Lesquereux, 1876, 7th Ann. Rep. Geol. Sur. Ind., Low. Carb. & Coal Meas. [Ety. proper name.]
ASTEROPHYLLITES, Brongniart, 1828, Prodr. Hist. Veg. Foss. [Ety. *aster*, a star; *phyllon*, a leaf.]
  acicularis, Dawson, 1862, Quar. Jour. Geol. Soc., vol. 18, Devonian. [Sig. full of small pins.]
  apertus, Lesquereux, 1858, Geo. Sur. Penn., vol. 2, Coal Meas. [Ety. from opening between leaves and stem.]
  brardi, Brongniart, 1828, Prodr. Hist. Veg. Foss., Coal Meas. [Ety. proper name.]
  crassicaulis, Lesquereux, 1858, Geo. Sur. Penn., vol. 2, Coal Meas. [Ety. *crassus*, thick; *caulis*, a stem.]
  curta, see Bechera curta.
  equisetiformis, Schlotheim, 1824, in Vers. Darst. Flora der Vorwelt, Coal Meas. [Ety. like unto *Equisetum*.]
  erectifolius, Andrews, 1875, Ohio Pal., vol. 2, Coal Meas. [Sig. leaves erect.]
  foliosus, Lindley & Hutton, 1833, Foss. Flora, Coal Meas. [Sig. full of leaves.]
  gracilis, Lesquereux, 1860, Geo. Sur. Ark., vol. 2, Coal Meas. [Sig. slender.]
  grandis, Lindley & Hutton, 1833, Foss. Flora, Coal Meas. [Sig. great.]
  lanceolatus, Lesquereux, 1858, Geo. Sur. Penn., vol. 2, Coal Meas. [Ety. from the lanceolate point of the leaf.]
  latifolius, Dawson, 1862, Quar. Jour. Geol. Soc., vol. 18, see *Annularia dawsoni*.
  laxus, Dawson, 1868, Acad. Geol., Devonian. [Sig. loose.]
  lentus, Dawson, 1871, Foss. Plants Can., Devonian. [Sig. pliant, tough.]
  longifolius, Brongniart, 1828, Prodr. Hist. Veg. Foss., Coal Meas. [Sig. long-leaved.]
  minutus, Andrews, 1875, Ohio Pal., vol. 2, Coal Meas. [Sig. small.]
  ovalis, Lesquereux, 1858, Geo. Sur. Penn., vol. 2, Coal Meas. [Ety. from the oval nuts.]
  parvulus, Dawson, 1862, Quar. Jour. Geol. Soc., vol. 18, Devonian. [Sig. small.]
  rigidus, Brongniart, 1828, Prodr. Hist. Veg. Foss., Coal Meas. [Sig. rigid.]
  scutigerus, Dawson, 1862, Quar. Jour. Geol. Soc., vol. 18, Devonian. [Sig. shield-bearing.]
  sublævis, Lesquereux, 1858, Geo. Sur. Penn., vol. 2, Coal Meas. [Sig. nearly smooth.]
  trinervis, Dawson, 1863, Can. Nat., vol. 8, Coal Meas. [Sig. three-veined.]
  tuberculatus, Brongniart, 1828, Prodr. Hist. Veg. Foss., Coal Meas. [Sig. covered with tubercles.]
BECHERA, Sternberg, 1824, Vers. Darst. Flora der Vorwelt. [Ety. proper name.] Becheria would be better orthography.
  curta, Dawson, 1868, Acad. Geol., Coal Meas. [Sig. short.]
  grandis, Bunbury, 1847, Quar. Jour. Geo. Soc., vol. 3, Coal Meas. [Sig. great.]
  tenuis, Bunbury, 1846, Am. Jour. Sci., 2d series, vol. 2, Coal Meas. [Sig. thin, slender.]
BEINERTIA, Gœppert, 1836, Syst. Filic. Foss. [Ety. proper name.]
  gœpperti, Dawson, 1863, Can. Nat., vol. 8, Coal Meas. [Ety. proper name.]
BRACHYPHYLLUM, Brongniart, 1828, Prodr. Hist. Veg. Foss. [Ety. *brachys*, short; *phyllon*, a leaf.]
  obtusum, Lesquereux, 1858, Geo. Sur. Penn., vol. 2, Coal Meas. [Sig. obtuse.]
BUTHOTREPHIS, Hall, 1847, Pal. N. Y., vol. 1. [Sig. growing in the depths of the sea.]
  antiquata, Hall, 1847, Pal. N. Y., vol. 1, Calcif. Gr. [Sig. ancient.]
  (?) cæspitosa, Hall, 1850, 3rd Reg. Rep., Trenton Gr. [Sig. turf-like.]
  flexuosa, Emmons, 1844, (Fucoides flexuosa) Tac. Syst., Hud. Riv. Gr. [Sig. crooked.]
  gracilis, Hall, 1847, Pal. N. Y., vol. 1, Trenton to Clinton Gr. [Sig. slender.]
  gracilis *var.* crassa, Hall, 1847, Pal. N. Y., vol. 2, Clinton Gr. [Ety. *crassus*, thick.]
  gracilis *var.* intermedia, Hall, 1852, Pal. N. Y., vol. 2, Trenton to Clinton Gr. [Ety. intermediate in size between the *gracilis* and *crassa*.]
  impudica, Hall, 1852, Pal. N. Y., vol. 2, Clinton Gr. [Sig. shameless, immodest.]
  lesquereuxi, Grote & Pitt, 1876, Buff. Soc. Nat. Hist., Water Lime Gr. [Ety. proper name.]
  palmata, Hall, 1852, Pal. N. Y., vol. 2, Clinton Gr. [Sig. palmate, having lobes like the fingers of the hand.]

PLANTÆ.

ramosa, Hall, 1852, Pal. N. Y., vol. 2, Clinton Gr. [Sig. branching.]
ramulosa, S. A. Miller, 1874, Cin. Quar. Jour. Sci., Cin'ti Gr. [Sig. full of little branches.]
subnodosa, Hall, 1847, Pal. N. Y., vol. 1, Hud.Riv.Gr. [Sig. somewhat nodose.]
succulens, Hall, 1847, Pal. N. Y., vol. 1, Trenton Gr. [Ety. *succulens*, sappy, from the succulent stems.]
CALAMITES, Guettard, 1751, Mem. Ac. Sci. Paris, [Ety. *calamus*, a reed.]
approximatus, Schlotheim, 1820, Petrefactenkunde, Coal Meas. [Ety. from the closeness of the joints.]
arenaceus, Jager, 1827, Pflanzen-Versteinerungen, Coal Meas. [Sig. sandy.]
bistriatus, Lesquereux, 1858, Geo. Sur. Penn., vol. 2, Coal Meas. [Sig. doublestriated.] This name was preoccupied by Sternberg.
cannæformis, Schlotheim, 1820, Petrefactenkunde, Coal Meas. [Sig. like the plant, *canna*.]
cisti, Brongniart, 1828, Hist. Veg. Foss., Coal Meas. [Ety. proper name.]
cruciatus, Sternberg, 1824, Vers. Darst. Flora der Vorwelt, Coal Meas. [Sig. cross-shaped.]
decoratus, Steinhaur, 1818, (Phytolithus decoratus) Trans. Am. Phil. Assoc., Coal Meas. [Sig. ornamented.]
disjunctus, Lesquereux, 1858, Geo. Sur. Penn., vol. 2, Coal Meas. [Sig. separated into joints.]
dubius, Artis, 1838, Antedil. Phytology, Coal Meas. [Sig. doubtful.]
gigas, Brongniart, 1828, Hist. Veg. Foss., Coal Meas. [Sig. large.]
gracilis, Lesquereux, 1861, Geo. Sur. Ky., vol. 4, Coal Meas. [Sig. slender.]
inornatus, Dawson, 1862, Quar. Jour. Geo. Soc., vol. 18, Devonian. [Sig. unadorned.]
nodosus, Schlotheim, 1820, Petrefactenkunde, Coal Meas. [Sig. knotty, knobbed.]
nova-scoticus, Dawson, 1863, Can. Nat. & Geol., vol. 8, Coal Meas. [Ety. proper name.]
pachyderma, Brongniart, 1828, Hist. Veg. Foss., Coal Meas. [Sig. thick-barked.]
ramosus, Artis, 1838, Antedil. Phytology, Coal Meas. [Sig. branching.]
suckovi, Brongniart, 1828, Hist.Veg.Foss., Coal Meas. [Ety. proper name.]
sulcatus, Martin, 1809, (Phytolithus sulcatus) Petrif. Derb., Coal Meas. [Sig. furrowed.]
transitionis, Dawson, 1861, Can. Nat. & Geol., Devonian. [Sig. going across, passage.] This name was preoccupied by Gœppert in 1834.
undulatus, Sternberg, 1824, Vers. Darst. Flora der Vorelt, Coal Meas. [Sig. wavy.]
voltzi, Brongniart, 1828, Hist. Veg. Foss., Coal Meas. [Ety. proper name.]

CALAMODENDRON, Binney, 1868, Observ. on the Struct. of Foss. Plants, etc. [Ety. *calamus*, a reed; *dendron*, a tree.]
antiquum, Dawson, 1871, Foss. Plants Canada, Devonian. [Sig. ancient.]
obscurum, Dawson, 1863, Can. Nat., vol. 8, Coal Meas. [Sig. obscure, hidden.]
tenuistriatum, Dawson, 1871, Foss. Plants Canada, Devonian. [Sig. fine-lined.]
CALAMOCLADUS, Schimper, 1869, Palæontologie Vegetale. Proposed to include *Asterophyllites equisetiformis, A. foliosus, A. longifolius, A. rigidus* & *Bechera grandis*.
CALLIPTERIS, Brongniart, 1828,Tabl. des Veg. Foss. (Ety. *kallos*, beautiful; *pteris*, a fern.]
sullivanti, Lesquereux, 1858, Geo. Sur. Penn., vol. 2, Coal Meas. [Ety. proper name.]
CARDIOCARPON, Brongniart, 1828, Prodr. Hist. Veg. Foss. [Ety. *kardia*, a heart; *karpos*, fruit.]
affine, Lesquereux, 1860, Geo. of Ark., Coal Meas. [Sig. near to.]
annulatum, Newberry, 1853, Ann. of Sci., vol. 1, Coal Meas. [Sig. ringed.]
baileyi, Dawson, 1868, Acad. Geol., Devonian, Coal Meas. [Ety. proper name.]
bicuspidatum, Sternberg, 1820, (Carpolithes bicuspidatus) Flora derVorwelt, Coal Meas. [Sig. double-pointed.]
bisectum, Dawson, 1863, Can. Nat. & Geol., vol. 8, Coal Meas. [Sig. divided.]
cornutum, Dawson, 1862, Quar. Jour.Geo. Soc., vol. 18, Devonian. [Sig. horned.]
crampi, Hartt, 1868, Acad. Geol., Devonian. [Ety. proper name.]
elongatum, Newberry, 1853, Ann. of Sci., vol. 1, Coal Meas. [Sig. lengthened.]
fluitans, Dawson, 1863, Can. Nat. & Geo., vol. 8, Coal Meas. [Sig. floating.]
ingens, Lesquereux, 1860, Geo. of Ark., Coal Meas. [Sig. huge, large.]
latum, Newberry, 1853, Ann. of Sci., vol. 1, Coal Meas. [Sig. broad.]
marginatum, Artis, 1838, Antedil. Phytol., Coal Meas. [Sig. margined.]
minus, Newberry, 1853, Ann. of Sci., vol. 1, Coal Meas. [Sig. less.]
newberryi, Andrews, 1875, Ohio Pal.,vol. 2, Coal Meas. [Ety. proper name.]
obliquum, Dawson, 1862, Quar. Jour. Geo. Soc., vol. 18, Devonian. [Sig. oblique.]
orbiculare, Newberry, 1853, Ann. of Sci., vol. 1, Coal Meas. [Sig. orbicular.]
ovale, Dawson, 1871, Foss. Plants Can., Devonian. [Sig. egg-shaped.]
plicatum, Lesquereux. 1858, Geo. Sur. Penn., vol. 2, Coal Meas. [Sig. plaited.]
punctatum, Gœppert, 1836, Syst. Filic. Foss., Coal Meas. [Sig. punctate.]
retusum, Sternberg, 1820, (Carpolithes retusus) Flora der Vorwelt, Coal Meas. [Sig. blunt.]
samaræforme, Newberry, 1853, Ann. Sci., vol. 1, Coal Meas. [Sig. like elm seed.]
tenellum, Dawson, 1873, Rep. Foss.Plants, Low. Carb. [Sig. delicate.]

PLANTÆ.

trevortoni, Lesquereux. 1858, Geo. Sur. Penn., vol. 2, Coal Meas. [Ety. proper name.]
CAULOLITHES, Schlotheim, 1820, Petrefactenkunde. [Ety. *karpos*, fruit; *lithos*, stone.]
bicuspidatus, see Cardiocarpon bicuspidutum.
bifidus, Lesquereux, 1858, Geo. Sur. Penn., vol. 2, Coal Meas. [Sig. bifid, cloven.]
bullatus, Lesquereux, 1870, Geo. Sur. Ill., vol. 4, Coal Meas. [Sig. studded with knobs.]
cistula, Lesquereux, 1866, Geo. Sur. Ill., vol. 2, Coal Meas. [Sig. a little chest or coffer.]
clavatus, see Rhabdocarpus clavatus.
compactus, Dawson, 1871, Foss. Plants Canada, Devonian. [Sig. compact.]
corticosus, Lesquereux, 1870, Geo. Sur. Ill., vol. 4, Coal Meas. [Sig. having thick bark.]
disjunctus, Lesquereux, 1858, Geo. Sur. Penn., vol. 2, Coal Meas. [Sig. separated.]
fasciculatus, Lesquereux, 1866, Geo. Sur. Ill., vol. 2, Coal Meas. [Sig. a small bundle.]
fragarioides, Newberry, 1873, Ohio Pal., vol. 1, Coal Meas. [Sig. resembling a strawberry.]
jacksonensis, Lesquereux, 1866, Geo. Sur. Ill., vol. 2, Coal Meas. [Ety. proper name.]
limatus, Dawson, 1863, Quar. Jour. Geo. Soc., vol. 19, Devonian. [Sig. elegant.]
multistriatus, Presl, 1833, in Sternberg Flora der Vorwelt, Coal Meas. [Sig. many-lined.]
persicaria, Lesquereux, 1870, Geo. Sur. Ill., vol. 4, Coal Meas. [Ety. *Persicaria*, an existing genus of plants.]
platimarginatus, Lesquereux, 1860, Geo. Sur. Ark., Coal Meas. [Sig. flat-margined.]
retusus, see Cardiocarpon retusum.
siliqua, Dawson, 1863, Quar. Jour. Geo. Soc., vol. 19, Devonian. [Sig. a pod.]
spicatus, Dawson, 1863, Quar. Jour. Geo. Soc., vol. 19, Devonian. [Sig. spiked.]
trilocularis, see Trigonocarpon trilocularis.
umbonatus, Sternberg, 1820, Vers. Darst. Flora der Vorwelt, Coal Meas. [Sig. having a shield.]
vesicularis, Lesquereux, 1870, Geo. Sur. Ill., vol. 4, Coal Meas. [Sig. bladder-like.]
CAULERPITES, Sternb, 1833, Vers. Darst. Flora der Vorwelt. [Ety. *kaulos*, stem; *erpo*, to creep.]
marginatus, Lesquereux, 1866, Am. Phil. Soc.,vol.13,Coal Meas. [Sig. bordered.]
CAULOPTERIS, Lindley & Hutton, 1833, Foss. Flora. [Ety. *kaulos*, stem; *pteris*, fern.]
acanthophora, Lesquereux, 1870, Geo. Sur. Ill.,vol. 4, Coal Meas. [Sig. thorn-bearing.]

antiqua, Newberry, 1871, Quar. Jour. Geo. Soc.,vol. 27, Devonian. [Sig. ancient.]
cisti, Brongniart, 1828, (Sigillaria cisti) Prodr. Hist. Veg. Foss., Coal Meas. [Ety. proper name.]
gigantea, Lesquereux, 1858, Geo. Sur. Penn., vol. 2, Coal Meas. [Sig. large.]
insignis, Lesquereux, 1866, Geo. Sur. Ill., vol. 2, Coal Meas. [Sig. remarkable.]
intermedia, Lesquereux, 1870, Geo. Sur. Ill., vol. 4, Coal Meas. [Sig. intermediate between *Sigillaria macrodiscus* and *Caulopteris cisti*.]
lockwoodi, Dawson, 1871, Quar. Jour.Geo. Soc., vol. 27, Devonian. [Ety. proper name.]
obtecta, Lesquereux, 1870, Geo. Sur. Ill., vol. 4, Coal Meas. [Sig. covered.]
peregrina, Newberry, 1871, Quar. Jour. Geo. Soc., vol. 27, Devonian. [Sig. foreign, strange.]
punctata, Lesquereux, 1858, Geo. Sur. Penn., vol. 2, Coal Meas. [Sig. dotted.] The name was preoccupied by Gœppert in 1836.
wortheni, Lesquereux, 1866, Geo. Sur. Ill., vol. 2, Coal Meas. [Ety. proper name.]
CHONDRITES, Sternberg, 1833, Vers. Darst. Flora der Vorwelt. [Ety. from its resemblance to *Chondrus crispus*, or Irish moss.]
antiquus, Brongniart, 1828, (*Fucoides antiquus*) Hist. Veg. Foss., Devonian. [Sig. ancient.]
colletti, Lesquereux, 1870, Geo. Sur. Ill., vol. 4, Coal Meas. [Ety. proper name.]
targioni, Brongniart, 1828, (*Fucoides targioni*) Coal Meas. [Ety. proper name.]
CONOSTICHUS, Lesquereux, 1876, 7th Ann. Rep. Geo. Sur. Ind. [Ety. *konos*, a cone; *stichos*, a row.]
ornatus, Lesquereux, 1876, 7th Ann. Rep. Geo. Sur. Ind., Coal Meas. [Sig. adorned.]
CORDAITES, Unger, 1850, Gen. et. sp., p. 277. [Ety. proper name.]
angustifolius, Dawson, 1861,Can.Nat.,vol. 6, Ham. Gr. [Sig. narrow-leaved.]
angustifolius, Lesquereux, 1870, Geo. Sur. Ill., vol. 4, Coal Meas. The name was preoccupied.
borassifolius, Sternberg, 1820,(Flabellaria borassifolia) Vers. Darst. Flora der Vorwelt, Coal Meas. [Ety. leaved like *Borassus*.]
flexuosus, Dawson, 1863, Quar. Jour.Geo. Soc.,vol. 19, Devonian. [Sig. winding, full of turns.]
robbi, Dawson, 1861, Can. Nat., vol. 6, Ham. Gr. [Ety. proper name.]
simplex, Dawson, 1863, Can. Nat., vol. 8, Coal Meas. [Sig. simple.]
CREMATOPTERIS, Schimper, 1865, Monograph, Foss. Plants. [Sig. hanging-fern.]
pennsylvanica, Lesquereux, 1858, Geo. Sur. Penn., vol. 2, Coal Meas. [Ety. proper name.]

PLANTÆ.

CRUZIANA, D'Orbigny, 1842, Geo. du Voy. Amer. [Ety. proper name.]
  linnarrsoni, White, 1874, Rep. Invert. Foss.,Potsdam Gr. [Ety. proper name.]
  rustica, White, 1874, Rep. Invert. Foss., Potsdam Gr. [Sig. plain, simple.]
  similis, Billings, 1874, Pal. Foss., vol. 2, Potsdam Gr. [Sig. like in aspect.]
CYCLOPTERIS, Brongniart, 1828, Prodr. Hist. Veg. Foss. [Ety. *kuklos*, a circle; *pteris*, a fern.]
  acadica, Dawson, 1861, Quar. Jour. Geo. Soc., vol. 17, Coal Meas. [Ety. proper name.]
  alleghaniensis, Meek, 1876, Desc. Foss. Plants Va., Ponent or Vespertine Gr., near base of Low. Carb. [Ety. proper name.]
  antiqua, Dawson, 1863, Can. Nat. & Geo., vol. 8, Coal Meas. [Sig. ancient.]
  browni, Dawson, 1862, Quar. Jour. Geo. Soc., vol. 18, Devonian. [Ety. proper name.]
  bockschi, Gœppert, 1836, (Adiantites bochschii) Syst. Filic. Foss. See Nœggerathia bockschi.
  crispa, Germ. & Kaulf, 1831, (Filicites crispa) Nova. Acta. Acad., vol. 15, Coal Meas. [Sig. wavy.]
  elegans, Lesquereux, 1858, Geo. Sur. Penn.,vol. 2, Coal Meas. [Sig. elegant.]
  fimbriata, see Neuropteris fimbriata.
  flabellata, Brongniart, 1828, Prodr. Hist. Veg. Foss., Coal Meas. [Sig. like a fan.]
  germari, Gutbier, 1835, Verst. Zwick. Schwarzk., Coal Meas. [Ety. proper name.]
  hallana, Gœppert, 1836, Syst. Filic. Foss., Chemung Gr. [Ety. proper name.]
  hispida, Dawson, 1863, Can. Nat. & Geol., vol. 8, Coal Meas. [Sig. rough, hairy.]
  hirsuta, Lesquereux,1858,Geo.Sur.Penn., vol. 2, Coal Meas. [Sig. hairy.]
  incerta, Dawson, 1862, Quar. Jour. Geo. Soc.,vol. 18, Ham.Gr. [Sig. uncertain.]
  jacksoni, Dawson, 1861, Can. Nat., vol. 6, Catskill Gr. [Ety. proper name.]
  laciniata, Lesquereux, 1858, Geo. Sur. Penn.,vol. 2,Coal Meas. [Sig. fringed.]
  lescuriana, Meek, 1876, Desc. Foss. Plants Va., Ponent or Vespertine Gr., near base of Low.Carb. [Ety. proper name.]
  oblata, Lindley & Hutton, 1837, Foss. Flora, vol. 3, Coal Meas. [Sig. oblate.]
  obliqua, Brongniart, 1828, Prodr. Hist. Veg. Foss.,Coal Meas. [Sig. oblique.]
  obtusa, Lesquereux, 1858, (Nœggerathia obtusa) Geo. Sur. Penn., vol. 2, Devonian. See Archæopteris obtusa.
  orbicularis, Brongniart, 1828, Prodr. Hist. Veg.Foss.,Coal Meas. [Sig. orbicular.]
  rogersi, Dawson, 1863, Quar. Jour. Geol. Soc., vol. 19, Devonian. [Ety. proper name.]
  trichomanoides, Brongniart, 1828, Hist. Veg. Foss.,Coal Meas. [Ety. *Trichomanes*, the maiden-hair fern; *eidos*, form.]
  undans, Lesquereux, 1858, Geo. Sur. Penn., vol. 2, Coal Meas. [Sig. waved.]
  valida, Dawson, 1862,Quar. Jour.Geo.Soc., vol. 18, Devonian. [Sig. sound, strong.]
  varia, Dawson, 1862, Quar. Jour.Geo. Soc., vol. 18, Devonian. [Sig. changing.]
  virginiana, Meek, 1876, Desc. Foss. Plants Va., Ponent or Vespertine Gr., near base of Low.Carb. [Ety. proper name.]
  wilsoni, Wood, 1860, Proc. Acad. Nat. Sci., Coal Meas. [Ety. proper name.]
CYCLOSTIGMA, Haughton, 1860, Ann. & Mag. Nat. Hist., 3d series, vol. 5. [Ety. *kuklos*, a circle; *stigma*, a dot or puncture.]
  densifolium, Dawson, 1871, Foss. Plants Can., Devonian. [Sig. dense-leaved.]
DADOXYLON, Endlicher, 1840, Syn. Con. [Sig. pine or torch-wood.]
  acadianum, Dawson, 1863, Can. Nat., vol. 8, Coal Meas. [Ety. proper name.]
  annulatum, Dawson, 1863, Can. Nat., vol. 8, Coal Meas. [Sig. annulated.]
  antiquum, Dawson, 1863, Can. Nat., vol. 8, Coal Meas. [Sig. ancient.]
  halli, Dawson, 1862, Quar. Jour. Geo. Soc., vol. 18, Ham. Gr. [Ety. proper name.]
  materiarium, Dawson, 1863, Can. Nat., vol. 8, Up. Coal Meas. [Sig. belonging to wood.]
  newberryi, Dawson, 1871, Foss. Plants Can., Devonian. [Ety. proper name.]
  ouangondianum, Dawson,1861,Can. Nat., vol. 6, Devonian. [Ety. proper name.]
DANÆITES, Gœppert, 1836, Syst. Filic. Foss. [Ety. proper name.]
  asplenioides *var.* major, Bunbury, 1846, Quar. Jour. Geol. Soc., vol. 2, Coal Meas. [Sig. large.]
Dictyolites, see Dictyophyton.
  becki, see Dictyophyton becki.
DICTYOPHYTON, Hall, 1863, 16th Reg. Rep. [Ety. *dictyon*, a net; *phyton*, a plant.]
  annulatum, Hall, 1863, 16th Reg. Rep., Chemung Gr. [Sig. ringed.]
  becki, Conradi, 1837,(Lithodictuon becki) Ann. Rep. N. Y., Medina sandstone. [Ety. proper name.]
  conradi, Hall, 1863, 16th Reg. Rep., Chemung Gr. [Ety. proper name.]
  fenestratum, Hall, 1863, 16th Reg. Rep., Chemung Gr. [Sig. reticulated.]
  filitextile, Hall,1863, 16th Reg. Rep., Chemung Gr. [Sig. woven like threads.]
  newberryi, Hall, 1863, 16th Reg. Rep., Portage Gr. [Ety. proper name.]
  nodosum, Hall, 1863, 16th Reg. Rep., Chemung Gr. [Sig. knotted.]
  redfieldi, Hall, 1863, 16th Reg. Rep., Portage Gr. [Ety. proper name.]
  rude, Hall,1863,18th Rep. Rep.,Chemung Gr. [Sig. rude.]
  tuberosum, Conrad, 1842, (*Hydnoceras tuberosum*) Jour. Acad. Nat. Sci. Phil., vol. 8, Chemung Gr. [Sig. composed of tuber-like parts.]
DICTYOPTERIS, Gutbier, 1835, Verst. Zwick. Schwarzk. [Ety. *dictyon*, a net; *pteris*, a fern.]

neuropteroides, Gutbier, 1852, Verst. Stein Sachs., Coal Meas. [Sig. like *Neuropteris*.]
obliqua, Bunbury, 1847, Quar. Jour. Geo. Soc.,vol. 3, Coal Meas. [Sig. oblique.]
rubella, Lesquereux, 1870, Geo. Sur. Ill., vol. 4,Coal Meas. [Sig. somewhat red.]
DIDYMOPHYLLUM, Gœppert, 1841, Gatt. der Foss. Pflanzen. [Ety. *didymos*, double; *phyllon*, a leaf.]
reniforme, Dawson, 1862, Quar. Jour. Geo. Soc., vol. 18, Ham. Gr. [Sig. kidney-shaped.] [nata.
Diplazites emarginatus see Pecopteris emargi-
DIPLOSTEGIUM, Corda, 1845, Beiträge zur Flora der Vorwelt. [Ety. *diplos*, double; *stege*, a covering.]
brownanum, Corda, 1845, Beiträge zur Flora der Vorwelt, Coal Meas. [Ety. proper name.]
retusum, Dawson, 1863, Can. Nat., vol. 8, Coal Meas. [Sig. turned back.]
truncatum, Lesquereux, 1860, Geo. Sur. Ark., vol. 2, Coal Meas. [Sig. truncated, cut off.]
EOPHYTON, Torell. [Ety. *eos*, dawn; *phyton*, a plant.]
jukesi, Billings, 1874, Pal. Foss., vol. 2, Potsdam Gr. [Ety. proper name.]
linnæanum (?) Torell, Potsdam Gr. [Ety. proper name.]
explanatum, Dawson, 1870, Can. Nat. & Geol., Low. Arenig rocks. [Sig. spread out.]
EQUISETITES, Sternberg, 1833, Vers. Darst. Flora der Vorwelt. [Ety.*Equus*, a horse; *seta*, a hair or bristle; in allusion to the resemblance'to a horse-tail.]
columnaris, Brongniart, 1828, (Equisetum columnaris) Hist. Veg. Foss., Coal Meas. [Sig. columnar.]
curtus, Dawson, 1863, Syn. Carb. Flora in Can. Nat.,vol. 8,Coal Meas. [Sig. short.]
occidentalis, Lesquereux, 1870, Geo. Sur. Ill., vol. 4, Coal Meas. [Sig. western.]
stellifolius, Harlan, 1835, (Equisetum stellifolium) Trans. Geo. Soc. Penn., Coal Meas. [Ety. *stella*, a star, and *folium*, a leaf; in allusion to the whorled leaves.]
Equisetum, see Equisetites.
columnare, see Equisetites columnaris.
stellifolium, see Equisetites stellifolius.
EREMOPTERIS, Schimper, 1869, Traite de Palæontologie Vegetale. [Ety. *eremos*, solitary; *pteris*, a fern.]
artemisiæfolia, Sternberg, 1824, (Sphenopteris artemisiæfolia) Vers. Darst. Flora der Vorwelt, Coal Meas. [Sig. leaved like the plant *Artemisia*.]
marginata, Andrews, 1875, Ohio Pal., vol. 2, Coal Meas. [Sig. bordered.]
Filicites crispa, see Cyclopteris crispa.
lonchiticus, see Alethopteris lonchitidis.
pennæformis, see Pecopteris pennæformis.
pluckeneti, see Alethopteris pluckeneti.
Flabellaria borassifolia, see Cordaites borassifolius.

FUCOIDES, Brongniart, 1822, in Memoires de la Soc. d'Hist. Nat. de Paris. [Ety. *fucus*, sea weed; *eidos*, form.]
alleghanensis, Harlan, 1830, Jour. Acad. Nat. Sci., vol. 6, Carb. See Harlania halli.
auriformis, Hall, 1843, Geo. Rep. 4th Dist. N. Y., Medina and Clinton Gr. Prof. Hall says, in 1852, that this can scarcely be referred to the organic remains.
bilobata, see Rusophycus bilobatus.
brongniarti, see Harlania halli.
cauda-galli, see Spirophyton cauda-galli.
dennisus,Conrad, 1838, probably Phytopsis tubulosa.
dentatus, M. Brongniart; see Diplograptus pristiniformis.
flexuosa, see Buthotrephis flexuosa.
gracilis, see Buthotrephis gracilis.
graphica, Hall, 1843, Geo. Rep. 4th Dist. N. Y., Portage Gr. [Sig. written.]
harlani, see Arthrophycus harlani.
heterophyllus, Hall, 1843, Geo. Rep. 4th Dist. N. Y., Medina and Clinton Gr. Not likely the remains of an organic substance.
retort, Vanuxem, 1843, Geo. Rep. 3rd Dist. N. Y., Portage Gr. [Sig. a retort.]
rigidus, syn. for Buthotrephis flexuosa.
secalinus, see Graptolithus and Diplograptus secalinus.
serra, Brongniart, see Graptolithus bryonoides.
simplex, Emmons, see Graptolithus and Diplograptus secalinus.
velum, see Spirophyton velum.
verticalis, Hall, 1843, Geo. Rep. 4th Dist. N. Y., Portage Gr. [Sig. vertical.]
Galium sphenophylloides, see Annularia sphenophylloides.
Gordia marina, Hall, 1847, Pal. N. Y., vol. 1. Not the remains of any organic form; it is described as a track, see *Helminthoidichnites*.
HALONIA, Lindley & Hutton, 1835, Foss. Flora. [Ety. from its close affinity with *Halonia*.]
pulchella, Lesquereux, 1860, Geo. Sur. Ark., vol. 2,Coal Meas. [Sig. beautiful.]
tuberculata, Brongniart, 1828, Hist. Veg. Foss., Coal Meas. [Sig. tuberculated.]
HARLANIA, Gœppert, 1852, Foss..Flora des Ueberg. [Ety. proper name.] Syn. for Arthrophycus.
halli, Gœppert, 1852, Foss. Flora des Ueberg, (Fucoides alleghaniensis, F. brongniarti, and F. harlani) Medina Gr. to Low. Carb. [Ety. proper name.] Syn. for Arthrophycus harlani.
HIPPODOPHYCUS, Hall, 1872, 24th Reg. Rep. [Ety. *hippodos*, horse-foot; *phukos*, a sea plant.]
cowlesi, Hall, 1872, 24th Reg. Rep., Chemung Gr. [Ety. proper name.]
HYMENOPHYLLITES, Gœppert, 1836, Syst. Filic. Foss. [Ety.*hymen*, a membrane; *phyllon*, a leaf.]

adnascens, Lindley & Hutton, 1835, Foss. Flora, vol. 2, Coal Meas. [Sig. growing upon.]
alatus, Brongniart, 1828, Hist. Veg. Foss. (Sphenopteris alata) Coal Meas. [Sig. winged.]
affinis, Lesquereux, 1858, Geo. Sur. Penn., vol. 2, Coal Meas. [Sig. near to.]
arborescens, Lesquereux, 1870, Geo. Sur. Ill., vol. 4, Coal Meas. [Sig. branching like a tree.]
ballantini, Andrews, 1875, Ohio Pal., vol. 2, Coal Meas. [Ety. proper name.]
capillaris, Lesquereux, 1858, Geo. Sur. Penn., vol. 2, Coal Meas. [Ety. *capillus*, hair.]
clarki, Lesquereux, 1866, Geo. Sur., Ill. vol. 2, Coal Meas. [Ety. proper name.]
curtilobus, Dawson, 1862, Quar. Jour. Geol. Soc., vol. 18, Devonian. [Sig. short-lobed.]
delicatulus, Brongniart, 1828, Hist. Veg. Foss., Coal Meas. [Sig. delicate.]
fimbriatus, Lesquereux, 1858, Geo. Sur. Penn., vol. 2, Coal Meas. [Sig. fringed.]
flexicaulis, Lesquereux, 1860, Geo. Sur. Ark., vol. 2, Coal Meas. [Sig. flexible.]
furcatus, Brongniart, 1828, (Sphenopteris furcata) Hist. Veg. Foss., Coal Meas. [Sig. forked.]
gersdorfi, Gœppert, 1836, Syst. Filic. Foss. Devonian. [Ety. proper name.]
giganteus, see Hymenophyllites lactuca.
gutbieranus, Unger, 1850, Gen. et. sp., Coal Meas. [Ety. proper name.]
hildrethi, Lesquereux, 1861,Geo. Sur. Ky., vol. 4, Coal Meas. [Ety. proper name.]
inflatus, Lesquereux, 1870, Geo. Sur. Ill., vol. 4, Coal Meas. [Sig. inflated.]
lactuca, Sternberg, 1833, (Schizopteris lactuca) Vers. Darst. Flora der Vorwelt, Coal Meas. [Sig. the plant Lettuce.]
mollis, Lesquereux, 1870, Geo. Sur. Ill., vol. 4, Coal Meas. [Sig. flexible.]
myriophyllus, Brongniart, 1828, (Sphenopteris myriophylla) Prodr. Hist. Veg. Foss., Coal Meas. [Sig. many-leaved.]
pentadactylus, Dawson, 1863, Can. Nat. & Geol., vol. 8, Coal Meas. [Sig. five-fingered.]
pinnatifidus, Lesquereux, 1866, Geo. Sur. Ill., vol. 2, Coal Meas. [Ety. *pinnatus*, winged; *fidus*, cleft.]
schlotheimi, Brongniart, 1828, (Sphenopteris schlotheimi) Hist. Veg. Foss., Coal Meas. [Ety. proper name.]
spinosus, Gœppert, 1841, (Sphenopteris spinosa) Gatt. Foss. Pflanzen, Coal Meas. [Sig. spiny.]
splendens, Lesquereux, 1870, Geo. Sur. Ill., vol. 4, Coal Meas. [Sig. splendid.]
strongi, Lesquereux, 1870, Geo. Sur. Ill., vol. 4, Coal Meas. [Ety. proper name.]
subfurcatus, Dawson, 1868, Acad. Geol., Devonian. [Ety. from resemblance to H. *furcatus*.]
tenuifolius, Brongniart, 1828, (Sphenopteris tenuifolia) Hist. Veg. Foss., Coal Meas. [Sig. slender-leaved.]
thallyformis, Lesquereux, 1870, Geo. Sur. Ill., vol. 4, Coal Meas. [Sig. frond-like.]
trichomanoides, Brongniart, 1828,(Sphenopteris trichomanoides) Hist. Veg. Foss., Coal Meas. [Ety. from resemblance to *Trichomanes*, or maiden-hair.]
tridactylites, Brongniart, 1828, (Sphenopteris tridactylites) Hist. Veg. Foss., Coal Meas. [Sig. three-fingered.]
ICHNOPHYCUS, Hall, 1852, Pal. N. Y., vol. 2. [Ety. *ichnos*, a foot print; *phukos*, a sea weed.]
tridactylus, Hall, 1852, Pal. N. Y., vol. 2, Clinton Gr. [Sig. three-fingered.]
KNORRIA, Sternberg, 1825, Vers. Darst. Flora der Vorwelt. [Ety. proper name.]
imbricata, Sternberg, 1825, Vers. Darst. Flora der Vorwelt, Chester Gr. [Sig. imbricated.]
selloni, Sternberg, 1825, Vers. Darst. Flora der Vorwelt, Coal Meas. [Ety. proper name.]
taxina, Lindley & Hutton, 1833-5, Foss. Flora, Coal Meas. [Sig. like the Yew tree.]
LEPIDODENDRON, Sternberg, 1821,Vers. Darst. Flora der Vorwelt. [Ety. *lepis*, a scale; *dendron*, a tree.]
aculeatum, Sternberg, 1825, Vers. Darst. Flora der Vorwelt, Coal Meas. [Sig. very sharp or needle-like.]
binerve, Bunbury, 1847, Quar. Jour. Geo. Soc., vol. 3, Coal Meas. [Sig. double nerved.]
bordæ, Wood, 1860, Proc. Acad. Nat. Sci., Coal Meas. [Ety. proper name.]
carinatum, Lesquereux, 1858, Geo. Sur. Penn., vol. 2, Coal Meas. [Sig. the margins of the scars are keeled.]
chemungense, Hall, 1843, Geo. Rep. 4th Dist. N. Y., Chemung Gr. [Ety. proper name.]
chilallœum, Wood, 1860, Proc. Acad. Nat. Sci., Coal Meas. [Ety. *chilos*, fodder; *alloios*, of another kind.]
clypeatum, Lesquereux, 1858, Geo. Sur. Penn.,vol. 2, Coal Meas. [Sig. a shield.]
conicum, Lesquereux, 1858, Geo. Sur. Penn., vol. 2, Coal Meas. [Ety. from the conical scars.]
corrugatum, Dawson, 1862, Quar. Jour. Geo. Soc., vol. 18, Devonian and Low. Carb. [Sig. corrugated.]
costatum, Lesquereux, 1866, Geo. Sur. Ill., vol. 2, Chester Gr. [Sig. ribbed.]
crenatum, Sternberg, 1820, Vers. Darst. Flora der Vorwelt, Coal Meas. [Sig. crenate.]
crucinatum, Lesquereux, 1870, Geo. Sur. Ill., vol. 4, Coal Meas. [Sig. tormented, twisted.]
decurtatum, Dawson, 1863, Can. Nat., vol. 8, Coal Meas. [Sig. curtailed.]

dichotomum, Sternberg, 1820, Vers. Darst.
Flora der Vorwelt, Coal Meas. [Sig. dividing into two.]
dikrocheilum, Wood, 1860, Proc. Acad. Nat. Sci., Coal Meas. [Ety. two-edged.]
dilatatum, Lindley & Hutton, 1833, Foss. Flora, Coal Meas. [Sig. widened.]
diplostegiodes, Lesquereux, 1860, Geo. Sur. Ark., vol. 2, Coal Meas. [Sig. resembling *Diplostegium*.]
distans, Lesquereux, 1858, Geo. Sur. Penn., vol. 2, Coal Meas. [Ety. from the distant scars.]
elegans, Brongniart, 1828, Hist. Veg. Foss., Coal Meas. [Sig. elegant.]
forulatum, Lesquereux, 1870, Geo. Sur. Ill., vol. 4, Coal Meas. [Sig. having long, narrow furrows.]
gaspanum, Dawson, 1860, Can. Nat. & Geo., vol. 5, Catskill Gr. [Ety. proper name.]
giganteum, Lesquereux, 1858, Geo. Sur. Penn., vol. 2, Coal Meas. [Sig. very large.]
gracile, Lindley & Hutton, 1833, Foss. Flora, Coal Meas. [Sig. slender.]
groeni, Lesquereux, 1870, Geo. Sur. Ill., vol. 4, Coal Meas. [Ety. proper name.]
harcourti, Witham, 1832, Trans. Nat. Hist. Soc., New, upon Tyne, Coal Meas. [Ety. proper name.]
mammillatum, Lesquereux, 1870, Geo. Sur. Ill., vol. 4, Coal Meas. [Sig. mammillated.]
mielecki, Gœppert, 1836, Syst. Filic. Foss., Coal Meas. [Ety. proper name.]
modulatum, Lesquereux, 1860, Geo. Sur. Ark., vol. 2, Coal Meas. [Sig. modulated.]
morrisanum, Lesquereux, 1870, Geo. Sur. Ill., vol. 4, Coal Meas. [Ety. proper name.]
obovatum, Sternberg, 1820, Vers. Darst. Flora der Vorwelt, Coal Meas. [Ety. from the obovate scars.]
obscurum, Lesquereux, 1866, Geo. Sur. Ill., vol. 2, Coal Meas. [Sig. obscure, not distinct.]
obtusum, Lesquereux, 1858, Geo. Sur. Penn., vol. 2, Coal Meas. [Ety. from the obtuse bases of the scars.]
oculatum, Lesquereux, 1858, Geo. Sur. Penn., vol. 2, Coal Meas. [Sig. having eyes.]
personatum, Dawson, 1863, Can. Nat. & Geo.,vol. 8, Coal Meas. [Sig. masked.]
pictoense, Dawson, 1863,Can. Nat. & Geo., vol. 8, Coal Meas. [Ety. proper name.]
plicatum, Dawson, 1863, Can. Nat. & Geo., vol. 8, Coal Meas. [Sig. folded.]
plumarium, Lindley & Hutton, 1835, Foss. Flora, vol. 2, Coal Meas. [Sig. embroidered with feathers.]
primævum, Rogers, 1858, Geo. Sur. Penn., vol. 2, Ham. Gr. [Sig. first formed.]
radiato-plicatum, Dawson, 1873, Rep. Foss. Plants, Low. Carb. [Ety. *radiatus*, radiated; *plicatus*, folded.]

radicans, Lesquereux, 1866, Geo. Sur. Ill., vol. 2, Coal Meas. [Sig. rooted.]
rectangulum, Wood, 1860, Proc. Acad. Nat.Sci.,Coal Meas. [Sig. rectangular.]
rigens, Lesquereux, 1870, Geo. Sur. Ill., vol. 4, Coal Meas. [Sig. stiff.]
rimosum, Sternberg, 1820, Vers. Darst. Flora der Vorwelt, Coal Meas. [Sig. full of clefts.]
rugosum, Brongniart, 1828, Prodr. Hist. Veg. Foss., Coal Meas. [Sig.wrinkled.]
rushvillense, Andrews, 1875, Ohio Pal., vol. 2, Coal Meas. [Ety. proper name.]
salebrosum, Wood, 1860, Proc. Acad. Nat. Sci., Coal Meas. [Sig. rough.]
scobiniforme, Meek, 1870, Desc. Foss. Plants Va., Ponent or Vespertine Gr., near the base of the Low. Carb. [Sig. rasp-like.]
selaginoides, Sternberg, 1824, Vers. Darst. Flora der Vorwelt, Coal Meas. [Ety. from the plant *Selago*.]
sigillarioides, Lesquereux, 1858, Geo. Sur. Penn., vol. 2, Coal Meas. [Ety. from its resemblance to *Sigillaria*.]
simplex, Lesquereux, 1860, Geo. Sur. Ill., vol. 2, Coal Meas. [Sig. simple.]
sternbergi, Lindley & Hutton, 1835, Foss. Flora, Coal Meas. [Ety. proper name.]
tijoui, Lesquereux, 1870, Geo. Sur. Ill., vol. 4, Coal Meas. [Ety. proper name.]
tumidum, Bunbury, 1847, (Lepidophloios tumidum) Quar. Jour. Geo. Soc., vol. 3, Coal Meas. [Sig. tumid.]
turbinatum, Lesquereux, 1866, Geo. Sur. Ill., vol. 2, Coal Meas. [Sig. top-shaped.]
undulatum, Gutbier, 1843, in Gæa von Sachsen, Coal Meas. [Sig. wavy.]
ureum,Wood, 1860, Proc. Acad. Nat. Sci., Coal Meas. [Sig. belonging to the tail.]
veltheimanum, Sternberg, 1823, Vers. Darst. Flora der Vorwelt, Chester Gr. [Ety. proper name.]
vestitum, Lesquereux, 1858, Geo. Sur. Penn., vol. 2, Coal Meas. [Sig. clothed, covered.]
wortheni, Lesquereux, 1866, Geo. Sur. Ill., vol. 2, Coal Meas. [Ety. proper name.]
LEPTOPHLÆUM, Dawson, 1862, Quar. Jour. Geo. Soc., vol. 18. [Ety. *leptos*, slender; *phlois*, the bark of a tree.]
rhombicum, Dawson, 1862, Quar. Jour. Geo. Soc., vol. 18, Devonian. [Sig. rhomb-like.]
LEPIDOPHLOIOS, Sternberg, 1823, Vers. Darst. Flora der Vorwelt. [Ety. *lepis*, a scale; *phloios*, the bark.]
acadianus, Dawson, 1863, Can. Nat. & Geo., vol. 8, Coal Meas. [Ety. proper name.]
antiquus, Dawson, 1871, Foss. Plants Canada, Devonian. [Sig. ancient.]
auriculatus, Lesquereux, 1870, Geo. Sur. Ill., vol. 4, Coal Meas. [Sig. ear-like.]
crassicaulis, Corda, 1833, in Flora der Vorwelt, vol. 2, Coal Meas. [Sig. thick-stemmed.]

PLANTÆ.

ichthyolepis, Wood, 1860, Proc. Acad. Nat. Sci., Coal Meas. [Ety. *ichthys*, a fish; *lepis*, a scale.]
irregularis, Lesquereux, 1860, Geo. Sur. Ark., vol. 2, Coal Meas. [Sig. irregular.]
laricinus, Sternberg, 1823, Vers. Darst. Flora der Vorwelt, Coal Meas. [Ety. from resemblance to the larch tree.]
lesquereuxi, Andrews, 1875, Ohio Pal. vol. 2, Coal Meas. [Ety. proper name.]
obcordatus, Lesquereux, 1860, Geo. Sur. Ill., vol. 2, Coal Meas. [Sig. inversely, heart-shaped.]
parvus, Dawson, 1863, Can. Nat. & Geo., vol. 8, Coal Meas. [Sig. small.]
platystigma, Dawson, 1863, Can. Nat. & Geo., vol. 8, Coal Meas. [Sig. flat-scarred.]
prominulus, Dawson, 1863, Can. Nat. & Geo., vol. 8, Coal Meas. [Sig. projecting a little.]
protuberans, Lesquereux, 1870, Geo. Sur. Ill., vol. 4, Coal Meas. [Sig. protuberant.]
tetragonus, Dawson, 1863, Can. Nat. & Geo., vol. 8, Coal Meas. [Sig. quadrangular.]
tumidus, see Lepidodendron tumidum.
LEPIDOPHYLLUM, Brongniart, 1828, Prodr. Hist. Veg. Foss. [Ety. *lepis*, a scale; *phyllon*, a leaf.]
acuminatum, Lesquereux, 1858, Geo. Sur. Penn., vol. 2, Coal Meas. [Sig. sharp-pointed.] The name was preoccupied by Gutbier in 1843.
affine, Lesquereux, 1858, Geo. Sur. Penn., vol. 2, Coal Meas. [Sig. closely related.]
auriculatum, Lesquereux, 1866, Geo. Sur. Ill., vol. 2, Coal Meas. [Sig. ear-shaped.]
brevifolium, Lesquereux, 1858, Geo. Sur. Penn., Coal Meas. [Ety. *brevis*, short; *folium*, a leaf.]
foliaceum, Lesquereux, 1870, Geo. Sur. Ill., vol. 4, Coal Meas. [Sig. leaf-like.]
hastatum, Lesquereux, 1858, Geo. Sur. Penn., vol. 2, Coal Meas. [Sig. halbert-shaped.]
intermedium, Lindley & Hutton, 1833, Foss. Flora, vol. 1, Coal Meas. [Sig. intermediate.]
lanceolatum, Lindley & Hutton, 1831-33, Foss. Flora, Coal Meas. [Sig. lanceolate.]
majum, Brongniart, 1828, Prodrome d'une Histoire de Vegetaux, Fossiles, Coal Meas. [Sig. large.]
obtusum, Lesquereux, 1858, Geo. Sur. Penn., vol. 2, Coal Meas. [Sig. obtuse.]
plicatum, Lesquereux, 1858, Geo. Sur. Penn., vol. 2, Coal Meas. [Sig. folded.]
rostellatum, Lesquereux, 1870, Geo. Sur. Ill., vol. 4, Coal Meas. [Sig. little-beaked.]
striatum, Lesquereux, 1870, Geo. Sur. Ill., vol. 4, Coal Meas. [Sig. striated.]
trinerve, Lindley & Hutton, 1835, Foss. Flora, vol. 2, Coal Meas. [Sig. three-veined.]

LEPIDOSTROBUS, Brongniart, 1828, Prodr. Hist. Veg. Foss. [Ety. *lepis*, a scale; *strobus*, a cone.]
connivens, Lesquereux, 1870, Geo. Sur. Ill., vol. 4, Coal Meas. [Sig. dissembling, closing.]
globosus, Dawson, 1861, Can. Nat. & Geo., vol. 6, Devonian. [Sig. globose.]
hastifolius, Lesquereux, 1866, Geo. Sur. Ill., vol. 2, Coal Meas. [Sig. spear-leaved.]
lancifolius, Lesquereux, 1870, Geo. Sur. Ill., vol. 4, Coal Meas. [Sig. lance-leaved.]
longifolius, Dawson, 1863, Can. Nat. & Geo., vol. 8, Coal Meas. [Sig. long-leaved.]
oblongifolius, Lesquereux, 1870, Geo. Sur. Ill., vol. 4, Coal Meas. [Sig. oblong-leaved.]
ornatus, Parkinson, 1811, Organic Remains, Coal Meas. [Sig. ornate.]
ovatifolius, Lesquereux, 1870, Geo. Sur. Ill., vol. 4, Coal Meas. [Sig. ovate-leaved.]
pinaster, Lindley & Hutton, 1837, Foss. Flora, vol. 3, Coal Meas. [Sig. like the cone of a *Pinaster*.]
princeps, Lesquereux, 1866, Geo. Sur. Ill., vol. 2, Coal Meas. [Sig. original, principal.]
richardsoni, Dawson, 1861, Can. Nat. & Geo., vol. 6, Devonian. [Ety. proper name.]
squamosus, Dawson, 1863, Can. Nat. & Geo., vol. 8, Coal Meas. [Sig. scaly.]
trigonolepis, Bunbury, 1847, Quar. Jour. Geo. Soc., vol. 3, Coal Meas. [Ety. from the triangular scars.]
truncatus, Lesquereux, 1870, Geo. Sur. Ill., vol. 4, Coal Meas. [Sig. truncated.]
variabilis, Lindley & Hutton, 1833, Foss. Flora, Coal Meas. [Sig. variable.]
LESCUROPTERIS, Schimper, 1869, Palæontologie Vegetale. [Ety. proper name; *pteris*, a fern.]
moori, Lesquereux, 1858, (Neuropteris moori) Geo. Sur. Penn., vol. 2, Coal Meas. [Ety. proper name.]
LICROPHYCUS, Billings, 1862, Pal. Foss., vol. 1. [Ety. *likros*, a fan; *phykos*, sea weed.]
formosus, Billings, 1866, Cntal. Sil. Foss. Antic., Hud. Riv. Gr. [Sig. beautiful.]
hiltonensis, Billings, 1862, Pal. Foss., vol. 1, Black Riv. & Trenton Gr. [Ety. proper name.]
hudsonicus, Billings, 1862, Pal. Foss., vol. 1, Hud. Riv. Gr. [Ety. proper name.]
minor, Billings, 1862, Pal. Foss., vol. 2, Trenton Gr. [Ety. *minor*, less; it is smaller than *ottawensis*.]
ottawensis, Billings, 1862, Pal. Foss., vol. 1, Trenton Gr. [Ety. proper name.]
robustus, Billings, 1866, Catal. Sil. Foss. Antic., Hud. Riv. Gr. [Sig. robust.]
vagans, Billings, 1866, Catal. Sil. Foss. Antic., Hud. Riv. Gr. [Sig. wandering.]
LONCHOPTERIS, Brongniart, 1828, Prodr. Hist. Veg. Foss. [Ety. *lonche*, a spear; *pteris*, fern.]

tenuis, Dawson, 1863, Can. Nat. & Geol., vol. 8, Coal Meas. [Sig. slender.]
LYCOPODITES, Brongniart, 1822, Mem. du Mus. d'Hist. Nat. de Paris. [Ety. from *Lycopodium*, the club moss.]
annulariæfolius, Lesquereux, 1870, Geo. Sur. Ill., vol. 4, Coal Meas. [Sig. with ring-shaped leaves.]
asterophyllitæfolius, Lesquereux, 1866, Geo. Sur. Ill., vol. 2, Coal Meas. [Ety. leaved like the *Asterophyllites*.]
carifolius, see Selaginites cavifolius.
comosus, Dawson, 1863, Quar. Jour. Geo. Soc., vol. 19, Devonian. [Sig. hairy.]
matthewi, Dawson, 1861, Can. Nat. & Geo., vol. 6, Devonian. [Ety. proper name.]
mecki, Lesquereux, 1870, Geo. Sur. Ill. vol. 4, Coal Meas. [Ety. proper name.]
plumulus, Dawson, 1873,Rep.Foss. Plants, Low. Carb. [Sig. a little feather.]
richardsoni, Dawson, 1868, Quar. Jour. Geo. Soc., vol. 19, Devonian. [Ety. proper name.]
vanuxemi, Dawson, 1862, Quar. Jour. Geo. Soc., vol. 18, Chemung Gr. [Ety. proper name.]
MEGALOPTERIS, Dawson, 1875. Not defined. [Ety. *megale*, great; *pteris*, a fern.]
dawsoni, Hartt, 1868, (Neuropteris dawsoni) Acad. Geol., Devonian. [Ety. proper name.]
hartti, Andrews, 1875, Ohio Pal., vol. 2, Coal Meas. [Ety. proper name.]
lata, Andrews, 1875, Ohio Pal., vol. 2, Coal Meas. [Sig. wide.]
minima, Andrews, 1875, Ohio Pal., vol. 2, Coal Meas. [Sig. very small.]
ovata, Andrews, 1875, Ohio Pal., vol. 2, Coal Meas. [Sig. egg-shaped.]
MEGAPHYTON, Artis, 1838, Antedil. Phytol. [Ety. *megas*, great; *phyton*, a shoot.]
humile, Dawson, 1863, Can. Nat. & Geol., vol. 8,Coal Meas. [Sig. small, humble.]
maclayi, Lesquereux, 1860, Geo. Sur. Ill., vol. 2, Coal Meas. [Ety. proper name.]
magnificum, Dawson, 1863, Can. Nat. & Geol.,vol. 8,Coal Meas. [Sig. splendid.]
protuberans, Lesquereux, 1866, Geo. Sur. Ill., vol. 2, Chester Gr. [Sig. protuberant.]
MYRIAXITES, Murchison, 1839, Sil. Syst. [Ety. *myrias*, innumerable.]
murchisoni, Emmons, 1844,Taconic. Syst., Potsdam Gr. [Ety. proper name.]
sillimani, Emmons, 1844, Taconic. Syst., Potsdam Gr. [Ety. proper name.]
NEMAPODIA, Emmons, 1844, Taconic. Syst., [Ety. *nema*, thread ; *podion*, foot.]
tenuissima, Emmons, 1844, Taconic. Syst., Potsdam Gr. [Sig. fine-lined.] Prof. Hall regards this as a recent track.]
NEMATOPHYCUS, Carruthers, 1872, Month. Micro. Jour. [Ety. *nematos*, a thread ; *phukos*, sea weed.] This is a syn. for Prototaxites.
logani, Carruthers, 1872, Month. Micro. Jour., Devonian. [Ety. proper name.] Syn. for Prototaxites logani.

NEMATOXYLON, Dawson, 1863, Quar. Jour. Geo. Soc., vol. 19. [Ety. *nema*, a thread; *xylon*, wood.] Carruthers says this is a syn. for Prototaxites and belongs to the *Algæ*.
crassum, Dawson, 1863, Quar. Jour. Geo. Soc., vol. 19, Devonian. [Sig. thick.] Carruthers says this is a syn. for Prototaxites logani.]
tenue, Dawson, 1863, Quar. Jour. Geo. Soc., vol. 19, Devonian. [Sig. slender.] Carruthers says this is a syn. for Prototaxites logani.]
Nephropteris, Brongniart,1828, Tab.des gener.
elegans, see Cyclopteris elegans.
fimbriata, see Neuropteris fimbriata.
germari, see Cyclopteris germari.
hirsuta, see Cyclopteris hirsuta.
laciniata, see Cyclopteris laciniata.
orbicularis, see Cyclopteris orbicularis.
trichomanoides, see Cyclopteris trichomanoides.
undans, see Cyclopteris undans.
NEREITES, Murchison, 1839, Sil. Syst. [Ety. from a resemblance to the track of the *Nereis*.] Prof. Hall says these species are of Devonian age.
deweyi, Emmons, 1844, Taconic. Syst., Potsdam Gr. [Ety. proper name.]
gracilis, Emmons, 1844, Taconic. Syst., Potsdam Gr. [Sig. slender.]
jacksoni, Emmons, 1844, Taconic. Syst., Potsdam Gr. [Ety. proper name.]
lanceolatus, Emmons,1844,Taconic. Syst., Potsdam Gr. [Sig. sword-like.]
loomisi, Emmons, 1844, Taconic Syst., Potsdam Gr. [Ety. proper name.]
pugnus, Emmons, 1844, Taconic. Syst., Potsdam Gr. [Sig. a hand full.]
NERIOPTERIS, Newberry,1873, Ohio Pal., vol. 1. [Ety. *nerion*, the Oleander ; *pteris*, a fern.]
lanceolata, Newberry, 1873, Ohio Pal.,vol. 1, Coal Meas. [Sig. spear-shaped.]
NEUROPTERIS, Brongniart, 1822, Mem. du Mus. d'Hist. Nat. de Paris. [Ety. *neuron*, a nerve; *pteris*, a fern.]
acutifolia, Brongniart, 1828, Hist. Veg. Foss., Coal Meas. [Sig. acute-leaved.]
adiantites, Lesquereux, 1858, Geo. Sur. Penn., vol. 2, Coal Meas. [Ety. from resemblance to *Adiantum*, the maiden hair fern.]
angustifolia, Brongniart, 1828, Hist. Veg. Foss., Coal Meas. [Sig.narrow-leaved.]
attenuata, Lindley & Hutton, 1837, Foss. Flora, vol. 3, Coal Meas. [Sig. attenuated.]
auriculata, Brongniart, 1828, Hist. Veg. Foss., Coal Meas. [Sig. eared.]
capitata, Lesquereux, 1870, Geo. Sur. Ill., vol. 4, Coal Meas. [Sig. capitate.]
cisti, Brongniart, 1828, Hist. Veg. Foss., Coal Meas. [Ety. proper name.]
clarksoni, Lesquereux, 1858, Geo. Sur. Penn., vol. 2, Coal Meas. [Ety. proper name.]

collinsi, Lesquereux, 1870, Geo. Sur. Ill., vol. 4, Coal Meas. [Ety. proper name.]
cordata, Brongniart, 1828, Hist. Veg. Foss., Coal Meas. [Sig. heart-shaped.]
coriacea, Lesquereux, 1870, Geo. Sur. Ill., vol. 4, Coal Meas. [Sig. having the texture of rough skin.]
crassa, Dawson, 1868, Acad. Geol., Devonian. [Sig. thick.]
crenulata, Brongniart, 1828, Hist. Veg. Foss., Coal Meas. [Sig. crenulated.]
cyclopteroides, Dawson, 1863, Can. Nat. & Geol., vol. 8, Coal Meas. [Sig. like Cyclopteris.]
dawsoni, see Megalopteris dawsoni.
delicatula, Lesquereux, 1858, Geo. Sur. Penn., vol. 2, Coal Meas. [Sig. small and delicate.]
dentata, Lesquereux, 1858, Geo. Sur. Penn.,vol. 2, Coal Meas. [Sig.toothed.]
desori, Lesquereux, 1858, Geo. Sur. Penn., vol. 2, Coal Meas. [Ety. proper name.]
eveni, Lesquereux, 1866, Geo. Sur. Ill., vol. 2, Coal Meas. [Ety proper name.]
fasciculata, Lesquereux, 1870, Geo. Sur. Ill., vol. 4, Coal Meas. [Sig. faggot-like, in bundles.]
fimbriata, Lesquereux, 1854, Jour. Bost. Soc.Nat.Hist.,Coal Meas. [Sig.fringed.]
fissa, Lesquereux, 1858, Geo. Sur. Penn., vol. 2, Coal Meas. [Sig. split.]
flexuosa, Sternberg, 1825, Vers. Darst. Flora der Vorwelt, Coal Meas. [Sig. wavy.]
gibbosa, Lesquereux, 1858, Geo. Sur. Penn.,vol.2, Coal Meas. [Sig. gibbous.]
gigantea, Sternberg, 1825, Vers. Darst. Flora der Vorwelt, Coal Meas. [Sig. very large.]
grangeri, Brongniart, 1828,Hist.Veg.Foss., Coal Meas. [Ety. proper name.]
heterophylla, Brongniart, 1822, (Filicites heterophylla) Mem. du Mus. d'Hist. Nat. de Paris, Coal Meas. [Ety. heteros, different; phyllon, a leaf.]
hirsuta, Lesquereux, 1854, Bost. Jour. Nat. Hist., Coal Meas. [Sig. hairy.]
inflata, Lesquereux, 1866, Geo. Sur. Ill., vol. 2, Coal Meas. [Sig. inflated.]
ingens, Lindley & Hutton, 1833, Foss. Flora, Coal Meas. [Sig. huge.]
lacerata, syn. for Neuropteris fimbriata.
linnæifolia, Bunbury, 1847, Quar. Jour. Geo. Soc., vol. 3, Coal Meas. [Ety. from a resemblance to the leaves of Linnæa borealis.]
loshi, Brongniart, 1828, Hist. Veg. Foss., Coal Meas. [Ety. proper name.]
marginata, see Alethopteris marginata.
microphylla, Brongniart, 1828, Hist. Veg. Foss., Coal Meas. [Sig. small-leaved.]
minor, Lesquereux, 1858, Geo. Sur. Penn., vol. 2, Coal Meas. [Sig. less.]
moori, see Lescuropteris moori.
pachyderma, Lesquereux, 1866, Geo. Sur. Ill., vol. 2, Coal Meas. [Sig. thick-barked.]

perelegans, Dawson, 1863, Can. Nat. & Geol., vol. 8, Coal Meas. [Sig. very elegant.]
plicata, Sternberg, 1825, Vers. Darst.Flora der Vorwelt, Coal Meas. [Sig. folded.]
polymorpha, Dawson, 1862, Quar. Jour. Geo. Soc., vol. 18, Devonian. [Sig. many-formed.]
rarinervis, Bunbury, 1847, Quar. Jour. Geo. Soc., vol. 3, Coal Meas. [Ety. rarus, few; nervus, a vein.]
retorquata, Dawson, 1871, Foss. Plants Canada, Devonian. [Sig.turned back.]
rogersi, Lesquereux, 1858,Geo.Sur.Penn., vol. 2, Coal Meas. [Ety. proper name.]
rotundifolia, Brongniart, 1828, Hist. Veg. Foss., Coal Meas. [Sig. round-leaved.]
selwyni, Dawson, 1871, Foss. Plants Canada, Devonian. [Ety. proper name.]
serrulata, Dawson, 1862, Quar. Jour. Geo. Soc., vol. 18, Devonian. [Sig. like a little saw.]
smilacifolia, Sternberg, 1824, Vers. Darst. Flora der Vorwelt, Coal Meas. [Ety. Smilax, an existing genus; folium, a leaf.]
soreti, Brongniart, 1828, Hist. Veg. Foss., Coal Meas. [Ety. proper name.]
speciosa, Lesquereux, see N. rogersi.
tenuifolia, Sternberg, 1825, Vers. Darst. Flora der Vorwelt, Coal Meas. [Ety. tenuis, narrow; folium, a leaf.]
tenuinervis, Lesquereux, 1858, Geo. Sur. Penn., vol. 2, Coal Meas. [Sig. fine-veined.]
undans, Lesquereux, 1858, Geo. Sur. Penn., vol. 2, Coal Meas. [Sig. wavy.]
verbenæfolia, Lesquereux, 1866, Geo. Sur. Ill., vol. 2, Coal Meas. [Sig. leaved like the Verbena.]
vermicularis, Lesquereux, 1861, Geo. Sur. Ky., vol. 4, Coal Meas. [Sig. worm-shaped.]
villiersi, Brongniart, 1828,Prodr.Hist.Veg. Foss., Coal Meas. [Ety. proper name.]
voltzi, Brongniart, 1828, Prodr. Hist.Veg. Foss., Coal Meas. [Ety. proper name.]
NŒGGERATHIA, Sternberg, 1822, Vers. Darst. Flora der Vorwelt. [Ety. proper name.]
beinertiana, Gœppert, 1842, Gatt. d. Foss. Pflanzen, Coal Meas. [Ety. proper name.]
bockschi, Gœppert, 1836, (Cyclopteris bockschi) Syst. Filic. Foss., Coal Meas. [Ety. proper name.]
bockschiana, syn. for N. bockschi.
dispar, Dawson, 1863, Can. Nat. & Geol., vol. 8, Coal Meas. [Sig. different.]
flabellata, Lindley & Hutton, 1833, Foss. Flora, vol. 1, Coal Meas. [Sig. spread out like a fan.]
gilboensis, Dawson, 1871, Quar. Jour Geo. Soc., vol. 27, Devonian. [Ety. proper name.]
minor, Lesquereux,1858, Geo. Sur. Penn., vol.2, Coal Meas. [Ety. from the small leaves.]

PLANTÆ.

obliqua, Gœppert, 1841, Gatt. Foss. Pflanzen, Coal Meas. [Ety. from the oblique attachment of the leaves.].
obtusa, Lesquereux, 1858, Geo.Sur.Penn., vol. 2, Coal Meas. [Ety. *obtusus*, from the blunt termination of the leaf.] Prof. Dawson refers this species to the genus Cyclopteris.
ODONTOPTERIS, Brongniart, 1822, Mem. du Mus. d'Hist. Nat. de Paris. [Ety. *odous*, a tooth; *pteris*, a fern.]
æqualis, Lesquereux, 1866, Geo. Sur. Ill., vol. 2, Coal Meas. [Sig. equal, of the same shape as another.]
alata, Lesquereux, 1858, Catal. Pottsville Foss., Coal Meas. [Sig. winged.]
antiqua, Dawson, 1863, Can. Nat. & Geo., Coal Meas. [Sig. ancient.]
bradleyi, Lesquereux, 1870, Geo. Sur. Ill., vol. 4, Coal Meas. [Ety. proper name.]
brardi, Brongniart, 1828, Prodr. Hist.Veg. Foss., Coal Meas. [Ety. proper name.]
crenulata, Brongniart, 1828, Hist. Veg. Foss., Coal Meas. [Sig. crenulated.]
dubia, Lesquereux, 1858, Geo. Sur. Penn., vol. 2, Coal Meas. [Sig. doubtful.]
gracillima, Newberry, 1873, Ohio Pal., vol. 1, Coal Meas. [Sig. very slender.]
heterophylla, Lesquereux, 1866, Geo. Sur. Ill., vol. 2, Coal Meas. [Ety. *heteros*, different; *phyllon*, a leaf.]
intermedia, Lesquereux, 1860, Geo. Sur. Ark., vol. 2, Coal Meas. [Sig. intermediate.]
neuropteroides, Newberry, 1873, Ohio Pal., vol. 1, Lower, Coal Meas. [Ety. from its resemblance to *Neuropteris*.]
schlotheimi, Brongniart, 1828, Hist. Veg. Foss., Coal Meas. [Ety. proper name.]
squamosa, Lesquereux, 1858, Geo. Sur. Penn., vol. 2, Coal Meas. [Sig. scaly.]
subcuneata, Bunbury, 1847, Quar. Geo. Jour., vol. 3, Coal Meas. [Sig. somewhat wedge-shaped.]
wortheni, Lesquereux, 1866, Geo. Sur. Ill., vol. 2, Coal Meas. [Ety. proper name.]
ORMOXYLON, Dawson, 1871, Foss. Plants Canada. [Ety. *ormos*, a chain, a cord; *xylon*, wood.]
erianum, Dawson, 1871, Foss. Plants Canada, Devonian. [Ety. proper name.]
ORTHOGONIOPTERIS, Andrews, 1875, Ohio Pal., vol. 2. [Ety. *orthogoniopteris*, rectangular-fern.]
clara, Andrews, 1875, Ohio Pal., vol. 2, Coal Meas. [Sig. clear, distinct.]
gilberti, Andrews, 1875, Ohio Pal., vol. 2, Coal Meas. [Ety. proper name.]
Pachyphyllum, Lesquereux, 1858, Geo. Sur. Penn., vol. 2. [Ety. *pachys*, thick; *phyllon*, a leaf.] This name was preoccupied in the class Polypi. See Rhacophyllum.
affine, see Rhacophyllum affine.
fimbriatum, see Rhacophyllum fimbriatum.
hirsutum, see Rhacophyllum hirsutum.
laceratum, see Rhacophyllum laceratum.
lactuca, see Rhacophyllum lactuca.

PACHYPTERIS, Brongniart, 1828, Prodr. Hist, Veg. Foss. [Ety. *pachys*, thick; *pteris*, fern.]
gracillima, Lesquereux, 1870,Geo. Sur. Ill., vol. 4, Coal Meas. [Sig. very slender.]
PALÆOCHORDA, McCoy, 1849, Quar. Jour. Geol. Soc., vol. 4. [Ety. *palaios*, ancient; *chorde*, an intestine.]
marina, Fitch, 1856, Am. Geol., Potsdam Gr. [Sig. pertaining to the sea.]
tenuis, Fitch, 1856, Am. Geol., Potsdam Gr. [Sig. slender.]
PALÆOPHYCUS, Hall, 1847, Pal. N. Y., vol. 1. [Ety. *palaios*, ancient; *phykos*, seaweed.]
articulatus, Winchell, 1864, Am. Jour. Sci. & Arts, 2d series, vol. 37, Potsdam Gr. [Ety. from the articulated stems.]
beauharnoisensis, Billings,1862, Pal.Foss., vol.1, Calcif. Gr. [Ety. proper name.]
beverleyensis, Billings,1862,Pal. Foss.,vol. 1, Potsdam Gr. [Ety. proper name.]
congregatus, Billings, 1861, Pal. Foss., vol. 1, Potsdam Gr. [Sig. assembled together.]
divaricatus, Lesquereux, 1876, 7th Ann. Rep. Geol. Sur. Ind., Coal Meas. [Sig. wide apart.]
funiculus, Billings, 1862, Pal. Foss., vol. 1, Calcif. Gr. [Sig. a rope.]
gracilis, Lesquereux, 1876, 7th Ann. Rep. Geol. Sur. Ind., Coal Meas. [Sig. slender.]
incipiens, Billings, 1861, Pal. Foss., vol. 1, Potsdam Gr. [Sig. the beginning.]
informis, Winchell, 1864, Am. Jour. Sci. & Arts, vol. 87, Potsdam Gr. [Sig. shapeless.]
irregularis, Hall, 1847, Pal. N. Y., vol. 1, Calcif. Gr. [Sig. irregular.]
milleri, Lesquereux, 1876, 7th Ann. Rep. Geol. Sur. Ind., Coal Meas. [Ety. proper name.]
obscurus, Billings, 1862, Pal. Foss., vol. 1, Trenton Gr. [Sig. obscure.]
rogosus, Hall, 1847, Pal. N. Y., vol. 1, Trenton Gr. [Sig. full of wrinkles.]
simplex, Hall, 1847, Pal. N. Y., vol. 1, Trenton Gr. [Sig. simple.]
striatus, Hall, 1852, Pal. N. Y., vol. 2, Clinton Gr. [Sig. striated.]
tortuosus, Hall, 1852, Pal. N. Y., vol. 2, Medina sandstone. [Ety. from the tortuous branches.]
tubularis, Hall, 1847, Pal. N. Y., vol. 1, Calcif. Gr. [Sig. hollow like a pipe.]
virgatus, Hall, 1847, Pal. N. Y., vol. 1, Hud. Riv. Gr. [Sig. twig-like.]
Palæopteris, being preoccupied, see Archæopteris.
acadica, see Archæopteris acadica.
hartii, see Archæopteris harti.
PALÆOXYRIS, Brongniart,1828, Ann. Sc. Nat., vol. 15. [Ety. *palaios*, ancient; *Xyris*, a plant.]
appendiculata, Lesquereux, 1870, Geo. Sur. Ill., vol. 4, Coal Meas. [Sig. having lateral appendages.]

corrugata, Lesquereux, 1870, Geo. Sur. Ill., vol. 4, Coal Meas. [Sig. corrugated.]
prendeli, Lesquereux, 1870, Geo. Sur. Ill., vol. 4, Coal Meas. [Ety. proper name.]

PECOPTERIS, Brongniart, 1822, Mem. du Mus. d'Hist. Nat. de Paris. [Ety. *peko*, to comb; *pteris*, a fern.]
abbreviata, Brongniart, 1828, Hist. Veg. Foss., Coal Meas. [Sig. from the short lobes of the pinnules.]
acuta, Brongniart, 1828, Prodr. Hist. Veg. Foss., Coal Meas. [Sig. acute.]
æqualis, Brongniart, 1828, Prodr. Hist. Veg. Foss., Coal Meas. [Sig. equal.]
aquilina, see Alethopteris aquilina.
arborescens, Brongniart, 1828, Prodr. Hist. Veg. Foss., Coal Meas. [Sig. tree-like.]
arguta, Brongniart, 1828, Hist. Veg. Foss., Coal Meas. [Sig. sharply defined.]
aspidioides, Brongniart, 1828, Hist. Veg. Foss., Coal Meas. [Sig. like the *Aspidium*.]
bucklandi, Brongniart, 1828, Prodr. Hist. Veg. Foss., Coal Meas. [Ety. proper name.]
bullata, Bunbury, 1847, Quar. Jour. Geo. Soc., vol. 3, Coal Meas. [Sig. bossed.]
callosa, Lesquereux, 1866, Geo. Sur. Ill., vol. 2, Coal Meas. [Sig. having a thick covering.]
candollana, Brongniart, 1828, Prodr. Hist. Veg. Foss., Coal Meas. [Ety. proper name.]
chærophylloides, Brongniart, 1828, Hist. Veg. Foss., Coal Meas. [Sig. like the *Chærophyllum*.]
cisti, Brongniart, 1828, Hist. Veg. Foss., Coal Meas. [Ety. proper name.]
concinna, Lesquereux, 1858, Geo. Sur. Penn., vol. 2, Coal Meas. [Sig. neat.] The name was preoccupied by Presl in 1833.
crenulata, see Alethopteris crenulata.
cristata, see Alethopteris cristata.
cyathea, Brongniart, 1828, Prodr. Hist. Veg. Foss., Coal Meas. [Ety. *kyathos*, a cup.]
decurrens, Lesquereux, 1858, Geo. Sur. Penn., vol. 2, Coal Meas. [Ety. from the decurrent nature of the leaves.]
decurrens, Dawson, 1862. The name was preoccupied, see P. discrepans.
densifolia, Dawson, 1874, Foss. Plants, Can., Devonian. [Sig. dense-leaved.]
dentata, Brongniart, 1828, Hist. Veg. Foss., Coal Meas. [Sig. toothed.]
discrepans, Dawson, 1863, Quar. Jour. Geo. Soc., vol. 19, Devonian. [Sig. different.]
distans, Lesquereux, 1858, Geo. Sur. Penn., vol. 2, Coal Meas. [Ety. from the distant pinules.] The name was preoccupied by Rost in 1839.
dubia, Gutbier, 1843, in Gæa von Sachsen, Coal Meas. [Sig. doubtful.]
elegans, Gœppert, 1836, Syst. Filic. Foss., Coal Meas. [Sig. elegant.]

elliptica, Bunbury, 1846, Quar. Jour. Geo. Soc., vol. 2, Coal Meas. [Sig. elliptical.]
emarginata, Gœppert, 1836, (Diplazites emarginata) Syst. Filic. Foss., Coal Meas. [Sig. notched.]
erosa, see Alethopteris erosa.
flavicans, Presl, 1833, in Sternberg, Vers. Darst. Flora der Vorwelt, Coal Meas. [Sig. yellow (?).]
hemiteloides, Brongniart, 1828, Hist. Veg. Foss., Coal Meas. [Sig. like *Hemitelites*.]
heterophylla, Lindley & Hutton, see Alethopteris heterophylla.
heterophylla, Dawson, syn. for P. mantelli.
incompleta, Lesquereux, 1858, Geo. Sur. Penn., vol. 2, Coal Meas. [Ety. from the incomplete condition of the specimen.]
ingens, Dawson, 1862, Quar. Jour. Geo. Soc., vol. 18, Devonian. [Sig. huge, enormous.]
lepidorhachis, Brongniart, 1828, Hist. Veg. Foss., Coal Meas. [Ety. *lepis*, a scale; *rachis*, a ridge.]
longifolia, Brongniart, 1828, Hist. Veg. Foss., Coal Meas. [Sig. long-leaved.] Alethopteris longifolia. (?)
loshi, Brongniart, 1828, Prodr. Hist. Veg. Foss., Coal Meas. [Ety. proper name.]
mantelli, Brongniart, 1828, Prodr. Hist. Veg. Foss., Coal Meas. [Ety. proper name.]
milleri, Harlan, 1835, Trans. Geo. Soc. Penn., Coal Meas. [Ety. proper name.]
muricata, see Alethopteris muricata.
murrayana, Brongniart, 1828, Hist. Veg. Foss., Coal Meas. [Ety. proper name.]
nervosa, see Alethopteris nervosa.
newberryi, Lesquereux, 1858, Geo. Sur. Penn., vol. 2, Coal Meas. [Ety. proper name.]
notata, Lesquereux, 1858, Geo. Sur. Penn., vol. 2, Coal Meas. [Sig. dotted.]
obsoleta, Harlan, 1835, Trans. Geo. Soc. Penn., Coal Meas. [Sig. obsolete.]
oreopteroides, Brongniart, 1828, Hist. Veg. Foss., Coal Meas. [Sig. like the fern *Oreopteris*.]
ovata, Brongniart, 1828, Prodr. Hist. Veg. Foss., Coal Meas. [Sig. ovate.]
ovata, Gutbier, 1843, in Gæa von Sachsen, Coal Meas. The name was preoccupied.
pennæformis, Brongniart, 1822, (Filicites pennæformis) Mem. du Mus. d'Hist. Nat. de Paris, Coal Meas. [Sig. featherformed.]
plukeneti, see Alethopteris plukeneti.
plumosa, Brongniart, 1828, Hist. Veg. Foss., Coal Meas. [Sig. feathery.]
polymorpha, Brongniart, 1828, Prodr. Hist. Veg. Foss., Coal Meas. [Sig. many-formed.]
preciosa, Hartt, 1868, Acad. Geo., Devonian. [Sig. precious, splendid.]
pteroides, see Alethopteris pteroides.

PLANTÆ.

pusilla, Lesquereux, 1858, Geo. Sur. Penn., vol. 2, Coal Meas. [Sig. very small.]
rigida, Dawson, 1863, Can. Nat. & Geo., vol. 8, Coal Meas. [Sig. inflexible.]
serlii, see Alethopteris serlii.
serrulata, Hartt, 1868, Acad. Geo., Devonian. [Sig. like a small saw.]
shæfferi, Lesquereux, 1858, Catal. Potts. Foss., Coal Meas. [Ety. proper name.]
sillimani, Brongniart, 1828, Hist. Veg. Foss., Coal Meas. [Ety. proper name.]
squamosa, Lesquereux, 1870, Geo. Sur. Ill., vol. 4, Coal Meas. [Sig. scaly.]
strongi, Lesquereux, 1870, Geo. Sur. Ill., vol. 4, Coal Meas. [Ety. proper name.]
tæniopteroides, Bunbury, 1847, Quar. Jour. Geo. Soc., vol. 3, Coal Meas. [Sig. like *Tæniopteris.*]
tenuis, Brongniart, 1828, Prodr. Hist. Veg. Foss., Coal Meas. [Sig. slender.]
unita, Brongniart, 1828, Prodr. Hist. Veg. Foss., Coal Meas. [Sig. from the united pinnules.]
urophylla, see Alethopteris urophylla.
velutina, Lesquereux, 1858, Geo. Sur. Penn., vol. 2, Coal Meas. [Sig. velvety.]
villosa, Brongniart, 1828, Hist. Veg. Foss., Coal Meas. [Sig. covered with short hair-like projections.]
PHYLLOPTERIS, Brongniart, 1828, Tab. d. Gen. etc. [Ety. *phyllon*, a leaf; *pteris*, a fern.]
antiqua, Dawson, 1863, Can. Nat., vol. 8, Coal Meas. [Sig. ancient.]
Phytolithus, Martin, 1809, Petrificata Derbiensia. [Ety. *phyton*, a plant; *lithos*, stone.]
cancellatus, see Sigillaria cancellata.
dawsoni, see Sigillaria dawsoni.
decoratus, see Calamites decoratus.
martini, see Sigillaria martini.
notatus, see Sigillaria notata.
reticulatus, see Stigmaria reticulata.
sulcatus, see Calamites sulcatus.
tessellatus, see Sigillaria tessellata.
transversus, see Artisia transversa.
verrucosus, see Stigmaria verrucosa.
PHYSOPHYCUS, Schimper, 1869, Pal. Veg. [Ety. *phyton*, a plant; *phykos*, a sea weed.]
marginatus, Lesquereux, 1866, (Caulerpites marginatus) Trans. Am. Phil. Soc., vol. 13, Coal Meas. [Sig. margined.]
PHYTOPSIS, Hall, 1847, Pal. N.Y., vol. 1. [Ety. *phyton*, a plant; *opsis*, resemblance.]
cellulosum, Hall, 1847, Pal. N. Y., vol. 1, Birdseye Gr. [Sig. from the cellular substance.]
tubulosum, Hall, 1847, Pal. N. Y., vol. 1, Birdseye Gr. [Sig. abounding in tubes.]
PINNULARIA, Lindley & Hutton, 1835, Foss. Flora, vol. 2. [Ety. *pinna*, a feather.]
calamitarum, Lesquereux, 1858, Geo. Sur. Penn., vol. 2, Coal Meas. [Sig. like the genus *Calamites.*]
capillacea, Lindley & Hutton, 1835, Foss. Flora, Coal Meas. [Sig. stringy, as the roots of herbs.]
confervoides, Lesquereux, 1858, Geo. Sur. Penn., vol. 2, Coal Meas. [Sig. like *Conferva.*]

crassa, Dawson, 1863, Can. Nat., vol. 8, Coal Meas. [Sig. thick.]
dispalans, Dawson, 1852, Quar. Jour. Geo. Soc., vol. 18, Devonian. [Sig. straggling, stray.]
elongata, Dawson, 1871, Foss. Plants Can., Devonian. [Sig. lengthened.]
fucoides, Lesquereux, 1868, Geo. Sur. Penn., vol. 2, Coal Meas. [Ety. *fucus*, sea weed; *eidos*, likeness.]
horizontalis, Lesquereux, 1858, Geo. Sur. Penn., vol. 2, Coal Meas. [Sig. horizontal.]
nodosa, Dawson, 1871, Foss. Plants Can., Devonian. [Sig. knotty.]
pinnata, Lesquereux, 1858, Geo. Sur. Penn., vol. 2, Coal Meas. [Sig. winged.]
ramosissima, Dawson, 1863, Can. Nat., vol. 8, Coal Meas. [Sig. very branchy.]
POLYPORITES, Lindley & Hutton, 1833, Foss. Flora. [Ety. from its resemblance to the *Polyporus versicolor*.]
bowmanni, Lindley & Hutton, 1833, Foss. Flora, Coal Meas. [Ety. proper name.]
polysporus, Newberry, 1873, Ohio Pal., vol. 1. [Ety. *polys*, many; *sporos*, seed.]
mirabilis, Newberry, 1873, Ohio Pal., vol. 1, Coal Meas. [Sig. wonderful, strange.]
PROTOTAXITES, Dawson, 1859, Quar. Jour. Geo. Soc., vol. 15. [Ety. *protos*, first; *taxus*, Yew tree; so named from the spirally marked cells characteristic of the genus *Taxites*.] Carruthers says it is an Alga and has therefore called it Nematophycus.
logani, Dawson, 1859, Quar. Jour. Geo. Soc., vol. 15, Devonian. [Ety. proper name.] This is the oldest known tree in America, according to Dawson, but Carruthers says it is a huge sea-weed and has named it Nematophycus logani.
PSARONIUS, Cotta, 1832, Dendrol. in Beziehung. [Ety. *psaros*, speckled.]
erianus, Dawson, 1871, Foss. Plants Can., Devonian. [Ety. proper name.]
textilis, Dawson, 1871, Foss. Plants Can., Devonian. [Sig. woven, like a web.]
PSILOPHYTON, Dawson, 1859, Quar. Jour. Geo. Soc., vol. 15. [Ety. *psilon*, smooth; *phyton*, stem.]
elegans, Dawson, 1862, Quar. Jour. Geo. Soc., vol. 18, Devonian. [Sig. elegant.]
glabrum, Dawson, 1862, Quar. Jour. Geo. Soc., vol. 18, Devonian. [Sig. without hair, smooth.]
princeps, Dawson, 1859, Quar. Jour. Geo. Soc., Upper Silurian and Devonian. [Sig. original, principal.] This is the oldest known plant in America. It is supposed to have grown in a marsh.
robustius, Dawson, 1859, Quar. Jour. Geo. Soc., vol. 15, Devonian. [Sig. strong, like oak.]
PTILOCARPUS, Lesquereux, 1870, Geo. Sur. Ill., vol. 4. [Ety. *ptilon*, a wing; *karpos*, fruit.]

bicornutus, Lesquereux, 1870, Geo. Sur. Ill., vol. 4, Coal Meas. [Sig. from two short horns at the base.]
RHABDOCARPUS, Gœppert & Berger, 1848, de Fruct. et Sem. [Ety. *rhabdos*, stria; *karpos*, fruit.]
acuminatus, Newberry, 1873, Ohio Pal., vol. 1, Coal Meas. [Sig. pointed.]
amygdaliformis, Gœppert & Berger, 1848, de Fruct et Sem., Coal Meas. [Sig. like the almond, *Amygdalus*.]
apiculatus, Newberry, 1873, Ohio Pal., vol. 1, Coal Meas. [Sig. pointed.]
arcuatus, Lesquereux, 1861, Geo. Sur. Ky., Coal Meas. [Sig. bent.]
carinatus, Newberry, 1873, Ohio Pal., vol. 1, Coal Meas. [Sig. keeled.]
clavatus, Sternberg, 1820, (Carpolithes clavatus) Vers. Darst. Flora der Vorwelt, Coal Meas. [Sig. club-shaped.]
costatus, Newberry, 1873, Ohio Pal., vol. 1, Coal Meas. [Sig. ribbed.]
danai, Foster, 1854, Ann. of Sci., vol. 1, Coal Meas. [Ety. proper name.]
insignis, Dawson, 1863, Can. Nat. & Geo., vol. 8, Coal Meas. [Sig. remarkable.]
lævis, Newberry, 1873, Ohio Pal., vol. 1, Coal Meas. [Sig. smooth.]
mammillatus, Lesquereux, 1870, Geo. Sur. Ill., vol. 4, Coal Meas. Sig. mammillated.]
minutus, Lesquereux, 1860, Geo.Sur.Ark. vol. 2, Coal Meas. [Sig. very small.]
venosus, Lesquereux, 1858, Geo. Sur. Penn., vol. 2, Coal Meas. [Sig. full of veins.]
RHACHIOPTERIS, Dawson, 1862, Quar. Jour. Geo. Soc., vol. 18. [Ety. *rachis*, a ridge; *pteris*, a fern.]
cyclopteroides, Dawson, 1862, Quar. Jour. Geo. Soc., vol. 18, Catskill Gr. [Ety. like the genus *Cyclopteris*.]
gigantea, Dawson, 1871, Foss. Plants Can., Devonian. [Sig. very large.]
palmata, Dawson, 1870, Proc. Royal Soc., Devonian. [Sig. like the palm tree.]
pinnata, Dawson, 1862, Quar. Jour. Geo. Soc., vol. 18, Ham.Gr. [Sig. feathered.]
punctata, Dawson, 1862, Quar. Jour.Geo. Soc., vol. 18, Catskill Gr. [Sig. punctured.]
striata, Dawson, 1862, Quar. Jour. Geo. Soc., vol. 18, Ham. Gr. [Sig. striated.]
tenuistriata, Dawson, 1862, Quar. Jour. Geo. Soc., vol. 18, Ham. Gr. [Sig. fine-lined.]
RHACOPHYLLUM, Schimper, 1869, Palæontologie Vegetale. [Ety. *rakos*, rugged; *phyllon*, a leaf.]
affine, Lesquereux, 1858, (Pachyphyllum affine) Geo.Sur.Penn., vol.2,Coal Meas. [Sig. near to; related to R. fimbriatum.]
fimbriatum, Lesquereux, 1858, (Pachyphyllum fimbriatum) Geo.Sur. Penn., vol. 2, Coal Meas. [Sig. fringed.]
hirsutum, Lesquereux, 1858, (Pachyphyllum hirsutum) Geo. Sur. Penn., vol. 2, Coal Meas. [Sig. hairy.]

laceratum, Lesquereux, 1858, (Pachyphyllum laceratum) Geo. Sur. Penn., vol. 2, Coal Meas. [Sig. lacerated.]
lactuca, Brongniart, 1828, Hist.Veg. Foss., Coal Meas. [Ety. *lactuca*, Lettuce.]
Rhizolithes, F. Braun, 1847, in Flora, etc. [Ety. *rhiza*, a root; *lithos*, stone.]
palmatifidus, Lesquereux, 1860, Geo. Sur. Ark., vol. 2, Coal Meas. [Sig. divided like a hand.]
RHIZOMOPTERIS, Schimper, 1869, Traite de Palæontologie Vegetale. [Sig. the rhizomas of ferns.] This was formed to include *Selaginites uncinatus* and *S. erdmanni*.]
Rotularia longifolia, see Sphenophyllum longifolium.
RUSOPHYCUS, Hall, 1852, Pal. N. Y., vol. 2. [Ety. *rusos*, rugose; *phykos*, sea-plant.]
bilobatus, Vanuxem, 1842, (Fucoides bilobatus) Geo. Rep. N. Y., Cin'ti & Clinton Gr. [Sig. two-lobed.]
clavatus, Hall, 1852, Pal. N. Y., vol. 2, Clinton Gr. [Sig. club-shaped.]
grenvillensis, Billings, 1862, Pal. Foss., vol. 1, Chazy Gr. [Ety. proper name.]
pudicus, Hall, 1852, Pal. N. Y., vol. 2, Hud. Riv. & Clinton Gr. [Sig. shamefaced, modest.]
subangulatus, Hall, 1852, Pal. N. Y., vol. 2, Clinton Gr. [Sig. somewhat angulated.]
Sagenaria, Brongniart, 1822, Memoires du Museum d'Histoire Naturelle.
veltheimiana, see Lepidodendron veltheimanum.
Schizopteris, Brongniart, 1828, Hist.Veg. Foss. [Ety. *schizo*, to cleave; *pteris*, a fern.]
lactuca, see Hymenophyllites lactuca.
SCHUTZIA, Gœppert, 1848, Permian Flora. [Ety. proper name.]
bracteata Lesquereux, 1870, Geo. Sur. Ill., vol. 4, Coal Meas. [Sig. covered with leaves or plates.]
SCOLITHUS, Haldeman, 1840, Supp. to Monograph of Limniades. [Ety. *scolex*, a worm; *lithos*, a stone.]
canadensis, Billings, 1862, Pal. Foss., vol. 1, Potsdam Gr. [Ety. proper name.]
linearis, Hall, 1847, Pal. N. Y., vol. 1, Potsdam to Hud. Riv. Gr. [Sig. drawn out in lines.]
verticalis, Hall, 1852, Pal. N. Y., vol. 2, Medina sandstone. [Ety. from penetrating the strata vertically.]
SCOLOPENDRITES, Lesquereux, 1858, Geo.Sur. Penn., vol. 2. [Ety. from the living fern Scolopendrium.]
dentatus, Lesquereux, 1858, Geo. Sur. Penn., vol. 2, Coal Meas. [Ety. *dentatus*, toothed.] In the text the specific name is written *grosse-dentata*, a Franco-Latin hybrid, but on plate 8, fig. 7, the name is corrected by leaving off the French.
SELAGINITES, Brongniart, 1828, Prodr. Hist. Veg. Foss. [Ety. from the plant *Selago*.]

cavifolius, Lesquereux, 1861, (Lycopodites cavifolius) Geo. Rep. Ky., vol. 4, Coal Meas. [Sig. hollow-leaved.]
crassus, Lesquereux, 1866, Geo. Sur. Ill., vol. 2, Coal Meas. [Sig. thick, stout.]
formosus, Dawson, 1861, Can. Nat., vol. 6, Devonian. [Sig. beautiful.]
uncinatus, Lesquereux, 1866, Geo. Sur. Ill., vol. 2, Coal Meas. [Sig. hooked.]

SIGILLARIA, Brongniart, 1822, Mem. du Mus. d'Hist. Nat. de Paris. [Ety. *sigillum*, a seal; from the seal-like scars of fallen leaves stamped upon the bark.]
alternans, Lindley & Hutton, 1831-33, Foss. Flora, Coal Meas. [Sig. alternating.]
alveolaris, Brongniart, 1828, Prodr. Hist. Veg. Foss., Coal Meas. [Ety. *alveolus*, a small channel.]
augusta, Brongniart, 1828, Hist. Veg. Foss., Coal Meas. [Sig. narrow.]
attenuata, Lesquereux, 1858, Catal. Potts. Foss., Coal Meas. [Sig. attenuated.]
brardi, Brongniart, 1828, Prodr. Hist. Veg. Foss., Coal Meas. [Ety. proper name.]
bretonensis Dawson, 1868, Can. Nat. & Geo., 2nd series, vol. 3, Coal Meas. [Ety. proper name.]
brochanti, Brongniart, 1828, Hist. Veg. Foss., Coal Meas. [Ety. proper name.]
browni, Dawson, 1861, Quar. Jour. Geo. Soc., vol. 17, Coal Meas. [Ety. proper name.]
cancellata, Martin, 1809, (Phytolithus cancellatus) Petrif. Derb., Coal Meas. [Sig. cancellated.]
catenoides, Dawson, 1868, Can. Nat. & Geo., 2nd series, vol. 3, Coal Meas. [Sig. chain-like.]
catenulata, Lindley & Hutton, 1831-33, Foss. Flora, Coal Meas. [Sig. linked together in a chain.] Probably syn. for *S. alternans*.
chemungensis, Hall, 1843, Geo. Rep. 4th Dist. N. Y., Chemung Gr. [Ety. proper name.]
cisti, see Caulopteris cisti.
corrugata, Lesquereux, 1861, Geo. Sur. Ky., vol. 4; redefined 1870, Geo. Sur. Ill., vol. 4, Coal Meas. [Sig. corrugated.]
cymtoides, Wood, 1860, Proc. Acad. Nat. Sci., Coal Meas. [Ety. wave-like.]
dawsoni, Steinhaur, 1818, (Phytolithus dawsoni) Trans. Am. Phil. Assoc., Coal Meas. [Ety. proper name.]
defrancii, Brongniart, 1828, Prodr. Hist. Veg. Foss., Coal Meas. [Ety. proper name.]
dilatata, Lesquereux, 1858, Geo. Sur. Penn., vol. 2, Coal Meas. [Sig. expanded.]
discoidea, Lesquereux, 1858, Geo. Sur. Penn., vol. 2, Coal Meas. [Ety. from the discoidal shape of the scars.]
dournaisi, Brongniart, 1828, Prodr. Hist. Veg. Foss., Coal Meas. [Ety. proper name.]

dubia, Lesquereux, 1858, Geo. Sur. Penn., vol. 2, Coal Meas. [Sig. doubtful.]
elegans, Brongniart, 1828, Prodr. Hist. Veg. Foss., Coal Meas. [Sig. elegant.]
elongata, Brongniart, 1828, Prodr. Hist. Veg. Foss., Coal Meas. [Ety. from the elongated scars.]
eminens, Dawson, 1863, Can. Nat., vol. 8, Coal Meas. [Sig. standing out in relief, prominent.
fissa, Lesquereux, 1858, Geo. Sur. Penn., vol. 2, Coal Meas. [Sig. divivided, from the deeply emarginate scars.]
flexuosa, Lindley & Hutton, 1837, Foss. Flora, vol. 3, Coal Meas. [Sig. full of flexures.]
intermedia, Brongniart, 1828, Hist. Veg. Foss., Coal Meas. [Sig. intermediate.]
knorri, Brongniart, 1828, Prodr. Hist. Veg. Foss., Coal Meas. [Ety. proper name.]
laevigata, Brongniart, 1828, Prodr. Hist. Veg. Foss., Coal Meas. [Sig. smoothed.]
lepidodendrifolia, Brongniart, 1828, Hist. Veg. Foss., Coal Meas. [Ety. leaved like the Lepidodendron.]
lescurrei, Schimper, 1869, Traite de Palæontologie Vegetale, Coal Meas. [Ety. proper name.]
lorwayana, Dawson, 1873, Foss. Plants, Low. Carb. [Ety. proper name.]
martini, Steinhaur, 1818, (Phytolithus martini) Trans. Amer. Phil. Assoc., Coal Meas. [Ety. proper name.]
massiliensis, Lesquereux, 1870, Geo. Sur. Ill., vol. 4, Coal Meas. [Ety. proper name.]
menardi, Brongniart, 1828, Prodr. Hist. Veg. Foss., Coal Meas. [Ety. proper name.]
monostigma, Lesquereux, 1866, Geo. Sur. Ill., vol. 2, Coal Meas. [Sig. single-dotted.]
notata, Steinhaur, 1818, (Phytolithus notatus) Trans. Am. Phil. Assoc., Coal Meas. [Sig. marked.]
obliqua, Brongniart, 1828, Hist. Veg. Foss., Coal Meas. [Ety. from the oblique scars.]
obovata, Lesquereux, 1858, Geo. Sur. Penn., vol. 2, Coal Meas. [Sig. obovate.]
oculata, Brongniart, 1828, Prodr. Hist. Veg. Foss., Coal Meas. [Ety. *oculus*, an eye.]
organum, Lindley & Hutton, 1833, Foss. Flora, vol. 1, Coal Meas. [Sig. an instrument or pipe.]
pachyderma, Brongniart, Prodr. Hist. Veg. Foss., Coal Meas. [Sig. thick-barked.]
palpebra, Dawson, 1860, Quar. Jour. Geo. Soc., vol. 8, Devon. [Sig. an eye-lid.]
planicosta, Dawson, 1863, Can. Nat. & Geo., vol. 8, Coal Meas. [Sig. smooth-ribbed.]
polita, Lesquereux, 1858, Geo. Sur. Penn., vol. 2, Coal Meas. [Sig. smoothed.]
reniformis, Brongniart, 1828, Prodr. Hist. Veg. Foss., Coal Meas. [Sig. kidney-shaped.]

reticulata, Steinhaur, 1818, (Phytolithus reticulatus) Trans. Am. Phil. Assoc., Coal Meas. [Sig. reticulated.]
reticulata, Lesquereux, 1860, Geo. Sur. Ark., vol. 2, Coal Meas. [Sig. reticulated.]
rugosa, Brongniart, 1828, Prodr. Hist. Veg. Foss., Coal Meas. [Sig. wrinkled.]
saulli, Brongniart, 1828, Hist. Veg. Foss., Coal Meas. [Ety. proper name.]
schimperi, Lesquereux, 1858, Geo. Sur. Penn., vol. 2, Coal Meas. [Ety. proper name.]
schlotheimana, Brongniart, 1828, Hist. Veg. Foss., Coal Meas. [Ety. proper name.] American Sp. (?)
sculpta, Lesquereux, 1858, Geo. Sur. Penn., vol. 2, Coal Meas. [Sig. engraved.]
scutellata, Brongniart, 1828, Hist. Veg. Foss., Coal. Meas. [Sig. a little shield.]
semina, Lesquereux, 1870, Geo. Sur. Ill., vol. 4, Coal Meas. [Sig. seeds of Sigillaria.]
serlii, Brongniart, 1828, Prodr. Hist. Veg. Foss., Coal Meas. [Ety. proper name.]
sillimani, Brongniart, 1828, Hist. Veg. Foss., Coal Meas. [Ety. proper name.]
simplicitas, Vanuxem, 1843, Geo. Rep. N. Y., Catskill Gr. [Sig. from the straightness of grain.]
spinulosa, Germ., 1853, Vers. v. Wett, etc., Coal Meas. [Sig. full of spines.]
stellata, Lesquereux, 1858, Geo. Sur. Penn., vol. 2, Coal Meas. [Ety. from the radiating wrinkles around the scars.]
striata, Dawson, 1863, Can. Nat. & Geol., vol. 8, Coal Meas. [Sig. *striated*.] The name was preoccupied by Brongniart in 1828.
sydenensis, Dawson, 1863, Can. Nat. & Geol., vol. 8, Coal Meas. [Ety. proper name.]
tessellata, Steinhaur, 1818, (Phytolithus tessellatus) Trans. Am. Phil. Assoc. [Sig. from the square scars.]
tessellata, Brongniart, 1828, Hist. Veg. Foss., Coal Meas. The name was preoccupied.
vanuxemi, Gœppert, 1836, Flora Silurisch, Chemung Gr. [Ety. proper name.]
yardleyi, Lesquereux, 1866, Geo. Sur. Ill., vol. 2, Coal Meas. [Ety. proper name.]
SIGILLARIOIDES, Lesquereux, 1870, Geo. Sur. Ill., vol. 4. [Ety. from its resemblance to the genus *Sigillaria*.]
radicans, Lesquereux, 1870, Geo. Sur. Ill. vol. 4, Coal Meas. [Sig. rooted.]
stellaris, Lesquereux, 1870, Geo. Sur. Ill., vol. 4, Coal Meas. [Sig. starred.]
SPHENOPHYLLUM, Brongniart, 1828, Prodr. Hist. Veg. Foss. [Ety. *sphen*, a wedge; *phyllon*, a leaf.] This genus was called *Sphenophyllites* by Brongniart in 1822.
antiquum, Dawson, 1861, Can. Nat., vol. 6, Devonian. [Sig. ancient.]
bifurcatum, Lesquereux, 1860, Geo. Sur. Ark., vol. 2, Coal Meas. [Sig. bifurcated.]

brevifolium, Newberry, 1854, Ann. of Sci., vol. 1, Coal Meas. [Sig. short-leaved.]
cornutum, Lesquereux, 1870, Geo. Sur. Ill., vol. 4, Coal Meas. [Sig. horned.]
emarginatum, Brongniart, 1828, Prodr. Hist. Veg. Foss., Coal Meas. [Ety. from the peculiar notch at the apex of the leaf.]
erosum, Lindley & Hutton, 1833, Foss. Flora, Coal Meas. [Sig. gnawed, bitten away.]
filiculme, Lesquereux, 1858, Geo. Rep. Penn., vol. 2, Coal Meas. [Ety. *filum*, a thread; *culmus*, a straw.]
longifolium, Germar, 1828, (Rotularia longifolia) Act. Ac. Caes. Leop. Nat. Cur., vol. 15, Coal Meas. [Sig. long-leaved.]
oblongifolium, Germ. & Kaulf., 1828, Act. Ac. Nat. Cur., vol. 15, Coal Meas. [Ety. from the oblong leaf.]
saxifragifolium, Sternberg, 1825, (Rotularia saxifragifolia) Vers. Darst. Flora der Vorwelt, Coal Meas. [Sig. leaved like *Saxifraga*.]
schlotheimi, Brongniart, 1828, Prodr. Hist. Veg. Foss., Coal Meas. [Ety. proper name.]
simplicitas, Vanuxem, 1842, Geo. Rep. N. Y., Catskill Gr. [Sig. plainness, simplicity.]
trifoliatum, Lesquereux, 1858, Geo. Sur. Penn., vol. 2, Coal Meas. [Ety. *tres*, three; *folium*, a leaf.]
SPHENOPTERIS, Brongniart, 1822, Mem. du Mus. d'Hist. Nat. de Paris. [Ety. *sphen*, a wedge; *pteris*, a fern.]
abbreviata, Lesquereux, 1858, Geo. Sur. Penn., vol. 2, Coal Meas. [Sig. shortened.]
acuta, Brongniart, 1828, Prodr. Hist. Veg. Foss., Coal Meas. [Sig. sharp.]
adiantoides, Lindley & Hutton, 1833, Foss. Flora, Coal Meas. [Sig. like *Adiantum*.]
alata, see Hymenophyllites alatus.
artemisiæfolia, see Eremopteris artemisiæfolia.
canadensis, Dawson, 1863, Can. Nat. & Geol., Coal Meas. [Ety. proper name.]
davallana, Gœppert, 1841, Gatt. d. Foss. Pflanzen, Coal Meas. [Ety. proper name.]
decipiens, Lesquereux, 1860, Geo. Sur. Ark., Coal Meas. [Sig. deceiving.]
delicatula, see Hymenophyllites delicatulus.
dilatata, Lesquereux, 1860, Geo. Sur. Ark., Coal Meas. [Sig. spread out.]
dubuissoni, Brongniart, 1828, Prodr. Hist. Veg. Foss., Coal Meas. [Ety. proper name.]
elegans, Brongniart, 1828, Prodr. Hist. Veg. Foss., Coal Meas. [Sig. elegant.]
flagellaris, Lesquereux, 1858, Geo. Sur. Penn., Coal Meas. [Sig. like a whip.]
furcata, see Hymenophyllites furcatus.
gersdorfi, see Hymenophyllites gersdorfi.

PLANTÆ.

glandulosa, Lesquereux, 1858, Geo. Sur. Penn., vol. 2, Coal Meas. [Sig. glandular.]
gracilis, Brongniart, 1828, Prodr. Hist. Veg. Foss., Coal Meas. [Sig. slender.]
gravenhorsti, Brongniart, 1828, Hist.Veg. Foss., Coal Meas. [Ety. proper name.]
harti, Dawson, 1862, Quar. Jour. Geol. Soc., vol. 18, Devonian. [Ety. proper name.]
hitchcockana, Dawson, 1862, Quar. Jour. Geol. Soc., vol. 18, Devonian. [Ety. proper name.]
hœninghausi, Brongniart, 1828, Hist.Veg. Foss., Coal Meas. [Ety. proper name.]
hymenophylloides, Brongniart, 1828, Hist. Veg. Foss., Coal Meas. [Ety. from the resemblance to *Hymenophyllites*.]
intermedia, Lesquereux, 1858, Geo. Sur. Penn., vol. 2, Coal Meas. [Sig. intermediate between *S. chærophylloides* and *Pecopteris athyrioides*.]
irregularis, Sternberg, 1833, Vers. Darst. Flora der Vorwelt, Coal Meas. [Sig. irregular.]
latifolia, Brongniart, 1828, Prodr. Hist. Veg. Foss., Coal Meas. [Sig. broad-leaved.]
latior, Dawson, 1863, Can. Nat., vol. 8, Coal Meas. [Sig. broader.]
laxa, Hall, 1843, Geo. Rep. 4th Dist. N. Y., Chemung Gr. [Sig. loosely arranged.] This name was preoccupied by Sternberg.
lesquereuxi, Newberry, 1858, Geo. Sur. Penn., vol. 2, Coal Meas. [Ety. proper name.]
macilenta, Lindley & Hutton, 1833, Foss. Flora, Coal Meas. [Sig. poor, lean.]
marginata, Dawson, 1862, Quar. Jour. Geo. Soc., vol. 18, Devonian. [Sig. bordered.]
microloba, Gœppert, 1836, Syst. Filic. Foss., Coal Meas. [Sig. small-lobed.]
mixta, Schimper, 1869, Traite de Paleontologie Vegetale, Coal Meas. [Sig. mixed.]
munda, Dawson, 1863, Can. Nat. & Geo. vol. 8, Coal Meas. [Sig. neat, elegant.]
myriophylla, see Hymenophyllites myriophyllus.
newberryi, Lesquereux, 1858, Geo. Sur. Penn., vol. 2, Coal Meas. [Ety. proper name.]
obtusiloba, Brongniart, 1828, Hist. Veg. Foss., Coal Meas. [Sig. obtuse-lobed.]
paupercula, Lesquereux, 1866, Geo. Sur. Ill., vol. 2, Coal Meas. [Sig. poor.]
pilosa, Dawson, 1868, Acad. Geol., Devonian. [Sig. hairy, shaggy.]
plicata, Lesquereux, 1858, Geo. Sur. Penn., vol. 2, Coal Meas. [Sig. folded.]
polyphylla, Lindley & Hutton, 1835, Foss. Flora, Coal Meas. [Sig. many-leaved.]
recurva, Dawson, 1863, Quar. Jour. Geo. Soc., vol. 19, Devonian. [Sig. turned back.]
rigida, Brongniart, 1828, Hist. Veg. Foss., Coal Meas. [Sig. rigid.]
scaberrima, Lesquereux, 1828, Hist. Veg. Foss., Coal Meas. [Sig. very rough.]
schlotheimi, see Hymenophyllites schlotheimi.
spinosa, see Hymenophyllites spinosus.
splendens, Dawson, 1871, Foss. Plants Canada, Devonian. [Sig. splendid.]
squamosa, Lesquereux, 1858, Geo. Sur. Penn., vol. 2, Coal Meas. [Sig. scaly.]
tenella, Brongniart, 1828, Hist. Veg. Foss., Coal Meas. [Sig. delicate.]
tenuifolia, see Hymenophyllites tenuifolius.
trichomanoides, see Hymenophyllites trichomanoides.
tridactylites, see Hymenophyllites tridactylites.
trifoliata, Brongniart, 1828, Hist. Veg. Foss., Coal Meas. [Sig. three-leaved.]
SPHENOTHALLUS, Hall, 1847, Pal. N. Y., vol. 1, [Ety. *sphen*, a wedge; *thallos*, a branch or frond.]
angustifolius, Hall, 1847, Pal. N. Y., vol. 1, Hud. Riv. Gr. [Sig. narrow-leaved.]
latifolius, Hall, 1847, Pal. N. Y., vol. 1, Hud. Riv. Gr. [Sig. broad-leaved.]
SPIROPHYTON, Hall, 1863, 16th Reg. Rep. [Ety. *speira*, a coil; *phyton*, a plant.]
cauda-galli, Vanuxem, 1842, (*Fucoides cauda-galli*) Geo. Rep. N.Y., Devonian. [Sig. like the tail of a cock.]
crassum, Hall, 1863, 16th Reg. Rep., Carb. Conglomerate. [Sig. thick.]
typus, Hall, 1863, 16th Reg. Rep., Chemung Gr. [Ety. type of the genus.]
velum, Vanuxem, 1842, (Fucoides velum) Geo.Rep.N.Y. [Ety. *velum*, a curtain.]
SPORANGITES, Dawson, 1863, Can. Nat. and Geol., vol. 8. [Sig. seed-vessel.]
glaber, Dawson, 1863, Can. Nat., vol. 8, Coal Meas. [Sig. smooth.]
papillatus, Dawson, 1863, Can. Nat., vol. 8, Coal Meas. [Sig. covered with papilli.]
STAPHYLOPTERIS, 1838, Presl, in Sternb. Vers. Darst. Flora der Vorwelt. [Ety. *staphyle*, a bunch of grapes; *pteris*, a fern.]
asteroides, Lesquereux, 1870, Geo. Sur. Ill., vol. 4, Coal Meas. [Sig. star-like.]
sagittata, Lesquereux, 1870, Geo. Sur. Ill., vol. 4, Coal Meas. [Sig. barbed like an arrow.]
stellata, Lesquereux, 1860, Geo. Sur. Ark., vol. 2, Coal Meas. [Sig. starred.]
wortheni, Lesquereux, 1870, Geo. Sur. Ill., vol. 4, Coal Meas. [Ety. proper name.]
STERNBERGIA, Artis, 1826, Antediluvian Phytology. [Ety. proper name.]
var. angularis, Dawson, 1868, Acad. Geol., Coal Meas. [Sig. angular.]
var. approximata, Dawson, 1868, Acad. Geol., Coal Meas. [Sig. approximate.]
var. distans, Dawson, 1868, Acad. Geol., Coal Meas. [Sig. distant.]
var. obscura, Dawson, 1868, Acad. Geol., Coal Meas. [Sig. obscure.]

STIGMARIA, Brongniart. 1822, Mem. du Mus. d'Hist. Nat. de Paris. [Ety. *stigma*, a dot or puncture.] This genus is regarded as representing the roots of Sigillaria.
anabathra, Corda, 1845, Beiträge zur Flora der Vorwelt, Coal Meas. [Sig. a ladder.]
areolata, Dawson, 1871, Foss. Plants Canada, Devonian. [Sig. areolate, divided into irregular squares, or small angular spaces.]
costata, Lesquereux, 1858, Geo. Sur. Penn., vol. 2, Coal Meas. [Sig. ribbed.]
elliptica, Lesquereux, 1870, Geo. Sur. Ill., vol. 4, Coal Meas. [Sig. in the form of an ellipse.]
eveni, see Stigmarioides eveni.
exigua, Dawson, 1862, Quar. Jour. Geo. Soc., vol. 18, Chemung Gr. [Sig. small.]
ficoides, Brongniart, 1822, Mem. du Mus. d'Hist. Nat. de Paris, Coal Meas. [Sig. like a fig.]
ficoides *var*. a, b, c, d, e, f, g, h, i, k, l, Dawson, 1868, Acad. Geol., Coal Meas.
ficoides *var*. reticulata, Gœppert, 1841, Gatt. d. Foss. Pflanzen, Coal Meas. [Sig. reticulated.]
ficoides *var*. stellata, Gœppert, 1841, Gatt. d. Foss. Pflanzen, Coal Meas. [Sig. starred.]
ficoides *var*. undulata, Gœppert, 1841, Gatt. d. Foss. Pflanzen, Coal Meas. [Sig. wavy.]
irregularis, Lesquereux, 1858, Geo. Sur. Penn., vol. 2, Coal Meas. [Ety. from the irregularity of the scars.]
minor, Gœppert, 1841, Gatt. d. Foss. Pflanzen, Coal Meas. [Sig. less.]
minuta, Lesquereux, 1858, Geo. Sur. Penn., vol. 2, Coal Meas. [Ety. in allusion to the small size of the lower scars.]
minutissima, Dawson, 1871, Foss. Plants Can., Devonian. [Sig. very minute.]
perlata, Dawson, 1871, Foss. Plants Canada, Devonian. [Sig. very wide.]
pusilla, Dawson, 1863, Quar. Jour. Geo. Soc., vol. 19, Devonian. [Sig. very small.]
radicans, Lesquereux, 1858, Geo. Sur. Penn., vol. 2, Coal Meas. [Sig. throwing out roots.]
umbonata, Lesquereux, 1858, Geo. Sur. Penn., vol. 2, Coal Meas. [Sig. embossed.]
verrucosa, Martin, 1809, (Phytolithus verrucosus) Petrif. Derb., Coal Meas. [Sig. warty.]
STIGMARIOIDES, Lesquereux, 1870, Geo. Sur. Ill., vol. 4. [Ety. from its resemblance to *Stigmaria*.]
affinis, Lesquereux, 1870, Geo. Sur. Ill., vol. 4, Coal Meas. [Sig. contiguous.]
eveni, Lesquereux, 1866, (Stigmaria eveni) Geo. Sur. Ill., vol. 4, Coal Meas. [Ety. proper name.]
linearis, Lesquereux, 1870, Geo. Sur. Ill., vol. 4, Coal Meas. [Sig. drawn out in lines.]
rugosus, Lesquereux, 1870, Geo. Sur. Ill., vol. 4, Coal Meas. [Sig. wrinkled.]
selago, Lesquereux, 1870, Geo. Sur. Ill., vol. 4, Coal Meas. [Sig. like the plant *Selago*.]
truncatus, Lesquereux, 1870, Geo. Sur. Ill., vol. 4, Coal Meas. [Sig. truncated, cut short.]
tuberosus, Lesquereux, 1870, Geo. Sur. Ill., vol. 4, Coal Meas. [Sig. tuberose, composed of tuber-like parts.]
villosus, Lesquereux, 1870, Geo. Sur. Ill., vol. 4, Coal Meas. [Sig. villous, covered with short hair-like projections.]
SYRINGODENDRON, Sternberg, 1820, Vers. Darst. Flora der Vorwelt. [Ety. *syrinx*, a pipe; *dendron*, a tree.]
bistriatum, Wood, 1860, Proc. Acad. Nat. Sci., Coal Meas. [Sig. double-striated.]
cyclostegium, Brongniart, 1828, Hist. Veg. Foss., Coal Meas. [Sig. circular covering.]
cyclostigma, Brongniart, 1828, Hist. Veg. Foss., Coal Meas. [Sig. circular puncture.]
gracile, Dawson, 1862, Quar. Jour. Geo. Soc., vol. 18, Chemung Gr. [Sig. slender.]
pachyderma, see Sigillaria pachyderma.
pes-capreoli, Sternb., 1828, Vers. Darst. Flora der Vorwelt, Coal Meas. [Ety. *pes*, stalk; *capreolus*, small tendril which supports it.]
porteri, Lesquereux, 1870, Geo. Sur. Ill., vol. 4, Coal Meas. [Ety. proper name.]
SYRINGOXYLON, Dawson, 1862, Quar. Jour. Geo. Soc., vol. 18. [Ety. *syrinx*, a pipe, *xylon*, wood.]
mirabile, Dawson, 1862, Quar. Jour. Geo. Soc., vol. 18, Ham. Gr. [Sig. extraordinary.]
TÆNIOPTERIS, Brongniart, 1828, Prodr. Hist. Veg. Foss. [Ety. *tainia*, a ribbon; *pteris*, a fern.]
magnifolia, Bunbury, 1847, Quar. Jour. Geo. Soc. Lond., vol. 3, Coal Meas. [Sig. large-leaved.]
TRICHOMANITES, Gœppert, 1836, Syst. Filic. Foss. [Ety. from the plant *Trichomanes*.]
filicula, Dawson, 1863, Quar. Jour. Geo. Soc., vol. 19, Devonian. [Ety. *filicula*, a small fern.]
TRIGONOCARPON, Brongniart, 1828, Prodr. Hist. Veg. Foss. [Ety. *trigon*, a triangle; *karpos*, a fruit.]
avellanum, Dawson, 1863, Can. Nat. & Geo., vol. 8, Coal Meas. [Sig. a filbert.]
bertholletiforme, Foster, 1853, Ann. of Sci., vol. 1, Coal Meas. [Sig. like *Bertholletia*.]
carbonarium, King, 1856, Proc. Acad. Nat. Sci., vol. 7, Coal Meas. [Sig. pertaining to coal.]

dawsi, Lindley & Hutton, 1837, Foss. Flora, Coal Meas. [Ety. proper name.]
hildrethi, Dawson, syn. (?) for Trigonocarpon triloculare.
hookeri, Dawson, 1861, Quar. Jour. Geol. Soc., vol. 17, Coal Meas. [Ety. proper name.]
intermedium, Dawson, 1863, Can. Nat., vol. 8, Coal Meas. [Sig. intermediate.]
juglans, Lesquereux, 1866, Geo. Sur. Ill., vol. 2, Low. Coal Meas. [Sig. a walnut.]
magnum, Newberry, 1873, Ohio Pal., vol. 1, Coal Meas. [Sig. large.]
mentzelianus, Gœppert & Berger, 1848, de Fruct. et Sem., Coal Meas. [Ety. proper name.]
minus, Dawson, 1863, Can. Nat. & Geol., vol. 8, Coal Meas. [Sig. less.]
multicarinatum, Newberry, 1873, Ohio Pal., vol. 1, Carb. Conglomerate. [Sig. many-carinated.]
multistriatum, see Carpolithes multistriatus.
nœggerathi, Brongniart, 1828, Prodr. Hist. Veg. Foss., Coal Meas. [Ety. proper name.]
oblongum, Lindley & Hutton, 1837, Foss. Flora, vol. 3, Coal Meas. [Sig. oblong.]
olivæforme, Lindley & Hutton, 1837, Foss. Flora, vol. 3, Coal Meas. [Sig. like an olive.]
ornatum, Newberry, 1873, Ohio Pal., vol. 1, Carb. Conglomerate. [Sig. ornate, adorned.]
parkinsoni, Brongniart, 1828, Prodr. Hist. Veg. Foss., Coal Meas. [Ety. proper name.]
perantiquum, Dawson, 1871, Foss. Plants Canada, Devonian. [Sig. very ancient.]
racemosum, Dawson, 1862, Quar. Jour. Geo. Soc., vol. 18, Devonian. [Sig. clustering.]
rostellatum, Lesquereux, 1866, Geo. Sur. Ill., vol. 2, Up. Coal Meas. [Sig. little-beaked.]
rotundum, Dawson, 1863, Can. Nat., vol. 8, Coal Meas. [Sig. wheel-shaped.]
schultzanum, Gœppert & Berger, 1848, de Fruct., etc., Coal Meas. [Ety. proper name.]
sigillariæ, Dawson, 1863, Can. Nat., vol. 8, Coal Meas. [Ety. from the genus *Sigillaria*.]
tricuspidatum, Newberry, 1873, Ohio Pal., vol. 1, Coal Meas. [Sig. three-pointed.]
triloculare, Hildreth, 1835, (Carpolithes trilocularis) Am. Jour. Sci., vol. 31, Conglomerate and Low. Coal Meas. [Sig. three-chambered.]
woodruffi, Moss, 1852, Proc. Acad. Nat. Sci., vol. 5, Coal Meas. [Ety. proper name.]
TROCHOPHYLLUM, Wood, 1860, Proc. Acad. Nat. Sci. This name was proposed as a substitute for *Annularia*, Sternb., because the latter was preoccupied as a generic name in the sub-kingdom Mollusca. [Ety. *trochos*, a wheel; *phyllon*, a leaf.]
ULODENDRON, Lindley & Hutton, 1831, Foss. Flora. [Ety. *ule*, wood; *dendron*, a tree.]
ellipticum, Sternberg, 1838, Vers. Darst. Flora der Vorwelt, Coal Meas. [Sig. elliptical.]
elongatum, Lesquereux, 1870, Geo. Sur. Ill., vol. 4, Coal Meas. [Sig. drawn out.]
lindleyanum, Presl, 1833, in Sternberg, Vers. Darst. Flora der Vorwelt, Coal Meas. [Ety. proper name.]
majum, Lindley & Hutton, 1831-33, Foss. Flora, Coal Meas. [Sig. great.]
minus, Lindley & Hutton, 1833, Foss. Flora, Coal Meas. [Sig. less.]
punctatum, Presl, 1833, in Sternberg, Vers. Flora der Vorwelt, Coal Meas. [Sig. dotted.]
UPHANTÆNIA, Vanuxem, 1843, Geo. Rep. N. Y. [Ety. *uphantos*, woven; *tainia*, ribbon.]
chemungensis, Vanuxem, 1843, Geo. Rep. N. Y., Chemung Gr. [Ety. proper name.]
WALCHIA, Sternberg, 1825, Vers. Darst. Flora der Vorwelt. [Ety. proper name.]
gracilis, Dawson, Coal Meas. [Sig. slender.]
robusta, Dawson, Coal Meas. [Sig. robust.]
WHITTLESEYA, Newberry, 1853, Ann. of Sci., vol. 1. [Ety. proper name.]
elegans, Newberry, 1853, Ann. of Sci., vol. 1, Cuyahoga shale. [Sig. elegant.]
ZAMITES, Brongiart, 1825, in Annales des Sciences Naturelles, vol. 4. [Ety. from its resemblance to the existing genus *Zamia*.]
gramineus, Bunbury, 1847, Quar. Jour. Geo. Soc. Lond., vol. 3, Coal Meas. [Sig. grassy.]
obtusifolius, Rogers, 1844, Rep. Ass'n Am. Geol., Coal Meas. [Sig. obtuse leaved.]

ANIMAL KINGDOM.

SUB-KINGDOM PROTISTA.

CLASS RHIZOPODA.—[Ety. *rhiza*, a root; *podes*, feet.]
CLASS PORIFERA.—[Ety. *porus*, a pore; *fero*, to bear.]

CLASS RHIZOPODA.—Dentalina, Eozoon, Fusulina, Nullipora, Receptaculites, Rotalia.
CLASS PORIFERA.—Archeocyathus, Astræospongia, Astylospongia, Aulocopina, Brachiospongia, Calathium, Cnemidium, Conopterium, Eospongia, Palæacis, Palæomanon, Rhabdaria, Trachyum, Trichospongia.
INCERTÆ SEDIS.—Pasceolus, Ribeiria. The latter genus is placed by Woodward in the class Lamellibranchiata. Salter referred it to the Crustacea. It is very doubtful whether it is American.

ARCHEOCYATHUS, Billings, 1861, Pal. Foss., vol. 1, Potsdam Gr. [Ety. *arche*, beginning; *cyathus*, a cup.]
atlanticus, Billings, 1861, Pal. Foss., vol. 1, Potsdam Gr. [Ety. mythological name.]
minganensis, Billings, 1859, (Petraia minganensis) Can. Nat. & Geol., vol. 4, Calcif. Gr. [Ety. proper name.]
profundus, Billings, 1861, Pal. Foss., vol. 1, Potsdam Gr. [Sig. deep.]
rensselæricus, Ford, 1873, Am. Jour. Sci. & Arts, 3rd ser., vol. 5, Low. Potsdam Gr. [Ety. proper name.]

ASTRÆOSPONGIA, Roemer, 1860, Sil. Fauna. des West Tenn. [Ety. *astræa*, from *aster*, a star; *spongia*, sponge.]
hamiltonensis, Meek & Worthen, 1866, Proc. Chi. Acad. Sci., vol. 1, Ham. Gr. [Ety. proper name.]
meniscus, Roemer, 1848, (Blumenbachium miniscus) Leonh. & Bronn's Jahrb., Niagara Gr. [Ety. *meniskos*, a little moon.]

ASTYLOSPONGIA, Roemer, 1860, Sil. Fauna. des West Tenn. [Ety. *astylos*, without a pillar or prop; *spongia*, sponge.]
christiana, Meek & Worthen, 1868, Geo. Sur. Ill., vol. 3, Niagara Gr. [Ety. proper name.]
imbricato-articulata, Roemer, 1848, (Siphonia imbricato-articulata) Leonh. & Bronn's Jahrb., Niagara Gr. [Ety. *imbricatus*, imbricated; *articulatus*, articulated.]
inciso-lobata, Roemer, 1848, (Spongia inciso-lobata) Leonh. & Bronn's Jahrb., Up. Sil. [Sig. cut into lobes.]
bursa, Hall, 1876, 28th Reg. Rep., Niagara Gr. [Sig. a purse.]
parvula, Billings, 1861, Pal. Foss., vol. 1, Trenton Gr. [Sig. small.]
perryi, Billings, 1861, Geol. Vermont, Black Riv. Gr. [Ety. proper name.]
præmorsa, Goldfuss, 1826, (Siphonia præmorsa) Petref. Germ., Niagara Gr. [Sig. jagged as if bitten off.]
stellatim-sulcata, Roemer, 1848, (Spongia stellatim-sulcata) Leonh. & Bronn's Jahrb., Up. Sil. [Sig. star-furrowed; in allusion to the star-like depressions on the outer surface.]

AULOCOPINA, Billings, 1875, Can. Nat. & Geol. [Ety. *aulokopeo*, cut into pipes.]
granti, Billings, 1875, Can. Nat. & Geol., Niagara Gr. [Ety. proper name.]
Blumenbachium, Konig, 1820, Icones, fossiles, sectiles,
meniscus, see Astræospongia meniscus.
BRACHIOSPONGIA, Marsh, 1867, Am. Jour. Sci. & Arts, 2nd series, vol. 44. [Ety. *brachium*, an arm; *spongia*, sponge.]
digitata, Owen, 1857, (Scyphia digitata) Geo. of Ky., vol. 2, Cin'ti Gr. [Sig. having fingers or toes.]
roemerana, Marsh, 1867, Am. Jour. Sci. and Arts, 2nd series, vol. 44, Cin'ti Gr. [Ety. proper name.]
lyoni, Marsh, 1867, Am. Jour. Sci. & Arts, 2nd series, vol. 44, Cin'ti Gr. [Ety. proper name.]

PROTISTA. 43

CALATHIUM, Billings, 1865, Pal. Foss., vol. 1. [Ety. *kalathos*, a small wicker basket.]
affine, Billings, 1865, Pal. Foss., vol. 1, Quebec Gr. [Sig. contiguous.]
anstedi, Billings, 1865, Pal. Foss., vol. 1, Quebec Gr. [Ety. proper name.]
canadense, Billings, 1865, Pal. Foss., vol. 1, Chazy Gr. [Ety. proper name.]
fittoni, Billings, 1865, Pal. Foss., vol. 1, Quebec Gr. [Ety. proper name.]
formosum, Billings, 1865, Pal. Foss., vol. 1, Quebec Gr. [Sig. beautiful.]
pannosum, Billings, 1865, Pal. Foss., vol. 1, Quebec Gr. [Sig. ragged.]
paradoxicum, Billings, 1865, Pal. Foss., vol. 1, Calcif. Gr. [Sig. puzzling, questionable.]
CNEMIDIUM, Goldfuss, 1826, Petref. Germ. [Ety. *knemidos*, armor for the legs, a sort of boot.]
trentonensis, Worthen, 1875, Geo. Sur. Ill., vol. 6, Trenton Gr. [Ety. proper name.]
CONOPTERIUM, Winchell, 1865, Proc. Acad. Nat. Sci. [Ety. *konos*, a cone; *poterion*, a cup.]
effusum, Winchell, 1865, Proc. Acad. Nat. Sci., Lithographic limestone. [Sig. spread abroad.]
Coscinopora sulcata, see Receptaculites oweni.
DENTALINA, D'Orbigny, 1826, Ann. Sci. Nat., vol. 7. [Ety. *dentale*, tooth; *inus*, resemblance.]
priscilla, Dawson, 1868, Acad.Geol., Carb. [Ety. proper name.]
FOSPONGIA, Billings, 1861, Pal. Foss., vol. 1. [Ety. *eos*, dawn; *spongia*, sponge.]
roemeri, Billings, 1861, Pal. Foss., vol. 1, Chazy Gr. [Ety. proper name.]
varians, Billings, 1861, Pal. Foss., vol. 1, Chazy Gr. [Ety. *varians*, variable.]
EOZOON, Dawson, 1865, Can. Nat. & Geo., 2d series, vol. 2. [Ety. *eos*, dawn; *zoon*, animal.]
canadense, Dawson, 1865, Can. Nat. & Geo., 2d series, vol. 2, Laurentian Gr. [Ety. proper name.]
FUSULINA, Fischer, 1837, Oryct. du Gouv. de Moscou. [Ety. *fusus*, a spindle; *inus*, little.]
cylindrica, Fischer, 1837, Oryct. du Gouv. de Moscou., Coal Meas. [Sig. cylindrical.]
cylindrica *var.* ventricosa,Meek & Hayden, 1859, Proc. Acad. Nat. Sci., vol. 10, Coal Meas. [Sig. ventricose.]
elongata, Shumard, 1858, Trans. St. Louis Acad. Sci., Permian Gr. [Sig. elongated.]
gracilis, Meek, 1864, Pal. of California, vol. 1, pt. 4, Coal Meas. [Sig. slender.]
robusta, Meek, 1864, Pal. California, vol. 1, Coal Meas. [Sig. robust.]
ventricosa, Meek & Hayden, 1864, Pal. Upper Mo., Coal Meas. See F. cylindrica *var.* ventricosa.]
Ischadites tesselatus, Winchell & Marcy, Syn. for Receptaculites infundibulus.

Lunulites, (?) *dactioloides*, see Receptaculites dactioloides.
NULLIPORA, Lamarck, 1801, Systeme des Animaux sans Vertebres. [Ety. *nullus*, no; *porus*, pore.]
(?) obtexta, White, 1862, Proc. Bost. Soc. Nat. Hist., vol. 9, Burlington Gr. [Sig. woven over.]
Orbitulites, (?) *reticulata*, see Receptaculites reticulatus.
PALÆACIS, Haime, 1860, Hist. Nat. des Coralliaires. [Ety. *palaios*, ancient; *akis*, a barb.]
compressus, Meek & Worthen, 1866, (Sphenopoterium compressum) Proc. Acad. Nat. Sci. Phil., Keokuk Gr. [Sig. flattened.]
cuneatus, Meek & Worthen, 1860, (Sphenopoterium cuneatum) Proc. Acad. Nat. Sci. Phil., St. Louis Gr. [Sig. wedge-shaped.]
enormis, Meek & Worthen, 1866, (Sphenopoterium enorme) Geo. Sur. Ill., vol. 2, Kinderhook Gr. [Sig. very large.]
enormis *var.* depressus, Meek & Worthen, 1866, (Sphenopoterium enorme *var.* depressum) Geo. Sur. Ill., vol. 2, Kinderhook Gr. [Sig. depressed.]
obtusus, Meek & Worthen, 1860, (Sphenopoterium obtusum) Proc. Acad. Nat. Sci., Keokuk Gr. [Sig. obtuse.]
PALÆOMANON, Roemer, 1860, Sil. Fauna. West Tenn. [Ety. *palaios*, ancient; *Manon*, a recent genus of sponges.]
cratera, Roemer, 1848, (Siphonia cratera) Leonh. und Bronn's Jahrb., Up. Sil. [Sig. a cup or goblet.]
PASCEOLUS, Billings, 1857, Rep. of Progr. [Ety. *pasceolus*, a leather money bag.]
claudii, S. A. Miller, 1874, Cin. Quar. Jour. Sci., Cin.Gr. [Ety. proper name.]
darwini, S. A. Miller, 1874, Cin. Quar. Jour. Sci., Cin. Gr. [Ety. proper name.]
globosus, Billings, 1857, Rep. of Progr., Trenton Gr. [Sig. globular.]
gregarius, Billings, 1866, Catal. Sil. Foss., Antic., Anticosti Gr. [Sig. found in flocks.]
halli, Billings, 1857, Rep. of Progr., Anticosti Gr. [Ety. proper name.]
intermedius, Billings, 1866, Catal. Sil. Foss. Antic., Anticosti Gr. [Sig. intermediate.]
RECEPTACULITES, DeFrance, 1827, Dict. Sci. Nat. vol. 45. [Ety. *receptaculum*, a receptacle; *lithos*, stone.]
calciferus, Billings, 1865, Pal. Foss., vol. 1, Calcif. Gr. [Ety. from the Calciferous Group.]
canadensis, Billings, 1863, (Ischadites canadensis) Geo. of Canada, Anticosti Gr. [Ety. proper name.]
dactioloides, Owen, 1840, (Lunulites dactioloides) Rep. on Min. Lands. Up. Sil. [Ety. from the thimble-like punctures. The correct orthography is *dactyloides*.]

elegantulus, Billings, 1865, Pal. Foss., vol. 1, Calcif. Gr. [Sig. elegant.]
formosus, Meek & Worthen, 1870, Proc. Acad. Nat. Sci., Niagara Gr. [Sig. beautiful.]
fungosus, Hall, 1861, Geo. Rep. Wis., Galena Gr. [Sig. spongy.]
globularis, Hall, 1861, Supp. Geo. Rep. Wis., Galena Gr. [Sig. globular.]
hemisphericus, Hall, 1861, Geo. Rep. Wis. Niagara Gr. [Sig. hemispherical.]
infundibulus, Hall, 1861, Geo. Rep. Wis., Niagara Gr. [Sig. a funnel.]
insularis, Billings, 1866, Catal. Sil. Foss. Antic., Anticosti Gr. [Sig. upon an island.]
iowensis, Owen, 1852, (Selenoides iowensis) Geo. Rep. Wis., Iowa and Minn. [Ety. proper name.]
jonesi, Billings, 1865, Pal. Foss., vol. 1, Low. Held. Gr. [Ety. proper name.]
neptuni, De France, 1827, Dict. des Sci. Nat., vol. 45, Trenton Gr. [Ety. proper name.]
occidentalis, Salter, 1859, Can. Org. Rem. Decade 1, Trenton Gr. [Sig. Western.]
ohioensis, Hall & Whitfield, 1875, Ohio Pal., vol. 2, Niagara Gr. [Ety proper name.]
oweni, Hall, 1861, Geo. Rep. Wis., Galena Gr. [Ety. proper name.]
reticulatus, Owen, 1840, (Orbituloides reticulata) Rep. on Min. lands, Niagara Gr. [Sig. reticulated.]
subturbinatus, Hall, 1863, Trans. Alb. Inst., vol. 4, Niagara Gr. [Sig. somewhat top-shaped.]
sulcatus, Owen, 1844. This name was preoccupied and the species is now named R. oweni.

RHABDARIA, Billings, 1865, Pal. Foss., vol. 1. [Ety. rhabdos, a rod.]
fragilis, Billings, 1865, Pal. Foss., vol. 1, Calcif. Gr. [Sig. frail, brittle.]
furcata, Billings, 1865, Pal. Foss., vol. 1, Calcif. Gr. [Sig. forked.]

RIBEIRIA, Sharpe, 1853, Jour. Geo. Soc., vol. 9. [Ety. proper name (?).]
(?) calcifera, Billings, 1865, Pal. Foss., vol. 1, Calcif. Gr. [Ety. from the Calciferous Group.]
(?) longiuscula, Billings, 1865, Pal. Foss., vol. 1, Calcif. Gr. [Sig. somewhat the longest.]

ROTALIA, Lamarck, 1804, Ann. Mus. [Ety. rota, a wheel.]
baileyi, Hall, 1858, Trans. Alb. Inst., vol. 4, Warsaw Gr. [Ety. proper name.]

Scyphia, Oken, 1815, Lehrb. Naturg.
digitata, see Brachiospongia digitata.
stellata, Troost, 1840. Not clearly defined.
Selenoides, Owen, 1852. Syn. for Receptaculites.
iowensis, Owen, 1852, see Receptaculites iowensis.
Siphonia, Parkinson, 1820, Organ. Rem.
cratera, see Palæomanon cratera.
imbricato-articulata, see Astylospongia imbricato-articulata.
præmorsa, see Astylospongia præmorsa.
Sphenopoterium, Meek & Worthen, 1866. A Syn. for Palæacis.
compressum, see Palæacis compressus.
cuneatum, see Palæacis cuneatus.
enorme, see Palæacis enormis.
enorme var. depressum, see Palæacis enormis var. depressus.
obtusum, see Palæacis obtusus.
Spongia, Linnæus, 1789, Systema Naturæ.
inciso-lobata, see Astylospongia incisolobata.
stellatim-sulcata, see Astylospongia stellatim-sulcata.

TRACHYUM, Billings, 1865, Pal. Foss., vol. 1. [Ety. trachus, rough, rugged.]
cyathiforme, Billings, 1865, Pal. Foss., vol. 1, Quebec Gr. [Sig. cup-shaped.]
rugosum, Billings, 1865, Pal. Foss., vol. 1, Quebec Gr. [Sig. wrinkled.]

TRICHOSPONGIA, Billings, 1865, Pal. Foss., vol. 1. [Ety. trichius, to show hairs; spongia, a sponge.]
sericea, Billings, 1865, Pal. Foss., vol. 1, Calciferous Gr. [Sig. silky.]

SUB-KINGDOM RADIATA.

CLASS POLYPI.

ORDER ZOANTHARIA.

FAMILY ASTRÆIDÆ.—Astræa, Sarcinula.
FAMILY CYATHOPHYLLIDÆ.—Acervularia, Acrophyllum, Amplexus, Anthophyllum, Aulophyllum, Axophyllum, Baryphyllum, Blothrophyllum, Campophyllum, Chonophyllum, Clisiophyllum, Combophyllum, Cyathophyllum, Cyclolites, (?) Cystiphyllum, Diphyphyllum, Duncanella, Eridophyllum, Ethmophyllum, Hadrophyllum, Heliophyllum, Heterophrentis, Lithostrotion, Lonsdalcia, Lophophyllum, Microcyclus, Omphyma, Pachyphyllum, Palaeocyclus, Palaeophyllum, Petraia, Phillipsastrea, Ptycophyllum, Smithia, Streptelasma, Strombodes, Vesicularia, Zaphrentis.
FAMILY CYATHAXONIDÆ.—Cyathaxonia.
FAMILY FAVOSITIDÆ.—Alveolites, Astrocerium, Bolboporites, Chetetes, Cladopora, Columnaria, Columnopora, Dendropora, Emmonsia, Faviphyllum, Favistella, Favosites, Leptopora, Limaria, Lunatipora, Michelinia, Monticulipora, Nebulipora, Sphærolites, Stellipora, Stenopora, Striatopora, Tetradium, Thecia, Trachypora, Vermipora.
FAMILY HALYSITIDÆ.—Calapœcia, Haimeophyllum, Halysites, Syringopora.
FAMILY MILLEPORIDÆ.—Fistulipora, Heliolites, Lyellia, Plasmopora, Rhombopora, Thecostegites.
FAMILY PORITIDÆ.—Pleurodictyum, Protarea.

ORDER ALCYONARIA.

FAMILY ALCYONIDÆ.—Aulopora, Heliopora, Quenstedtia.
FAMILY TUBIPORIDÆ.—Cannopora, Caunopora, Dictyostroma, Lamellopora, Stromatocerium, Stromatopora, Syringostroma, Tubipora.
FAMILY GRAPTOLITIDÆ.—Buthograptus, Callograptus, Chonograptus, Cladograptus, Climacograptus, Dawsonia, Dendrograptus, Dicranograptus, Dictyonema, Didymograptus, Diplograptus, Discophyllum, Graptolithus, Inocaulis, Megalograptus, Monograptus, Nemagraptus, Nereograptus, Phyllograptus, Plumalina, Ptilograptus, Rastrites, Retiograptus, Retiolites, Staurograptus, Tetragraptus, Thamnograptus.

ACERVULARIA, Schweigger, 1820, Handb. der Naturg. [Ety. *acervus*, a heap, considered as a body.]
clintonensis, Nicholson, 1875, Ohio Pal., vol. 2, Clinton Gr. [Ety. proper name.]
davidsoni, Edwards & Haime, 1851, Monograph, Corniferous & Ham. Gr. [Ety. proper name.]
inequalis, Hall, 1873, 23rd Reg. Rep. Chemung Gr. [Sig. unequal.]
profunda, Hall, 1858, Geo. of Iowa, Corniferous and Ham. Gr. [Sig. deep.]
rugosa, Hall, 1843, (Astrea rugosa) Geo. Rep. 4th Dist. N. Y., Onondaga Gr. [Sig. wrinkled.] Is this Cyathophyllum rugosum?

ACROPHYLLUM, Thomson & Nicholson, 1876, Ann. Mag. Nat. Hist., 4th series, vol. 17. '[Ety. *akros*, the point or summit; *phyllon*, a leaf.]
oneidaense, Billings, (Clisiophyllum oneidaense) Can. Jour. Corniferous Gr. [Ety. proper name.]

Agaricia, Lamarck, 1801, Syst. des Anim. sans. Vert.

swinderniana, see Thecia swinderniana.

ALVEOLITES, Lamarck, 1801, Syst. des An. sans. Vert. [Ety. *alveus*, a cavity; *lithos*, stone.]
billingsi, Nicholson, 1874, Geo. Mag. N. S., vol. 1, Corniferous Gr. [Ety. proper name.]
confertus, Nicholson, 1874, Geo. Mag. N. S., vol. 1, Cornif. Gr. [Sig. thick or crammed in close together.]
cryptodens, Billings, 1859, Can. Jour., vol. 4, Up. Held. Gr. [Sig. hidden-toothed.]
distans, Nicholson, 1874, Geo. Mag. N. S., vol. 1, Cornif. Gr. [Ety. in allusion to the distance between the calices.]
dubia, see Favosites dubius.
exsul, Hall, 1876, 28th Reg. Rep. Niagara Gr. [Sig. a wanderer.]
fisheri, Billings, 1859, Can. Jour. Up. Held. Gr. [Ety. proper name.]
frondosus, Nicholson. 1874, Geo. Mag. N. S., vol. 1, Ham. Gr. [Sig. full of branches.]
goldfussi, Billings, 1859, Can. Jour. Ham. Gr. [Ety. proper name.]
granulosus, James, 1875, Catal. Cin. Foss., Cincinnati Gr. [Sig. full of granules.]
labechi, Edwards & Haime, 1851, Pal. Foss. des Terr., Palæoz., Anticosti Gr. [Ety. proper name.]
labiosus, Billings, 1859, Can. Jour., Corniferous Gr. [Sig. large-lipped.]
megastoma, Winchell, 1866, Rep. Low. Peninsula, Mich., Ham. Gr. [Sig. large-mouthed; from the large oblique cell mouths.]
niagarensis, Rominger, 1876, Foss. Corals, Niagara Gr. [Ety. proper name.]
ramulosus, Nicholson, 1874, Geo. Mag. Corniferous Gr. [Sig. full of little sprigs.].
repens, Fought, 1749, (Millepora repens) Amœn. Acad., vol. 1, Niagara Gr. [Sig. creeping.]
reticulata, see Favosites reticulatus.
rockfordensis, Hall, 1873, Chemung Gr., 26th Reg. Rep. [Ety. proper name.]
roemeri, Billings, 1859, Can. Jour., Ham. Gr. [Ety. proper name.]
selwyni, Nicholson, 1874, Geo. Mag. N. S., vol. 1, Corniferous Gr. [Ety. proper name.]
squamosus, Billings, 1859, Can. Jour., Corniferous Gr. [Sig. scaly, rough.]
strigillatus, Winchell, 1866, Rep. Low. Peninsula, Mich., Ham. Gr. [Sig. wide-furrowed.]
subramosus, Rominger, 1876, Foss. Corals, Ham. Gr. [Sig. somewhat ramose.]
vallorum, Meek, 1868, Trans. Chi. Acad. Sci., Devonian. [Sig. intrenched.]

AMPLEXUS, Sowerby, 1812, Mineral Conchology, vol. 1. [Ety. *amplexus*, encircling, surrounding.]
cingulatus, Billings, 1862, Pal. Foss., vol. 1, Mid. Sil. [Sig. banded, from the sharp-edged rings of growth.]
coralloides, Sowerby, 1812, Min. Conch., vol. 1, Warsaw Gr. [Sig. like a coral.]
exilis, Billings, 1875. Can. Nat. & Geol. Corniferous Gr. [Sig. small.]
fragilis, White & St. John, 1868, Trans. Chi. Acad. Sci., Keokuk Gr. [Sig. frail.]
mirabilis, Billings, 1875. Can. Nat. & Geol., Coal Meas. [Sig. wonderful.]
shumardi, M. Edwards, 1851, (Cyathophyllum shumardi) Mon. des Polyp. Foss., Niagara Gr. [Ety. proper name.]
yandelli, Edwards & Haime, 1851, Pal. Foss. des Ter. Palæoz., Devonian. [Ety. proper name.]
zaphrentiformis, White, 1876, Geo. Uinta Mountains, Low. Aubrey Gr. [Sig. in form like the genus *Zaphrentis*.]

ANTHOPHYLLUM, Schweigger, 1820, Handb. der Naturg. [Ety. *anthos*, a flower; *phyllon*, a leaf.]
denticulatum, Goldfuss, 1826, Petref. Germ., Niagara Gr. [Sig. denticulated.]
expansum, Owen, 1840, Rep. on Mineral Lands, Devonian. [Sig. spread out.]

Astræa, Lamarck, 1815, Histoire Naturelle des Animaux sans Vertebres.
gigas, see Phillipsastrea gigas.
helianthoides, as identified by D'Archiac & Verneuil, see Heliophyllum Halli.
mammillaris, Fischer as identified by Castlenau, Syn. for Lithostrotion canadense.
mammillaris, see Phillipsastrea mammillaris.
rugosa, see Acervularia rugosa.
tesselata, Troost. 1840, 5th Geo. Rep. Tenn., Low. Carb. Not satisfactorily defined.

POLYPI.

ASTROCERIUM, Hall, 1852, Pal. N. Y., vol. 2. [Ety. *aster*, a star; *herium*, honey-comb.] This genus is generally considered a synonym for Favosites and hence I have placed the species under that head also.
   constrictum, Hall, 1852, Pal. N. Y., vol. 2, Niagara Gr. [Sig. constricted.]
   parasiticum, Hall, 1852, Pal. N. Y., vol. 2, Niagara Gr. [Sig. parasitic.]
   pyriforme, Hall, 1852, Pal. N. Y., vol. 2, Niagara Gr. [Sig. pear-shaped.]
   venustum, Hall, 1852, Pal. N. Y., vol. 2, Niagara Gr. [Sig. elegant.]
AULOPHYLLUM, Edwards & Haime, 1850, Brit. Pal. Foss. [Ety. *aulos*, a pipe; *phyllon*, a plant.]
   richardsoni, Meek, 1868, Trans. Chi. Acad. Sci., Devonian. [Ety. proper name.]
AULOPORA, Goldfuss, 1826, Germ. Petref. [Ety. *aulos*, a pipe; *poros*, a pore.]
   aperta, Winchell, 1866, Rep. Low. Peninsula, Mich., Ham. Gr. [Sig. open, wide.]
   arachnoidea, Hall, 1847, Pal. N. Y. vol. 1, Trenton & Hud. Riv. Gr. [Sig. in the form of a cob-web.]
   conferta, Winchell, 1866, Rep. Low. Peninsula, Mich. Ham. Gr. [Sig. thick or crammed in close together.]
   cornuta, see Quenstedtia cornuta.
   cyclopora, Winchell, 1866, Rep. Low. Peninsula, Mich., Ham. Gr. [Ety. *kuklos*, a circle; *poros*, a pore.]
   erecta, Rominger, 1876, Foss. Corals, Ham. Gr. [Sig. erect.]
   filiformis, Billings, 1859, Can. Jour., vol. 4, Corniferous Gr. [Sig. thread-like.]
   iowensis, Hall, 1873, 23rd Reg. Rep., Chemung Gr. [Ety. proper name.]
   precius, Hall, 1876, 28th Reg. Rep., Niagara Gr. [Sig. a kind of grapevine.]
   repens, Walch et Knorr, 1775, (Milleporites repens) Samulung von Merk., Niagara Gr. [Sig. creeping.]
   saxivadum, Hall, 1873, 23rd Reg. Rep., Chemung Gr. [Sig. stone-bottomed.]
   schohariæ, Hall, 1874, 26th Reg. Rep., Low. Held. Gr. [Ety. proper name.]
   serpens, Goldfuss, 1826, Germ. Petref., Hamilton Gr. [Sig. crawling.]
   serpuloidea, Winchell, 1866, Rep. Low. Peninsula, Mich., Ham. Gr. [Sig. worm-formed.]
   tubæformis, Goldfuss, 1826, Germ. Petref.? Lonsdale, 1839, Murch. Sil. Syst., Ham. Gr. [Sig. trumpet formed.]
   umbellifera, see Quenstedtia *umbellifera*.
Arinura, Castelnau, syn. for Lithostrotion.
   canadense, see Lithostrotion canadense.
AXOPHYLLUM, Edwards & Haime, 1850, Brit. Pal. Foss. [Ety. *axon*, axis; *phyllon*, a plant.]
   infundibulum, Worthen, 1875, Geo. Sur. Ill., vol. 6, Coal Meas. [Sig. a funnel.]
   rudis, White & St. John, 1868, Trans. Chi. Acad. Sci., Coal Meas. [Sig. rude, rustic.]

BARYPHYLLUM, Edwards & Haime, 1850, Brit. Foss. Corals. [Ety. *barys*, heavy; *phyllon*, a plant.]
   arenarium, Meek & Worthen, 1868, Geo. Sur. Ill., vol. 3, Onondaga Gr. [Sig. sandy.]
BLOTHROPHYLLUM, Billings, 1859, Can. Jour., vol. 4. [Ety. *blothros*, tall-growing; *phyllon*, a plant.]
   approximatum, Nicholson, 1875, Can. Nat. & Geol., Corniferous Gr. [Sig. approximate.]
   cæspitosum, Rominger, 1876, Foss. Corals, Niagara Gr. [Sig. turfy.]
   decorticatum, Billings, 1859, Can. Jour., vol. 4, Corniferous Gr. [Sig. peeled, barked.]
BOLBOPORITES, Billings, 1859, Can. Nat. & Geol., vol. 4. [Ety. *bolbos*, a bulb; *poros*, a pore.]
   americanus, Billings, 1859, Can. Nat. & Geo., vol. 4, Chazy Gr. [Ety. proper name.]
BYTHOGRAPTUS, Hall, 1861, Geo. Rep. Wis. [Ety. *buthos*, in the deep; *grapho*, to write.]
   laxus, Hall, 1861, Geo. Rep. Wis., Trenton Gr. [Sig. loose.]
Calamopora, Goldfuss, syn. for Favosites.
   basaltica, see Favosites basalticus.
   cristata, see Favosites cristatus.
   cumberlandica, see Favosites cumberlandicus.
   favosa, see Favosites favosus.
   fibrosa, see Monticulipora fibrosa.
   forbesi var. discoidea, see Favosites forbesi var. discoidea.
   gothlandica, see Favosites gothlandicus.
   heliolitiformis, see Favosites heliolitiformis.
   hemispherica, see Favosites hemisphericus.
   infundibuliformis, Goldfuss, as identified by D'Archiac & Verneuil.
   mackrothi, Geinitz, see Chetetes mackrothi.
   maxima, see Favosites maximus.
   tumida, see Chetetes tumidus.
   winchelli, see Favosites winchelli.
CALAPŒCIA, Billings, 1865, Can. Nat. & Geo., 2d ser., vol. 2, Mid. Sil. [Ety. *kalos*, beautiful; *poikilos*, spotted.]
   anticostiensis, Billings, 1865, Can. Nat. & Geo., 2d series, vol. 2, Mid. Sil. [Ety. proper name.]
   canadensis, Billings, 1865, Can. Nat. & Geo., 2d series, vol. 2, Black Riv. Gr. [Ety. proper name.]
   huronensis, Billings, 1865, Can. Nat. & Geo., 2d series, vol. 2, Hud. Riv. Gr. [Ety. proper name.]
CALLOGRAPTUS, Hall, 1865, Can. Org. Rem., Decade 2. [Ety. *kallos*, beautiful; *grapho*, to write.]
   (?) diffusus, see Dendrograptus diffusus.
   elegans, Hall, 1865, Can. Org. Rem., Decade 2, Quebec Gr. [Sig. elegant.]
   salteri, Hall, 1865, Can. Org. Rem., Decade 2, Quebec Gr. [Ety. proper name.]

CAMPOPHYLLUM, Edwards & Haime, 1850, British Foss. Corals. [Ety. *kampos*, a sea animal; *phyllon*, a plant.]
   nanum, Hall, 1873, 23d Reg. Rep. Chemung Gr. [Sig. dwarfish.]
   texanum, Shumard, 1858, Trans. St. Louis Acad. Sci., Permian Gr. [Ety. proper name.]
   torquium, Owen, 1852, (Cyathophyllum torquium) Geo. Rep. Wis., Iowa and Minn., Coal Meas. [Sig. twisted.]
Caninia, Syn. for Zaphrentis.
   bilateralis, see Zaphrentis bilateralis.
CANNAPORA, Hall, 1852, Pal. N. Y., vol. 2. [Ety. *kanna*, reed; *poros*, pore.]
   junciformis, Hall, 1852, Pal. N. Y., vol. 2, Clinton Gr. [Sig. like rush stems in form.]
Catenipora, Lamarck, 1816, Anim. sans. Vert. Syn. for Halysites.
CAUNOPORA, Phillips, 1841, Palaeozoic fossils of Cornwall, Devon and West Somerset. [Ety. *kaunos*, loose; *poros*, perforation.]
   planulata, Hall, 1873, 23rd Reg. Rep., Chemung Gr. [Sig. somewhat flat.]
CHETETES, Fisher, 1837, Oryct. du Gouv. Moscou. [Ety. *chaite*, hair.]
   approximatus, see Monticulipora approximata.
   attritus, see Monticulipora attrita.
   barrandi, see Monticulipora barrandi.
   briareus, see Monticulipora briareus.
   calicula, see Monticulipora calicula.
   carbonarius, Worthen, 1875, Geo. Sur. Ill., vol. 6, Coal Meas. [Sig. pertaining to coal.]
   cincinnatiensis, see Monticulipora cincinnatiensis.
   clathratulus, syn. for Cyclopora jamesi.
   clavacoideus, see Monticulipora clavacoidea.
   columnaris, see Tetradium columnare.
   consimilis, Hall, 1876, 28th Reg. Rep., Niagara Gr. [Sig. similar in all parts.]
   corticans, syn. for Monticulipora tuberculata.
   dalei, see Monticulipora dalli.
   decipiens, see Monticulipora decipiens.
   delicatulus, see Monticulipora delicatula.
   discoideus, see Monticulipora discoidea.
   fibrosus, see Monticulipora fibrosa.
   fletcheri, see Monticulipora fletcheri.
   frondosus, see Monticulipora frondosa.
   gracilis, see Monticulipora gracilis.
   hamiltonensis, Winchell, 1866, Rep. Low. Pen. Mich., Ham. Gr. [Ety. proper name.]
   helderbergiae, Hall, 1874, 26th Reg. Rep. Low., Held. Gr. [Ety. proper name.]
   jamesi, see Monticulipora jamesi.
   lycoperdon, see Monticulipora lycoperdon.
   mackrothi, Geinitz, 1846, (Calamopora mackrothi) Grund., p. 586, Permian Gr. [Ety. proper name.]
   mammulatus, see Monticulipora mammulata.
   milleporaceus, Troost. Where?
   microscopica, Winchell, 1866, Rep. Low. Pen. Mich., Ham. Gr. [Ety. *mikros*, small; *skopeo*, to view—from the small branches.]
   moniliformis, Nicholson, 1874, Geo. Mag. N. S., vol. 1, Ham. Gr. [Sig. bead-like.]
   muscatinensis, White 1876, Proc. Acad. Nat. Sci., Devonian. [Ety. proper name.]
   newberryi, see Monticulipora newberryi.
   nodulosus, see Monticulipora nodulosa.
   onealli, see Monticulipora onealli.
   ortoni, see Monticulipora ortoni.
   papillatus, see Monticulipora papillata.
   pavonia, see Monticulipora pavonia.
   petechialis, see Monticulipora petechialis.
   petropolitanus, Pander. It is not evident that this species is found in America.
   pulchellus, Edwards & Haime, as identified by Nicholson, but I think it is not found in this country.
   quadrangularis, Nicholson, 1874, Geo. Mag. N. S., vol. 1, Ham. Gr. [Sig. four cornered; quadrangular.]
   quadratus, see Monticulipora quadrata.
   rhombicus, Nicholson, 1875, syn. for Monticulipora quadrata.
   rugosus, Hall, see Monticulipora rugosa.
   rugosus, Edwards & Haime, 1851. This name was preoccupied. Moreover it is merely a form of Monticulipora dalii.
   sigillarioides, see Monticulipora sigillarioidea.
   sphaericus, Hall, 1874, 26th Reg. Rep. Low. Held. Gr. [Sig. spherical.]
   subpulchellus, see Monticulipora subpulchella.
   tuberculatus, see Monticulipora tuberculata.
   tumidus, Phillips, 1836, (Calamopora tumida) Geol. Yorkshire, Low. Carboniferous. [Ety. *tumidus*, swollen.]
CHONOGRAPTUS, Hall, 1873, Ann. Mag. Nat. Hist., 4th series, vol. 13. [Ety. *konos*, a twig; *grapho*, to write.] This subgenus is proposed to include the two species, Graptolithus flexilis and G. rigidus.
CHONOPHYLLUM, Edwards & Haime, 1850, Brit. Corals. [Ety. *konos*, a cone; *phyllon*, a plant.]
   belli, Billings, 1865, Can. Nat. & Geo., vol. 2, Clinton Gr. [Ety. proper name.]
   ellipticum, Hall, 1873, 23d Reg. Rep. Chemung Gr. [Sig. in the form of an ellipse.]
   magnificum, Billings, 1859, Can. Jour. Corniferous Gr. [Sig. magnificent.]
   niagarense, Hall, 1852, Pal. N. Y., vol. 2, Niagara Gr. [Ety. proper name.]
   ponderosum, Rominger, 1876, Foss. Corals Ham. Gr. [Sig. bulky, heavy.]
Chonostegites clappi, see Michelinia clappi.

POLYPI.

CLADOGRAPTUS, Geinitz, 1852, (Cladograpsus) Verst. Grauw. Sachs., etc. [Ety. *klados*, a twig; *grapho*, to write.]
dissimilaris, Emmons, 1856, Am. Geol., Quebec Gr. [Sig. dissimilar.]
inequalis, Emmons, 1856, Am. Geol., Quebec Gr. [Sig. unequal.]

CLADOPORA, Hall, 1852, Pal. N. Y., vol. 2. [Ety. *klados*, a twig; *poros*, a pore.]
alpenensis, Rominger, 1876, Foss. Corals, Ham. Gr. [Ety. proper name.]
aspera, Rominger, 1876, Foss. Corals, Up. Held. Gr. [Sig. rough.]
cæspitosa, Hall, 1852, Pal. N. Y., vol. 2, Niagara Gr. [Sig. turf-like.]
canadensis, Rominger, 1876, Foss. Corals, Ham. Gr. [Ety. proper name.]
cervicornis, Hall, 1852, Pal. N. Y., vol. 2, Niagara Gr. [Sig. shaped like the horns of a stag.]
dichotoma, Hall, 1858, Geo. Sur. Iowa, Ham. Gr. [Sig. dividing into two.]
expatiata, Rominger, 1876, Foss. Corals, Up. Held. Gr. [Sig. spread out.]
fibrosa, Hall, 1852, Pal. N. Y., vol. 2, Niagara Gr. [Sig. fibrous.]
imbricata, Rominger, 1876, Foss. Corals, Up. Held. Gr. [Sig. imbricated.]
labiosa, Billings, 1859, Can. Jour., Corniferous Gr. [Sig. full lipped.]
laqueata, Rominger, 1876, Foss. Corals, Niagara Gr. [Sig. paneled.]
lichenoidea, Winchell & Marcy, 1865, Bost. Soc. Nat. Hist., Niagara Gr. [Sig. resembling a lichen.]
macrophora, Hall, 1852, Pal. N. Y., vol. 2, Niagara Gr. [Sig. from the wide meshes.]
magna, Hall, 1873, 23rd Reg. Rep., Up. Held. Gr. [Sig. great.]
multipora, Hall 1852, Pal. N. Y., vol. 2, Niagara Gr. [Sig. having many pores.]
palmata, Hall, 1873, 23rd Reg. Rep., Up. Held. Gr. [Sig. having five lobes.]
pinguis, Rominger, 1876, Foss. Corals, Up. Held. Gr. [Sig. plump.]
prolifica, Hall, 1873, 23rd Reg. Rep., Up. Held. Gr. [Sig. prolific.]
pulchra, Rominger, 1876, Foss. Corals, Up. Held. Gr. [Sig. beautiful.]
reticulata, Hall, 1852, Pal. N. Y., vol. 2, Niagara Gr. [Sig. reticulated.]
rimosa, Rominger, 1876, Foss. Corals. Up. Held. Gr. [Sig. full of fissures.]
robusta, Rominger, 1876, Foss. Corals, Corniferous & Ham. Gr. [Sig. robust.]
seriata, Hall, 1852, Pal. N. Y., vol. 2, Niagara Gr. [Sig. from the regular alternating series of openings.]
turgida, Rominger, 1876, Foss. Corals, Up. Held. Gr. [Sig. swollen.]
verticillata, Winchell & Marcy, Bost. Soc. Nat. Hist., Niagara Gr. [Sig. whorled.]

CLIMACOGRAPTUS, Hall, 1865, Can. Org. Rem. Decade 2. [Ety. *klimax*, a small ladder; *grapho*, to write.]
antennarius, Hall, 1863, (Graptolithus antennarius) Geo. of Can., Quebec Gr. [Ety. from the *antennæ*.]
bicornis, Hall, 1847, (Graptolithus bicornis) Pal. N. Y., vol. 1, Hud. Riv. Gr. [Ety. having two horns.] This is the type of the genus.
parvus, Hall, 1865, Can. Org. Rem., Decade 2, Hud. Riv. Gr. [Sig. small.]
typicalis Hall, 1865, Can. Org. Rem., Decade 2, Hud. Riv. Gr. [Ety. type of the genus, though the genus was founded on *C. bicornis*.]

CLISIOPHYLLUM, Dana, 1846, Explor. Exped., vol. 8. [Ety. *klision*, a tent; from the conical central boss; *phyllon*, a plant.]
gabbi, Meek, 1864, Pal. California, Carb. [Ety. proper name.]
oneidænse, Billings, 1859, Can. Jour. Corniferous Gr. [Ety. proper name.] This species is placed in the new genus acrophyllum, by Thomson & Nicholson.

COLUMNARIA, Goldfuss, 1826, Germ. Petref. [Ety. *Columnarius*, formed of columns.]
alveolata, Goldfuss, 1826, Germ. Petref. Black Riv. Gr. [Sig. hollowed out.]
blainvilli, Billings, 1858, Can. Nat. & Geo., vol. 3, Hud. Riv. Gr. [Ety. proper name.]
carterensis, Safford, 1869, Geo. of Tenn., Trenton Gr. [Ety proper name.]
divergens, Troost, 1840, 5th Geo. Rep. Tenn., Devonian. [Sig. divergent.]
erratica, Billings, 1858, Can. Nat. & Geo., Trenton Gr. [Sig. straying abroad.]
goldfussi, Billings, 1858, Can. Nat. & Geo., vol. 3, Hud. Riv. Gr. [Ety. proper name.]
herzeri, syn. for Favistella stellata.
incerta, Billings, 1859, Can. Nat. & Geo., vol. 4, Chazy Gr. [Sig. uncertain.]
inequalis, Hall, 1852, Pal. N. Y., vol. 2, Coralline limestone. [Sig. unequal.]
intermedia, Eaton, 1832, Geo. Text Book, (?) Gr. [Sig. intermediate.]
parva, Billings, 1859, Can. Nat. & Geo., vol. 4, Chazy Gr. [Sig. small.]
rigida, Billings, 1858, Can. Nat. & Geo., vol. 3, Hud. Riv. Gr. [Sig. rigid.]
troosti, Castelnau, 1843, syn. for Lonsdaleia papillata.

COLUMNOPORA, Nicholson, 1874, London Geo. Mag., N. S., vol. 1. [Ety. *columna*, a column; *pora*, a pore.]
cribriformis, Nicholson,1874, London Geo. Mag., vol. 1, Cin'ti Gr. [Sig. having the form of a seive.]
huronica, Rominger, 1876, (Houghtonia huronica) Foss. Corals, Hud. Riv. Gr. [Ety. proper name.]

COMBOPHYLLUM, Edwards & Haime, 1858. [Ety. *kombos*, a strip of cloth; *phyllon*, a plant.]

multiradiatum, Meek, 1868, Trans. Chi. Acad. Sci., Devonian. [Sig. many rayed.]
Conophyllum, Hall, 1852. Syn. for Chonophyllum.
niagarense, see Chonophyllum Niagarense.
Constellaria, Dana, 1848, Zoophytes, syn. for Stellipora.
constellata, syn. for Stellipora antheloidea.
polystomella, Nicholson, syn. for Stellipora antheloidea.
CYATHAXONIA, Michelin, 1846, Icon. Zooph. [Ety. kuathus, a cup; axones, a tablet made to turn on its axis.]
distorta, Worthen, 1875, Geo. Sur. Ill., vol. 6, Coal Meas. [Sig. distorted.]
prolifera, see Lophophyllum proliferum.
CYATHOPHYLLUM, Goldfuss, 1826, Petref. Germ. [Ety. kuathus, a cup; phyllon, a plant.]
arcticum, Meek, 1868, Trans. Chi. Acad. Sci., Devonian. [Sig. from the arctic regions of the north.]
anticostiense, Billings, 1862, Pal. Foss., vol. 1, Mid. Sil. [Ety. proper name.]
billingsi, Dawson, 1868, Acad. Geol. Low. Carb. [Ety. proper name.]
caespitosum, (?) Goldfuss, 1826, Petref. Germ., Corniferous Gr. [Sig. turfy.]
calyculare, Owen, 1840, Rep. on Min. Lands, Devonian. [Sig. like a flower bud.]
ceratites, as identified by D'Archiac & Verneuil. Not American.
coalitum, Rominger, 1876, Foss. Corals, Corniferous Gr. [Sig. grown together.]
corinthium, Owen, 1840, Rep. on Min. Lands, Devonian. [Ety. from resemblance to a Corinthian column.]
cristatum, Rominger, 1876, Foss. Corals, Ham. Gr. [Sig. tufted, crested.]
dianthus, (?) Goldfuss, 1826, Germ. Petref., Onondaga Gr. [Sig. the flower of Jove.]
eriphele, Billings, 1862, Pal. Foss., vol. 1, Mid. Sil. [Ety. mythological name.]
euryone, Billings, 1862, Pal. Foss., vol. 1, Mid. Sil. [Ety. mythological name.]
excentricum, as identified by D'Archiac & Verneuil. Not American.
flexuosum, (?) Owen, syn. for Campophyllum torquium.
geniculatum, Rominger, 1876, Foss. Corals, Ham. Gr. [Sig. knotted, geniculated.]
gigas, Yandell & Shumard, 1847, Contrib. to Geol. Ky., Devonian. [Sig. large.]
gracile, Troost, 1840, 5th Geo. Rep. Tenn. Low. Carb. [Sig. slender.]
helianthoides, Goldfuss, 1826, Germ. Petref., Devonian. [Sig. rayed like the sunflower.] See Heliophyllum halli.
houghtoni, Rominger, 1876, Foss. Corals, Devonian. [Ety. proper name.]
interruptum, Billings, 1862, Pal. Foss., vol. 1, Mid. Sil. [Sig. interrupted.]
uvenis, Rominger, 1876, Foss. Corals, Up. Held. Gr. [Sig. young.]
lesueuri, Hall, 1859, figured without specific name in 1843, 4th Dist. Rep. N. Y., Onondaga Gr. [Ety. proper name.]
nymphale, Billings, 1862, Pal. Foss., vol. 1, Mid. Sil. [Ety. a fountain.]
panicum, Winchell, 1866, Rep. Low. Pen. Mich., Ham. Gr. [Sig. panic, or stragling appearance.]
partitum, Winchell, 1866, Rep. Low. Pen. Mich., Ham. Gr. [Ety. from the vertical partitions in the internal cavity.]
pasithea, Billings, 1862, Pal. Foss., vol. 1, Mid. Sil. [Ety. mythological name.]
pelagicum, Billings, 1862, Pal. Foss., vol. 1, Mid. Sil. [Sig. belonging to the deep sea.]
pennanti, Billings, 1862, Pal. Foss., vol. 1, Mid. Sil. [Ety. proper name.]
profundum, Conrad, see Streptelasma profunda.
quadrigeminum, as identified by d'Archiac & Verneuil. Not American.
radicula, Rominger, 1876, Foss. Corals, Niagara Gr. [Sig. a little root.]
rugosum, Edwards & Haime, 1851, Pal. Foss. des Terr Palaeoz., Corniferous Gr. [Sig. wrinkled.]
scyphus, Rominger, 1876, Foss. Corals, Ham. Gr. [Sig. a cup.]
shumardi, see Amplexus shumardi.
solitarium, Billings, 1866, Catal. Sil. Foss. Antic., Clinton & Niagara Gr. [Sig. alone.]
torquium, see Campophyllum torquium.
turbinatum, of the N. Y. Rep's, see Heliophyllum halli.
undulatum et multiplicatum, Owen, 1840, Rep. on Min. lands, Devonian. [Sig. many waved lines.]
vanuxemi, Hall, 1859, figured without specific name in 1843, 4th Dist. Rep. N. Y., Ham. Gr. [Ety. proper name.]
vermiculare, (?) Owen, syn. for Campophyllum torquium.
wahlenbergi, Billings, 1862, Pal. Foss., vol. 1, Mid. Sil. [Ety. proper name.]
Cyathopora iowensis, see Striatopora iowensis. There is no genus Cyathopora, and if Dr. Owen did not intend to refer his species to Cyathophora, then he failed to establish a genus by neglecting to define it.
CYCLOLITES, Lamarck, 1801, Syst. Anim. sans. Verteb. [Ety. kuklos, a circle; lithos, a stone.]
rotuloides, Hall, 1852, Pal. N. Y., vol. 2, Clinton Gr. [Ety. rotula, a little wheel; eidos, a form.]
CYSTIPHYLLUM, Lonsdale, 1839, Murch. Sil. Syst. [Ety. kustis, a vesicle or small cavity; phyllon, a plant.]
aggregatum, Billings, 1859, Can. Jour. Ham. Gr. [Sig. aggregated.]
americanum, Edwards & Haime, 1851, Monogr. Pal. Foss., Ham. Gr. [Ety. proper name.]

americanum *var.* arcticum, Meek, 1868, Trans. Chi. Acad. Sci., Ham. Gr. [Ety. from the arctic or northern region.]
cylindricum, (not Lonsdale) see Cystiphyllum americanum.
fruticosum, Nicholson, 1875, Geo. Mag., vol. 2, N. S., Corniferous Gr. [Sig. full of stems.]
grande, Billings, 1859, Can. Jour., Corniferous Gr. [Sig. large.]
huronense, Billings, 1866, Catal. Sil. Foss. Antic., Clinton & Niagara Gr. [Ety. proper name.]
maritima, Billings, 1862, Pal. Foss., vol. 1, Mid. Sil. [Sig. of the sea.]
mundulum, Hall, 1873, 23d Reg. Rep., Chemung Gr. [Sig. fine, neat.]
ohioense, Nicholson, 1875, Ohio Pal., vol. 2, Corniferous Gr. [Ety. proper name.]
senecaense, Billings, 1859, Can. Jour., Devonian. [Ety. proper name.]
squamosum, Nicholson, 1875, Geo. Mag. N. S., vol. 2, Corniferous Gr. [Sig. rough or scaley.]
sulcatum, Billings, 1859, Can. Nat. & Geo. vol.4, Corniferous Gr. [Sig. furrowed.]
superbum, Nicholson, 1875, Geo. Mag., vol. 2, N. S., Ham. Gr. [Sig. stately, nice.]
vesiculosum, Goldfuss, 1826, Petref. Germ. Devonian. [Sig. full of vesicles.]
DAWSONIA, Nicholson, 1873, Ann. Mag. Nat. Hist. 4th ser., vol, 12. [Ety. proper name.] Supposed to be the ovarian vesicles of Graptolites.
acuminata, Nicholson, 1873, Ann. Mag. Nat. Hist., 4th ser., vol. 12, Quebec Gr. [Sig. pointed.]
campanulata, Nicholson, 1873, Ann. Mag. Nat. Hist., Quebec Gr. [Sig. bell-shaped.]
rotunda, Nicholson, 1873, Ann. Mag. Nat. Hist., Quebec Gr. [Sig. round.]
tenuistriata, Nicholson, 1873, Ann. Mag. Nat. Hist., Quebec Gr. [Sig. finely striated.]
DENDROGRAPTUS, Hall, 1865, Can. Org. Rem., Decade 2. [Ety. *dendron*, a tree; *grapho*, to write.]
diffusus, Hall, 1865, Can. Org. Rem., Decade 2, Quebec Gr. [Sig. diffused, hanging loose.]
divergens, Hall, 1865, Can. Org. Rem., Decade 2, Quebec Gr. [Ety. *divergo*, to extend from a point in different directions.]
erectus, Hall, 1865, Can. Org. Rem., Decade 2, Quebec Gr. [Sig. standing upright.]
flexuosus, Hall, 1865, Can. Org. Rem., Decade 2, Quebec Gr. [Sig. crooked.]
fruticosus, Hall, 1865, Can. Org. Rem., Decade 2, Quebec Gr. [Sig. full of shoots or stems.]
gracilis, Hall, 1865, Can. Org. Rem., Decade 2. Quebec Gr. [Sig. slender.]
hallianus, Prout, 1851, (Graptolithus hallianus) Am. Jour. Sci. 2d ser., vol. 11, Potsdam sandstone. [Ety. proper name.]
striatus, Hall, 1865, Can. Org. Rem., Decade 2, Quebec Gr. [Sig. striated.]
DENDROPORA, Michelin, 1846, Icon. Zooph. [Ety. *dendron*, a tree; *pora*, a pore.]
alternans, Rominger, 1876, Foss. Corals Ham. Gr. [Sig. alternately.]
neglecta, Rominger, 1876, Foss. Corals, Up. Held. Gr. [Sig. overlooked.]
ornata, Rominger, 1876, Foss. Corals, Ham. Gr. [Sig. ornamented.]
proboscidialis, Rominger, 1876, Foss. Corals, Ham. Gr. [Sig. having a proboscis.]
reticulata, Rominger, 1876, Foss. Corals, Ham. Gr. [Sig. reticulated.]
Dichograptus, syn. for Graptolithus.
DICRANOGRAPTUS, Hall, 1865, Can. Org. Rem. Decade 2. [Ety. *dikranos*, two pointed; *grapho*, to write.]
divaricatus, Hall, 1859, (Graptolithus divaricatus) Pal. N. Y., vol. 3, Hud. Riv. Gr. [Sig. wide apart.]
furcatus, Hall, 1847, (Graptolithus furcatus) Pal. N. Y., vol. 1, Hud. Riv. Gr. [Sig. forked.]
ramosus, Hall, 1847, (Graptolithus ramosus) Pal. N. Y., vol. 1, Hud. Riv. Gr. [Sig. branching.]
sextans, Hall, 1847, (Graptolithus sextans) Pal. N. Y., vol. 1, Hud. Riv. Gr. [Ety. *sextans*, a sixth part; probably named from the regular diverging bifurcations at an angle of sixty degrees.]
DICTYONEMA, Hall, 1852, Pal. N. Y., vol. 2. [Ety. *dictyon*, a net; *nema*, a thread.]
cadens, Hall, 1865, Can. Org. Rem., Decade 2, Ham. Gr. [Sig. drooping.]
fenestrata, Hall, 1868, 20th Reg. Rep., Up. Held. Gr. [Sig. having openings.]
gracilis, Hall, 1852, Pal. N. Y., vol. 2, Niagara Gr. [Sig. slender.]
grandis, Nicholson, 1873, Ann. Mag. Nat. Hist., 4th ser., vol. 12, Quebec Gr. [Sig. grand.]
hamiltoniæ, Hall, 1865, Can. Org. Rem., Decade 2, Ham. Gr. [Ety. proper name.]
irregularis, Hall, 1865, Can. Org. Rem., Decade 2, Quebec Gr. [Sig. irregular.]
murrayi, Hall, 1865, Can. Org. Rem., Decade 2. Quebec Gr. [Ety. proper name.]
neenah, Hall, 1861, Geo. Rep. Wis. Trenton Gr. [Ety. proper name.]
quadrangularis, Hall, 1865, Can. Org. Rem., Decade 2, Quebec Gr. [Sig. four sided.)
retiformis, Hall, 1843, (Gorgonia retiformis) Geo. Rep. 4th Dist. N. Y., Niagara Gr. [Sig. net-formed.]
robusta, Hall, 1865, Can. Org. Rem., Decade 2, Quebec Gr. [Sig. hardy, strong.]

splendens, Billings, 1874, Pal. Foss. vol. 2, Up. Sil. [Sig. splendid.]
websteri, Dawson, 1868, Acad. Geol., Niagara Gr. [Ety. proper name.]
DICTYOSTROMA, Nicholson, 1875, O. Pal., vol. 2. [Ety. *dictyon*, a net; *stroma*, a layer.]
undulata, Nicholson, 1875, Ohio Pal. vol. 2, Niagara Gr. [Ety. from the undulating layers.]
DIDYMOGRAPTUS, McCoy, 1851, Pal. Foss. [Ety. *didymos*, twin or double; *grapho*, to write.]
caduceus, Salter, 1853, (Graptolithus caduceus) Quar. Jour. Geo. Soc., vol. 9, Quebec Gr. [Sig. falling down, frail.]
DIPHYPHYLLUM, Lonsdale, 1845, Russ. & Ural. Mts. [Ety. *diphyia*, a division; *phyllon*, a leaf.]
archiaci, Billings, 1859, Can. Jour., Up. Held. & Ham. Gr. [Ety. proper name.]
arundinaceum, Billings, 1859, Can. Jour. Corniferous Gr. [Sig. like a reed.]
cæspitosum, Hall, 1852, (Diplophyllum cæspitosum) Pal. N. Y., vol. 2, Niagara Gr. [Sig. turf-like.]
coralliferum, Hall, 1852, (Diplophyllum coralliferum) Pal. N. Y., vol. 2, Coralline limestone. [Ety. *coralium*, a coral; *fero*, to bear; in allusion to the fact that the specimens were found enclosed in a species of *Stromatopora*.]
gigas, Rominger, 1876, Foss. Corals., Niagara Gr. [Sig. great.]
huronicum, Rominger, 1876, Foss. Corals, Niagara Gr. [Ety. proper name.]
rectiseptatum, Rominger, 1876, Foss. Corals, Ham. Gr. [Ety. from the straight septæ.]
stramineum, Billings, 1859, Can. Jour., Corniferous Gr. [Sig. made of straw.]
DIPLOGRAPTUS, McCoy, 1854, (Diplograpsus) Brit. Pal. Rocks. [Ety. *diploos*, duplex; *grapho*, to write.]
amplexicaulis, Hall, 1847, (Graptolithus amplexicaule) Pal. N. Y., vol. 1, Trent. Gr. [Ety. *amplexus*, embracing; *caulis*, a stalk or stem.; in allusion to the triangular scales surrounding the central stipe.]
angustifolius, Hall, 1859, (Graptolithus angustifolius) Pal. N. Y., vol. 3, Hud. Riv. Gr. [Sig. narrow-leaved.]
ciliatus, Emmons, 1856, Am. Geol., Trent. Gr. [Sig. fringed.]
dissimilaris, Emmons, 1856, Am. Geol., Quebec Gr. [Sig. dissimilar.]
foliaceus, (?) Murch, 1839, (*Graptolites foliaceus*) Murch, Sil. Syst., Hud. Riv. Gr. [Sig. leaf-like.]
foliosus, Emmons, 1856, Am. Geol., Quebec Gr. [Sig. full of leaves.]
folium, (?) Hisinger, 1837, (Prionotus folium) Leth. Suec., Hud. Riv. Gr. [Sig. a leaf or thin plate.]
inutilis, Hall, 1865, Can. Org. Rem., Dec. 2, Quebec Gr. [Sig. not useful.]
laciniatus, Emmons, 1856, Am. Geol., Quebec Gr. [Sig. pointed, jagged.]

marcidus, Hall, 1859, (Graptolithus marcidus) Pal. N. Y., vol. 3, Hud. Riv. Gr. [Sig. rotten, flagging, from the shrunken stipe.]
mucronatus, Hall, 1847, (Graptolithus mucronatus) Pal. N. Y., vol. 1, Hud. Riv. Gr. [Sig. sharp-pointed.]
obliquus, Emmons, 1856, Am. Geol., Quebec Gr. [Sig. oblique.]
peosta, Hall, 1861, (Graptolithus peosta) Geo. Rep. Wis., Trenton Gr. [Ety. proper name. (?)]
pristiniformis, Hall, 1858, (Graptolithus pristiniformis) Rep. of Progr. Can. Sur., Quebec Gr. [Sig. an ancient form.] This species is probably identical with *Fucoides dentata*, Brongiart, 1828, Hist. Veg. Foss., vol. 1.
pristis, (?) Hisinger, 1837, (Prionotus pristis) Leth. Suec., Hud. Riv. Gr. [Sig. a saw fish.]
putillus, Hall, 1865, Can. Org. Rem., Decade 2, Hud. Riv. Gr. [Sig. a dwarf.]
rugosus, Emmons, 1856, Am. Geol., Quebec Gr. [Sig. rugose.]
rectangularis, McCoy, 1854, Brit. Pal. Rocks, Low. Sil. [Sig. rectangular.]
secalinus, Eaton, 1847, (Fucoides secalinus) Pal. N. Y., vol. 1, Hud. Riv. Gr. [Sig. resembling a small grain.]
setaceus, Emmons, 1856, (Glossograpsus setaceus) Am. Geol., Quebec Gr. [Sig. hairy.]
spinulosus, Hall, 1859, (Graptolithus spinulosus) Pal. N. Y., vol. 3, Hud. Riv. Gr. [Sig. full of little spines.]
whitfieldi, Hall, 1859, (Graptolithus whitfieldi) Pal. N. Y., vol. 3, Hud. Riv. Gr. [Ety. proper name.]
Diplophyllum, Hall, 1852, Pal. N. Y., vol. 2, syn. for Diphyphyllum.
cæspitosum, see Diphyphyllum caespitosum.
coralliferum, see Diphyphyllum coralliferum.
DISCOPHYLLUM, Hall, 1847, Pal. N. Y.. vol. 1. [Ety. *diskos*, a disc; *phyllon*, a leaf.]
peltatum, Hall, 1847, Pal. N. Y., vol. 1, Hud. Riv.Gr. [Sig. half-moon-shaped.]
DUNCANELLA, Nicholson, 1874, Ann. Mag. Nat. Hist., 4th ser., vol. 13. [Ety. proper name.]
borealis, Nicholson, 1874, Ann. Mag. Nat. Hist., 4th ser., vol. 13, Niagara Gr. [Sig. belonging to northern latitudes.]
EMMONSIA, Edwards & Haime, 1851, Monographie des Polyp., Foss. des Terr. Palaeoz. [Ety. proper name.]
hemispherica, Troost, 1840, (Calamopora hemispherica) 5th Geo. Rep. Tenn., Corniferous Gr. [Sig. hemispherical.]
hemispherica, Yandell & Shumard, 1847, (Favosites hemispherica) Contrib. to Geol. of Ky., Corniferous Gr. This species was first described by Troost, who gave the Falls of the Ohio as one of its localities.

ERIDOPHYLLUM, Edwards & Haime, 1850, Brit. Foss. Corals. [Ety. *eridos*, in dispute; *phyllon*, a plant.]
rugosum, Edwards & Haime, 1851, Pal. Foss. des Terr. Pal., Niagara Gr. [Sig. wrinkled.]
simcoense, Billings, 1859, Can. Jour., new ser., vol. 4., Clinton Gr. [Ety. proper name.]
strictum, Edwards & Haime, 1851, Pol. Foss. des Terr. Pal., Corniferous Gr. [Sig. gathered, narrow.]
vennori, Billings, 1865, Can. Nat. & Geo., 2d ser., vol. 2, Clinton Gr. [Ety. proper name.]
verneuilianum, Edwards & Haime, 1851, Pol. Foss. des Terr. Pal., Corniferous Gr. [Ety. proper name.]

ETHMOPHYLLUM, Meek, 1868, Am. Jour. Sci. & Arts, 2d ser., vol. 45. [Ety. *ethmos*, a seive; *phyllon*, a plant.]
gracile, Meek, 1868, Am. Jour. Sci. & Arts, 2d ser., vol. 45, Up. Sil. [Sig. slender.]
whitneyi, Meek, 1868, Am. Jour. Sci. & Arts, 2d ser., vol. 45, Up. Sil. [Ety. proper name.]

FAVIPHYLLUM, as used by Hall, 1852, Stans. Exped. to Great Salt Lake. [Ety. *favus*, honey-comb; *phyllon*, a plant.]
rugosum, Hall, 1852, Stansbury's Exped. to Great Salt Lake, Coal Meas. [Sig. wrinkled.]

FAVISTELLA, Hall, 1847, Pal. N. Y., vol. 1. [Ety. *favus*, honey-comb; *stella*, a star.]
favosidea, Hall, 1852, Pal. N. Y., vol. 2, Clinton Gr. [Sig. honey-comb-like.]
stellata, Hall, 1847, Pal. N. Y., Hudson Riv. Gr. [Sig. starred.]

FAVOSITES, Lamarck, 1812, Cours. de Zool. du Mus. d'Hist. Nat. [Ety. *favus*, honey-comb.]
alpenensis, Winchell, 1866, Rep. Low. Pen. Mich., Ham. Gr. [Ety. proper name.]
alveolaris, DeBlainville, 1834, Man. d' Actinol., as identified by Hall, in 1843, Onondaga Gr. [Sig. full of cells.]
asper, D'Orbigny, 1849, Prodr. de Palaeont., Clinton Gr. [Sig. rough.]
basalticus, Goldfuss, 1826, Germ. Petref., (Calamopora basaltica) Devonian. [Sig. basaltic.]
billingsi, Rominger, 1876, Foss. Corals, Ham. Gr. [Ety. proper name.]
capax, Billings, 1866, Catal. Sil. Foss. Antic., Hud. Riv. Gr. [Sig. large.]
cervicornis, DeBlainville, as identified by Billings, 1859, Can. Jour., vol. 4, Devonian. [Sig. deer-horned.]
clausus, Rominger, 1876, Foss. Corals, Devonian. [Sig. closed up.]
conicus, Hall, 1874, 26th Reg. Rep., Low. Held. Gr. [Sig. cone-shaped.]
constrictus, Hall, 1852, (Astrocerium constrictum) Pal. N. Y., vol. 2, Niagara Gr. [Sig. constricted.]
cristatus, Edwards & Haime, 1851, Pol. Foss. Terr. Palaeoz., Niagara Gr. [Sig. peaked.]
cumberlandicus, Troost, 1840, (Calamopora cumberlandica) 5th Geo. Rep. Tenn., Low. Carb. [Ety. proper name.]
digitatus, Rominger, 1876, Foss. Corals, Ham. Gr. [Sig. fingered.]
dubius, Blainville, 1839,(Alveolites dubia) Corniferous Gr. [Sig. doubtful.]
dumosus, Winchell, 1866, Rep. Low. Pen. Mich., Ham. Gr. [Sig. bushy, clustered.]
emmonsi, Rominger, 1876, Foss. Corals, Up. Held. Gr. [Ety. proper name.]
epidermatus, Rominger, 1862, Am. Jour. Sci. & Arts, Corniferous Gr. [Sig. covered with a crust or skin.]
excretus, Hall, 1876, 28th Reg. Rep., Niagara Gr. [Sig. separated.]
favosus, Goldfuss, 1826, Germ. Petref., (Calamopora favosa) Niagara Gr. [Sig. honey-combed.]
forbesi, Edwards & Haime, 1854, Brit. Foss. Corals, Niagara Gr. [Ety. proper name.]
forbesi, *var*. discoideus, Roemer, 1860, (Calamopora forbesi, *var*. discoidea) Sil. Fauna W. Tenn., Niagara Gr. [Sig. like a dish.]
goldfussi, D'Orbigny, 1850, Prodr. de Palaeont, Devonian. [Ety. proper name.]
gothlandicus, Lamarck, 1816, Hist. An. sans. Vert., Up. Sil. [Ety. proper name.]
hamiltonensis, Rominger, 1876, Foss. Corals, Ham. Gr. [Ety. proper name.]
helderbergiæ, Hall, 1874, 26th Reg. Rep. Low. Held. Gr. [Ety. proper name.]
heliolitiformis, Rominger, 1862, (Calamopora heliolitiformis) Am. Jour. Sci., vol. 34, 2nd series, Devonian. [Sig. from its resemblance to *Heliolites*.]
hemisphericus, see Emmonsia hemispherica.
hisingeri, Edwards & Haime, 1851, Pal. Foss. des Terr. Palæoz., Niagara Gr. [Ety. proper name.]
hispidus, Rominger, 1876, Foss. Corals, Niagara Gr. [Sig. bristly.]
infundibuliformis, as identified by d'Archiac & Verneuil. Not America.
intertextus, Rominger, 1876, Foss. Corals, Ham. Gr. [Sig. interlaced.]
invaginatus, Nicholson, 1875, Ohio Pal., vol. 2, Corniferous Gr. [Sig. sheathed, enwrapped.]
limitaris, Rominger, 1876, Foss. Corals, Corniferous Gr. [Sig. on the border, from its resemblance to *Cladopora*.]
lycoperdon, see Monticulipora lycoperdon.
manus, Winchell, 1863, Proc. Acad. Nat. Sci., Kinderhook Gr. [Sig. the hand.]
maximus, Troost, 1840, (Calamopora maxima) 5th Geo. Rep. Tenn., Devonian. [Sig. largest.]

minimus, Hall, 1874, 26th Reg. Rep., Low. Held. Gr. [Sig. smallest.]
niagarensis, Hall, 1852, Pal. N. Y., vol. 2, Niagara Gr. [Ety. proper name.]
niagarensis, var. spinigerus, Hall, 1876, 28th Reg. Rep., Niagara Gr. [Sig. spiny.]
nitellus, Winchell, 1866, Rep. Low. Pen. Mich., Ham. Gr. [Sig. delicate, smooth.]
obliquus, Rominger, 1876, Foss. Corals, Niagara Gr. [Sig. oblique.]
parasiticus, Hall, 1852, (Astrocerium parasiticum) Pal. N. Y., vol. 2, Niagara Gr. [Sig. parasitic.] This name was preoccupied by Phillips in his Geol. of Yorkshire.
placenta, Rominger, 1876, Foss. Corals, Ham. Gr. [Sig. a cake.]
pleurodictyloides, Nicholson, 1875, Ohio Pal., vol. 2, Corniferous Gr. [Sig. resembling *Pleurodictyum*.]
polymorphus, Goldfuss, 1826, Germ. Petref., Corniferous Gr. [Sig. having many-forms.]
prolificus, Billings, 1865, Can. Nat. & Geol., 2d series, vol. 2, Mid. Sil. [Sig. abundant.]
pyriformis, Hall, 1852, (Astrocerium pyriforme) Pal. N. Y., vol. 2, Niagara Gr. [Sig. pear-shaped.]
radiatus, Rominger, 1876, Foss. Corals, Ham. Gr. [Sig. radiated.]
radiciformis, Rominger, 1876, Foss. Corals, Devonian. [Sig. root-like.]
reticulatus, DeBlainville, 1830, (Alveolites reticulata) Dict., vol. 60, Niagara Gr. [Sig. reticulated.]
spongilla, Rominger, 1876, Foss. Corals, Niagara Gr. [Sig. a little sponge.]
striatus, Say, 1818, Am. Jour. Sci., vol. 1, Niagara Gr. [Sig. striated.]
tuberosus, Rominger, 1876, Foss. Corals, Corniferous Gr. [Sig. composed of tubes.]
turbinatus, Billings, 1859, Can. Jour., vol. 4, Corniferous Gr. [Sig. top-shaped.]
venustus, Hall, 1852, (Astrocerium venustum) Pal. N. Y., vol. 2, Niagara Gr. [Sig. elegant.]
whitfieldi, White, 1874, Rep. Invert. Foss., Low. Carb. [Ety. proper name.]
winchelli, Rominger, 1862, (Calamopora winchelli) Am. Jour. Sci., vol. 34, 2d ser., Devonian. [Ety. proper name.]

Filicites gracilis, Shumard, see Plumalina gracilis.

FISTULIPORA, McCoy, 1849, Ann. & Mag. Nat. Hist., 2d ser., vol. 3. [Ety. *fistula*, a pipe; *pora*, a pore.]
acervulosa, Rominger, 1866, Proc. Acad. Nat. Sci., Ham. Gr. [Sig. many clustered together.]
canadensis, Billings, 1859, Can. Nat. & Geo., vol. 4, Devonian. [Ety. proper name.]
compressa, Rominger, 1866, Proc. Acad. Nat. Sci., Keokuk Gr. [Sig. compressed.]
crassa, Rominger, 1866, Proc. Acad. Nat. Sci., Ham. Gr. [Sig. thick.]
elegans, Rominger, 1866, Proc. Acad. Nat. Sci., Ham. Gr. [Sig. elegant.]
eriensis, Rominger, 1866, Proc. Acad. Nat. Sci., Ham. Gr. [Ety. proper name.]
flabellum, Rominger, 1866, Proc. Acad. Nat. Sci., Warsaw Gr. [Sig. fan-like.]
halli, Rominger, 1866, Proc. Acad. Nat. Sci., Niagara Gr. [Ety. proper name.]
helios, Rominger, 1866, Proc. Acad. Nat. Sci., Corniferous Gr. [Sig. the sun.]
labiosa, Winchell, 1866, Rep. Low. Pen. Mich., Ham. Gr. [Sig. full-lipped.]
lunata, Rominger, 1866, Proc. Acad. Nat. Sci., Up. Held. Gr. [Sig. crescent-shaped.]
minuta, Rominger, 1866, Proc. Acad. Nat. Sci., Ham. Gr. [Sig. very small.]
neglecta, Rominger, 1866, Proc. Acad. Nat. Sci., Niagara Gr. [Sig. overlooked.]
nodulifera, Meek, 1872, Pal. E. Neb., Coal Meas. [Sig. bearing knots.]
occidens, Hall, 1873, 23d Reg. Rep., Chemung Gr. [Sig. western.]
peculiaris, Rominger, 1866, Proc. Acad. Nat. Sci., Keokuk Gr. [Sig. peculiar.]
saffordi, Winchell, 1866, Rep. Low. Pen. Mich., Ham. Gr. [Ety. proper name.]
spergenensis, Rominger, 1866, Proc. Acad. Nat. Sci., Warsaw Gr. [Ety. proper name.]
spinulifera, Rominger, 1866, Proc. Acad. Nat. Sci., Ham. Gr. [Sig. spine-bearing.]
stellifera, Rominger, 1866, Proc. Acad. Nat. Sci., Ham. Gr. [Sig. star-bearing.]
sulcata, Rominger, 1866, Proc. Acad. Nat. Sci., Ham. Gr. [Sig. furrowed.]
trifolia, Rominger, 1866, Proc. Acad. Nat. Sci., Keokuk Gr. [Sig. three-leaved.]
utriculus, Rominger, 1866, Proc. Acad. Nat. Sci., Ham. Gr. [Sig. a little bag.]

(*Glossograptus*, Emmons, (Glossograpsus) 1856, Am. Geol., pt. 2. Prof. Hall uses the termination *graptus*, instead of *grapsus*, because the latter termination is used in the nomenclature of crustacea. This genus is a synonym for Diplograptus.
ciliatus, Emmons, 1856, Am. Geol. This name will be found preoccupied by Emmons, under the genus *Diplograptus*.
setaceus, Emmons, 1856, Am. Geol. See Diplograptus setaceus.

GRAPTOLITHUS, Linnæus, 1736, Syst. Naturæ. [Ety. *grapho*, to write; *lithos*, stone.]
abnormis, Hall, 1858, Rep. of Progr. Can. Sur., Quebec Gr. [Sig. abnormal.]
alatus, Hall, 1858, Rep. of Progr. Can. Sur., Quebec Gr. [Sig. winged.]
amplexicaule, see Diplograptus amplexicaulis.
angustifolius, see Diplograptus angustifolius.

antennarius, see Climacograptus antennarius.
arcuatus, Hall, 1865, Can. Org. Rem., Decade 2, Quebec Gr. [Sig. bent, bow-shaped.]
belmontensis, White, conditional name.
bicornis, see Climacograptus bicornis.
bifidus, Hall, 18 8, Can. Nat. & Geo., vol. 3, Quebec Gr. [Ety. *bis*, twice; *fidi*, to split.]
bigsbyi, Hall, 1865, Can. Org. Rem., Decade 2, Quebec Gr. [Ety. proper name.]
bryonoides, Hall, 1858, Rep. of Progr. Can. Sur., Quebec Gr. [Sig. like moss.] Same species called Fucoides serra, Brongiart, 1828, Veg. Foss., vol. 1. See *Tetragraptus bryonoides*.
caduceus, see Didymograptus caduceus.
clintonensis, Hall, 1843, Geo. Rep., 4th Dist. N. Y., Clinton Gr. [Ety. proper name.]
constrictus, Hall, 1865, Can. Org. Rem., Decade 2, Quebec Gr. [Sig. constricted.]
crucifer, Hall, 1858, Rep. of Progr. Can. Sur., Quebec Gr. [Sig. the cross-bearer.]
dentatus, Emmons, 1842, Geo. Rep. N. Y., Utica slate. [Sig. toothed.] This may be Fucoides dentatus of Brongiart.
denticulatus, Hall, 1858, Rep. of Progr. Can. Sur., Quebec Gr. [Sig. small-toothed.]
divaricatus, Hall, 1859, Pal. N. Y., vol. 3, Hud. Riv. Gr. [Sig. wide apart.] See Dicranograptus divaricatus.
divergens, Hall, 1859, Pal. N. Y., vol. 3, Hud. Riv. Gr. [Sig. to extend from a point in different directions.]
ensiformis, see Retiolites ensiformis.
extensus, Hall, 1858, Rep. of Progr. Can. Sur., Quebec Gr. [Sig. extended.]
extenuatus, Hall, 1865, Can. Org. Rem. Decade 2, Quebec Gr. [Sig. thinned away.]
flaccidus, Hall, 1865, Can. Org. Rem. Decade 2, Utica shales. [Sig. withered.]
flexilis, Hall, 1858, Rep. of Progr. Can. Sur., Quebec Gr. [Sig. pliant, flexible.] See Chonograptus flexilis.
foliaceus, see Diplograptus foliaceus.
folium, (?) see Diplograptus folium.
fruticosus, Hall, 1858, Rep. of Progr., Quebec Gr. [Sig. bushy.]
furcatus, Hall, 1847, Pal. N. Y., vol. 1, Hud. Riv. Gr. [Sig. forked.] See Dicranograptus furcatus.
gracilis, Hall, 1847, Pal. N. Y., vol. 1, Hud. Riv. Gr. [Sig. slender.]
hallianus, see Dendrograptus hallianus.
headi, Hall, 1858, Rep. of Progr. Sur. Can., Quebec Gr. [Ety. proper name.]
hypniformis, see Diplograptus hypniformis.
indentus, Hall, 1858, Rep. of Prog. Can. Sur., Quebec Gr. [Sig. indented.]
lævis, Hall, 1847, Pal. N. Y., vol. 1, Hud. Riv. Gr. [Sig. smooth.]

logani, Hall, 1858, Rep. of Progr. Can. Sur., Quebec Gr. [Ety. proper name.]
marcidus, see Diplograptus marcidus.
milesi, Hall, 1861, Geo. Rep. Vermont, Quebec Gr. [Ety. proper name.]
mucronatus, see Diplograptus mucronatus.
multifasciatus, Hall, 1865, Decade 2, Can. Org. Rem., Hud. Riv. Gr. [Sig. many swathed or banded.]
nitidus, Hall, 1858, Rep. of Progr. Can. Sur., Quebec Gr. [Sig. neat.]
octobrachiatus, Hall, 1858, Rep. of Progr. Can. Sur., Quebec Gr. [Sig. having eight arms.]
octonarius, Hall, 1858, Rep. of Progr. Can. Sur., Quebec Gr. [Sig. belonging to the number eight.]
patulus, Hall, 1858, Rep. of Progr. Can. Sur., Quebec Gr. [Sig. spread out.]
pennatulus, Hall, 1865, Can. Org. Rem., Decade 2, Quebec Gr. [Sig. having small wings.]
peosta, see Diplograptus peosta.
pristis, see Diplograptus pristis.
putillus, see Diplograptus putillus.
pristiniformis, see Diplograptus pristiniformis.
quadribrachiatus, Hall, 1858, Rep. of Progr. Sur. Can., Quebec Gr. [Sig. having four arms.] See *Tetragraptus quadribrachiatus*.
quadrimucronatus, Hall, 1865, Can. Org. Rem., Decade 2, Utica shales. [Sig. having four sharp points or spines.]
ramosus, see Dicranograptus ramosus.
ramulus, Hall, 1865, Can. Org. Rem. Decade 2, Quebec Gr. [Sig. a little branch.]
ramulus, White, 1874, Rep. Invertebrate Foss., Trenton Gr. This name was preoccupied by Hall in 1865.
richardsoni, Hall, 1865, Can. Org. Rem., Decade 2, Quebec Gr. [Ety. proper name.]
rigidus, see Chonograptus rigidus.
scalaris, Linnæus as identified by Vanuxem, 1842, Geo. Rep. 3rd Dist. N. Y., Utica slate. [Sig. a ladder.]
secalinus, see Diplograptus secalinus.
serratulus, Hall, 1847, Pal. N. Y., vol. 1, Hud. Riv. Gr. [Sig. like a small saw.]
sagittarius, Hisinger, 1837, (Prionotus sagittarius) Leth. Suec. Supp., Hud. Riv. Gr. [Sig. an archer.]
sextans, see Dicranograptus sextans.
similis, Hall, 1865, Can. Org. Rem., Decade 2, Quebec Gr. [Sig. like in aspect.]
spinulosus, see Diplograptus spinulosus.
tentaculatus, see Retiograptus tentaculatus.
tenuis, (?) Portlock, 1843, Geo. Rep. Londonderry, Hud. Riv. Gr. [Sig. slender.]
tenuis, Hall. This name was preoccupied by Portlock, in 1843.
venosus, see Retiolites venosus.
whitfieldi, see Diplograptus whitfieldi.

POLYPI.

HADROPHYLLUM, Edwards & Haime, 1850, Brit. Foss. Corals. [Ety. *hadros*, mighty; *phyllon*, a plant.]
orbignyi, Edwards & Haime, 1850, Brit. Foss. Corals, Devonian. [Ety. proper name.]
HAIMEOPHYLLUM, Billings, 1859, Can. Jour., vol. 4. [Ety. proper name; *phyllon*, a plant.]
ordinatum, Billings, 1859, Can. Jour., vol. 4, Corniferous Gr. [Sig. well-arranged.]
HALYSITES, Fischer, 1813, Zoognosia, vol. 1. [Ety. *halysion*, a small chain or necklace.]
agglomeratus, Hall, 1843, (Catenipora agglomerata) Geo. Rep., 4th Dist. N. Y., Niagara Gr. [Sig. to gather into a mass.]
catenulatus, Linnæus, 1767, Syst. Nat., Niagara Gr. [Ety. *catena*, a chain; *latus*, wide.]
compactus, Rominger, 1876, Foss. Corals, Niagara Gr. Syn. for H. agglomerata.
escharoides, Lamarck, 1816, Hist. des Anim. sans. Vert., Niagara Gr. [Sig. grate-shaped.]
gracilis, Hall, 1851, Geo. Lake Sup. Land Dist., vol. 2, Hud. Riv. Gr. [Sig. slender.]
meandrina, Troost, 1840,(Catenipora meandrina) 5th Geo. Rep. Tenn., Niagara Gr. The definition is too meagre for identification.
HELIOLITES, Dana, 1846, Zooph. [Ety. *helios*, sun; *lithos*, stone.]
affinis, Billings, 1865, Can. Nat. & Geo., 2d ser., vol. 2, Hud. Riv. & Mid. Sil. [Sig. contiguous.]
elegans, Hall, 1852, Pal. N. Y., vol. 2, Niagara Gr. [Sig. elegant.]
exiguus, Billings, 1865, Can. Nat. & Geo., 2d ser., vol. 2, Mid. Sil. [Sig. little.]
interstinctus, Linne, 1767, (Madrepora interstincta) Syst. Nat., Niagara Gr. [Sig. divided.]
macrostylus, Hall, 1852, Pal. N. Y., vol. 2, Niagara Gr. [Ety. *makros*, long; *stylos*, a column.]
megastoma, McCoy, 1851, Brit. Pal. Foss., Niagara Gr. [Ety. *megas*, great; *stoma*, mouth.]
pyriformis, Guett., 1770, Mem. 3, Niagara Gr. [Sig. pear-shaped.]
sparsus. Billings, 1865, Can. Nat. & Geo., 2d ser., vol. 2, Mid. Sil. [Sig. scattered.]
speciosus, Billings, 1865, Can. Nat. & Geo., 2d ser., vol. 2, Mid. Sil. [Sig. beautiful.]
spinoporus, Hall, 1852, Pal. N.Y., vol. 2, Niagara Gr. [Ety. *spina*, a spine; *porus*, a pore.] In allusion to the spiniform rays in the interior of the tubes.
subtubulatus, McCoy, as identified by Rominger, 1876, Foss. Corals, Niagara Gr. [Sig. somewhat like *H. tubulatus*.]
tenuis, Billings, 1865, Can. Nat. & Geo., 2d ser., vol. 2, Mid. Sil. [Sig. slender.]

HELIOPHYLLUM, Hall, 1846, in Dana Zooph. [Ety. *helios*, the sun; *phyllon*, a plant.]
colbornense, Nicholson, 1875, Can. Nat. & Geol., Corniferous Gr. [Ety. proper name.]
colligatum, Billings, 1859, Can. Jour., Up. Held. Gr. [Sig. collected together.]
exiguum, Billings, 1859, Can. Jour., Corniferous Gr. [Sig. little.]
eriense, Billings, 1859, Can. Jour., vol. 4, Corniferous Gr. [Ety. proper name.]
halli, Edwards & Haime, 1850, Brit. Foss. Corals, Ham. Gr. [Ety. proper name.]
canadense, Billings, 1859, Can. Jour., vol. 4, Corniferous Gr. [Ety. proper name.]
cayugaense, Billings, 1859, Can. Jour., vol. 4, Corniferous Gr. [Ety. proper name]
prolificum, Nicholson, 1874, Rep. Pal. Ont. Can., Corniferous Gr. [Sig. prolific.]
sub-cæspitosum, Nicholson, 1874, Geo. Mag. Lond. N. S., vol. 1, Ham. Gr. [Sig. somewhat like *Cyathophyllum cæspitosum*.]
tenuiseptum, Billings, 1859, Can. Jour., Ham. Gr. [Ety. having slender septæ.]
HETEROPHRENTIS, Billings, 1875, Can. Nat. & Geol. [Ety. *heteros*, irregular; *phren*, the midriff or the lamella.]
compta, Billings, 1875, Can. Nat. & Geol., Corniferous Gr. [Sig. elegant.]
excellens, Billings, 1875, Can. Nat. & Geol., Corniferous Gr. [Sig. excellent.]
prolifica, Billings, 1875, Can. Nat. & Geol., Corniferous Gr. [Sig. prolific.]
spatiosa, Billings, 1859, (Zaphrentis spatiosa) Can. Nat. & Geol., vol.4, Onondaga & Corniferous Gr. [Sig. large.]
Houghtonia, syn. for Columnopora.
huronica, see Columnopora huronica.
INOCAULIS, Hall, 1852, Pal. N. Y., vol. 2. [Ety. *inos*, small sprouts, like the roots of herbs; *kaulos*, the stock or stem.]
bella, Hall & Whitfield, 1875, Ohio Pal., vol. 2, Niagara Gr. [Sig. beautiful.]
plumulosa, Hall, 1852, Pal. N. Y., vol. 2, Niagara Gr. [Sig. full of feathers.]
LAMELLOPORA, Owen, 1840, Rep. on Mineral lands. [Ety. *lamella*, a small plate; *poros*, a perforation.] This genus is a synonym for *Stromatopora*, or very closely related to it.]
infundibularis, Owen, 1840, Rep. on Mineral lands, Devonian. [Sig. funnel-shaped.]
LEPTOPORA, Winchell, 1863, Proc. Acad. Nat. Sci. [Ety. *leptos*, shallow; *pora*, a cell.]
typa, Winchell, 1863, Proc. Acad. Nat. Sci., Chemung Gr. [Ety. the type of the genus.]
LIMARIA, Steininger, 1833, Bul. Soc. Geol. France. [Ety. *Limarius*, belonging to slime.]
crassa, Rominger, 1876, Foss. Corals, Niagara Gr. [Sig. thick.]

POLYPI. 57

falcata, Prout, 1859, Trans. St. Louis Acad. Sci., Up. Held. Gr. [Sig.sickle-shaped.]
fruticosa, Steininger, 1833, Bul. Soc. Geol. France, Niagara Gr. [Sig. full of shoots or stems.]
laminata, Hall, 1852, Pal. N. Y., vol. 2, Niagara, Gr. [Sig. laminated.]
ramulosa, Hall, 1852, Pal. N. Y., vol. 2, Niagara Gr. [Sig. full of little branches.]

Linipora rotunda, one of Troost's names.

LITHOSTROTION, Lhwyd, 1699, Lithophyl. Britann. Ichnographia. [Ety. *lithos*, stone; *strotion*, little rafter.]
  basaltiforme, Conybeare & Phillips, 1822, as identified by Owen, 1852, see Lithostrotion canadense.
  californiense, Meek, 1864, Pal. California, Carb. [Ety. proper name.]
  canadense, Castelnau, 1843, (Axinura canadensis)Terr. Sil. Amerique, Lower Carb. [Ety. proper name.]
  mamillare, Castelnau, syn. for Lithostrotion canadense.
  pictoense, Billings, 1868, Acad.Geol., Low. Carb. [Ety. proper name.]
  proliferum, Hall, 1858, Geo. Rep. Iowa., St. Louis Gr. [Sig. putting forth a new shoot.]

LONSDALEIA, McCoy, 1849, Ann. & Mag. Nat. Hist., 2d ser., vol. 3. [Ety. proper name.]
  papillata, Fischer, 1837, (Cyathophyllum papillatum) Oryct. de Moscou. [Sig. covered with papilli.]

LOPHOPHYLLUM, Edwards & Haime, 1850, Brit. Foss. Corals. [Ety. *lophos*, a ridge; *phyllon*, a plant.]
  expansum, White, 1876, Proc. Acad. Nat. Sci. Phil. Keokuk Gr. [Sig. expanded.]
  proliferum, McChesney, 1860, (Cyathaxonia prolifera) New. Pal. Foss., Coal Meas. [Sig. abundant.]

LUNATIPORA, Winchell, 1866, Rep. Low. Peninsula Mich. [Ety. *lunatus*, crescent-formed ; *poros*, a pore.]
  michiganensis, Winchell, 1866, Rep. Low. Peninsula Mich., Ham. Gr. [Ety. proper name.]

LYELLIA, Edwards & Haime, 1851, Mon. Pol. Foss. Terr. Palæoz. [Ety. proper name.]
  americana, Edwards & Haime, 1851, Mon. Pol. Foss. Terr. Palæoz., Corniferous Gr. [Ety. proper name.]
  decipiens, Rominger, 1876, Foss. Corals, Niagara Gr. (Sig. deceiving.]
  papillata, Rominger, 1876, Foss. Corals, Niagara Gr. [Sig. covered with papilli.]
  parvituba, Rominger, 1876, Pal. Foss. Corals, Niagara Gr. [Sig. small-tubed.]

Madrepora, Linnaeus, 1748, Syst. Nat.
  repens, Troost, 1840, 5th Geo. Rep. Tenn. Not satisfactorily defined.

MEGALOGRAPTUS, S. A. Miller, 1874, Cincin'ti Quar. Jour. Sci., vol. 1. [Ety. *megale*, large ; *grapho*, to write.]
  welchi, S. A. Miller, 1874, Cin'ti Quar. Jour. Sci., Cin'ti Gr. [Ety. proper name.]

MICHELINIA, De Koninck, 1842, Descr. des Anim. Foss. Belg. [Ety. proper name.]
  clappi, Edwards & Haime, (Chonostegites clappi) Corniferous Gr. [Ety. proper name.]
  convexa, 1850, D'Orbigny, Prodr. de Palæont., Onondaga & Corniferous Gr. [Sig. convex.]
  cylindrica, Edwards & Haime, 1851, Mon. Pol. Foss., Corniferous Gr. [Sig. cylindrical.]
  favositoidea, Billings, 1859, Can. Jour., vol. 4, Corniferous Gr. [Sig. like *Favosites.*]
  insignis, Rominger, 1876, Foss. Corals, Up. Held. & Ham. Gr. [Sig. remarkable.]
  intermittens, Billings, 1859, Can. Nat. & Geol., vol. 4, Corniferous Gr. [Sig. ceasing for a time.]
  lenticularis, Hall, 1874, 26th Reg. Rep. Low. Held. Gr. [Sig. lens-shaped.]
  trochiscus. Rominger, 1876, Foss. Corals. Ham. Gr. [Sig. a small round ball.]

MICROCYCLUS, Meek & Worthen, 1868, Geo. Sur. Ill., vol. 3. [Ety. *mikros*, small; *kuklos*, circle.]
  discus, Meek & Worthen, 1868, Geo. Sur. Ill.; vol. 3, Ham. Gr. [Sig. a quoit.]

Millepora, Linnaeus, 1748, Syst. Nat.
  repens, see Alveolites repens.

MONOGRAPTUS, Emmons,1856,(Monograpsus) Am. Geol. [Ety. *monos*, one; *grapho*, to write.]
  elegans, Emmons, 1856, Am. Geol., Quebec Gr. [Sig. elegant.]
  rectus, Emmons, 1856, Am. Geol. Quebec Gr. [Sig. straight.]

MONTICULIPORA, D'Orb., 1850, Prodr. de Palæont., vol. 1. [Ety. *monticulus*, a hillock; *poros*, a pore.]
  approximata, Nicholson, 1874, (Chetetes approximatus) Quar. Jour. Geol. Soc., vol. 30, Cin. Gr. [Sig. near to—from its near approach to M. dalii.]
  attrita, Nicholson, 1874, (Chetetes attritus) Quar. Jour. Geol. Soc. vol. 30, Cint'i Gr. I believe this is not a good species, but Nicholson says now that it should be called *Dekayia attrita.*
  barrandi, Nicholson, 1874, (Chetetes barrandii) Quar. Jour. Geol. Soc., vol. 30, Cin. Gr. [Ety. proper name.]
  briareus, Nicholson, 1875, (Chetetes briareus) Ohio Pal., vol. 2, Cin'ti Gr. [Ety. mythological name.]
  caliculus, James, 1875, (Chetetes calicula) Int. Catal. Cin. Foss., Cin'ti Gr. [Sig. a little cup.]

cincinnatiensis, James, 1875, (Chetetes cincinnatiensis) Int. Catal. Cin. Foss. Cin'ti Gr. [Ety. proper name.]
clavicoides, James, 1875, (Chetetes clavicoides) Int. Catal. Cin. Foss., Cin'ti Gr. [Sig. club-shaped.] A very doubtful species.
dalii, Edwards & Haime, 1851, (Chetetes dalii) Pol. Foss. des Terr. Palæoz., Cin'ti Gr. [Ety. proper name.]
decipiens, Rominger, 1866, (Chetetes decipiens) Proc. Acad. Nat. Sci. Phil., Cin'ti Gr. [Sig. doubtful.]
delicatula, Nicholson, 1874, (Chetetes delicatulus) Quar. Jour. Geol. Soc., vol. 30, Cin'ti Gr. [Sig. delicate.]
discoidea, Nicholson, 1875, (Chetetes discoideus) Ohio Pal., vol. 2, Cin'ti Gr. [Sig. disc-like.]
fibrosa, Goldfuss, 1826, (Calamopora fibrosa) Germ. Petref., Cin'ti to Clinton Gr. [Sig. fibrous.]
fletcheri, Edwards & Haime, 1851, (Chetetes fletcheri) Pol. Foss. des Terr. Palæoz.,Cin'ti Gr. [Ety. proper name.]
frondosa, D'Orbigny, 1850, Prodr. des Palæont, Cin'ti Gr. [Sig. branchy.]
gracilis, Nicholson, 1874, (Chetetes gracilis) Quar. Jour. Geo. Soc., vol. 30, Cin'ti Gr. [Sig. slender.]
jamesi, Nicholson, 1874, (Chetetes jamesi) Quar. Jour. Geo. Soc., vol. 30, Cin'ti Gr. [Ety. proper name.]
lycoperdon, Say, 1847, (Favosites lycoperdon) Hall, Pal. N. Y., vol. 1, Trenton & Hudson Riv. Gr. [Sig. puff-ball-shaped.]
mammillata, D'Orbigny, 1850, Prodr. de Palæont, Cin'ti Gr. [Sig. covered with nipples—mammillated.]
monticula, White, 1876, Proc. Acad. Nat. Sci., Devonian. [Sig. small conical projections.]
newberryi, Nicholson, 1875, (Chetetes newberryi) Ohio Pal. vol. 2, Cin'ti Gr. [Ety. proper name.]
nodulosa, Nicholson, 1874, (Chetetes nodulosa) Quar. Jour. Geol. Soc., vol. 30, Cin'ti Gr. [Sig. covered with small knots.]
o'nealli, James, 1875, (Cbetetes o'nealli) Int. Catal. Cin. Foss., Cin'ti Gr. [Ety. proper name.]
ortoni, Nicholson, 1874, (Chetetes ortoni) Quar. Jour. Geol. Soc., vol. 30, Cin'ti Gr. [Ety. proper name.]
pavonia, D'Orbigny, 1850, (Ptylodictya pavonia) Prodr. de Palæont., Cin'ti Gr. [Ety. *Pavonia*, a genus of polyps.]
petechialis, Nicholson, 1875, (Chetetes petechialis) Ohio Pal., vol. 2, Cin'ti Gr. [Sig. spotted.]
quadrata, Rominger, 1866, (Chetetes quadratus) Proc. Acad. Nat. Sci. Phil., Cin'ti Gr. [Sig. four cornered.]
rugosa, Hall, 1847, (Chetetes rugosus) Pal. N. Y., vol. 1, Cin'ti Gr. [Sig. wrinkled.]

rugosa, Edwards & Haime, 1851. The name was preoccupied.
sigillarioides, Nicholson, 1875, (Chetetes sigillarioides) Pal. Ohio, vol. 2, Cin'ti Gr. [Ety. from its resemblance to *Sigillaria*.]
subpulchella, Nicholson, 1875, (Chetetes subpulchellus) Ohio Pal., vol. 2, Cin'ti Gr. [Sig. somewhat like *M. pulchella*.]
tuberculata, Edwards & Haime, 1851, (Chetetes tuberculatus) Pol. Foss. des Terr. Palæoz., Cin'ti Gr. [Sig. tuberculated.]

NEBULIPORA, McCoy, 1850, Ann. & Mag. Nat. Hist., 2d ser., vol. 6. [Ety. *nebula*, thick-mist; *pora*, a pore.]
papillata, McCoy, 1850, Ann. & Mag. Nat. Hist., 2d ser., vol. 6, Cin'ti Gr. [Sig. covered with papilli.]

NEMAGRAPTUS, Emmons, (Nemagrapsus) 1856, Am. Geol., pt. 2. Prof. Hall uses the termination *graptus*, because *grapsus* is used in the nomenclature of crustacea. [Ety. *nema*, a thread; *grapho*, to write.]
capillaris, Emmons, 1856, Am. Geol., pt. 2, Quebec Gr. Prof. Hall says this species is apparently part of a *Graptolithus gracilis* or of some similar species. [Sig. like a hair.]
elegans, Emmons, 1856, Am. Geol., pt. 2. Quebec Gr. [Sig. elegant.]

NEREOGRAPTUS, Geinitz, 1852, Die Versteinerungen der Grauwacken—formation, etc. [Ety. *Nereis*, existing annelids; *grapho*, to write.]
deweyi, Emmons, 1856, Am. Geol., Quebec Gr. [Ety. proper name.]
gracilis, Emmons, 1856, Am. Geol., Quebec Gr. [Sig. slender.]
jacksoni, Emmons, 1856, Am. Geol., Quebec Gr. [Ety. proper name.]
lanceolatus, Emmons, 1856, Am. Geol., Quebec Gr. [Sig. lanceolate.]
loomisi, Emmons, 1856, Am. Geol., Quebec Gr. [Ety. proper name.]
pugnus, Emmons, 1856, Am. Geol., Quebec Gr. [Sig. a fist.]
robustus, Emmons, 1856, Am. Geol., Quebec Gr. [Sig. robust.]

OLDHAMIA, Forbes, 1850, Dublin Geo. Jour. [Ety. proper name.]
antiqua, Forbes, 1850, Dublin Geo. Jour., Potsdam Gr. [Sig. ancient.]
fruticosa, Hall, 1865, Can. Org. Rem., Decade 2, Trenton Gr. [Sig. branchy.]

OMPHYMA, Rafinesque, 1820, Ann. des Sci. Phys. de Bruxelles, vol. 5. [Ety. *omphax*, (?) a precious stone.]
congregata, Billings, 1866, Catal. Sil. Foss. Antic., Clinton and Niagara Gr. [Sig. congregated together.]
drummondi, Billings, 1866, Catal. Sil. Foss. Antic., Clinton & Niagara Gr. [Ety. proper name.]

stokesi, M. Edwards, (Ptychophyllum stokesi) as identified by Rominger, 1876, Foss. Corals, Niagara Gr. [Ety. proper name.]
verrucosa, Edwards & Haime, 1851, Mon. Pol. Foss. Terr. Palæoz., Niagara Gr. [Sig. warty.]

PACHYPHYLLUM, Edwards & Haime, 1850, Brit. Foss. Corals. [Ety. *pachys*, thick; *phyllon*, a leaf.]
solitarium, Hall, 1873, 23d Reg. Rep. Chemung Gr. [Sig. alone, solitary.]
woodmani, White, 1870, (Smithia woodmani) Geo. Rep. Iowa, vol. 1, Ham. Gr. [Ety. proper name.]

PALÆOCYCLUS, Edwards & Haime, 1849, Polypiers Fossiles. [Ety. *palaios*, ancient; *kuklos*, a circle.]
kirbyi, Meek, 1868, Trans. Chi. Acad. Sci., Devonian. [Ety. proper name.]

PALÆOPHYLLUM, Billings, 1858, Rep. of Prog. Can. Sur. [Ety. *palaios*, ancient; *phyllon*, a leaf.]
divaricans, Nicholson, 1875, Pal. Ohio, vol. 2, Cin'ti Gr. [Sig. wide-apart.]
rugosum, Billings, 1858, Rep. of Progr., Trenton Gr. [Sig. wrinkled.]

Palaeotrochis, Emmons, 1856, Geo. Rep. Midland counties of North Carolina. This form Prof. Hall, and later Prof. Marsh, ascertained to be a peculiar concretion.
major, Emmons. A concretion.
minor, Emmons. A concretion.

PETRAIA, Munster, 1839, Beitrage zur Petrefactenkunde, etc. [Ety. *petraios*, that grows among rocks.] Streptelasma is regarded by some as a sub-genus and by others as a synonym.
angulata, Billings, 1862, Pal. Foss., vol. 1, Hud. Riv. Gr. [Sig. angular.]
aperta, Billings, 1862, Pal. Foss., vol. 1, Black Riv. Gr. [Sig. standing open.]
fanningana, Safford, 1869, Geo. of Tenn., Low. Held. Gr. [Ety. proper name.]
forresteri, Honeyman, 1868, Acadian Geology. Catalogue name.
latuscula, Billings, 1862, Pal. Foss., vol. 1, Mid. Sil. [Sig. a little waist.]
logani, Nicholson, 1875, Can. Nat. & Geo., Corniferous Gr. [Ety. proper name.]
minganensis, see Archeocyathus minganensis.
ottawaensis, Billings, 1865, Can. Nat. & Geo., 2d series, vol. 2, Trenton Gr. [Ety. proper name.]
pygmea, Billings, 1862, Pal. Foss., vol. 1, Mid. Sil. [Sig. diminutive in size.]
pulchella, Billings, 1865, Can. Nat. & Geo., 2d series, vol. 2, Mid. Sil. [Sig. lovely, very pretty.]
rustica, Billings, 1858, Rep. of Progr., Hud. Riv. Gr. [Sig. rustic, uncouth.]
selecta, Billings, 1865, Can. Nat. & Geo., 2d series, vol. 2, Mid. Sil. [Sig. select, choice.]
waynensis, Safford, 1869, Geo. of Tenn., Low. Held. Gr. [Ety. proper name.]

PHILLIPSASTREA, D'Orbigny, 1849, Note Sur des Polypiers Fossiles. [Ety. proper name; *astrea*, from *aster*, a star.]
affinis, Billings, 1874, Pal. Foss., vol. 2, Devonian. [Sig. adjoining.]
gigas, Owen, 1840, (Astræa gigas) Rep. on Mineral lands, Devonian. [Sig. large.]
mammillaris, Owen, 1840, (Astrea mammillaris) Rep. on Mineral lands, Devonian. [Sig. mammillated.]
verneuili, Edwards & Haime, 1850, Polypiers Foss. des Terr., Ham. Gr. [Ety. proper name.]
yandelli, Rominger, 1876, Foss. Corals, Up. Held. Gr. [Ety. proper name.]

PHYLLOGRAPTUS, Hall, 1858, Rep. of Progr. Can. Sur. [Ety. *phyllon*, a leaf; *grapho*, to write.]
angustifolius, Hall, 1858, Rep. of Progr. Can. Sur., Quebec Gr. [Sig. narrow-leaved.]
anna, Hall, 1865, Can. Org. Rem., Decade 2, Quebec Gr. [Ety. proper name.]
ilicifolius, Hall, 1858, Rep. of Progr. Can. Sur., Quebec Gr. [Sig. oak-leaved.]
loringi, White, 1874, Rep. Invertebrate Foss., Quebec Gr. [Ety. proper name.]
similis, Hall, 1858, Can. Nat. & Geo., vol. 4, Quebec Gr. [Sig. similar.]
typus, Hall, 1858, Rep. of Progr. Can. Sur., Quebec Gr. [Ety. type of the genus.]

PLASMOPORA, Edwards & Haime, 1849, Comptes rend., t. 29. [Ety. *plasma*, a cast; *poros*, a pore.]
follis, Edwards & Haime, 1850, Mon. Pol. Foss., Niagara Gr. [Sig. a bag of leather.]

PLEURODICTYUM, Goldfuss, 1826, Petref. Germ., vol. 1. [Ety. *pleura*, side; *dictyon*, a net.]
problematicum, Goldfuss, 1826, Petref. Germ., vol. 1, Onondaga Gr. [Sig. doubtful.]

PLUMALINA, Hall, 1858, Can. Nat. & Geo., vol. 3. [Sig. like a feather.]
gracilis, Shumard, 1855, (Filicites gracilis) Geo. Rep. Mo., Lithographic Gr. [Sig. slender.]
plumaria, Hall, 1843, (Filicites ?) Geo. Rep. 4th Dist. N. Y., Chemung Gr. [Sig. plume-like.]

Polydilasma, Hall, 1852, Pal. N. Y., vol. 2, syn. for Zaphrentis.
turbinatum, see Zaphrentis turbinata.

PORITES, Lamarck, 1816, Hist. des Anim. sans Vert. [Ety. *poros*, a pore.]
astræiformis, Owen, 1840, Rep. on Mineral lands, Devonian. This may be the same species subsequently described as Pachyphyllum woodmani.
pyriformis, as identified by d'Archiac & Verneuil. Not American.
vetustus, see Protarea vetusta.

Prionotus, Hisinger, Leth. Suec.
folium, see Diplograptus folium.
pristis, see Diplograptus pristis.

POLYPI.

PROTAREA, Edwards & Haime, 1849, Pol. Foss. des Terr. Palæoz. [Ety. *protos*, first, formerly ; *araios*, porous, spongy.]
  vetusta, Hall, 1847, (Porites vetusta) Pal. N. Y., vol. 1, Trenton & Hud. Riv. Gr. [Sig. ancient.]
  verneuili, Edwards & Haime, 1851, Pol. Foss. des Ter. Palæoz., Silurian. (?) [Ety. proper name.]

PTILOGRAPTUS, Hall, 1865, Can. Org. Rem., Decade 2. [Ety. *ptilon*, a feather; *grapho*, to write.]
  geinitzianus, Hall, 1865, Can. Org. Rem., Decade 2, Quebec Gr. [Ety. proper name.]
  plumosus, Hall, 1865, Can. Org. Rem., Decade 2, Quebec Gr. [Sig. plume-like.]

PTYCHOPHYLLUM, Lonsdale, 1839, Sil. Syst. [Ety. *ptyche*, a ridge or wrinkle; *phyllon*, a plant.]
  canadense, Billings, 1862, Pal. Foss., vol. 1, Mid. Sil. [Ety. proper name.]
  stokesi, Edwards & Haime, 1851, Mon. Pol. Foss. des Terr. Palæoz., Niagara Gr. [Ety. proper name.]

QUENSTEDTIA, Rominger, 1876, Foss. Corals. [Ety. proper name.]
  cornuta, Billings, 1859, (Aulopora cornuta) Can. Jour., vol. 4, Corniferous Gr. [Sig. horned.]
  niagarensis, Rominger, 1876, Foss. Corals, Niagara Gr. [Ety. proper name.]
  umbellifera, Billings, 1859, (Aulopora umbellifera) Can. Jour., vol. 4, Corniferous Gr. [Sig. umbrella-bearing.]

RASTRITES, Barrande, 1850, Graptolites de Boheme. [Sig. a rake.]
  barrandi, Hall, 1859, Pal. N. Y., vol. 3, Hud. Riv. Gr. [Ety. proper name.] This may be only part of *Graptolithus gracilis*.

RETIOGRAPTUS, Suess, 1851, Ueber Bomischen Graptolithen. [Ety. *rete*, a net; *grapho*, to write.]
  barrandi, Hall, 1860, 13th Reg. Rep., Hud. Riv. Gr. [Ety. proper name.]
  eucharis, Hall, 1865, Can. Org. Rem., Decade 2, Utica slate. [Sig. very graceful.]
  geinitzianus, Hall, 1859, Pal. N. Y., vol. 3, Hud. Riv. Gr. [Ety. proper name.]
  tentaculatus, Hall, 1858, (Graptolithus tentaculatus) Rep. of Progr. Can. Sur., Quebec Gr. [Sig. having feelers.]

RETIOLITES, Barrande, 1850, Graptolites de Boheme. [Ety. *rete*, a net; *lithos*, stone.]
  ensiformis, Hall, 1858, (Graptolithus ensiformis) Rep. of Progr. Can. Sur., Quebec Gr. [Sig. sword-shaped.]
  venosus, Hall, 1852, (Graptolithus venosus) Pal. N. Y., vol. 2, Clinton Gr. [Sig. reticulated or venose.]

RHOMBOPORA, Meek, 1872, Pal. Eastern Nebraska. [Ety. *rhombus*, a rhomb; *pora*, a pore.]
  lepidodendroidea, Meek, 1872, Pal. Eastern Nebraska, Upper Coal Meas. [Sig. like the *Lepidodendron*.]

SARCINULA, Lamarck, 1816, Hist. des Anim. sans. Vert. [Ety. *sarcinula*, a little bundle.]
  glabra, Owen, 1840, Rep. on Mineral lands, Devonian. [Sig. smooth.]
  (?) obsoleta, Hall, 1857, Geo. Lake Sup. Land Dist., vol. 2, Hud. Riv. Gr. [Sig. obsolete.]
  ramosa, Eaton, 1832, Geo. Text Book, (?) Gr. [Sig. full of branches.]

SMITHIA, Edwards & Haime, 1851, Pol. Foss. des Terr. Palæoz. [Ety. proper name.] This name was preoccupied for a genus in Botany.
  johanna, Hall, 1873, 23d Reg. Rep., Chemung Gr. [Ety. proper name.]
  multiradiata, Hall, 1873, 23d Reg. Rep., Chemung Gr. [Sig. many-rayed.]
  woodmani, see Pachyphyllum woodmani.
  verrilli, Meek, 1868, Trans. Chi. Acad. Sci., Devonian. [Ety. proper name.]

SPHÆROLITES, Hinde, 1875, Proc. Geo. Soc. Lond. [Ety. from forming a sphæroidal body.]
  nicholsoni, Hinde, 1875, Proc. Geo. Soc. Lond., Low. Held. Gr. [Ety. proper name.]

STAUROGRAPTUS, Emmons, 1856, (Staurograpsus) Am. Geol., pt. 2. The termination *graptus*, is used because *grapsus* is in use in the nomenclature of crustacea, [Ety. *stauros*, a cross; *grapho*, to write.]
  dichotomus, Emmons, 1856, Am. Geol., Quebec Gr. [Sig. divided in two parts.]

STELLIPORA, Hall, 1847, Pal. N. Y., vol. 1. [Ety. *stella*, a star; *pora*, pore.]
  antheloidea, Hall, 1847, Pal. N. Y., vol. 1, Trenton & Hud. Riv. Gr. [Sig. like a coral of the genus *Anthelia*.]

STENOPORA, Lonsdale, 1845, Geol. Russ. & Ural. Mts., vol. 1. [Ety. *stenos*, narrow; *poros*, a pore.] Most of the species placed in this genus belong to *Monticulipora*.
  bulbosa, Billings, 1865, Can. Nat. & Geo., 2d series, vol. 2, Mid. Sil. [Sig. bulbous.]
  adherens, Billings, 1859, Can. Nat. & Geo., vol. 4, Chazy Gr. [Sig. adhering.]
  columnaris, Geinitz, 1866,Carb. und Dyas. in Neb., Coal Meas. [Sig. columnar.]
  crassa, Lonsdale, 1845, Geo. Rus., vol. 1, Coal Meas. [Sig. thick.]
  fibrosa, see Monticulipora fibrosa.
  exilis, Dawson, 1868, Acad. Geol., Low. Carb. [Sig. slender.]
  huronensis, Billings, 1865, Pal. Foss., vol. 1, Hudson River Gr. [Ety. proper name.]
  libana, Safford, 1869, Geo. of Tenn. Not defined.
  patula, Billings, 1859, Can. Nat. & Geo., vol. 4, Chazy Gr. [Sig. broad.]

spinigera, Lonsdale, 1845, Geo. Rus., vol. 1, Coal Meas. [Sig. bearing-spines.]
STREPTELASMA, Hall, 1847, Pal. N. Y., vol. 1. [Ety. *streptos*, twisted; *elasma*, lamella.]
calyculus, Hall, 1852, Pal. N. Y., vol. 2, Niagara Gr. [Sig. a little cup.]
conulus, Rominger, 1876, Foss, Corals, Niagara Gr. [Sig. a little cone.]
corniculum, Hall, 1847, Pal. N. Y., vol. 1, Trenton & Hud. Riv. Gr. [Sig. a little horn.]
crassum, Hall, 1847, Pal. N. Y., vol. 1, Trenton Gr. [Sig. thick.]
expansum, Hall, 1847, Pal. N. Y., vol. 1, Chazy Gr. [Sig. expanded.]
minimum, Hall, 1876, 28th Reg. Rep., Niagara Gr. [Sig. smallest.]
multilamellosum, Hall, 1847, Pal. N. Y., vol. 1, Trenton Gr. [Sig. having many lamellæ.]
parvulum, Hall, 1847, Pal. N. Y., vol. 1, Trenton Gr. [Sig. small.]
patulum, Rominger, 1876, Foss Corals, Niagara Gr. [Sig. standing open.]
profundum, Conrad, 1843, Proc.Acad.Nat. Sci., (Cyathophyllum profundum) Black Riv. & Trenton Grs. Was this name preoccupied by Germar in 1840? [Sig. deep.]
profunda, Hall, 1847, Pal. N. Y., vol. 1, Birdseye & Black Riv. Grs. If, as some claim, Streptelasma is a synonym for Petraia then this name was preoccupied by Germar in 1840, as well as by Conrad in 1843.
radicans, Hall, 1876, 28th Reg. Rep., Niagara Gr. [Sig. a root.]
rectum, Hall, 1843, (Strombodes rectus) Geo. Rep., 4th Dist. N. Y., Ham. Gr. [Sig. straight.]
spongiaxis, Rominger, 1876, Foss. Corals., Niagara Gr. [Sig. from the sponge-like axis.]
strictum, Hall, 1874, 26th Reg. Rep., Niagara Gr. [Sig. narrow, close.]
STRIATOPORA, Hall, 1852, Pal. N. Y., vol. 2. [Ety. *striatus*, striated; *pora*, a pore.]
carbonaria, White, 1862, Proc. Bost. Soc., Nat. Hist., vol. 9, Burlington Gr. [Sig. pertaining to coal.]
cavernosa, Rominger, 1876, Foss. Corals, Corniferous Gr. [Sig. full of cavities.]
flexuosa, Hall, 1852, Pal. N. Y., vol. 2, Niagara Gr. [Sig. bent.]
huronensis, Rominger, 1876, Foss. Corals, Niagara Gr. [Ety. proper name.]
iowensis, Owen, 1840, (Cyathopora iowensis) Rep. on Min. lands of Iowa, etc., Ham. Gr. [Ety. proper name.]
issa, Hall, 1874, 26th Reg. Rep. Low. Held. Gr. [Ety. proper name.]
linnæana, Billings, 1859, Can. Jour., Ham. Gr. [Ety. proper name.]
missouriensis, Meek & Worthen, 1868, Geo. Sur. Ill., vol. 3, Low. Held. Gr. [Ety. proper name.]

rugosa, Hall, 1858, Geo. of Iowa, Ham. Gr. [Sig. wrinkled.] Syn. for *Striatopora iowensis*.
STROMATOCERIUM, Hall, 1847, Pal. N. Y., vol. 1. [Ety. *stroma*, layer; *kerion*, a honey-comb.]
rugosum, Hall, 1847, Pal. N. Y., vol. 1, Birdseye & Black River Gr. [Sig. wrinkled.]
STROMATOPORA, Goldfuss, 1826, Germ. Petref. [Ety. *stroma*, stratum; *poros*, a pore.] This genus is regarded by some authors as a sponge.
alternata, Hall, 1873, 23rd Reg. Rep., Chemung Gr. [Sig. alternating.]
cæspitosa, Winchell, 1866, Rep. Low. Penin. Mich., Ham. Gr. [Sig. turf-like.]
compacta, Billings, 1862, Pal. Foss., vol. 1, Black River Gr. [Sig. compact.]
concentrica, Goldfuss, 1826, Germ. Petref., Niagara Gr. [Sig. concentric.]
constellata, Hall, 1852, Pal. N. Y., vol. 2, Coralline limestone. [Sig. starry.]
erratica, Hall, 1873, 23d Reg. Rep., Chemung Gr. [Sig. erratic.]
expansa, Hall, 1873, 23d Reg. Rep., Chemung Gr. [Sig. expanded.]
granulata, Nicholson, 1873, Ann. & Mag. Nat. Hist., 4th series, vol. 12, Corniferous Gr. [Sig. granulated.]
hindi, Nicholson, 1874, Ann. & Mag. Nat. Hist., 4th series, vol. 13. [Ety. proper name.]
incrustans, Hall, 1873, 23d Reg. Rep., Chemung Gr. [Sig. incrusting.]
mammillata, Nicholson, 1873, Ann. & Mag. Nat. Hist., 4th series, vol. 12, Corniferous Gr. [Sig. mammillated.]
monticulifera, Winchell, 1866, Rep. Low. Penin. Mich., Ham. Gr. [Sig. bearing conical projections.]
nodulata, Nicholson, 1875, Ohio Pal., vol. 2, Corniferous Gr. [Sig. from the elevations that cover it.]
nux, Winchell, 1866, Rep. Low. Penin. Mich., Ham. Gr. [Sig. a nut.]
ostiolata, Nicholson, 1873, Ann. & Mag. Nat. Hist. 4th ser., vol. 12, Guelph Gr. [Sig. a little door.]
perforata, Nicholson, 1874, Ann. & Mag. Nat. Hist., 4th ser., vol. 13, Corniferous Gr. [Sig. perforated.]
ponderosa, Nicholson, 1875, Ohio Pal. vol. 2, Corniferous Gr. [Sig. from its ponderous appearance.]
pustulifera, Winchell, 1866, Rep. Low. Penin. Mich., Ham. Gr. [Sig. bearing pustules.]
pustulosa, Safford. Not defined.
solidula, Hall, 1873, 23rd Reg. Rep., Chemung Gr. [Sig. solid.]
substriatella, Nicholson, 1875, Ohio Pal., vol. 2, Corniferous Gr. [Sig. somewhat finely striated.]
tuberculata, Nicholson, 1873, Ann. & Mag. Nat. Hist., 4th ser., vol. 12, Corniferous Gr. [Sig. tuberculated.]

verrucosa, Troost, 1840, 5th Geo. Rep. Tenn., Devonian. [Sig. warty.]
STROMBODES, Schweigger, 1820, Handb. der Naturg. [Ety. *strombos*, twisting; in allusion to the lamellæ spirally twisting about the center.]
alpenensis, Rominger, 1876, Foss. Corals, Ham. Gr. [Ety. proper name.] Is this a syn. for Phillipsastrea mammillaris?]
diffluens, Edwards & Haime, 1851, Pol. Foss. des Terr. Palæoz., Anticosti Gr. [Sig. flowing in different directions.]
distortus, Hall, 1843, Geo. Rep. 4th Dist. N. Y., Ham. Gr. [Sig. distorted, irregular in shape.]
eximius, Billings, 1866, Catal. Sil. Foss., Antic., Clinton & Niagara Gr. [Sig. uncommon, excellent.]
gracilis, Billings, 1862, Pal. Foss., vol. 1, Mid. Sil. [Sig. slender.]
helianthoides, (?) Heliophyllum halli.
mammillatus, Owen, as defined by Rominger, 1876, Foss. Corals, Niagara Gr. [Sig. mammillated.]
pentagonus, Goldfuss, 1826, Germ. Petref. Niagara Gr. [Sig. pentagonal.]
pygmæus, Rominger, 1876, Foss. Corals, Niagara Gr. [Sig. dwarfed.]
(?) *rectus*, see Streptelasma rectum.
simplex, see Zaphrentis simplex.
striatus, D'Orbigny, Niagara Gr. [Sig. striated.]
vermicularis, Lonsdale, 1849, Trans. Geo. Soc., vol. 5, Devonian. [Sig. vermicular.]
SYRINGOPORA, Goldfuss, 1826, Germ. Petref. [Ety. *syrinx*, a pipe; *pora*, a pore.]
alectiformis, Winchell, 1866, Rep. Low. Penin. Mich., Ham. Gr. [Ety. like a coral of the genus Alecto.]
annulata, Rominger, 1876, Foss. Corals, Niagara Gr. [Sig. ringed.]
cleviana, Edwards & Haime, 1851, Polyp. Foss., Corniferous Gr. [Ety. proper name.]
compacta, Billings, 1858, Can. Nat. & Geo., vol. 3, Upp. Sil. [Sig. compact.]
crassata, Winchell, 1866. Rep. Low. Penin. Mich., Ham. Gr. [Sig. thick; in allusion to the thickness of the tube walls.]
dalmani, Billings, 1858, Can. Nat. & Geo., vol. 3, Upp. Sil. [Ety proper name.]
debilis, Billings, 1858, Can. Nat. & Geo. vol. 3, Upp. Sil. [Sig. weak, feeble.]
elegans, Billings, 1858, Can. Nat. & Geo., vol. 3, Corniferous Gr. [Sig. elegant.]
fenestrata, Winchell, 1866, Rep. Low. Penin. Mich., Ham. Gr. [Sig. reticulated.]
fibrata, Rominger, 1876, Foss. Corals, Niagara Gr. [Sig. fibrous.]
harveyi, White, 1862, Proc. Bost. Soc. Nat. Hist. vol. 9, Chemung Gr. [Ety. proper name.]
hisingeri, Billings, 1858, Can. Nat. & Geo., vol. 4, Corniferous Gr. [Ety. proper name.]
intermedia, Nicholson, 1874, Rep. Pal. Prov. Ont., Can., Devonian. [Sig. intermediate.]
laxata, Billings, 1859, Can. Jour., vol. 4, Corniferous Gr. [Sig. loose.]
macluri, Billings, 1860, Can. Jour., vol. 5, Corniferous Gr. [Ety. proper name.]
multattenuata, McChesney, 1860, New Pal. Foss., Coal Meas. [Ety. *multum*, much; *attenuatus*, slender.]
multicaulis, Hall, 1852, Pal. N. Y., vol. 2, Niagara Gr. [Sig. having many stems.]
nobilis, Billings, 1858, Can. Nat. & Geo., vol. 4, Corniferous Gr. [Sig. remarkable.]
perelegans, Billings, 1859, Can. Jour., vol. 4, Corniferous Gr. [Sig. very elegant.]
retiformis, Billings, 1858, Can. Nat. & Geo., vol. 3, Upp. Sil. [Sig. net-formed.]
reticulata, Goldfuss, 1826, Petref. Germ., Devonian. [Sig. reticulated.]
tabulata, Edwards & Haime, 1851, Pol. Foss. Terr. Palæoz., Up. Held. Gr. [Sig. tabulated.]
tenella, Rominger, 1876, Foss. Corals, Niagara Gr. [Sig. delicate.]
tubiporoides, Yandell & Shumard, 1847, Contributions to Geo. of Ky., Corniferous Gr. [Ety. resembling corals of the genus Tubipora.]
tubiporoides, Billings. The name was pre-occupied in 1847.
verneuili, Edwards & Haime, 1851, Polyp. Foss., Corniferous Gr. [Ety. proper name.]
verticillata, Goldfuss, 1826, Petref. Germ., Up. Sil. [Sig. verticillate, whorled.]
SYRINGOSTROMA, Nicholson, 1875, Ohio Pal., vol. 2. [Ety. *syrinx*, a pipe; *stroma*, a layer.]
columnare, Nicholson, 1875, Ohio Pal., vol. 2, Corniferous Gr. [Sig. columnar.]
densum, Nicholson, 1875, Ohio Pal., vol. 2, Corniferous Gr. [Ety. from the apparently dense calcareous structure.]
TETRADIUM, Dana, 1846, Zooph, vol. 8. [Ety. *tetras*, four.]
columnare, Hall, 1847, (Chetetes columnaris) Pal. N. Y., vol. 1, Trenton Gr. [Sig. columnar.]
fibratum, Safford, 1856, Am. Jour. Sci., vol. 22, Cin'ti Gr. [Sig. threaded.]
fibratum *var.* apertum, Safford, 1856, Am. Jour. Sci., vol. 22, Cin'ti Gr. [Sig. open.]
fibratum *var.* minus, Safford, 1856, Am. Jour. Sci. & Arts, vol. 22, Cin'ti Gr. [Sig. less.]
TETRAGRAPTUS, Salter, 1863, Quar. Jour. Geo. Soc., vol. 19. [Ety. *tetras*, four; *grapho*, to write]. Prof. Hall does not regard this sub-genus of Graptolithus, considering the present knowledge of the subject, with much favor. *G. bryonoides* is made the typical species. *G. quadribrachiatus*, also placed in it.

approximatus, Nicholson, 1873, Ann. Mag. Nat. Hist., 4th series, vol. 12, Quebec Gr. [Sig. near to.]
THAMNOGRAPTUS, Hall, 1859, Pal. N. Y., vol. 3. [Ety. *thamnus*, a shrub; *grapho*, to write.]
anna, Hall, 1865, Can. Org. Rem., Decade 2, Quebec Gr. [Ety. proper name.]
capillaris, Hall, 1859, Pal. N. Y., vol. 3, Hud. Riv. Gr. [Sig. hair-like.]
typus, Hall, 1859, Pal. N. Y., vol. 3, Hud. Riv. Gr. [Ety. type of the genus.]
THECIA, Edwards & Haime, 1849, Comptes rend., t. 29. [Ety. proper name. (?)]
major, Rominger, 1876, Foss. Corals, Niagara Gr. [Sig. large.]
minor, Rominger, 1876, Foss. Corals, Niagara Gr. [Sig. less.]
ramosa, Rominger, 1876, Foss. Corals, Up. Held. Gr. [Sig. branching.]
swinderniana, Goldfuss, 1826, (Agaricia swinderniana) Petref. Germ., Niagara Gr. [Ety. proper name.]
THECOSTEGITES, Edwards & Haime, 1849, Comptes rend., t. 29. [Ety. *theke*, a sheath; *stege*, a covering.]
bouchardi, Edwards & Haime, 1854, Brit. Foss. Corals, Up. Held. Gr. [Ety. proper name.]
hemisphæricus, Roemer, 1860, Sil. Fauna. West Tenn., Niagara Gr. [Sig. hemispherical.]
TRACHYPORA, Edwards & Haime, 1851, Pol. Foss. Terr. Palæoz. [Ety. *trachys*, rough; *poros*, a pore.]
elegantula, Billings, 1859, Can. Jour., Ham. Gr. [Sig. beautiful.]
TUBIPORA, Lamarck, 1815, Hist. des Anim. sans Verteb. [Ety. *tubus*, a tube; *porus*, a pore.]
lamellosa, Owen, 1840, Rep. on Mineral lands, Devonian. [Sig. in very thin plates.]
VERMIPORA, Hall, 1874, 26th Reg. Rep. [Ety. *vermis*, a worm; *pora*, a pore.]
serpuloides, Hall, 1874, 26th Reg. Rep., Low. Held. Gr. [Sig. like the *Serpula*.]
fasciculata, Rominger, 1876, Foss. Corals, Ham. Gr. [Sig. bundled.]
niagarensis, Rominger, 1876, Foss. Corals, Niagara Gr. [Ety. proper name.]
VESICULARIA, Rominger, 1876, Foss. Corals. [Ety. *vesicularia*, from the vesiculose structure.]
major, Rominger, 1876, Foss. Corals, Niagara Gr. [Sig. large.]
minor, Rominger, 1876, Foss. Corals, Niagara Gr. [Sig. less.]
variolosa, Rominger, 1876, Foss. Corals, Niagara Gr. [Sig. full of cell pits.]
ZAPHRENTIS, Rafinesque, 1820, Ann. des. Sci. Phys. Brux., vol. 5. [Ety. *za*, very; *phrentis*, a diaphragm.]
affinis, Billings, 1865, Can. Nat. & Geo., 2d series, vol. 2, Mid. Sil. [Sig. contiguous.]
bellistriata, Billings, 1865, Can. Nat. & Geo., 2d series, vol. 2, Hud. Riv. Gr. [Sig. beautifully striated.]
bigsbyi, Billings, 1866, Catal. Sil. Foss. Antic., Clinton & Niagara Gr. [Ety. proper name.
bilateralis, Hall, 1852, (Caninia bilateralis) Pal. N. Y., vol. 2, Clinton & Niagara Gr. [Sig. two-sided.]
canadensis, Billings, 1862, Pal. Foss., vol. 1, Hud. Riv. Gr. [Ety. proper name.]
celator, Hall, 1876, 28th Reg. Rep., Niagara Gr. [Sig. a concealer.]
cinctosa, Billings, 1866, Catal. Sil. Foss., Antic., Clinton & Niagara Gr. [Sig. full of circles or belts.]
cingulosa, Billings, 1874, Pal. Foss., vol. 2, Devonian. [Sig. circled with lines.]
compressa, Rominger, 1876, Foss. Corals, Up. Held. Gr. [Sig. compressed.]
conigera, Rominger, 1876, Foss. Corals Up. Held. Gr. [Sig. cone-bearing.]
corniculum, Edwards & Haime, 1851, Pol. Foss. Terr. Palæoz., Up. Held. Gr. [Sig. a little horn.]
corticata, Billings, 1874, Pal. Foss., vol. 2, Devonian. [Sig. coated.]
cystica, Winchell, 1866, Rep. Low. Penin. Mich., Ham. Gr. [Sig. like a pouch.]
edwardsi, Nicholson, 1875, Ohio Pal., vol. 2, Corniferous Gr. [Ety. proper name.]
egeria, Billings, 1875, Can. Nat. & Geol., Corniferous Gr. [Ety. mythological name.]
elliptica, White, 1862, Proc. Bost. Soc. Nat. Hist., vol. 9, Burlington Gr. [Sig. elliptical.]
eriphyle, Billings, 1875, Can. Nat. & Geo., Corniferous Gr. [Ety. mythological name.]
fenestrata, Nicholson, 1875, Can. Nat. & Geo., Corniferous Gr. [Sig. full of openings.]
gigantea, Rafinesque, 1820, as identified by Edwards & Haime, 1851, Polyp. Foss., Corniferous Gr. [Sig. very large.]
glans, White, 1862, Proc. Bost. Soc. Nat. Hist., vol. 9, Burlington Gr. [Sig. an acorn.]
gregaria, Rominger, 1876, Foss. Corals, Niagara Gr. [Sig. in flocks.]
hecuba, Billings, 1875, Can. Nat. & Geo., Corniferous Gr. [Ety. mythological name.]
ida, Winchell, 1863, Proc. Acad. Nat. Sci., Lithographic limestone. [Ety. proper name.]
incondita, Billings, 1874, Pal. Foss., vol. 2, Devonian. [Sig. confused.]
invenusta, Billings, 1875, Can. Nat. & Geo., Corniferous Gr. [Sig. not elegant.]
macfarlani, Meek, 1868, Trans. Chi. Acad. Sci., Devonian. [Ety. proper name.]
minas, Dawson, 1868, Acad. Geo. Low. Carb. [Sig. proper name.]

multilamella, Hall, 1852, Stans. Ex. to Gt. Salt Lake, Coal Meas. [Sig. many lamellæ.]

multilamellata, Nicholson, 1875, Ohio Pal., vol. 2, Corniferous Gr. Preoccupied above.

nodulosa, Rominger, 1876, Foss. Corals, Corniferous Gr. [Sig. full of knots.]

patens, Billings, 1865, Can. Nat. & Geo., 2d ser., vol. 2, Mid. Sil. [Sig. spreading.]

prolifica, Billings, 1859, Can. Nat. & Geo., vol. 4, Corniferous Gr. [Sig. prolific.]

recta, Meek, 1868, Trans. Chi. Acad. Sci., Devonian. [Sig. straight.]

rugatula, Billings, 1874, Pal. Foss., vol. 2, Up. Sil. [Sig. covered with little wrinkles.]

simplex, Hall, 1843, (Strombodes simplex) Geo. Rep., 4th Dist. N. Y., Ham. Gr. [Sig. simple.]

solida, Hall, 1873, 23d Reg. Rep., Chemung Gr. [Sig. solid.]

spatiosa, see Heterophrentis spatiosa.

spinulifera, Hall, 1858, Geo. Sur. Iowa, Warsaw Gr. [Sig. bearing little spines.]

spinulosa, Edwards & Haime, 1851, Pol. Foss. Terr. Palæoz., Low. Carb. [Sig. covered with spines.]

stansburyi, Hall, 1852, Stans. Ex. to Gt. Salt Lake, Coal Meas. [Ety. proper name.]

stokesi, Edwards & Haime, 1851, Pol. Foss. Terr. Palæoz., Niagara Gr. [Ety. proper name.]

subrecta, Billings, 1875, Can. Nat. & Geo., Corniferous Gr. [Sig. somewhat like Z. recta.]

traversensis, Winchell, 1866, Rep. Low. Penin. Mich., Ham. Gr. [Ety. proper name.]

turbinata, Hall, 1852, (Polydilasma turbinatum) Pal. N. Y., vol. 2, Niagara Gr. [Sig. top-shaped.]

umbonata, Rominger, 1876, Foss. Corals, Ham. Gr. [Sig. protuberant.]

ungula, Rominger, 1876, Foss. Corals, Up. Held. Gr. [Sig. a talon.]

wortheni, Nicholson, 1875, Ohio Pal., vol. 2, Corniferous Gr. [Ety. proper name.]

CORRIGENDA.

Brachiospongia, page 42, read, [Ety. *brachium*, an arm.]

Cnemidium trentonensis, page 43, read, *C. trentonense*.

Receptaculites infundibulus, page 44, read, *R. infundibulum*.

Aulopora saxivadum, page 47, read, *A. saxivada*. [Sig. creeping over stone.]

Axophyllum rudis, page 47, read, *A. rude*.

Campophyllum, page 48, read, [Ety. *kampto*, to bend.]

Chonophyllum, page 48, read, [Sig. *chonos*, a funnel.]

Chonograptus, page 48, read, *Clonograptus*. [Ety. *klon*, a twig.]

Cladopora lichenoidea, page 49, read, *C. lichenoides*.

Cyathophyllum juvenis, page 50, read, *C. juvene*.

Cyathophyllum maritima, page 51, read, *C. maritimum*.

Dictyonema fenestrata, page 51, read, *D. fenestratum*. —

Dictyonema gracilis, page 51, read, *D. gracile*.

Dictyonema grandis, page 51, read, *D. grande*.

Dictyonema irregularis, page 51, read, *D. irregulare*.

Dictyonema quadrangularis, page 51, read, *D. quadrangulare*.

Dictyonema retiformis, page 51, read, *D. retiforme*.

Dictyonema robusta, page 51, read, *D. robustum*.

Dictyostroma undulata, page 52, read, *D. undulatum*.

Favosites pleurodictyloides, page 54, read, *F. pleurodictyoides*.

Halysites catenulatus, page 56, read, [Ety. *catenula*, a little chain.]

Leptopora typa, page 56, read, *L. typus*.

ECHINODERMATA.

ORDER CRINOIDEA.

FAMILY CYATHOCRINIDÆ.—Agassizocrinus, Anomalocrinus, Barycrinus, Belemnocrinus, Bursacrinus, Carabocrinus, Catillocrinus, Cleiocrinus, Closterocrinus, Coccocrinus, Cœliocrinus, Cotyledonocrinus, Ctenocrinus, Cyathocrinus, Dendrocrinus, Dichocrinus, Erisocrinus, Eucalyptocrinus, Eupachycrinus, Forbesocrinus, Glyptaster, Glyptocrinus, Graphiocrinus, Hadlocrinus, Haplocrinus, Heterocrinus, Homocrinus, Hybocrinus, Ichthyocrinus, Lecanocrinus, Macrostylocrinus, Mariscrinus, Myelodactylus, Myrtillocrinus, Nipterocrinus, Onychocrinus, Pachycrinus, Palæocrinus, Platocrinus, Porocrinus, Poteriocrinus, Pterotocrinus, Retiocrinus, Rhodocrinus, Scaphiocrinus, Schizocrinus, Scyphocrinus, Stephanocrinus, Synbathocrinus, Taxocrinus, Technocrinus, Thysanocrinus, Zeacrinus.

FAMILY ACTINOCRINIDÆ.—Acrocrinus, Actinocrinus, Agaricocrinus, Alloprosallocrinus, Amphoracrinus, Batocrinus, Cœlocrinus, Coronocrinus, Dolatocrinus, Dorycrinus, Eretmocrinus, Gilbertsocrinus, Goniasteroidocrinus, Hadrocrinus, Lampterocrinus, Lyriocrinus, Megistocrinus, Melocrinus, Mespilocrinus, Physetocrinus, Pygorhynchus, (?) Saccocrinus, Steganocrinus, Strotocrinus, Vasocrinus.

FAMILY CALCEOCRINIDÆ.—Calceocrinus.

FAMILY ANCYROCRINIDÆ. (?)—Ancyrocrinus.

FAMILY EDRIOCRINIDÆ. (?)—Aspidocrinus, Edriocrinus.

FAMILY BRACHIOCRINIDÆ. (?)—Brachiocrinus, Syringocrinus. (?)

ORDER PERISCHO-ECHINOIDEA.

FAMILY PALÆCHINIDÆ.—Lepidechinus, Lepidesthes, Melonites, Oligoporus, Palæchinus.

FAMILY ARCHÆOCIDARIDÆ.—Archæocidaris, Eocidaris, Lepidocidaris, Pholidocidaris.

ORDER CYSTOIDEA.—Amygdalocystites, Anomalocystites, Apiocystites, Ateleocystites, Callocystites, Caryocrinus, Codaster, Comarocystites, Crinocystites, Cyclocystoides, Cystocrinus, Dictyocrinus, Echinocystites, Echino-encrinites, Eocystites, Glyptocystites, Gomphocystites, Hemicosmites, Heterocystites, Holocystites, Lepadocrinus, Lichenocrinus, Malocystites, Palæocystites, Pleurocystites, Sphærocystites, Strobilocystites.

ORDER BLASTOIDEA.—Blastoidocrinus, Codonites, Eleutherocrinus, Granatocrinus, Nucleocrinus, Pentremites.

ORDER OPHIUROIDEA.—Protaster, Onychaster.

ORDER ASTEROIDEA.—Codaster, Eugaster, Palæaster, Palæasterina, Palæocoma, Petraster, Schœnaster, Stenaster, Taeniaster.

ORDER AGELACRINIDÆ.—Agelacrinus, Edrioaster, Hemicystites.

ACROCRINUS, Yandell, 1855, Am. Jour. Sci., vol. 20. [Ety. *akros*, the summit; *krinon*, lily.]
shumardi, Yandell, 1855, Am. Jour. Sci., vol. 20, Kaskaskia Gr. [Ety. proper name.]
urniformis, Hall, 1858, Geo. Rep. Iowa, Kaskaskia Gr. [Sig. urn-shaped.]
ACTINOCRINUS, Miller, 1821, Nat. Hist. Crinoidea. [Ety. *aktin*, a ray or thorn; *krinon*, a lily.]
abnormis, see Megistocrinus abnormis.
ægilops, see Strotocrinus ægilops.
æqualis, see Batocrinus æqualis.
æquibrachiatus, see Batocrinus æquibrachiatus.
æquibrachiatus var. alatus, see Batocrinus æquibrachiatus var. alatus.
agassizi, Troost, 1850, Catal. Not defined.
althea, Hall, 1861, Desc. New Crinoidea, Burlington Gr. [Ety. mythological name.]
amplus, Meek & Worthen, 1861, Proc. Acad. Nat. Sci. Phil., Burlington Gr. [Sig. full sized.]
andrewsianus, syn. for Eretmocrinus verneuilianus.
araneolus, see Steganocrinus araneolus.
asterias, McChesney, 1860, Desc. New Pal. Foss., Burlington Gr. [Sig. in the fashion of a star.]
asteriscus, see Batocrinus asteriscus.
biturbinatus, Hall, 1858, Geo. Rep. Iowa, Keokuk Gr. [Sig. double cone-shaped.]
brevicornis, Hall, 1858, Geo. Rep. Iowa, Burlington Gr. [Sig. short-horned.]
brevis, Hall, 1858, Geo. Rep. Iowa, Burlington Gr. [Sig. short.]
brontes, Hall, 1860, Supp. to Geo. Sur. Iowa, Warsaw Gr. [Ety. mythological name.]
cælatus, Hall, 1858, Geo. Sur. Iowa, Burlington Gr. [Sig. carved in relief, sculptured.]
calyculoides, see Eretmocrinus calyculoides.
calyculus, Hall, 1860, Supp. to Geo. Iowa, Warsaw Gr. [Sig. a little cup or flower-bud.]
calypso, Hall, 1862, 15th Reg. Rep. N. Y., Ham. Gr. [Ety. mythological name.]
cassedayi, Lyon, 1861, Proc. Acad. Nat. Sci. Phil., Up. Held. Gr. [Ety. proper name.]
carica, see Batocrinus carica.
caroli, Hall, 1860, Supp. to Geo. Sur. Iowa, Warsaw Gr. [Ety. proper name.]
cauliculus, Hall, 1862, 15th Reg. Rep. N. Y., Ham. Gr. [Sig. the small stalk or stem of a plant.]
chloris, Hall, 1861, Desc. New Crinoidea, Burlington Gr. [Ety. mythological name.]
christyi, Shumard, 1855, see Batocrinus christyi.

christyi, Hall, 1863, see Saccocrinus christyi.
clarus, Hall, 1861, Desc. New Crinoidea, Burlington Gr. [Sig. brilliant, illustrious.]
clavigerus, see Batocrinus clavigerus.
clio, see Eretmocrinus clio.
clivosus, Hall, 1861, Jour. Bost. Soc. Nat. Hist., vol. 7, Burlington Gr. [Sig. hilly.]
clœlia, see Eretmocrinus clœlia.
clypeatus, Hall, 1860, Supp. Geo. Sur. Iowa, Burlington Gr. [Sig. shield-like.]
concavus, see Cœlocrinus concavus.
concinnus, Shumard, 1855, Geo. Rep. Mo., Burlington Gr. [Sig. handsome.]
corbulis, see Batocrinus corbulis.
coreyi, Lyon & Casseday, 1859, Am. Jour. Sci., Burlington Gr. [Ety. proper name.]
corniculum, Hall, 1858, Geo. Rep. Iowa, Burlington Gr. [Ety. *corniculum*, a little horn—from the urn-shaped body below.]
cornigerus, Hall, 1858, see Dorycrinus cornigerus.
cornigerus, Lyon & Casseday, 1859, Am. Jour. Sci., vol. 28. This name being preoccupied, Shumard proposed for this species *A. kentuckiensis*.
cornutus, Troost, 1850, Catal. Not defined.
coronatus, Hall, 1860, Supp. to Geo. Sur. Iowa, Burlington Gr. [Sig. crowned.]
daphne, Hall, 1863, Crin. Wav. Sands. Ohio, Waverly Gr. [Ety. mythological name.]
decornis, Hall, 1860, Supp. to Geo. Sur. Iowa, Burlington Gr. [Sig. without horns.]
delicatulus, Meek & Worthen, 1869, Proc. Acad. Nat. Sci. Phil. Burlington Gr. [Sig. rather delicate.]
desideratus, see Dorycrinus desideratus.
discoideus, see Batocrinus discoideus.
divaricatus, see Dorycrinus divaricatus.
divergens, see Amphoracrinus divergens.
dodecadactylus, see Batocrinus dodecadactylus.
doris, see Batocrinus doris.
eicosidactylus, see Batocrinus icosidactylus.
erodus, Hall, 1861, Desc. New Crinoidea, Burlington Gr. [Ety. *erosus*, eroded, having the parts jagged as if gnawed.]
eryx, Hall 1861, Desc. New Crinoidea, Burlington Gr. [Ety. mythological name.]
eucharis, Hall, 1862. 15th Reg. Rep. N. Y., Ham. Gr. [Ety. *eu*, very; *charis*, beauty.]
evansi, Owen & Shumard, 1850, Jour. Acad. Nat. Sci., 2d ser., vol. 2, Burlington Gr. [Ety. proper name.]
excerptus, Hall, 1861, Desc. New Crinoidea, Burlington Gr. [Sig. picked out, selected.]
fibula, Troost, 1850, Catal. Not defined.

ECHINODERMATA.

fiscellus, Hall, 1861, Desc. New Crinoidea, Burlington Gr. [Sig. a small basket for fruit, woven of slender twigs.]
fosteri, McChesney, 1860, syn. for A. crelatus.
formosus, see Batocrinus fromosus.
gemmiformis, see Batocrinus gemmiformis.
gibbosus, Troost, 1850. Not defined.
glans, Hall, 1860, Supp. to Geo. Sur. Iowa, Burlington Gr. [Sig. an acorn.]
glyptus, see Strotocrinus glyptus.
gouldi, see Dorycrinus gouldi.
hageri, McChesney, 1860, New Pal. Foss., Burlington Gr. [Ety. proper name.]
helice, Hall, 1863, Crin. Wav. Sands. Ohio, Waverly Gr. [Ety. mythological name.]
helice var. eris, Hall, 1864, 17th Reg. Rep. N. Y., Waverly Gr. [Ety. proper name.]
humboldti, Troost, 1850, Catal. Not defined.
hurdianus, McChesney, 1860, New Pal. Foss., Burlington Gr. [Ety. proper name.]
indianensis, Lyon & Casseday, 1859, Am. Jour. Sci., vol. 28, Warsaw Gr. [Ety. proper name.]
inflatus, see Amphoracrinus inflatus.
infrequens, Hall, 1861, Desc. New Crinoidea, Burlington Gr. [Sig. not frequent, rare.]
insculptus, see Strotocrinus insculptus.
inornatus, see Batocrinus inornatus.
irregularis, see Batocrinus irregularis.
jugosus, Hall, 1860, Supp. Geo. Sur. Iowa, Keokuk Gr. [Sig. yoked together.]
kentuckiensis, Shumard, Trans. St. Louis Acad. Sci., Up. Held. Gr. [Ety. proper name.]
konincki, see Batocrinus konincki.
lagena, Hall, 1861, Desc. New Crinoidea, Burlington Gr. [Sig. a flask or bottle.]
lagunculus, see Batocrinus laguncplus.
laura, see Batocrinus laura.
lepidus, see Batocrinus lepidus.
leucosia, Hall, 1861, Desc. New Crinoidea, Burlington Gr. [Ety. mythological name.]
limabrachiatus, Hall, 1861, Desc. New Crinoidea, Burlington Gr. [Ety. *lima*, a file; *brachiatus*, armed.]
liratus, see Strotocrinus liratus.
lobatus, Hall, 1860, Supp. to Geo. Sur. Iowa, Keokuk Gr. [Sig. lobate.]
locellus, Hall, 1861, Desc. New Crinoidea, Burlington Gr. [Sig. a chest or casket.]
longirostris, see Batocrinus longirostris.
longus, Meek & Worthen, Proc. Acad. Nat. Sci. Phil., Burlington Gr. [Ety. *longus*, long.]
lowii, Hall, 1852, Geo. Rep. Iowa, Keokuk Gr. [Ety. proper name.]
lucina, Hall, 1861, Desc. New Crinoidea, Burlington Gr. [Ety. mythological name.]
matuta, see Eretmocrinus matuta.

matuta var. attenuata, see Eretmocrinus matuta var. attenuata.
Meeki, Lyon, 1861, Proc Acad. Nat. Sci. Phil., Niagara Gr. [Ety. proper name.]
minor, Hall, 1858, Geo. Rep Iowa, Burlington Gr. [Sig. less.]
mississippiensis, see Dorycrinus mississippiensis.
mississippiensis var. spiniger, see Dorycrinas mississipiensis var. spiniger.
missouriensis, see Dorycrinus missouriensis.
moniliformis, Miller, cited by Troost, but the species does not exist in American strata.
mortoni, Troost, 1850. Not defined.
multibrachiatus, Hall, 1858, Geo. Rep. Iowa, Burlington Gr. [Sig. many-armed.]
multibrachiatus var. echinatus, Hall, 1861, Desc. New Crinoidea, Warsaw Gr. [Ety. *echinatus*, prickly.]
multicornis, Lyon, 1860, Trans. Am. Phil. Soc., vol. 13, Devonian. [Sig. many-horned.]
mundulus, see Batocrinus mundulus.
multiradiatus, Shumard, 1857, Trans. Acad. Sci. St. Louis, Burlington Gr. [Sig. many-rayed.]
nashvillæ, Troost, 1850, Catal. in Proc. Am. Ass'n, Ad. Sci., Keokuk Gr. [Ety. proper name.]
nashvillæ var. subtractus, White, 1862, Proc. Bost. Soc. Nat. Hist., vol. 9, Burlington Gr. [Sig. taken away from.]
nyssa, Hall, 1862, 15th Reg. Rep. N. Y., Ham. Gr. [Ety. mythological name.]
oblatus, Hall, 1860, Supp. to Geo. Sur. Iowa, Burlington Gr. [Sig. broader than long.]
obpyramidalis, Winchell & Marcy, syn. for Melocrinus verneuili.
olliculus, syn. for Megistocrinus whitii.
olla, Hall, 1861, Desc. New Crinoidea, Burlington Gr. [Sig. a pot or jar.]
opusculus, Hall, 1860, Supp. to Geo. Sur. Iowa, Burlington Gr. [Sig. a little fortification.]
ornatus, Hall, 1858, Geo. Rep. Iowa, Burlington Gr. [Sig. ornamented.]
ovatus, Hall, 1861, Desc. New Crinoidea, Burlington Gr. [Sig. egg-shaped.]
papillatus, see Batocrinus papillatus.
parvus, Shumard, 1855, Geo. Rep. Mo., St. Louis Gr. [Sig. small.]
pendens, Hall, 1860, Supp. to Geo. Sur. Iowa, Burlington Gr. [Sig. hanging down.]
penicillus, Meek & Worthen, 1869, Proc. Acad. Nat. Sci. Phil., Burlington Gr. [Sig. a painter's brush or pencil.]
pentagonus, see Steganocrinus pentagonus.
pentaspinus, Lyon, 1860, Trans. Am. Phil. Soc., vol. 14, Devonian. [Sig. five-spined.]
pernodosus, Hall, 1858, Geo. Rep. Iowa, Keokuk Gr. [Sig. very knotty.]

perumbrosus, see Strotocrinus perumbrosus.
pistilliformis, see Batocrinus pistilliformis.
pistillus, see Batocrinus pistillus.
planobasalis, see Amphoracrinus planobasalis.
planodiscus, see Batocrinus planodiscus.
plumosus, see Glyptocrinus plumosus.
pocillum, Hall, 1862, 15th Reg. Rep. N. Y., Ham. Gr. [Sig. a little cup.]
polydactylus, Schenectady Reflector, 1835, see Mariacrinus pachydactylus.
præcursor, Hall, 1862, 15th Reg. Rep. N. Y., Ham. Gr. [Sig. a forerunner.]
proboscidialis, Hall, 1858, Geo. Rep. Iowa, Burlington Gr. [Sig. having a proboscis].
pyriformis, Shumard, see Batocrinus pyriformis.
pyriformis var. rudis, Meek & Worthen, 1861, Proc. Acad. Nat. Sci. Phil., Burlington Gr. [Sig. rough.]
pyramidatus, Hall, 1858, Geo. Rep. Iowa, Burlington Gr. [Ety. from pyramidal summit.]
quadrispinus, see Amphoracrinus quadrispinus.
quaternarius, Hall, 1860, Supp. to Geo. Sur. Iowa, Burlington Gr. [Sig. containing four, quaternary.]
quaternarius var. spiniferus, Hall, 1861, Desc. New Crinoides, Burlington Gr. [Sig. spine-bearing.]
quinquelobus, see Dorycrinus quinquelobus.
ramulosus, Hall, 1858, Geo. Rep. Iowa, Keokuk Gr. [Sig. branchy.]
regalis, see Strotocrinus regalis.
remibrachiatus, see Batocrinus remibrachiatus.
reticulatus, Hall, 1861, Desc. New Crinoidea, Burlington Gr. [Sig. reticulated like net-work.]
rotundus, Yandell & Shumard, 1855, Geo. Rep. Mo., Burlington Gr. [Sig. round.]
rudis, see Strotocrinus radis.
rusticus, Hall, 1861, Desc. New Crinoidea, Burlington Gr. [Sig. coarse, rough.]
scitulus, Meek & Worthen, 1860, Proc. Acad. Nat. Sci. Phil., Burlington Gr. [Sig. handsome, pretty.]
sculptus, Hall, 1858, Geo. Rep. Iowa, Burlington Gr. [Sig. engraved, sculptured.]
securis, Hall, 1861, Desc. New Crinoides, Burlington Gr. [Sig. an axe or hatchet with a broad edge.]
semiradiatus, see Saccocrinus semiradiatus.
senarius, Hall, 1860, Supp. to Geo. Sur. Iowa, Burlington Gr. [Sig. consisting of six.]
sexarmatus, Hall, 1860, Supp. to Geo. Sur. Iowa, Burlington Gr. [Sig. six-armed.]
sillimani, syn. for A. rusticus.
similis, Hall, 1860, Supp. to Geo. Sur. Iowa, Warsaw Gr. [Sig. similar.]

sinousus, see Batocrinus sinuosus.
speciosus, syn. for Strotocrinus regalis.
spinobrachiatus, see Amphoracrinus spinobrachiatus.
spinotentaculus, Hall, 1860, Supp. to Geo. Sur. Iowa, Burlington Gr. [Ety. *spina*, a spine; *tentaculus*, a feeler.]
spinulosus, Hall, 1860, Supp. to Geo. Sur. Iowa, Keokuk Gr. [Sig. full of little spines.]
steropes, Hall, 1860, Supp. to Geo. Sur. Iowa, Warsaw Gr. [Ety. mythological name.]
subaculeatus, see Dorycrinus subaculeatus.
subæqualis, see Batocrinus subæqualis.
subturbinatus, see Dorycrinus subturbinatus.
subumbrosus, Hall, 1860, Supp. to Geo. Sur. Iowa, Burlington Gr. [Sig. somewhat umbrella-like.] Syn. (?) for Strotocrinus liratus.
subventricosus, see Physetocrinus subventricosus.
superlatus, Hall, 1858, Geo. Rep. Iowa, Burlington Gr. [Sig. very wide.]
symmetricus, see Dorycrinus symmetricus.
tenuidiscus, Hall, 1861, Desc. New Crinoidea, Burlington Gr. [Sig. slender disc.]
tenuiradiatus, Hall, 1847, see Palæocystites tenuiradiatus.
tenuiradiatus, Hall, 1861, Desc. New Crin. Burlington Gr. [Sig. slender-rayed.]
tenuisculptus, syn. for Actinocrinus sculptus.
thalia, Hall, 1861, Desc. New Crinoidea, Burlington Gr. [Ety. mythological name.]
themis, Hall, 1861, Desc. New Crinoidea, Burlington Gr. [Ety. mythological name.]
thetis, Hall, 1861, Desc. New Crinoidea, Burlington Gr. [Ety. mythological name.]
thoas, Hall, 1861, Desc. New Crinoidea, Burlington Gr. [Ety. mythological name.]
tholus, see Strotocrinus tholus.
tricornis, Hall, 1858, Geo. Rep. Iowa, Burlington Gr. [Sig. three-horned.]
trinodus, see Dorycrinus trinodus.
turbinatus, see Batocrinus turbinatus.
turbinatus var. elegans, see Batocrinus turbinatus var. elegans.
umbrosus, Hall, see Strotocrinus umbrosus.
unicarinatus, Hall, 1860, Supp. to Geo. Sur. Iowa, Keokuk Gr. [Sig. single keeled.]
unicornis, see Dorycrinus unicornis.
unispinus, see Dorycrinus unispinus.
urna, Troost, 1850. Not defined.
urniformis, McChesney, 1860, New Pal. Foss., Burlington Gr. [Sig. urn-formed.]
validus, Meek & Worthen, 1860, Proc. Acad. Nat. Sci., Phil., Burlington Gr. [Sig. stoutly built.]

ECHINODERMATA. 69

ventricosus, Hall, 1858, Geo. Rep. Iowa, Burlington Gr. [Sig. bulging out.]
ventricosus var. cancellatus, Hall, 1861, Desc. New Crinoidea, Burlington Gr. [Sig. cancellated.]
ventricosus var. internodus, Hall, 1861, Desc. New Crinoidea, Burlington Gr. [Ety. *inter*, between; *nodus*, a joint.]
verneuili, see Melocrinus verneuili.
verneuilianus, see Eretmocrinus verneuilianus.
verrucosus, Hall, 1858, Geo. Rep. Iowa, Burlington Gr. [Sig. full of warts, warty.]
viaticus, White, 1874, Rep. Invert. Foss., Lower Carb. [Ety. *viaticus*, a parting repast—application not evident.]
viminalis, Hall, 1863, Crinoidea in Wav. Sandstone Ohio, Waverly Sandstone. [Sig. bearing twigs for plaiting.]
wachsmuthi, White, 1862, Proc. Bost. Soc. Nat. Hist., vol. 9, Burlington Gr. [Ety. proper name.]
whitfieldi, see sub-genus Saccocrinus whitfieldi.
whitii, see Megistocrinus whitii.
yandelli, Shumard, 1857, Trans. St. Louis Acad. Sci., Lower Carb. [Ety. proper name.]

AGARICOCRINUS, Troost, 1850, Catal. in Proc. Am. Ass'n. [Ety. *agarikon*, a sort of tree-fungus; *krinon*, a lily.]
americanus, Roemer, 1854, (Amphoracrinus americanus) Bronn's Leth. Geog., vol. 2, Warsaw Gr. [Ety. proper name.]
bellatrema, see Amphoracrinus bellatrema.
bullatus, Hall, 1858, Geo. Rep. Iowa, Burlington Gr. [Sig. studded with round knobs; bossed.]
calyculus, Hall, 1860, Supp. to Geo. Iowa, Warsaw Gr. [Sig. the cup of a flower.]
corrugatus, Hall, 1861, Desc. New Crinoidea, Burlington Gr. [Sig. corrugated.]
excavatus, see Amphoracrinus excavatus.
geometricus, Hall, 1860, Supp. to Geo. Sur. Iowa, Burlington Gr. [Sig. geometrical.]
gracilis, Meek & Worthen, 1861, Proc. Acad. Nat. Sci. Phil., Burlington Gr. [Sig. slender.]
inflatus, Hall, 1861, Desc. New Crinoidea, Burlington Gr. [Sig. inflated.]
nodosus, Meek & Worthen, 1869, Proc. Acad. Nat. Sci. Phil., Burlington Gr. [Sig. knobbed.]
ornotrema, Hall, 1861, Desc. New Crinoidea, Burlington Gr. [Sig. not evident—the word may be a printer's mistake for *ornatotrema*, having a beautiful opening.]
planoconvexus, Hall, 1861, Desc. New Crinoidea, Burlington Gr. [Ety. *planus*, flat; *convexus*, convex.]

pentagonus, Hall, 1860, Supp. to Geo. Sur. Iowa, Burlington Gr. [Sig. pentagonal.]
pentagonus var. convexus, Hall, 1860, Supp. to Geo. Sur. Iowa, Burlington Gr. [Sig. convex.]
stellatus, Hall, 1858, Geo. Rep. Iowa, Burlington Gr. [Sig. covered with stars.]
tuberosus, Troost, 1849, Catal. This is the same species described by Roemer as Amphoracrinus americanus in Bronn's Leth. Geognostica, vol. 2, 1852-54, and by the rules of priority. it must bear Roemer's name, for Troost did not figure or describe it.
whitfieldi, Hall, 1858, Geo. Rep. Iowa, Keokuk Gr. [Ety. proper name.]
wortheni, Hall, 1858, Geo. Rep. Iowa, Keokuk Gr. [Ety. proper name.]

AGASSIZOCRINUS, Troost, 1850, Ms. of Monograph Crinoidea. [Ety. *Agassiz*, the eminent naturalist; *krinon*, a lily.]
carbonarius, Worthen, 1873, Geo. Sur. Ill., vol. 5, Up. Coal Meas. [Ety. from the *Coal Measures*.]
conicus, Owen & Shumard, 1851, Jour. Acad. Nat. Sci. Phil., 2d ser., vol. 2, Chester Gr. [Sig. like a cone.]
dactyliformis, Troost, 1850, as identified by Shumard, 1853, Marcy's Rep. Red. Riv., Chester Gr. [Sig. finger-shaped.]
chesterensis, Worthen, 1873, Geo. Sur. Ill., vol. 5, Chester Gr. [Ety. proper name.]
constrictus, Hall, 1858, Geo. Rep. Iowa, Kaskaskia Gr. [Sig. constricted.]
gibbosus, Hall, 1858, Geo. Rep. Iowa, Chester Gr. [Sig. gibbous.]
globosus, Worthen, 1873, Geo. Sur. Ill., vol. 5, Chester Gr. [Sig. globose.]
gracilis, Troost, 1850. Not defined.]
occidentalis, Owen & Shumard, 1852, (Poteriocrinus occidentalis) Jour. Acad. Nat. Sci. Phil., Chester Gr. [Sig. western.]
pentagonus. Worthen, 1873, Geo. Sur. Ill., vol. 5, Chester Gr. [Sig. pentagonal.]
tumidus, Owen & Shumard, 1852, (Poteriocrinus tumidus) Jour. Acad. Nat. Sci. Phil., Chester Gr. [Sig. tumid or swollen.]

AGELACRINUS, Vanuxem, 1842, (Agelacrinites) Geo. Rep. 3rd Dist. N. Y. [Ety. *agele*, a herd; *krinon*, a lily.]
billingsi, Chapman, 1860, Can. Jour., vol. 5, Trenton Gr. [Ety. proper name.]
cincinnatiensis, Roemer, 1851, Verh. Naturh. Rhein. Westph., vol. 8, Cincinnati Gr. [Ety. proper name.]
dicksoni, Billings, 1857, Rep. of Progr., Trenton Gr. [Ety. proper name.]
hamiltonensis, Vanuxem, 1842, Geo. Rep. 3rd Dist. N. Y., Ham. Gr. [Ety. proper name.]
kaskaskiensis, Hall, 1858, Geo. Rep. Iowa, Kaskaskia Gr. [Ety. proper name.]

pileus, Hall, 1866, Pamphlet, Cin'ti Gr. [Ety. *pileus*, a felt cap or hat made to fit close.]
squamosus, Meek & Worthen, 1868, Proc. Acad. Nat. Sci. Phil., Keokuk Gr. [Sig. covered with scales.]
stellatus, see Hemicystites stellatus.
vorticellatus, Hall, 1866, Pamphlet, Cincinnati Gr. [Sig. whorled.]

ALLOPROSALLOCRINUS, Casseday & Lyon, 1860, Proc. Am. Acad. Arts & Sci., vol. 5. [Ety. *alloprosallos*, inclining first to one side and then to another; *krinon*, a lily.]
conicus, Casseday & Lyon, 1860, Proc. Am. Acad. Arts & Sci., vol. 5, Warsaw Gr. [Sig. conical.]
euconus, Meek & Worthen, 1865, Proc. Acad. Nat. Sci. Phil., Burlington Gr. [Ety. *eu*, perfect; *konos*, cone.]
depressus, Casseday & Lyon, 1860, Proc. Am. Acad. Arts & Sci., vol. 5, Warsaw Gr. [Sig. depressed.]
lex, Troost, 1850, (Conocrinus lex). Not defined.
tuberculosus, Troost, 1850, (Conocrinus tuberculosus). Not defined.

AMPHORACRINUS, Austin, 1848, Quar. Jour. Geo. Soc. Lond., vol. 4. Sub-genus of Actinocrinus. [Ety. *amphora*, a cup or goblet; *krinon*, a lily.]
americanus, see Agaricocrinus americanus.
bellatrema, Hall, 1861, (Actinocrinus bellatrema) Jour. Bost. Soc. Nat. Hist., vol. 7, Burlington Gr. [Sig. beautiful opening.]
divergens, Hall, 1860, (Actinocrinus divergens) Supp. Geo. Rep. Iowa, Burlington Gr. [Sig. separating.]
excavatus, Hall, 1861, (Actinocrinus excavatus) Desc. New Crinoidea, Burlington Gr. [Sig. hollowed out.]
inflatus, Hall, 1861, (Actinocrinus inflatus) Desc. New Crinoidea, Burlington Gr. [Sig. inflated.]
planobasalis, Hall, 1858, (Actinocrinus planobasalis) Geo. Rep. Iowa, Burlington Gr. [Sig. flat-based.]
quadrispinus, White, 1862, (Actinocrinus quadrispinus) Proc. Bost. Soc. Nat. Hist., vol. 9, Burlington Gr. [Sig. four-spined.]
spinobrachiatus, Hall, 1860, (Actinocrinus spinobrachiatus) Supp. Geo. Rep. Iowa, Burlington Gr. [Sig. spiny armed.]

AMYGDALOCYSTITES, Billings, 1854, Can. Jour., vol. 2. [Ety. *amygdalos*, an almond; *kustis*, a bladder.]
florealis, Billings, 1854, Can. Jour., vol. 2, Trenton Gr. [Sig. flower-like.]
radiatus, Billings, 1854, Can. Jour., vol. 2, Trenton Gr. [Sig. radiating from a point.]
tenuistriatus, Billings, 1854, Can. Jour., vol. 2, Trenton Gr. [Sig. finely striated.]

ANCYROCRINUS, Hall, 1862, 15th Reg. Rep. N. Y. [Ety. *ankura*, a grapnel; *krinon*, a lily.]
bulbosus, Hall, 1862, 15th Reg. Rep. N. Y., Ham. Gr. [Sig. bulbous.]
spinosus, Hall, 1862, 15th Reg. Rep. N. Y., Up. Held. Gr. [Sig. covered with many spines.]

ANOMALOCRINUS, Meek & Worthen, 1868, Geo. of Ill., vol. 3. [Ety. *anomos*, irregular; *krinon*, a lily.]
incurvus, Meek & Worthen, 1865, (Heterocrinus incurvus) Proc. Acad. Nat. Sci. Phil. [Ety. from one incurved arm.]

ANOMALOCYSTITES, Hall, 1859, Pal. N. Y., vol. 3. [Ety. *anomos*, irregular; *kustis*, a bladder.] Prof. Billings regards this genus as a synonym for *Ateleocystites*.
balanoides, Meek, 1872, Am. Jour. Sci., 3rd ser., vol. 3. [Sig. resembling the shell of the *Balanus*.]
cornutus, Hall, 1859, Pal. N. Y., vol. 3, Low. Held. Gr. [Sig. horned.]
disparilis, Hall, 1859, Pal. N. Y., vol. 3, Oriskany sandstone. [Sig. different.]

APIOCYSTITES, Forbes, 1848, Mem. Geo. Sur. Great Brit. [Ety. *apion*, a pear; *kustis*, a bladder.] A. syn (?) for Lepadocrinus.
canadensis, Billings, 1866, Catal. Sil. Foss. Antic., Niagara Gr. [Ety. proper name.]
elegans, Hall, 1852, Pal. N. Y., vol. 2, Niagara Gr. [Sig. elegant.]
huronensis, Billings, 1866, Catal. Sil. Foss. Antic., Niagara Gr. [Ety. proper name.]
imago, Hall, 1867, 20th Reg. Rep., Niagara Gr. [Ety. *imago*, an image, likeness, perfect state.]
tecumseth, Billings, 1866, Catal. Sil. Foss. Antic., Niagara Gr. [Ety. proper name.]

ARCHÆOCIDARIS, McCoy, 1844, Carb. Foss., Ireland. [Ety. *archaios*, ancient; *cidaris*, a turban.]
aculeatus, Shumard, 1858, Trans. St. Louis Acad. Sci., Permian Gr. [Sig. armed with sharp points.]
agassizi, Hall, 1858, Geo. Rep. Iowa, Kaskaskia Gr. [Ety. proper name.]
biangulatus, Shumard, 1858, Trans. St. Louis, Acad. Sci., Coal Meas. [Sig. double-angled.]
cratis, White, 1876, Geol. of Uinta Mountains, Lower Aubrey Gr. [Sig. like a basket.]
gracilis, Newberry, 1861, Ives, Col. Ex. Ex., Up. Carb. [Sig. slender.]
keokuk, Hall, 1858, Geo. Rep. Iowa, Kaskaskia Gr. [Ety. proper name.]
longispinus, Newberry, 1861, Ives, Col. Ex. Ex., Up. Carb. [Sig. long-spined.]
megastylus, Shumard, 1858, Trans. St. Louis, Acad. Sci., Up. Coal Meas. [Sig. having large spines.]

ECHINODERMATA.

mucronatus, Meek & Worthen, 1860, Proc. Acad. Nat. Sci. Phil., Chester Gr. [Sig. sharp-pointed.]
norwoodi, Hall, 1858, Geo. Rep. Iowa, Kaskaskia Gr. [Ety. proper name.]
ornatus, Newberry, 1861, Ives, Col. Ex. Ex., Up. Carb. [Sig. ornamented.]
shumardiana, Hall, 1858, Geo. Rep. Iowa, Kaskaskia Gr. [Ety. proper name.]
triserrata, Meek, 1872, Pal. E. Neb., Up. Coal Meas. [Sig. three-notched.]
trudifera, White, 1874, Rep. Invert. Foss., Carb. [Sig. bearing a pike.]
verneuiliana, Swallow, 1858, Trans. St. Louis Acad. Sci. This name was preoccupied by King, and the species is now known as A. aculeatus.
wortheni, Hall, 1858, Geo. Rep. Iowa, Kaskaskia Gr. [Ety. proper name.]

ASPIDOCRINUS, Hall, 1859, Pal. N. Y., vol. 3. [Ety. *aspis*, shield; *krinon*, a lily.]
callosus, Hall, 1859, Pal. N. Y., vol. 3, Low. Held. Gr. [Sig. thick-skinned.]
digitatus, Hall, 1859, Pal. N. Y., vol. 3, Low. Held. Gr. [Sig. fingered.]
scutelliformis, Hall, 1859, Pal. N. Y., vol. 3, Low. Held. Gr. [Sig. from the scutelliform base.]

Asterias, Lamarck, 1815, Hist. Nat. Anim. sans. Vert.
anthoni, see Palæaster jamesi.
antiqua, see Palæaster antiqua.
antiquata, see Palæaster antiquata.
matutina, see Palæaster matutina.

Asterocrinus, Lyon, 1857, Geo. Sur. Ky., vol. 3. This name was preoccupied by Munster. See Pterotocrinus.
capitalis, see Pterotocrinus capitalis.
coronarius, see Pterotocrinus coronarius.

Astrias, Troost, 1850, Catalogue. Not defined.
tennesseew, Troost, 1850. Not defined.

Astrocrinites, Conrad in Catalogue Ann. Geo. Rep., 1840–'41. This name was proposed but not defined, moreover the name was preoccupied.
pachydactylus, see Mariacrinus pachydactylus.

Ataxocrinus, Lyon, 1869, syn. for Anomalocrinus.
caponiformis, syn. for Anomalocrinus incurvus.

ATELEOCYSTITES, Billings, 1858, Can. Org. Rem., Decade 3. [Ety. *ateles*, defective or incomplete; *kustis*, a b l a d d e r.] Professor Billings regards Hall's genus *Anomalocystites*, and DeKoninck's genus *Placocystites* as congeneric with this genus and therefore synonyms.
huxleyi, Billings, 1858, Can. Org. Rem., Decade 3, Trenton Gr. [Ety. proper name.]

Balanocrinus, Troost, 1850. This name was pre-occupied. See Lampterocrinus.
inflatus, see Lampterocrinus inflatus.

BARYCRINUS, Wachsmuth, 1868, Proc. Acad. Nat. Sci. [Ety. *barus*, heavy; *krinon*, a lily.]

geometricus, Meek & Worthen, 1873, Geo. Sur. Ill., vol. 5, Keokuk Gr. [Sig. geometrical.]
hercules, Meek & Worthen, 1868, Proc. Acad. Nat. Sci. Phil., Keokuk Gr. [Ety. mythological name.]
hoveyi, Hall, 1861, (Cyathocrinus hoveyi) Desc. New Crin., Keokuk Gr. [Ety. proper name.]
magnificus, Meek & Worthen, 1868, Proc. Acad. Nat. Sci. Phil., Keokuk Gr. [Sig. large-sized.]
mammatus, Worthen, 1873, Geo. Sur. Ill., vol. 5, Keokuk Gr. [Sig. covered with protuberances.]
pentagonus, Worthen, 1873, Geo. Sur. Ill., vol. 5, Keokuk Gr. [Sig. pentagonal.]
spectabilis, Meek & Worthen, 1869, Proc. Acad. Nat. Sci. Phil., St. Louis Gr. [Sig. notable or worth seeing.]
striatus, Worthen, 1875, Geo. Sur. Ill., vol. 6, Keokuk Gr. [Sig. striated.]
subtumidus, Meek & Worthen, 1869, Proc. Acad. Nat. Sci. Phil., Keokuk Gr. [Sig. somewhat tumid or as if blown up.]
thomæ, Hall, 1860, (Cyathocrinus thomæ) Supp. Geo. Sur. Iowa, St. Louis Gr. [Ety. proper name.]

BATOCRINUS, Casseday, 1869, Proc. Acad. Nat. Sci. Phil. [Ety. *batos*, a prickly bush; *krinon*, a lily.]
æqualis, Hall, 1858, (Actinocrinus æqualis) Geo. Rep. Iowa, Burlington Gr. [Sig. equal.]
æquibrachiatus, McChesney, 1860, (Actinocrinus æquibrachiatus) New Pal. Foss., Burlington Gr. [Sig. equal-armed.]
æquibrachiatus var. alatus, Hall, 1861, (Actinocrinus æquibrachiatus var. alatus) Bost. Jour. Nat. Hist., vol. 7, Burlington Gr. [Sig. winged.]
asteriscus, Meek & Worthen, 1860,(Actinocrinus asteriscus) Proc. Acad. Nat. Sci. Phil., Burlington Gr. [Sig. little star.]
calyculoides, see Eretmocrinus calyculoides.
carica, Hall, 1861, (Actinocrinus carica) Desc. New Crinoidea, Burlington Gr. [Sig. a kind of dried fig.]
cassedayanus, Meek & Worthen, 1868, Proc. Acad. Nat. Sci. Phil., Burlington Gr. [Ety. proper name.]
christyi, Shumard, 1855, (Actinocrinus christyi) Geo. Sur. Mo., Burlington Gr. [Ety. proper name.]
claviger, Hall, 1860, (Actinocrinus clavigerus) Supp. Geo. Sur. Iowa, Burlington Gr. [Sig. club-bearing.]
cloelia, see Eretmocrinus cloelia.
corbulis, Hall, 1861, (Actinocrinus corbulis) Desc. New Crinoidea, Burlington Gr. [Sig. a little basket.]
discoideus, Hall, 1858, (Actinocrinus discoideus) Geo. Rep. Iowa, Burlington Gr. [Sig. quoit-shaped.]

dodecadactylus, Meek & Worthen, 1861, (Actinocrinus dodecadactylus) Proc. Nat. Sci. Phil., Burlington Gr. [Sig. twelve-fingered.]
doris, Hall, 1861, (Actinocrinus doris) Desc. New Crinoidea, Burlington Gr. [Ety. a mythological name.]
formosus, Hall, 1860, (Actinocrinus formosus) Supp. to Geo. Sur. Iowa, Burlington Gr. [Sig. beautiful.]
gemmiformis, Hall, 1860, (Actinocrinus gemmiformis) Supp. to Geo. Sur. Iowa, Burlington Gr. [Sig. bud-like.]
icosidactylus, Casseday, 1854 (Actinocrinus icosidactylus) Zeitsch. Deutsch. Geol. Gesellsch, Warsaw Gr. [Sig. twenty-fingered.]
inornatus, Hall, 1860, (Actinocrinus inornatus) Supp. to Geo. Sur. Iowa, Burlington Gr. [Sig. not ornamented.]
irregularis, Casseday, 1854, Zeitsch. Deutsch Geol. Gesell., Warsaw Gr. [Sig. irregular.]
konincki, Shumard, 1855, (Actinocrinus konincki) Geo. Rep. Mo., Burlington Gr. [Ety. proper name.]
lagunculus, Hall, 1860, (Actinocrinus lagunculus) Supp. to Geo. Sur. Iowa, Warsaw Gr. [Sig. small flask, small bottle.]
laura, Hall, 1861, (Actinocrinus laura) Desc. New Crinoidea, Burlington Gr. [Ety. proper name.]
lepidus, Hall, 1860, (Actinocrinus lepidus) Supp. to Geo. Sur. Iowa, Burlington Gr. [Sig. elegant.]
longirostris, Hall, 1858, (Actinocrinus longirostris) Geo. Rep. Iowa, Burlington Gr. [Sig. having a long proboscis.]
magnificus, see Eretmocrinus magnificus.
mundulus, Hall, 1860, (Actinocrinus mundulus) Supp. to Geo. Sur. Iowa, Warsaw Gr. [Sig. neat, trim, nice.]
neglectus, Meek & Worthen, 1868, Proc. Acad. Nat. Sci. Phil., Burlington Gr. [Sig. overlooked.]
papillatus, Hall, 1860, (Actinocrinus papillatus) Supp. to Geo. Sur. Iowa, Burlington Gr. [Sig. shaped like a bud.]
pistilliformis, Meek & Worthen, 1865, Proc. Acad. Nat. Sci. Phil., Kindershook Gr. [Sig. in the form of a pestle.]
pistillus, Meek & Worthen, 1865, Proc. Acad. Nat. Sci. Phil., Burlington Gr. [Sig. a pestle.]
planodiscus, Hall, 1860, (Actinocrinus planodiscus) Supp. to Geo. Sur. Iowa, Warsaw Gr. [Sig. a plain disc.]
pyriformis, Shumard, 1855, (Actinocrinus pyriformis) Geo. Sur. Mo., Burlington Gr. [Sig. pear-shaped.]
quasillus, Meek & Worthen, 1869, Proc. Acad. Nat. Sci. Phil., Burlington Gr. [Sig. a small basket.]
remibrachiatus, see sub-genus Eretmocrinus remibrachiatus.
sinuosus, Hall, 1860, (Actinocrinus sinuosus) Supp. to Geo. Sur. Iowa, Burlington Gr. [Sig. full of curves.]
subaequalis, McChesney, 1860, (Actinocrinus subaequalis) New Pal. Foss., Burlington Gr. [Sig. somewhat equal.]
trochiscus, Meek & Worthen, 1868, Proc. Acad. Nat. Sci. Phil., Burlington Gr. [Sig. a little wheel.]
turbinatus, Hall, 1858, (Actinocrinus turbinatus) Geo. Rep. Iowa, Burlington Gr. [Sig. top-shaped.]
turbinatus var. elegans, (Actinocrinus turbinatus var. elegans) Geo. Rep. Iowa, Burlington Gr. [Sig. elegant.]
verneuilianus, see Eretmocrinus verneuilianus.

BELEMNOCRINUS, White, 1862, Proc. Bost. Soc. Nat. Hist., vol. 9. [Ety. *belemnon*, a dart, javelin; *krinon*, a lily.]
typus, White, 1862, Proc. Bost. Soc. Nat. Hist., vol. 9, Burlington Gr. [Ety. type of the genus.]
whitii, Meek & Worthen, 1866, Proc. Acad. Nat. Sci. Phil., Burlington Gr. [Ety. proper name.]

BLASTOIDOCRINUS, Billings, 1859, Decade 4. [Ety. *blastos*, bud; *eidos*, form; *krinon*, lily.]
carcharidens, Billings, 1859, Geo. Sur. of Can., Decade 4, Chazy Gr. [Ety. *carcharus*, a shark; *dens*, a tooth.]

BRACHIOCRINUS, Hall, 1859, Pal. N. Y., vol. 3. [Ety. *brachium*, an arm; *krinon*, a lily.]
nodosarius, Hall, 1859, Pal. N. Y., vol. 3, Low. Held. Gr. [Sig. knotty.]

BURSACRINUS, Meek & Worthen, 1861, Proc. Acad. Nat. Sci. Phil. [Ety. *bursa*, a purse; *krinon*, a lily.]
confirmatus, White, 1862, Proc. Bost. Soc. Nat. Hist., vol. 9, Burlington Gr. [Sig. very thick or firm.]
wachsmuthi, Meek & Worthen, 1861, Proc. Acad. Nat. Sci. Phil., Burlington Gr. [Ety. proper name.]

Cacabocrinus, Troost, 1850. It was never described. The fossils that have been referred to it will be found under Dolatocrinus.

CALATHOCRINUS, Hall, 1861. This name was preoccupied by Von Meyer, in 1848.

CALCEOCRINUS, Hall, 1852, Pal. N. Y., vol. 2. [Ety. *calceus*, a shoe; *krinon*, a lily.]
barrisi, Worthen, 1875, Geo. Sur. Ill., vol. 6, Devonian. [Ety. proper name.]
bradleyi, Meek & Worthen, 1869, Proc. Acad. Nat. Sci. Phil., Keokuk Gr. [Ety. proper name.]
chrysalis, Hall, 1859, (Cheirocrinus chrysalis) 13th Reg. Rep. N. Y., Niagara Gr. [Sig. chrysalis.]
clarus, Hall, 1862, (Cheirocrinus clarus) 15th Reg. Rep. N. Y., Hamilton Gr. [Sig. distinct, remarkable.]

ECHINODERMATA. 73

dactylus, Hall, 1859, (Cheirocrinus dactylus) 13th Reg. Rep. N. Y., Burlington Gr. [Sig. a finger.]
lamellosus, Hall, 1859, (Cheirocrinus lamellosus) 13th Reg. Rep. N. Y., Burlington Gr. [Sig. in very thin plates.]
nodosus, Hall, 1859, (Cheirocrinus nodosus) 13th Reg. Rep. N. Y., Warsaw Gr. [Sig. knobbed.]
perplexus, Shumard, 1866, (Cheirocrinus perplexus) Trans. St. Louis Acad. Sci., Keokuk Gr. [Sig. intricate, obscure.]
stigmatus, Hall, 1863, (Cheirocrinus stigmatus) Trans. Alb. Inst., vol. 4, Niagara Gr. [Sig. marked, branded.]
tunicatus, Hall, 1859, (Cheirocrinus tunicatus) 13th Reg. Rep. N. Y., Warsaw Gr. [Sig. coated.]
ventricosus, Hall, 1859, (Cheirocrinus ventricosus) 13th Reg. Rep. N. Y., Burlington Gr. [Sig. bulging out.]
wachsmuthi, Meek & Worthen, 1869, Proc. Acad. Nat. Sci. Phil., Burlington Gr. [Ety. proper name.]
CALLOCYSTITES, Hall, 1852, Pal. N. Y., vol. 2. [Ety. *kallos*, beautiful; *kustis*, bladder.]
jewetti, Hall, 1852, Pal. N. Y., vol. 2, Niagara Gr. [Ety. proper name.]
Campanulites, Troost, 1850. Not defined.
tesselatus, Troost, 1850. Not defined.
CARABOCRINUS, Billings, 1857, Rep. of Progr. [Ety. *karabos*, a crab; *krinon*, a lily.]
radiatus, Billings, 1857, Rep. of Progr., Trenton Gr. [Sig. radiating from a point.]
tuberculatus, Billings, 1859, Decade 4, Hud. Riv. Gr. [Sig. covered with tubercles.]
vancortlandti, Billings, 1859, Decade 4, Trenton Gr. [Ety. proper name.]
CARYOCRINUS, Say, 1825, Jour. Acad. Nat. Sci., vol. 4. [Ety. *karyon*, a nut; *krinon*, a lily.]
globosus, Troost, 1850. Not defined.
granulatus, Troost, 1850. Not defined.
hexagonus, Troost. Not defined.
insculptus, Troost. Not defined.
loricatus, Say, 1825, Jour. Acad. Nat. Sci., vol. 4, Clinton & Niagara Gr. Syn. for C. ornatus.
meconoideus, Troost. Not defined.
ornatus, Say, 1825, Jour. Acad. Nat. Sci., vol. 4, Clinton & Niagara Gr. [Sig. adorned.]
Caryocystites, Von Buch, as cited by Hall in 1861, in Geo. Rep. Wis. See Holocystites.
alternatus, see Holocystites alternatus.
cylindricus, see Holocystites cylindricus.
CATILLOCRINUS, Troost, 1850, Cat. Foss., described by Shumard, 1866, Trans. St. Louis Acad. Sci. [Ety. *catillus*, a small bowl; *krinon*, a lily.]
bradleyi, Meek & Worthen, 1868, Proc. Acad. Nat. Sci. Phil., Keokuk Gr. [Ety. proper name.]

tennesseæ, Troost, 1850, Catalogue, but described by Shumard, in 1866, in Trans. St. Louis Acad. Sci., Warsaw Gr. [Ety. proper name.]
wachsmuthi, Meek & Worthen, 1866, (Synbathocrinus wachsmuthi) Proc. Acad. Nat. Sci. Phil., Burlington Gr. [Ety. proper name.]
CHEIROCRINUS, Hall, 1860, 13th Reg. Rep. This name seems to have been pre-occupied by Eichwald, in 1856, and therefore cannot stand. Shumard suggests the close affinity with Calceocrinus, and I have therefore referred the species to that genus.
chrysalis, see Calceocrinus chrysalis.
clarus, see Calceocrinus clarus.
dactylus, see Calceocrinus dactylus.
lamellosus, see Calceocrinus lamellosus.
nodosus, see Calceocrinus nodosus.
perplexus, see Calceocrinus perplexus.
stigmatus, see Calceocrinus stigmatus.
tunicatus, see Calceocrinus tunicatus.
ventricosus, see Calceocrinus ventricosus.
CLEIOCRINUS, Billings, 1857, Rep. of Progr. [Ety. *kleio*, I close; *krinon*, a lily.]
grandis, Billings, 1859, Can. Org. Rem., Decade 4, Trenton Gr. [Sig. great.]
libanus, Safford, 1869, Geo. of Tenn. Not defined.
magnificus, Billings, 1859, Can. Org. Rem., Decade 4, Trenton Gr. [Sig. magnificent.]
regius, Billings, 1857, Rep. of Progr., Trenton Gr. [Sig. royal, magnificent.]
CLOSTEROCRINUS, Hall, 1852, Pal. N. Y., vol. 2. [Ety. *kloster*, a spindle; *krinon*, a lily.]
elongatus, Hall, 1852, Pal. N. Y., vol. 2, Clinton Gr. [Sig. lengthened.]
COCCOCRINUS, Muller, 1855, Verhand. Naturhist. Vereins Rhein und Westph., Jahr. 12. [Ety. *kokkos*, a berry; *krinon*, a lily.]
bacca, Roemer, 1860, Sil. Fauna. West Tenn., Niagara Gr. [Sig. a small round fruit, a berry.]
CODASTER, McCoy, 1849, Ann. & Mag. Nat. Hist. 2d ser., vol. 3. [Ety. *kodon*, a bell; *aster*, a star.]
alternatus, Lyon, 1857, Geo. Sur. Ky., vol. 3, Corniferous Gr. [Sig. alternating.]
americanus, Shumard, 1858, Trans. St. Louis Acad. Sci., Up. Held. Gr. [Ety. proper name.]
kentuckiensis, Shumard, 1858, Trans. St. Louis Acad. Sci., Low. Carb. [Ety. proper name.]
pyramidatus, Shumard, 1858, Trans. St. Louis Acad. Sci., Up. Held. Gr. [Sig. in the form of a pyramid.]
whitii, Hall, 1861, Desc. New Crinoidea, Bost. Jour. Nat. Hist., vol. 7, Burlington Gr. [Ety. proper name.]

ECHINODERMATA.

CODONITES, Meek & Worthen, 1869, Proc. Acad. Nat. Sci. Phil. [Ety. *kodon*, a bell; *lithos*, stone.]
  gracilis, Meek & Worthen, 1873, Geo. Sur. Ill., vol. 5, Burlington Gr. [Sig. slender.]
  stelliformis, Owen & Shumard, 1850, (Pentremites stelliformis), Jour. Acad. Nat. Sci. Phil., 2d series, vol. 2, Burlington Gr. [Sig. star-shaped.]
CŒLIOCRINUS, White, 1863, Jour. Bost. Soc. Nat. Hist., vol. 7. [Ety. *koilia*, the belly; *krinon*, a lily.]
  dilatatus, Hall, 1861, (Poteriocrinus dilatatus) Desc. New Crinoidea, Burlington Gr. [Sig. widened, spread out.]
  subspinosus, White, 1863, Jour. Bost. Soc. Nat. Hist., vol. 7, Burlington Gr. [Sig. somewhat covered with spines.]
  ventricosus, Hall, 1861, (Poteriocrinus ventricosus) Desc. New Crinoidea, Burlington Gr. [Sig. bulging out.]
CŒLOCRINUS, Meek & Worthen, 1865, Proc. Acad. Nat. Sci. Phil. [Ety. *koilos*, hollow; *krinon*, a lily.]
  concavus, Meek & Worthen, 1861, (Actinocrinus concavus) Proc. Acad. Nat. Sci. Phil., Burlington Gr. [Sig. depressed, concave.]
COMAROCYSTITES, Billings, 1854, Can. Jour., vol. 2. [Ety. *komaron*, a strawberry; *kustis*, a bladder.]
  obconicus, Meek & Worthen, 1865, Proc. Acad. Nat. Sci. Phil., Trenton Gr. [Sig. inversely conical.]
  punctatus, Billings, 1854, Can. Jour., vol. 2, Trenton Gr. [Sig. punctate.]
  shumardi, Meek & Worthen, 1865, Proc. Acad. Nat. Sci. Phil., Trenton Gr. [Ety. proper name.]
Conocrinus, Troost. Not defined.
CORONOCRINUS, Hall, 1859, Pal. N. Y., vol. 3. [Ety. *korone*, a crown; *krinon*, a lily.]
  polydactylus, Hall, 1859, Pal. N. Y., vol. 3, Low. Held. Gr. [Sig. many-fingered.]
COTYLEDONOCRINUS, Casseday & Lyon, 1860, Proc. Am. Acad. Arts & Sci., vol. 5. [Ety. *kotyledon*, any cup-shaped hollow or cavity; *krinon*, a lily.]
  pentalobus, Casseday & Lyon, 1860, Proc. Am. Acad. Arts & Sci., vol. 5, Kaskaskia Gr. [Sig. five-lobed.]
CRINOCYSTITES, Hall, 1864, 20th Reg. Rep. [Ety. *krinon*, a lily; *kustis*, a bladder.]
  chrysalis, Hall, 1864, 20th Reg. Rep., Niagara Gr. [Sig. chrysalis.]
  (?) *rectus*, Hall, 1864, see Rhodocrinus (?) rectus.
Crumenæcrinites, Troost, 1850. Not defined.
  ovalis, Troost, 1850. Not defined.
CTENOCRINUS, Bronn, 1840, Leonh. und Bronn, Jahrb. [Ety. *kten*, comb; *krinon*, a lily.] This genus may include the species referred to Macrostylocrinus, if so, it would have priority.
  bainbridgensis, Hall & Whitfield, 1875, Ohio Pal., vol. 2, Portage Gr. [Ety proper name.]
  breviradiatus, Hall & Whitfield, 1872, Ham. Gr. [Sig. having short rays.]
Cupellæcrinus. This name was proposed by Troost in his catalogue published in 1850, in Proc. Am. Ass'n, with the following species, *C. buchii*, *C. corrugatus*, *C. inflatus*, *C. lævis*, *C. magnificus*, *C. pentagonalis*, *C. rosæformis*, *C. stellatus*, and *C. striatus*. As the words have never been illustrated or defined, they must be regarded as out of the list.
CYATHOCRINUS, Miller, 1821, Nat. Hist. Crinoidea. [Ety. *cyathos*, a cup or goblet; *krinon*, a lily.]
  angulatus, Meek & Worthen, 1860, Proc. Acad. Nat. Sci. Phil., Keokuk Gr. [Sig. having angles, cornered.]
  arboreus, Meek & Worthen, 1865, Proc. Acad. Nat. Sci. Phil., Keokuk Gr. [Sig. tree-like.]
  bulbosus, Hall, 1862, 15th Reg. Rep. N. Y., Up. Held. Gr. [Sig. bulbous.]
  bullatus, Hall, 1858, Geo. Rep. Iowa, Keokuk Gr. [Sig. bossed, studded with small round knobs.]
  conglobatus, Troost. Not defined.
  cora, Hall, 1864, 20th Reg. Rep. Niagara Gr. [Ety. a Roman proper name.]
  cornutus, Owen & Shumard, 1850, Jour. Acad. Nat. Sci., 2d ser., vol. 2, Burlington Gr. [Sig. horned.]
  corrugatus, Troost. Not defined.
  crassibrachiatus, Hall, 1860, Supp. to Geo. Iowa, Burlington Gr. [Sig. thick-armed.]
  crassus, see Zeacrinus crassus.
  crateriformis, Troost. Not defined.
  decadactylus, Lyon & Casseday, 1859, Am. Jour. Sci. & Arts, vol. 28, Low. Carb. [Sig. ten-fingered.]
  decadactylus, see Scaphiocrinus decadactylus. This name was preoccupied.
  depressus, Troost, see Zeacrinus depressus.
  divaricatus, Hall, 1858, Geo. Rep. Iowa, Burlington Gr. [Sig. wide apart, diverging.]
  enormis, Meek & Worthen, 1865, (Poteriocrinus enormis) Proc. Acad. Nat. Sci. Phil., Burlington Gr. [Sig. unusually large.]
  farleyi, Meek & Worthen, 1866, Proc. Acad. Nat. Sci. Phil., Keokuk Gr. [Ety. proper name.]
  fasciatus, Hall, 1876, 28th Reg. Rep. Niagara Gr. [Sig. striped.]
  florealis, see Zeacrinus florealis.
  fragilis, Meek & Worthen, 1868, Proc. Acad. Nat. Sci. Phil., Burlington Gr. [Sig. frail, easily broken.]
  globosus, Troost. Not defined.
  granuliferus, Shumard, 1852, Red. Riv. Expl. Louisiana, Kaskaskia Gr. [Sig. bearing granules.]
  hexadactylus, Lyon & Casseday, 1859, Am. Jour. Sci., vol. 29, Kaskaskia Gr. [Sig. six-fingered.]
  hoveyi, see Barycrinus hoveyi.

incipiens, Hall, 1861, Desc. New Crinoidea, Burlington Gr. [Sig. a beginning.]
inflatus, Troost. Not defined.
inflexus, Geinitz, 1866, syn. for Poteriocrinus hemisphericus.
insporatus, Lyon, 1860, Trans. Am. Phil. Soc., vol. 13, Low. Carb. [Sig. unexpected.]
intermedius, Hall, 1858, Geo. Rep. Iowa, Keokuk Gr. [Sig. intermediate.]
iowensis, Owen & Shumard, 1850, Jour. Acad. Nat. Sci., 2d ser., vol. 2, Burlington Gr. [Ety. proper name.]
kelloggi, White, 1862, Proc. Bost. Soc. Nat. Hist., vol. 9, Keokuk Gr. [Ety. proper name.]
laeviculus, Lyon, 1861, Proc. Acad. Nat. Sci. Phil., Up. Held. Gr. [Sig. nearly smooth.]
latus, Hall, 1861, Desc. New Crinoidea, Burlington Gr. [Sig. wide.]
lamellosus, White, 1863, Jour. Bost. Soc. Nat. Hist., vol. 7, Burlington Gr. [Sig. in very thin plates.]
lyoni, Hall, 1861, Desc. New Crinoidea, Warsaw Gr. [Ety. proper name.]
macropleurus, Hall, 1861, Desc. New Crinoidea, Burlington Gr. [Sig. long-sided.]
magister, Hall, 1858, Geo. Rep. Iowa, Keokuk Gr. [Sig. the master, chief.]
magnoliiformis, see Zeacrinus magnoliiformis.
malvaceus, Hall, 1858, Geo. Rep. Iowa, Burlington Gr. [Sig. mallow-shaped.]
maniformis, see Zeacrinus maniformis.
multibrachiatus, Lyon & Casseday, 1859, Am. Jour. Sci., vol. 28, Warsaw Gr. [Sig. many-armed.]
ornatissimus, Hall, 1843, Geo. Rep. 4th Dist. N. Y., Portage Gr. [Sig. highly ornamented.]
parvibrachiatus, Hall, 1861, Desc. New Crinoidea, Keokuk Gr. [Sig. small-armed.]
pentalobus, Hall, 1858, Geo. Rep. Iowa, Kaskaskia Gr. [Sig. five-lobed.]
planus, Troost. Not defined.
polyxo, Hall, 1863, Trans. Alb. Inst., vol. 4, Niagara Gr. [Ety. mythological name.]
poterium, Meek & Worthen, 1870, Proc. Acad. Nat. Sci. Phil., Keokuk Gr. [Sig. a drinking vessel, a goblet,]
protuberans, Hall, 1858, Geo. Rep. Iowa, Keokuk Gr. [Sig. protuberant.]
pusillus, Hall, 1863, Trans. Alb. Inst. vol. 4, Niagara Gr. [Sig. very small.]
pyriformis, Murchison, Sil. Researches, 1839, as identified by Hall, but described by Conrad as Ichthyocrinus laevis.
quinquelobus, Meek & Worthen, 1865, Proc. Acad. Nat. Sci. Phil., Keokuk Gr. [Sig. five-lobed.]
rarus, Lyon, 1860, Trans. Am. Phil. Soc., vol. 13, Low. Carb. [Sig. rare.]

rigidus, White, 1862, Proc. Bost. Soc. Nat. Hist., vol. 9, Burlington Gr. [Sig. rigid, not flexible.]
robustus, Troost. Not defined.
roemeri, Troost. Not defined.
rotundatus, Hall, 1858, Geo. Rep. Iowa, Burlington Gr. [Sig. rounded.]
saffordi, Meek & Worthen, 1860, Proc. Acad. Nat. Sci. Phil., Keokuk Gr. [Ety. proper name.]
sangamonensis, Meek & Worthen, 1860, Proc. Acad. Nat. Sci. Phil., Up. Coal Meas. [Ety. proper name.]
scitulus, syn. for C. sculptilis.
sculptilis, Hall, 1860, Supp. to Geo. Rep. Iowa, Burlington Gr. [Sig. carved, engraved.]
sculptus, Troost. Not defined.
solidus, Hall, 1861, Desc. New Crinoidea, Burlington Gr. [Sig. solid, dense, compact.]
spurius, Hall, 1858, Geo. Rep. Iowa, Keokuk Gr. [Sig. spurious.]
stellatus, Troost, 1850, Cat. Crin. Proc. Am. Assoc., Ad. Sci., Keokuk Gr. [Sig. glittering with stars.]
subtumidus, Meek & Worthen, 1865, Proc. Acad. Nat. Sci. Phil., Warsaw Gr. [Sig. somewhat tumid.]
tennesseee, Troost. Not defined.
tenuidactylus, Meek & Worthen, 1868, Proc. Acad. Nat. Sci. Phil., Burlington Gr. [Sig. slender-fingered.]
thomæ, see Barycrinus thomæ.
tiariæformis, see Ichthyocrinus tiariformis.
tumidus, Hall, 1858, Geo. Rep. Iowa, Keokuk Gr. [Sig. tumid, or as if swollen.]
viminalis, Hall, 1861, Desc. New Crinoidea, Burlington Gr. [Sig. rod-like.]
waukoma, Hall, 1864, 20th Reg. Rep., Niagara Gr. [Ety. proper name.]
wortheni, Lyon, 1861, Proc. Acad. Nat. Sci. Phil., Up. Held. Gr. [Ety. proper name.]
wachsmuthi, Meek & Worthen, 1861, Proc. Acad. Nat. Sci. Phil., Burlington Gr. [Ety. proper name.]
Cyclaster, Billings, 1857, Rep. of Progr. This name was preoccupied. See Edrioaster.
bigsbyi, see Edrioaster bigsbyi.
CYCLOCYSTOIDES, Billings & Salter, 1858, Can. Org. Rem., Decade 3. [Ety. *kuklos*, a circle; *kustis*, a bladder; *eidos*, form.]
anteceptus, Hall, 1866, Pamphlet, Trenton & Hud. Riv. Gr. [Sig. anticipated, comprehended before.]
halli, Billings, 1858, Can. Org. Rem., Decade 3, Trenton Gr. [Ety. proper name.]
huronensis, Billings, 1865, Pal. Foss. vol. 1, Hud. Riv. Gr. [Ety. proper name.]
salteri, Hall, 1866, Pamphlet, Trenton Gr. [Ety. proper name.]
CYSTOCRINUS, Roemer, 1860, Sil. Fauna. West Tenn. [Ety. *kustis*, a bladder; *krinon*, a lily.]

ECHINODERMATA.

tennesseensis, Roemer, 1860, Sil. Fauna.
West Tenn., Niagara Gr. [Ety. proper name.]
Cytocrinus, Roemer, 1860, Sil. Fauna. West Tenn. Syn. for Macrostylocrinus.
lævis, see Macrostylocrinus lævis.
Dæmonocrinites, Troost. Not defined.
Decadactylocrinites, Owen. Not defined.
DENDROCRINUS, Hall, 1852, Pal. N. Y., vol. 2. [Ety. *dendron*, a tree; *krinon*, a lily.]
acutidactylus, Billings, 1857, Rep. of Progr., Trenton Gr. [Sig. sharp-fingered.]
angulatus, see Palæocrinus angulatus.
caduceus, Hall, 1866, Pamphlet, Cin'ti Gr. [Sig. the herald's staff.]
casii, Meek, 1871, Am. Jour. Sci., 3d ser., vol. 2, Cin'ti Gr. [Ety. proper name.]
cincinnatiensis, Meek, 1872, Proc. Acad. Nat. Sci. Phil., Cin'ti Gr. [Ety. proper name.]
conjugans, Billings, 1857, Rep. of Progr., Trenton Gr. [Sig. joined, united.]
cylindricus, Billings, 1859, Can. Org. Rem., Decade 4, Trenton Gr. [Sig. cylindrical.]
dyeri, Meek, 1872, Proc. Acad. Nat. Sci. Phil., Cin'ti Gr. [Ety. proper name.]
gregarius, Billings, 1857, Rep. of Progr., Trenton Gr. [Sig. occuring in flocks or masses.]
humilis, Billings, 1857, Rep. of Progr., Trenton Gr. [Sig. small.]
jewetti, Billings, 1859, Can. Org. Rem., Decade 4, Trenton Gr. [Ety. proper name.]
latibrachiatus, Billings,1857, Rep. of Prog., Hud. Riv. Gr. [Sig. wide-armed.]
longidactylus, Hall, 1852, Pal. N. Y., vol. 2, Niagara Gr. [Sig. long-fingered.]
modestus, Safford, 1869, Geo. of Tenn. Not defined.
nucleus, Hall, 1876, 28th Reg. Rep., Niagara Gr. [Sig. a kernel.]
oswegoensis, Meek & Worthen, 1868, Geo. Sur. Ill., vol. 3, Cin'ti Gr. [Ety. proper name.]
polydactylus, Shumard, 1857, (Homocrinus polydactylus) Trans. St. Louis Acad. Sci., Cin'ti Gr. [Sig. many-fingered.]
posticus, Hall, 1866, Pamphlet, Cin'ti Gr. [Sig. behind, posterior.]
proboscidiatus, Billings, 1857, Rep. of Progr., Trenton Gr. [Sig. having a proboscis.]
rusticus, Billings, 1857, Rep. of Progr., Trenton Gr. [Sig. rustic, coarse, rough.]
similis, Billings, 1857, Rep. of Progr., Trenton Gr. [Sig. similar to D. acutidactylus, and D. proboscidiatus.]
tener, Billings, 1866, Catal. Sil. Foss. Antic., Hud. Riv. Gr. [Sig. delicate.]
DICHOCRINUS, Munster, 1839, Beitrag. Zur. Petref. [Ety. *dicha*, in two parts; *krinon*, a lily.]

angustus, White, 1862, Proc. Bost. Soc. Nat. Hist., vol. 9, Burlington Gr. [Sig. narrow.]
chesterensis, see Pterocrinus chesterensis.
constrictus, Meek & Worthen, 1860, Proc. Acad. Nat. Sci. Phil., St. Louis Gr. [Sig. constricted.]
conus, Meek & Worthen, 1860, Proc. Acad. Sci. Phil., Burlington Gr. [Sig. a cone.]
cornigerus, see Pterocrinus cornigerus.
crassitestus, White, 1862, Proc. Bost. Soc. Nat. Hist., vol. 9, Burlington Gr. [Sig. like a thick vessel or pot lid.]
crassus, see Pterocrinus crassus.
dichotomus, Hall, 1860, Supp. to Geo. Rep. Iowa, Warsaw Gr. [Sig. dividing into two.]
elegans, Casseday & Lyon, 1860, Proc. Am. Acad. Arts & Sci., vol. 5, Kaskaskia Gr. [Sig. elegant.]
expansus, Meek & Worthen, 1868, Proc. Acad. Nat. Sci. Phil., Keokuk Gr. [Sig. spread out.]
ficus, Casseday & Lyon, 1860, Proc. Am. Acad. Arts & Sci., vol. 5, Keokuk Gr. [Sig. a fig.]
lachrymosus, Hall, 1860, Supp. to Geo. Rep. Iowa, Burlington Gr. [Sig. full of tears.]
lævis, Hall, 1860, Supp. Geo. Rep. Iowa, Burlington Gr. [Sig. smooth.]
lineatus, Meek & Worthen, 1869, Proc. Acad. Nat. Sci. Phil., Burlington Gr. [Sig. marked with lines.]
liratus, Hall, 1861, Desc. New Crinoidea, Burlington Gr. [Sig. lined, furrowed.]
ovatus, Owen & Shumard, 1850, Jour. Acad. Nat. Sci., 2d ser., vol. 2, Burlington Gr. [Sig. egg-shaped.]
pisum, Meek & Worthen, 1869, Proc. Acad. Nat. Sci. Phil., Burlington Gr. [Sig. a pea.]
plicatus, Hall, 1861, Desc. New Crinoidea, Burlington Gr. [Sig. plaited.]
pocillum, Hall, 1861, Desc. New Crinoidea, Burlington Gr. [Sig. a little cup.]
polydactylus, Casseday & Lyon, 1860, Proc. Am. Acad. Arts & Sci., vol. 5, Keokuk Gr. [Sig. many-fingered.]
protuberans, Hall, 1858, Geo. Rep. Iowa, Kaskaskia Gr. [Sig. protuberant.]
scitulus, Hall, 1861, Desc. New Crinoidea, Burlington Gr. [Sig. handsome, neat.]
sculptus, Casseday & Lyon, 1860, Proc. Am. Acad. Arts & Sci., vol. 5, Keokuk Gr. [Sig. engraved.]
sexlobatus, see Pterocrinus sexlobatus.
simplex, Shumard, 1857, Trans. St. Louis Acad. Sci., Warsaw Gr. [Sig. simple.]
striatus, Owen & Shumard, 1850, Jour. Acad. Nat. Sci., 2d ser., vol. 2, Burlington Gr. [Sig. striated.]
symmetricus, Casseday & Lyon, 1860, Proc. Am. Acad. Arts & Sci., vol. 5, Kaskaskia Gr. [Sig. symmetrical.]

ECHINODERMATA.

DICTYOCRINUS, Conrad, 1841, (Dictuocrinites) Ann. Rep. N. Y. [Ety. *dictyon*, a net; *krinon*, a lily.]
squamifer, Hall, 1859, Pal. N. Y., vol. 3, Low. Held. Gr. [Sig. scale-bearing.]
DOLATOCRINUS, Lyon, 1857, Geo. Sur. Ky., vol. 3. [Ety. *dolatus*, hewn or tooled; *krinon*, a lily.]
glyptus, Hall, 1862, (Cacabocrinus glyptus) 15th Reg. Rep., Ham. Gr. [Sig. sculptured.]
glyptus var. intermedius, Hall, 1862, 15th Reg. Rep. N. Y., Ham. Gr. [Sig. intermediate.]
lacus, Lyon, 1857, Geo. Sur. Ky., vol. 3, Corniferous Gr. [Sig. a tub, a vat, a basin.]
lamellosus, Hall, 1862, (Cacabocrinus lamellosus) 15th Reg. Rep. N. Y., Up. Held. Gr. [Sig. in very thin plates.]
liratus, Hall, 1862, (Cacabocrinus liratus) 15th Reg. Rep. N. Y., Ham. Gr. [Sig. furrowed.]
liratus var. multilira, Hall, 1862, Cacabocrinus liratus var. multilira) 13th Reg. Rep. N. Y., Ham. Gr. [Sig. many-furrowed.]
marshi, Lyon, 1860, Trans. Am. Phil. Soc., Devonian. [Ety. proper name.]
speciosus, Hall, 1862, (Cacabocrinus speciosus) 15th Reg. Rep. N. Y., Up. Held. Gr. [Sig. beautiful.]
troosti, Hall, 1862, (Cacabocrinus troosti) 15th Reg. Rep. N. Y., Ham. Gr. [Ety. proper name.]
Donacicrinites, Troost. Not defined.
simplex, Troost. Not defined.
DORYCRINUS, Roemer, 1853, Wiegm. Arch. [Ety. *dory*, a spear; *krinon*, a lily.]
canaliculatus, Meek & Worthen, 1869, Proc. Acad. Nat. Sci., Burlington Gr. [Sig. channeled.]
cornigerus, Hall, 1858, (Actinocrinus cornigerus) Geo. Rep. Iowa, Burlington Gr. [Sig. horned.]
desideratus, Hall, 1861, (Actinocrinus desideratus) Desc. New Crinoidea, Burlington Gr. [Sig. wished for, desirable.]
divaricatus, Hall, 1860, (Actinocrinus divaricatus) Supp. to Geo. Sur. Iowa, Burlington Gr. [Sig. wide apart.]
gouldi, Hall, 1858, (Actinocrinus gouldi) Geo. Rep. Iowa, Keokuk Gr. [Ety. proper name.]
kelloggi, Worthen, 1875, Geo. Sur. Ill., vol. 6, Keokuk Gr. [Ety. proper name.]
mississippiensis, Roemer, 1853, Archiv. fur Nat., Jahr. 19, Keokuk Gr. [Ety. proper name.]
mississippiensis var. spiniger, Hall, 1860, Supp. to Geo. Sur. Iowa, Keokuk Gr. [Sig. spiny.]
missouriensis, Shumard, 1858, (Actinocrinus missouriensis) Geo. Rep. Mo., Burlington Gr. [Ety. proper name.]
quinquelobus, Hall, 1860 (Actinocrinus quinquelobus) Supp. to Geo. Rep. Iowa, Burlington Gr. [Sig. five-lobed.]
quinquelobus var. intermedius, Meek & Worthen, 1868, Proc. Acad. Nat. Sci., Burlington Gr. [Sig. intermediate.]
roemeri, Meek & Worthen, 1868, Proc. Acad. Nat. Sci. Phil., Burlington Gr. [Ety. proper name.]
subaculeatus, Hall, 1858, (Actinocrinus subaculeatus) Geo. Rep. Iowa, Burlington Gr. [Sig. armed with points somewhat sharp.]
subturbinatus, Meek & Worthen, 1860, (Actinocrinus subturbinatus) Proc. Acad. Nat. Sci. Phil., Burlington Gr. [Sig. somewhat top-shaped.]
symmetricus, Hall, 1858, (Actinocrinus symmetricus) Geo. Rep. Iowa, Burlington Gr. [Sig. symmetrical.]
trinodus, Hall, 1858, (Actinocrinus trinodus) Geo. Rep. Iowa, Burlington Gr. [Sig. three-knobbed.]
unicornis, Owen & Shumard, 1850, (Actinocrinus unicornis) Jour. Acad. Nat. Sci. Phil., vol. 2, new ser., Burlington Gr. [Sig. single-horned.]
unispinus, Hall, 1861, (Actinocrinus unispinus) Desc. New Crinoidea, Burlington Gr. [Sig. one-spined.]
ECHINOCYSTITES, Hall, 1864, 20th Reg. Rep. N. Y. [Ety. *echinos*, the sea urchin; *kustis*, a bladder.]
nodosus, Hall, 1864, 20th Reg. Rep. Niagara Gr. [Sig. knobbed.]
ECHINO-ENCRINITES, Meyer, 1826, Karst. Archiv. Nat., vol. 7. [Ety. *echinos*, the sea urchin; *krinon*, a lily.]
anatiformis, Hall, 1847, Pal. N. Y., vol. 1, Trenton Gr. [Sig. resembling the barnacle *Anatifa*.] I think this species is not generically distinct from *Lepadocrinus moori*.
fenestratus, Troost. Not defined.
Echinus drydenensis, see Eocidaris drydenensis.
gyracanthus, see Tentaculites gyracanthus.
EDRIOASTER, Billings, 1858, Can. Org., Rem. [Ety. *edrion*, a seat; *aster*, a star; in allusion to the sessile condition of the species. This name is a substitute for Cyclaster, proposed in 1857, the latter name having been preoccupied.]
bigsbyi, Billings, 1857, (Cyclaster bigsbyi) Rep. of Progr., Trenton Gr. [Ety. proper name.]
EDRIOCRINUS, Hall, 1859, Pal. N. Y., vol. 3. [Ety. *edrion*, a seat; *krinon*, a lily.]
pocilliformis, Hall, 1859, Pal. N. Y., vol. 3, Low. Held. Gr. [Sig. like a little cup.]
pyriformis, Hall, 1862, 15th Reg. Rep. N. Y., Up. Held. Gr. [Sig. pear-shaped.]
sacculus, Hall, 1859, Pal. N. Y., vol. 3, Oriskany sandstone. [Sig. a little bag.]

Elæacrinus, Roemer, 1852, syn. for Nucleocrinus.]
  kirkwoodensis, see Nucleocrinus kirkwoodensis.
  verneuili, see Nucleocrinus verneuili.
ELEUTHEROCRINUS, Shumard & Yandell, 1856, Proc. Acad. Nat. Sci. Phil., vol. 8. [Ety. *eleutheros*, free; *krinon*, a lily.]
  cassedayi, Shumard & Yandell, 1856, Proc. Acad. Nat. Sci. Phil., vol. 8, Up. Held. Gr. [Ety. proper name.]
  whitfieldi, Hall, 1862, 15th Reg. Rep. N. Y., Ham. Gr. [Ety. proper name.]
EOCIDARIS, Desor, 1858, Synopsis des Echinides Fossiles. [Ety. *eos*, the dawn; *cidaris*, a turban.]
  drydenensis, Vanuxem, 1842, (Echinus drydenensis) Geo. Rep. 3rd Dist. N. Y., Chemung Gr. [Ety. proper name.]
  hallianus, Geinitz, 1866, Carb. und Dyas. in Neb., Up. Coal Meas. [Ety. proper name.]
  squamosus, Meek & Worthen, 1869, Proc. Acad. Nat. Sci. Phil., Burlington Gr. [Sig. scaly.]
EOCYSTITES, Billings, 1868, Acad. Geol. [Ety. *eos*, dawn; *kustis*, a bladder.]
  primævus, Billings, 1868, Acad. Geol., St. John's Gr. [Sig. in the first period of life.]
ERETMOCRINUS, Lyon & Casseday, 1859, Am. Jour. Sci., vol. 28. [Ety. *eretmos*, an oar; *krinon*, a lily.]
  calyculoides, Hall, 1860, (Actinocrinus calyculoides) Supp. to Geo. Sur. Iowa, Burlington Gr. [Sig. like a little cup.]
  clio, Hall, 1861, (Actinocrinus clio) Desc. New Crinoidea, Burlington Gr. [Ety. mythological name.]
  clœlia, Hall, 1861, (Actinocrinus clœlia) Desc. New Crinoidea, Burlington Gr. [Ety. mythological name.]
  magnificus, Lyon & Casseday, 1859, Am. Jour. Sci., vol. 28, Low. Carb. [Sig. magnificent.]
  matuta, Hall, 1861, (Actinocrinus matuta) Desc. New Crinoidea, Burlington Gr. [Ety. *Matuta*, the name of a goddess of the morning.]
  matuta *var.* attenuata, Hall, 1861, (Actinocrinus matuta *var.* attenuata) Desc. New Crinoidea, Burlington Gr. [Sig. drawn out, attenuated.]
  remibrachiatus, Hall, 1861, (Actinocrinus remibrachiatus) Desc. New Crinoidea, Burlington Gr. [Sig. paddle-armed.]
  verneuilianus, Shumard, 1855, (Actinocrinus verneuilianus) Geo. Sur. Mo., Burlington Gr. [Ety. proper name.]
ERISOCRINUS, Meek & Worthen, 1865, Am. Jour. Sci., vol. 89. [Ety. *eris*, contention; *krinon*, a lily.] The authors have expressed their doubts about this genus, because it is so closely allied to Dr. De Koninck's genus, *Philocrinus*.

  antiquus, Meek & Worthen, 1869, Proc. Acad. Nat. Sci. Phil., Burlington Gr. [Sig. ancient.]
  conoideus, Meek & Worthen, 1865, Proc. Acad. Nat. Sci. Phil., Up. Coal Meas. [Sig. somewhat conical.]
  nebrascensis, Meek & Worthen, 1865, Am. Jour. Sci., vol. 89, Up. Coal Meas. [Ety. proper name.] This is regarded as merely a variety of *E. typus*.
  typus, Meek & Worthen, 1865, Am. Jour. Sci., vol. 89, Up. Coal Meas. [Ety. type of the genus.]
  tuberculatus, see Eupachycrinus tuberculatus.
  whitii, Meek & Worthen, 1869, Proc. Acad. Nat. Sci. Phil., Burlington Gr. [Ety. proper name.]
EUCALYPTOCRINUS, Goldfuss, 1826, Petref. Germ. [Ety. *eu*, well; *kalyptos*, covered; *krinon*, a lily.]
  armosus, see Glyptocrinus armosus.
  cælatus, Hall, 1843, (Hypanthocrinites cælatus) Geo. Rep., 4th Dist., N. Y., Niagara Gr. [Sig. carved in relief, sculptured.]
  chicagoensis, Winchell & Marcy, 1865, Mem. Bos. Soc. Nat. Hist., Niagara Gr. [Ety. proper name.]
  conicus, Troost. Not defined.
  cornutus, Hall, 1864, 20th Reg. Rep., Niagara Gr. [Sig. horned.]
  cornutus *var.* excavatus, Hall, 1864, 20th Reg. Rep., N. Y., Niagara Gr. [Sig. hollowed out.]
  crassus, Hall, 1863, Trans. Alb. Inst., vol. 4, Niagara Gr. [Sig. thick, stout.]
  decorus, Phillips, 1839, (Hypanthocrinites decorus) Murch. Sil. Syst., Niagara Gr. [Sig. ornamented in relief.]
  extensus, Troost. Not defined.
  gibbosus, Troost. Not defined.
  goldfussi, Troost. Not defined.
  laevis, Troost. Not defined.
  magnus, Worthen, 1875, Geo. Sur. Ill., vol. 6, Up. Sil. [Sig. large.]
  nashvillæ, Troost. Not defined.
  obconicus, Hall, 1864, 20th Reg. Rep., Niagara Gr. [Sig. inversely conical.]
  ornatus, Hall, 1861, Rep. of Progr. Sur. of Wis., Niagara Gr. [Sig. adorned.]
  ovatus, Troost, as figured by Hall, 1876, 28th Reg. Rep., Niagara Gr. [Sig. ovate.]
  papulosus, Hall, 1852, Pal. N. Y., vol. 2, Niagara Gr. [Sig. covered with pimples.]
  phillipsii, Troost. Not defined.
  ramifer, Roemer, 1860, Sil. Fauna. West Tenn., Niagara Gr. [Sig. bearing branches.]
  splendidus, Troost, Catal. Hall & Whitfield, 1875, Ohio Pal., vol. 2, Niagara Gr. [Sig. splendid.]
  tennesseæ, Troost. Not defined.
EUGASTER, Hall, 1868, 20th Reg. Rep. [Ety. *euge*, pre-eminent, remarkable; *aster*, a star.]

ECHINODERMATA. 79

logani, Hall, 1858, 20th Reg. Rep., Ham. Gr. [Ety. proper name.]
EUPACHYCRINUS, Meek & Worthen, 1865, Proc. Acad. Nat. Sci. [Ety. *eu*, very; *pachys*, thick; *krinon*, a lily.]
bassetti, Worthen, 1875, Geo. Sur. Ill., vol. 6, Coal Meas. [Ety. proper name.]
boydi, Meek & Worthen, 1870, Proc. Acad. Nat. Sci. Phil., Chester Gr. [Ety. proper name.]
craigi, Worthen, 1875, Geo. Sur. Ill., vol. 6, Coal Meas. [Ety. proper name.]
fayettensis, Worthen, 1873, Geo. Sur. Ill., vol. 5, Up. Coal Meas. [Ety. proper name.]
platybasis, White, 1876, Geo. Uinta Mountains, Low. Aubrey Gr. [Sig. having a flat base.]
tuberculatus, Meek & Worthen, 1865, (Erisocrinus tuberculatus) Proc. Acad. Nat. Sci. Phil., Coal Meas. [Sig. tuberculated.]
verrucosus, White & St. John, 1869, Trans. Chi. Acad. Sci., Coal Meas. [Sig. covered with tubercles.]
FORBESOCRINUS, DeKoninck & LeHon, 1854, Resch. Crin. Carb. Belg. [Ety. proper name; *krinon*, a lily.]
agassizi, Hall, 1858 & 1860, Geo. Sur. of Iowa & Supp., Burlington Gr. [Ety. proper name.]
agassizi *var.* giganteus, Meek & Worthen, 1861, Proc. Acad. Nat. Sci. Phil., Burlington Gr. [Sig. very large.]
asteriformis, Hall, 1861, Desc. New Crinoidea, Burlington Gr. [Sig. star-shaped.]
cestriensis, Hall, 1860, Supp. to Geo. Iowa, Burlington Gr. [Ety. proper name.]
communis, Hall, 1863, Crin. Waverly sandstone Ohio, Waverly Gr. [Sig. of frequent occurrence.]
giddingi, Hall, 1858, Geo. Rep. Iowa, Keokuk Gr. [Ety. proper name.]
juvenis, Hall, 1861, Desc. New Crinoidea, Burlington Gr. [Sig. young.]
kelloggi, Hall, 1863, Crin. Wav. Sands. Ohio, Waverly Gr. [Ety. proper name.]
lobatus, Hall, 1862, 15th Reg. Rep. N. Y., Ham. Gr. [Sig. lobate.]
lobatus *var.* tardus, Hall, 1863, Crin. Wav. Sands. Ohio, Waverly Gr. [Sig. sluggish.]
meeki, see Onychocrinus meeki.
monroensis, see Onychocrinus monroensis.
multibrachiatus, Lyon & Casseday, 1859, Am. Jour. Sci., vol. 28, Kaskaskia Gr. [Sig. many-armed.]
norwoodi, see Onychocrinus norwoodi.
nuntius, see Taxocrinus nuntius.
pratteni, see Melocrinus pratteni.
ramulosus, Lyon & Casseday, 1859, Am. Jour. Sci., vol. 28, Kaskaskia Gr. [Sig. full of branches.]
ramulosus, Hall, 1860, see F. subramulosus.
saffordi, Hall, 1860, Supp. to Geo. Sur. Iowa, Burlington Gr. [Ety. proper name.]
semioratus, see Taxocrinus semiovatus.
shumardianus, Hall, 1858, Geo. Rep. Iowa, St. Louis Gr. [Ety. proper name.]
spiniger, Hall, 1861, Desc. New Crinoidea, Burlington Gr. [Sig. thorny.]
subramulosus, Shumard, 1866, (Hall, 1860, F. ramulosus) Supp. to Geo. Sur. Iowa, Burlington Gr. The specific name *ramulosus* being preoccupied Shumard proposed subramulosus, in Trans. St. Louis. Acad. Sci. [Sig. somewhat branchy.]
thiemei, see Taxocrinus thiemii.
whitfieldi, see Onychocrinus whitfieldi.
wortheni, Hall, 1858, Geo. Rep. Iowa, Keokuk Gr. [Ety. proper name.]
GLYPTASTER, Hall, 1852, Pal. N. Y., vol. 2. [Ety. *glyptos*, sculptured; *aster*, a star.]
brachiatus, Hall, 1852, Pal. N. Y., vol. 2, Niagara Gr. [Sig. having arms.]
inornatus, Hall, 1863, Trans. Alb. Inst., vol. 4, Niagara Gr. [Sig. not adorned.]
occidentalis, Hall, 1863, Trans. Alb. Inst., vol. 4, Niagara Gr. [Sig. western.]
pentangularis, Hall, 1867, 20th Reg. Rep., Niagara Gr. [Sig. five-cornered; pentagonal.]
GLYPTOCRINUS, Hall, 1847, Pal. N. Y., vol. 1. [Ety. *glyptos*, sculptured; *krinon*, a lily.]
armosus, McChesney, 1861, (Eucalyptocrinus armosus) New Pal. Foss., Niagara Gr. [Sig. many-armed.]
baeri, Meek, 1872, Am. Jour. Sci., 3rd series, vol. 3, Cin'ti Gr. [Ety. proper name.]
carleyi, Hall, 1862, Trans. Alb. Inst., vol. 4, Niagara Gr. [Ety. proper name.]
decadactylus, Hall, 1847, Pal. N. Y., vol. 1, Cin'ti Gr. [Ety. ten-fingered, but the fossil has twenty fingers.]
dyeri, Meek, 1872, Proc. Acad. Nat. Sci. Phil., Cin'ti Gr. [Ety. proper name.]
fimbriatus, Shumard, 1855, Geo. Rep. Mo., Trenton Gr. [Sig. fringed.]
fornshelli, S. A. Miller, 1874, Cin. Quar. Jour. Sci., Cin'ti Gr. [Ety. proper name.]
lacunosus, Billings, 1857, Rep. of Progr., Trenton Gr. [Sig. having deep depressions.]
libanus, Safford, 1869, Geo. of Tenn. Not defined.
marginatus, Billings, 1857, Rep. of Progr., Trenton Gr. [Sig. margined—from the border on the margin of the plates.]
nealli, Hall, 1866, Pamphlet, Cin'ti Gr. [Ety. proper name.]
nobilis, Hall, 1861, Rep. of Progr. Sur of Wis., Niagara Gr. [Sig. remarkable.]
ornatus, Billings, 1857, Rep. of Progr., Trenton Gr. [Sig. adorned.]

parvus, Hall, 1866, Pamphlet, Cin'ti Gr. [Sig. small.]
plumosus, Hall, 1843, (Actinocrinus plumosus) Geo. Rep., 4th Dist. N. Y., Clinton Gr. [Sig. feathery.]
priscus, Billings, 1857, Rep. of Progr. Geo. Sur. Can., Black Riv. & Trenton Gr. [Sig. ancient.]
quinquepartitus, Billings, 1859, Can. Org. Rem., Decade 4, Trenton Gr. [Sig. five-parted.]
ramulosus, Billings, 1856, Can. Nat. Geo., vol. 1, Trenton Gr. [Sig. full of little branches.]
shafferi, S. A. Miller, 1875, Cin. Quar. Jour. Sci., vol. 2, Cin'ti Gr. [Ety. proper name.]
siphonatus, Hall, 1861, syn. for Glyptocrinus armosus.
subglobosus, Meek, 1873, Pal. Ohio, vol. 1, Cin'ti Gr. [Sig. somewhat globose.]
GLYPTOCYSTITES, Billings, 1854, Can. Jour., vol. 2. [Ety. glyptos, sculptured; kustis, a bladder.]
forbesi, Billings, 1857, Rep. of Progr. Chazy Gr. [Ety. proper name.]
gracilis, Billings, 1858, Can. Org. Rem., Decade 3, Trenton Gr. [Sig. slender.]
logani, Billings, 1857, Rep. of Progress, Trenton Gr. [Ety. proper name.]
logani var. gracilis, Billings, 1858, Can. Org. Rem., Decade 3, Trenton Gr. [Sig. slender.]
multiporus, Billings, 1854, Can. Jour., vol. 2, Trenton Gr. [Sig. having many passages.]
GOMPHOCYSTITES, Hall, 1864, 20th Reg. Rep. [Ety. gomphos, nail or rudder; kustis, a bladder.]
clavus, Hall, 1864, 20th Reg. Rep., Niagara Gr. [Sig. a club.]
glans, Hall, 1864, 20th Reg. Rep., Niagara Gr. [Sig. an acorn.]
tenax, Hall, 1864, 20th Reg. Rep., Niagara Gr. [Sig. holding on—the arm plates appear to have been fringed with tentacles.]
GONIASTEROIDOCRINUS, Lyon & Casseday, 1859, Am. Jour. Sci., vol. 28. [Ety. like the recent genus Goniaster; krinon, a lily.]
fiscellus, Meek & Worthen, 1861, (Trematocrinus fiscellus) Proc. Acad. Nat. Sci. Phil., Burlington Gr. [Sig. a small basket made of twigs.]
obovatus, Meek & Worthen, 1869, Proc. Acad. Nat. Sci. Phil., Burlington Gr. [Sig. egg-shaped, with large end uppermost.]
papillatus, Hall, 1860, (Trematocrinus papillatus) Supp. to Geo. Rep. Iowa, Burlington Gr. [Sig. bud-shaped.]
reticulatus, Hall, 1861, (Trematocrinus reticulatus) Desc. New Crinoidea, Burlington Gr. [Sig. reticulated.]
robustus, Hall, 1860, (Trematocrinus robustus) Supp. to Geo. Rep. Iowa, Keokuk Gr. [Sig. robust.]

spinigerus, Hall, 1862, (Trematocrinus spinigerus) 15th Reg. Rep. N. Y., Ham. Gr. [Sig. bearing spines.]
tenuiradiatus, Meek & Worthen, 1869, Proc. Acad. Nat. Sci. Phil., Burlington Gr. [Sig. slenderly rayed.]
tuberculosus, Hall, 1860, (Trematocrinus tuberculosus) Supp. to Geo. Rep. Iowa, Burlington Gr. [Sig. covered with tubercles.]
tuberosus, Lyon & Casseday, 1859, Am. Jour. Sci., vol. 28, Kaskaskia Gr. [Sig. knobby.]
typus, Hall, 1860, (Trematocrinus typus) Supp. to Geo. Rep. Iowa, Burlington Gr. [Sig. the type—the type of the genus Trematocrinus.]
GRANATOCRINUS, Troost, 1850, Cat. Foss., and described by Hall, 1862, 15th Reg. Rep. N. Y. [Ety. granatos, granular; krinon, a lily.]
cidariformis, Troost. Not defined.
cornutus, Meek & Worthen, 1861, (Pentremites cornutus) Proc. Acad. Nat. Sci. Phil., St. Louis Gr. [Ety. from the horn-like interradial pieces.]
curtus, Shumard, 1855, (Pentremites curtus) Geo. Rep. Mo., Warsaw Gr. [Sig. short.]
glaber, Meek & Worthen, 1869, Proc. Acad. Nat. Sci. Phil., St. Louis Gr. [Sig. smooth.]
granulatus, Roemer, 1852, (Pentremites granulatus) Monog. Blast., Warsaw Gr. [Sig. granulated.]
granulosus, Meek & Worthen, 1865, Proc. Acad. Nat. Sci. Phil., Keokuk Gr. [Sig. covered with small granules.]
lotoblastus, White, 1874, Rep. Invert. Foss., Low. Carb. [Ety. lotos, the Lotos plant; blastos, a bud.]
melo, Owen & Shumard, 1850, (Pentremites melo) Jour. Acad. Nat. Sci. Phil., 2d ser., vol. 2, Burlington Gr. [Sig. a melon.]
melo var. projectus, see Granatocrinus projectus.
melonoides, Meek & Worthen, 1869, Proc. Acad. Nat. Sci. Phil., Burlington, Gr. [Sig. resembling a melon.]
missouriensis, Shumard, 1866, Trans. St. Louis Acad. Sci., Chemung Gr. [Ety. proper name.]
neglectus, Meek & Worthen, 1869, Proc. Acad. Nat. Sci. Phil., Burlington Gr. [Sig. overlooked.]
norwoodi, Owen & Shumard, 1850, (Pentremites norwoodi) Jour. Acad. Nat. Sci. Phil., 2d series, vol. 2, Burlington Gr. [Ety. proper name.]
pisum, Meek & Worthen, 1869, Proc. Acad. Nat. Sci. Phil., Burlington Gr. [Sig. a pea.]
projectus, Meek & Worthen, 1861, Proc. Acad. Nat. Sci. Phil., Burlington Gr. [Sig. projected, thrown out.]

ECHINODERMATA. 81

roemeri, Shumard, 1855, (Pentremites roemeri) Geo. Rep. Mo., Chemung Gr. [Ety. proper name.]
sayi, Shumard, 1855, (Pentremites sayi) Geo. Rep. Mo., Burlington Gr. [Ety. proper name.]
shumardi, Meek & Worthen, 1866, Proc. Acad. Nat. Sci. Phil., Burlington Gr. [Ety. proper name.]

GRAPHIOCRINUS, De Koninck & Le Hon, 1854, Rech. Crin. Carb. Belg. [Ety. *graphion*, writing instrument; *krinon*, a lily.]
brachialis, Lyon, 1857, Geo. Sur. Ky., vol. 3, Low. Carb. [Sig. having arms.]
dactylus, Hall, 1860, Supp. to Geo. Rep. Iowa, St. Louis Gr. [Sig. fingered.]
quatuordecembrachialis, Lyon, 1857, Geo. Sur. Ky., vol. 3, Burlington Gr. [Sig. having fourteen arms.]

HADROCRINUS, Lyon, 1869, Trans. Am. Phil. Soc., vol. 13. [Ety. *adros*, full grown; *krinon*, a lily.]
discus, Lyon, 1869, Trans. Am. Phil. Soc., vol. 13, Devonian. [Sig. a dish.]
pentagonus, Lyon, 1869, Trans. Phil. Soc., vol. 13, Devonian. [Sig. pentagonal.]
plenissimus, Lyon, 1869, Trans. Phil. Soc., vol. 13, Devonian. [Sig. the largest.]

HAPLOCRINUS, Steininger, Bul. Soc. Geol. France t. 8, 1st series. [Ety. *haploos*, simple; *krinon*, a lily.]
clio, Hall, 1862, 15th Reg. Rep. N. Y., Marcellus shale. [Ety. mythological name.]
granulatus, Troost. Not defined.
hemisphericus, Troost. Not defined.
maximus, Troost. Not defined.
ovalis, Troost. Not defined.

HEMICOSMITES, Von Buch, 1840, Monatsber d. Berlin Akad. [Ety. *hemi*, half; *cosmites*, sphere.]
subglobosus, Hall, 1864, in 20th Reg. Rep., Niagara Gr. [Sig. somewhat globose.]

HEMICYSTITES, Hall, 1852, Pal. N. Y., vol. 2. [Ety. *hemi*, half; *kustis*, a bladder.]
altus, a synonym for H. granulatus.
granulatus, Hall, 1872, Pamphlet, Cin'ti Gr. [Ety. from the granulated appearance of the upper surface.]
parasiticus, Hall, 1852, Pal. N. Y., vol. 2, Niagara Gr. [Sig. parasitic.]
stellatus, Hall, 1866, Pamphlet, Cin'ti Gr. [Sig. star-shaped.]

HETEROCRINUS, Hall, 1847, Pal. N. Y., vol. 1. [Ety. *heteros*, irregular; *krinon*, a lily.]
articulosus, Billings, 1859, Can. Org. Rem., Decade 4, Trenton Gr. [Sig. full of joints.]
canadensis, Billings, 1859, Can. Org. Rem., Decade 4, Trenton Gr. [Ety. proper name.] This species can hardly be separated from *H. simplex*, by specific differences, and it is doubtful whether it constitutes more than a variety.
constrictus, Hall, 1866, Pamphlet, Cin'ti Gr. [Sig. constricted.]

constrictus var. compactus, Meek, 1873, Ohio Pal., vol. 1, Cin'ti Gr. [Sig. compact.]
crassus, Meek & Worthen, 1865, Proc. Acad. Nat. Sci. Phil., Cin'ti Gr. [Ety. from the thick arm plates.]
exilis, Hall, 1866, Pamphlet, Trenton & Hud. Riv. Gr. [Sig. small, slender.]
exiguus, syn. for H. exilis.
gracilis, Hall, 1847, Pal. N. Y., vol. 1. Hud. Riv. Gr. [Sig. slender.]
heterodactylus, Hall, 1847, Pal. N. Y., vol. 1, Hud. Riv. Gr. [Sig. irregularly fingered.]
inæqualis, Billings, 1859, Can. Org. Rem., Decade 4, Trenton Gr. [Sig. unequal.]
incurvus, see Anomalocrinus incurvus.
isodactylus, syn. for Heterocrinus constrictus var. compactus.
juvenis, Hall, 1866, Pamphlet, Cin'ti Gr. [Sig. young.]
laxus, Hall, 1866, Pamphlet, Trenton & Hud. Riv. Gr. [Sig. loose.]
polyxo, syn. for H. subcrassus.
simplex, Hall, 1847, Pal. N. Y., vol. 1, Trenton & Hud. Riv. Gr. [Sig. simple.]
simplex var. grandis, Meek, 1873, Pal. Ohio, vol. 1, Cin'ti Gr. [Sig. great.]
subcrassus, Meek & Worthen, 1865, Proc. Acad. Nat. Sci. Phil., Cin'ti Gr. [Sig. somewhat like *H. crassus*.]
tenuis, Billings, 1857, Rep. of Progr., Trenton Gr. [Sig. slender.]

HETEROCYSTITES, Hall, 1852, Pal. N. Y., vol. 2. [Ety. *heteros*, irregular; *kustis*, a bladder.]
armatus, Hall, 1852, Pal. N. Y., vol. 2, Niagara Gr. [Sig. having arms.]

HOLOCYSTITES, Hall, 1864, in 20th Reg. Rep. [Ety. *holos*, entire; *kustis*, a bladder.]
abnormis, Hall, 1864, in 20th Reg. Rep., Niagara Gr. [Sig. abnormal, out of the usual order.]
alternatus, Hall, 1861, (Caryocystites alternatus) Rep. of Prog. Geo. Sur. Wis., Niagara Gr. [Ety. from the alternating plates.]
cylindricus, Hall, 1861, (Caryocystites cylindricus) Ann. Rep. Geo. Wis., Niagara Gr. [Sig. cylindrical.]
ovatus, Hall, 1864, in 20th Reg. Rep., Niagara Gr. [Sig. egg-shaped.]
scutellatus, Hall, 1864, in 20th Reg. Rep., Niagara Gr. [Sig. a salver or waiter of a nearly square form.]
sphæricus, Winchell & Marcy, 1865, Mem. Bost. Soc. Nat. Hist., vol. 1, Niagara Gr. [Sig. round-bodied.]
winchelli, Hall, 1864, in 20th Reg. Rep., Niagara Gr. [Ety. proper name.]

HOMOCRINUS, Hall, 1852, Pal. N. Y., vol. 2. [Ety. *homos*, like; *krinon*, a lily.]
alternatus, Hall, 1847, (Poteriocrinus alternatus) Pal. N. Y., vol. 1, Trenton Gr. [Sig. from the alternating plates.]

angustus, Meek & Worthen, 1870, Proc. Acad. Nat. Sci. Phil., Cin'ti Gr. [Sig. narrowed, contracted.]

cylindricus, Hall, 1852, Pal. N. Y., vol. 2, Niagara Gr. [Sig. cylindrical.]

gracilis, Hall, 1847, (Poteriocrinus gracilis) Pal. N. Y., vol. 1, Trenton Gr. [Sig. slender.]

parvus, Hall, 1852, Pal. N. Y., vol. 2, Niagara Gr. [Sig. small.]

polydactylus, see Dendrocrinus polydactylus.

proboscidialis, Hall, 1859, Pal. N. Y., vol. 3, Oriskany sandstone. [Ety. from the proboscis.]

scoparius, Hall, 1859, Pal. N. Y., vol. 3, Low. Held. Gr. [Sig. a sweeper.]

HYBOCRINUS, Billings, 1857, Rep. of Progr. [Ety. *hubos*, hump-backed; *krinon*, a lily.]

conicus, Billings, 1857, Rep. of Progr., Trenton Gr. [Sig. conical.]

pristinus, Billings, 1859, Decade 4, Chazy Gr. [Sig. primitive, early.]

tumidus, Billings, 1857, Rep. of Progr., Trenton Gr. [Sig. tumid, swollen out.]

Hydreionocrinus (?) *verrucosus*, White & St. John, 1868, Trans. Chi. Acad. Sci. See Eupachycrinus verrucosus.

Hypanthocrinites, Phillips, 1839, Murch. Sil. Syst. [Ety. *upo*, under; *anthos*, a flower; *krinon*, a lily.]

cælatus, see Eucalyptocrinus cælatus.

decorus, see Euclayptocrinus decorus.

ICHTHYOCRINUS, Conrad, 1842, Jour. Acad. Nat. Sci. Phil., vol. 8. [Ety. *ichthys*, a fish; *krinon*, a lily.]

burlingtonensis, Hall, 1858, Geo. Rep. Iowa, Burlington Gr. [Ety. proper name.]

clintonensis, Hall, 1852, Pal. N. Y., vol. 2, Clinton Gr. [Ety. proper name.]

corbis, Winchell & Marcy, 1865, Mem. Bost. Soc. Nat. Hist., Niagara Gr. [Sig. a basket.] Prof. Hall regards this as a syn. for I. subangularis.

laevis, Conrad, 1842, Jour. Acad. Nat. Sci., vol. 8, Niagara Gr. [Sig. smooth.]

subangularis, Hall, 1862, Trans. Alb. Inst., vol. 4, Niagara Gr. [Sig. somewhat angular.]

tiariformis, Troost, 1850, (Cyathocrinus tiariformis) Catal. Low. Carb. [Sig. like a turban.]

Icosidactylocrinites. Not defined.

LAMPTEROCRINUS, Roemer, 1860, Sil. Fauna West Tenn. [Ety. *lampter*, a lamp; *krinon*, a lily.]

inflatus, Hall, 1861, (Balanocrinus inflatus) Rep. of Progr. Sur. of Wis., Niagara Gr. [Sig. inflated.]

sculptus, syn. for L. tennesseensis.

tennesseensis, Roemer, 1860, Sil. Fauna West Tenn., Niagara Gr. [Ety. proper name.]

LECANOCRINUS, Hall, 1852, Pal. N. Y., vol. 2. [Ety. *lekane*, a basin, from the bowl-shaped form of the calyx in the typical species; *krinon*, a lily.]

caliculus, Hall, 1852, Pal. N. Y., vol. 2, Niagara Gr. [Sig. a little cup.]

elegans, Billings, 1857, Rep of Progr., Trenton Gr. [Sig. elegant.]

lævis, Billings, 1857, Rep. of Progress, Trenton Gr. [Sig. smooth.]

macropetalus, Hall, 1852, Pal. N. Y., vol. 2, Niagara Gr. [Sig. having long-flower leaves; from the wide scapular plates.]

ornatus, Hall, 1852, Pal. N. Y., vol. 2, Niagara Gr. [Sig. adorned.]

pusillus, syn. for Cyathocrinus pusillus.

simplex, Hall, 1852, Pal. N. Y., vol. 2, Niagara Gr. [Sig. simple.]

LEPADOCRINUS, Conrad, 1840, Ann. Rep. N. Y. [Ety. from the resemblance to the Lepas or Barnacle *Anatifa*; *krinon*, a lily.]

gebhardi, Conrad, 1840, Ann. Rep. N. Y., Low. Held. Gr. [Ety. proper name.]

moorii, Meek, 1871, (Lepocrinites moorii) Am. Jour. Sci., 3rd ser., vol. 2, Cin'ti Gr. [Ety. proper name.] I think this species is not generically distinct from Echino-encrinites anatiformis.

LEPIDECHINUS, Hall, 1861, Desc. New Spec. Crinoidea. [Ety. *lepis*, scale; *echinus*, the sea urchin.]

imbricatus, Hall, 1861, Desc. New Crinoidea, Burlington Gr. [Sig. imbricated.]

rarispinus, Hall, 1867, 20th Reg. Rep., Chemung Gr. [Sig. having few spines.]

LEPIDESTHES, Meek & Worthen, 1868, Geo. Sur. Ill., vol. 3. [Ety. *lepis*, a scale; *esthes*, a garment.]

coreyi, Meek & Worthen, 1868, Geo. Sur. Ill., vol. 3, Keokuk Gr. [Ety. proper name.]

Lepidocidaris, Meek & Worthen, 1873, Geo. Sur. Ill., vol. 5. A proposed subgenus of Eocidaris.

Lepidodiscus, Meek & Worthen, 1873, Geo. Sur. Ill., vol. 3. [Ety. *lepis*, a scale; *diskos*, a quoit.] A proposed subgenus of Agelacrinus.

Lepocrinites, Conrad, 1840. The correct orthography seems to be Lepadocrinus.

moorei, Meek, see Lepadocrinus moorii.

LICHENOCRINUS, Hall, 1866, Pamphlet. [Ety. *lichen*, a tree moss; *krinon*, a lily.]

crateriformis, Hall, 1866, Pamphlet, Trenton & Hud. Riv. Gr. [Sig. having the form of a cup.]

dyeri, Hall, 1866, Phamphlet, Cin'ti Gr. [Ety. proper name.]

tuberculatus, S. A. Miller, 1874, Cin. Quar. Jour. Sci., vol. 1, Cin'ti Gr. [Ety. from the tuberculated plates.]

ECHINODERMATA.

LYRIOCRINUS, Hall, 1852, Pal. N. Y., vol. 2. [Ety. *lyrion*, small lyre, a musical instrument; *krinon*, a lily.]
dactylus, Hall, 1843, (Marsupiocrinites (?) dactylus) Geo. Rep. 4th Dist. N. Y., Niagara Gr. [Sig. fingered.]
sculptilis, Hall, 1864, 20th Reg. Rep., Niagara Gr. [Sig. produced by carving.]

MACROSTYLOCRINUS, Hall, 1852, Pal. N. Y., vol. 2. [Ety. *makros*, long; *stylos*, an arm; *krinon*, a lily.]
lævis, Roemer, 1860, (Cytocrinus lævis) Sil. Faun. West Tenn., Niagara Gr. [Sig. smooth.]
ornatus, Hall, 1852, Pal. N. Y., vol. 2, Niagara Gr. [Sig. adorned.]
striatus, Hall, 1868, Trans. Alb. Inst., vol. 4, Niagara Gr. [Sig. striated.]

MALOCYSTITES, Billings, 1858, Can. Org. Rem. Decade 3. [Ety. *malum*, an apple; *kustis*, a bladder.)
barrandi, Billings, 1858, Can. Org. Rem., Decade 3, Chazy Gr. [Ety. proper name.]
murchisoni, Billings, 1858, Can. Org. Rem., Decade 3, Chazy Gr. [Ety. proper name.]

MARIACRINUS, Hall, 1859, Pal. N. Y., vol. 3. [Ety. *Maria*, Mary—a proper name; *krinon*, a lily.] See also subgenus Technocrinus.
macropetalus, Hall, 1859, Pal. N. Y., vol. 3, Low. Held. Gr. [Sig. having long flower leaves; from the wide basal and radial plates.]
nobilissimus, Hall, 1859, Pal. N. Y., vol. 3, Low. Held. Gr. [Sig. most remarkable.]
pachydactylus, Conrad, 1841, (Astrocrinites pachydactylus) Ann. Rep. N. Y., Low. Held. Gr. [Sig. thick-fingered.] Syn. (?) for M. polydactylus.
paucidactylus, Hall, 1859, Pal. N. Y., vol. 3, Low. Held. Gr. [Sig. few-fingered.]
plumosus, Hall, 1859, Pal. N. Y., vol. 3, Low. Held. Gr. [Sig. feathery.]
polydactylus, Bonny, 1837, (Actinocrinus polydactylus) Am. Jour., vol. 31, Up. Sil. [Sig. many-fingered.]
ramosus, Hall, 1859, Pal. N. Y., vol. 3, Low. Held. Gr. [Sig. branching.]
stoloniferus, Hall, 1859, Pal. N. Y., vol. 3, Low. Held. Gr. [Sig. bearing branches.]

Marsupiocrinites, Phillips, 1839, Murch., Sil. Syst. [Sig. a purse or bag; *krinon*, a lily.]
dactylus, see Lyriocrinus dactylus.

MEGISTOCRINUS, Owen & Shumard, 1852, Geo. Sur. Wis., Iowa and Minn. [Ety. *megistos*, very great; *krinon*, a lily.]
abnormis, Lyon, 1857, (Actinocrinus abnormis) Geo. Sur. Ky., vol. 3, Up. Held. Gr. [Sig. abnormal.]
crassus, White, 1862, Proc. Bost. Soc. Nat. Hist., vol. 9, Burlington Gr. [Sig. thick.]
depressus, Hall, 1862, 15th Reg. Rep. N. Y., Hamilton Gr. [Sig. depressed.]
evansi, Owen & Shumard, 1852, Geo. Sur. Wis., Iowa and Minn., Burlington Gr. [Ety. proper name.]
farnsworthi, White, 1876, Proc. Acad. Nat. Sci., Devonian. [Ety. proper name.]
infelix, Winchell & Marcy, Mem. Bost. Soc. Nat. Hist., Niagara Gr. Prof. Hall regards this species as the young of Saccocrinus christyi. [Sig. miserable.]
knappi, Lyon & Casseday, 1861, Proc. Acad. Nat. Sci. Phil., Up. Held. Gr. [Ety. proper name.]
latus, Hall, 1858, Geo. of Iowa, Ham. Gr. [Sig. broad.]
marcouanus, Winchell & Marcy, see Saccocrinus christyi.
necis, Winchell & Marcy, 1865, Mem. Bost. Soc. Nat. Hist., Niagara Gr. [Sig. doubtful.] Prof. Hall regards this species as a synonym for Saccocrinus christyi.
olliculus, Hall, 1861, Desc. New Crinoidea, Burlington Gr. [Sig. a little pot.]
ontario, Hall, 1862, 15th Reg. Rep. N. Y., Ham. Gr. [Ety. proper name.]
parvirostris, Meek & Worthen, 1869, Proc. Acad. Nat. Sci. Phil., Burlington Gr. [Sig. small-beaked.]
plenus, White, 1862, Proc. Bost. Soc. Nat. Hist., vol. 9, Burlington Gr. [Sig. full, large.]
rugosus, Lyon & Casseday, 1859, Am. Jour. Sci., vol. 28, Up. Held. Gr. [Sig. rugose.]
spinosus, Lyon, 1861, Proc. Acad. Nat. Sci. Phil., Up. Held. Gr. [Sig. covered with spines.]
spinulosus, Lyon, 1861, Jour. Acad. Nat. Sci. Phil., Up. Held. Gr. [Sig. covered with little spines.]
whitii, Hall, 1861, Jour. Bost. Soc. Nat. Hist., vol. 7, Burlington Gr. [Ety. proper name.]

MELOCRINUS, Goldfuss, 1826, Petref. Germ. [Ety. *melo*, a melon; *krinon*, a lily.]
bainbridgensis, see Ctenocrinus bainbridgensis.
breviradiatus, see Ctenocrinus breviradiatus.
nodosus, Hall, 1861, Geo. Rep. Wis., Devonian. [Ety. *nodosus*, knobbed.]
obconicus, Hall, 1863, Trans. Alb. Inst., vol. 4, Niagara Gr. [Sig. inversely conical.]
pratteni, McChesney, 1860, (Forbesocrinus pratteni) New Pal. Foss., Warsaw Gr. [Ety. proper name.]
sculptus, Hall, 1852, Pal. N. Y., vol. 2, Niagara Gr. [Sig. engraved, sculptured.]
verneuili, Troost, 1850, (Actinocrinus verneuili) Proc. Am. Assoc. Sci., Niagara Gr. [Ety. proper name.]

MELONITES, Owen & Norwood, 1846, Am. Jour. Sci., 2d series, vol. 2. [Ety. *melon*, a melon; *lithos*, stone.]
danæ, see Oligoporus danæ.
multiporus, Owen & Norwood, 1846, Am. Jour. Sci., 2d series, vol. 2, St. Louis Gr. [Sig. having many pores.]
stewarti, Safford, 1869, Geo. of Tennessee, Low. Carb. [Ety. proper name.]
MESPILOCRINUS, DeKoninck & LeHon, 1854, Rech. Crin., Terr. Carb. Belg. [Ety. *mespilum*, a medlar; *krinon*, a lily.]
konincki, Hall, 1860, Supp. to Geo. Rep. Iowa, Burlington Gr. [Ety. proper name.]
scitulus, Hall, 1861, Desc. New Crinoidea, Burlington Gr. [Sig. handsome, pretty.]
MYELODACTYLUS, Hall, 1852, Pal. N. Y., vol. 2. [Ety. *myelos*, the inside pith, from the foramen or medullary canal penetrating the column of joints; *dactylus*, a finger.]
brachiatus, Hall, 1852, Pal. N. Y., vol. 2, Niagara Gr. [Sig. having arms.]
convolutus, Hall, 1852, Pal. N. Y., vol. 2, Niagara Gr. [Sig. convoluted, rolled as it were together.]
MYRTILLOCRINUS, Sandberger, 1856, Verst. der Rhein. Schi. Syst. in Nassau. [Ety. *myrtillus*, a myrtle; *krinon*, a lily.]
americanus, Hall, 1862, 15th Reg. Rep. N. Y., Up. Held. [Ety. proper name.]
NIPTEROCRINUS, Wachsmuth, 1868, Proc. Acad. Nat. Sci. Phil. [Ety. *nipter*, a washing vessel; *krinon*, a lily.]
arboreus, Worthen, 1863, Geo. Sur., Ill., vol. 5, Burlington Gr. [Sig. branched like a tree.]
wachsmuthi, Meek & Worthen, 1868, Proc. Acad. Nat. Sci. Phil., Burlington Gr. [Ety. proper name.]
NUCLEOCRINUS, Conrad, 1842, Jour. Acad. Nat. Sci. Phil., vol. 8. [Ety. *nucleus*, a little nut; *krinon*, a lily.]
angularis, Lyon, 1857, (Olivanites angularis) Geo. Sur. Ky., vol. 3, Corniferous Gr. [Sig. angular.]
conradi, Hall, 1862, 15th Reg. Rep. N. Y., Up. Held. Gr. [Ety. proper name.]
elegans, Conrad, 1842, Jour. Acad. Nat. Sci. Phil., Ham. Gr. [Sig. elegant.]
hallii, syn. for Nucleocrinus elegans.
kirkwoodensis, Shumard, 1863, Trans. St. Louis Acad. Sci., St. Louis Gr. [Ety. proper name.]
lucina, Hall, 1862, 15th Reg. Rep. N. Y., Ham. Gr. [Ety. mythological name.]
verneuili, Troost, 1841, (Pentremites verneuili) 6th Rep. on the Geo. of Tenn., Corniferous Gr. [Ety. proper name.]
OLIGOPORUS, Meek & Worthen, 1860, Proc. Acad. Nat. Sci. Phil. [Ety. *oligos*, few; *poros*, a passage.]
danæ, Meek & Worthen, 1860, (Melonites danæ) Proc. Acad. Nat. Sci. Phil., Keokuk Gr. [Ety. proper name.]

nobilis, Meek & Worthen, 1869, Proc. Acad. Nat. Sci. Phil., Burlington Gr. [Sig. remarkable.]
Olivanites, syn. for Nucleocrinus.
angularis, see Nucleocrinus angularis.
verneuili, see Nucleocrinus verneuili.
ONYCHASTER, Meek & Worthen, 1868, Geo. Sur. Ill., vol. 3. [Ety. *onyx*, a claw; *aster*, a star.]
barrisi, Hall, 1861, (Protaster barrisi) Desc. New Crinoidea, Burlington Gr. [Ety. proper name.]
flexilis, Meek & Worthen, 1868, Geo. Sur. Ill., vol. 3, Keokuk Gr. [Sig. pliant.]
ONYCHOCRINUS, Lyon & Casseday, 1859, Am. Jour. Sci., 2d series, vol. 29. [Ety. *onyx*, a claw; *krinon*, a lily.]
diversus, Meek & Worthen, 1860, Proc. Acad. Nat. Sci. Phil., Burlington Gr. [Sig. separated.]
exculptus, Lyon & Casseday, 1859, Am. Jour. Sci., Keokuk Gr. [Sig. carved out.]
magnus, Worthen, 1875, Geo. Sur. Ill., vol. 6, St. Louis Gr. [Sig. large.]
meeki, Hall, 1858, (Forbesiocrinus meeki) Geo. Sur. Iowa, Chester Gr. [Ety. proper name.]
monroensis, Meek & Worthen, 1861, (Forbesiocrinus monroensis) Proc. Acad. Nat. Sci., Keokuk Gr. [Ety. proper name.]
norwoodi, Meek & Worthen, 1860, (Forbesiocrinus norwoodi) Proc. Acad. Nat. Sci., Keokuk Gr. [Ety. proper name.]
whitfieldi, 1858, (Forbesiocrinus whitfieldi) Geo. Sur. Iowa, Chester Gr. [Ety. proper name.]
PACHYCRINUS, Billings, 1859, Decade 4. [Ety. *pachys*, thick; *krinon*, a lily.]
crassibasalis, Billings, 1859, Decade 4, Chazy Gr. [Ety. from the thick basal plates.]
PALÆASTER, Hall, 1852, Pal. N. Y., vol. 2. [Ety. *palaios*, ancient; *aster*, a star.]
antiquatus, Locke, 1846, (Asterias antiquata) Proc. Acad. Nat. Sci. Phil., vol. 3, Cin'ti Gr. [Sig. ancient.]
antiquus, Troost, 1835, (Asterias antiqua) Trans. Geo. Soc. Penn., vol. 1, Hud. Riv. Gr. [Sig. ancient.]
dyeri, Meek, 1872, Am. Jour. Sci., 3rd series, vol. 3, Cin'ti Gr. [Ety. proper name.]
eucharis, Hall, 1868, 20th Reg. Rep., Ham. Gr. [Sig. very graceful.]
granulosus, Hall, 1868, 20th Reg. Rep., Cin'ti Gr. [Sig. granular.]
incomptus, Meek, 1872, Am. Jour. Sci., 3rd ser., vol. 3, Cin'ti Gr. [Sig. unadorned.]
jamesi, Dana, 1863, (Palæasterina (?) jamesi) Am. Jour. of Sci., 2d series, vol. 35, Cin'ti Gr. [Ety. proper name.]
matutinus, Hall, 1847, (Asterias matutina) Pal. N. Y., vol. 1, Trenton Gr. [Sig. in the morning.]

parviusculus, Billings, 1860, Can. Nat. & Geo., vol. 5, Mid. Sil. [Sig. very small.]
niagarensis, Hall, 1852, Pal. N. Y., vol. 2, Niagara Gr. [Ety. proper name.]
pulchellus, see Stenaster pulchellus.
shafferi, Hall, 1868, 20th Reg. Rep., Cin'ti Gr. [Ety. proper name.]
wilberianus, Meek & Worthen, 1861, (Petraster wilberianus) Proc. Acad. Nat. Sci. Phil., Cin'ti Gr. [Ety. proper name.]

PALÆASTERINA, McCoy, 1851, Brit. Pal. Foss., but first defined by Salter, 1857, Ann. Mag. Nat. Hist. [Ety. *palaios*, ancient; *aster*, a star; *inus*, resemblance.]
fimbriata, see Schœnaster fimbriatus.
jamesi, see Palæaster jamesi.
rigida, see Petraster rigidus.
rugosa, Billings, 1857, Rep. of Progr., Hud. Riv. Gr. [Sig. rugose.]
stellata, Billings, 1857, Rep. of Prog., Trenton Gr. [Sig. star-shaped.]

PALÆCHINUS, McCoy, 1844, Carb. Foss., Ireland. [Ety. *palaios*, ancient; *echinus*, the sea urchin.]
burlingtonensis, Meek & Worthen, 1860, Proc. Acad. Nat. Sci. Phil., Burlington Gr. [Ety. proper name.]
gracilis, Meek & Worthen, 1869, Proc. Acad. Nat. Sci. Phil., Burlington Gr. [Sig. slender.]

PALÆOCOMA, Salter, 1857, Ann. & Mag. Nat. Hist., 2d series, vol. 17. [Ety. *palaios*, ancient; *coma*, hair.]
cylindrica, see Tæniaster cylindricus.
princeps, Hall, 1868,* (Ptilonaster princeps) 20th Reg. Rep. Chemung Gr. [Sig. early, ancient.]
spinosa, see Tæniaster spinosus.

PALÆOCRINUS, Billings, 1859, Decade 4. [Ety. *palaios*, ancient; *krinon*, a lily.]
angulatus, Billings, 1857, (Dendrocrinus angulatus) Rep. of Progr., Trenton Gr. [Sig. cornered, angular.]
pulchellus, Billings, 1859, Can. Org. Rem., Decade 4, Trenton Gr. [Sig. minutely pretty.]
rhombiferous, Billings, 1859, Can. Org. Rem., Decade 4, Trenton Gr. [Sig. bearing rhombs.]
striatus, Billings, 1859, Can. Org. Rem., Decade 4., Chazy Gr. [Sig. striated.]
sulcatus, Safford, 1869, Geo. of Tenn. Not defined.

PALÆOCYSTITES, Billings, 1858, Can. Org. Rem., Decade 3. [Ety. *palaios*, ancient; *kustis*, a bladder.]
chapmani, Billings, 1858, Can. Org. Rem., Decade 3, Chazy Gr. [Ety. proper name.]
dawsoni, Billings, 1858, Can. Org. Rem., Decade 3, Chazy Gr. [Ety. proper name.]
pulcher, Billings, 1859, Can. Nat. Geo., vol. 4, Chazy Gr. [Sig. beautiful.]

tenuiradiatus, Hall, 1847, (Actinocrinus tenuiradiatus) Pal. N. Y., vol. 1, Chazy Gr. [Sig. slender-rayed.]

Pentacrinites hamptoni, Emmons, 1842, Geo. Rep. N. Y., Trenton Gr. This is merely the plate of a crinoid column.

Pentagonites, proposed by Rafinesque for a crinoid column.

PENTREMITES, Say, 1820, Am. Jour. Sci., vol. 2. [Ety. *pente*, five; *remos*, a board or plate.]
angularis, Lyon, 1860, Trans. St. Louis Acad. Sci., Low. Carb. [Sig. angular.]
bipyramidalis, Hall, 1858, Geo. Rep. Iowa, Keokuk Gr. [Sig. doubly-pyramidal.]
burlingtonensis, Meek & Worthen, 1869, Proc. Acad. Nat. Sci. Phil., Burlington Gr. [Ety. proper name.]
calyce, Hall, 1862, 15th Reg. Rep. N. Y., Ham. Gr. [Sig. a cup.]
calycinus, Lyon, 1860, Trans. St. Louis Acad. Sci., Kaskaskia Gr. [Sig. a small cup.]
cervinus, Hall, 1858, Geo. Rep. Iowa, Kaskaskia Gr. [Sig. deer-like.]
cherokeus, Troost, 1850, Catal. Proc. Am. Assoc. Ad. Sci., Kaskaskia Gr. [Ety. proper name.]
conoideus, Hall, 1856, Trans. Alb. Inst., vol. 4, Warsaw Gr. [Sig. shaped like a cone.]
cornutus, see Granatocrinus cornutus.
curtus, see Granatocrinus curtus.
decussatus, Shumard, 1858, Trans. St. Louis Acad. Sci., Low. Carb. [Sig. cross marked—arranged in pairs that alternately cross each other.]
elegans, Lyon, 1860, Trans. St. Louis Acad. Sci., Kaskaskia Gr. [Sig. elegant.]
elongatus, Shumard, 1855, Geo. Rep. Mo., Encrinital limestone. [Sig. lengthened.]
florealis, Schlotheim, 1820, syn. for P. godoni.
globosus, Say, as identified by Troost, 1850, probably Pentremites sulcatus.
godoni, DeFrance, 1818, Dict. Sci. Nat., Kaskaskia Gr. [Ety. proper name.]
granulatus, see Granatocrinus granulatus.
grosvenori, Shumard, 1858, Trans. St. Louis Acad. Sci., Warsaw Gr. [Ety. proper name.]
koninckianus, Hall, 1856, Trans. Alb. Inst., vol. 4, Warsaw Gr. [Ety. proper name.]
laterniformis, Owen & Shumard, 1850, Jour. Acad. Nat. Sci., 2d series, vol. 2, Chester Gr. [Sig. lantern-formed.]
leda, Hall, 1862, 15th Reg. Rep. N. Y., Ham. Gr. [Ety. mythological name.]
lineatus, Shumard, 1858, Trans. St. Louis Acad. Sci., Burlington Gr. [Sig. marked with lines.]
longicostalis, Hall, 1860, Supp. to Geo. Iowa, Warsaw Gr. [Sig. long-ribbed.]
lycorias, Hall, 1862, 15th Reg. Rep. N. Y., Ham. Gr. [Ety. mythological name.]

maia, Hall, 1862, 15th Reg. Rep. N. Y., Ham. Gr. [Ety. mythological name.]
melo, see Granatocrinus melo.
missouriensis, Swallow, 1863, Trans. St. Louis Acad. Sci., Low. Carb. [Ety. proper name.]
norwoodi, see Granatocrinus norwoodi.
obesus, Lyon, 1857, Geo. Sur. Ky., vol. 3, Kaskaskia Gr. [Sig. fat, plump in form.]
obliquatus, syn. for P. laterniformis.
ovalis, Owen. Not defined.
pyriformis, Say, 1825, Jour. Acad. Nat. Sci. Phil., vol. 4, Kaskaskia Gr. [Sig. from the pyriform body.]
reinwardti, Troost, 1835, Trans. Geo. Soc. Penn., vol. 1, Niagara Gr. [Ety. proper name.]
robustus, Lyon, 1860, Trans. St. Louis Acad. Sci., Low. Carb. [Sig. robust.]
roemeri, see Granatocrinus roemeri.
sayi, see Granatocrinus sayi.
sirius, White, 1862, Proc. Bost. Soc. Nat. Hist., vol. 9, Burlington Gr. [Ety. sirius, the dog star.]
stelliformis, see Codonites stelliformis.
subconoideus, Meek, 1872, Hayden's Geo. Rep. Low. Carb. [Sig. somewhat cone-like.]
subcylindricus, Hall & Whitfield, 1875, Ohio Pal., vol. 2, Niagara Gr. [Sig. somewhat cylindrical.]
subtruncatus, Hall, 1858, Geo. of Iowa, Ham. Gr. [Sig. somewhat truncated.]
sulcatus, Roemer, 1852, Monog. Blastoid., Kaskaskia Gr. [Sig. furrowed.]
symmetricus, Hall, 1858, Geo. Rep. Iowa, Kaskaskia Gr. [Sig. symmetrical.]
tennesseex, Troost. Not defined.
troosti, Shumard, 1866, Trans. St. Louis Acad. Sci., Kaskaskia Gr. [Ety. proper name.]
truncatus, Conrad, 1843, Proc. Acad. Nat. Sci. Phil., Warsaw Gr. [Sig. truncated.]
versouviensis, Worthen, 1875, Geo. Sur. Ill., vol. 6, St. Louis Gr. [Ety. proper name.]
verneuili, see Nucleocrinus verneuili.
whitii, Hall, 1862, 15th Reg. Rep. N. Y., Ham. Gr. [Ety. proper name.]
woodmani, Meek & Worthen, 1868, Proc. Acad. Nat. Sci. Phil., Keokuk Gr. [Ety. proper name.]
wortheni, Hall, 1858, Geo. Rep. Iowa, Keokuk Gr. [Ety. proper name.]
PETRASTER, Billings, 1858, Can. Org. Rem., Decade 3. [Ety. petros, a stone; aster, a star.] Prof. Hall regards this genus as a synonym for Palæaster.
bellulus, Billings, 1865, Pal. Foss., vol. 1, Niagara Gr. [Sig. pretty.]
rigidus, Billings, 1857, (Palæasterina rigidus) Rep. of Progr., Trenton Gr. [Sig. rigid.]
wilberianus, see Palæaster wilberianus.
Philocrinus, Koninck, 1863. [Ety. philos, favorite; krinon, a lily.]

pelvis, Meek & Worthen, 1865, Am. Jour. Sci., 2nd series, vol. 39. Syn. for Erisocrinus typus.
PHOLIDOCIDARIS, Meek & Worthen, 1869, Proc. Acad. Nat. Sci. Phil. [Ety. pholidos, a scale; kideris, a turban.]
irregularis, Meek & Worthen, 1869, Proc. Acad. Nat. Sci. Phil., Keokuk Gr. [Sig. irregular.]
PHYSETOCRINUS, Meek & Worthen, 1869, Proc. Acad. Nat. Sci. Phil. [Ety. physetos, blown up, inflated; krinon, a lily.]
asper, Meek & Worthen, 1869, Proc. Acad. Nat. Sci. Phil., Burlington Gr. [Sig. rough.]
dilatatus, Meek & Worthen, 1869, (Strotocrinus dilatatus) Proc. Acad. Nat. Sci. Phil., Burlington Gr. [Sig. spread out.]
subventricosus, McChesney, 1860, (Actinocrinus subventricosus) Desc. New Pal. Foss., Burlington Gr. [Sig. somewhat ventricose.]
PLATYCRINUS, Miller, 1821, Nat. Hist., Crinoidea. [Ety. platys, flat; krinon, lily.]
æqualis, Hall, 1861, Desc. New Crin., Burlington Gr. [Sig. equal.]
americanus, Owen & Shumard, 1852, Jour. Acad. Nat. Sci., 2d ser., vol. 2, Burlington Gr. [Ety. proper name.]
anndixoni, Troost. Not defined.
asper, Meek & Worthen, 1861, Proc. Acad. Nat. Sci. Phil., Burlington Gr. [Sig. rough.]
bedfordensis, Hall & Whitfield, 1875, Ohio Pal., vol. 2, Erie shales. [Ety. proper name.]
brevinodus, Hall, 1861, Desc. New Crinoidea, Keokuk Gr. [Sig. shortknotted.]
burlingtonensis, Owen & Shumard, 1850, Jour. Acad. Nat. Sci., 2d ser., vol. 2, Burlington Gr. [Ety. proper name.]
calyculus, Hall, 1861, Desc. New Crinoidea, Burlington Gr. [Sig. a little cup.]
canaliculatus, Hall, 1858, Geo. Rep. Iowa, Burlington Gr. [Sig. channeled, grooved.]
cavus, Hall, 1858, Geo. Rep. Iowa, vol. 1, pt. 2, Burlington Gr. [Sig. hollow, concave.]
clytis, Hall, 1861, Desc. New Crinoidea, Burlington Gr. [Sig. remarkable.]
contritus, Hall, 1863, Crin. Wav. Sands. Ohio, Waverly Gr. [Sig. worn out.]
corrugatus, Owen & Shumard, 1850, Jour. Acad. Nat. Sci. 2d ser., vol. 2, Burlington Gr. [Sig. corrugated, wrinkled.]
depressus, Owen. Not defined.
discoideus, Owen & Shumard, 1850, Jour. Acad. Nat. Sci., 2d series, vol. 2, Burlington Gr. [Sig. quoit-shaped, disclike.]
eboraceus, Hall, 18 2, 15th Reg. Rep. N. Y., Ham. Gr. [Ety. eboracum, Latin name for York.]

elegans, Hall, 1861, Desc. New Crinoidea, Burlington, Gr. [Sig. elegant.]
eminulus, Hall, 1861, Desc. New Crinoidea, Burlington Gr. [Sig. projecting a little.]
eriensis, Hall, 1862, 15th Reg. Rep. N. Y., Ham. Gr. [Ety. proper name.]
excavatus, Hall, 1861, Desc. New Crinoidea, Burlington Gr. [Sig. hollowed out.]
exsertus, Hall, 1858, Geo. Rep. Iowa, Burlington Gr. [Sig. projecting, thrust forth.]
georgii, Hall, 1860, Supp. to Iowa Geo. Rep., Warsaw, Gr. [Ety. proper name.]
glyptus, Hall, 1861, Desc. New Crinoidea, Burlington, Gr. [Sig. sculptured.]
graphicus, Hall, 1863, Crin. Wav. Sands. Ohio, Waverly Gr. [Sig. written on.]
halli, Shumard, 1866, Trans. St. Louis Acad. Sci., vol. 2, Burlington Gr. [Ety. proper name.]
haydeni, Meek, 1872, Hayden's Geo. Rep., Low. Carb. [Ety. proper name.]
hemisphericus, Meek & Worthen, 1865, Proc. Acad. Nat. Sci., Phil., Keokuk Gr. [Sig. hemispherical.]
huntsvillæ, Troost. Not defined.
incomptus, White, 1863, Jour. Bost. Soc. Nat. Hist., vol. 7, Burlington Gr. [Sig. unadorned.]
inornatus, syn. for P. burlingtonensis.
insculptus, Troost. Not defined.
leai, Lyon, 1860, Trans. Am. Phil. Soc., Low. Held. Gr. [Ety. proper name.]
lodensis, Hall & Whitfield, 1875, Ohio Pal., vol. 2, Waverly Gr. [Ety. proper name.]
montanaensis, Meek, 1871, Hayden's Geo. Rep. Low. Carb. [Ety. proper name.]
multibrachiatus, Meek & Worthen, 1861, Proc. Acad. Nat. Sci. Phil., Warsaw Gr. [Sig. many-armed.]
niotensis, Meek & Worthen, 1865, Proc. Acad. Nat. Sci. Phil., Keokuk Gr. [Ety. proper name.]
nodobrachiatus, Hall, 1858, Geo. Rep. Iowa, Burlington Gr. [Sig. having knotty arms.]
nodulosus, Hall, 1858, Geo. Rep. Iowa, Burlington Gr. [Sig. covered with small knots.]
nucleiformis, Hall, 1858, Geo. Rep. Iowa, Burlington Gr. [Sig. in the shape of a kernel.]
olla, Hall, 1861, Desc. New Crinoidea, but the name was preoccupied and Shumard proposed for the species, P. halli.
ornogranulus, McChesney, Syn. for P. americanus.
oweni, Meek & Worthen, 1861, Proc. Acad. Nat. Sci., Phil., Burlington Gr. [Ety. proper name.]
papillatus, Hall, 1861, Desc. New Crinoidea, Burlington Gr. [Sig. shaped like a bud.]

parvinodus, Hall, 1861, Desc. New Crinoidea, Burlington Gr. [Sig. having small knots.]
parvulus, Meek & Worthen, 1865, Proc. Acad. Nat. Sci. Phil.; Chester Gr. [Sig. very small.]
parvus, Hall, 1859, Pal. N. Y., vol. 3, Low. Held. Gr. [Sig. small.]
penicillus, Meek & Worthen, 1860, Proc. Acad. Nat. Sci. Phil., St. Louis Gr. [Sig. a small brush.]
perasper, Meek & Worthen, 1865, Proc. Acad. Nat. Sci. Phil., Burlington Gr. [Sig. very rough.]
pileiformis, Hall, 1858, Geo. Rep. Iowa, vol. 1, pt. 2, Burlington Gr. [Sig. like a small cap.]
planus, Owen & Shumard, 1850, Jour. Acad. Nat. Sci. Phil., 2d ser., vol. 2, Burlington Gr. [Sig. plain, level.]
plenus, Meek & Worthen, 1860, Proc. Acad. Nat. Sci., Phil., St. Louis Gr. [Sig. full, large.]
pleurovimineus, White, 1862, Proc. Bost. Soc. Nat. Hist., vol. 9, Burlington Gr. [Ety. *pleuron*, the side; *vimineus*, made of wicker work.]
plumosus, Hall, 1859, Pal. N. Y., vol. 3, Low. Held. Gr. [Sig. feathery.]
pocilliformis, Hall, 1858, Geo. Rep. Iowa, vol. 1, pt. 2, Burlington Gr. [Sig. like a little cup.]
polydactylus, Troost. Not defined.
præmaturus, Hall & Whitfield, 1875, Ohio Pal., vol. 2, Niagara Gr. [Sig. præmature.]
prattenianus, Meek & Worthen, Sept. 1860, Proc. Acad. Nat. Sci., Phil., St. Louis Gr. [Ety. proper name.]
pumilis, Hall, 1860, Supp. to Geo. Iowa, Warsaw Gr. [Sig. a dwarf.]
quinquenodus, White, 1862, Proc. Bost. Soc. Nat. Hist., vol. 9, Burlington Gr. [Sig. five-noded.]
ramulosus, Hall, 1859, Pal. N. Y., vol. 3, Low. Held. Gr. [Sig. full of branches.]
regalis, Hall, 1861, Desc. New Crinoidea, Burlington Gr. [Sig. royal, splendid.]
richfieldensis, Hall & Whitfield, 1875, Ohio Pal., vol. 2, Waverly Gr. [Ety. proper name.]
saffordi, Troost, 1850, Hall, 1858, Geo. Rep. Iowa, Keokuk Gr. [Ety. proper name.]
saræ, Hall, 1858, Geo. Rep. Iowa, St. Louis Gr. [Ety. proper name.]
scobina, Meek & Worthen, 1861, Proc. Acad. Nat. Sci., Phil., Burlington Gr. [Sig. a rasp.]
sculptus, Hall, 1858, Geo. Rep. Iowa, Burlington Gr. [Sig. engraved.]
shumardianus, Hall, 1858, Geo. Rep. Iowa, vol. 1, pt. 2, Burlington Gr. [Ety. proper name.]
striobrachiatus, Hall, 1861, Desc. New Crinoidea, Burlington Gr. [Sig. having striated arms.]

subspinosus, Hall, 1858, Geo. Rep. Iowa, Burlington Gr. [Sig. somewhat spiny.]
subspinulosus, Hall, 1860, Supp. to Geo. Iowa, Burlington Gr. [Sig. somewhat full of little spines.]
tennesseensis, Roemer, 1860, Sil. Fauna West Tenn., Niagara Gr. [Ety. proper name.]
tentaculatus, Hall, 1859, Pal. N. Y., vol. 3, Low. Held. Gr. [Sig. having feelers.]
tenuibrachiatus, Meek & Worthen, 1869, Proc. Acad. Nat. Sci. Phil., Burlington Gr. [Sig. slender-armed.]
truncatulus, Hall, 1858, Geo. Rep. Iowa, Burlington Gr. [Sig. somewhat truncated.]
truncatus, Hall, 1858, Geo. Rep. Iowa, Burlington Gr. [Sig. truncated, cut short.]
tuberosus, Hall, 1858, Geo. Rep. Iowa, Burlington Gr. [Sig. covered with protuberances.]
verrucosus, White, 1863, Jour. Bost. Soc. Nat. Hist., vol. 7, Burlington Gr. [Sig. covered with wart-like projections.]
wortheni, Hall, 1858, Geo. Rep. Iowa, vol. 1, pt. 2, Burlington Gr. [Ety. proper name.]
yandelli, Owen & Shumard, 1850, Jour. Acad. Nat. Sci., 2d ser., vol. 2, Burlington Gr. [Ety. proper name.]

PLEUROCYSTITES, Billings, 1854, Can. Jour., vol. 2. [Ety. *pleuron*, the side; *kustis*, a bladder.]
anticostiensis, Billings, 1857, Rep. of Progr. Hud. Riv. Gr. [Ety. proper name.]
elegans, Billings, 1857, Rep. of Progr., Trenton Gr. [Sig. elegant.]
exornatus, Billings, 1857, Rep. of Progr., Trenton Gr. [Sig. adorned.]
filitextus, Billings, 1854, Can. Jour., vol. 2, Trenton Gr. [Sig. woven like thread.]
robustus, Billings, 1854, Can. Jour., vol. 2, Trenton Gr. [Sig. robust.]
squamosus, Billings, 1854, Can. Jour., vol. 2, Trenton Gr. [Sig. scaly.]

POROCRINUS, Billings, 1857, Rep. of Progr. [Ety. from the poriferous areas similar to the pectinated rhombs of the cystidea.]
conicus, Billings, 1857, Rep. of Progr., Trenton Gr. [Sig. conical.]
crassus, Meek & Worthen, 1868, Geo. Sur. Ill., vol. 3, Cin'ti Gr. [Sig. thick.]
pentagonus, Meek & Worthen, 1865, Proc. Acad. Nat. Sci., Phil., Cin'ti Gr. [Sig. five-cornered.]

POTERIOCRINUS, Miller, 1821, Nat. Hist. Crinoidea. [Ety. *poterion*, a goblet; *krinon*, a lily.]
aequalis, Hall, 1860, Supp. to Geo. Iowa, Burlington Gr. [Sig. equal.]
alternatus, see Homocrinus alternatus.

barrisi, Hall, 1861, Desc. New Crinoidea, Burlington Gr. [Ety. proper name.]
bayensis, see Scaphiocrinus bayensis.
bisselli, Worthen, 1873, Geo. Sur. Ill., vol. 5, Chester Gr. [Ety. proper name.]
bursiformis, White, 1862, Proc. Bost. Soc. Nat. Hist., vol. 9, Burlington Gr. [Sig. purse-shaped.]
caduceus, see Dendrocrinus caduceus.
calyculus, Hall, 1858, Geo. Rep. Iowa, Burlington Gr. [Sig. a little cup.]
carbonarius, see Scaphiocrinus carbonarius.
carinatus, Meek & Worthen, 1861, Proc. Acad. Nat. Sci. Phil., Burlington Gr. [Sig. keeled.]
concinnus, Meek & Worthen, 1870, Proc. Acad. Nat. Sci. Phil., Keokuk Gr. [Sig. handsome.]
coreyi, Worthen, 1875, Geo. Sur. Ill., vol. 6, Keokuk Gr. [Ety. proper name.]
corycia, Hall, 1863, Crin. Wav. Sands. Ohio, Waverly Gr. [Ety. mythological name.]
crineus, Hall, 1863, Crin. Wav. Sands. Ohio, Waverly Gr. [Sig. hairy.]
cultidactylus, Hall, 1860, Supp. to Geo. Iowa, Burlington Gr. [Sig. elegantly fingered.]
cylindricus, Lyon, 1860, Trans. Am. Phil. Soc., vol. 13, Devonian. [Sig. cylindrical.]
decadactylus, see Scaphiocrinus decadactylus.
diffusus, Hall, 1862, 15th Reg. Rep. N. Y., Ham. Gr. [Sig. spread-out.]
dilatatus, see Coeliocrinus dilatatus.
divaricatus, Hall, 1860, Supp. to Geo. Iowa, Warsaw Gr. [Sig. spread-out.]
enormis, see Cyathocrinus enormis.
florealis, see Zeacrinus florealis.
fusiformis, Hall, 1861, Desc. New Crinoidea, Burlington Gr. [Sig. spindle-shaped.]
gracilis, see Homocrinus gracilis.
hardinensis, Worthen, 1873, Geo. Sur. Ill., vol. 5, St. Louis Gr. [Ety. proper name.]
hemisphericus, Shumard, 1858, Trans. St. Louis Acad. Sci., Coal Meas. [Sig. hemispherical.]
hoveyi, Worthen, 1875, Geo. Sur. Ill., vol. 6, Keokuk Gr. [Ety. proper name.]
indentus, Hall, 1862, 15th Reg. Rep. N. Y., Ham. Gr. [Sig. indented.]
indianensis, Meek & Worthen, 1865, Proc. Acad. Nat. Sci. Phil., Keokuk Gr. [Ety. proper name.]
keokuk, Hall, 1860, Supp. to Geo. Rep. Iowa, Keokuk Gr. [Ety. proper name.]
lasallensis, Worthen, 1875, Geo. Sur. Ill., vol. 6, Coal Meas. [Ety. proper name.]
lepidus, Hall, 1861, Desc. New Crinoidea, Burlington Gr. [Sig. elegant.]

longidactylus, Shumard, 1855. This name was preoccupied, and the species is now named P. missouriensis.

macoupinensis, Worthen, 1873, Geo. Sur. Ill., vol. 5, Up. Coal Meas. [Ety. proper name.]

maniformis, see Zeacrinus maniformis.

meekianus, Shumard, 1855, Geo. Rep. Mo., Encrinital limestone. [Ety. proper name.]

missouriensis, Shumard, 1857, Trans. St. Louis Acad. Sci., St. Louis Gr. [Ety. proper name.]

montanaensis, Meek, 1872, Hayden's Geo. Rep., Low. Carb. [Ety. proper name.]

municipalis, Troost. Not defined.

nassa, Hall, 1862, 15th Reg. Rep. N. Y., Ham. Gr. [Sig. a wicker basket with a narrow neck.]

nereus, Hall, 1862, 15th Reg. Rep. N. Y., Ham. Gr. [Ety. mythological name.]

norwoodi, Meek & Worthen, 1865, Proc. Acad. Nat. Sci. Phil., Kaskaskia Gr. [Ety. proper name.]

nycteus, Hall, 1862, 15th Reg. Rep. N. Y., Ham. Gr. [Ety. mythological name.]

obuncus, White, 1862, Proc. Bost. Soc. Nat. Hist., vol. 9, Burlington Gr. [Sig. bent in, hooked.]

occidentalis, see Agassizocrinus occidentalis.

perplexus, Meek & Worthen, 1869, Proc. Acad. Nat. Sci., Phil., Burlington Gr. [Sig. intricate.]

pisiformis, Roemer, 1860, Sil. Fauna West Tenn., Niagara Gr. [Sig. pea-shaped.]

pleias, Hall, 1863, Crin. Wav. Sands. Ohio; Waverly Gr. [Ety. *Pleias*, one of the seven stars, a Pleiad.]

posticus, Hall, 1866, Pamphlet, Cin'ti Gr. [Sig. posterior, the back part.]

proboscidialis, Worthen, 1875, Geo. Sur. Ill., vol. 6, St. Louis Gr. [Sig. having a proboscis.]

rhombiferus, Owen & Shumard, 1852, Geo. Sur. Wis., Iowa & Minn., Burlington Gr. [Sig. bearing rhombs.]

rugosus, Shumard, 1858, Trans. St. Louis Acad. Sci., Coal Meas. [Sig. wrinkled.]

salignoides, White, 1862, Proc. Bost. Soc. Nat. Hist., vol. 9, Burlington Gr. [Sig. like willow-work.]

simplex, Lyon, 1860, Trans. Am. Phil. Soc. vol. 13, Devonian. [Sig. simple.]

solidus, Meek & Worthen, 1861, Proc. Acad. Nat. Sci., Phil., Burlington Gr. [Sig. solid, compact.]

spinosus, see Zeacrinus spinosus.

subimpressus, Meek & Worthen, 1861, Proc. Acad. Nat. Sci., Phil., Burlington Gr. [Sig. slightly indented.]

subtumidus, Meek & Worthen, 1865, Proc. Acad. Nat. Sci., Phil., Kaskaskia Gr. [Sig. somewhat swollen or tumid.]

swallovi, Meek & Worthen, 1860, Proc. Acad. Nat. Sci., Phil., Burlington Gr. [Ety. proper name.]

tenuibrachiatus, Meek & Worthen, 1861, Proc. Acad. Nat. Sci., Phil., Burlington Gr. [Sig. slender-armed.]

tenuidactylus, see Scaphiocrinus tenuidactylus.

tumidus, see Agassizocrinus tumidus.

vanhornei, Worthen, 1875, Geo. Sur. Ill., vol. 6, St. Louis Gr. [Ety. proper name.]

ventricosus, see Cœliocrinus ventricosus.

verticillus, Hall, 1862, 15th Reg. Rep. N. Y., Ham. Gr. [Sig. whorled, verticillate.]

wachsmuthi, see Scaphiocrinus wachsmuthi.

PROTASTER, Forbes, 1849, Mem. Geo. Sur. Great Britain, Decade 1. [Ety. *protos*, first; *aster*, star.]

barrisi, see Onychaster barrisi.

forbesi, Hall, 1859, Pal. N. Y., vol. 3, Low. Held. Gr. [Ety. proper name.]

granuliferus, Meek, 1872, Am. Jour. Sci., 3rd ser., vol. 3, Cin'ti Gr. [Sig. bearing granules.]

gregarius, Meek & Worthen, 1869, Proc. Acad. Nat. Sci. Phil., Keokuk Gr. [Sig. found in flocks.]

PTEROTOCRINUS, Lyon & Casseday, 1860, Am. Jour. Sci., vol. 29. [Ety. *pterotos*, feathered; *krinon*, a lily.]

capitalis, Lyon, 1857, (Astrocrinus capitalis) Geo. Sur. Ky., vol. 3, Chester Gr. [Sig. relating to the head.]

chesterensis, Meek & Worthen, 1860, (Actinocrinus chesterensis) Proc. Acad. Nat. Sci. Phil., Chester Gr. [Ety. proper name.]

cornigerus, Shumard, 1857, (Dichocrinus cornigerus) Trans. St. Louis, Acad. Sci. Kaskaskia Gr. [Sig. horned.]

coronarius, Lyon, 1857, (Asterocrinus coronarius) Geo. Sur. Ky., vol. 3, Chester Gr. [Sig. wreathed, crowned.]

crassus, Meek & Worthen, 1860, (Dichocrinus crassus) Proc. Acad. Nat. Sci. Phil., Chester Gr. [Sig. thick.]

depressus, Lyon & Casseday, 1860, Am. Jour. Sci., vol. 29, Chester Gr. [Sig. depressed.]

pyramidalis, Lyon & Casseday, 1860, Am. Jour. Sci., vol. 29, Kaskaskia Gr. [Sig. pointed like a pyramid.]

rugosus, Lyon & Casseday, 1860, Am. Jour. Sci., vol. 29, Kaskaskia Gr. [Sig. wrinkled.]

sexlobatus, Shumard, 1857, (Dichocrinus sexlobatus) Trans. St. Louis Acad. Sci., Kaskaskia, Gr. [Sig. six-lobed.]

Ptilonaster, Hall, 1868, syn. for Palæocoma.

princeps, see Palæocoma princeps.

PYGORHYNCHUS, Agassiz, 1839. [Ety. *pygos*, solid; *rhynchos*, beak.]

gouldi, Bonve, 1846, Proc. Bost. Soc. Nat. Hist., Kaskaskia (?) Gr. [Ety. proper name.]

ECHINODERMATA.

RETIOCRINUS, Billings, 1858, Can. Org. Rem. Decade 4. [Ety. *retium*, net; *krinon*, a lily.]
- fimbriatus, Billings, 1859, Can. Org. Rem. Decade 4, Hud. Riv. Gr. [Sig. fringed.]
- stellaris, Billings, 1859, Can. Org. Rem. Decade 4, Trenton Gr. [Sig. rayed like a star.]

RHODOCRINUS, Miller, 1821, Nat. Hist. Crinoidea. [Ety. *rhodon*, a rose; *krinon*, a lily.]
- asperatus, Billings, 1859, Can. Org. Rem. Decade 4, Chazy Gr. [Sig. made rough, uneven.]
- barrisi, Hall, 1861, Desc. New Crinoidea, Burlington Gr. [Ety. proper name.]
- barrisi *var.* divergens, Hall, 1861, Desc. New Crinoidea, Burlington Gr. [Sig. divergent, separating.]
- gracilis, Hall, 1862, 15th Reg. Rep. N. Y., Ham. Gr. [Sig. slender.]
- halli, Lyon, 1861, Proc. Acad. Nat. Sci., Phil., Low. Held. Gr. [Ety. proper name.]
- melissa, Hall, 1863, Trans. Alb. Inst., vol. 4, Niagara Gr. [Ety. mythological name.]
- *microbasalis*, see Thysanocrinus microbasalis.
- nanus, Meek & Worthen, 1866, Proc. Acad. Nat. Sci., Phil., Burlington Gr. [Sig. dwarfish.]
- nodulosus, Hall, 1862, 15th Reg. Rep. N. Y., Ham. Gr. [Sig. covered with small knots.]
- *pyriformis*, see Thysanocrinus pyriformis.
- rectus, Hall, 1867, 20th Reg. Rep., Niagara Gr. [Sig. straight.]
- spinosus, Hall, 1862, 15th Reg. Rep. N. Y., Ham. Gr. [Sig. covered with many spines.]
- varsoviensis, Hall, 1860, Supp. Geo. Rep. Iowa, Warsaw Gr. [Ety. proper name.]
- wachsmuthi, Hall, 1861, Desc. New Crinoidea, Burlington Gr. [Sig. proper name.]
- whitii, Hall, 1861, Desc. New Crinoidea, Burlington Gr. [Ety. proper name.]
- whitii *var.* burlingtonensis, Hall, 1861, Desc. New Crinoidea, Burlington Gr. [Ety. proper name.]
- wortheni, Hall, 1858, Geo. Rep. Iowa, Burlington Gr. [Ety. proper name.]

SACCOCRINUS, Hall, 1852, Pal. N. Y., vol. 2. [Ety. *sakkos*, a bag or sack; *krinon*, a lily.]
- christyi, Hall, 1863, (Actinocrinus christyi) Trans. Alb. Inst., vol. 4, Niagara Gr. This species was called by Winchell & Marcy, 1865, Megistocrinus marcouanus, and again by Hall, in 1865, Saccocrinus whitfieldi.
- ornatus, Hall & Whitfield, 1875, Ohio Pal., vol. 2, Niagara Gr. [Sig. adorned.]
- semiradiatus, Hall, 1867, 20th Reg. Rep., Niagara Gr. [Sig. half-rayed.]
- speciosus, Hall, 1852, Pal. N. Y., vol. 2, Niagara Gr. [Sig. beautiful.]
- tennesseensis, Troost, Ms., Hall & Whitfield, 1875, Ohio Pal., vol. 2, Niagara Gr. [Ety. proper name.]
- *whitfieldi*, Hall, 1867, (Actinocrinus christyi, 1863) 20th Reg. Rep., Niagara Gr. This name must be regarded as a synonym for Saccocrinus christyi as well as Megistocrinus marcouanus, unless the fossil should be referred to the genus Actinocrinus, in which case, the specific name *christyi* being preoccupied, *marcouanus* would have the preference.

SCAPHOCRINUS, Hall, 1858, Geo. Rep. Iowa. [Ety. *scaphion*, a skiff; *krinon*, a lily.]
- abnormis, Worthen, 1875, Geo. Sur. Ill., vol. 6, St. Louis Gr. [Sig. abnormal.]
- ægina, Hall, 1863, Desc. New Crin. Waverly sandstone, Waverly Gr. [Ety. mythological name.]
- æqualis, Hall, 1861, Desc. New Crin., Keokuk Gr. [Sig. equal.]
- bayensis, Meek & Worthen, 1865, Proc. Acad. Nat. Sci. Phil., Chester Gr. [Ety. proper name.]
- carbonarius, Meek & Worthen, 1861, Proc. Acad. Nat. Sci. Phil., Up. Coal Meas. [Sig. from the coal measures.]
- carinatus, Hall, 1861, Desc. New Crin., Burlington Gr. [Sig. keeled.]
- clio, Meek & Worthen, 1869, Proc. Acad. Nat. Sci. Phil., Burlington Gr. [Ety. mythological name.]
- coreyi, Meek & Worthen, 1869, Proc. Acad. Nat. Sci. Phil., Keokuk Gr. [Ety. proper name.]
- dactyliformis, Hall, 1858, Geo. Rep. Iowa, St. Louis Gr. [Sig. finger-shaped.]
- decabrachiatus, Hall, 1858, Geo. Rep. Iowa, Kaskaskia Gr. [Sig. ten-armed.]
- decadactylus, Meek & Worthen, 1860, (Poteriocrinus decadactylus) Proc. Acad. Nat. Sci. Phil., Keokuk Gr. [Ety. ten-fingered.] P. decadactylus was preoccupied.
- delicatus, Meek & Worthen, 1869, Proc. Acad. Nat. Sci. Phil., Burlington Gr. [Sig. delicate.]
- depressus, Meek & Worthen, 1870, Proc. Acad. Nat. Sci. Phil., Keokuk Gr. [Sig. depressed.]
- dichotomus, Hall, 1858, Geo. Rep. Iowa, Burlington Gr. [Sig. dividing into two.]
- divaricatus, Hall, 1860, Supp. to Geol. Iowa, Burlington Gr. [Sig. wide apart.]
- doris, Hall, 1861, Desc. New Crinoidea, Burlington Gr. [Ety. mythological name.]
- fiscellus, Meek & Worthen, 1869, Proc. Acad. Nat. Sci. Phil., Burlington Gr. [Sig. a small basket of woven slender twigs.]
- halli, Hall, 1861, Desc. New Crinoidea, Burlington Gr. [Ety. proper name.]

huntsvillæ, Worthen, 1873, Geo. Sur. Ill., vol. 5, St. Louis Gr. [Ety. proper name.]
internodius, Hall, 1858, Geo. Rep. Iowa, Kaskaskia Gr. [Sig. from the length of the joints.]
juvenis, Meek & Worthen, 1869, Proc. Acad. Nat. Sci. Phil., Burlington Gr. [Sig. young.]
liriope, Hall, 1863, (spelled lyriope as published,) Crin. Wav. Sands. Ohio, Waverly Gr. [Ety. mythological name.]
longidactylus, McChesney, 1860, New Pal. Foss., Kaskaskia Gr. [Sig. long-fingered.]
macadamsi, Worthen, 1873, Geo. Sur. Ill., vol. 5, Keokuk Gr. [Ety. proper name.]
macrodactylus, Meek & Worthen, 1869, Proc. Acad. Nat. Sci. Phil., Burlington Gr. [Sig. long-fingered.]
nanus, Meek & Worthen, 1869, Proc. Acad. Nat. Sci. Phil., Burlington Gr. [Sig. dwarfish.]
nodobrachiatus, Hall, 1861, Desc. New Crinoidea, Warsaw Gr. [Sig. arms nodose.]
notabilis, Meek & Worthen, 1869, Proc. Acad. Nat. Sci. Phil., Burlington Gr. [Sig. noteworthy, extraordinary.]
orbicularis, Hall, 1861, Desc. New Crinoidea, Keokuk Gr. [Sig. orb-shaped.]
penicillus, Meek & Worthen, 1869, Proc. Acad. Nat. Sci. Phil., Burlington Gr. [Sig. a little brush.]
ramulosus, Hall, 1861, Desc. New Crinoidea, Burlington Gr. [Sig. full of little branches.]
randolphensis, Worthen, 1873, Geo. Sur. Ill., vol. 5, Chester Gr. [Ety. proper name.]
robustus, Hall, 1861, Desc. New Crinoidea, Keokuk Gr. [Sig. robust.]
rudis, Meek & Worthen, 1869, Proc. Acad. Nat. Sci. Phil., Burlington Gr. [Sig. rude, rough, coarse.]
rusticellus, White, 1863, Proc. Bost. Soc. Nat. Hist., vol. 7, Burlington Gr. [Sig. a little rough, rustic.]
scalaris, Meek & Worthen, 1869, Proc. Acad. Nat. Sci. Phil., Burlington Gr. [Sig. resembling a ladder.]
scoparius, Hall, 1858, Geo. Rep. Iowa, Kaskaskia Gr. [Sig. a sweeper.]
simplex, Hall, 1858, Geo. Rep. Iowa, Burlington Gr. [Sig. simple.]
spinobrachiatus, Hall, 1861, Desc. New Crinoidea, Burlington Gr. [Sig. spiny-armed.]
striatus, Meek & Worthen, 1869, Proc. Acad. Nat. Sci. Phil., Burlington Gr. [Sig. striated.]
subcarinatus, Hall, 1863, Crin. Wav. Sands. Ohio, Waverly Gr. [Sig. somewhat keeled.]
subtortuosus, Hall, 1863, Crin. Wav. Sands., Waverly Gr. [Sig. somewhat twisted.]
tenuidactylus, Meek & Worthen, 1865, Proc. Acad. Nat. Sci. Phil., Burlington Gr. [Sig. slender-fingered.]
tethys, Meek & Worthen, 1869, Proc. Acad. Nat. Sci. Phil., Burlington Gr. [Ety. mythological name.]
tortuosus, Hall, 1861, Desc. New Crinoidea, Burlington Gr. [Sig. twisted.]
unicus, Hall, 1861, Desc. New Crinoidea, Keokuk Gr. [Sig. unique.]
wachsmuthi, Meek & Worthen, 1861, (Poteriocrinus wachsmuthi) Proc. Acad. Nat. Sci. Phil., Burlington Gr. [Ety. proper name.]
whitii, Hall, 1861, Desc. New Crinoidea, Burlington Gr. [Ety. proper name.]

SCHIZOCRINUS, Hall, 1847, Pal. N. Y., vol. 1. [Ety. *schiza*, a cleft, in allusion to the cleft or double interscapular plates; *krinon*, a lily.]
nodosus, Hall, 1847, Pal. N. Y., vol. 1, Trenton Gr. [Sig. knobbed.]
striatus, Hall, 1847, Pal. N. Y., vol. 1, Trenton Gr. [Sig. striated.]

SCHÆNASTER, Meek & Worthen, 1866, Geo. Sur. Ill., vol. 2. [Ety. *schoinos*, a rope; *aster*, a star.]
fimbriatus, Meek & Worthen, 1860, (Palæasterina fimbriatus,) Proc. Acad. Nat. Sci. Phil., St. Louis Gr. [Sig. fringed.]
wachsmuthi, Meek & Worthen, 1866, Proc. Acad. Nat. Sci. Phil., Burlington Gr. [Ety. proper name.]

SCYPHOCRINUS, Hall, 1847, Pal. N. Y., vol. 1. [Ety. *scyphos*, a little cup; *krinon*, a lily.]
heterocostalis, Hall, 1847, Pal. N. Y., vol. 1, Trenton Gr. [Ety. *heteros*, irregalar; *costalis*, ribbed.]

Sphærocrinus, Meek & Worthen, 1866. The name was preoccupied by Roemer, and the authors substituted *Coelocrinus* for it.

SPHÆROCYSTITES, Hall, 1859, Pal. N. Y., vol. 3. [Ety. *sphaira*, a sphere from the spheroidal form of the body; *kustis*, a bladder.]
multifasciatus, Hall, 1859, Pal. N. Y., vol. 3, Low. Held. Gr. [Sig. many banded.]

STEGANOCRINUS, Meek & Worthen, 1866, Geo. Sur. Ill., vol. 2. [Ety. *steganos*, covered; *krinon*, a lily.]
araneolus, Meek & Worthen, 1860, (Actinocrinus araneolus) Proc. Acad. Nat. Sci. Phil., Burlington Gr. [Sig. a small spider.]
pentagonus, Hall, 1858, (Actinocrinus pentagonus) Geo. Sur. Iowa, vol. 1, pt. 2, Burlington Gr. [Sig. pentagonal.]

STENASTER, Billings, 1858, Can. Org. Rem., Decade 3. [Ety. *stenos*, narrow, in allusion to the contracted body; *aster*, a star.]

ECHINODERMATA.

grandis, Meek, 1872, Am. Jour. Sci., 3rd series, vol. 3, Cin'ti Gr. [Sig. great.]
buxleyi, Billings, 1865, Pal. Foss., vol. 1, Quebec Gr. [Ety. proper name.]
pulchellus, Billings, 1857, (Palæaster pulchella) Rep. of Progr. Trenton Gr. [Sig. minutely beautiful.]
salteri, Billings, 1858, Can. Org. Rem., Decade 3, Trenton Gr. [Ety. proper name.]
STEPHANOCRINUS, Conrad, 1842, Jour. Acad. Nat. Sci.,vol. 8. [Ety. stephanos, a coronet; krinon, a lily.]
angulatus, Conrad, 1842, Jour. Acad. Nat. Sci., vol. 8, Niagara Gr. [Sig. angular.]
gemmiformis, Hall, 1852, Pal. N. Y., vol. 2, Niagara Gr. [Sig. bud-shaped.]
STROBILOCYSTITES, White, 1876, Proc. Acad. Nat. Sci.. [Ety. strobilos, a pine cone; kustis, a bladder.]
calvini, White, 1876, Proc. Acad. Nat. Sci., Devonian. [Ety. proper name.]
STROTOCRINUS, Meek & Worthen, 1866, Geo. Sur. Ill., vol. 2. [Ety. strotos, spread; krinon, a lily.]
ægilops, Hall, 1860, (Actinocrinus ægilops) Supp. to Geo. Sur. Iowa, Burlington Gr. [Sig. a sweet acorn.]
asperrimus, Meek & Worthen, 1869, Proc. Acad. Nat. Sci. Phil., Burlington Gr. [Sig. extremely rough.]
dilatatus, see Physetocrinus dilatatus.
ectypus, Meek & Worthen, 1869, Proc. Acad. Nat. Sci. Phil., Burlington Gr. [Sig. engraved in relief, embossed.]
glyptus, Hall, 1860, (Actinocrinus glyptus) Supp. to Geo. Sur. Iowa, Burlington Gr. [Sig. sculptured.]
insculptus, Hall, 1861, (Actinocrinus insculptus) Desc. New Crinoidea, Burlington Gr. [Sig. engraved.]
liratus, Hall, 1860, (Actinocrinus liratus) Supp. Iowa Geo. Rep., vol. 1, pt. 2, Burlington Gr. [Sig. furrowed, ridged.]
perumbrosus, Hall, 1860, (Actinocrinus perumbrosus) Supp. to vol. 1, pt. 2, Iowa Geo. Rep., Burlington Gr. [Sig. very shady.]
regalis, Hall, 1860, (Actinocrinus regalis) Supp. to Geo. Rep. Iowa, Burlington Gr. [Sig. regal, splendid.]
rudis, Hall, 1860, (Actinocrinus rudis) Supp. to Geo. Sur. Iowa, Burlington Gr. [Sig. rude.]
tenuiradiatus, Hall, 1861, (Actinocrinus tenuiradiatus) Desc. New Crinoidea, Burlington Gr. [Sig. slender-armed.]
tholus, Hall, 1860, (Actinocrinus tholus) Supp. to Geo. Iowa, Burlington Gr. [Sig. a dome, a rotunda.]
umbrosus, Hall, 1858, (Actinocrinus umbrosus) Geo. Rep. Iowa, vol. 1, pt. 2, Burlington Gr. [Sig. shady, umbrella-like.]
SYNBATHOCRINUS, Phillips, 1836, Geol. Yorkshire. [Ety. syn, together; baden, walking; krinon, a lily.]

brevis, Meek & Worthen, 1869, Proc. Acad. Nat. Sci. Phil., Burlington Gr. [Sig. short.]
dentatus, Owen & Shumard, 1852, Geo. Sur. Wis., Iowa & Minn., Burlington Gr. [Sig. toothed.]
granulatus, Troost. Not defined.
matutinus, Hall, 1858, Geo. of Iowa, Ham. Gr. [Sig. in the morning.]
oweni, Hall, 1860, 13th Reg. Rep. N. Y., Ham. Gr. [Ety. proper name.]
papillatus, Hall, 1861, Desc. New Crinoidea, Burlington Gr. [Sig. bud-like.]
swallovi, Hall, 1858, Geo. Rep. Iowa, St. Louis Gr. [Ety. proper name.]
tennesseæ, Troost. Not defined.
tennesseensis, Roemer, 1860, Sil. Fauna West Tenn., Niagara Gr. [Ety. proper name.]
robustus, Shumard, 1866, Trans. St. Louis Acad. Sci., Warsaw Gr. [Sig. robust.]
wachsmuthi, Meek & Worthen, 1866, see Catillocrinus wachsmuthi.
wachsmuthi, Meek & Worthen, 1869, Proc. Acad. Nat. Sci. Phil., Burlington Gr. [Ety. proper name.]
wortheni, Hall, 1858, Geo. Rep. Iowa, Burlington Gr. [Ety. proper name.]
SYRINGOCRINUS, Billings, 1859, Can. Org. Rem., Decade 4. [Ety. syrinx, a pipe; krinon, a lily.]
paradoxicus, Billings, 1859, Can. Org. Rem., Decade 4, Trenton Gr. [Sig. puzzling, questionable.]
TÆNIASTER, Billings, 1858, Can. Org. Rem., Decade 3. [Ety. tainia, a ribbon; aster, a star.]
cylindricus, Billings, 1857, (Palæocrinus cylindrica) Rep. of Progr., Trenton Gr. [Sig. cylindrical.]
spinosus, Billings, 1857, (Palæocrinus spinosa) Rep. of Progr., Trenton Gr. [Sig. covered with many spines.]
TAXOCRINUS, Phillips, 1843, Morris Cat. Brit. Foss. [Ety. taxus, the yew tree; krinon, a lily.]
gracilis, Meek & Worthen, 1865, Proc. Acad. Nat. Sci. Phil., Ham. Gr. [Sig. slender.]
interscapularis, Hall, 1858, Geo. of Iowa, Ham. Gr. [Sig. spaced between the shoulder pieces.]
nuntius, Hall, 1862, (Forbesiocrinus nuntius) 15th Reg. Rep. N. Y., Ham. Gr. [Sig. a messenger, an interpreter.]
semiovatus, Meek & Worthen, 1860, (Forbesiocrinus semiovatus) Proc. Acad. Nat. Sci. Phil., St. Louis Gr. [Sig. half-ovate.]
thiemii, Hall, 1861, (Forbesiocrinus thiemei) Desc. New Crinoidea, Burlington Gr. [Ety. proper name.]
TECHNOCRINUS, Hall, 1859, Pal. N. Y., vol. 3, proposed as a subgenus of Mariacrinus. [Ety. techne, art; krinon, a lily.]
andrewsi, Hall, 1859, Pal. N. Y., vol. 3, Oriskany sandstone. [Ety. proper name.]

ECHINODERMATA. 93

sculptus, Hall, 1859, Pal. N. Y., vol. 3, Oriskany sandstone. [Sig. engraved, sculptured.]
spinulosus, Hall, 1859, Pal. N. Y., vol. 3, Oriskany sandstone. [Sig. covered with little spines.]
striatus, Hall, 1859, Pal. N. Y., vol. 3, Oriskany sandstone. [Sig. striated.]

THYSANOCRINUS, Hall, 1852, Pal. N. Y., vol. 2. [Ety. *thysanos*, fringed; from the fimbriated arms or fingers of the species; *krinon*, a lily.]
aculeatus, Hall, 1852, Pal. N. Y., vol. 2, Niagara Gr. [Sig. armed with sharp points.]
canaliculatus, Hall, 1852, Pal. N. Y., vol. 2, Niagara Gr. [Sig. channeled, grooved.]
immaturus, Hall, 1852, Pal. N. Y., vol. 2, Niagara Gr. [Sig. not mature, not full grown.]
liliiformis, Hall, 1852, Pal. N. Y., vol. 2, Niagara Gr. [Sig. lily-shaped.]
microbasalis, Billings, 1857, (Rhodocrinus microbasalis) Rep. of Progr., Trenton Gr. [Sig. having a small base.]
pyriformis, Billings, 1857, (Rhodocrinus pyriformis) Rep. of Progr., Trenton Gr. [Sig. pear-shaped.] —

Trematocrinus, syn. for Goniasteroidocrinus.
fiscellus, see Goniasteroidocrinus fiscellus.
papillatus, see G. papillatus.
reticulatus, see G. reticulatus.
robustus, see G. robustus.
spinigerus, see G. spinigerus.
tuberculatus, see G. tuberculatus.
typus, see G. typus.

VASOCRINUS, Lyon, 1857, Geo. Sur. Ky., vol. 3. [Ety. *vas*, a vessel ; *krinon*, a lily.]
sculptus, Lyon, 1857,Geo. Sur. Ky.,vol. 3, Corniferous Gr. [Sig. sculptured.]
valens, Lyon, 1857, Geo. Sur. Ky., vol. 3, Corniferous Gr. [Sig. strong, vigorous.]

ZEACRINUS, Troost, Catal. Foss., 1850, and described by Hall, 1858, Geo. Iowa. [Ety. *zea*, indian corn; *krinon*, a lily.]
acanthophorus, Meek & Worthen, 1870, Proc. Acad. Nat. Sci. Phil., Coal Meas. [Sig. spine-bearing.]
arboreus, Worthen, 1873, Geo. Sur. Ill., vol. 5, St. Louis Gr. [Sig. tree-like.]
armiger, Meek & Worthen, 1870, Proc. Acad. Nat. Sci. Phil., Chester Gr. [Sig. armed.]
asper, Meek & Worthen, 1869, Proc. Acad. Nat. Sci. Phil., Burlington Gr. [Sig. rough.]
bifurcatus, McChesney, 1860, New Pal. Foss., Kaskaskia Gr. [Sig. bifurcated.]
cariniferus, Worthen, 1873, Geo. Sur. Ill., vol. 5, St. Louis Gr. [Sig. keel-bearing.]
compactilis, Worthen, 1873, Geo. Sur. Ill., vol. 5, St. Louis Gr. [Sig. compact, pressed together.]

crassus, Meek & Worthen, 1860, (Cyathocrinus crassus) Proc. Acad. Nat. Sci. Phil., Low. Coal Meas. [Sig. thick.]
crateriformis, Troost. Not defined.
depressus, Troost, as defined by Hall, 1858, Geo. Rep. Iowa, Chester Gr. [Sig. depressed.]
discus, Meek & Worthen, 1860, Proc. Acad. Nat. Sci. Phil., Up. Coal Meas. [Sig. a quoit.]
elegans, Hall, 1858, Geo. Rep. Iowa, Burlington Gr. [Sig. elegant.]
florealis, Yandell & Shumard, 1847, (Cyathocrinus florealis) Cont. to Geol. Ky. Kaskaskia Gr. [Sig. flower-like.]
formosus, Worthen, 1873, Geo. Sur. Ill., vol. 5, Chester Gr. [Sig. beautiful.]
intermedius, Hall, 1858, Geo. Rep. Iowa, Kaskaskia Gr. [Sig. intermediate.]
lyra, Meek & Worthen, 1869, Proc. Acad. Nat. Sci. Phil., Burlington Gr. [Sig. a lute, a stringed instrument.]
magnoliaformis, Owen & Norwood, 1846, (Cyathocrinus magnoliaformis) Research Pot. Carb. Rocks Ky., Kaskaskia Gr. [Sig. magnolia-like.]
maniformis, Yandell & Shumard, 1847, (Poteriocrinus maniformis) Contributions to the Geo. of Ky., Kaskaskia Gr. [Sig. like a hand.]
merope, Hall, 1863, Extract from 17th Reg. Rep. N. Y., Waverly Gr. [Ety. mythological name.]
mucrospinus, McChesney, 1859, New Pal. Foss., Coal Meas. [Sig. sharp-spined.]
ovalis, Lyon & Casseday, 1859, Am. Jour. Sci., vol. 28, Kaskaskia Gr. [Sig. egg-shaped.]
paternus, Hall, 1863, Crin. Wav. Sands. Ohio, Waverly Gr. [Sig. paternal.]
perangulatus, White, 1862, Proc. Bost. Soc. Nat. Hist., vol. 9, Burlington Gr. [Sig. very angular.]
planobrachiatus, Meek & Worthen, 1860, Proc. Acad. Nat. Sci. Phil., Keokuk Gr. [Sig. smooth-armed.]
ramosus, Hall, 1858, Geo. Rep. Iowa, Burlington Gr. [Sig. branching.]
sacculus, White, 1862, Proc. Bost. Soc. Nat. Hist., vol. 9, Burlington Gr. [Sig. a little bag.]
sacculus var concinnus, White, 1862, Proc. Bost. Soc. Nat. Hist., vol. 9, Burlington Gr. [Sig. beautiful.]
scobina, Meek & Worthen, 1869, Proc. Acad. Nat. Sci. Phil., Burlington Gr. [Sig. a rasp.]
scoparius, Hall, 1861, Desc. New Crinoidea, Burlington Gr. [Sig. a sweeper.]
serratus, Meek & Worthen, 1861, Proc. Acad. Nat. Sci. Phil., Burlington Gr. [Sig. serrated.]
spinosus, Owen & Shumard, 1852, (Poteriocrinus spinosus,) Jour. Acad. Nat. Sci. Phil., Kaskaskia Gr. [Sig. spiny.]

stimpsoni, Lyon, 1869, Trans. Am. Phil. Soc., vol. 13, Low. Carb. [Ety. proper name.]
subtumidus, Worthen, 1873, Geo. Sur. Ill., vol. 5, Chester Gr. [Sig. somewhat tumid or swollen.]

troostianus, Meek & Worthen, 1860, Proc. Acad. Nat. Sci. Phil., Burlington Gr. [Ety. proper name.]
wortheni, Hall, 1858, Geo. Rep. Iowa, Kaskaskia, Gr. [Ety. proper name.]

ADDENDA.

BELEMNOCRINUS florifer, Wachsmuth & Springer, 1877, Am. Jour. Sci. & Arts, 3rd series, vol. 13, Burlington Gr. [Sig. flower-bearing—the appearance is that of a bouquet of flowers in a conical vase.]
pourtalesi, Wachsmuth & Springer, 1877, Am. Jour. Sci. & Arts, 3d series, vol. 13, Burlington Gr. [Ety. proper name.] These authors have pointed out the family relationship of the genus Belemnocrinus with Apiocrinus and the genus should therefore be classed with the Apiocrinidæ.
EUCLADOCRINUS, Meek, 1871, proposed as a subgenus of Platycrinus and founded upon P. montanaensis.

SUB-KINGDOM MOLLUSCA.

First Class,	. . . .	BRYOZOA.
Second Class,	. . . .	BRACHIOPODA.
Third Class,	. . . .	PTEROPODA.
Fourth Class,	. . . .	GASTEROPODA.
Fifth Class,	. . . .	CEPHALOPODA.
Sixth Class,	. . . .	LAMELLIBRANCHIATA.

CLASS BRYOZOA.

FAMILY CELLEPORIDÆ.—Flustra, Paleschara, Sagenella.
FAMILY ESCHARIDÆ.—Arthroclema, Clathropora, Coscinium, Eschara, Escharopora, Glauconome, Helopora, Intricaria, Phænopora, Ptilodictya, Semicoscinium, Stictopora.
FAMILY CRISIDÆ.—Alecto, Hippothoa.
FAMILY RETEPORIDÆ.—Archimedes, Botryllopora, Carinopora, Cryptopora, Evactinopora, Fenestella, Fenestralia, Gorgonia, (?) Hemitrypa, Ichthyorachis, Lyropora, Phyllopora, Polypora, Ptilopora, Retepora, Septopora, Synocladia, Tæniopora.
FAMILY THAMNISCIDÆ.—Thamniscus.
FAMILY TUBULIPORIDÆ.—Callopora, Ceramopora, Ceriopora, Cyclopora, Hornera, Lichenalia, Rhinopora, Trematopora.

ALECTO, Lamouroux, 1821, Exposi. Method. [Ety. mythological name.]
  auloporoides, Nicholson, 1875, Ohio Pal., vol. 2, Cin'ti Gr. [Sig. like *Aulopora*.]
  canadensis, Nicholson, 1875, Can. Nat. & Geol., Corniferous Gr. [Ety. proper name.]
  confusa, Nicholson, 1875, Ohio Pal., vol. 2, Cin'ti Gr. [Sig. confused.]
  frondosa, James, 1875, Ohio Pal., vol. 2, Cin'ti Gr. [Sig. branchy.]
  inflata, Hall, 1847, Pal. N. Y., vol. 1, Trenton & Hud. Riv. Gr. [Sig. inflated.]
  nexilis, James, 1875, Int. to Catal. Cin. Foss., Cin'ti Gr. [Sig. wreathed together.]
ARCHIMEDES, LeSueur, 1842, (Retepora archimedes) Am. Jour. Sci., vol. 43. [Ety. from its resemblance to the machine for raising water, consisting of a tube rolled in a spiral form round a cylinder, invented by Archimedes, a Greek mathematician.]
  laxa, Hall, 1857, Proc. Am. Ass'n Ad. Sci., vol. 10, Kaskaskia Gr. [Sig. loose.]
  meekiana, Hall, 1857, Proc. Am. Ass'n Ad. Sci., vol. 10, Kaskaskia Gr. [Ety. proper name.]
  oweniana, Hall, 1857, Proc. Am. Ass'n Ad. Sci., vol. 10, Keokuk Gr. [Ety. proper name.]
  reversa, Hall, 1858, Geo. Rep. Iowa, Warsaw Gr. [Ety. from the reversed direction of the spiral frond.]
  swalloviana, Hall, 1857, Proc. Am. Ass'n Ad. Sci., vol. 10, Kaskaskia Gr. [Ety. proper name.]
  wortheni, Hall, 1857, Proc. Am. Ass'n Ad. Sci., vol. 10, Warsaw Gr. [Ety. proper name.]
Archimedipora, D'Orb., 1849, Prod. de Pal. Syn. for Archimedes.
  archimedes, see Archimedes.
ARTHROCLEMA, Billings, 1862, Pal. Foss., vol. 1. [Ety. *arthron*, a joint; *klema*, a twig.]

pulchella, Billings, 1862. Pal. Foss. vol., 1, Trenton Gr. [Sig. beautiful.]

BOTRYLLOPORA, Nicholson, 1874, Geo. Mag. Lond., n. s., vol. 1. [Ety. *botryllos*, a cluster; *poros*, a pore.]
socialis, Nicholson, 1874, Geo. Mag. Lond., n. s., vol. 1, Ham. Gr. [Sig. living in groups or flocks.]

CALLOPORA, Hall, 1852, Pal. N. Y., vol. 2. [Ety. *kallos*, beautiful; *poros*, a pore.]
aspera, Hall, 1852, Pal. N. Y., vol. 2, Niagara Gr. [Sig. rough.]
elegantula, Hall, 1852, Pal. N. Y., vol. 2, Niagara Gr. [Sig. elegant.]
florida, Hall, 1852, Pal. N. Y., vol. 2, Niagara Gr. [Sig. adorned.]
heteropora, Hall, 1874, 26th Reg. Rep., Low. Held. Gr. [Ety. *heteros*, irregular; *poros*, a pore.]
hyale, Hall, 1874, 26th Reg. Rep., Low. Held. Gr. [Ety. mythological name.]
incrassata, Nicholson, 1874, Geo. Mag. Lond., n. s., vol. 1, Corniferous and Ham. Gr. [Sig. stout, thickened.]
laminata, Hall, 1852, Pal. N. Y., vol. 2, Niagara Gr. [Sig. composed of thin plates.]
macropora, Hall, 1874, 26th Reg. Rep., Devonian. [Sig. having long pores.]
missouriensis, Rominger, 1866, Proc. Acad. Nat. Sci. Phil., Keokuk Gr. [Ety. proper name.]
nummiformis, Hall, 1852, Pal. N.Y., vol. 2, Niagara Gr. [Sig. in the form of small coins.]
perelegans, Hall, 1874, 26th Reg. Rep. Low. Held. Gr. [Sig. very elegant.]
ponderosa, Hall, 1874, 26th Reg. Rep., Low. Held. Gr. [Sig. heavy, bulky.]
punctata, Hall, 1858, Geo. Rep. Iowa, Warsaw Gr. [Sig. punctated.]
punctillata, Winchell, 1866, Rep. Low. Peninsula Mich., Ham. Gr. [Sig. a little dot.]
singularis, Hall, 1876, 28th Reg. Rep., Niagara Gr. [Sig. singular.]
unispina, Hall, 1874, 26th Reg. Rep., Low. Held. Gr. [Sig. one-spined.]
venusta, Hall, 1874, 26th Reg. Rep., Low. Held. Gr. [Sig. elegant.]

CARINOPORA, Nicholson, 1874, Ann. Mag. Nat. Hist., 4th series, vol. 13. [Ety. *carina*, a keel; *poros*, a pore.]
hindi, Nicholson, 1874, Ann. Mag. Nat. Hist., 4th series, vol. 13, Corniferous Gr. [Ety. proper name.]

CERAMOPORA, Hall, 1852, Pal. N. Y., vol. 2. [Ety. *keramis*, imbricated like roof tile, *poros*, a pore.]
agellus, Hall, 1876, 28th Reg. Rep., Niagara Gr. [Sig. a small field.]
confluens, Hall, 1876, 28th Reg. Rep., Niagara Gr. [Sig. colonies blended together.]
foliacea, Hall, 1852, Pal. N. Y., vol. 2, Niagara Gr. [Sig. leaf-like.]

huronensis, Nicholson, 1875, Geo. Mag., n. s., vol. 2, Ham. Gr. [Ety proper name.]
imbricata, Hall, 1852, Pal. N. Y., vol. 2, Niagara Gr. [Sig. imbricated, like shingles on a roof.]
incrustans, Hall, 1852, Pal. N. Y., vol. 2, Niagara Gr. [Sig. incrusting other substances.]
labecula, Hall, 1876, 28th Reg. Rep., Niagara Gr. [Sig. a little spot.]
maculata, Hall, 1874, 26th Reg. Rep., Low. Held. Gr. [Sig. forming maculæ or small clusters.]
maxima, Hall, 1874, 26th Reg. Rep., Low. Held. Gr. [Sig. largest.]
nicholsoni, James, 1875, Int. to Catal. Cin. Foss., Cin'ti Gr. [Ety. proper name.]
ohioensis, Nicholson, 1875, Ohio Pal., vol. 2, Cin'ti Gr. [Ety. proper name.]

CERIOPORA, Goldfuss, 1826, Germ. Petref. [Ety. *kerion*, a honey-comb; *poros*, a pore.]
hamiltonensis, Nicholson, 1874, Geo. Mag. Lond. n. s., vol. 1, Ham Gr. [Ety. proper name.]

CLATHROPORA, Hall, 1852, Pal. N. Y., vol. 2. [Ety. *clathrum*, a lattice; *pora*, a pore.] This genus is regarded by some as a synonym for Coscinium.
alcicornis, Hall, 1852, Pal. N. Y., vol. 2, Niagara Gr. [Sig. elk-horned.]
clintonensis, Hall & Whitfield, 1875, Ohio Pal., vol. 2, Clinton Gr. [Ety. proper name.]
flabellata, Hall, 1851, Foster & Whitney's Rep., vol. 2, Trenton Gr. [Sig. spread out like a fan.]
frondosa, Hall, 1852, Pal. N. Y., vol. 2, [Sig. branchy.]
intertexta, Nicholson, 1874, Geo. Mag. Lond., n. s., vol. 1, Corniferous Gr. [Sig. inter-woven.]

COSCINIUM, Keyserling, 1846. [Ety. *koskinon*, a sieve.]
asterium, Prout, 1860, Trans. St. Louis Acad. Sci., Keokuk Gr. [Sig. like a star.]
cribriforme, Prout, 1858, Trans. St. Louis Acad. Sci., Up. Held. Gr. [Ety. full of openings like a sieve.]
cyclops, Keyserling, 1846, Up. Held. Gr. [Ety. mythological name.]
elegans, Prout, 1860, Trans. St. Louis Acad. Sci., Lower beds of St. Louis Gr. [Sig. elegant.]
escharoides, Prout, 1860, Trans. St. Louis Acad. Sci., erroneously written *escharense*, Keokuk Gr. [Sig. like the genus *eschara*.]
keyserlingi, Prout, 1858, Trans. St. Louis Acad. Sci., Warsaw Gr. [Ety. proper name.]
michelinia, Prout, 1860, Trans. St. Louis Acad. Sci., St. Louis Gr. [Ety. proper name.]

plumosum, Prout, 1860, Trans. St. Louis Acad. Sci., St. Louis Gr. [Sig. feathery.]
sagenella, Prout, 1860, Trans. St. Louis Acad. Sci., St. Louis Gr. [Sig. from its resemblance to the genus *Sagenella*.]
tuberculatum, Prout, 1860, Trans. St. Louis Acad. Sci., Keokuk Gr. [Sig. covered with tubercles.]
wortheni, Prout, 1860, Trans. St. Louis Acad. Sci., Keokuk Gr. [Ety. proper name.]

CRYPTOPORA, Nicholson, 1874, Ann. Mag. Nat. Hist., 4th ser., vol. 13. [Ety. *krypton*, concealed; *poros*, a pore.]
mirabilis, Nicholson, 1874, Ann. Mag. Nat. Hist., 4th ser., vol. 13, Corniferous Gr. [Sig. wonderful.]

CYCLOPORA, Prout, 1860, Trans. St. Louis Acad. Sci. [Ety. *kuklos*, circle; *poros*, a pore.]
discoidea, Prout, 1860, Trans. St. Louis Acad. Sci., Keokuk Gr. [Sig. disc-like, quoit-shaped.]
fungia, Prout, 1860, Trans. St. Louis Acad. Sci., Keokuk Gr. [Ety. from its resemblance to the genus *Fungia*.]
jamesi, Prout, 1860, Trans. St. Louis Acad. Sci., Cin'ti Gr. [Ety. proper name.]
polymorpha, Prout, 1860, Trans. St. Louis Acad. Sci., Chester Gr. [Sig. many-formed.]

Dianæsopora, Hall, 1852. Not defined.
dichotoma, see Trematopora dichotoma.

ESCHARA, Lamarck, 1801, Syst. An. sans. Vert. [Ety. *eschar*, a scar.]
(?) concentrica, Prout, Trans. St. Louis Acad. Sci., Carb. [Sig. in concentric lines.]
ovatipora, Troost, 1840, 5th Geo. Rep. Tenn., Low. Sil. [Ety. *ovatus*, ovate; *pora*, a pore.]
reticulata, Troost, 1840, 5th Geo. Rep. Tenn., Low. Sil. [Sig. reticulated.]
tuberculata, Prout, 1858, Trans. St. Louis Acad. Sci., Carb. [Sig. tuberculated.]

ESCHAROPORA, Hall, 1847, Pal. N. Y. vol. 1. [Ety. *eschar*, a scar; *poros*, a pore or cell.]
lirata, Hall, 1874, 26th Reg. Rep., Low. Held. Gr. [Sig. lined.]
nebulosa, Hall, 1874, 26th Reg. Rep., Low. Held. Gr. [Sig. misty, foggy.]
recta, Hall, 1847, Pal. N. Y., vol. 1, Trenton Gr. [Sig. straight.]
recta var. nodosa, Hall, 1847, Pal. N. Y., vol. 1, Trenton Gr. [Sig. nodose, knotty.]
tenuis, Hall, 1874, 26th Reg. Rep., N. Y., Low. Held. Gr. [Sig. slender.]

EVACTINOPORA, Meek & Worthen, 1865, Proc. Acad. Nat. Sci. Phil. [Ety. *evactinos*, with beautiful rays; *poros*, a pore.]
grandis, Meek & Worthen, 1868, Geo. Sur. Ill., vol. 3, Burlington Gr. [Sig. grand, large.]

radiata, Meek & Worthen, 1865, Proc. Acad. Nat. Sci. Phil., Burlington Gr. [Sig. rayed.]
sexradiata, Meek & Worthen, 1868, Geo. Sur. Ill., vol. 3, Burlington Gr. [Sig. six-rayed.]

FENESTELLA, Lonsdale, 1839, Murch. Sil. Syst. [Ety. *fenestella*, a little window.]
acmea, Hall, 1876, 28th Reg. Rep., Niagara Gr. [Sig. pointed.]
acuticosta, Roemer, 1860, Sil. Fauna West Tenn., Niagara Gr. [Sig. sharp-ribbed.]
aspera, Hall, 1847, (Gorgonia aspera) Pal. N. Y., vol. 1, Chazy Gr. [Sig. rough.]
banyana, Prout, 1859, Trans. St. Louis Acad. Sci., Low. Carb. [Ety. from its resemblance to the rootlets of the banyan-tree of India.]
bifurcata, Prout, 1866, Trans. St. Louis Acad. Sci., Ham. Gr. [Sig. bifurcated, forked.]
corticata, Prout, 1858, Trans. St. Louis Acad. Sci., Low. Carb. [Sig. having rind or bark.]
crebripora, Hall, 1874, 26th Reg. Rep., Low. Held. Gr. [Sig. having the openings very close.]
cribrosa, Hall, 1852, Pal. N. Y., vol. 2, Niagara Gr. [Sig. like a sieve.]
dawsoni, Nicholson, 1875, Geo. Mag. vol. 2, n. s. Ham. Gr. [Ety. proper name.]
delicata, Meek, 1871, Proc. Acad. Nat. Sci. Phil., vol. 23, Waverly Gr. [Sig. delicate, thin.]
dilata, Prout, 1866, Trans. St. Louis Acad. Sci., Ham. Gr. [Sig. spread out.]
elegans, Hall, 1852, Pal. N. Y., vol. 2, Niagara Gr. [Sig. elegant.]
eximia, Winchell, 1866, Rep. Low. Peninsula Mich., Ham. Gr. [Sig. excellent, select.]
filiformis, Nicholson, 1874, Geo. Mag., vol. 1, n. s., Corniferous Gr. [Sig. thread-like.]
filitexta, Winchell, 1866, Rep. Low. Peninsula Mich., Ham. Gr. [Sig. woven like threads.]
flabellata, Phillips, 1836, Geo. York. pt. 2, Coal Meas. or Permian. [Sig. fan-like.]
gracilis, Hall, 1847, (Retepora gracilis) Pal. N. Y., vol. 1, Chazy Gr. [Sig. slender.]
hemitrypa, Prout, 1859, Trans. St. Louis Acad. Low. Carb. [Sig. from half-closed fenestrules.]
idalia, Hall, 1874, 26th Reg. Rep., Low. Held. Gr. [Ety. mythological name.]
incepta, Hall, 1847, (Retepora incepta) Pal. N. Y., vol. 1, Chazy Gr. [Sig. incipient.]
intermedia, Prout, 1858, Trans. St. Louis Acad. Sci., Carb. [Sig. intermediate.]
lyelli, Dawson, 1868, Acad. Geol., Low. Carb. [Ety. proper name.]

magnifica, Nicholson, 1874, Geo. Mag., vol. 1, n. s., Corniferous Gr. [Sig. magnificent.]
marginalis, Nicholson, 1874, Geo. Mag., vol. 1, n. s., Corniferous Gr. [Sig. pertaining to the margin.]
multiporata, *var.* lodiensis, Meek, 1875, Ohio Pal., vol. 2, Waverly Gr. [Ety. proper name.]
nervata, Nicholson, 1875, Ohio Pal., vol. 2, Niag. Gr. [Sig. full of nerves or fibers.]
nervia, Hall, 1874, 26th Reg. Rep., Low. Held. Gr. [Ety. proper name.]
nodosa, Prout, 1866, Trans. St. Louis Acad. Sci., Ham. Gr. [Sig. nodose, knotty.]
norwoodiana, Prout, 1858, Trans. St. Louis Acad. Sci., Coal Meas. [Ety. proper name.]
parvulipora, Hall, 1876, 28th Reg. Rep., Niagara Gr. [Sig. having very small pores.]
plebeia, Geinitz, 1866, Carb. und Dyas. Neb., Upper Coal Meas. This name was preoccupied by McCoy, and the species is distinct.
plumosa, Prout, 1858, Trans. St. Louis Acad. Sci., Warsaw Gr. [Sig. feathery.]
popeana, Prout, 1858, Trans. St. Louis Acad. Sci., Permian Gr. [Ety. proper name.]
præcursor, Hall, 1874, 26th Reg. Rep., Low. Held. Gr. [Sig. forerunner.]
prisca, Lonsdale, 1839, Murch. Sil. Syst., Clinton Gr. [Sig. ancient.]
puncto-striata, Hall, 1876, 28th Reg. Rep., Niagara Gr. [Sig. dotted and striated.]
shumardi, Prout, 1858, Trans. St. Louis Acad. Sci., vol. 1, Up. Coal Meas. [Ety. proper name.]
subretiformis, Prout, 1858, Trans. St. Louis Acad. Sci., Carb. [Sig. somewhat net-shaped.]
sylvia, Hall, 1874, 26th Reg. Rep., Low. Held. Gr. [Ety. mythological name.]
tenuiceps, Hall, 1852, Pal. N. Y., vol. 2, Niagara Gr. [Sig. slender-headed.]
tenuis, Hall, 1852, Pal. N. Y., vol. 2, Clinton Gr. [Sig. thin, slender.]
trituberculata, Prout, 1858, Trans. St. Louis Acad. Sci., Carb. [Sig. having three tubercles.]
variabilis, Prout, 1858, Trans. St. Louis Acad. Sci., Carb. [Sig. variable, changing.]
FENESTRALIA, Prout, 1858, Trans. St. Louis Acad. Sci. [Ety. from the genus *Fenestella*.]
st. ludovici, Prout, 1858, Trans. St. Louis Acad. Sci., St. Louis Gr. [Ety. proper name.] *Sanctus Ludovicus*, Latin for St. Louis.
FLUSTRA, Linnæus, 1745, Amænitates acadæmicæ. [Sig. calm.]
carbaseoides, Eaton, 1832, Geo. Text Book, Devonian. [Ety. from a supposed resemblance to the recent *Flustra carbasea*.]

spatulata, Prout, 1859, Trans. St. Louis Acad. Sci., Low. Carb. [Sig. spatula-shaped.]
tuberculata, Prout. 1859, Trans. St. Louis Acad. Sci., Low. Carb. [Sig. tuberculated.]
GLAUCONOME, Goldfuss, 1826, Germ. Petref. [Ety. mythological name.]
nereidis, White, 1874, Rep. Invert. Foss., Carb. [Ety. *nerium*, a tree; *eidos*, like.]
trilineata, Meek, 1872, Pal. E. Neb., Coal Meas. [Sig. three lined or striated.]
GORGONIA, Linnaeus, 1745, Amænitates academicæ. [Ety. mythological name.]
(?) *aspera*, see Fenestella aspera.
dubia, Goldfuss, 1826, Petref. Germ. Permian. [Sig. doubtful.]
perantiqua, see Retepora perantiqua.
HELOPORA, Hall, 1852, Pal. N. Y., vol. 2. [Ety. *helos*, a club; *poros*, a perforation.]
armata, Billings, 1866, Catal. Sil. Foss. Antic., Anticosti Gr. [Sig. furnished with defenses.]
bellula, Billings, 1866, Catal. Sil. Foss. Antic., Anticosti Gr. [Sig. pretty, neat.]
circe, Billings, 1866, Catal. Sil. Foss. Antic., Anticosti Gr. [Ety. mythological name.]
concava, Billings, 1866, Catal. Sil. Foss. Antic., Anticosti Gr. [Sig. concave.]
formosa, Billings, 1865, Catal. Sil. Foss. Antic., Anticosti Gr. [Sig. beautiful.]
fragilis, Hall, 1852, Pal. N. Y., vol. 2, Clinton Gr. [Sig. frail, easily broken.]
fragilis *var.* acadiensis, Hall, 1860, Can. Nat. & Geo., vol. 5, Anticosti Gr. [Ety. proper name.]
irregularis, Billings, 1866, Catal. Sil. Foss. Antic., Anticosti Gr. [Sig. irregular.]
lineata, Billings, 1866, Catal. Sil. Foss. Antic., Anticosti Gr. [Sig. lined, striated.]
lineopora, Billings, 1866, Catal. Sil. Foss. Antic., Anticosti Gr. [Sig. lined with perforations.]
nodosa, Billings, 1866, Catal. Sil. Foss. Antic., Anticosti Gr. [Sig. full of knots.]
striatopora, Billings, 1866, Catal. Sil. Foss. Antic., Anticosti Gr. [Sig. striated coral.]
strigosa, Billings, 1866, Catal. Sil. Foss. Antic., Anticosti Gr. [Sig. thin, meagre.]
varipora, Billings, 1866, Catal. Sil. Foss. Antic., Anticosti Gr. [Sig. cells varying in size.]
HEMITRYPA, Phillips, 1841, Pal. Foss. of Cornwall, Devon. and West Somerset. [Ety. *hemitrypa*, half-foramen.]
dubia, Hall, 1876, 28th Reg. Rep., Niagara Gr. [Sig. doubtful.]
prima, Hall, 1874, 26th Reg. Rep., Low. Held. Gr. [Sig. first.]

HETERODICTYA, Nicholson, 1875, Geo. Mag., vol. 2, n. s. [Ety. *heteros*, irregular; *dictyon*, a net.] The correct orthography is *Heterodictyon*.
  gigantea, Nicholson, 1875, Geo. Mag., vol. 2, Low. Carb. [Sig. large, gigantic.]
Hippothoa, Lamouroux, 1821, Expos. method. *inflata*, see Alecto inflata.
HORNERA, Lamouroux, 1821, Expos. method. des genres de l'Ordre des Pol. [Ety. proper name.]
  dichotoma, Hall, 1852, Pal. N. Y., vol. 2, Niagara Gr. [Sig. divided into two; forked.]
ICHTHYORACHIS, McCoy, 1844, Carb. Foss. Ireland. [Ety. *ichthys*, a fish; *rachis*, a backbone.]
  nereis, Hall, 1874, 26th Reg. Rep., Low. Held. Gr. [Ety. mythological name.]
INTRICARIA, Defrance, 1823, Dictionnaire des Sciences Naturelles. [Ety. *intrico*, to entangle.]
  reticulata, Hall, 1847, Pal. N. Y., vol. 1, Trenton & Cin'ti Gr. [Sig. networked.]
LICHENALIA, Hall, 1852, Pal. N. Y., vol. 2. [Sig. from resemblance to a marine lichen.]
  concentrica, Hall, 1852, Pal. N. Y., vol. 2, Niagara Gr. [Sig. concentrically wrinkled.]
  concentrica *var.* parvula, Hall, 1876, 28th Reg. Rep. N.Y., Niag. Gr. [Sig. small.]
LYROPORA, Hall, 1857, Proc. Am. Assoc. Ad. Sci., vol. 10. [Ety. *lyra*, a lute; *pora*, a pore.]
  lyra, Hall, 1857, Proc. Am. Ass. Ad. Sci., vol. 10, Kaskaskia Gr. [Sig. a lute.]
  quincuncialis, Hall, 1857, Proc. Am. Ass. Ad. Sci., vol. 10, Kaskaskia Gr. [Sig. in the form of a quincunx; an oblique arrangement.]
  retrorsa, Meek & Worthen, 1868, Geo.Sur. Ill., vol. 3, Burlington Gr. [Sig. turned back.]
  subquadrans, Hall, 1857, Proc. Am. Ass. Ad. Sci., vol. 10, Kaskaskia Gr. [Sig. somewhat square.]
PALESCHARA, Hall, 1874, 26th Reg. Rep. [Sig. ancient *Eschara*.]
  aspera, Hall, 1876, 28th Reg. Rep., Niagara Gr. [Sig. rough.]
  bifoliata, Hall, 1874, 26th Reg. Rep., Low. Held. Gr. [Sig. two-leaved.]
  incrustans, Hall, 1874, 26th Reg. Rep., Low. Held. Gr. [Sig. incrusting other objects.]
  maculata, Hall, 1876, 28th Reg. Rep., Niagara Gr. [Sig. spotted.]
  offula, Hall, 1876, 28th Reg. Rep., Niagara Gr. [Sig. a small piece.]
  sphærion, Hall, 1876, 28th Reg. Rep., Niagara Gr. [Sig. a little ball.]
PHÆNOPORA, Hall, 1852, Pal. N. Y., vol. 2. [Ety. *phaino*, to open or make a window; *poros*, a pore.]
  constellata, Hall, 1852, Pal. N. Y., vol. 2, Clinton Gr. [Sig. starred.]
  ensiformis, Hall, 1852, Pal. N. Y., vol. 2, Clinton Gr. [Ety. like a sword.]
  expansa, Hall & Whitfield, 1875, Ohio Pal., vol. 2, Clinton Gr. [Sig. spread out.]
  explanata, Hall, 1852, Pal. N. Y., vol. 2, Clinton Gr. [Sig. spread out.]
  multipora, Hall, 1851, Geo. Lake Sup. Land. Dist., vol. 2, Trenton Gr. [Sig. having many pores.]
PHYLLOPORA, King, 1849, Ann. Nat. Hist., 2d ser., vol. 3. [Ety. *phyllon*, a leaf; *poros*, a pore.]
  ehrenbergi, Geinitz, 1846, (Gorgonia ehrenbergi) Grundriss, Permian Gr. [Ety. proper name.]
POLYPORA, McCoy, 1845, Carb. Foss. Ireland. [Ety. *polys*, many; *poros*, a pore.]
  biarmica, (?) Keyserling, 1846, Russia and the Ural Mts., Geinitz referred a form from the Coal Meas. to this species, and Prout referred a form from the Chester Gr. to it. [Sig. strong-jointed.]
  elegans, Hall, 1874, 26th Reg. Rep., Low. Held. Gr. [Sig. elegant.]
  gracilis, Prout, 1860, Trans. St. Louis Acad. Sci., Keokuk Gr. [Sig. slender.]
  halliana, Prout, 1860, Trans. St. Louis Acad. Sci., St. Louis Gr. [Ety. proper name.]
  hamiltonensis, Prout, 1866, Geo. Sur. Ill., vol. 2, Ham. Gr. [Ety. proper name.]
  imbricata, Prout, 1866, Trans. St. Louis Acad. Sci., Devonian. [Sig. imbricated.]
  incepta, Hall, 1852, Pal. N. Y., vol. 2, Niagara Gr. [Sig. incipient.]
  intermedia, Prout, 1858, Trans. St. Louis Acad. Sci., Up. Held. Gr. [Sig. intermediate.]
  lilia, Hall, 1874, 26th Reg. Rep., Low. Held. Gr. [Ety. proper name.]
  mexicana, Prout, 1858, Trans. St. Louis Acad. Sci., Permian Gr. [Ety. proper name.]
  (?) psyche, Billings, 1874, Pal. Foss., vol. 2, Devonian. [Ety. mythological name.]
  pulchella, Nicholson, 1874, Geo. Mag. Lond., n. s., vol. 1, Corniferous Gr. [Sig. beautiful.]
  rigida, Prout, 1866, Trans. St. Louis Acad. Sci., Ham. Gr. [Sig. rigid.]
  shumardi, Prout, 1858, Trans. St. Louis Acad. Sci., Up. Held. Gr. [Ety. proper name.]
  stragula, White, 1874, Rep. Invert. Foss., Carb. [Sig. outward covering or garment.]
  submarginata, Meek, 1872, Pal. E. Neb., Coal Meas. [Sig. somewhat like *P. marginata*.] This species was referred by Geinitz to P. marginata of McCoy.
  tenella, Nicholson, 1874, Geo. Mag. Lond. n. s., vol. 1, Corniferous Gr. [Sig. delicate.]

tuberculata, Prout, 1859, Trans. St. Louis Acad. Sci., Low. Carb. [Sig. tuberculated.]
tuberculata, Nicholson, 1874, Geo. Mag. Lond., vol. 1, Corniferous Gr. This name was preoccupied by Prout in 1859.
varsoviensis, Prout, 1858, Trans. St. Louis Acad. Sci., Warsaw Gr. [Ety. proper name.]

PTILODICTYA, Lonsdale, 1839, Murch. Sil. Syst. [Ety. *ptilon*, a wing; *dictyon*, a net.] The correct orthography is Ptilodictyon.

acuminata, James, 1876, Int. Catal. Cin. Foss., Cin'ti Gr. [Sig. sharp-pointed.]
alcyone, Billings, 1866, Catal. Sil. Foss. Antic., Anticosti Gr. [Ety. mythological name.]
arctipora, Nicholson, 1875, Ohio Pal., vol. 2, Cin'ti Gr. [Sig. having narrow pores.]
arguta, Billings, 1866, Catal. Sil. Foss. Antic., Anticosti Gr. [Sig. very distinct.]
canadensis, Billings, 1866, Catal. Sil. Foss. Antic., Hud. Riv. Gr. [Ety. proper name.]
carbonaria, see Stictopora carbonaria.
cosciniformis, Nicholson, 1875, Geo. Mag., n. s., vol. 2, Ham. Gr. [Sig. resembling Coscinium.]
dictyota, Meek, 1872, Hayden's Geo. Rep., Low. Carb. [Sig. net-worked.]
emacerata, Nicholson, 1875, Ohio Pal., vol. 2, Cin'ti Gr. [Sig. thin, lean.]
excellens, Billings, 1866, Catal. Sil. Foss. Antic., Anticosti Gr. [Sig. excellent.]
explicans, Safford, 1869, Geo. of Tenn. Not defined.
falciformis, Nicholson, 1875, Ohio Pal., vol. 2, Cin'ti Gr. [Sig. sword-shaped.]
fenestelliformis, Nicholson, 1875, Ohio Pal., vol. 2, Cin'ti Gr. [Ety. from resemblance to *Fenestella*.]
flagellum, Nicholson, 1875, Ohio Pal., vol. 2, Cin'ti Gr. [Sig. a small whip.]
fragilis, Billings, 1866, Catal. Sil. Foss. Antic. Hud. Riv. Gr. [Sig. frail.]
gladiola, Billings, 1866, Catal. Sil. Foss. Antic., Anticosti Gr. [Sig. sword-grass.]
libana, Safford, 1869, Geo. of Tenn., Trenton Gr. [Ety. proper name.]
meeki, Nicholson, 1874, Geo. Mag., n. s., vol. 1, Corniferous & Ham. Gr. [Ety. proper name.]
multiramis, Safford. Not defined.
nitidula, Billings, 1866, Catal. Sil. Foss. Antic., Hud. Riv. Gr. [Sig. neat.]
pavonia, see Monticulipora pavonia.
rustica, Billings, 1865, Catal. Sil. Foss. Antic., Anticosti Gr. [Sig. rough.]
sereata, see Stictopora sereata.
sulcata, Billings, 1866, Catal. Sil. Foss. Antic., Anticosti Gr. [Sig. furrowed.]

superba, Billings, 1866, Catal. Sil. Foss. Antic., Anticosti Gr. [Sig. superb.]
symmetra, Safford. Not defined.
tarda, Billings, 1874, Pal. Foss., vol. 2, Devonian. [Sig. thick.]
tenera, Billings, 1866, Catal. Sil. Foss. Antic., Anticosti Gr. [Sig. tender.]

PTILOPORA, McCoy, 1844, Syn. Carb. Foss., Ireland. [Ety. *ptilon*, a plume; *poros*, a pore.]

prouti, Hall, 1858, Geo. Rep. Iowa, Warsaw Gr. [Ety. proper name.]

RETEPORA, Lamarck, 1801, Syst. An. sans. Vert. [Ety. *rete*, a net; *poros*, a pore.]

angulata, Hall, 1852, Pal. N. Y., vol. 2, Clinton Gr. [Sig. angular.]
antiqua, as identified by d'Archiac & Verneuil. Not American.
archimedes, Lesueur, see Archimedes.
asperato-striata, Hall, 1852, Pal. N. Y., vol. 2, Niagara Gr. [Sig. rough and striated.]
clintoni, Vanuxem, 1842, Geo. Rep. 3d Dist. N. Y., Clinton Gr. [Ety. proper name.]
diffusa, Hall, 1852, Pal. N. Y., vol. 2, Niagara Gr. [Sig. diffused, extended.]
fenestrata, Hall, 1850, 3d Reg. Rep., Clinton Gr. [Sig. having open windows.]
foliacea, Hall, 1847. This name Prof. Hall says may be erased from the list of fossils.
gracilis, see Fenestella gracilis.
hamiltonensis, Prout, 1866, Trans. St. Louis Acad. Sci., Ham. Gr. [Ety. proper name.]
incepta, see Fenestella incepta.
perantiqua, Hall, 1847, (Gorgonia (?) perantiqua) Pal. N. Y., vol. 1, Trenton Gr. [Sig. very ancient.]
phillipsi, Nicholson, 1874, Geo. Mag., n. s., vol. 1, Corniferous Gr. [Ety. proper name.]
trentonensis, Nicholson, 1875, Geo. Mag., vol. 2, n. s., Trenton Gr. [Ety. proper name.]

RHINOPORA, Hall, 1852, Pal. N. Y., vol. 2. [Ety. *rhine*, a file; *poros*, a pore.] Better orthography would be Rhinipora.

frondosa, Hall & Whitfield, 1875, Ohio Pal., vol. 2, Clinton Gr. [Sig. branchy.]
tuberculosa, Hall, 1852, Pal. N. Y., vol. 2, Niagara Gr. [Sig. abounding in tubercles.]
tubulosa, Hall, 1852, Pal. N. Y., vol. 2, Clinton Gr. [Sig. abounding in tubes.]
verrucosa, Hall, 1852, Pal. N. Y., vol. 2, Clinton Gr. [Sig. covered with wart-like projections.]

SAGENELLA, Hall, 1852, Pal. N. Y., vol. 2, Niagara Gr. [Ety. *sagenella*, a little drag net.]

elegans, Hall, 1876, 28th Reg. Rep., Niagara Gr. [Sig. elegant.]
membranacea, Hall, 1852, Pal. N. Y., vol. 2, Niagara Gr. [Sig. membrane-like.]

SEMICOSCINIUM, Prout, 1859, Trans. St. Louis Acad. Sci. [Sig. somewhat like fossils of the genus *Coscinium*.]
eriense, Prout, 1860, Trans. St. Louis Acad. Sci., Devonian. [Ety. proper name.]
rhomboideum, Prout, 1859, Trans. St. Louis Acad. Sci., Up. Held. Gr. [Sig. rhomboidal.]
tuberculatum, Prout, 1860, Trans. St. Louis Acad. Sci., Up. Held. Gr. [Sig. tuberculated.]

SEPTOPORA, Prout, 1859, Trans. St. Louis Acad. Sci. [Ety. *septum*, partition or division; *pora*, a pore.]
cestriensis, Prout, 1850, Trans. St. Louis Acad. Sci., Chester Gr. [Ety. proper name.]

STICTOPORA, Hall, 1847, Pal. N. Y., vol. 1. [Ety. *stictos*, spotted or punctured; *poros*, a pore.]
acuta, Hall, 1847, Pal. N. Y., vol. 1, Trenton Gr. [Sig. sharp.]
carbonaria, Meek, 1871, Proc. Acad. Nat. Sci. Phil., Coal Meas. [Sig. from the Coal Measures.]
crassa, Hall, 1852, Pal. N. Y., vol. 2, Clinton Gr. [Sig. thick.]
elegantula, Hall, 1847, Pal. N. Y., vol. 1, Trenton Gr. [Sig. quite elegant.]
fenestrata, Hall, 1847, Pal. N. Y., vol. 1, Chazy Gr. [Sig. open, having windows.]
gilberti, Meek, 1871, Proc. Acad. Nat. Sci. Phil., Corniferous Gr. [Ety. proper name.]
glomerata, Hall, 1847, Pal. N. Y., vol. 1, Chazy Gr. [Sig. confused, out of order.]
labyrinthica, Hall, 1847, Pal. N. Y., vol. 1, Birdseye Gr. [Sig. intricate.]
lichenoides, Meek, 1873, Pal. Ohio, vol. 1, Corniferous Gr. [Sig. resembling a *Lichen*.]
magna, Hall & Whitfield, 1875, Pal. Ohio, vol. 2, Clinton Gr. [Sig. great, large.]
punctipora, Hall, 1852, Pal. N. Y., vol. 2, Niagara Gr. [Sig. *punctus*, a puncture; *pora*, a pore.]
ramosa, Hall, 1847, Pal. N. Y., vol. 1, Birdseye Gr. [Sig. branching.]
raripora, Hall, 1852, Pal. N. Y., vol. 2, Clinton Gr. [Sig. cells distant.]
serrata, Meek, 1875, Pal. Ohio, vol. 2, Coal Meas. [Sig. serrated.]
shafferi, Meek, 1872, Proc. Acad. Nat. Sci. Phil., Cin'ti Gr. [Ety. proper name.]
similis, Hall, 1876, 28th Reg. Rep., Niagara Gr. [Sig. similar.]
sulcata, Winchell, 1866, Rep. Low. Penin. Mich., Ham. Gr. [Sig. furrowed.]
variabilis, Prout, 1866, Trans. St. Louis Acad. Sci., Up. Sil. [Sig. variable, not always the same.]

SYNOCLADIA, King, 1849, Trans. Geol. Soc. Lond., 2d Ser., vol. 3. [Ety. *syn*. together; *klados*, a young branch.]

biserialis, Swallow, 1858, Trans. St. Louis Acad. Sci., Up. Coal Meas. [Sig. in allusion to the two rows of cellules.]
TÆNIOPORA, Nicholson, 1874, Geo. Mag. Lond., n. s., vol. 1. [Ety. *tainia*, a ribbon; *poros*, a pore.]
exigua, Nicholson, 1874, Geo. Mag. Lond., n. s., vol. 1, Ham. Gr. [Sig. little, small.]
penniformis, Nicholson, 1874, Geo. Mag. Lond. n. s., vol. 1, Ham. Gr. [Sig. resembling a feather.]

THAMNISCUS, King, 1849, Perm. Foss. [Ety. *thamniskos*, a little shrub.]
niagarensis, Hall, 1876, 28th Reg. Rep., Niagara Gr. [Ety. proper name.]

TREMATOPORA, Hall, 1852, Pal. N. Y., vol. 2. [Ety. *trema*, a hole; *poros*, a pore.]
aspera, Hall, 1852, Pal. N. Y., vol. 2, Niagara Gr. [Sig. rough.]
coalescens, Hall, 1852, Pal. N. Y., vol. 2, Niagara Gr. [Sig. uniting together.]
constricta, Hall, 1874, 26th Reg Rep., Low. Held. Gr. [Sig. constricted.]
corticosa, Hall, 1874, 26th Reg. Rep., Low. Held. Gr. [Sig. like thick bark.]
densa, Hall, 1874. 26th Reg. Rep., Low. Held. Gr. [Sig. thick, close.]
dichotoma, Hall, 1852, (Diamesopora dichotoma) Pal. N. Y., vol. 2. Niagara Gr. [Sig. dividing into two.]
echinata, Hall, 1876, 28th Reg. Rep., Niagara Gr. [Sig. set with spines.]
fragilis, Winchell, 1863, Proc. Acad. Nat. Sci., Chemung Gr. [Sig. frail.]
granulifera, Hall, 1852, Pal. N. Y., vol. 2, Niagara Gr. [Sig. bearing granules.] The same species is marked "n. sp." in 28th Reg. Rep., probably by mistake.
infrequens, Hall, 1876, 28th Reg. Rep., Niagara Gr. [Sig. not frequent.]
maculosa, Hall, 1874, 26th Reg. Rep., Low. Held. Gr. [Sig. spotted.]
minuta, Hall, 1876, 28th Reg. Rep., Niagara Gr. [Sig. very small.]
osculum, Hall, 1876, 28th Reg. Rep., Niagara Gr. [Sig. pretty mouth.]
ostiolata, Hall, 1852, Pal. N. Y., vol. 2, Niagara Gr. [Sig. having small openings.]
ponderosa, Hall, 1874, 26th Reg. Rep., Low. Held. Gr. [Sig. heavy.]
punctata, Hall, 1874, Pal. N. Y., vol. 2, Niagara Gr. [Sig. dotted.]
regularis, Hall. 1874, 26th Reg. Rep., Low. Held. Gr. [Sig. in allusion to the regular arrangement of the cells.]
rhombifera, Hall, 1874, 26th Reg. Rep. Low. Held. Gr. [Sig. rhomb-bearing.]
signatus, Hall, 1874, 26th Reg. Rep., Low. Held. Gr. [Sig. marked.]
solida, Hall, 1852, Pal. N. Y., vol. 2, Niagara Gr. [Sig. solid.]
sparsa, Hall, 1852, Pal. N. Y., vol. 2, Niagara Gr. [Sig. scattered.]
spinulosa, Hall, 1852, Pal. N. Y., vol. 2, Niagara Gr. [Sig. full of little spines.]

spinulosa, Hall, 1876, 28th Reg. Rep., Niagara Gr. The name was preoccupied. Possibly the species are identical.

striata, Hall, 1852, Pal. N. Y., vol. 2, Niagara Gr. [Sig. striated.]

superba, Billings, 1866, Catal. Sil. Foss. Antic., Clinton & Niagara Gr. [Sig. splendid.]

tuberculosa, Hall, 1852, Pal. N. Y., vol. 2, Niagara Gr. [Sig. full of tubercles.]

tubulosa, Hall, 1852, Pal. N. Y., vol. 2, Niagara Gr. [Sig. full of tubes.]

varia, Hall, 1876, 28th Reg. Rep., Niagara Gr. [Sig. different; from the variable cell mouths.]

variolata, Hall, 1876, 28th Reg. Rep., Niagara Gr. [Sig. from the distant and variable cell mouths.]

vesiculosa, Winchell, 1863, Proc. Acad. Nat. Sci., Chemung Gr. [Sig. full of vesicles.]

CLASS BRACHIOPODA.

FAMILY CALCEOLIDÆ.—Calceola.
FAMILY CRANIIDÆ.—Crania, Pholidops, Pseudocrania, Schizocrania.
FAMILY DISCINIDÆ.—Discina, Trematis.
FAMILY KONINCKIIDÆ.—Koninckia.
FAMILY LINGULIDÆ.—Kutorgina, Leptobolus, Lingula, Lingulella, Lingulepis, Lingulops, Obolus, Obolella.
FAMILY ORBICULIDÆ.—Acrotreta, Iphidea, Orbicula.
FAMILY ORTHIDÆ.—Ægilops, Meekella, Orthis, Orthisina, Skenidium, Tropidoleptus, Vitulina.
FAMILY PENTAMERIDÆ. — Amphigenia, Anastrophia, Camarella, Camarophoria, Gypidula, Pentamerella, Pentamerus, Stricklandinia.
FAMILY PORAMBONITIDÆ.—Porambonites.
FAMILY PRODUCTIDÆ.—Aulosteges, Chonetes, Productella, Productus, Strophalosia.
FAMILY RHYNCHONELLIDÆ.—Acambona, Eatonia, Leiorhynchus, Rhynchonella, Stenoschisma.
FAMILY SPIRIFERIDÆ. — Ambocœlia, Athyris, Atrypa, Camarium, Cryptonella, Cyrtia, Cyrtina, Eichwaldia, Martinia, Merista, Meristella, Meristina, Nucleospira, Pentagonia, Retzia, Rhynchospira, Spirifera, Spiriferina, Syntrielasma, Syringothyris, Trematospira, Trigonotreta, Waldheimia, Zygospira.
FAMILY STROPHOMENIDÆ. — Leptæna, Streptorhynchus, Strophodonta, Strophomena.
FAMILY TEREBRATULIDÆ. — Centronella, Cœlospira, Leptocœlia, Rensselæria, Terebratula.
FAMILY TRIMERELLIDÆ.—Dinobolus, Monomerella, Trimerella.

ACAMBONA, White, 1862, Proc. Bost. Soc. Nat. Hist., vol. 9. [Ety. *aka*, a point; *umbona*, umbo.]
   prima, White, 1862, Proc. Bost. Soc. Nat. Hist., vol. 9, Burlington Gr. [Sig. first.]
ACROTRETA, Kutorga, 1848, Uber die Siphonotretæ aus den Verhandlungen der Kaiserlichen Mineralogischen Gesellscaft fur Jahr. [Ety. *akros*, the top or summit; *tretos*, perforated.]
   attenuata, Meek, 1872, Hayden's Geo. Rep., Potsdam Gr. [Sig. drawn out.]
   gemma, Billings, 1865, Pal. Foss., vol. 1, Quebec Gr. [Sig. a young bud.]
   pyxidicula, W h i t e, 1874, Rep. Invert. Foss., Quebec Gr. [Sig. a little box.]
   subsidua, White, 1874, Rep. Invert. Foss., Potsdam Gr. [Sig. sinking down.]
ÆGILOPS, Hall, 1850, 3rd Reg. Rep. [Ety. *ægilops*, an acorn.]
   subcarinata, Hall, 1850, 3rd Reg. Rep., Trenton Gr. [Sig. somewhat keeled.]

AMBOCŒLIA, Hall, 1860, 13th Reg. Rep. [Ety. *ambon*, umbo; *koilos*, the belly.]
   gemmula, syn. for Spirifera planoconvexa.
   gregaria, see Ambocœlia umbonata *var.* gregaria.
   minuta, White, 1862, Proc. Bost. Soc. Nat. Hist., vol. 9, Chemung Gr. [Sig. very small.]
   nucleus, syn. for Ambocœlia umbonata.
   præumbona, Hall, 1857, (Orthis præumbona) 10th Reg. Rep., Ham. Gr. [Sig. very protuberant.]
   subumbona, see Spirifera, subumbona.
   umbonata, Conrad, 1842, (Orthis umbonata) Jour. Acad. Nat. Sci., vol. 8, Marcellus shale & Ham. Gr. [Sig. protuberant, bossed.]
   umbonata *var.* gregaria, Hall, 1860, (Ambocœlia gregaria) 13th R e g. Rep., Chemung Gr. [Sig. in a common flock.]

AMPHIGENIA, Hall, 1867, Pal. N. Y., vol. 4. [Ety. *amphi*, on both sides; *genea*, growth.]
curta, Meek & Worthen, 1868, (Stricklandinia elongata *var.* curta) Geo. Sur. Ill., vol. 3, Oriskany sandstone. [Sig. short.]
elongata, Vanuxem, 1842, (Pentamerus elongatus) Geo. 3rd Dist. N. Y., Schoharie grit & Up. Held. Gr. [Sig. elongated.]
elongata *var.* undulata, Hall, 1867, Pal. N. Y., vol. 4, Up. Held. Gr. [Sig. waved.]
elongata *var.* subtrigonalis, Hall, 1857, (Meganteris subtrigonalis) 10th Reg. Rep., Up. Held. Gr. [Sig. somewhat triangular.]
ANASTROPHIA, Hall, 1867, Pal. N. Y., vol. 4. [Ety. *ana*, with; *strophe*, a turning round; the relation of the valves is the reverse of that of *Pentamerus*.]
interplicata, Hall, 1852, (Atrypa interplicata) Pal. N. Y., vol. 2, Niagara Gr. [Sig. from the interplications.]
reversa, Billings, 1857, (Pentamerus reversus) Rep. of Prog. Mid. Sil. [Sig. reversed; the dorsal valve being largest.]
verneuili, Hall, 1859, Pal. N. Y., vol. 3, Low. Held. Gr. [Ety. proper name.]
Anomia, Linnæus, 1767, Syst. Nat., 12th Ed. [Ety. *anomios*, unequal.]
biloba, see Orthis biloba.
pecten, see Strophomena pecten.
reticularis, see Atrypa reticularis.
Anomites, Wahlenberg, 1821, Act., Upsal.
exporrectus, see Cyrtia exporrecta.
glaber, see Spirifera glabra.
punctatus, see Productus punctatus.
resupinatus, see Orthis resupinatus.
reticularis, see Atrypa reticularis.
rhomboidalis, see Strophomena rhomboidalis.
scabriculus, see Productus scabriculus.
semireticulatus, see Productus semireticulatus.
ATHYRIS, McCoy, 1844, Carb. Foss. Ireland. [Ety. *a*, without; *thuris*, a small door; in allusion to the absence of a deltidium or door. But the name is erroneous.]
americana, Swallow, 1863, Trans. St. Louis Acad. Sci., Low. Carb. [Ety. proper name.]
angelica, Hall, 1861, 14th Reg. Rep., Chemung Gr. [Ety. proper name.]
argentea, Shepard, 1838, Am. Jour. Sci., Up. Coal Meas. [Sig. pertaining to silver.]
biloba, Winchell, 1865, (Spirigera biloba) Proc. Acad. Nat. Sci., Kinderhook Gr. [Sig. double-lobed.]
caput-serpentis, Swallow, 1863, Trans. St. Louis Acad. Sci., Up. Coal Meas. [Ety. *caput*, head; *serpens*, a serpent.]
charitonensis, Swallow, 1860, (Spirigera charitonensis) Trans. St. Louis Acad. Sci., Coal Meas. [Ety. proper name.]
chloe, Billings, 1860, Can. Jour., vol. 5. Devonian. [Ety. proper name.]
clara, syn. for Meristella nasuta.
clintonensis, Swallow, 1863, Trans. St. Louis Acad. Sci., Low. Carb. [Ety. proper name.]
cora, Hall, 1860, 13th Reg. Rep., Ham. & Chemung Gr. [Ety. mythological name.]
corpulenta, Winchell, 1863, (Spirigera corpulenta) Proc. Acad. Nat. Sci., Chemung Gr. [Sig. corpulent.]
crassicardinalis, White, 1860, Bost. Jour. Nat. Hist., vol. 8, Waverly Gr. [Sig. thick on the hinge.]
eborea, Winchell, 1866, Rep. Low. Peninsula Mich., Ham. Gr. [Sig. made of ivory.]
euzona, Swallow, 1863, Trans. St. Louis Acad. Sci., Low. Carb. [Sig. beautifully girdled.]
differentis, McChesney, 1860, New Pal. Foss., Carb. [Sig. different.]
formosa, Swallow, 1863, Trans. St. Louis Acad. Sci., Low. Carb. [Sig. beautiful.]
fultonensis, Swallow, 1860, (Spirigera fultonensis) Trans. St. Louis Acad. Sci., Ham. Gr. [Ety. proper name.]
hannibalensis, Swallow, 1860, (Spirigera hannibalensis) Trans. St. Louis Acad. Sci., Chemung Gr. [Ety. proper name.]
hawni, Swallow, 1860, (Spirigera hawnii) Trans. St. Louis Acad. Sci., Coal Meas. [Ety. proper name.]
headi, see Zygospira headi.
headi var. anticostiensis, see Zygospira headi *var.* anticostiensis.
headi var. borealis, see Zygospira headi *var.* borealis.
hirsuta, Hall, 1858, Trans. Alb. Inst., vol. 4, Warsaw Gr. [Sig. rough, prickly.]
incrassata, Hall, 1858, Geo. Rep. Iowa, Burlington Gr. [Sig. thickened.]
jacksoni, Swallow, 1860, (Spirigera jacksoni) Trans. St. Louis Acad. Sci., Coal Meas. [Ety. proper name.]
julia, see Meristella julia.
junia, Billings, 1866, Catal. Sil. Foss. Antic., Anticosti Gr. [Ety. proper name.]
lamellosa, Leveille, 1835, (Spirifer lamellosus) Mem. Geol. Soc. France, Waverly Gr. [Sig. in very thin plates.]
lara, Billings, 1866, Catal. Sil. Foss. Antic., Anticosti Gr. [Ety. mythological name.]
maconensis, Swallow, 1860, (Spirigera maconensis) Trans. St. Louis Acad. Sci., Coal Meas. [Ety. proper name.]
maia, see Spirifera maia.
minima, Swallow, 1860, (Spirigera minima) Trans. St. Louis Acad. Sci., Ham. Gr. [Sig. the least.]
missouriensis, Swallow, 1860, (Spirigera missouriensis) Trans. St. Louis Acad. Sci., Coal Meas. [Ety. proper name.]

BRACHIOPODA. 105

missouriensis, Winchell, 1865, (Spirigera missouriensis) Proc. Acad. Nat. Sci., Lithographic limestone. This name was preoccupied.
monticola, White, 1874, (Spirigera monticola) Rep. Invert. Foss., Low. Carb. [Sig. inhabiting a mountain.]
papilioniformis, McChesney, 1865, Desc. New Pal. Foss., Chester Gr. [Sig. resembling a butterfly.]
parvirostris, Meek & Worthen, 1860, Proc. Acad. Nat. Sci. Phil., Keokuk Gr. [Sig. little-beaked.]
pectinifera, Swallow, 1863, Trans. St. Louis Acad. Sci., Low. Carb. [Sig. comb-bearing.]
ohioensis, Winchell, 1865, Proc. Acad. Nat. Sci., Chemung Gr. [Ety. proper name.]
orbicularis, McChesney, 1860, New Pal. Foss., Coal Meas. [Sig. orbicular.]
planosulcata, Phillips, 1836, Geol. York., vol. 2, Keokuk Gr. [Sig. plain-furrowed.]
plattensis, Swallow, 1863, Trans. St. Louis Acad. Sci., Up. Coal Meas. [Ety. proper name.]
polita, Hall, 1843, (Atrypa polita) Geo. 4th Dist. N. Y., Chemung Gr. [Sig. smoothed.]
prinstana, see Meristella prinstana.
prouti, Swallow, 1860, (Spirigera Proutii) Trans. St. Louis Acad. Sci., Chemung Gr. [Ety. proper name.]
reflexa, Swallow, 1863, Trans. St. Louis Acad. Sci., Low. Carb. [Sig. turned back.]
singletoni, Swallow, 1863, Trans. St. Louis Acad. Sci., Low. Coal Meas. [Ety. proper name.]
solitaria, Billings, 1866, Catal. Sil. Foss. Antic., Anticosti Gr. [Sig. alone, solitary.]
spiriferoides, Eaton, 1831, (Terebratula spiriferoides) Am. Jour. Sci., vol. 21, Cornif. & Ham. Gr. [Sig. resembling a Spirifera.]
sublamellosa, Hall, 1858, Geo. Rep. Iowa, Kaskaskia Gr. [Sig. somewhat like *A. lamellosa.*]
subquadrata, Hall, 1858, Geo. Rep. Iowa, Kaskaskia Gr. [Sig. somewhat squared.]
subtilita, Hall, 1852, Stansbury's Exped. to Great Salt Lake, Coal Meas. [Sig. fine, thin.]
tumida, Dalman, 1827, (Atrypa tumida) Niagara Gr. [Sig. tumid, swollen out.] The fossil usually referred to this species is Meristella maria.
tumidula, Billings, 1866, Catal. Sil. Foss. Antic., Anticosti Gr. [Sig. diminutive of *tumidus*; from its resemblance to *A. tumida* and smaller size.]
umbonata, see Meristella umbonata.
vittata, Hall, 1860, 13th Reg. Rep., Cornif. & Ham. Gr. [Sig. banded.]

turgida, Shaler, 1865, Bulletin No. 4, M. C. Z., Anticosti Gr. [Sig. swollen, inflated.]
ATRYPA, Dalman, 1827, Vet. Acad. [Ety. *a*, without; *trypa*, a hole or perforation. It was supposed the shells had no foramen under the beak. The name is erroneous.]
acutiplicata, see Leptocœlia acutiplicata.
acutirostra, see Rhynchonella acutirostra.
aequiradiata, Conrad, 1842, Jour. Acad. Nat. Sci., vol. 8, Low. Held. Gr. [Sig. equal-rayed.]
æquiradiata, see Rhynchonella æquiradiata.
affinis, syn. for Atrypa reticularis.
altilis, see Rhynchonella altilis.
ambigua, see Camarella ambigua.
aprinis, see Rhynchonella aprinis.
arata, see Pentamerella arata.
aspera, Schlotheim, 1813, (Terebratula aspera) Petrefactenkunde, Ham. Gr. [Sig. rough.]
aspera var. occidentalis, Hall, 1858, Geo. Rep. Iowa, vol. 1, pt. 2, Ham. Gr. [Sig. western.]
bidens, see Rhynchonella bidens.
bisulcata, see Camarella bisulcata.
borealis, Schlotheim, as identified by d'Archiac & Verneuil. Not American.
brevirostris, as identified by Hall. See Pentamerus brevirostris and Anastrophia verneuili.
camura, see Trematospira camura.
capax, see Rhynchonella capax.
cassidea, as identified by d'Archiac & Verneuil. Not American.
chemungensis, Conrad, 1842, Jour. Acad. Nat. Sci., vol. 8, Chemung Gr. [Ety. proper name.]
circulus, see Camarella circulus.
concinna, see Nucleospira concinna.
comis, see Pentamerus comis.
concentrica, syn. for Athyris spiriferoides.
congesta, see Camarella congesta.
congregata, see Stenoschisma congregata.
contracta, see Stenoschisma contracta.
corallifera, see Eichwaldia corallifera.
crassirostra, Hall, 1852, Pal. N. Y., vol. 2, Niagara Gr. [Ety. *crassus*, thick; *rostra*, beak.]
crenulata, see Terebratula crenulata.
cuboides, as identified by Hall and others, see Rhynchonella venustula.
cuneata, see Rhynchonella cuneata.
cuspidata, see Camarella cuspidata.
cylindrica, see Meristella cylindrica.
deflecta, Hall, 1847, Pal. N. Y., vol. 1, Trenton Gr. [Sig. bent or turned aside.]
dentata, see Rhynchonella dentata.
disparilis, see Cœlospira disparilis.
dubia, see Rhynchonella dubia.
dumosa, Hall, 1843, Geo. Rep., 4th Dist. N. Y., Chemung Gr. [Sig. bushy.]
duplicata, see Stenoschisma duplicata.
elongata, syn. for Rensselæria ovoides.
emacerata, see Rhynchonella emacerata.

14

exigua, Hall, 1847, Pal. N. Y., vol. 1, Trenton Gr. [Sig. small.]
eximia, see Stenoschisma eximia.
extans, see Camarella extans.
flabella, syn. for Leptocœlia hemispherica.
flabellites, see Leptocœlia flabellites.
galeata, see Pentamerus galeatus.
gibbosa, Hall, 1852, Pal. N. Y., vol. 2, Clinton Gr. [Sig. gibbous, tumid.]
globuliformis, see Leiorhynchus globuliformis.
hemiplicata, see Camarella hemiplicata.
hemispherica, see Leptocoelia hemispherica.
hirsuta, see Trematospira hirsuta.
hystrix, Hall, 1843, Geo. Rep. 4th Dist. N. Y., Chemung Gr. [Sig. covered with spines.]
impressa, Hall, 1857, 10th Reg. Rep., Schoharie grit. [Sig. impressed.]
impressa, Shaler. The name was preoccupied.
increbescens, syn. for Rhynchonella capax.
inflata, Conrad, 1843, Geo. Rep. 3rd Dist. N. Y., Catskill Gr. [Sig. inflated.]
intermedia, Hall, 1852, Pal. N. Y., vol. 2, Clinton Gr. [Sig. intermediate.]
interplicata, see Anastrophia interplicata.
laevis, see Merista lævis.
lamellata, see Rhynchonella lamellata.
laticosta, Phillips, 1841, (Terebratula laticosta) Pal. Foss. Devonian. [Sig. wide-ribbed.] This species is not clearly identified in America.
lentiformis, syn. for Atrypa reticularis.
limitaris, see Leiorhynchus limitaris.
marginalis (?), Dalman, 1827, (Terebratula marginalis) Vet. Acad., Niagara Gr. [Sig. bordered.]
medialis, see Eatonia medialis.
mesacostalis, see Leiorhynchus mesacostalis.
modesta, see Zygospira modesta.
nasuta, see Meristella nasuta.
naviformis, Hall, 1843, Geo. 4th Dist. N. Y., Clinton Gr. [Sig. resembling a boat.]
neglecta, see Rhynchonella neglecta.
nitida, see Meristina nitida.
nitida var. oblata, see Meristina nitida var. oblata.
nodostriata, Hall, 1852, Pal. N. Y., vol. 2, Niagara Gr. [Sig. rough and striated.]
nucleolata, Hall, 1852, Pal. N. Y., vol. 2, Coralline limestone. [Ety. nucleus, a kernel; latus, wide.]
nucleus, see Camarella nucleus.
oblata, Hall, 1852, Pal. N. Y., vol. 2, Medina Gr. [Sig. oblate.]
obtusiplicata, see Rhynchonella obtusiplicata.
octocostata, see Pentamerella arata.
peculiaris, see Eatonia peculiaris.
planoconvexa, see Leptocœlia planoconvexa.
plebeia, Conrad, 1843, Geo. Rep. 3rd Dist. N. Y., Ham. Gr. Preoccupied name.

pleiopleura, see Rhynchonella pleiopleura.
plena, see Rhynchonella plena.
plicata, see Rhynchonella plicata.
plicatella (?), see Rhynchonella plicatella.
plicatula, see Rhynchonella plicatula.
plicifera, see Rhynchonella plicifera.
polita, see Athyris polita.
prisca, syn. for Atrypa reticularis.
pseudomarginalis, Hall, 1860, 13th Reg. Rep., Up. Held. Gr. [Sig. falsemargined.]
quadricostata, Hall, 1843, see Rhynchonella quadricostata.
quadricostata, Hall, 1852, see Rhynchonella quadricostata.
rectiplicata, Conrad, 1842, Jour. Acad. Nat. Sci., vol. 8, Low. Held. Gr. [Sig. straight-plicated.]
recurvirostra, see Rhynchonella recurvirostra.
reticularis, Linnæus, 1767, (Anomia reticularis) Syst. Nat., ed. 12. It occurs with its varieties in all the Groups of the Upper Silurian and Devonian formations except the Oriskany sandstone. Some of its varieties or synonyms are *Atrypa affinis, A. lentiformis, A. prisca, A. tribulis, Hipparionyx consimilaris,* etc. [Sig. reticulated.]
robusta, see Rhynchonella robusta.
rostrata, see Meristella rostrata.
rugosa, see Rhynchonella rugosa.
scitula, see Meristella scitula.
semiplicata, see Rhynchonella semiplicata.
singularis, see Eatonia singularis.
sordida, see Rhynchonella sordida.
spinosa, Hall, 1843, Geo. 4th Dist. N. Y., Cornif., Ham., Tully & Chemung Gr. Equal to Atrypa aspera var. occidentalis.
subcuboides, D'Orbigny, see Rhynchonella venustula.
subtrigonalis, see Rhynchonella subtrigonalis.
sulcata, see Merista sulcata.
tenuilineata, Hall, 1843, Geo. 4th Dist. N. Y., Chemung Gr. [Ety. tenuis, fine; lineatus, lined.]
tribulis, syn. for Atrypa reticularis.
tumida, see Athyris tumida.
unguiformis, syn. for Orthis hipparionyx.
unisulcata, see Meristella unisulcata.

AULOSTEGES, Helmerson, 1847, Bull. de la Classe Physi. Math. Acad. Sci. St. Petersburg. [Ety. aulos, a tube; stege, a chamber.]
guadalupensis, Shumard, 1858, Trans. St. Louis Acad. Sci., Permian Gr. [Ety. proper name.]
spondyliformis, White & St. John, 1868, Trans. Chi. Acad. Sci., Up. Coal Meas. [Sig. vertebra-formed.]

Brachymerus, Shaler. The name was preoccupied for a genus of Coleoptera. See Anastrophia.

BRACHIOPODA.

Brachyprion geniculatum, see Strophomena geniculata.
leda, see Strophomena leda.
ventricosum, see Strophomena ventricosa.

CALCEOLA, Lamarck, 1801, Syst. des Anim. sans. Vert. [Ety. *calceola*, a slipper.]
 americana, Safford, syn. for C. tennesseensis.
 plicata, Conrad, 1840, Ann. Rep. N. Y., Held. Gr. [Ety. from the plications toward the aperture.]
 sandalina, Lamarck, as identified by Troost, see C. tennesseensis.
 tennesseensis, Roemer, 1852, Lethæ. Geognost., Niagara Gr. [Ety. proper name.]

CAMARELLA, Billings, August, 1859, Can. Nat. & Geol., vol. 4. [Ety. *kamara*, arching chamber; *ella*, diminutive.]
 ambigua, Hall, 1847, (Atrypa ambigua) Pal. N. Y., vol. 1, Trenton Gr. [Sig. doubtful.]
 antiquata, Billings, 1861, Pal. Foss., vol. 1, Potsdam Gr. [Sig. very ancient.]
 bisulcata, Emmons, 1842, (Orthis bisulcata) Geo. Rep. N. Y., Trenton Gr. [Sig. double-furrowed.]
 breviplicata, Billings, 1865, Pal. Foss., vol. 1, Quebec Gr. [Sig. short-plicated.]
 calcifera, Billings, 1861, Can. Nat. & Geo., vol. 6, Calcif. Gr. [Ety. from the Calciferous Group.]
 circulus, Hall, 1847, (Atrypa circulus) Pal. N. Y., vol. 1, Trenton Gr. [Sig. circular.]
 congesta, Conrad, 1842, (Atrypa congesta) Jour. Acad. Nat. Sci., vol. 8, Clinton Gr. [Sig. a heap.]
 costata, Billings, 1865, Pal. Foss., vol. 1, Quebec Gr. [Sig. ribbed.]
 cuspidata, Hall, 1847, (Atrypa cuspidata) Pal. N. Y., vol. 1, Trenton Gr. [Sig. pointed.]
 extans, Emmons, 1842, (Atrypa extans) Geo. Rep. N. Y., Trenton Gr. [Sig. standing out.]
 hemiplicata, Hall, 1847, (Atrypa hemiplicata) Pal. N. Y., vol. 1, Trenton Gr. [Sig. half-plicated.]
 lenticularis, Billings, 1866, Catal. Sil. Foss. Antic., Anticosti Gr. [Sig. lenticular.]
 longirostra, Billings, 1859, Can. Nat. and Geo., vol. 4, Chazy Gr. [Sig. long-beaked.]
 nucleus, Hall, 1847, (Atrypa nucleus) Pal. N. Y., vol. 1, Trenton Gr. [Sig. a kernel.]
 ops, Billings, 1862, Pal. Foss., vol. 1, Mid. Sil. [Ety. mythological name.]
 ortoni, Meek, 1872, (Dicraniscus ortoni) Am. Jour. Sci., 3rd ser., vol. 4, Clinton Gr. [Ety. proper name.]
 panderi, Billings, 1859, Can. Nat. and Geo., vol. 4, Black Riv. Gr. [Ety. proper name.]
 parva, Billings, 1865, Pal. Foss., vol. 1, Quebec Gr. [Sig. small.]
 polita, Billings, 1865, Pal. Foss., vol. 1, Quebec Gr. [Sig. smoothed.]
 reversa, see Anastrophia reversa.
 varians, Billings, 1859, Can. Nat. and Geo., vol. 4, Chazy Gr. [Sig. variable.]
 volborthi, Billings, 1859, Can. Nat. and Geo., vol. 4, Black Riv. Gr. [Ety. proper name.]

CAMARIUM, Hall, 1859, Pal. N. Y., vol. 3. [Ety. *kamara*, arching septum.]
 elongatum, Hall, 1859, Pal. N. Y., vol. 3, Low. Held. Gr. [Sig. elongated.]
 typum, Hall, 1859, Pal. N. Y., vol. 3, Low. Held. Gr. [Sig. type of the genus.]

CAMEROPHORIA, King, 1844, Ann. & Mag. Nat. Hist., vol. 14. [Ety. *kamara*, an arched chamber; *phoreo*, I carry.]
 bisulcata, Shumard, 1858, Trans. St. Louis Acad. Sci., Permian Gr. [Sig. double furrowed.]
 eucharis, Hall, 1867, Pal. N. Y., vol. 4, Corniferous Gr. [Sig. graceful.]
 globulina, Phillips, 1844, as identified by Geinitz, is Rhynchonella uta.
 schlotheimi, Von Buch, 1834, (Terebratulites schlotheimi) Ueber Terebratel. Permian Gr. [Ety. proper name.] This is the type of the genus.
 subtrigona, Meek & Worthen, 1860, (Rhynchonella subtrigona) Proc. Acad. Nat. Sci. Phil., Keokuk Gr. [Sig. somewhat three-cornered.]
 swalloviana, Shumard, 1859, Trans. St. Louis Acad. Sci., Permian Gr. [Ety. proper name.]

CENTRONELLA, Billings, 1859, Can. Nat. & Geo., vol. 4. [Ety. a little point.]
 allii, Winchell, 1865, Proc. Acad. Nat. Sci., Chemung Gr. [Ety. proper name.]
 alveata, Hall, 1857, (Rhynchonella alveata) 10th Reg. Rep., Onondaga Gr. [Sig. channeled.]
 anna, Hartt, 1868, Acad. Geol., Low. Carb. [Ety. proper name.]
 billingsiana, Meek & Worthen, 1868, Geo. Sur. Ill., vol. 3, Niagara Gr. [Ety. proper name.]
 glansfagea, Hall, 1857, (Rhynchonella glansfagea) 10th Reg. Rep., Schoharie grit, Cornif. Gr. and Oriskany sandstone. [Ety. *glans*, an acorn; *fagea*, beech tree.]
 glaucia, Hall, 1867, Pal. N. Y., vol. 4, Ham. Gr. [Ety. proper name.]
 hecate, Billings, 1861, Can. Jour. vol. 6, Up. Held. Gr. [Ety. mythological name.]
 impressa, Hall, 1861, 14th Reg. Rep., Ham. Gr. [Sig. impressed.] Prof. Billings says this is a syn. for *C. hecate*.
 julia, Winchell, 1862, Proc. Acad. Nat. Sci., vol. 14, Marshall Gr. [Ety. proper name.]
 ovata, Hall, 1867, Pal. N. Y., vol. 4, Up. Held. Gr. [Sig. egg-shaped.]

Charionella, Billings, 1861, Can. Jour. Ind. Sci. and Art. Prof. Hall regards this name as a syn. for Meristella.
circe, syn. for Meristella scitula.
doris, see Meristella doris.
(?) *hyale*, see Meristella hyale.
CHONETES, Fischer, 1837, Oryckt. Moscou. [Ety. *chone*, a little cup.]
acutiradiata, Hall, 1843, (Strophomena acutiradiata) Geo. Rep. 4th Dist. N. Y., Cornif. Gr. [Sig. sharp-radiated.]
antiope, Billings, 1874, Pal. Foss., vol. 2, Low. Devonian. [Ety. mythological name.]
arcuata, Hall, 1857, 10th Reg. Rep., Cornif. Gr. [Sig. bent.]
armata, DeKoninck, the specimens referred to this species belong to *C. pusilla*.]
canadensis, Billings, 1874, Pal. Foss., vol. 2, Devonian. [Ety. proper name.]
carinata, Conrad, 1842, (Strophomena carinata) Jour. Acad. Nat. Sci., vol. 8, Ham. Gr. [Sig. carinated.]
complanata, Hall, 1857, 10th Reg. Rep., Oriskany sandstone. [Sig. smoothed.]
cornuta, Hall, 1843, (Strophomena cornuta) Geo. Rep. 4th Dist. N. Y., Clinton Gr. [Sig. horned.]
dawsoni, Billings, 1874, Pal. Foss., vol. 2, Low. Devonian. [Ety. proper name.]
deflecta, Hall, 1857, 10th Reg. Rep., Ham. Gr. [Sig. bent down.]
emmetensis, Winchell, 1866, Rep. Low. Peninsula Mich., Ham. Gr. [Ety. proper name.]
fischeri, Norwood & Pratten, 1854, Jour. Acad. Nat. Sci., vol. 3, Kinderhook Gr. [Ety. proper name.]
flemingi, Norwood & Pratten, 1854, Jour. Acad. Nat. Sci., vol. 3, Permian Gr. [Ety. proper name.]
geniculata, White, 1862, Proc. Bost. Soc. Nat. Hist., vol. 9, Chemung Gr. [Sig. bent, geniculated.]
gibbosa, syn. for C. deflecta.
glabra, Hall, 1857, 10th Reg. Rep., Up. Hold. Gr. [Sig. smooth.]
glabra, Geinitz, 1866, Carb. und Dyas. The name was preoccupied.
granulifera, Owen, 1852, Geo. Rep. Wis., Iowa and Minn., Coal Meas. [Sig. granule-bearing.]
hemispherica, Hall, 1857, 10th Reg. Rep., Schoharie grit and Cornif. Gr. [Sig. hemispherical.]
illinoisensis, Worthen, 1860, Trans. St. Louis Acad. Sci., Chester Gr. [Ety. proper name.]
iowensis, Owen, 1852, Geo. Rep. Iowa, Wis. and Minn., Carb. [Ety. proper name.]
konincklana, Norwood & Pratten, 1854, Jour. Acad. Nat. Sci., vol. 3, Devonian. [Ety. proper name.]
laticosta, syn. for C. mucronata.
lepida, Hall, 1857, 10th Reg. Rep. Marcellus shale & Ham. Gr. [Sig. pretty.]
lineata, Conrad, 1839, (Strophomena lineata) Ann. Geo. Rep. N. Y., Corniferous Gr. [Sig. lined.]
littoni, Norwood & Pratten, 1854, Jour. Acad. Nat. Sci., vol. 3, Ham. Gr. [Ety. proper name.]
logani, Norwood & Pratten, 1854, Jour. Acad. Nat. Sci., vol. 3, Burlington Gr. [Ety. proper name.]
logani *var.* aurora, Hall, 1867, Pal. N. Y., vol. 4, Tully limestone & Ham. Gr. [Sig. the morning.]
maclurii, Norwood & Pratten, 1854, Jour. Acad. Nat. Sci., vol. 3, Ham. Gr. [Ety. proper name.]
martini, Norwood & Pratten, 1854, Jour. Acad. Nat. Sci., vol. 3, Ham. Gr. [Ety. proper name.]
melonica, Billings, 1874, Pal. Foss., vol. 2, Devonian. [Sig. a small melon.]
mesoloba, Norwood & Pratten, 1854, Jour. Acad. Nat. Sci., vol. 3, Coal Meas. [Sig. middle-lobed.]
michiganensis, Stevens, 1858, Am. Jour. Sci., vol. 25, Low. Carb. [Ety. proper name.]
millepunctata, Meek & Worthen, 1870, Proc. Acad. Nat. Sci., Coal Meas. [Sig. many-dotted.]
minima, Hall, 1876, 28th Reg. Rep. Niagara Gr. [Sig. least.]
mucronata, Hall, 1843, Geo. Rep. 4th Dist. N. Y., Corniferous & Ham. Gr. [Sig. sharp-pointed.]
mucronata, Meek & Hayden, 1858, Proc. Acad. Nat. Sci., Coal Meas. This name was preoccupied, moreover it is a syn. for *C. granulifera*.
multicosta, Winchell, 1863, Proc. Acad. Nat. Sci., Chemung Gr. [Sig. many-ribbed.]
muricata, Hall, 1867, Pal. N. Y., vol. 4, Chemung Gr. [Sig. like the shell *Murex*.]
nova-scotica, Hall, 1860, Can. Nat. & Geo., vol. 5, Up. Sil. [Ety. proper name.]
ornata, Shumard, 1855, Geo. of Mo., Chemung Gr. [Sig. ornamented.]
parva, Shumard, 1855, Geo. of Mo., Coal Meas. [Sig. small.]
permiana, Shumard, 1858, Trans. St. Louis Acad. Sci., Permian Gr. [Ety. proper name.]
planumbona, Meek & Worthen, 1860, Proc. Acad. Nat. Sci., Keokuk Gr. [Sig. smooth on the *umbo*.]
platynota, White, 1874, Rep. Invert. Foss., Low. Carb. [Ety. *platys*, broad; *notos*, ridge.]
pulchella, Winchell, 1862, Proc. Acad. Nat. Sci., Marshall Gr. [Sig. beautiful.]
pusilla, Hall, 1857, 10th Reg. Rep., Ham. Gr. [Sig. very small.]
scitula, Hall, 1857, 10th Reg. Rep., Ham. Gr. [Sig. pretty.]

setigera, Hall, 1843, (Strophomena setigera) Geo. Rep. 4th Dist. N. Y., Ham. & Chemung Gr. [Sig. having bristles.]
shumardiana, DeKoninck, 1847, Recherches sur les Anim. Foss., Low. Carb. [Ety. proper name.]
smithi, Norwood & Pratten, 1854, Jour. Acad. Nat. Sci., vol. 3, Coal Meas. [Ety. proper name.]
syrtalis, syn. for C. carinata.
tenuistriata, Hall, 1860, Can. Nat. and Geo., vol. 5, Up. Sil. [Sig. fine-lined.]
tuomeyi, Norwood & Pratten, 1854, Jour. Acad. Nat. Sci., vol. 3, Ham. Gr. [Ety. proper name.]
variolata, DeKoninck, 1847, Recher. Anim. Foss., Coal Meas. [Sig. spotted with pimples.]
verneuiliana, Norwood & Pratten, 1854, Jour. Acad. Nat. Sci., vol. 3, Coal Meas. [Ety. proper name.]
yandelliana, Hall, 1857, 10th Reg. Rep., Corniferous Gr. [Ety. proper name.]
COELOSPIRA, Hall, 1858, Trans. Alb. Inst., vol. 4. [Ety. *koilos*, hollow; *spira*, a spire.]
concava, Hall, 1859, (Leptocoelia concava) Pal. N. Y., vol. 3, Corniferous Gr. [Sig. concave.]
dichotoma, Hall, 1859, Pal. N. Y., vol. 3, Oriskany sandstone. [Sig. dividing into two.]
disparilis, Hall, 1852, (Atrypa disparilis) Pal. N. Y., vol. 2, Niagara Gr. [Sig. different, unequal.]
CRANIA, Retzius, 1781, Schriften der Berliner Gesellschaft Naturforschende Freund. [Ety. *kranion*, the upper part of a skull.]
acadiensis, Hall, 1860, Can. Nat. & Geo., vol. 5, Up. Sil. [Ety. proper name.]
aurora, Hall, 1863, 16th Reg. Rep. Schoharie Grit. [Sig. morning.]
bella, Billings, 1874, Pal. Foss., vol. 2, passage beds between Upper Sil. & Devonian. [Sig. beautiful.]
bordeni, Hall, 1872, 24th Reg. Reg., Devonian. [Ety. proper name.]
corrugata, Hall, 1843, (Orbicula corrugatus) Geo. Rep. N. Y., Niagara Gr. [Sig. corrugated.]
crenistriata, Hall, 1860, 13th Reg. Rep., Ham. Gr. [Sig. convex-lined.]
deformata, Hall, 1847, (Orbicula deformata) Pal. N. Y., vol. 1, Chazy Gr. [Sig. déformed.] Is it a Crania?
dyeri, S. A. Miller, 1875, Cin. Quar. Jour. Sci., vol. 2, Cin'ti Gr. [Ety. proper name.]
excentrica, Emmons, 1856, (Orbicula excentrica) Am. Geol., Quebec Gr. [Sig. from the center.]
famelica, Hall, 1873, 23rd Reg. Rep., Chemung Gr. [Sig. famished.]
gregaria, Hall, 1863, 16th Reg. Rep., Ham. Gr. [Sig. occurring in flocks.]
hamiltoniæ, Hall, 1860, 13th Reg. Rep., Ham. Gr. [Ety. proper name.]

lælia, Hall, 1866, Pamphlet, Cin'ti Gr. [Ety. proper name.]
leoni, Hall, 1860, 13th Reg. Rep. N. Y., Chemung Gr. [Ety. proper name.]
modesta, White & St. John, 1868, Trans. Chi. Acad. Sci., Coal Meas. [Sig. not large, modest.]
multipunctata, S. A. Miller, 1875, Cin. Quar. Jour. Sci.,vol. 2, Cin'ti Gr. [Sig. many-dotted.]
permiana, Shumard, 1859, Trans. St. Louis Acad. Sci., Permian Gr. [Ety. proper name.]
prima, Owen, 1852, (Orbicula prima) Geo. Sur. Iowa, Wis. & Minn., Potsdam Gr. [Sig. first.]
quadricostata, Vanuxem, 1842, (Orbicula quadricostata) Geo. Rep. 3rd Dist. N. Y., Genessee slate. [Sig. four-ribbed.]
radicans, Winchell, 1866, Rep. Low. Peninsula Mich., Ham. Gr. [Sig. rooting; in allusion to the spines on the exterior.]
reposita, White, 1866, Proc. Bost. Soc. Nat. Hist., vol. 9, Ham. Gr. [Sig. remote, distant.]
reticularis, S. A. Miller, 1875, Cin. Quar. Jour. Sci., vol. 2, Cin'ti Gr. [Sig. reticulated.]
scabiosa, Hall, 1866, Pamphlet, Cin'ti Gr. [Sig. scabby.]
setifera, Hall, 1863, Trans. Alb. Inst., vol. 4, Niagara Gr. [Sig. bearing bristles.]
setigera, Hall, 1866, Pamphlet, Trenton Gr. [Sig. having bristles.]
sheldoni, White, 1862, Proc. Bost. Soc. Nat. Hist., vol. 9, Ham. Gr. [Ety. proper name.]
siluriana, Hall, 1863, Trans. Alb. Inst., vol. 4, Niagara Gr. [Ety. proper name.]
tenuilamellata, Hall, 1852, (Orbicula tenuilamellata) Pal. N. Y., vol. 2, Niagara Gr. [Sig. having very thin plates.] This species is also classed with the *Discina*.
trentonensis, Hall, 1866, Pamphlet, Trenton Gr. [Ety. proper name.]
truncata, Emmons, 1856, (Orbicula truncata) Am. Geol., Trenton Gr. [Sig. truncated.]
CRYPTONELLA, Hall, 1861, 14th Reg. Rep. [Sig. a little cavity.] Prof. Billings regarded this as a synonym for Charionella.]
calvini, Hall, 1870, 23rd Reg. Rep. N. Y., Chemung Gr. [Ety. proper name.]
eudora, Hall, 1867, Pal. N. Y., vol. 4, Chemung Gr. [Ety. proper name.]
iphis, Hall, 1867, Pal. N. Y., vol. 4, Corniferous Gr. [Ety. mythological name.]
lens, see Terebratula lens.
lincklæni, see Terebratula lincklæni.
planirostra, Hall, 1860, (Terebratula planirostra) 13th Reg. Rep., Ham. Gr. [Sig. smooth-beaked.]

rectirostra, Hall, 1860, (Terebratula rectirostra) 13th Reg. Rep. Ham. Gr. [Sig. straight-beaked.]
CYRTIA, Dalman, 1827, Kongl. Vet. Acad. Handl. [Ety. *kyrtia*, a fishing basket.]
acutirostris, see Cyrtina acutirostris.
biplicata, see Cyrtina biplicata.
curvilineata, see Cyrtina curvilineata.
dalmani, see Cyrtina dalmani.
exporrecta, Wahlenberg, 1821, Nova. Acta. Regiæ. Soc. Sci., vol. 8, Niagara Gr., as identified by Hall in 24th Reg. Rep. [Sig. smooth, without wrinkles.]
exporrecta *var.* arrecta, Niagara Gr., as identified by Hall in 24th Reg. Rep. [Sig. erected, steep.]
hamiltonensis, see Cyrtina hamiltonensis.
missouriensis, see Cyrtina missouriensis.
myrtea, Billings, 1862, Pal. Foss., vol. 1, Mid. Sil. [Ety. proper name.]
occidentalis, see Cyrtina occidentalis.
rostrata, see Cyrtina rostrata.
triquetra, see Cyrtina triquetra.
umbonata, see Cyrtina umbonata.
CYRTINA, Davidson, 1858, Monog. Brit. Carb. Brach. [Ety. the dimniutive of *Cyrtia* is *Cyrtidium*, but the author said he preferred bad Greek to a long name.]
acutirostris, Shumard, 1855, (Cyrtia acutirostris) Geo. Rep. Mo., Chemung Gr. [Sig. sharp-beaked.]
affinis, Billings, 1874, Pal. Foss., vol. 2, Devonian. [Sig. near to.]
billingsi, Meek, 1868, Trans. Chi. Acad. Sci., Ham. Gr. [Ety. proper name.]
biplicata, Hall, 1857, (Cyrtia biplicata) 10th Reg. Rep., Schoharie grit and Cornif. Gr. [Sig. double-plicated.]
crassa, Hall, 1867, Pal. N. Y., vol. 4, Cornif. Gr. [Sig. thick.]
curvilineata, White, 1865, (Cyrtia curvilineata) Proc. Bost. Soc. Nat. Hist., vol. 9, Ham. Gr. [Sig. having bent lines.]
dalmani, Hall, 1857, (Cyrtia dalmani) 10th Reg. Rep., Low. Held. Gr. [Sig. proper name.]
hamiltonensis, Hall, 1857, (Cyrtia hamiltonensis) 10th Reg. Rep., Schoharie grit, Cornif. and Ham. Gr. [Ety. from the Hamilton group.]
hamiltonensis *var.* recta, Hall, 1867, Pal. N. Y., vol. 4, Ham. Gr. [Sig. straight.]
missouriensis, Swallow, 1860, (Cyrtia missouriensis) Trans. St. Louis Acad. Sci., Ham. Gr. [Ety. proper name.]
occidentalis, Swallow, 1860, (Cyrtia occidentalis) Trans. St. Louis Acad. Sci., Ham. Gr. [Sig. western.]
panda, Meek, 1868, Trans. Chi. Acad. Sci., Ham. Gr. [Sig. bent downwards.]
pyramidalis, Hall, 1852, (Spirifer pyramidalis) Pal. N. Y., vol. 2, Niagara Gr. [Sig. pointed like a pyramid.]
rostrata, Hall, 1857, (Cyrtia rostrata) 10th Reg. Rep., Oriskany sandstone. [Sig. beaked.]
triquetra, Hall, 1858, (Cyrtia triquetra) Geo. Rep. Iowa, vol. 1, pt. 2, Ham. Gr. [Sig. a triangle.]
umbonata, Hall, 1858, (Cyrtia umbonata) Geo. Rep. Iowa, vol. 1, pt. 2, Ham. Gr. [Sig. protuberant, bossed.]
DELTHYRIS, Dalman, 1827, syn. for Spirifera.
acanthoptera, syn. for Spirifera disjuncta.
acuminata, Conrad, see Spirifera acuminata.
acuminata, Hall, syn. for Spirifera mesacostalis.
acutilirata, see Orthis acutilirata.
arenosa, see Spirifera arenosa.
audacula, see Spirifera audacula.
bialveata, see Spirifera bialveata.
bilobata, see Orthis bilobata.
brachynota, see Spirifera brachynota.
chemungensis, syn. for Spirifera disjuncta.
congesta, see Spirifera congesta.
cuspidata, syn. for Spirifera disjuncta.
decemplicata, see Spirifera decemplicata.
deltoidea, syn. for Orthis lynx.
disjuncta, see Spirifera disjuncta.
duodenaria, see Spirifera duodenaria.
duplicata, see Spirifera duplicata.
euruteines, see Spirifera euruteines.
expansa, see Pterotheca expansa.
fimbriata, see Spirifera fimbriata.
granulifera, see Spirifera granulifera.
granulosa, see Spirifera granulosa.
inermis, see Spirifera disjuncta.
lævis, see Spirifera lævis.
macronota, see Spirifera macronota.
macropleura, see Spirifera macropleura.
medialis, see Spirifera medialis.
mesacostalis, see Spirifera mesacostalis.
mesastrialis, see Spirifera mesastrialis.
microptera, syn. for Orthis lynx.
mucronata, see Spirifera mucronata.
niagarensis, see Spirifera niagarensis.
pachyoptera, see Spirifera pachyoptera.
perlata, see Spirifera disjuncta.
prolata, see Spirifera prolata.
prora, see Spirifera prora.
radiatus, see Spirifera radiata.
raricosta, see Spirifera raricosta.
rugatina, see Spirifera rugatina.
sculptilis, see Spirifera sculptilis.
staminea, see Spirifera staminea.
triloba, see Spirifera triloba.
undulatus, see Spirifera undulata.
varica, see Orthis varica.
ziczac, see Spirifera ziczac.
DICRANISCUS, Meek, syn. for Camarella.
ortoni, see Camarella ortoni.
DINOBOLUS, Hall, March, 1871, notes on Brachiopoda. [Ety. *dis*, twice; *Obolus*, a genus of shells.]
canadensis, Billings, 1857, (Obolus canadensis) Rep. of Progr. Can. Sur., Black Riv. Gr. [Ety. proper name.]
conradi, Hall, 1868, (Obolus conradi) 20th Reg. Rep. N. Y., Niagara Gr. [Ety. proper name.]
galtensis, Billings, 1872, (Obolellina galtensis) Guelph Gr. [Ety. proper name.]

magnificus, Billings, 1872, (Obolellina magnifica) Canadian Naturalist, vol. 7, Black Riv. Gr. [Sig. magnificent.]
DISCINA, Lamarck, 1819, Hist. Nat. Anim. sans. Vert. [Ety. *discus*, a flat round plate; the termination *inus*, implying resemblance.]
acadica, Hartt, 1868, Acad. Geol., St. John's Gr. [Ety. proper name.]
alleghania, Hall, 1860, 13th Reg. Rep., Chemung Gr. [Ety. proper name.]
ampla, Hall, 1867, Pal. N. Y., vol. 4, Oriskany sandstone. [Sig. full sized.] Proposed instead of *D. grandis* of Hall.
capax, White, 1862, Proc. Bost. Soc. Nat. Hist., vol. 9, Chemung Gr. [Sig. capacious.]
caputiformis, syn. for D. nitida.
circe, Billings, 1862, Pal. Foss., vol. 1, Trenton Gr. [Ety. mythological name.] See remarks on D. lamellosa.
conradi, Hall, 1859, Pal. N. Y., vol. 3, Low. Held. Gr. [Ety. proper name.]
convexa, Shumard, 1858, Trans. St. Louis Acad. Sci., Coal Meas. [Sig. convex.]
discus, Hall, 1859, Pal. N. Y., vol. 3, Low. Held. Gr. [Sig. a quoit.]
doria, Hall, 1863, 16th Reg. Rep., Ham. Gr. [Ety. mythological name.]
elmira, Hall, 1863, 16th Reg. Rep., Chemung Gr. [Ety. proper name.]
gallaheri, Winchell, 1865, Proc. Acad. Nat. Sci., Marshall Gr. [Ety. proper name.]
grandis, Vanuxem, 1842, Geo. Rep. 3rd Dist. N. Y., Cornif. & Ham. Gr. [Sig. great.]
grandis, Hall, 1859, Pal. N. Y., vol. 3. The name was preoccupied. See *D. ampla*.
humilis, Hall, 1863, 16th Reg. Rep., Marcellus slate & Ham. Gr. [Sig. dwarfish.]
inutilis, Hall, 1863, 16th Reg. Rep., Potsdam Gr. [Sig. trifling.]
lamellosa, Hall, 1847, (Orbicula lamellosa). The name was preoccupied by Broderick in 1833. Billings has described it as *D. circe*.
lodensis, Vanuxem, 1842, (Orbicula lodensis) Geo. Rep. 3rd Dist. N. Y., Genessee slate. [Ety. proper name.]
media, Hall, 1863, 16th Reg. Rep., Ham. and Chemung Gr. [Sig. intermediate.]
minuta, Hall, 1843, (Orbicula minuta) Geo. Rep. 4th Dist. N. Y., Marcellus shale. [Sig. very small.]
missouriensis, Shumard, 1858, Trans. St. Louis Acad. Sci., Coal Meas. Syn. for D. nitida.
neglecta, Hall, 1863, 16th Reg. Rep., Chemung Gr. [Sig. overlooked.]
newberryi, Hall, 1863, 16th Reg. Rep., Chemung Gr. [Ety. proper name.]
nitida, Phillips, 1836, (Orbicula nitida) Geo. of York., Coal Meas. [Sig. neat, smooth.]

patellaris, Winchell, 1863, Proc. Acad. Nat. Sci., Chemung Gr. [Sig. a small plate.]
pelopea, Billings, 1862, Pal. Foss., vol. 1, Trenton Gr. [Ety. mythological name.]
pleuritis, Meek, 1875, Ohio Pal., vol. 2, Waverly Gr. [Ety. from *pleuron*, the side.]
randalli, Hall, 1863, 16th Reg. Rep., Ham. Gr. [Ety. proper name.]
seneca, Hall, 1863, 16th Reg. Rep., Ham. Gr. [Ety. proper name.]
subtrigonalis, McChesney, 1865, Desc. New Pal. Foss., Coal Meas. [Sig. somewhat triangular.]
tenuilamellata, Hall, 1852, (Orbicula tenuilamellata) Pal. N. Y., vol. 2, Niagara Gr. [Sig. having very thin plates.]
tenuilamellata, var. subplana, Hall, 1860, Can. Nat. and Geol., vol. 5, Up. Sil. [Sig. somewhat smooth.]
trigonalis, syn. for D. subtrigonalis.
truncata, Hall, 1862, 16th Reg. Rep., Chemung Gr. and Genessee slate. [Sig. truncated.]
tullia, Hall, 1863, 16th Reg. Rep., Tully limestone. [Ety. proper name.]
vanuxemi, Hall, 1859, Pal. N. Y., vol. 3, Low. Held. Gr. [Ety. proper name.]
EATONIA, Hall, 1857, 10th Reg. Rep., [Ety. proper name.]
eminens, Hall, 1859, Pal. N. Y., vol. 3, Low. Held. Gr. [Sig. eminent, standing out.]
medialis, Vanuxem, 1843, (Atrypa medialis) Geo. Rep. 3rd Dist. N. Y., Low. Held. Gr. [Sig. middle.]
peculiaris, Conrad, 1841, (Atrypa peculiaris) Ann. Rep. N. Y., Oriskany and Low. Held. Gr. [Sig. peculiar.]
pumila, Hall, 1859, Pal. N. Y., vol. 3, Oriskany sandstone. [Sig. a dwarf or pigmy.]
singularis, Vanuxem, 1843, (Atrypa singularis) Geo. Rep. 3rd Dist. N. Y., Low. Held. Gr. [Sig. singular.]
sinuata, Hall, 1859, Pal. N. Y., vol. 3, Oriskany sandstone. [Sig. marked with depressions.]
whitfieldi, Hall, 1859, Pal. N. Y., vol. 3, Oriskany sandstone. [Ety. proper name.]
EICHWALDIA, Billings, 1858, Rep. of Progr. [Ety. proper name.]
anticostiensis, Billings, 1866, Catal. Sil. Foss. Antic., Hud. Riv. Gr. [Ety. proper name.]
concinna, Hall, 1868, 20th Reg. Rep., Niagara Gr. [Sig. beautiful.]
coralifera, Hall, 1852, (Atrypa coralifera) Pal. N. Y., vol. 2, Niagara Gr. Prof. Davidson regards this shell as identical with *E. Capewelli*, which was described in 1848, in Bull. Soc. Geol. France, vol. 3. [Sig. coral-bearing.]
gibbosa, Hall, 1868, 20th Reg. Rep., Niagara Gr. [Sig. gibbous.]

BRACHIOPODA.

reticulata, Hall, 1862, (Rhynchonella (?) reticulata) Trans. Alb. Inst., vol. 4, Niagara Gr. [Sig. reticulated.]
subtrigonalis, Billings,1858, Rep. of Prog., Black Riv. Gr. [Sig. somewhat three-cornered.]

Goniocælia, syn. for Pentagonia.

GYPIDULA, Hall, 1867, Pal. N. Y., vol. 4. [Ety. *gyps*, vulture; in allusion to the strongly incurved beak.]
laeviuscula, Hall, 1867, Pal. N. Y., vol. 4, Devonian. [Sig. slightly smooth.]
obsolescens, see Pentamerella obsolescens.
occidentalis, Hall, 1858, (Pentamerus occidentalis) Geo. Rep. Iowa, vol. 1, pt. 2, Ham. Gr. [Sig. western.]

Hemipronites, Pander,1830. This name, never having been defined, has been superseded by *Streptorhynchus*, if the two names refer to the same form.

Hipparionyx, Vanuxem, 1842, Geo. 3rd Dist. N. Y. [Sig. colts hoof.] Syn. for Orthis. The genus was founded on a cast.
consimilis, Vanuxem, syn. for Atrypa reticularis.
proximus, syn. for Orthis hipparionyx.
similaris, Vanuxem, 1842, Geo. Rep. 3rd Dist. N. Y., Oriskany sandstone. [Sig. similar.]

IPHIDEA, Billings, 1874, Pal. Foss., vol. 2. [Ety. proper name.]
bella, Billings,˙1874, Pal. Foss., vol. 2, Potsdam Gr. [Sig. beautiful.]
labradoricus, Billings, 1861, (Obolus labradoricus) Pal. Foss., vol. 1, Potsdam Gr. [Ety. proper name.]
sculptilis, Meek, 1872, Hayden's Geo. Rep., Quebec Gr. [Sig. carved.]

KONINCKIA, Suess, 1853. MS. published by Woodward, 1854, in Manual of Mollusca. [Ety. proper name.]
americana, Swallow, 1863, Trans. St. Louis Acad. Sci., Low. Carb. [Ety. proper name.]

KUTORGINA, Billings, 1861, Pal. Foss., vol. 1. [Ety. proper name.]
cingulata, Billings, 1861, Pal. Foss, vol. 1, Potsdam Gr. [Sig. encircled with lines.]

LEIORHYNCHUS, Hall, 1860, 13th Reg. Rep. [Ety. *leios*, smooth; *rhynchos*, a beak.]
dubius, Hall, 1867, Pal. N. Y., vol. 4, Marcellus shale. [Sig. doubtful.]
globuliformis, Vanuxem, 1842, (Atrypa globuliformis) Geo., 3rd Dist. N. Y., Chemung Gr. [Sig. in the form of a globe.]
huronensis, Nicholson, 1874, Geo. Mag. Lond., n. s., vol. 1, Ham. Gr. [Ety. proper name.]
iris, Hall, 1867, Pal. N. Y., vol. 4, Chemung Gr. [Ety. proper name.]
kelloggi, Hall, 1867, Pal. N. Y., vol. 4, Chemung Gr. [Ety. proper name.]
limitaris, Vanuxem, 1842, (Orthis limitaris) Geo. 3rd Dist. N. Y. (Atrypa limitaris, 4th Dist. N. Y.) Marcellus shale. [Ety. supposed to be limited to, and to characterize, the Marcellus shale.]
mesacostalis, Hall, 1843, (Atrypa mesacostalis) Geo. 4th Dist. N. Y., Chemung Gr. [Sig. middle-ribbed.]
multicosta, Hall, 1860, 13th Reg. Rep., Ham. Gr. [Sig. many-ribbed.]
mysia, Hall, 1867, Pal. N. Y., vol. 4, Marcellus shale. [Ety. proper name.]
newberryi, Hall, 1873, 23rd Reg. Rep., Chemung Gr. [Ety. proper name.]
quadricostatus, Vanuxem, 1842, (Orthis quadricostata) Geo. 3rd Dist. N. Y., Genessee slate. [Sig. four-ribbed.]
.sesquiplicatus, Winchell, 1866, Rep. Low. Penin. Mich., Ham. Gr. [Sig. a plication and a half.]
sinuatus, Hall, 1867, Pal. N. Y., vol. 4, Chemung Gr. [Sig. wavy.]

LEPTÆNA, Dalman, 1827, Kongl. Vet. Acad. Handl. [Ety. *leptos*, thin.]
alternata, see Strophomena alternata.
alternistriata, see Strophomena alternistriata.
analoga, see Strophomena analoga.
aspera, James, syn. for L. sericea.
bipartita, see Strophomena bipartita.
camerata, see Strophomena camerata.
concava, Hall, 1859, Pal. N. Y., vol. 3, Low. Held. Gr. [Sig. concave.]
decipiens, Billings, 1862, Pal. Foss., vol. 1, Quebec Gr. [Sig. doubtful.]
deflecta, see Streptorhynchus deflecta.
deltoidea, see Strophomena deltoidea.
depressa, see Strophomena depressa.
fasciata, see Strophomena fasciata.
filitexta, see Strophomena filitexta.
fragaria, syn. for Productella subaculeata.
incrassata, see Strophomena incrassata.
indenta, see Strophodonta indenta.
laticosta, syn. for Tropidoleptus carinatus.
membranacea, see Productella hirsuta.
mesacosta, Shumard, 1855, Geo. Rep. Mo., Trenton Gr. [Sig. middle-ribbed.]
nasuta, see Strophomena nasuta.
nucleata, Hall, 1857, 10th Reg. Rep., Oriskany sandstone. [Sig. kerneled.]
obscura, see Strophomena obscura.
orthididea, see Strophomena orthididea.
planoconvexa, see Streptorhynchus planoconvexa.
planumbona, see Streptorhynchus planumbona.
plicifera, see Strophomena plicifera.
profunda, see Strophodonta profunda.
punctulifera, see Strophodonta punctulifera.
quadrilatera, syn. for Strophomena rhomboidalis.
recta, see Streptorhynchus recta.
rugosa, see Strophomena rugosa.
semiovalis, syn. for L. sericea.
sericea, Sowerby, 1839, Murch. Sil. Syst., Trenton to Clinton Gr. [Sig. silky.]
sordida, Billings, 1862, Pal. Foss., vol. 1, Quebec Gr. [Sig. paltry.]

subtenta, see Streptorhynchus subtentus.
tenuilineata, see Strophomena tenuilineata.
tenuistriata, see Strophomena tenuistriata.
transversalis, Dalman, 1827, Vet. Acad. Handl., Anticosti Gr. [Sig. crosswise.]
trilobata, see Strophomena trilobata.

LEPTOBOLUS, Hall, 1871, Pamphlet. [Sig. minute Obolus.]
   insignis, Hall, Pamphlet, Utica slate. [Sig. marked naturally.]
   lepis, Hall, 1871, Pamphlet, Cin'ti Gr. [Sig. a scale.]
   occidentalis, Hall, 1871, Pamphlet, Hud. Riv. Gr. [Sig. western.]

LEPTOCŒLIA, Hall, 1857, 10th Reg. Rep., and 1859, 12th Reg. Rep. [Ety. *leptos*, minute; *koilia*, belly; in allusion to the shallow visceral cavity.]
   acutiplicata, Conrad, 1841, (Atrypa acutiplicata) Ann. Rep. N. Y., Up. Held. Gr. [Sig. sharply-plicated.]
   concava, see Cœlospira concava.
   dichotoma, see Cœlospira dichotoma.
   disparilis, see Cœlospira disparilis.
   fimbriata, Hall, 1859, Pal. N. Y., vol. 3, Oriskany sandstone. [Sig. fringed.]
   flabellites, Conrad, 1841, (Atrypa flabellites) Ann. Rep. N. Y., Oriskany sandstone. [Sig. a fan.]
   hemispherica, Sowerby, 1839, (Atrypa hemispherica) Murch. Sil. Syst., Clinton Gr. [Sig. hemispherical.]
   imbricata, Hall, 1857, 10th Reg. Rep., Low. Held. Gr. [Sig. imbricating lamellæ of growth.]
   intermedia, Hall, 1860, Can. Nat. & Geo., vol. 5, Up. Sil. [Sig. intermediate.]
   planoconvexa, Hall, 1852, (Atrypa planoconvexa) Pal. N. Y., vol. 2, Clinton Gr. [Ety. *planus*, level; *convexus*, convex.]
   propria, syn. for Leptocœlia flabellites.

LINGULA, Bruguiere, 1792, Encyc. Meth. [Ety. *lingula*, a little tongue.]
   acuminata, Conrad, 1839, Ann. Rep. N. Y., Potsdam and Calcif. Gr. [Sig. terminating sharply.]
   acutirostra, Hall, 1843, Geo. Rep. 4th Dist. N. Y., Clinton Gr. [Sig. sharp-beaked.]
   aequalis, Hall, 1847, Pal. N. Y., vol. 1, Trenton Gr. [Sig. equal.]
   alveata, Hall, 1863, 16th Reg. Rep., Ham. Gr. [Sig. channeled.]
   ampla, Owen, 1852, Geo. Sur. Wis., Iowa and Minn., Potsdam Gr. [Sig. full-sized.]
   antiqua, Emmons, 1842, Geo. Rep. N. Y., Potsdam Gr. [Sig. ancient.]
   antiquata, Emmons, 1856, Am. Geol., Trenton Gr. [Sig. ancient.]
   artemus, Billings, 1874, Pal. Foss., vol. 2, Passage beds between Up. Sil. and Devonian. [Ety. proper name.]
   attenuata, Sowerby. The fossil referred by Hall to this species, is described by Billings under the name of L. Daphne.
   aurora, see Lingulella aurora.

   belli, Billings, 1859, Can. Nat. Geo., vol. 4, Chazy Gr. [Ety. proper name.]
   briseis, Billings, 1862, Pal. Foss., vol. 4, Trenton Gr. [Ety. proper name.]
   canadensis, Billings, 1862, Pal. Foss., vol. 1, Hud. Riv. Gr. [Ety. proper name.]
   carbonaria, Shumard, 1858, Trans. St. Louis Acad. Sci., Coal Meas. [Sig. pertaining to the Coal Measures.]
   centrilineata, Hall, 1859, Pal. N. Y., vol. 3, Low. Held. Gr. [Sig. the central line from beak to base.]
   ceryx, Hall, 1863, 16th Reg. Rep., Scoharie grit. [Sig. a herald.]
   clintoni, Vanuxem, 1842, Geo. Rep. N. Y., Clinton Gr. [Ety. proper name.] Syn. for L. oblonga.
   cobourgensis, Billings, 1862, Pal. Foss., vol. 1, Trenton Gr. [Ety. proper name.]
   concentrica, Conrad, 1839, Ann. Rep. N. Y., Genessee slate. [Sig. arranged in concentric lines.]
   covingtonensis, Hall & Whitfield, 1875, Ohio Pal., vol. 2, Cin'ti Gr. [Ety. proper name.]
   crassa, Hall, 1847, Pal. N. Y., vol. 1, Trenton Gr. [Sig. thick.]
   cuneata, Conrad, 1839, Geo. Rep. N. Y., Medina sandstone. [Sig. wedge-shaped.]
   curta, Conrad, 1842, Jour. Acad. Nat. Sci., vol. 8, Trenton Gr. [Sig. short.]
   cuyahoga, Hall, 1863, 16th Reg. Rep., Green shales, upper part of Chemung Gr. [Ety. proper name.]
   cyane, Billings, 1865, Pal. Foss., vol. 1, Quebec Gr. [Ety. mythological name.]
   daphne, Billings, 1862, Pal. Foss., vol. 1, Trenton Gr. [Ety. mythological name.] See *L. attenuata*.
   delia, Hall, 1863, 16th Reg. Rep., upper part of Ham. Gr. [Ety. mythological name.]
   densa, Hall, 1867, Pal. N. Y., vol. 4, upper part of Ham. Gr. [Sig. thick.]
   desiderata, Hall, 1863, 16th Reg. Rep., Corniferous Gr. [Sig. longed for, rare.]
   elegantula, syn. for Lingula quadrata.
   elliptica, Hall, 1843, Geo. Rep. 4th Dist. N. Y., Clinton Gr. The name was preoccupied by Phillips in 1836.
   elliptica, Emmons, 1856, Am. Geol. The name was preoccupied.
   elongata, Hall, 1847, Pal. N. Y., vol. 1, Trenton Gr. [Sig. elongated.]
   exilis, Hall, 1860, 13th Reg. Rep., Marcellus shale. [Sig. thin, fine.]
   eva, Billings, 1861, Can. Nat. Geo., vol. 6, Black Riv. Gr. [Ety. proper name.]
   forbesi, Billings, 1862, Pal. Foss. vol. 1, Hud. Riv. & Mid. Sil. Gr. [Ety. proper name.]
   halli, White, 1862, Proc. Bost. Soc. Nat. Hist., vol. 9, Burlington Gr. [Ety. proper name.]

huronensis, Billings, 1859, Can. Nat. Geo., vol. 4, Chazy & Black Riv. Gr. [Ety. proper name.]
insularis, Billings, 1866, Catal. Sil. Foss. Antic., Anticosti Gr. [Sig. upon an island.]
iole, Billings, 1865, Pal. Foss., vol. 1, Quebec Gr. [Ety. mythological name.]
irene, Billings, 1862, Pal. Foss., vol. 1, Quebec Gr. [Ety. proper name.]
iris, Billings, 1865, Pal. Foss., vol. 1, Quebec Gr. [Ety. proper name.]
kingstonensis, Billings, 1862, Pal. Foss., vol. 1, Black Riv. Gr. [Ety. proper name.]
lamellata, Hall, 1843, Geo. Rep. 4th Dist. N. Y., Clinton & Niagara Gr. [Sig. composed of thin plates.]
leaena, Hall, 1863, 16th Reg. Rep., Ham. Gr. [Ety. proper name.]
ligea, Hall, 1860, 13th Reg. Rep., upper part of Ham. Gr. [Ety. mythological name.]
ligea var. Hall, 1867, Pal. N. Y., vol. 4, Portage Gr.
lucretia, Billings, 1874, Pal. Foss., vol. 2, Passage beds between Up. Sil. and Devonian. [Ety. proper name.]
lyelli, Billings, 1859, Can. Nat. Geo., vol. 4, Calcif. & Chazy Gr. [Ety. proper name.]
maida, Hall, 1863, 16th Reg. Rep., Ham. Gr. [Ety. proper name.]
manni, Hall, 1863, 16th Reg. Rep., Up. Held. Gr. [Ety. proper name.]
mantelli, Billings, 1859, Can. Nat. Geo., vol. 4, Calcif. Gr. [Ety. proper name.]
manticula, White, 1864, Rep. Invert. Foss., Quebec Gr. [Sig. a small wallet.]
mathewi, Hartt, 1868, Acad. Geol., St. Johns Gr. [Ety. proper name.]
melie, Hall, 1867, Pal. N. Y., vol. 4, Chemung Gr. [Ety. mythological name.]
membranacea, Winchell, 1863, Proc. Acad. Nat. Sci. Phil., vol. 15, Low. Carb. [Sig. like a membrane.]
minuta, Meek, 1868, Trans. Chi. Acad. Sci., Devonian. [Sig. very small.]
mosia, Hall, 1863, 16th Reg. Rep., Potsdam Gr. [Ety. proper name.]
murrayi, Billings, 1874, Pal. Foss., vol. 2, Potsdam Gr. [Ety. proper name.]
mytiloides, Sowerby, 1812, Min. Conch., Tab. 19, Coal Meas. [Sig. like the *Mytilus*, or mussel shell.]
nebrascensis, Meek, 1872, Pal. E. Neb., Coal Meas. [Ety. proper name.]
norwoodi, James, 1875, Cin. Quar. Jour. Sci., vol. 2, Cin'ti Gr. [Ety. proper name.]
nuda, Hall, 1863, 16th Reg. Rep., Ham. Gr. [Sig. naked.]
nympha, Billings, 1865, Pal. Foss., vol. 1, Quebec Gr. [Ety. mythological name.]

oblata, Hall, 1843, Geo. Rep. 4th Dist. N. Y., Clinton Gr. [Sig. oblate.]
oblonga, Conrad, 1839, Ann. Rep. N. Y., Clinton Gr. [Sig. rather oblong.]
obtusa, Hall, 1847, Pal. N. Y., vol. 1, Trenton Gr. [Sig. obtuse.]
ovata, McCoy, 1844, Syn. Sil. Foss., Ireland. Not clearly identified in America.
paliformis, Hall, 1860, 13th Reg. Rep., Ham. Gr. [Sig. shovel-like.]
papillosa, Emmons, 1856, Am. Geol., Trenton Gr. [Sig. covered with pimples.]
perlata, Hall, 1859, Pal. N. Y., vol. 3, Low. Held. Gr. [Sig. very wide.]
perovata, Hall, 1852, Pal. N. Y., vol. 2, Clinton Gr. [Sig. very ovate or nearly round.]
perryi, Billings, 1861, Pal. Foss., vol. 1, Black Riv. Gr. [Ety. proper name.]
philomela, Billings, 1862, Pal. Foss., vol. 1, Trenton Gr. [Ety. mythological name.]
pinniformis, see Lingulepis pinniformis.
polita, see Obolella polita.
prima, see Lingulepis prima.
prima, Emmons, 1856, Am. Geol. This name was preoccupied.
progne, Billings, 1862, Pal. Foss., vol. 1, Trenton Gr. [Ety. mythological name.]
punctata, Hall, 1863, 16th Reg. Rep., Ham. Gr. [Sig. dotted.]
quadrata, Eichwald, 1829, Zool. Specialis., vol. 1, Trenton to Mid. Sil. [Ety. in allusion to the somewhat four-sided shape.]
quebecensis, Billings, 1862, Pal. Foss., vol. 1, Quebec Gr. [Ety. proper name.]
rectilatera, Hall, 1859, Pal. N. Y., vol. 3, Low. Held. Gr. [Sig. straight-sided.] This name was preoccupied.
rectilateralis, Emmons, 1842, Geo. Rep. N. Y., Utica slate. [Sig. straight-sided.]
riciniformis, Hall, 1847, Pal. N. Y., vol. 1, Trenton Gr. [Sig. like a tike or tick.]
scotica, Davidson, 1860, Monogr. Scot. Carb. Brach., Waverly Gr. [Ety. proper name.]
spathata, Hall, 1859, Pal. N. Y., vol. 3, Low. Held. Gr. [Sig. spatula-shaped.]
spatiosa, Hall, 1859, Pal. N. Y., vol. 3, Low. Held. Gr. [Sig. large.]
spatulata, Vanuxem, 1842, Geo. Rep. 3rd and 4th Dist. N. Y., Genessee slate. [Sig. spatula-shaped.]
subspatulata, Meek & Worthen, 1868, Geo. Sur. Ill., vol. 3, Ham. Gr. [Sig. somewhat spatula-shaped.]
trentonensis, Conrad, 1842, Jour. Acad. Nat. Sci., vol. 8, Trenton Gr. [Ety. proper name.]
umbonata, Cox, 1857, Geo. Sur. Ky., vol. 3, Coal Meas. [Sig. protuberant.]

vanhorni, S. A. Miller, 1875, Cin. Quar. Jour. Sci., vol. 2, Cin'ti Gr. [Ety. proper name.]
winona, Hall, 1863, 16th Reg. Rep., Potsdam Gr. [Ety. proper name.]

LINGULELLA, Salter, 1861, Mem. Geo., North Wales. [Ety. dinimutive of *Lingula*.]
(?) affinis, Billings, 1874, Pal. Foss., vol. 2, Potsdam Gr. [Sig. near to.]
aurora, Hall, 1861, (Lingula aurora) Geo. Rep. Wis., Potsdam sandstone. [Sig. morning.]
cincinnatiensis, Hall & Whitfield, 1875, Ohio Pal., vol. 2, Cin'ti Gr. [Ety. proper name.]
lamborni, Meek, 1871, Proc. Acad. Nat. Sci., Potsdam Gr. [Ety. proper name.]
(?) spissa, Billings, 1874, Pal. Foss., vol. 2, Potsdam Gr. [Sig. thick.]

LINGULEPIS, Hall, 1863, 16th Reg. Rep. [Ety. *lingula*, a little tongue; *lepis*, a scale.]
morsii, N. H. Winchell, 1876, (Lingula morsensis) Geol. Fillmore Co., Minn., St. Peters sandstone. [Ety. proper name—in honor of Mr. Morse.]
pinniformis, Owen, 1852, (Lingula pinniformis) Geo. Rep. Iowa, Wis. & Minn. Potsdam Gr. [Sig. like the *Pinna*.]
prima, Conrad, 1847, (Lingula prima) Pal. N. Y., vol. 1, Potsdam Gr. [Sig. first.]

LINGULOPS, Hall, 1871 notes on Brachiopoda. [Ety. *lingula*, a genus of shells; *opsis*, appearance.]
whitfieldi, Hall, 1871, notes on Brachiopoda, Low. Sil. [Ety. proper name.]

MARTINIA, McCoy, 1844, syn. Carb. Foss., Ireland. [Ety. proper name.]
athyroides, Winchell, 1866, Rep. Low. Peninsula Mich., Ham. Gr. [Ety. resembling an *athyris*.]
planoconvexa, syn. for Spirifera planoconvexa.

MEEKELLA, White & St. John, 1868, Trans. Chi. Acad. Sci. [Ety. proper name.]
striato-costata, Cox, 1857, (Plicatula striato-costata) Geo. Rep. Ky., vol. 3, Coal Meas. [Ety. *striatus*, lined; *costatus*, ribbed.]

Meganteris aequiradiata, see Rensselæria aequiradiata.
cumberlandiæ, see Rensselæria cumberlandiæ.
elliptica, see Rensselæria elliptica.
elongata, see Amphigenia elongata.
lævis, see Rensselæria lævis.
mutabilis, see Rensselæria mutabilis.
ovalis, see Rensselæria ovalis.
ovoides, see Rensselæria ovoides.
subtrigonalis, see Amphigenia elongata var. subtrigonalis.
suessiana, see Rensselæria suessiana.

MERISTA, Suess, 1851, Jahrb. Geol. Reichs. Austalt. [Ety. *meros*, a part.]
arcuata, Hall, 1859, Pal. N. Y., vol. 3, Low. Held. Gr. [Sig. bent or arched.]
barrisi, Hall, 1862, 15th Reg. Rep., Marcellus shale. [Ety. proper name.]
bella, Hall, 1859, Pal. N. Y., vol. 3, Low. Held. Gr. [Sig. beautiful.]
bisulcata, Vanuxem, 1843, (Atrypa bisulcata) Geo. Rep. 3rd Dist. N. Y., Low. Held. Gr. [Sig. double-depressed.]
doris, Hall, 1862, 15th Reg. Rep., Marcellus shale. [Ety. mythological name.]
haskinsi, Hall, 1862, 15th Reg. Rep., Ham. Gr. [Ety. proper name.]
houghtoni, Winchell, 1862, Proc. Acad. Nat. Sci., Portage Gr. [Ety. proper name.]
lævis, Vanuxem, 1843, (Atrypa lævis) Geo. Rep. 3rd Dist. N. Y., Low. Held. Gr. [Sig. smooth.]
lata, Hall, 1859, Pal. N. Y., vol. 3, Oriskany sandstone. [Sig. broad.]
lens, Winchell, 1866, Rep. Low. Peninsula Mich., Ham. Gr. [Sig. a lentil, here signifying a double convex *Merista*.]
meeki, Hall, 1859, Pal. N. Y., vol. 3, Low. Held. Gr. [Ety. proper name.]
princeps, Hall, 1859, Pal. N. Y., vol. 3, Low. Held. Gr. [Sig. primitive.]
subquadrata, Hall, 1859, Pal. N. Y., vol. 3, Low. Held. Gr. [Sig. somewhat four-cornered.]
sulcata, Vanuxem, 1842, (Atrypa sulcata) Geo. Rep. N. Y., Water Lime Gr. [Sig. wavy.]

MERISTELLA, Hall, 1860, 13th Reg. Rep. [Ety. diminutive of *Merista*.]
barrisi, Hall, 1860, 13th Reg. Rep., Marcellus shale & Ham. Gr. [Ety. proper name.]
cylindrica, Hall, 1852, (Atrypa cylindrica) Pal. N. Y., vol. 2, Clinton & Niagara Gr. [Sig. cylindrical.]
doris, 1860, Hall, 13th Reg. Rep., Schoharie Grit & Corniferous Gr. [Ety. mythological name.]
elissa, syn. for Meristella nasuta.
haskinsi, Hall, 1860, 13th Reg. Rep., Ham. Gr. [Ety. proper name.]
(?) hyale, Billings, 1862, (Charionella (?) hyale) Pal. Foss., vol. 1, Guelph Gr. [Ety. proper name.]
julia, Billings, 1862, (Athyris julia) Pal. Foss., vol. 1, Mid. Sil. [Ety. proper name.]
lenta, Hall, 1867, Pal. N. Y., vol. 4, Oriskany sandstone. [Sig. heavy.]
maria, see Meristina maria.
meta, Hall, 1867, Pal. N. Y., vol. 4, Ham. Gr. [Sig. having a conical form.]
multicosta, Hall, 1862, 15th Reg. Rep., Ham. Gr. [Sig. having many costæ.]
nasuta, Conrad, 1840, (Atrypa nasuta) Ann. Rep. N. Y., Schoharie Grit, Up. Held. Corniferous & Ham. Gr. [Sig. having a prominent nose.]
prinstana, Billings, 1862, (Athyris prinstana) Pal. Foss., vol. 1, Mid. Sil. [Ety. proper name.]

rostrata, Hall, 1843, (Atrypa rostrata) Geo. Rep. 4th Dist. N. Y., Ham. Gr. & Tully limestone. [Sig. beaked.]
scitula, Hall, 1843, (Atrypa scitula) Geo. 4th Dist. N. Y., Corniferous Gr. Prof. Hall regards *Charionella circe* as a syn. for this species. [Sig. pretty.]
umbonata, Billings, 1862, (Athyris umbonata) Pal. Foss., vol. 1, Mid. Sil. [Sig. protuberant.]
unisulcata, Conrad, 1841, (Atrypa unisulcata) Ann. Rep. N. Y., Up. Held. & Ham. Gr. [Sig. having one depression.]
MERISTINA, Hall, 1867, 20th Reg. Rep. N. Y. [Ety. *Merista*, a genus of shells; *inus*, implying resemblance.]
maria, Hall, 1863, (Meristella maria) Trans. Alb. Inst., vol. 4, Niagara Gr. [Ety. proper name.]
nitida, Hall, 1852, (Atrypa nitida) Pal. N. Y., vol. 2, Niagara Gr. [Sig. smooth.]
nitida *var*. oblata, Hall, 1852, (Atrypa nidita *var*. oblata) Pal. N. Y., vol. 2, Niagara Gr. [Sig. oblate.]
MONOMERELLA, Billings, 1861, Can. Nat. & Geo., vol. 6. [Ety. *monos*, one; *meros*, a part; *ella*, diminutive termination.]
newberryi, Hall & Whitfield, 1875, Ohio Pal., vol. 2, Niagara Gr. [Ety. proper name.]
orbicularis, Billings, 1871, Can. Nat., vol. 6, Guelph Gr. [Sig. *orbicular*.]
prisca, Billings, 1871, Can. Nat. & Geol., vol. 6, Guelph Gr. [Sig. ancient.]
NUCLEOSPIRA, Hall, 1859, Pal. N. Y., vol. 3. [Ety. *nucleus*, a kernel ; *spira*, a spire.]
barrisi, White, 1860, Bost. Jour. Nat. Hist., Kinderhook Gr. [Ety. proper name.]
concentrica, Hall, 1859, Pal. N. Y., vol. 3, Low. Held. Gr. [Sig. concentrically lined.]
concinna, Hall, 1843, (Atrypa concinna) Geo. 4th Dist. N. Y., Hamilton Gr. [Sig. handsome.]
elegans, Hall, 1859, Pal. N. Y., vol. 3, Low. Held. Gr. [Sig. elegant.]
pisiformis, Hall, 1859, (Orthis pisum, 1851,) Pal. N. Y., vol. 2,) Pal. N. Y., vol. 3, Niagara Gr. [Ety. from its resemblance to *Spirifer pisum* of Murchison.]
ventricosa, Hall, 1859, Pal. N. Y., vol. 3, Low. Held. Gr. This species was first described in 1856, in 9th Reg. Rep., as *Spirifera ventricosa*. [Sig. bulging out.]
OBOLELLA, Billings, 1861, Pal. Foss., vol. 1. [Ety. diminutive of Obolus, a small Greek coin.]
cælata, Hall, 1847, (Orbicula cælata) Pal. N. Y., vol. 1, Hud. Riv. Gr. [Sig. sculptured.]
chromatica, Billings, 1861, Pal. Foss., vol. 1, Potsdam Gr. [Sig. colored.]
cingulata, Billings, 1861, Pal. Foss., vol. 1, Potsdam Gr. [Sig. encircled with lines.]

circe, Billings, 1871, Can. Nat. and Geol., Potsdam Gr. [Ety. mythological name.]
crassa, Hall, 1847, (Orbicula crassa) Pal. N. Y., vol. 1, Hud. Riv. Gr. [Sig. thick.]
desiderata, Billings, 1862, Pal. Foss., vol. 1, Quebec Gr. [Sig. rare.]
desquamata, Hall, 1847, (Avicula desquamata) Pal. N. Y., vol. 1, Hud. Riv. Gr. [Sig. scaled off.]
gemma, Billings, 1871, Can. Nat. & Geol., Potsdam Gr. [Sig. a young bud.]
ida, Billings, 1862, Pal. Foss., vol. 1, Quebec Gr. [Ety. mythological name.]
miser, Billings, 1874, Pal. Foss., vol. 2, Low. Potsdam Gr. [Sig. paltry.]
nana, Meek & Hayden, 1861, Proc. Acad. Nat. Sci. Phil., Potsdam Gr. [Sig. dwarfish.]
nitida, Ford, 1863, Am. Jour. Sci. & Arts, 3rd ser., vol. 5, Low. Potsdam Gr. [Sig. neat.]
polita, Hall, 1861, Geo. Rep. Wis., (Lingula polita) Potsdam Gr. [Sig. smoothed.]
pretiosa, Billings, 1862, Pal. Foss., vol. 1, Quebec Gr. [Sig. valuable, precious.]
transversa, Hartt, 1868, Acad. Geol., St. John's Gr. [Ety. from the transversely oval form.]
Obolellina, Billings, Dec., 1871, syn. for Dinobolus.
  canadensis, see Dinobolus canadensis.
  galtensis, see Dinobolus galtensis.
  magnifica, see Dinobolus magnificus.
OBOLUS, Eichwald, 1829, Zoologia Specialis, vol. 1. [Ety. *obolus*, a small coin.]
  canadensis, see Dinobolus canadensis.
  conradi, see Dinobolus conradi.
  galtensis, see Trimerella galtensis.
  labradoricus, see Iphidea labradorica.
  (?) murrayi, Billings, 1865, Pal. Foss., vol. 1, Quebec Gr. [Ety. proper name.]
Orbicula, Cuvier, 1808, syn. for Crania.
  cælata, see Obolella cælata.
  cancellata, see Trematis cancellata.
  corrugata, see Crania corrugata.
  crassa, see Obolella crassa.
  deformata, see Crania deformata.
  excentrica, see Crania excentrica.
  filosa, see Schizocrania filosa.
  grandis, see Discina grandis.
  lamellosa, see Discina lamellosa.
  lodensis, see Discina lodensis.
  minuta, see Discina minuta.
  nitida, see Discina nitida.
  prima, see Crania prima.
  quadricostata, see Crania quadricostata.
  squamiformis, see Pholidops squamiformis.
  subtruncata, see Pholidops subtruncatus.
  tenuilamellata, see Crania tenuilamellata.
  terminalis, see Trematis terminalis.
  truncata, see Crania truncata.
ORTHIS, Dalman, 1827, Kongl. Vet. Acad. Handl. [Ety. *orthos*, straight, in allusion to the straight hinge line.]

BRACHIOPODA. 117

acuminata, Billings, 1859, Can. Nat. Geo., vol. 4, Chazy Gr. [Sig. sharp-pointed.]
acutilirata, Conrad, 1842, (Delthyris acutilirata) Jour. Acad. Nat. Sci., vol. 8, Cin'ti Gr. [Sig. sharply ridged.]
æquivalvis, Hall, 1847, Pal. N. Y., vol. 1, Trenton Gr. [Sig. equal-valved.]
æquivalva, Shaler. The name was preoccupied.
æquivalvis, Hall, 1857, syn. for *Orthis eryna*. Moreover the name was preoccupied.
alata, Shaler. The name was preoccupied.
alsus, Hall, 1863, 16th Reg. Rep., Schoharie Grit. [Sig. cold. (?)]
anticostiensis, syn. for Orthis porcata.
apicalis, Billings, 1865, Pal. Foss., vol. 1, Quebec Gr. [Sig. sharp-pointed.]
arctostriata, Hall, 1860, 13th Reg. Rep., Ham. Gr. [Sig. closely striated.]
armanda, Billings, 1865, Pal. Foss., vol. 1, Quebec Gr. [Ety. proper name.]
assimilis, Hall, 1859, Pal. N. Y., vol. 3, Low. Held. Gr. [Sig. very-like.]
aurelia, Billings, 1874, Pal. Foss., vol. 2, Devonian. [Ety. proper name.]
barabuensis, Winchell, 1864, Am. Jour. Sci. & Arts, 2nd ser., vol. 37, Potsdam Gr. [Ety. proper name.]
battis, Billings, 1865, Pal. Foss., vol. 1, Quebec Gr. [Ety. mythological name.]
bellarugosa, Conrad, 1843, Proc. Acad. Nat. Sci. Phil., vol. 1, Trenton Gr. [Sig. beautifully wrinkled.]
bellula, James, 1873, Ohio Pal., vol. 1, Cin'ti Gr. [Sig. pretty.]
biforata, Schlotheim, 1820, (Terebratulites biforatus) Petrefact, Trenton & Hud. Riv. Gr. [Sig. two-holed or double-doored.]
billingsi, Hartt, 1868, Acad. Geol., St. John's Gr. [Ety. proper name.]
biloba, Linnæus, 1749, (Anomia biloba) Linne. Syst., Niagara Gr. [Sig. two-lobed.]
bilobata, Conrad, 1838, (Delthyris bilobata) Ann. Rep. N. Y., Low. Held. Gr. The name was preoccupied by Sowerby.
bisulcata, see Camarella bisulcata.
borealis, Billings, 1859, Can. Nat. Geo., vol. 4, Chazy & Trenton Gr. [Sig. northern.]
carbonaria, Swallow, 1858, syn. for Orthis pecosi.
carinata, Hall, 1843, Geo. Rep. 4th Dist. N. Y., Portage & Chemung Gr. [Sig. keeled.]
carleyi, syn. for Orthis retrorsa.
centrilineata, Hall, 1847, Pal. N. Y., vol. 1, Hud. Riv. Gr. [Ety. *centrum*, center; *lineatus*, striated.]
circulus, Hall, 1843, Geo. Rep. 4th Dist. N. Y., Clinton Gr. [Sig. a circle.]
clarkensis, Swallow, 1863, Trans. St. Louis Acad. Sci., Keokuk Gr. [Ety. proper name.]

cleobis, Hall, 1863, 16th Reg. Rep., Onondaga limestone & Up. Held. Gr. [Ety. mythological name.]
clytie, Hall, 1861, 14th Reg. Rep., Cin'ti Gr. [Ety. mythological name.]
coloradoensis, Shumard, 1860, Trans. St. Louis Acad. Sci., Potsdam Gr. [Ety. proper name.]
coloradoensis, Meek, 1870, see *O. desmopleura*.
concinna, Hall, 1859, Pal. N. Y., vol. 3, Low. Held. Gr. [Sig. handsome.]
cooperensis, Swallow, 1863, Trans. St. Louis Acad. Sci., Low. Carb. [Ety. proper name.]
corinna, Billings, 1865, Pal. Foss., vol. 1, Quebec Gr. [Ety. mythological name.]
costalis, Hall, 1847, Pal. N. Y., vol. 1, Chazy Gr. [Sig. ribbed.]
costata, Hall, 1845. This name was preoccupied by Sowerby in 1839.
crassa, Meek, 1874, Cin. Quar. Jour. Sci., vol. 1, Cin'ti Gr. [Sig. thick.]
crispata, Emmons, 1842, Geo. Rep. N. Y., Trenton Gr. [Sig. curled.]
cumberlandia, Hall, 1859, Pal. N. Y., vol. 3, Oriskany sandstone. [Ety. proper name.]
cuneata, Owen, 1852, Geo. Sur. Wis., Iowa and Minn., Devonian. [Sig. wedge-shaped.]
cyclas, Hall, 1860, 13th Reg. Rep., Ham. Gr. [Sig. of a round form.]
cyclus, syn. for Orthis multisecta.
cypha, syn. for Orthis crassa.
davidsoni, Verneuil, 1840, Bull. Geol. Soc. France, vol. 5, Up. Sil. [Ety. proper name.]
deflecta, see Streptorhynchus deflectus.
deformis, Hall, 1859, Pal. N. Y., vol. 3, Low. Held. Gr. [Sig. deformed.]
delicatula, Billings, 1865, Pal. Foss., vol. 1, Quebec Gr. [Sig. quite delicate.]
dentata, Pander, 1830, (Porambonites dentatus) Bietr. Geogn. Russl., Trent. and Hud. Riv. Gr. [Sig. toothed.]
desmopleura, Meek, 1870, Hayden's Geo. Rep., Silurian. [Ety. *desmos*, a band; *pleura*, the side.]
dichotoma, syn. for Orthis fissicosta.
discus, Hall, 1859, Pal. N. Y., vol. 3, Low. Held. Gr. [Sig. a quoit.]
disparilis, Conrad, 1843, Proc. Acad. Nat. Sci., vol. 1, Black Riv. & Trenton Gr. [Sig. different.]
dubia, Hall, 1858, Trans. Alb. Inst., vol. 4, Warsaw Gr. [Sig. doubtful.]
electra, Billings, 1862, Pal. Foss., vol. 1, Quebec Gr. [Ety. mythological name.]
elegantula, Dalman, 1827, Kongl. Vet. Acad. Handl., Clinton & Niagara Gr. [Sig. quite elegant.]
ella, Hall, 1861, 13th Reg. Rep., Cin'ti Gr. [Ety. proper name.]
emacerata, Hall, 1860, 13th Reg. Rep., Cin'ti Gr. [Sig. made-lean.]

emarginata, see Orthis oblata *var.* emarginata.]
eminens, Hall, 1859, Pal. N. Y., vol. 3, Low. Held. Gr. [Sig. conspicuous.]
erratica, Hall, 1847, Pal. N. Y., vol. 1, Hud. Riv. Gr. [Sig. wandering.]
eryna, Hall, 1867,(Corrigenda, eryna) Pal. N. Y., vol. 4, Cornif. Gr. [Ety. mythological name.] Named instead of O. æquivalvis in 10th Reg. Rep.
eudocia, Billings, 1862, Pal. Foss., vol. 1, Quebec Gr. [Ety. proper name.]
euryone, Billings, 1862, Pal. Foss., vol. 1, Quebec Gr. [Ety. proper name.]
evadne, Billings, 1862, Pal. Foss., vol. 1, Quebec Gr. [Ety. mythological name.]
fasciata, Hall, 1852, Pal. N. Y., vol. 2, Niagara Gr. [Sig. striped.]
fissicosta, Hall, 1847, Pal. N. Y., vol. 1, Cin'ti Gr. [Sig. having divided costæ.]
fissiplica, Roemer, 1860, Sil. Fauna West Tenn., Niagara Gr. [Sig. having divided plications.]
flabellum, Hall, 1843. This name was preoccupied by Sowerby in 1839.
flava, Winchell, 1865, Proc. Acad. Nat. Sci., Chemung Gr. [Sig. yellow.]
gemmicula, Billings, 1862, Pal. Foss., vol. 1, Quebec Gr. [Sig. a little bud.]
gibbosa, Billings, 1857, Rep. of Progr., Black Riv. Gr. [Sig. gibbous, tumid.]
hipparionyx, Vanuxem, 1843, (Hipparionyx proximus) Geo. Rep. 3rd Dist. N. Y., Oriskany sandstone. [Sig. a colt's hoof.]
hippolyte, Billings, 1862, Pal. Foss., vol. 1, Quebec Gr. [Ety. mythological name.]
hybrida, Sowerby, 1839, Murch. Sil. Syst., Niagara Gr. [Sig. a hybrid.]
idonea, Hall, 1867, Pal. N. Y., vol. 4, Ham. Gr. [Sig. suitable.]
imperator, Billings, 1859, Can. Nat. Geo., vol. 4, Chazy Gr. [Sig. chief.]
impressa, Hall, 1843, Geo. Rep. 4th Dist. N. Y., Chemung Gr. [Sig. impressed.]
inæqualis, Hall, 1858, Geo. of Iowa, Ham. Gr. [Sig. unequal.]
insculpta, Hall, 1847, Pal. N. Y., vol. 1, Trenton & Cin'ti Gr. [Sig. engraved.]
insignis, see Skenidium insignis.
interlineata, Sowerby, see Orthis tioga.
interstrialis, Phillips, 1841, Pal. Foss., Devonian. [Sig. interstriated.] This species is probably foreign to America.
iowensis, Hall, 1858, Geo. of Iowa, Ham. Gr. [Ety. proper name.]
iowensis *var.* lurnarius, Hall, 1858, Geo. of Iowa, Ham. Gr. [Sig. (?).]
iphigenia, Billings, 1862, Pal. Foss., vol. 1, Trenton Gr. [Ety. mythological name.]
jamesi, Hall, 1861, 14th Reg. Rep., Cin'ti Gr. [Ety. proper name.]
kankakensis, McChesney, 1860, Desc. New Pal. Foss., Hud. Riv. Gr. [Ety. proper name.]

keokuk, Hall, 1858, Geo. Rep. Iowa, Keokuk Gr. [Ety. proper name.] This species was referred to Orthis umbraculum of DeKoninck by Owen.
lasallensis, McChesney, 1860, New Pal. Foss. Prof. Meek regards this as a syn. for Streptorhynchus crassus.
laticosta, James, 1873, Pal. Ohio, vol. 1, Cin'ti Gr. [Sig. broad-ribbed.]
laurentina, Billings, 1857, Rep. of Geo. Sur. Can., Mid. Sil. [Ety. proper name.]
lenticularis, Vanuxem, 1842, Geo. Rep. 3rd Dist. N. Y., Cornif. Gr. [Sig. lenticular.]
lentiformis, Vanuxem, 1842, Geo. Rep. 3rd Dist. N. Y., Cornif. Gr. [Sig. lens-shaped.]
leonensis, Hall, 1867, Pal. N. Y., vol. 4, Chemung Gr. [Ety. proper name.]
lepida, Hall, 1860, 13th Reg. Rep., Ham. Gr. [Sig. pretty.]
lepis, as identified by d'Archiac & Verneuil. Not American.
leptænoides, Emmons, 1842, Geo. Rep. N. Y., Trenton Gr. [Sig. in the form of a Leptæna.]
leucosia, Hall, 1860, 13th Reg. Rep., Ham. Gr. [Ety. proper name.]
limitaris, see Leiorhynchus limitaris.
livia, Billings, 1860, Can. Jour. Ind. Sci. and Art, Cornif. Gr. [Ety. Roman proper name.]
lucia, Billings, 1874, Pal. Foss., vol. 2, Devonian. [Ety. proper name.]
lynx, Eichwald, 1830, (Terebratula lynx) Nat. Kizze von Podol., Trenton and Hud. Riv. Gr. [Ety. the name of a quadruped of the genus *Felis*.]
maria, Billings, 1862, Pal. Foss., vol. 1, Mid. Sil. [Ety. *Maria*, Mary a proper name.]
macfarlani, Meek, 1868, Trans. Chi. Acad. Sci., vol. 1, Ham. Gr. [Ety. proper name.]
media, Shaler, 1865, Bul. No. 4, M. C. Z., Anticosti Gr. This is probably only a variety of *O. elegantula*.
meeki, S. A. Miller, 1875, Cin. Quar. Jour. Sci., vol. 2, Cin'ti Gr. [Ety. proper name.] A variety of O. testudinaria.
merope, Billings, 1862, Pal. Foss., vol. 1, Trenton Gr. [Ety. mythological name.]
michelini, (Terebratula michelini) L'Eveille, 1835, Mem. Soc. Geol. France, vol. 2, Low. Carb. [Ety. proper name.]
michelini *var.* burlingtonensis, Hall, 1858, Geo. Rep. Iowa, Burlington Gr. [Ety. proper name.]
minna, Billings, 1865, Pal. Foss., vol. 1, Quebec Gr. [Ety. proper name.]
missouriensis, Shumard, 1855, Geo. Rep. Mo., Trenton Gr. [Ety. proper name.]
missouriensis, Swallow, 1860, Trans. St. Louis Acad. Sci., Chemung Gr. This name was preoccupied.

mitis, Hall, 1863, 16th Reg. Rep., Schoharie grit. [Sig. moderate.]
morrowensis, syn. for Orthis ella (?)
multisecta, James, 1873, Ohio Pal., vol. 1, Cin'ti Gr. [Sig. having many paths.]
multistriata, Hall, 1859, Pal. N. Y., vol. 3, Low Held. Gr. [Sig. many-striated.]
musculosa, Hall, 1857, Pal. Foss., Oriskany sandstone. [Sig. full of muscles.]
mycale, Billings, 1862, Pal. Foss., vol. 1, Quebec Gr. [Ety. proper name.]
nisis, Hall, 1872, 24th Reg. Rep., Niagara Gr. [Ety. proper name.]
nucleus, syn. for Ambocœlia umbonata.
oblata, Hall, 1859, Pal. N. Y., vol. 3, Low. Held. Gr. [Sig. broader than long.]
oblata var. emarginata, Hall, 1859, Pal. N. Y., vol. 3, Low. Held. Gr. [Sig. notched at the end.]
occasus, Hall, 1860, 13th Reg. Rep., Ham. Gr. [Sig. the west.]
occidentalis, Hall, 1847, Pal. N. Y., vol. 1, Trenton to Hud. Riv. Gr. [Sig. western.]
orbicularis, Sowerby, 1839, Murch. Sil. Syst., Up. Sil. [Sig. orbicular.]
orthambonites, Pander, as figured by Murchison & Verneuil, 1845, Russia and Ural mountains, Quebec Gr. [Sig. straight-umbo.]
pecosi, Marcou, 1858, Geo. N. America, Coal Meas. This species was subsequently described by Swallow under the name of Orthis carbonaria. [Ety. proper name.]
pecten, as identified by d'Archiac & Verneuil. Not American.
pectinella, Conrad, 1840, Ann. Rep. N. Y., Black Riv. & Trenton Gr. [Sig. a little comb.]
pectinella var. semiovalis, Hall, 1847, Pal. N. Y., vol 1, Trenton Gr. [Sig. half-oval.]
peduncularis, Hall, 1859, Pal. N. Y., vol. 3, Low. Held. Gr. [Sig. a little foot.]
peloris, Hall, 1863, 16th Reg. Rep., Schoharie grit. [Sig. *Peloris*; an existing shell fish.]
penelope, Hall, 1860, 13th Reg. Rep., Ham. Gr. [Ety. mythological name.]
pepina, Hall, 1863, 16th Reg. Rep., Potsdam Gr. [Ety. proper name.]
perelegans, Hall, 1859, Pal. N. Y., vol. 3, Low. Held. Gr. [Sig. very elegant.]
perversa, see Streptorhynchus perversus.
pervetus, Conrad, 1843, Proc. Acad. Nat. Sci., vol. 1, Black Riv. & Trenton Gr. [Sig. very old.]
pigra, Billings, 1859, Can. Nat. Geo., vol. 4, Chazy Gr. [Sig. sluggish.]
pisum, see Nucleospira pisiformis.
planoconvexa, Hall, 1859, Pal. N. Y., vol. 3, Low. Held. Gr. [Ety. *planus*, level; *convexus*, convex.]
platys, Billings, 1859, Can. Nat. Geo., vol. 4, Chazy Gr. [Sig. broad.]
plicata, Vanuxem, see Spirifera vanuxemi.

plicatella, Hall, 1847, Pal. N. Y., vol. 1, Trenton & Hud. Riv. Gr. [Sig. a little fold or plait.]
porcata, McCoy, 1846, Sil. Foss. of Ireland, Trenton, Hud. Riv. & Mid. Sil. [Sig. ridged.]
porcia, Billings, 1859, Can. Nat. Geo., vol. 4, Chazy Gr. [Ety. proper name.]
præumbona, see Ambocoelia præumbona.
pratteni, McChesney, 1860, New Pal. Foss., Coal Meas. [Ety. proper name.]
prava, Hall, 1858, Geo. of Iowa, Ham. Gr. [Sig. crooked.]
propinqua, Hall, 1857, 10th Reg. Rep. N. Y., Up. Held. Gr. [Sig. related to.]
punctostriata, Hall, 1852, Pal. N. Y., vol. 2, Niagara Gr. [Sig. *punctus*, pricked; *striatus*, striated.]
pyramidalis, see Skenidium pyramidalis.
quadricostata, see Leiorhynchus quadricostatus.
resupinata, Martin, 1809, Petref. Derb. Carb. [Sig. upside down.]
resupinoides, Cox, 1857, Geo. Sur. Ky., vol. 3, Coal Meas. [Sig. like *O. resupinata*.]
retrorsa, Salter, 1858, Geo. Sur. of G. B., Trenton & Hud. Riv. Gr. [Sig. turned backwards.]
rhynchonelliformis, Shaler, 1865, Bul. No. 4, M. C. Z., Anticosti Gr. [Sig. like a shell of the genus Rhynchonella.]
richmonda, syn. for Streptorhynchus crassus.
robusta, Hall, 1858, Geo. Rep. Iowa, Coal Meas. [Sig. robust.]
rugiplicata, Hall, 1872, 24th Reg. Rep., Niagara Gr. [Sig. wrinkled and plicated.]
ruida, Billings, 1866, Catal. Sil. Foss. Antic., Anticosti Gr. [Sig. rough.]
semele, Hall, 1863, 16th Reg. Rep., Onondaga & Up. Held. Gr. [Ety. mythological name.]
sinuata, Hall, 1847, Pal. N. Y., vol. 1, Hud. Riv. Gr. [Sig. waved.]
sola, Billings, 1866, Catal. Sil. Foss. Antic., Hud. Riv. Gr. [Sig. alone.]
solitaria, Hall, 1860, 13th Reg. Rep., N. Y., Ham. Gr. [Sig. alone.]
stonensis, Safford, 1860, Geo. of Tenn., Trenton & Nashville Gr. [Ety. proper name.]
striatula, Emmons, 1842, Geo. Rep. N. Y. This name was preoccupied by Schlotheim.
strophomenoides, Hall, 1859, Pal. N. Y., vol. 3, Low. Held. Gr. [Sig. like *Strophomena*.
subcarinata, Hall, 1859, Pal. N. Y., vol. 3, Low. Held. Gr. [Sig. somewhat keeled.]
subæquata, Conrad, 1843, Proc. Acad. Nat. Sci., vol. 1, Chazy to Trenton Gr. [Sig. somewhat equal.]
subjugata, syn. for Orthis occidentalis.

suborbicularis, Hall, 1858, Geo. of Iowa, Ham. Gr. [Sig. somewhat orbicular.]
subquadrata, Hall, 1847, Pal. N. Y., vol. 1, Trenton to Hud. Riv. Gr. [Sig. somewhat quadrate.]
subumbona, see Spirifera subumbona.
swallovi, Hall, 1858, Geo. Rep. Iowa, Burlington Gr. [Ety. proper name.]
tennidens, Hall, 1852, Pal. N. Y., vol. 2, Clinton Gr. [Sig. slender-toothed.]
tenuistriata, Hall, 1843, Geo. Rep. 4th Dist. N. Y., Portage Gr. [Sig. fine-striated.]
testudinaria, Dalman, 1827, Vet. Acad. Hand., Trenton & Hud. Riv.Gr. [Sig. arched like a tortoise shell.]
thiemii, White, 1860, Jour. Bost. Soc. Nat. Hist., vol. 7, Kinderhook Gr. [Ety. proper name.]
tioga, Hall, 1867, Pal. N. Y., vol. 4, (O. interlineata, Sow., Geo. Rep. 4th Dist. N. Y.) Portage & Chemung Gr. [Ety. proper name.]
tricenaria, Conrad, 1843, Proc. Acad. Nat. Sci., vol. 1, Trenton Gr. [Sig. of or belonging to thirty.]
trinucleus, Hall, 1852, Pal. N. Y., vol. 2, Clinton Gr. [Sig. a triple-nut.]
triplicatella, Meek, 1873, Ohio Pal., vol. 1, Cin'ti Gr. [Sig. having three plications in one-fold.
tritonia, Billings, 1862, Pal. Foss., vol. 1, Quebec Gr. [Ety. mythological name.]
tubulostriata, Hall, 1859, Pal. N. Y., vol. 3, Low. Held. Gr. [Sig. having tube-like striæ.]
tulliensis, Vanuxem, 1843, Geo. Rep. 3rd Dist. N. Y., Tully limestone. [Ety. proper name.]
uberi, Billings, 1866, Catal. Sil. Foss. Antic., Anticosti Gr. [Sig. abundant, fruitful.]
umbonata, see Ambocœlia umbonata.
umbraculum, DeKoninck, see Orthis Keokuk and Streptorhynchus umbraculum.
unguiculus, Phillips, as identified by Hall in 1843, see Ambocœlia umbonata var. gregaria.
unguiformis, syn. for Orthis hipparionyx.
vanuxemi, Hall, 1857, 10th Reg. Rep., Ham. Gr. [Ety. proper name.]
vanuxemi, Winchell, 1862, Proc. Acad. Nat. Sci., Portage Gr. The name was preoccupied.
varica, Conrad, 1842, (Delthyris varica) Jour. Acad. Nat. Sci., vol. 8, Low. Held. Gr. [Sig. straddling.]
ORTHISINA, D'Orbigny, 1849. [Ety. *orthis*, a genus of shells; *inus*, implying resembling to.]
alternata, see Streptorhynchus perversus.
arctostriata, see Streptorhynchus arctostriatus.
crassa, see Streptorhynchus crassus.
diversa, Shaler, syn. for Orthisina verneuili.

festinata, Billings, 1861, Pal. Foss., vol. 1, Potsdam Gr. [Sig. hasty.]
grandæva, Billings, 1859, Can. Nat. Geo., vol. 4, Calcif. Gr. [Sig. primeval.]
missouriensis, Swallow, 1858. Syn. for Meekella striato-costata.
occidentalis, Swallow, 1863, Trans. St. Louis Acad. Sci., Permian Gr. [Sig. western.]
shumardiana, Swallow, 1858, Trans. St. Louis Acad. Sci., Permian Gr. [Ety. proper name.]
verneuili, Eichwald, as identified by Billings, Trenton & Anticosti Gr. [Ety. proper name.]
PENTAGONIA, Cozzens, 1846, Ann. N. Y. Lyceum, vol. 3. [Ety. *pente*, five; *gonia*, an angle.]
peersi, Cozzens, 1846, Ann. N. Y. Lyceum, vol. 3, Devonian. [Ety. proper name.]
PENTAMERELLA, Hall, 1867, Pal. N. Y., vol. 4. [Ety. diminutive of *Pentamerus*.]
arata, Conrad, 1841, (Atrypa arata and Atrypa octo-costata) Ann. Rep. N. Y., Schoharie grit and Up. Held. Gr. [Sig. plowed, furrowed.]
dubia, Hall, 1860, (Spirifer dubius) 13th Reg. Rep., Ham. Gr. [Sig. doubtful.]
micula, Hall, 1867, Pal. N. Y., vol. 4, Ham. Gr. [Sig. a very small crumb.]
obsolescens, Hall, 1867, Pal. N. Y., vol. 4, Devonian. [Sig. obsolete.]
papilionensis, Hall, 1858, (Pentamerus papilionensis) Geo. Rep. Iowa, vol. 1, pt. 2, Ham. Gr. [Ety. proper name.]
PENTAMERUS, Sowerby, 1814, Min. Conch., vol. 1. [Ety. *penta*, five; *meros*, apartments.]
aratus, see Pentamerella arata.
arcuosus, McChesney, 1861, New Pal. Foss., Niagara Gr. [Sig. arched, bent over.]
barrandi, Billings, 1857, Rep. of Progr., Mid. Sil. [Ety. proper name.]
bisinuatus, McChesney, 1861, New Pal. Foss., Niagara Gr. [Sig. having two depressions.]
borealis, Meek, 1868, Trans. Chi. Acad. Sci., Ham. Gr. [Sig. northern.] This name was preoccupied by Eichwald in 1840.
brevirostris, Sowerby, 1839, (Terebratula brevirostris) Murch. Sil. Syst., Niagara Gr. [Sig. short-beaked.]
chicagoensis, Winchell & Marcy, 1865, Mem. Bost. Soc. Nat. Hist., Niagara Gr. [Ety. proper name.]
comis, Owen, 1852, (Atrypa comis) Geo. Sur. Wis., Iowa and Minn., Ham. Gr. [Sig. nice, delicate.]
crassiradiatus, McChesney, 1861, New Pal. Foss., Niagara Gr. [Sig. thick-rayed.]
elongatus, see Amphigenia elongata.
fornicatus, Hall, 1852, Pal. N. Y., vol. 2, Clinton Gr. [Sig. arched or vaulted over.]

galeatiformis, Meek & Worthen, syn. for *P. galeatus*.
galeatus, Dalman, 1827, (Atrypa galeatus) Vet. Acad. Handl., Low. Held. Gr. [Sig. helmet-shaped.]
intralineatus, Winchell, 1866, Rep. Low. Penin. Mich., Ham. Gr. [Sig. marked with lines on the inside.]
knappi, Hall, 1872, 24th Reg. Rep., Niagara Gr. [Ety. proper name.]
knighti, Sowerby, 1812, Min. Conch., vol. 1, Devonian. [Ety. proper name.]
littoni, Hall, 1859, Pal. N. Y., vol. 3, Low. Held. & Niagara. Gr. [Ety. proper name.]
multicostatus, Hall, 1861, Rep. of Progr. Wis. Sur., Niagara Gr. [Sig. many-ribbed.]
nucleus, Hall, 1872, 24th Reg. Rep., Clinton Gr. [Sig. a kernel.]
nysius, Hall, 1872, 24th Reg. Rep., Niagara Gr. [Ety. mythological name.] There are two varieties, one having coarse and the other finer radii. These are designated *P. nysius var. crassicostus* and *P. nysius var. tenuicostus*.
oblongus, Sowerby, 1839, Murch. Sil. Syst., Clinton & Niagara Gr. [Sig. longer than broad.]
oblongus var. cylindricus, Hall, 1872, 24th Reg. Rep., Niag. Gr. [Sig. cylindrical.]
occidentalis, Hall, 1852, Pal. N. Y., vol. 2, Guelph Gr. [Sig. western.]
occidentalis, Hall, 1858, Geo. Rep. Iowa, Ham. Gr. This name was preoccupied. The species is now referred to the genus *Gypidula*, by the author.
ovalis, Hall, 1852, Pal. N. Y., vol. 2, Clinton Gr. [Sig. egg-shaped.]
papilionensis, see Pentamerella papilionensis.
pergibbosus, Hall & Whitfield, 1875, Ohio Pal., vol. 2, Niagara Gr. [Sig. very gibbous.]
pseudogaleatus, Hall, 1859, Pal. N. Y., vol. 3, Low. Held. Gr. [Sig. false *galeatus*.]
reversus, see Anastrophia reversa.
salinensis, Swallow, 1860, Trans. St. Louis Acad. Sci., Devonian. [Ety. proper name.]
similior, Winchell & Marcy, 1865, (Spirifera similior) Mem. Bost. Soc. Nat. Hist., Niagara Gr. [Ety. from its resemblance to *Spirifera bicostata*.]
subglobosus, Meek & Worthen, 1868, Geo. Sur. Ill., vol. 3, Ham. Gr. [Sig. somewhat globose.]
trisinuatus, McChesney, 1861, Desc. New Pal. Foss., Niagara Gr. [Sig. marked with three depressions.]
ventricosus, Hall, 1861, Rep. Progr. Wis. Sur., Niagara Gr. [Sig. bulging out.]
verneuili, see Anastrophia verneuili.
PHOLIDOPS, Hall, 1859, Pal. N. Y., vol. 3, [Ety. *pholis, pholidos*, a scale.]
arenaria, Hall, 1867, Pal. N. Y., vol. 4, Oriskany sandstone. [Sig. sandy.]

areolata, Hall, 1863, 16th Reg. Rep., Schoharie grit. [Sig. divided into irregular squares or angular spaces.]
cincinnatiensis, Hall, 1872, Pamphlet, Cin'ti Gr. [Ety. proper name.]
hamiltoniæ, Hall, 1860, 13th Reg. Rep. N. Y., Ham. Gr. [Ety. proper name.]
linguloides, Hall, 1867, Pal. N. Y., vol. 4, Ham. Gr. [Sig. like a *Lingula*.]
oblata, Hall, 1867, Pal. N. Y., vol. 4, Ham. Gr. [Sig. oblate.]
ovalis, Hall, 1863, Trans. Alb. Inst., vol. 4, Niagara Gr. [Sig. oval.]
ovata, Hall, 1859, Pal. N. Y., vol. 3, Low. Held. Gr. [Sig. egg-shaped.]
squamiformis, Hall, 1843, (Orbicula squamiformis) Geo. Rep. 4th Dist. N. Y., Niagara Gr. [Sig. in the form of a scale.]
subtruncata, Hall, 1847, (Orbicula subtruncata) Pal. N. Y., vol. 1, Hud. Riv. Gr. [Sig. somewhat shortened.]
terminalis, Hall, 1859, Pal. N. Y., vol. 3, Oriskany sandstone. [Sig. terminating.]
trentonensis, Hall, 1866, Pamphlet, Trenton Gr. [Ety. proper name.]
Platystrophia, syn. for Orthis.
regularis, syn. for Orthis lynx.
Plectambonites arca, syn. for Leptæna transversalis.
glabra, syn. for Leptæna sericea.
tenera, syn. for Leptæna transversalis.
Plicatula, Lamarck, 1809.
striatocostata, see Meekella striatocostata.
PORAMBONITES, Pander, 1830, Beitrage zur Geog. des Russichen Reiches. [Ety. *poros*, opening; *ambon*, umbone.]
dentatus, see Orthis dentata.
ottawænsis, Billings, 1862, Pal. Foss., vol. 1, Black Riv. Gr. [Ety. proper name.]
PRODUCTELLA, Hall, 1867, Pal. N. Y., vol. 4, [Sig. diminutive of *Productus*.]
arctirostrata, Hall, 1857, (Productus arctirostratus) 10th Reg. Rep., Chemung Gr. [Sig. narrow-beaked.]
bialveata, Hall, 1867, Pal. N. Y., vol. 4, Chemung Gr. [Sig. double-channeled.]
boydi, Hall, 1857, (Productus boydii) 10th Reg. Rep., Chemung Gr. [Ety. proper name.]
concentrica, Hall, 1857, (Productus concentricus) 10th Reg. Rep., Kinderhook Gr. [Sig. concentrically lined.]
costatula, Hall, 1867, Pal. N. Y., vol. 4, Chemung Gr. [Sig. small-ribbed.]
costatula var. strigata, Hall, 1867, Pal. N. Y., vol. 4, Chemung Gr. [Sig. furrowed.]
dumosa, Hall, 1861, (Productus dumosus) 14th Reg. Rep., Ham. Gr. [Sig. bushy.]
eriensis, Nicholson, 1874, Geo. Mag., n. s.; vol. 1, Cornif. Gr. [Ety. proper name.]

exanthemata, Hall, 1857, (Productus exanthematus) 10th Reg. Rep., Ham. Gr. [Sig. covered with eruptions.]
hirsuta, Hall, 1857, (Productus hirsutus) 10th Reg. Rep., Chemung Gr. [Sig. rough, hairy.]
hirsuta *var.* rectispina, Hall, 1867, Pal. N. Y., vol. 4, Chemung Gr. [Sig. straight-spined.]
hystricula, Hall, 1867, Pal. N. Y., vol. 4, Chemung Gr. [Sig. somewhat covered with spines.]
lachrymosa, Conrad, 1842, (Strophomena lachrymosa) Jour. Acad. Nat. Sci., vol. 8, Chemung Gr. [Sig. full of tears.]
lachrymosa *var.* lima, Conrad, 1842, (Strophomena lima) Jour. Acad. Nat. Sci., vol. 8, Chemung Gr. [Sig. crooked or rough.]
laychrymosa *var.* stigmata, Hall, 1867, Pal. N. Y., vol. 4, Chemung Gr. [Sig. marked, branded.]
navicella, Hall, 1857, (Productus navicella) 10th Reg. Rep., Cornif. & Ham. Gr. [Sig. a small boat.]
newberryi, Hall, 1857, (Productus newberryi) 10th Reg. Rep., Chemung Gr. [Ety. proper name.]
onusta, Hall, 1867, Pal. N. Y., vol. 4, Chemung Gr. [Sig. filled.]
papulosa, Hall, 1867, Pal. N. Y., vol. 4, Ham. Gr. [Sig. pimpled.]
pyxidata, Hall, 1858, (Productus pyxidatus) Geo. of Iowa, Ham. Gr. [Sig. made like a box.]
rarispina, Hall, 1857, (Productus rarispinus) 10th Reg. Rep., Chemung Gr. [Sig. having few spines.]
shumardiana, Hall, 1858, (Productus shumardianus) Geo. Rep. of Iowa, vol. 1, pt. 2, Cornif. Gr., Marcellus shale, Ham. Gr., Burlington and Tully limestone. [Ety. proper name.]
speciosa, Hall, 1857, (Productus speciosus) 10th Reg. Rep., Chemung Gr. [Sig. beautiful.]
spinulicosta, Hall, 1857, (Productus spinulicostus) 10th Reg. Rep., Marcellus shales & Ham. Gr. [Sig. spined and ribbed.]
striatula, Hall, 1867, Pal. N. Y., vol. 4, Chemung Gr. [Sig. somewhat striated.]
subaculeata, Murchison, 1840, (Productus subaculeatus) Bul. Soc.Geo. de France, vol. 5, Cornif. Gr. [Sig. somewhat prickly.]
subalata, Hall, 1857, (Productus subalatus) 10th Reg. Rep., Ham. Gr. [Sig. somewhat winged.]
truncata, Hall, 1857, (Productus truncatus) 10th Reg. Rep., Marcellus shales & Ham. Gr. [Sig. cut short.]
tullia, Hall, 1867, Pal. N. Y., vol. 4, Ham. Gr. [Ety. proper name.]

PRODUCTUS, Sowerby, 1814, Min. Conch., vol. 1. [Ety. *productus*, produced—so named from one valve of the shell being prolonged beyond the other, and often to a great extent.]
æquicostatus, Shumard, 1855, Geo. Rep. Mo., Coal Meas. [Sig. equal-ribbed.]
alternatus, Norwood & Pratten, 1854, Jour. Acad. Nat. Sci., 2d series, vol. 3, Keokuk Gr. [Sig. alternating.]
altonensis, Norwood & Pratten, 1854, Jour. Acad. Nat. Sci., 2d series, vol. 3, Chester Gr. [Ety. proper name.]
americanus, Swallow, 1863, Trans. St. Louis Acad. Sci., Up. Coal Meas. [Ety. proper name.]
arctirostratus, see Productella arctirostrata.
arcuatus, Hall, 1858, Geo. Rep. Iowa, Kinderhook Gr. [Sig. arched, bent over.]
asper, McChesney, syn. for P. nebrascensis.
auriculatus, Swallow, 1863, Trans. St. Louis Acad. Sci., Coal Meas. [Sig. having ear-like appendages.]
biseriatus, Hall, 1858, Trans. Alb. Inst., vol. 4, Warsaw Gr. [Sig. in double rows or series.]
boonensis, Swallow, 1858, Trans. St. Louis Acad. Sci., Coal Meas. [Ety. proper name.]
boydi, see Productella boydi.
calhounianus, Swallow, 1858, Trans. St. Louis Acad. Sci., Coal Meas. [Ety. proper name.] Prof. Meek regarded this name as a synonym for *P. semireticulatus*.
callawayensis, Swallow, 1860, Trans. St. Louis Acad. Sci., Low. Devonian. [Ety. proper name.]
cancrini, as identified by Geinitz, is P. pertenuis of Meek.
capaci, D'Orbigny, 1843, as identified by early authors, is referred to *P. longispinus*.
cestriensis, Worthen, 1860, Trans. St. Louis Acad. Sci., Chester Gr. [Ety. proper name.]
clavus, Norwood & Pratten, 1854, Jour. Acad. Nat. Sci., 2nd series, vol. 3, Coal Meas. [Sig. club-shaped.]
comoides, as identified by d'Archiac & Veneuil. Not American.
concentricus, see Productella concentrica.
confragosus, Harlan, 1835, Trans. Geo. Soc. Penn., Carb. [Sig. rough, uneven.]
cooperensis, Swallow, 1860, Trans. St. Louis Acad. Sci., Chemung Gr. [Ety. proper name.]
cora, d'Orbigny, 1842, Geol.Voy. Amer., Coal Meas. [Ety. mythological name.]
cora *var.* mogoyoni, Marcou, 1858, Geo. N. Amer., Low. Carb. [Ety. proper name.]
coriformis, Swallow, 1863, Trans. St. Louis Acad. Sci., Low. Carb. [Sig. like *Productus cora.*]

costatoides, Swallow, 1858, Trans. St. Louis Acad. Sci., Up. Coal Meas. [Sig. resembling *P. costatus*.]
costatus, Sowerby, 1827, Min. Conch. vol. 6, Coal Meas. [Sig. ribbed.] It is doubtful whether this species has been identified in America.
curtirostratus, Winchell, 1865, Proc. Acad. Nat. Sci., Chemung Gr. [Sig. short-beaked.]
delawari, Marcou, 1858, Geol. N. Amer., Low. Carb. [Ety. proper name.]
depressus, Sowerby, 1825, see Strophomena depressa.
depressus, Swallow, 1863, Trans. St. Louis Acad. Sci., Low. Carb. [Sig. depressed.]
dissimilis, Hall, 1858, Geo. of Iowa, Ham. Gr. [Sig. unlike, various.]
dolorosus, Winchell, 1865, Proc. Acad. Nat. Sci., Chemung Gr. [Sig. wretched.]
dumosus, see Productella dumosa.
duplicostatus, Winchell, 1865, Proc. Acad. Nat. Sci., Marshall Gr. [Sig. double-plicated.]
elegans, Norwood & Pratten, 1854. This name was preoccupied, and the fossil is now named *P. cestriensis*.
exanthematus, see Productella exanthemata.
fasciculatus, McChesney, 1860, New Pal. Foss., Coal Meas. [Sig. bundled.]
fentonensis, Swallow, 1863, Trans. St. Louis Acad. Sci., Low. Carb. [Ety. proper name.]
flemingi, Sowerby, 1814, Min. Conch., vol. 1, Low. Carb. [Ety. proper name.]
flemingi var. burlingtonensis, Hall, 1858, Geo. Rep. Iowa, Burlington Gr. [Ety. proper name.]
gracilis, Winchell, 1865, Proc. Acad. Nat. Sci., Cuyahoga shale. [Sig. slender.]
gradatus, Swallow, 1863, Trans. St. Louis Acad. Sci., Keokuk Gr. [Sig. made with steps.]
hildrethianus, Norwood & Pratten, 1854, Jour. Acad. Nat. Sci., 2nd ser., vol. 3, Coal Meas. [Ety. proper name.]
hirsutus, see Productella hirsuta.
horridus, Geinitz, 1866. This name was preoccupied; moreover Prof. Meek regarded the fossil as *P. longispinus*.
indianensis, Hall, 1858, Trans. Alb. Inst., vol. 4, Warsaw Gr. [Ety. proper name.]
inflatus, syn. for *P. semireticulatus*.
ivesi, Newberry, 1861, Ives' Col. Ex. Exped., Mid. Carb. [Ety. proper name.]
lasallensis, Worthen, 1873, Geo. Sur. Ill., vol. 5, Up. Coal Meas. [Ety. proper name.]
lævicostus, White, 1860, Bost. Jour. Nat. Hist., Kinderhook Gr. [Sig. smooth-ribbed.]

latissimus, Sowerby, 1822, Min. Conch., Carb. [Sig. very wide.]
longispinus, Sowerby, 1814, Min. Conch., vol. 1, Coal Meas. [Sig. long-spined.]
lobatus, as identified by d'Archiac & Verneuil. Not American.
magnicostatus, Swallow, 1860, Trans. St. Louis Acad. Sci., Coal Meas. [Sig. large-ribbed.]
magnus, Meek & Worthen, 1861, Proc. Acad. Nat. Sci. Phil., Keokuk Gr. [Sig. large.]
marginicinctus, Prout, 1857, Trans. St. Louis Acad. Sci., St. Louis Gr. [Sig. encircled with a depression near the margin.]
mesialis, Hall, 1858, Geo. Rep. Iowa, Keokuk Gr. [Sig. middle-parted.]
mexicanus, Shumard, 1858, Trans. St. Louis Acad. Sci., Permian Gr. [Ety. proper name.]
morbillianus, Winchell, 1865, Proc. Acad. Nat. Sci., Burlington Gr. [Sig. measly, spotted.]
multistriatus, Meek, 1860, Proc. Acad. Nat. Sci., Coal Meas. [Sig. many-lined.]
muricatus, Norwood & Pratten, 1854, Jour. Acad. Nat. Sci. Phil., Coal Meas. Prof. Meek regarded this as a syn. for *P. longispinus*. [Sig. full of sharp points.]
navicella, see Productella navicella.
nebrascensis, Owen, 1852, Geo. Rep. Wis., Iowa & Minn., Coal Meas. [Ety. proper name.]
nodosus, Newberry, 1861, Ives' Col. Ex. Exped. Carb. [Sig. knotty.]
norwoodi, Swallow, 1858, Trans. St. Louis Acad. Sci., Permian Gr. [Ety. proper name.]
occidentalis, Newberry, 1861, Ives' Col. Ex. Exped., Up. Carb. [Sig. western.]
orbignyanus, Geinitz, 1866. This name was preoccupied by Sowerby in 1822.
ovatus, Hall, 1858, Geo. Rep. Iowa, St. Louis Gr. [Sig. egg-shaped.]
parvulus, Winchell, 1863, Proc. Acad. Nat. Sci., Chemung Gr. [Sig. very small.]
parvus, Meek & Worthen, 1860, Proc. Acad. Nat. Sci. Phil., Chester Gr. [Sig. small.]
pertenuis, Meek, 1872, Pal. E. Neb., Coal Meas. [Sig. very thin.]
phillipsi, Norwood & Pratten, 1854, Jour. Acad. Nat. Sci., vol. 3, 2nd series, Low. Carb. [Ety. proper name.]
pileiformis, syn. for Productus cora.
pileolus, Shumard, 1858, Trans. St. Louis Acad. Sci., Permian Gr. [Sig. a little bonnet or cap.]
popii, Shumard, 1858, Trans. St. Louis Acad. Sci., Permian Gr. [Ety. proper name.]
portlockianus, Norwood & Pratten, 1854, Jour. Acad. Nat. Sci., 2nd series, vol. 3, Coal Meas. [Ety. proper name.]

prattenianus, Norwood, 1854, Jour. Acad. Nat. Sci. Phil., 2nd series, vol. 3, Coal Meas. [Ety. proper name.]
punctatus, Martin, 1809, Petrif. Derb., Low. Carb. and Coal Meas. [Sig. covered with points, dotted.]
pyxidatus see Productella pyxidata.
rarispinus, see Productella rarispina.
rogersi, Norwood & Pratten, 1854, Jour. Acad. Nat. Sci., Coal Meas. [Ety. proper name.] Prof. Meek regarded this as a synonym for *P. nebrascensis*.
scabriculus, (Conchyliolithus Anomites scabriculus) Martin, 1809, Petrif. Derb., Carb. [Sig. rough.]
scitulus, Meek & Worthen, 1860; Proc. Acad. Nat. Sci., St. Louis Gr. [Sig. neat, pretty.]
semipunctatus, Hildreth, 1838, syn. for *P. punctatus*.
semireticulatus, Martin, 1809, (*Conchyliolithus Anomites semireticulatus*) Petrif. Derb., Keokuk Gr. [Sig. half-like a net or lattice work.]
semistriatus, Meek, 1860, Proc. Acad. Nat. Sci., Coal Meas. [Sig. half-striated.]
setigerus, Hall, 1858, Geo. Rep. Iowa, Keokuk Gr. [Sig. bearing bristles on the back.]
setigerus var. keokuk, Hall, 1858, Geo. Rep. Iowa, Keokuk Gr. [Ety. proper name.]
shumardianus, see Productella shumardiana.
speciosus, see Productella speciosa.
spinulicostus, see Productella spinulicosta.
spinulosus, Sowerby, 1814, Min. Conch., vol. 1, Carb. [Sig. full of spines.]
splendens, Norwood & Pratten, 1854, Jour. Acad. Nat. Sci. Phil., vol. 3, Coal Meas. [Sig. splendid.] Prof. Meek regarded this as a synonym for *P. longispinus*.
subaculeatus, see Productella subaculeata.
subalatus, see Productella subalata.
symmetricus, McChesney, 1860, Desc. New Pal. Foss., Coal Meas. [Sig. symmetrical.]
tenuicostus, Hall, 1858, Geo. Rep. Iowa, St. Louis Gr. [Sig. slender-ribbed.]
tenuistriatus, Verneuil, 1845, Geol. Russia & Ural Mountains, Carb. [Sig. fine-lined.]
truncatus, see Productella truncata.
tubulospinus, McChesney. Syn. for *P. punctatus*.
viminalis, White, 1862, Proc. Bost. Soc. Nat. Hist., vol. 9, Burlington Gr. [Sig. bearing twigs.]
vittatus, Hall, 1858, Geo. Rep. Iowa, Keokuk Gr. [Sig. bound in a fillet or hair-lace.]
wabashensis, Norwood & Pratten, syn. for *P. longispinus*.
wilberianus, McChesney, syn. for *P. nebrascensis*.
wortheni, Hall, 1858, Geo. Rep. Iowa, Keokuk Gr. [Ety. proper name.]

PSEUDOCRANIA, McCoy, 1851, Ann. & Mag. Nat. Hist., 2d series, vol. 8. [Ety. *pseudo*, false; *Crania*, a genus of shells.
anomala, Winchell, 1866, Rep. Low. Pen. Mich., Ham. Gr. [Sig. irregular.]

RENSSELÆRIA, Hall, 1859, Pal. N. Y., vol. 3. [Ety. proper name.]
æquiradiata, Conrad, 1842, (Atrypa æquiradiata) Jour. Acad. Nat. Sci., vol. 8, Low. Held. Gr. [Sig. equal-rayed.]
conradi, McChesney, 1861, New Pal. Foss., Oriskany sandstone. [Ety. proper name.]
cumberlandiæ, Hall, 1857, (Meganteris cumberlandiæ) 10th Reg. Rep., Oriskany sandstone. [Ety. proper name.]
elliptica, Hall, 1857, (Meganteris elliptica) 10th Reg. Rep., Low. Held. Gr. [Sig. elliptical.]
elongata, see Amphigenia elongata.
intermedia, Hall, 1859, Pal. N. Y., vol. 3, Oriskany sandstone. [Sig. intermediate.]
johanni, Hall, 1867, Pal. N. Y., vol. 4, Up. Held. Gr. [Ety. proper name.]
lævis, Hall, 1857, (Meganteris lævis) 10th Reg. Rep., Low. Held. Gr. [Sig. smooth.]
lævis, Meek, 1868. This name was preoccupied.
marylandica, Hall, 1859, Pal. N. Y., vol. 3, Oriskany sandstone. [Ety. proper name.]
mutabilis, Hall, 1857, (Meganteris mutabilis) 10th Reg. Rep., Low. Held. Gr. [Sig. changing, variable.]
ovalis, Hall, 1857, (Meganteris ovalis) 10th Reg. Rep., Oriskany sandstone. [Sig. oval.]
ovoides, Eaton, 1832, (Terebratula ovoides) Geo. Text-book, Oriskany sandstone. [Sig. ovoid.]
suessiana, Hall, 1857, (Meganteris suessiana) 10th Reg. Rep., Oriskany sandstone. [Ety. proper name.]

RETZIA, King, 1850, Monograph of Permian Foss. [Ety. proper name.]
altirostris, White, 1862, Proc. Bost. Soc. Nat. Hist., vol. 9, Chemung Gr. [Sig. high-beaked.]
compressa, Meek, 1864, Pal. California, Coal Meas. [Sig. compressed.]
eugenia, Billings, 1861, Can. Jour., Ham. Gr. [Ety. proper name.]
marcyi, Shumard, 1854, (Terebratula marcyi) Marcy's Exp. Red Riv., Coal Meas. [Ety. proper name.]
meekiana, Shumard, 1858, Trans. St. Louis Acad. Sci., Permian Gr. [Ety. proper name.]
mormoni, Marcou, 1858, (Terebratula mormonii) Geo. N. Amer., Coal Meas. [Ety. proper name.] This species was subsequently, though in the same year, described by Shumard under the name *R. punctilifera*.

osagensis, Swallow, 1860, Trans. St. Louis Acad. Sci., Chemung Gr. [Ety. proper name.]
papillata, Shumard, 1858, Trans. St.Louis Acad. Sci., Permian Gr. [Sig. pimpled.]
popiana, Swallow, 1860, Trans. St. Louis Acad. Sci., Chemung Gr. [Ety. proper name.]
punctilifera, Shumard, 1858, syn. for Retzia mormoni.
polypleura, Winchell, 1862, Proc. Acad. Nat. Sci., Portage Gr. [Sig. manysided.]
subglobosa, McChesney, syn. for Retzia mormoni.
vera, Hall, 1858, Geo. Rep. Iowa, Kaskaskia Gr. [Sig. true, natural.]
vera var. costata, Hall, 1858, Geo. Rep. Iowa, Kaskaskia Gr. [Sig. ribbed.]
verneuiliana, Hall, 1856,Trans. Alb. Inst., vol. 4, Warsaw Gr. [Ety. proper name.]
RHYNCHONELLA, Fischer, 1809, Mem. Soc. Imp. Mosc. [Ety. rhynchos, a beak; ella, little.]
abrupta, Hall, 1859, Pal. N. Y., vol. 3, Low. Held. Gr. [Sig. terminating abruptly.]
acadiensis, Davidson, 1863, Quar. Jour. Geo. Soc., vol. 19, Low. Carb. [Ety. proper name.]
acinus, Hall, 1863, Trans. Alb. Inst., vol. 4, Niagara Gr. [Sig. a cherry stone.]
acutiplicata, Hall, 1859, Pal. N. Y., vol. 3, Low. Held. Gr. [Sig. having plications acutely angular.]
acutirostris, Hall, 1847, (Atrypa acutirostra) Pal. N. Y., vol. 1, Chazy Gr. [Sig. sharp-beaked.]
æquivalvis, Hall, 1857, 10th Reg. Rep., Low. Held. Gr. [Sig. equal-valved.]
æquiradiata, Hall, 1852, (Atrypa æquiradiata) Pal. N. Y., vol. 2, Clinton Gr. [Sig. equal-rayed.]
algeri, McChesney, 1860, New Pal. Foss., Carb. [Ety. proper name.]
altilis, Hall, 1847, (Atrypa altilis) Pal. N. Y., vol. 1, Chazy Gr. [Sig. fat, fed.]
altiplicata, Hall, 1859, Pal. N. Y., vol. 3, Low. Held. Gr. [Sig. high plications.]
alveata, see Centronella alveata.
angulata, Linnæus, as identified by Geinitz, syn. for Syntrilasma hemiplicatum.
anticostiensis, Billings, 1862, Pal. Foss., vol. 1, Hud. Riv. Gr. [Ety. proper name.]
aprinis, DeVerneuil, 1845, (Terebratula aprinis) Geo. Russia & Ural Mts., Niagara Gr. [Sig. like a pig's head (?).]
arctirostrata, Swallow, 1863, Trans. St. Louis Acad. Sci., Low. Carb. [Sig. narrow-beaked.]
argentea, Billings, 1866, Catal. Sil. Foss. Antic., Anticosti Gr. [Sig. glittering.]

argenturbica, White, 1874, Rep. Invert. Foss., Cin'ti Gr. [Sig. from Silver City.]
barquensis, Winchell, 1862, Proc. Acad. Nat. Sci., Marshall Gr. [Ety. proper name.]
barrandi, Hall, 1857, 10th Reg. Rep., Oriskany sandstone. [Ety. proper name.]
bialveata, Hall, 1859, Pal. N. Y., vol. 3, Lower Held. Gr. [Sig. doublechanneled.]
bidens, Hall, 1852. (Atrypa bidens) Pal. N. Y., vol. 2, Clinton Gr. [Sig. having two teeth.]
bidentata, Hisinger, 1826, (Terebratula bidentata) Vet. Acad. Handl., Niagara Gr. [Sig. having two teeth.]
billingsi, see Stenoschisma billingsi.
boonensis, Shumard, 1855, Geo. Rep. Mo., Trenton Gr. [Ety. proper name.]
brevirostris, Sowerby, 1839, (Terebratula brevirostris) Murch. Sil. Syst., Niagara Gr. [Sig. short-beaked.] This species is probably Pentamerus brevirostris.
campbelliana, Hall, 1859, Pal. N. Y., vol. 3, Low. Held. Gr. [Ety. proper name.]
camerifera, Winchell, 1862, Proc. Acad. Nat. Sci., Marshall Gr. [Sig. chambered.]
capax, Conrad, 1842, (Atrypa capax) Jour. Acad. Nat. Sci., vol. 8, Hud. Riv. Gr. [Sig. large, capacious.]
caput-testudinis, White, 1862, Proc. Bost. Soc. Nat. Hist., vol. 9, Burlington Gr. [Sig. like a turtle's head.]
carica, see Stenoschisma carica.
carbonaria, McChesney, 1860, New Pal. Foss., Coal Meas. [Sig. from the Coal Measures.]
carolina, see Stenoschisma carolina.
castanea, Meek, 1868, Trans. Chi. Acad. Sci., Devonian. [Sig. a chestnut.]
congregata, see Stenoschisma congregatum.
contracta, see Stenoschisma contractum.
cooperensis, Shumard, 1855, Geo. Rep. Mo., Chem. Gr. [Ety. proper name.]
corinthia, Billings, 1865, Pal. Foss., vol. 1, Quebec Gr. [Sig. corinthian.]
cuboides, Sowerby, (Atrypa cuboides) see R. venustula.
cuneata, Dalman, 1827, (Terebratula cuneata) Vet. Acad. Handl., Niagara Gr. [Sig. wedge-shaped.]
dawsoniana, Davidson, 1863, Quar. Jour. Geo. Soc., vol. 19, Low. Carb. [Ety. proper name.]
dentata, Hall, 1847, (Atrypa dentata) Pal. N. Y., vol. 1, Hud. Riv. Gr. [Sig. having teeth.]
dotis, see Stenoschisma dotis.
dryope, Billings, 1874, Pal. Foss., vol. 2, Devonian. [Ety. mythological name.]
dubia, Hall, 1847, (Atrypa dubia) Pal. N. Y., vol. 1, Chazy Gr. [Sig. doubtful.]
duplicata, syn. for Stenoschisma contractum.

eatoniiformis, McChesney, 1860, New Pal. Foss., Carb. [Ety. from a resemblance to *Eatonia*.]
emacerata, Hall, 1852, (Atrypa emacerata) Pal. N. Y., vol. 2, Clinton Gr. [Sig. made lean.]
eminens, Hall, 1859, Pal. N. Y., vol. 3, Low. Held. Gr. [Sig. eminent, remarkable.]
endlichi, Meek, 1876, U. S. Geo. Sur. of Colorado, Up. Devonian. [Ety. proper name.]
eva, Billings, 1866, Catal. Sil. Foss., Antic., Anticosti Gr. [Ety. proper name.]
evangelina, Hartt, 1868, Acad. Geol., Low. Carb. [Ety. proper name.]
excellens, Billings, 1874, Pal. Foss., vol. 2, Devonian. [Sig. excellent.]
eximia, see Stenoschisma eximium.
explanata, McChesney, 1860, Desc. New Pal. Foss., Chester Gr. [Sig. made plane or smooth.]
fitchiana, Hall, 1857, 10th Reg. Rep.,Oriskany sandstone. [Ety. proper name.]
formosa, Hall, 1859, Pal. N. Y., vol. 3, Low. Held. Gr. [Sig. beautiful.]
fringilla, Billings, 1862, Pal. Foss., vol. 1, Mid. Sil. [Sig. a small bird.]
glacialis, Billings, 1862, Pal. Foss., vol. 1, Mid. Sil. [Sig. icy.]
glansfagea, see Centronella glansfagea.
grosvenori, Hall, 1858, Trans. Alb. Inst., vol. 4, Warsaw Gr. [Ety. proper name.]
guadalupæ, Shumard, 1858, Trans. St. Louis Acad. Sci., Permian Gr. [Ety. proper name.]
heteropsis, Winchell, 1865, Proc. Acad. Nat. Sci., Chemung Gr. [Sig. of irregular appearance.]
horsfordi, see Stenoschisma horsfordi.
hubbardi, Winchell, 1862, Proc. Acad. Nat. Sci., Marshall Gr. [Ety. proper name.]
huronensis, Winchell, 1862, Proc. Acad. Nat. Sci., Portage Gr. [Ety. proper name.]
ida, Hartt, 1868, Acad. Geol., Low. Carb. [Ety. proper name.]
increbescens, syn. for Rhynchonella capax.
indentata, Shumard, 1859, Trans. St. Louis Acad. Sci., Permian Gr. [Sig. indented.]
indianensis, Hall, 1863, Trans. Alb. Inst., vol. 4, Niagara Gr. [Ety. proper name.]
inæquiplicata, Hall, 1857, 10th Reg. Rep., Up. Held. Gr. [Sig. unequally plicated.]
interplicata, Sowerby, 1839, (Terebratula interplicata) Murch. Sil. Syst., Niagara Gr. [Sig. interplicated.]
inutilis, Hall, 1859, Pal. N. Y., vol. 3, Low. Held. Gr. [Sig. insignificant.]
janea, Billings, 1866, Catal. Sil. Foss. Antic., Anticosti Gr. [Ety. mythological name.]

lacunosa. Not an American species.
lamellata, Hall, 1852, (Atrypa lamellata) Pal. N. Y., vol. 2, Coralline limestone. [Sig. having thin plates.]
laura, Billings, syn. for Leiorhynchus multicostus.
macra, Hall, 1858, Trans. Alb. Inst., vol. 4, Warsaw Gr. [Sig. long.]
marshallensis, Winchell, 1862, Proc.Acad. Nat. Sci., Marshall Gr. [Ety. proper name.]
metallica, White, 1874, Rep. Invert.Foss., Carb. [Sig. metallic.]
mica, Billings,1866, Catal. Sil. Foss. Anti., Anticosti Gr. [Sig. a little crumb.]
micropleura, Winchell, 1865, Proc. Acad. Nat. Sci., Chemung Gr. [Sig. smallribbed.]
missouriensis, Shumard, 1855, Geo. of Mo., Chemung Gr. [Ety. proper name.]
multistriata, Hall, 1857, 10th Reg. Rep., Oriskany sandstone. [Sig. manystriated.]
mutabilis, Hall, 1857, 10th Reg. Rep., Low. Held. Gr. [Sig. variable.]
mutata, Hall, 1858, Trans. Alb. Inst., vol. 4, Warsaw Gr. [Sig. changing.]
neglecta, Hall, 1852, Pal. N. Y., vol. 2, Niagara Gr. [Sig. overlooked.]
neglecta *var.* scobina, Meek, 1872, Am. Jour. Sci. & Arts, 3rd series, vol. 4, Clinton Gr. [Sig. a file.]
nobilis, Hall, 1859, Pal. N. Y., vol. 3, Low. Held. Gr. [Sig. notable.]
nucleolata, Hall, 1857, 10th Reg. Rep., Low. Held. Gr. [Sig. like a small nut.]
nutrix, Billings, 1866, Catal. Sil. Foss. Antic., Anticosti Gr. [Sig. the breast, the pap.]
oblata, Hall, 1857, 10th Reg. Rep., Oriskany sandstone. [Sig. oblate, broader than long.]
obsolescens, Hall, 1860, 13th Reg. Rep., Ham. Gr. [Sig. old, obsolete.]
obtusiplicata, Hall, 1852, (Atrypa obtusiplicata) Pal. N. Y., vol. 2, Niagara Gr. [Sig. having obtuse plications.]
orbicularis, see Stenoschisma orbiculare.
orientalis, Billings, 1859, Can. Nat. Geo., vol. 4, Chazy Gr. [Sig. eastern, in the eastern provinces.]
osagensis, Swallow, 1858, syn. for Rhynchonella uta.
ottumwa, White, 1862, Proc. Bost. Soc. Nat. Hist., vol. 9, St. Louis Gr. [Ety. proper name.]
parvini, McChesney, syn. for Camerophoria subtrigona.
perrostellata, Swallow, 1863, Trans. St. Louis Acad. Sci., Low. Carb. [Sig. having a very little beak.]
persinuata, Winchell, 1865, Proc. Acad. Nat. Sci., Chemung Gr. [Sig. very sinuate.]
pisum, Hall & Whitfield, 1875, Ohio Pal., vol. 2, Niagara Gr. [Sig. a pea.]

planoconvexa, Hall, 1859, Pal. N. Y., vol. 3, Low. Held. Gr. [Ety. *planus*, level; *convexus*, convex.]
pleiopleura, Conrad, 1841, (Atrypa pleiopleura) Ann. Rep. N. Y., Oriskany sandstone. [Sig. wide-ribbed.]
plena, Hall, 1847, (Atrypa plena) Pal. N. Y., vol. 1, Chazy Gr. [Sig. full, large.]
plicata, Hall, 1852, (Atrypa plicata) Pal. N. Y., vol. 2, Medina Gr. [Sig. plaited, folded.]
plicatella, Hall, 1852, (Atrypa plicatella) Pal. N. Y., vol. 2, Niagara Gr. [Sig. having small plications.]
plicatula, Hall, 1843, (Atrypa plicatula) Geo. Rep. 4th Dist. N. Y., Clinton Gr. [Sig. having little plications.]
plicifera, Hall, 1847, (Atrypa plicifera) Pal. N. Y., vol. 1, Chazy Gr. [Sig. bearing plications.]
principalis, Hall, 1857, 10th Reg. Rep., Oriskany sandstone. [Sig. principal, chief.]
prolifica, see Stenoschisma prolificum.
pugnus, Martin, 1809, Petrif. Derb., Low. Carb. [Sig. the fist.]
pustulosa, White, 1860, Bost. Jour. Nat. Hist., vol. 7, Burlington Gr. [Sig. pustulose.]
pyramidata, Hall, 1859, Pal. N. Y., vol. 3, Low. Held. Gr. [Sig. made like a pyramid.]
pyrrha, Billings, 1866, Catal. Sil. Foss. Antic., Anticosti Gr. [Ety. mythological name.]
quadricostata, Hall, 1843, (Atrypa quadricostata) Geo. Rep. 4th Dist. N. Y., Genessee slate. [Sig. four-ribbed.]
quadricostata, Hall, 1852, (Atrypa quadricostata) Pal. N. Y., vol. 2, Clinton Gr. This name was preoccupied.
ramsayi, Hall, 1859, Pal. N. Y., vol. 3, Oriskany sandstone. [Ety. proper name.]
recurvirostra, Hall, 1847, (Atrypa recurvirostra) Pal. N. Y., vol. 1, Black Riv. to Hud. Riv. Gr. [Sig. bent-beaked.]
reticulata, see Eichwaldia reticulata.
ricinula, Hall, 1858, Trans. Alb. Inst., vol. 4, Warsaw Gr. [Sig. a little tick.]
ringens, Swallow, 1860, Trans. St. Louis Acad. Sci., Low. Carb. [Sig. gaping, having an open orifice.]
robusta, Hall, 1852, Pal. N. Y., vol. 2, (Atrypa robusta) Clinton Gr. [Sig. robust.]
royana, see Stenoschisma royanum.
ridleyana, Safford, 1869, Geo. of Tenn. Not defined.
rudis, Hall, 1859, Pal. N. Y., vol. 3, Low. Held. Gr. [Sig. rude, not fashioned.]
rugosa, Hall, 1852, (Atrypa rugosa) Pal. N. Y., vol. 2, Niagara Gr. [Sig. wrinkled.]
saffordi, Hall, 1860, Can. Nat. & Geo., vol. 5, Low. Held. Gr. [Ety. proper name.]

sageriana, Winchell, 1862, Proc. Acad. Nat. Sci., Marshall Gr. [Ety. proper name.]
sappho, see Stenoschisma sappho.
semiplicata, Conrad, 1841, (Atrypa semiplicata) Ann. Rep. N. Y., Low. Held. Gr. [Sig. half-folded.]
septata, Hall, 1859, Pal. N. Y., vol. 3, Oriskany sandstone. [Sig. divided by septa or partitions.]
sinuata, Hall, 1860, Can. Nat. & Geo., vol. 5, Up. Sil. [Sig. wavy.]
sordida, Hall, 1847, (Atrypa sordida) Pal. N. Y., vol. 1, Trenton Gr. [Sig. despicable, paltry.]
speciosa, Hall, 1857, 10th Reg. Rep., Oriskany sandstone. [Sig. beautiful.]
stephani, see Stenoschisma stephani.
subcircularis, Winchell, 1862, Proc.Acad. Nat. Sci., Marshall Gr. [Sig. somewhat circular.]
subcuboides. Not an American species.
subcuneata, Hall, 1856, Trans. Alb. Inst., vol. 4, Warsaw Gr. [Sig. somewhat wedge-shaped.]
subtrigona, see Camerophoria subtrigona.
subtrigonalis, Hall, 1847, (Atrypa subtrigonalis) Pal. N. Y., vol. 1, Trenton Gr. [Sig. somewhat triangular.]
sulcoplicata, Hall, 1859, Pal. N. Y., vol. 3, Low. Held.Gr. [Sig. grooved along the center of the plications.]
tennesseensis, Roemer, 1860, Sil. Fauna West Tenn., Niagara Gr. [Ety. proper name.]
tethys, see Stenoschisma tethys.
tetrantyx, Winchell, 1865, Proc. Acad. Nat. Sci., Kinderhook Gr. [Sig. having four-folds.]
texiana, Shumard, 1859, Trans. St. Louis Acad. Sci., Permian Gr. [Ety. proper name.]
thalia, see Stenoschisma billingsi.
transversa, Hall, 1859, Pal. N. Y., vol. 3, Low. Held. Gr. [Sig. wider than long.]
unica, Winchell, 1865, Proc. Acad. Nat. Sci., Chemung Gr. [Sig. single, alone.]
unisulcata, see Meristella unisulcata.
uta, Marcou, 1858, (Terebratula uta) Geo. N. Amer., Coal Meas. [Ety. proper name.] This was subsequently described by Swallow as R. osagensis.
vellicata, Hall, 1859, Pal. N. Y., vol. 3, Low. Held. Gr. [Sig. pinched.]
ventricosa, Hall, 1859, Pal. N. Y., vol. 3, Low. Held. Gr. [Sig. bulging out, bellying.]
venustula, Hall, 1867, Pal. N. Y., vol. 4, Tully limestone. This was identified by Vanuxem, 1842, Geo. 3rd Dist. N. Y., as *Atrypa cuboides* of Sowerby. [Sig. somewhat fair, handsome or pretty.]
vicina, Billings, 1866, Catal. Sil. Foss. Antic., Anticosti Gr. [Sig. neighboring, near to.]

warrenensis, Swallow, 1860, Trans. St. Louis Acad. Sci., Low. Devonian. [Ety. proper name.]
wasatchensis, White, 1874, Rep. Invert. Foss., Carb. [Ety. proper name.]
whitii, Winchell, 1862, Proc. Acad. Nat. Sci., Marshall Gr. [Ety. proper name.]
whitii, Hall, 1863, Trans. Alb. Inst., vol. 4, Niagara Gr. [Ety. proper name.] This name was preoccupied.
wilsoni, Sowerby, 1818, (Terebratula wilsoni) Min. Conch., vol. 2, Niagara Gr. [Ety. proper name.]
wortheni, Hall, 1858, Trans. Alb. Inst., vol. 4, Warsaw Gr. [Ety. proper name.]

RHYNCHOSPIRA, Hall, 1859, Pal. N. Y., vol. 3, [Ety. *rhynchos*, a beak; *spira*, a spire.]
deweyi, Hall, 1856, (Waldheimia deweyi) 9th Reg. Rep., Low. Held. Gr. [Ety. proper name.]
evax, Hall, 1863, Trans. Alb. Inst., vol. 4, Niagara Gr. [Sig. a word of joy, a hurra.]
formosa, Hall, 1856, (Waldheimia formosa) 9th Reg. Rep., Low. Held. Gr. [Sig. beautiful.]
lepida, Hall, 1860, 13th Reg. Rep., Ham. Gr. [Sig. pretty.]
nobilis, see Trematospira nobilis.
rectirostra, see Trematospira rectirostra.
subglobosa, Hall, 1867, Pal. N. Y., vol. 4, Schoharie grit. [Sig. somewhat globular.]
sinuata, Dawson, 1868, Acad. Geol., Up. Sil. [Sig. waved.]
Rhynobolus, Hall, 1871, syn. for Trimerella.
galtensis, Hall, see Trimerella galtensis.

SCHIZOCRANIA, Hall & Whitfield, 1875, Ohio Pal., vol. 2. [Ety. *schiza*, a cleft; *Crania*, a genus of fossil brachiopods.]
filosa, Hall, 1847, (Orbicula (?) filosa) Pal. N. Y., vol. 1, Cin'ti and Utica Gr. [Sig. thread-like, thready.]

SKENIDIUM, Hall, 1860, 13th Reg. Rep. [Ety. a little tent.]
halli, Safford, 1869, Geo. of Tenn. Not defined.
insignis, Hall, 1859, (Orthis insignis) Pal. N. Y., vol. 3, Low. Sil. [Sig. distinguished.]
pyramidalis, Hall, 1852, (Orthis pyramidalis) Pal. N. Y., vol. 2, Niagara Gr. [Sig. pyramidal.]

SPIRIFERA, Sowerby, 1815, Min. Conch., vol. 2. [Ety. *spira*, a spire; *fero*, to bear.]
acanthoptera Conrad, 1842, (Delthyris acanthoptera) Jour. Acad. Nat. Sci., vol. 8, Chemung Gr. [Ety. *akantha*, a spine; *pteron*, a wing.]
acuminata, Conrad, 1839, (Delthyris acuminata) Ann. Rep. N. Y., Cornif. and Ham. Gr. [Sig. sharp-pointed.]
aculicostata, DeKoninck, 1843, Desc. An. Foss. Terr. Carb. Belg., Low. Carb. [Sig. sharp-ribbed.]

agelaia, Meek, 1872, Hayden's Geo. Rep., Low. Carb. [Sig. belonging to a herd, common.]
alta, Hall, 1867, Pal. N. Y., vol. 4, Chemung Gr. [Sig. high, noble.]
amara, Swallow, 1860, Trans. St. Louis Acad. Sci., Chemung Gr. [Sig. brackish, salt.]
angusta, Hall, 1857, 10th Reg. Rep., Ham. Gr. [Sig. narrow, short.]
annæ, Swallow, 1860, Trans. St. Louis Acad. Sci., Ham. Gr. [Ety. proper name.]
arata, syn. for S. granulifera.
archiaci, see S. disjuncta.
arctisegmenta, Hall, 1857, 10th Reg. Rep., Up. Held. Gr. [Sig. having narrow segments or ribs.]
arenosa, Conrad, 1839, (Delthyris arenosa) Ann. Rep. N. Y., Oriskany sandstone. [Sig. sandy.]
arrecta, Hall, 1859, Pal. N. Y., vol. 3, Oriskany sandstone. [Sig. erect, steep.]
aspera, Hall, 1858, Geo. Rep. Iowa, Ham. Gr. [Sig. rough.]
audacula, Conrad, (Delthyris audacula) 1842, Jour. Acad. Nat. Sci., vol. 8, Ham. Gr. [Sig. rather bold.]
bialveata, Conrad, (Delthyris bialveata) 1842, Jour. Acad. Nat. Sci., vol. 8, Niagara Gr. [Sig. double-channeled.]
bicostata, Vanuxem, 1842, (Orthis bicostatus) Geol. Rep. 3rd Dist. N. Y., Niagara Gr. [Sig. double-ribbed.]
bidorsalis, Winchell, 1866, Rep. Low. Penin. Mich., Ham. Gr. [Sig. double-backed.]
bifurcata, Hall, 1858, Trans. Alb. Inst., vol. 4, Warsaw Gr. [Sig. bifurcated, forked.]
biloba, Linnæus, 1768, (Anomia biloba) Syst. Nat., Niagara Gr. [Sig. two-lobed.]
bimesialis, Hall, 1858, Geo. Rep. Iowa, vol. 1, pt. 2, Ham. Gr. [Sig. having two middle parts.]
biplicata, Hall, 1858, Geo. Rep. Iowa, vol. 1, pt. 2, Kinderhook Gr. [Sig. two-folded or plaited.]
boonensis, Swallow, 1860, Trans. St.Louis Acad. Sci., Low. Coal Meas. [Ety. proper name.]
brachynota, Hall, 1843, (Delthyris brachynota) Geo. 4th Dist. N. Y., Clinton Gr. [Sig. short-ridged.]
calcarata, syn. for S. disjuncta.
camerata, Morton, 1836, Am. Jour. Sci., vol. 29, Coal Meas. [Sig. vaulted or arched.]
camerata var. kansasensis, Swallow, 1866, Trans. St. Louis Acad. Sci., vol. 2, Coal Meas. [Ety. proper name.]
camerata var. percrassa, Swallow, 1866, Trans. St. Louis Acad. Sci., vol. 2, Coal Meas. [Sig. very thick.] This name was preoccupied as a species.

capax, Hall, 1858, Geo. Rep. Iowa, vol. 1, pt. 2, Kinderhook Gr. [Sig. large, capacious.]
carteri, Hall, 1857, 10th Reg. Rep., Waverly Gr. [Ety. proper name.]
cedarensis, Owen, 1852, Geo. Sur. Wis., Iowa and Minn., Ham. Gr. [Ety. proper name.]
centronata, Winchell, 1865, Proc. Acad. Nat. Sci., Cuyahoga shale. [Sig. having knots or points.]
clara, Swallow, 1853, Trans. St. Louis Acad. Sci., Low. Carb. [Sig. remarkable.]
clavatula, McChesney, 1861, Desc. New Pal. Foss., Burlington Gr. [Sig. little club.]
clintoni, syn. for S. granulifera.
clio, syn. for S. ziczac.
compacta, Meek, 1868, Trans. Chi. Acad. Sci., Ham. Gr. [Sig. joined, compact.]
concinna, Hall, 1857, 10th Reg. Rep., Low. Held. Gr. [Sig. handsome.]
congesta, syn. for S. granulifera.
consors, Winchell, 1866, Rep. Low. Peninsula Mich., Ham. Gr. [Sig. common.]
contracta, Meek & Worthen, 1866, Geol. Sur. Ill., vol. 2, Chester Gr. [Sig. contracted, gathered.]
cooperensis, Swallow, 1860, Trans. St. Louis Acad. Sci., Chemung Gr. [Ety. proper name.]
corticosa, Hall, 1857, 10th Reg. Rep., Ham. Gr. [Sig. covered with thick bark.]
crenistria, see Streptorhynchus crenistria.
crispa, Hisinger, 1826, (Terebratula crispa) Act. Acad. Sci. Holm., Niagara Gr. [Sig. curled.]
cumberlandiæ, Hall, 1857, 10th Reg. Rep., Oriskany sandstone. [Ety. proper name.]
cuspidata, Sowerby, 1812, Min. Conch., vol. 1, Devonian. [Sig. pointed.]
cycloptera, Hall, 1859, Pal. N. Y., vol. 3, Low. Held. Gr. [Ety. *kuklos*, a circle; *pteron*, a wing.]
cyrtiniformis, Hall, 1873, 23rd Reg. Rep., Chemung Gr. [Sig. like a shell of the genus *Cyrtina*.]
decemplicata, Hall, 1843, (Delthyris decemplicata) Geo. Rep. 4th Dist. N. Y., Niagara Gr. [Sig. having ten plications.]
disjuncta, Sowerby, 1840, Trans. Geo. Soc., 2nd ser., vol. 5, Chemung Gr. [Sig. divided.]
disparilis, Hall, 1857, 10th Reg. Rep., Up. Held Gr. [Sig. unequal.]
distans, syn. for S. disjuncta.
divaricata, Hall, 1857, 10th Reg. Rep., Cornif. & Ham. Gr. [Sig. straddling.]
dubia, see Pentamerella dubia.
duodenaria, Hall, 1843, (Delthyris duodenaria) Geol. 4th Dist. N. Y., Schoharie grit & Cornif. Gr. [Sig. twelve.]

duplicata, Conrad, 1842, (Delthyris duplicata) Jour. Acad. Nat. Sci., vol. 8, Ham. Gr. [Sig. double-plicated.]
eatoni, see S. medialis var. eatoni.
engelmanni, Meek, 1860, Proc. Acad. Nat. Sci., Oriskany sandstone. [Ety. proper name.]
eudora, Hall, 1861, Rep. of Prog. Wis. Sur., Niagara Gr. [Ety. proper name.]
eurutcines, Owen, 1840, (Delthyris eurutcines) Report on Min. Lands, Up. Held. Gr. [Sig. widely extended; from the long hinge line.]
euruteines var. fornacula, Hall, 1857, (S. fornacula) 10th Reg. Rep., Ham. Gr. [Sig. a little oven.]
exporrecta, see Cyrtia exporrecta.
exporrecta var. arrecta, see Cyrtia exporrecta var. arrecta.
extensa, syn. for S. disjuncta.
extenuata, Hall, 1858, Geo. Rep. Iowa, Kinderhook Gr. [Sig. drawn out to a thin edge.]
fasciger, Keyserling in Owen's report, see Spirifera camerata.
fastigiata, Meek & Worthen, 1870, Proc. Acad. Nat Sci. Phil., Keokuk Gr. [Sig. pointed, peaked like a roof.]
filicosta, Winchell, 1866, Rep. Low. Peninsula Mich., Ham. Gr. [Sig. having thread-like costæ.]
fimbriata, Conrad, 1842, (Delthyris fimbriata) Jour. Acad. Nat. Sci., vol. 8, Oriskany sandstone, Schoharie grit, Cornif. and Ham. Gr. [Sig. fringed.]
forbesi, Norwood & Pratten, 1854, Jour. Acad. Nat. Sci., vol. 3, Burlington & Chester Gr. [Ety. proper name.]
formosa, Hall, 1857, 10th Reg. Rep., Ham. Gr. [Sig. beautiful.]
fornax, Hall, 1857, 10th Reg. Rep., Ham. Gr. [Sig. a furnace.]
franklini, Meek, 1858, Trans. Chi. Acad. Sci., Ham. Gr. [Ety. proper name.]
fultonensis, Worthen, 1873, Geol. Rep. Ill., vol. 5, Low. Coal Meas. [Ety. proper name.]
gaspensis, Billings, 1874, Pal. Foss., vol. 2, Devonian. [Ety. proper name.]
gibbosa, Hall, 1861, Rep. of Progr. Wis. Sur., Niagara Gr. [Sig. gibbous, tumid.]
gigantea, syn. for S. disjuncta.
glabra var. contracta, Meek & Worthen, 1861, Proc. Acad. Nat. Sci., Chester Gr. [Sig. contracted.]
glabra, Martin, 1809, (Anomites glabra) Petrif. Derb., Low. Carb. [Sig. smooth.]
glanscerasi, White, 1862, Proc. Bost. Soc. Nat. Hist., Ham. Gr. [Sig. mast of the cherry tree.]
grandæva, syn. for S. disjuncta.
granulifera, Hall, 1843, (Delthyris granulifera) Geol. 4th Dist. N. Y., Ham. Gr. [Sig. bearing granules.]

granulosa, Conrad, 1839, (Delthyris granulosa) Ann. Rep. N. Y., Low. Held. Gr. [Sig. covered with small granules.]
gregaria, Clapp, 1857, 10th Reg. Rep. & Can. Jour., Up. Held. Gr. [Sig. occurring in flocks or masses.]
grieri, Hall, 1857, 10th Reg. Rep., Schoharie grit & Cornif. Gr. [Ety. proper name.]
grimesi, Hall, 1858, Geo. Rep. of Iowa, Burlington Gr. [Ety. proper name.]
guadalupensis, Shumard, 1858, Trans. St. Louis Acad. Sci., Permian Gr. [Ety. proper name.]
hannibalensis, Swallow, 1860, Trans. St. Louis Acad. Sci., Low. Carb. [Ety. proper name.]
hemicyclus, Meek & Worthen, 1868, Geo. Sur. Ill., vol. 3, Oriskany sandstone. [Sig. a half circle.]
hemiplicata, see Syntrielasma hemiplicatum.
heteroclitus, syn. for S. granulifera.
hungerfordi, Hall, 1858, Geo. Rep. Iowa, vol. 1, pt. 2, Ham. Gr. [Ety. proper name.]
huronensis, Winchell, 1862, Proc. Acad. Nat. Sci., Portage Gr. [Ety. proper name.]
imbrex, Hall, 1858, Geo. Rep. Iowa, Burlington Gr. [Sig. the gutter or roof tile.]
incerta, Hall, 1858, Geo. Rep. Iowa, Burlington Gr. [Sig. uncertain, doubtful.]
inconstans, syn. for Spirifera racinensis.
increbescens, Hall, 1858, Geo. Rep. Iowa, Kaskaskia Gr. [Sig. abundant.]
increbescens *var*. americana, Swallow, 1866, Trans. St. Louis Acad. Sci., vol. 2, Kaskaskia Gr. [Ety. proper name.]
increbescens *var*. transversalis, Hall, 1858, Geol. Rep. Iowa, Kaskaskia Gr. [Sig. transverse.]
inæquicostata, Owen, 1852, Geo. Rep. Wis., Iowa & Min., Carb. [Sig. having unequal costæ or ribs.]
inornata, syn. for S. disjuncta.
insolita, Winchell, 1862, Proc. Acad. Nat. Sci., Portage Gr. [Sig. unusual.]
intermedia, Hall, 1859, Pal. N. Y., vol. 3, Oriskany sandstone. [Sig. intermediate.] This name was preoccupied by Brongniart in 1829.
inutilis, Hall, 1858, Geo. Rep. Iowa, vol. 1, pt. 2, Ham. Gr. [Sig. insignificant.]
iowensis, Owen, 1852, Geo. Sur. Wis., Iowa & Min., Ham. Gr. [Ety. proper name.]
kelloggi, Swallow, 1863, Trans. St. Louis Acad. Sci., Keokuk Gr. [Ety. proper name.]
kennicotti, Meek, 1868, Trans. Chi. Acad. Sci., Ham. Gr. [Ety. proper name.]
kentuckiensis, see Spiriferina kentuckiensis.
kentuckiensis var. propatula, see Spiriferina kentuckiensis *var*. propatula.

keokuk, Hall, 1858, Geo. Rep. Iowa, Keokuk Gr. [Ety. proper name.]
keokuk *var*. shelbyensis, Swallow, 1866, Trans. St. Louis Acad. Sci., vol. 2, Keokuk Gr. [Ety. proper name.]
lævigata, Swallow, 1863, Trans. St. Louis Acad. Sci., Keokuk Gr. [Sig. smoothed.]
lævis, Hall, 1843, (Delthyris lævis) Geol. 4th Dist. N. Y., Portage Gr. [Sig. smooth.]
lamellosa, see Athyris lamellosa.
laminosus, McCoy, as identified by Geinitz, is Spiriferina kentuckiensis.
lateralis, Hall, 1858, Geo. Rep. Iowa, Warsaw Gr. [Sig. belonging to the side.]
latior, Swallow, 1863, Trans. St. Louis Acad. Sci., Low. Carb. [Sig. wider.]
leidyi, Norwood & Pratten, 1854, Jour. Acad. Nat. Sci., 2d series, vol. 3, Chester Gr. [Ety. proper name.]
leidyi *var*. chesterensis, Swallow, 1866, Trans. St. Louis Acad. Sci., vol. 2, Chester Gr. [Ety. proper name.]
leidyi *var*. merrimacensis, Swallow, 1866, Trans. St. Louis Acad. Sci., vol. 2, Warsaw Gr. [Ety. proper name.]
ligus, syn. for S. pinnata.
lineatoides, Swallow, 1860, Trans. St. Louis Acad. Sci., Low. Carb. [Sig. like *S. lineatus*.]
lineata, Martin, 1809, Petrif. Derb., Coal Meas. [Sig. marked with lines.]
lineatus *var*. striato-lineatus, Swallow, 1866, Trans. St. Louis Acad. Sci., vol. 2, Coal Meas. [Sig. striated and lined.]
littoni, Swallow, 1860, Trans. St. Louis Acad. Sci., St. Louis Gr. [Ety. proper name.]
logani, Hall, 1858, Geo. Rep. Iowa, Keokuk Gr. [Ety. proper name.]
lonsdalii, syn. for S. disjuncta.
macra, Hall, 1857, 10th Reg. Rep. Schoharie grit & Cornif. Gr. [Sig. lean.]
macra, Meek. This name was preoccupied.
macronota, Hall, 1843, (Delthyris macronota) Geo. 4th Dist. N. Y., Ham. Gr. [Sig. long-sided.]
macropleura, Conrad, 1840, (Delthyris macropleura) Ann. Rep. N. Y., Low. Held. Gr. [Sig. long-sided.]
macroptera, as identified by d'Archiac & Verneuil, is S. mucronota.
macrothyris, Hall, 1857, 10th Reg. Rep., Up. Held. Gr. [Sig. having a long foramen.]
maia, Billings, 1860, (Athyris maia) Can. Jour. Ind. Sci. & Arts, Cornif. Gr. [Ety. mythological name.]
manni, Hall, 1857, 10th Reg. Rep., Cornif. Gr. [Ety. proper name.]
marcyi, Hall, 1857, 10th Reg. Rep., Ham. Gr. [Ety. proper name.]
marionensis, Shumard, 1855, Geo. Rep. Mo., Ham. Gr. [Ety. proper name.]

medialis, Hall, 1843, (Delthyris medialis) Geo. 4th Dist. N. Y., Ham. Gr. [Sig. middle.]
medialis var. eatoni, Hall, 1857, (Spirifer eatoni) 10th Reg. Rep., Ham. Gr. [Ety. proper name.]
meeki, Swallow, 1860, Trans. St. Louis Acad. Sci., Low. Carb. [Ety. proper name.]
meristoides, Meek, 1868, Trans. St. Louis Acad. Sci., Ham. Gr. [Sig. in the form of a shell of the genus Merista.]
mesocostalis, Hall, 1843,.(Delthyris mesocostalis and D. acuminata) Geo. 4th Dist. N. Y., Chemung Gr. [Sig. having middle costæ.]
mesostrialis, Hall, 1843, (Delthyris mesostrialis) Geo. 4th Dist. N. Y., Ham. and Chemung Gr. [Sig. having middle striæ.]
meta, Hall, 1867, 20th Reg. Rep., Niagara Gr. [Sig. a pillar in the form of a cone.]
meusebachianus, syn. for Spirifera camerata.
mexicana, Shumard, 1858, Trans. St. Louis Acad. Sci., Permian Gr. [Ety. proper name.]
missouriensis, Swallow, 1860, Trans. St. Louis Acad. Sci., Chemung Gr. [Ety. proper name.]
modesta, Hall, 1859, Pal. N. Y., vol. 3, Low. Held. Gr. [Ety. in allusion to its small size.]
mucronata, Conrad, 1841, (Delthyris mucronata) Ann. Rep. N. Y., Marcellus shale, Ham. Gr. [Sig. sharp-pointed.]
multistriata, see Trematospira multistriata.
mysticensis, Meek, 1872, Hayden's Geo. Rep., Low. Carb. [Ety. proper name.]
neglecta, Hall, 1858, Geo. Rep. Iowa, Keokuk Gr. [Sig. overlooked, neglected.]
niagarensis, Conrad, 1842, Jour. Acad. Nat. Sci., vol. 8, Niagara Gr. [Ety. proper name.]
niagarensis var. oligoptycha, Roemer, 1860, Sil. Fauna West Tenn., Niagara Gr. [Ety. oligos, few; ptyche, a fold.]
norwoodiana, Hall, 1858, Trans. Alb. Inst., vol. 4, Warsaw Gr. [Ety. proper name.]
norwoodi, Meek, 1860, Proc. Acad. Nat. Sci., Devonian. [Ety. proper name.]
octocostata, Hall, 1859, Pal. N. Y., vol. 3, Low. Held. Gr. [Sig. having eight costæ or folds.]
octoplicata, Sowerby, 1827, as identified by Hall, syn. for Spiriferina kentuckiensis.
opima, Hall, 1858, Geo. Rep. Iowa, Coal Meas. [Sig. large.]
orestes, Hall, 1873, 23rd Reg. Rep., Chemung Gr. [Ety. mythological name.]
oregonensis, Shumard, 1863, Trans. St. Louis Acad. Sci., Coal Meas. [Ety. proper name.]
osagensis, Swallow, 1860, Trans. St. Louis Acad. Sci., Chemung Gr. [Ety. proper name.]

oweni, Hall, 1857, 10th Reg. Rep., Up. Held. Gr. [Ety. proper name.]
pachyptera, Goldfuss, as identified by Conrad in 1839, (Delthyris pachyptera). Not American.
parryana, Hall, 1858, Geo. Rep. Iowa, vol. 1, pt. 2, Ham. Gr. [Ety. proper name.]
peculiaris, Shumard, 1855, Geo. Rep. Mo., Chemung Gr. [Sig. peculiar, singular.]
perforata, see Trematospira perforata.
pinnata, Owen, 1852, Geo. Rep. Wis., Iowa & Min., Ham. Gr. [Sig. winged.]
percrassa, McCoy, 1855, Brit. Pal. Rocks., Sil. Not satisfactorily identified in America.
perextensa, Meek & Worthen, 1868, Geo. Sur. Ill., vol. 3, Carb. [Sig. very extended.]
(?) perforata, see Trematospira perforata.
perlamellosa, Hall, 1859, Pal. N. Y., vol. 3, Low. Held. Gr. [Sig. having very thin plates.]
perplexa, McChesney, 1860, New Pal. Foss., Coal Meas. [Sig. very perplexing.]
pertenuis, Hall, 1857, 10th Reg. Rep., Ham. Gr. [Sig. very slender.]
pharovicina, Winchell, 1862, Proc. Acad. Nat. Sci., Portage Gr. [Sig. close by the light house.]
pinonensis, Meek, 1870, Proc. Acad. Nat. Sci., Up. Held. Gr. [Ety. proper name.]
planoconvexa, Shumard, 1855, Geo. Rep. Mo., Coal Meas. [Ety. planus, plane; convexus, convex.]
plena, Hall, 1858, Geo. Rep. Iowa, Burlington Gr. [Sig. full, large.]
plicata, Vanuxem, 1843, see S. vanuxemi.
prœmatura, Hall, 1867, Pal. N. Y., vol. 4, Chemung Gr. [Sig. too early, ripe before its time.]
prolata, Vanuxem, 1842, (Delthyris prolata) Geo. Rep. N. Y., Chemung Gr. [Sig. prolonged.]
propinqua, Hall, 1858, Geo. Rep. Iowa, Keokuk Gr. [Ety. related to; from its resemblance to S. subcuspidata.]
prora, Conrad, 1842, (Delthyris prora) Jour. Acad. Nat. Sci., vol. 8, Ham. Gr. [Sig. the prow of a ship.]
protensa, syn. for S. disjuncta.
pseudolineata, Hall, 1858, Geo. Rep. Iowa, Keokuk Gr. [Sig. false-lined.]
pulchra, Meek, 1860, Proc. Acad. Nat. Sci., Coal Meas. [Sig. beautiful.]
pyramidalis, see Cyrtina pyramidalis.
pyxidata, Hall, 1859, Pal. N. Y., vol. 3, Oriskany sandstone. [Sig. made like a box.]
racinensis, McChesney, 1861, Pal. Foss., Niagara Gr. [Ety. proper name.]
radiata, Sowerby, 1825, Min. Conch., vol. 5, Niagara Gr. [Sig. radiated.]

raricosta, Conrad, 1842, (Delthyris raricosta) Jour. Acad. Nat. Sci., vol. 8, Schoharie grit and Cornif. Gr. [Sig. with few costæ.]
resupinata, as identified by d'Archiac & Verneuil. Not American.
richardsoni, Meek, 1868, Trans. Chi. Acad. Sci., Ham. Gr. [Ety. proper name.]
rockymontana, Marcou, 1858, Geo. N. Amer., Low. Carb. [Ety. proper name.]
rostellata, Hall, 1858, Geo. Rep. Iowa, Keokuk Gr. [Sig. having a little beak.]
rostellum, Hall, 1872, 24th Reg. Rep., Niagara Gr. [Sig. a little beak.]
rugicosta, Hall, 1860, Can. Nat. Geo., vol. 5, Up. Sil. [Sig. having furrowed or wrinkled plaits or folds.]
rugatina, Conrad, 1842, (Delthyris rugatina) Jour. Acad. Nat. Sci., vol. 8, Niagara Gr. [Sig. having little folds.]
saffordi, Hall, 1859, Pal. N. Y., vol. 3, Low. Held. Gr. [Ety. proper name.]
scobina, Meek, 1860, Proc. Acad. Nat. Sci., Coal Meas. [Sig. a rasp.]
sculptilis, Hall, 1843, (Delthyris sculptilis) Geo. Rep. 4th Dist. N. Y., Ham. Gr. [Sig. carved or graven.]
segmentata, Hall, 1857, 10th Reg. Rep., Up. Held. Gr. [Sig. made up of segments or pieces.]
semiplicata, Hall, 1860, 13th Reg. Rep., Ham. Gr. [Sig. half-plicated.]
setigera, Hall, 1858, Geo. Rep. Iowa, Kaskaskia Gr. [Sig. bristly.]
sillana, Winchell, 1865, Proc. Acad. Nat. Sci., Cuyahoga shale. [Ety. proper name.]
similior, see Pentamerus similior.
solidirostris, White, 1860, Bost. Jour. Nat. Hist., Kinderhook Gr. [Sig. solid-beaked.]
spinosa, see Spiriferina spinosa.
staminea, Hall, 1843, Geo. Rep. 4th Dist. N. Y., Niagara Gr. [Sig. made of threads, thready.]
striatiformis, Meek, 1875, Ohio Pal., vol. 2, Waverly Gr. [Sig. like S. striata.]
striata, Martin, 1809, (Anomites striatus) Petrif. Derb., Carb. [Sig. striated.]
striata var. triplicata, Marcou, 1858, Geol. North America, Low. Carb. [Sig. three-plicated.]
striatulus, as identified by d'Archiac & Verneuil. Not American.
subæqualis, Hall, 1858, Geo. Rep. Iowa, Warsaw Gr. [Sig. somewhat equal.]
subattenuata, Hall, 1858, Geo. Rep. Iowa, Ham. Gr. [Sig. somewhat attenuated.]
subcardiformis, Hall, 1858, Geo. Rep. Iowa, Warsaw Gr. [Sig. somewhat heart-shaped.]
subcuspidata, Hall, 1858, Geo. Rep. Iowa, Keokuk Gr. [Sig. somewhat-pointed.]

subelliptica, McChesney, 1860, New Pal. Foss., Coal Meas. [Sig. somewhat elliptical.]
sublineata, Meek, 1868, Trans. Chi. Acad. Sci., Ham. Gr. [Sig. somewhat striated.]
submucronata, Hall, 1857, 10th Reg. Rep., Oriskany sandstone. [Sig. somewhat sharp-pointed.]
submucronata, Hall, 1858, Geo. Rep. Iowa, vol. 1, pt. 2, Ham. Gr. This name was preoccupied. See S. subattenuata.
suborbicularis, Hall, 1858, Geo. Rep. Iowa, Keokuk Gr. [Sig. somewhat orb-shaped.]
subrotundata, Hall, 1858, Geo. Rep. Iowa, vol. 1, pt. 2, Kinderhook Gr. [Sig. somewhat rounded.]
subsulcata, Hall, 1860, Can. Nat. and Geol., vol. 5, Up. Sil. [Sig. somewhat furrowed.] This name was preoccupied by Dalman in 1828.
subumbonata, Hall, 1857, (Orthis subumbona) 10th Reg. Rep., (Ambocoelia subumbona) 13th Reg. Rep., Ham. Gr. and Tully limestone. [Sig. somewhat protuberant.]
subundifera, Meek & Worthen, 1868, Geo. Sur. Ill., vol. 3, Ham. Gr. [Sig. somewhat wavy.]
subvaricosa, Hall, 1873, 23rd Reg. Rep., Up. Held. Gr. [Sig. somewhat varicose.]
subventricosa, syn. S. opima.
sulcata, Hisinger, 1837, (Delthyris sulcatus) Petrif. Suecica, Niagara Gr. [Sig. furrowed.]
sulcifera, Shumard, 1858, Trans. St. Louis Acad. Sci., Permian Gr. [Sig. bearing furrows.]
superba, Billings, 1874, Pal. Foss., vol. 2, Devonian. [Sig. superb.] The name was preoccupied by Eichwald in 1842.
tenuicostata, Hall, 1858, Geo. Rep. Iowa, Warsaw Gr. [Sig. slender-ribbed.]
tenuimarginata, Hall, 1858, Geo. Rep. Iowa, Keokuk Gr. [Sig. thin-margined.]
tenuis, Hall, 1857, 10th Reg. Rep., Ham. Gr. [Sig. slender, thin.]
tenuistriata, Hall, 1859, Pal. N. Y., vol. 3, Low. Held. Gr. [Sig. having fine radiating striæ.]
tenuistriata, Shaler, 1865. The name was preoccupied.
texana, Meek, 1871, Proc. Acad. Nat. Sci., Coal Meas. [Ety. proper name.]
texta, Hall, 1857, 10th Reg. Rep., Chemung Gr. [Sig. woven.]
translata, Swallow, 1863, Trans. St. Louis Acad. Sci., Low. Carb. [Sig. transferred.]
transversa, McChesney, 1860, New Pal. Foss., Chester Gr. [Sig. transverse, crosswise.]
tribulis, Hall, 1859, Pal. N. Y., vol. 3, Oriskany sandstone. [Sig. one of the same flock.]

triplicata, syn. for Spirifera camerata.
tullia, Hall, 1867, Pal. N. Y., vol. 4, Ham. Gr. [Ety. proper name.]
undulata, Vanuxem, 1843, (Delthyris undulatus) Geo. 3rd Dist. N. Y., Onondaga Gr. [Sig. undulating.] The name was preoccupied.
unica, Hall, 1867, Pal. N. Y., vol. 4, Cornif. Gr. [Sig. excellent, chief.]
vanuxemi, Hall, 1859, Pal. N. Y., vol. 3, Low. Held. Gr., described as Orthis plicata by Vanuxem in the Geo. Rep. 3rd Dist. N. Y., but that name was preoccupied. [Ety. proper name.]
varicosa, Hall, 1857, 10th Reg. Rep., Up. Held. Gr. [Sig. varicose.]
ventricosa, see Nucleospira ventricosa.
venusta, syn. for Spirifera divaricata.
vernonensis, Swallow, 1860, Trans. St. Louis Acad. Sci., Chemung Gr. [Ety. proper name.]
verneuili, syn. for S. disjuncta.
whitneyi, Hall, 1858, Geo. Rep. Iowa, Ham. and Chemung Gr. [Ety. proper name.]
wortheni, Hall, 1857, 10th Reg. Rep., Ham. Gr. [Ety. proper name.]
ziczac, Hall, 1843, (Delthyris zigzag) Geo. Rep. 4th Dist. N. Y., Ham. Gr. [Sig. slanting in straight lines from side to side.]
SPIRIFERINA, D'Orbigny, 1847, Consid. Zool. et Geol. Sur. les Brachiopodes, Comptes rendus des Sciences de l'Academie des Sciences. [Ety. *Spirifera*, a genus of shells; *inus*, implying resemblance.]
billingsi, Shumard, 1858, Trans. St. Louis Acad. Sci., Permian Gr. [Ety. proper name.]
binacuta, Winchell, 1865, Proc. Acad. Nat. Sci., Burlington Gr. [Ety. *bis*, double; *acutus*, acute.]
clarksvillensis, Winchell, 1865, Proc. Acad. Nat. Sci., Low. Carb. [Ety. proper name.]
kentuckiensis, Shumard, 1855, (Spirifera kentuckiensis) Geo. Rep. Mo., Coal Meas. [Ety. proper name.]
kentuckiensis *var.* propatula, Swallow, 1860, (Spirifera kentuckiensis *var.* propatula) Trans. St. Louis Acad. Sci., vol. 2, Coal Meas. [Sig. open.]
spinosa, Norwood & Pratten, (Spirifera spinosa) 1855, Jour. Acad. Nat. Sci., vol. 3, 2d series, Warsaw Gr. [Sig. spiny.]
spinosa *var.* campestris, White, 1874, Rep. Invert. Foss., Carb. [Sig. of or belonging to the plain fields, rustic.]
subtexta, White, 1862, Proc. Bost. Soc. Nat. Hist., vol. 9, Burlington Gr. [Sig. somewhat woven.]

Spirigera, syn. for Athyris.
americana, syn. for A. americana.
biloba, syn. for A. biloba.
caput-serpentis, syn. for A. caput-serpentis.
charitonensis, syn. for A. charitonensis.
clintonensis, syn. for A. clintonensis.
concentrica, syn. for A. spiriferoides.
corpulenta, syn. for A. corpulenta.
eborea, syn. for A. eborea.
euzona, syn. for A. euzona.
formosa, syn. for A. formosa.
fultonensis, syn. for A. fultonensis.
hannibalensis, syn. for A. hannibalensis.
hawni, syn. for A. hawni.
jacksoni, syn. for A. jacksoni.
macouensis, syn. for A. macouensis.
minima, syn. for A. minima.
missouriensis, syn. for A. missouriensis.
monticola, syn. for A. monticola.
obmaxima, syn. for A. obmaxima.
ohioensis, syn. for A. ohioensis.
pectinifera, syn. for A. pectinifera.
plattensis, syn. for A. plattensis.
prouti, syn. for A. prouti.
reflexa, syn. for A. reflexa.
singletoni, syn. for A. singletoni.
spiriferoides, syn. for A. spiriferoides.

STENOSCHISMA, Conrad, 1839, Ann. Rep. N. Y. [Ety. *stenos*, narrow; *schisma*, a fissure.] Written *Stenocisma* by Conrad.
billingsi, Hall, 1867, Pal. N. Y., vol. 4, Cornif. Gr. The same that Billings called Rynchonella thalia, Can. Jour., 1860, but the name was preoccupied. [Ety. proper name.]
carica, Hall, 1867, Pal. N. Y., vol. 4, Ham. Gr. [Sig. a kind of dry fig.]
carolina, Hall, 1867, Pal. N. Y., vol. 4, Cornif. Gr. [Ety. proper name.]
congregatum, Conrad, 1841, (Atrypa congregata) Ann. Rep. N. Y., Ham. Gr. [Sig. gathered together.]
contractum, Hall, 1843, (Atrypa contracta) Geo. 4th Dist. N. Y., Chemung Gr. [Sig. contracted, drawn together.]
contractum *var.* saxatile, Hall, 1867, Pal. N. Y., vol. 4, Chemung Gr. [Sig. that lives among stone or rocks.]
dotis, Hall, 1867, Pal. N. Y., vol. 4, Ham. Gr. [Sig. an ornament.]
duplicatum, Hall, 1843, (Atrypa duplicata) Geo. 4th Dist. N. Y., Chemung Gr. [Sig. doubled.]
eximium, Hall, 1843, (Atrypa eximia) Geo. 4th Dist. N. Y., Chemung Gr. [Sig. choice, select.]
horsfordi, Hall, 1860, 13th Reg. Rep., Cornif. Gr., Marcellus shale & Ham. Gr. [Ety. proper name.]
orbiculare, Hall, 1860, (Rhynchonella orbicularis) 13th Reg. Rep., Chemung Gr. [Sig. orb-shaped.]
prolificum, Hall, 1867, Pal. N. Y., vol. 4, Ham. Gr. [Sig. fruitful, prolific.]
royanum, Hall, 1860, Pal. N. Y., vol. 4, Cornif. Gr. [Ety. proper name.]
sappho, Hall, 1860, (Rhynchonella sappho) 13th Reg. Rep., Marcellus shale & Ham. Gr. [Ety. proper name.]
stephani, Hall, 1867, Pal. N. Y., vol. 4, Chemung Gr. [Ety. proper name.]

tethys, Billings, 1860, (Rhynchonella tethys) Can. Jour., Cornif. Gr. [Ety. mythological name.]
STREPTORHYNCHUS, King, 1850, Monograph of Permian Fossils. [Ety. *strepto*, I bend or twist; *rhynchos*, a beak.]
antiquatus, Sowerby, 1839, (Strophomena antiquata) Murch. Sil. Syst., Mid. Sil. [Sig. ancient.]
arctostriatus, Hall, 1843, (Strophomena arctostriata) Geo. Rep. 4th Dist. N. Y., Chemung Gr. [Sig. closely striated.]
arctostriatus, Hall, 1860, 13th Reg. Rep., (Orthisina arctostriata) Ham. Gr. This name was preoccupied.
chemungensis, Conrad, 1843, (Strophomena chemungensis) Jour. Acad. Nat. Sci., Chemung Gr. [Ety. proper name.]
crassus, Meek & Hayden, 1858, (Orthisina crassa) Proc. Acad. Nat. Sci. Phil., Coal Meas. [Sig. thick.]
crenistriatus, Phillips, 1836, (Spirifera crenistria) Geo. York., vol. 2, Waverly Gr. [Sig. having crooked striæ.]
deflectus, Conrad, 1843, (Strophomena deflecta) Proc. Acad. Nat. Sci. Phil., Trenton Gr. [Sig. bent down.]
elongatus, James, 1874, Cin. Quar. Jour. Sci., Cin'ti. Gr. [Sig. from the long cardinal line.]
filitextus, Hall, 1847, (Strophomena filitexta) Pal. N. Y., vol. 1, Trenton and Hud. Riv. Gr. [Sig. woven like thread.]
hallianus, S. A. Miller, 1874, Cin. Quar. Jour. Sci., vol. 1, Cin'ti Gr. [Ety. proper name.]
hemiaster, syn. for S. subplanum.
lens, White, 1862, Proc. Bost. Soc. Nat. Hist., vol. 9, Chemung Gr. [Sig. a concavo-convex shell.]
nutans, James, 1873, (Hemipronites nutans) Pal. Ohio, vol. 1, Cin'ti Gr. [Sig. bent over.]
occidentalis, Newberry, 1861, Ives' Col. Ex. Exped., Up. Carb. [Sig. western.]
pandora, Billings, 1860, Can. Jour., Schoharie grit and Cornif. Gr. [Ety. mythological name.]
pectinaceus, Hall, 1843, (Strophomena pectinacea and S. bifurcata) Geo. Rep. 4th Dist. N. Y., Chemung Gr. and Waverly sandstone. [Sig. like a *Pecten*.]
perversus, Hall, 1857, (Orthis perversa) 10th Reg. Rep. (Orthisina alternata, 1860, 13th Reg. Rep.) Cornif. and Ham. Gr. [Sig. turned upside down.]
planoconvexus, Hall, 1847, (Leptæna planoconvexa) Pal. N. Y., vol. 1, Hud. River Gr. [Ety. *planus*, plane; *convexus*, convex.]
planumbonus, Hall, 1847, (Leptæna planumbona) Pal. N. Y., vol. 1, Trenton & Hud. Riv. Gr. [Sig. flat on the umbo.]
pyramidalis, Newberry, 1861, Ives' Col. Ex. Exped., Carb. [Sig. pointed like a pyramid.]
radiatus, Vanuxem, 1843, (Strophomena radiata) Geo. Rep. 3rd Dist. N. Y., Low. Held. Gr. [Sig. radiated.]
rectus, Conrad, 1843, (Strophomena recta) Proc. Acad. Nat. Sci., vol. 1, Black Riv. and Trenton Gr. [Sig. straight.]
sinuatus, Emmons, 1855, Am. Geol., Cin'ti Gr. [Sig. marked with depressions, wavy.]
subplanus, Conrad, 1842, Jour. Acad. Nat. Sci., vol. 8, (Strophomena subplana) Niagara Gr. [Sig. somewhat flat.]
subtentus, Conrad, 1847, (Strophomena subtenta) Pal. N. Y., vol. 1, Trenton & Hud. Riv. Gr. [Sig. somewhat bent.]
sulcatus, Verneuil, 1848, (Leptæna sulcata) Bull. Geol. Soc. France, vol. 5, Cin'ti Gr. [Sig. furrowed.]
tenuis, Hall, 1858, Trans. Alb. Inst., vol. 4, Niagara Gr. [Sig. slight, slender.]
umbraculum, V. Buch, (Orthis umbraculum) Ueber Delthyris, &c., Devonian to the Permian Gr. [Sig. an umbrella.]
vetustus, James, 1874, Cin. Quar. Jour. Sci., vol. 1, Cin'ti Gr. [Sig. old.]
woolworthianus, Hall, 1859, (Strophomena woolworthiana) Pal. N. Y., vol. 3, Low. Held. Gr. [Ety. proper name.]

Stricklandia, Billings, 1859, Can. Nat. Geo., vol. 4. This name having been previously applied to a genus of fossil plants the author abandoned it and proposed Stricklandinia.

STRICKLANDINIA, Billings, 1863, Can. Nat. & Geo., vol. 8. [Ety. proper name.]
anticostiensis, Billings, 1863, Can. Nat. Geo., vol. 8, Anticosti Gr. [Ety. proper name.]
(?) arachne, Billings, 1862, Pal. Foss., vol. 1, Quebec Gr. [Ety. mythological name.]
(?) arethusa, Billings, 1862, Pal. Foss., vol. 1, Quebec Gr. [Ety. mythological name.]
brevis, Billings, 1859, Can. Nat. Geo., vol. 4, Mid. Sil. [Sig. short.]
canadensis, Billings, 1859, Can. Nat. Geo., vol. 4, Clinton Gr. [Ety. proper name.]
castellana, White, 1876, Proc. Acad. Nat. Sci., Niagara Gr. [Ety. proper name.]
davidsoni, Billings, 1868, Lond. Geo. Mag., vol. 5, Up. Sil. [Ety. proper name.]
deformis, Meek & Worthen, 1870, Proc. Acad. Nat. Sci. Phil., Niagara Gr. [Sig. deformed, ugly shaped.]
elongata, see Amphigenia elongata.
elongata var. curta, see Amphigenia curta.
gaspensis, Billings, 1859, Can. Nat. Geo., vol. 4, Mid. Sil. [Ety. proper name.]

BRACHIOPODA. 135

lens. The fossil referred to this species was afterward described as S. davidsoni.
melissa, Billings, 1874, Pal. Foss., vol. 2, Mid. Sil. [Ety. mythological name.]
salteri, Billings, 1874, Pal. Foss., vol. 2, Mid. Sil. [Ety. proper name.]

STROPHALOSIA, King, 1844, Ann. & Mag. Nat. Hist., vol. 14. [Ety. *strophe*, a bending; *alos*, a disc.]
horrescens, Geinitz, 1866, Carb. und Dyas in Neb. Prof. Meek regarded this name as a syn. for Productus nebrascensis.
numularis, Winchell, 1863, Proc. Acad. Nat. Sci., Chemung Gr. [Sig. like a little coin.]
subaculeata, Murchison, 1845, Bul. Geol. Soc. de France, Coal Meas. [Sig. somewhat thorny or furnished with prickles.]

STROPHODONTA, Hall, 1852, Pal. N. Y., vol. 2. [Ety. *strophos*, bent; *odous*, tooth.]
æquicostata, Swallow, 1860, Trans. St. Louis Acad. Sci., Ham. Gr. [Sig. equal-ribbed.]
altidorsata, Swallow, 1860, Trans. St. Louis Acad. Sci., Devonian. [Sig. high backed.]
alveata, Hall, 1863, 16th Reg. Rep., Schoharie grit. [Sig. channeled.]
ampla, Hall, 1857, (Strophomena ampla) 10th Reg. Rep., Schoharie grit and Cornif. Gr. [Sig. full, large.]
arcuata, Hall, 1858, Geo. of Iowa, Ham. Gr. [Sig. bent, arched.]
becki, Hall, 1859, Pal. N. Y., vol. 3, Low. Held. Gr. [Ety. proper name.]
boonensis, Swallow, 1860, Trans. St. Louis Acad. Sci., Low. Devonian. [Ety. proper name.]
cælata, Hall, 1867, Pal. N. Y., vol. 4, Chemung Gr. [Sig. sculptured.]
callawayensis, Swallow, 1860, Trans. St. Louis Acad. Sci., Ham. Gr. [Ety. proper name.]
callosa, Hall, 1863, 16th Reg. Rep., Schoharie grit. [Sig. having a thick shell.]
canace, Hall, 1873, 23rd Reg. Rep., Chemung Gr. [Ety. mythological name.]
cavumbona, Hall, 1859, Pal. N. Y., vol. 3, Low. Held. Gr. [Sig. having a hollow umbo.]
cayuta, Hall, 1867, Pal. N. Y., vol. 4, Chemung Gr. [Ety. proper name.]
cincta, Winchell, 1866, Rep. Low. Peninsula Mich., Ham. Gr. [Ety. *cinctus*, girdled; in allusion to a ridge around the border of the inside of the ventral valve.]
concava, Hall, 1857, (Strophomena concava) 10th Reg. Rep., Cornif. and Ham. Gr. [Sig. concave.]
costata, Owen, 1852, Geo. Sur. Wis., Iowa and Minn., Devonian. [Sig. ribbed.]

crebristriata, Conrad, 1842, (Strophomena crebristriata) Jour. Acad. Nat. Sci., vol. 8, Schoharie grit. [Sig. thickly striated.]
cymbiformis, Swallow, 1860, Trans. St. Louis Acad. Sci., Devonian. [Sig. boat-shaped.]
demissa, Conrad, 1842, (Strophomena demissa) Jour. Acad. Nat. Sci., vol. 8, Schoharie grit, Cornif., Ham. and Chemung Gr. [Sig. hanging down.]
erratica, Winchell, 1866, Rep. Low. Peninsula Mich., Ham. Gr. [Sig. wandering, erratic.]
fragilis, syn. for Strophodonta perplana.
geniculata, Hall, 1859, Pal. N. Y., vol. 3, Low. Held. Gr. [Sig. geniculated.]
headleyana, Hall, 1859, Pal. N. Y., vol. 3, Low. Held. Gr. [Ety. proper name.]
hemispherica, Hall, 1857, (Strophomena hemispherica) 10th Reg. Rep., Schoharie Grit & Cornif. Gr. [Sig. hemispherical.]
hybrida, Hall, 1873, 23rd Reg. Rep., Chemung Gr. [Sig. a hybrid.]
imitata, Winchell, 1866, Rep. Low. Penin. Mich., Ham. Gr. [Sig. imitating, like *S. inæquistriata*.]
inæquiradiata, Hall, 1857, 10th Reg. Rep., Schoharie Grit & Cornif. Gr. [Sig. unequal-rayed.]
inæquistriata, Conrad, 1842,(Strophomena inæquistriata) Jour. Acad. Nat. Sci., vol. 8, Cornif. & Ham. Gr., Moscow shales. [Sig. having unequal striæ.]
indentata, Conrad, (Leptæna indentata) 1838, Ann. Rep. N. Y., Low. Held. Gr. [Sig. indented.]
inflexa, Swallow, 1860, Trans. St. Louis Acad. Sci., Devonian. [Sig. bowed, made crooked.]
intermedia, Hall, 1859, Pal. N. Y., vol. 3, Oriskany sandstone. [Sig. intermediate.]
iowensis, Owen, 1852, Geo. Sur. Wis., Iowa & Minn., Devonian. [Ety. proper name.]
junia, Hall, 1867, Pal. N. Y., vol. 4, Cornif., Ham. & Tully Gr., (changed from *textilis*, in the corrigenda and index). [Ety. proper name.]
kemperi, Swallow, 1860, Trans. St. Louis Acad. Sci., Devonian. [Ety. proper name.]
leavenworthiana, Hall, 1859, Pal. N. Y., vol. 3, Low. Held. Gr. [Ety. proper name.]
lepida, syn. for S. nacrea.
linckleni, Hall, 1857, 10th Reg. Rep., Oriskany sandstone. [Ety. proper name.]
magnifica, Hall, 1857, 10th Reg. Rep., Oriskany sandstone. [Sig. magnificent.]
magniventra, Hall, 1857, 10th Reg. Rep., Oriskany sandstone. [Sig. large bellied.]

mucronata, Conrad, 1842, (Strophomena mucronata) Jour. Acad. Nat. Sci., vol. 8, Chemung Gr. [Sig. sharp-pointed.]
nacrea, Hall, 1857, (Strophomena nacrea) 10th Reg. Rep., Cornif. and Ham. Gr. [Sig. iridescent, like mother of pearl.]
navalis, Swallow, 1860, Trans. St. Louis Acad. Sci., Devonian. [Sig. like a ship.]
parva, Owen, 1852, Geo. Sur. Wis., Iowa and Minn., Ham. Gr. [Sig. small.]
parva, Hall, 1863, 16th Reg. Rep., Schoharie grit. This name was preoccupied.
patersoni, Hall, 1857, (Strophomena patersoni) 10th Reg. Rep., Schoharie grit and Cornif. Gr. [Ety. proper name.]
perplana, Conrad, 1842, (Strophomena perplana) Jour. Acad. Nat. Sci., vol. 8, Onondaga, Schoharie, Cornif., Ham. and Chemung Gr. [Sig. very plain.]
perplana var. nervosa, Hall, 1843, (Strophomena nervosa) Geo. Rep. 4th Dist. N. Y., Chemung Gr. [Sig. full of sinews.]
planulata, Hall, 1859, Pal. N. Y., vol. 3, Low. Held. Gr. [Sig. flat.]
plicata, Hall, 1860, 13th Reg. Rep., Ham. Gr. [Sig. folded, plaited.]
prisca, Hall, 1852, Pal. N. Y., vol. 2, Clinton Gr. [Sig. ancient.]
profunda, Hall, 1852, (Leptæna profunda) Pal. N. Y., vol. 2, Niagara Gr. [Sig. deep.]
punctulifera, Conrad, 1838, (Leptæna punctulifera) Ann. Rep. N. Y., Low. Held. Gr. [Sig. bearing punctures.]
reversa, Hall, 1858, Geo. Rep. Iowa, Ham. Gr. [Sig. reversed.]
semifasciata, Hall, 1863, Trans. Alb. Inst., vol. 4, Niagara Gr. [Sig. half-banded.]
subcymbiformis, Swallow, 1860, Trans. St. Louis Acad. Sci., Devonian. [Sig. somewhat boat-shaped.]
varistriata, Conrad, 1842, (Strophomena varistriata) Jour. Acad. Nat. Sci., vol. 8, Low. Held. Gr. [Sig. differently striated.]
varistriata var. arata, Hall, 1859, Pal. N. Y., vol. 3, Low. Held. Gr. [Sig. furrowed.]
vascularia, Hall, 1859, Pal. N. Y., vol. 3, Oriskany sandstone. [Sig. vascular.]
STROPHOMENA, Rafinesque, 1825, Manuel de Malacologie of Blainville. [Ety. strophos, bent; mene, a crescent.]
acutirudiata, see Chonetes acutiradiata.
alternata, Conrad, 1838, (Leptæna alternata) Ann. Rep. N. Y., Trenton and Hud. Riv. Gr. [Sig. alternating.]
alterniradiata, Shaler, 1865, Bulletin, No. 4, M. C. Z., Anticosti Gr. [Sig. alternately radiated.]
alternistriata, Hall, 1847, Pal. N. Y., vol. 1, Trenton and Hud. Riv. Gr. [Sig. alternately striated.]

alternata var. loxorhytis, Meek, 1873, Ohio Pal., vol. 1, Cin'ti Gr. [Sig. cross-wrinkled.]
ampla, see Strophodonta ampla.
analoga, Phillips, 1836, Geol. Yorkshire, vol. 2, Low. Carb. [Sig. like analogous.]
anticostiensis, syn. for Strophomena alternata.
antiquata, see Streptorhynchus antiquatus.
arctostriata, see Streptorhynchus arctostriatus.
arcuata, Shaler, 1865. This name was preoccupied.
arethusa, Billings, 1862, Pal. Foss., vol. 1, Hud. Riv. Gr. [Ety. mythological name.]
aurora, Billings, 1865, Pal. Foss., vol. 1, Quebec Gr. [Sig. in the morning.]
bifurcata, syn. for Streptorhynchus pectinaceus.
bipartita, Hall, 1852, (Leptæna bipartita) Pal. N. Y., vol. 2, Coralline limestone. [Sig. double-partitioned.]
blainvilli, Billings, 1874, Pal. Foss., vol. 2, Up. Sil. [Ety. proper name.]
camerata, Conrad, 1842, Jour. Acad. Nat. Sci., vol. 8, Trenton Gr. [Sig. arched or vaulted.]
carinata, Conrad, 1838, see Tropidoleptus carinatus.
carinata, Conrad, 1842, see Chonetes carinata.
ceres, Billings, 1860, Can. Nat. & Geo., vol. 5, Hud. Riv. Gr. and Mid. Sil. [Ety. mythological name.]
chemungensis, Conrad, 1842, Jour. Acad. Nat. Sci., vol. 8, Chemung Gr. [Ety. proper name.]
concava, see Strophodonta concava.
conradi, Hall, 1859, Pal. N. Y., vol. 3, Low. Held. Gr. [Ety. proper name.]
convexa, Owen, 1840, Rep. on Mineral Lands, Calcif. Gr. [Sig. convex.]
cornuta, see Chonetes cornuta.
corrugata, Conrad, 1842, Jour. Acad. Nat. Sci., vol. 3, Clinton Gr. [Sig. wrinkled.]
crebristriata, see Strophodonta crebristriata.
crenistria, syn. for Strophodonta perplana.
declivis, syn. for Strophomena alternata.
deflecta, see Streptorhynchus deflectus.
deltoidea, Conrad, 1839, Ann. Rep. N. Y., Trenton Gr. [Sig. shaped like the Greek letter Delta.]
delthyris, syn. for Strophodonta perplana.
demissa, see Strophodonta demissa.
depressa, Sowerby, 1825, (Producta depressa) vol. 6, Min. Conchology, Up. Sil. [Sig. depressed.]
depressa var. ventricosa, see Strophomena rugosa var. ventricosa.
elegantula, Hall, 1843, Geo. Rep, 4th Dist. N. Y., Clinton Gr. [Sig. quite elegant.]

elongata, Conrad, 1842, Jour. Acad. Nat. Sci., vol. 8, Low. Held. Gr. [Sig. elongated.]
elliptica, Conrad, 1839, Ann. Rep. N. Y., Low. Held. Gr. [Sig. elliptical.]
euglypha, syn. for Strophodonta punctulifera.
fasciata, Hall, 1847, (Leptæna fasciata) Pal. N. Y., vol. 1, Chazy Gr. [Sig. swathed, banded.]
filitexta, see Streptorhynchus filitextus.
fluctuosa, Billings, 1860, Can. Nat. Geo., vol. 5, Trenton & Hud. Riv. Gr. [Sig. wavy.]
fontinalis, White, 1874, Rep. Invert. Foss., Quebec Gr. [Ety. fons, a fountain or spring.]
fracta, Meek, 1873, Pal. Ohio, vol. 1, Cin'ti Gr. [Sig. broken, weak.]
fragilis, syn. for Strophodonta perplana.
galatea, Billings, 1874, Pal. Foss., vol. 2, Devonian. [Ety. mythological name.]
geniculata, Shaler, (Brachyprion geniculatum). The name was preoccupied.
gibbosa, Conrad, 1841, Ann. Geo. Rep. N. Y., Onondaga Gr. [Sig. gibbous.]
hecuba, Billings, 1860, Can. Nat. Geo., vol. 5, Hud. Riv. Gr. [Ety. mythological name.]
hemispherica, see Strophodonta hemispherica.
imbecillis, Billings, 1865, Pal. Foss., vol. 1, Quebec Gr. [Sig. feeble.]
imbrex, Pander, 1845, in Russia & Ural Mountains, Hud. Riv. Gr. [Sig. imbricated.] The identification very doubtful in America.
impressa, syn. for Strophodonta varistriata.
incrassata, Hall, 1847, (Leptæna incrassata) Pal. N. Y., vol. 1, Chazy to Hud. Riv. Gr. [Sig. thickened.]
indentata, Conrad, 1842, Geo. Rep. 3rd Dist. N. Y., Held. Gr. [Sig. indented.]
inæquiradiata, see Strophodonta inæquiradiata.
inæquistriata, see Strophodonta inæquistriata.
interstrialis, Phillips, in Geo. 4th Dist. N. Y., see Strophodonta cayuta.
irene, Billings, 1874, Pal. Foss., vol. 2, Devonian. [Ety. proper name.]
ithacensis, Vanuxem, 1842, Geo. Rep. N. Y., Portage Gr. [Ety. proper name.]
julia, Billings, 1862, Pal. Foss., vol. 1, Mid. Sil. [Ety. proper name.]
lachrymosa, see Productella lachrymosa.
laevis, Emmons, 1842, Geo. Rep. N. Y., Birdseye Gr. [Sig. smooth.]
leda, Billings, 1860, Can. Nat. Geo., vol. 5, Mid. Sil. [Ety. mythological name.]
lepida, syn. for Strophodonta nacrea.
lima, see Productella lachrymosa var. lima.
lineata, see Chonetes lineata.
macra, syn. for Strophodonta semifasciata.
magniventra, see Strophodonta magniventra.

membranacea, of Phillips, as identified by Vanuxem, 1842, Geo. 3rd Dist. N. Y., see Productella hirsuta.
modesta, Conrad, 1839, Ann. Rep. N. Y., Low. Held. Gr. [Sig. not large.]
mucronata, see Strophodonta mucronata.
nacrea, see Strophodonta nacrea.
nasuta, Conrad, 1842, Jour. Acad. Nat. Sci., vol. 8, Trenton & Hud. Riv. Gr. [Sig. having a prominent nose.]
nassula, Conrad, 1848, Proc. Acad. Nat. Sci., vol. 3, Carb. [Sig. a little bag-net.]
nervosa, see Strophodonta perplana var. nervosa.
niagarensis, syn. for Strophodonta profunda.
nitens, Billings, 1860, Can. Nat. Geo., vol. 5, Hud. Riv. Gr. [Sig. neat.]
nutans, see Streptorhynchus nutans.
obscura, Hall, 1852, (Leptæna obscura) Pal. N. Y., vol. 2, Clinton Gr. [Sig. obscure.]
orthididæ, Hall, 1852, (Leptæna orthididæ) Pal. N. Y., vol. 1, Clinton Gr. [Sig. like a shell of the genus Orthis.]
patenta, Hall, 1852, (Leptæna patenta) Pal. N. Y., vol. 2, Clinton Gr. [Sig. spreading.] patens would be better orthography.
patersoni, see Strophodonta patersoni.
pecten, Linnæus, 1758, (Anomia pecten) Syst. Nat., Niagara Gr. [Sig. comblike.] Not American.
pectinacea, see Streptorhynchus pectinaceus.
perplana, see Strophodonta perplana.
philomela, Billings, 1860, Can. Nat. Geo., vol. 5, Mid. Sil. [Ety. mythological name.]
planoconvexa, see Streptorhynchus planoconvexus.
planumbona, see Streptorhynchus planumbonus.
plicata, syn. for Streptorhynchus subtentus.
plicifera, Hall, 1847, (Leptæna plicifera) Pal. N. Y., vol. 1, Chazy Gr. [Sig. bearing plications.]
pluristriata, syn. for Strophodonta perplana.
profunda, see Strophodonta profunda.
punctulifera, see Strophodonta punctulifera.
pustulosa, syn. for Productella truncata.
radiata, see Streptorhynchus radiatus.
recta, see Streptorhynchus rectus.
rectilateris, syn. for Strophodonta varistriata.
rhomboidalis, Wahlenberg, 1821, Acta. Soc. Sci., Upsaliensis. This species ranges from the Trenton Group to the Lower Silurian to the Hamilton and Chemung Group, regarding S. tenuistriata, S. depressa and S. rugosa as varieties only. The type however is the Devonian form. [Sig. rhomboidal.]

reticulata, Shaler, 1865, Bulletin No. 4, M. C. Z., Anticosti Gr. [Sig. reticulated.]
rugosa, Dalman, 1827, (Leptæna rugosa) Vet. Acad. Handlinger, Niagara & Low. Held. Gr. This form is supposed to be the type of Rafinesque's genus Strophomena. The species is usually regarded as merely a variety of S. rhomboidalis. [Sig. wrinkled.]
rugosa var. ventricosa, Hall, 1857, (S. depressa var. ventricosa) 10th Reg. Rep., Oriskany sandstone. [Sig. bulging-out.]
setigera, see Chonetes setigera.
semiovalis, Conrad, syn. for Leptæna sericea.
semiovalis, Shaler. The name had been twice preoccupied.
squamula, James, 1874, Cin. Quar. Jour. Sci., vol. 1, Cin'ti Gr. [Sig. a little scale.]
striata, Hall, 1843, Geo. Rep. 4th Dist. N. Y., Niagara Gr. [Sig. striated.]
subdemissa, syn. for Strophodonta demissa.
subplana, see Streptorhynchus subplanus.
subtenta, see Streptorhynchus subtentus.
syrtalis, syn. for Chonetes carinata.
tenuilineata, Conrad, 1842, Jour. Acad. Nat. Sci., vol. 8, Trenton Gr. [Sig. slender-lined.]
tenuistriata, Sowerby, 1839, (Leptæna tenuistriata) Murch. Sil. Syst., Low. Sil. [Sig. fine-lined.]
textilis, see Strophodonta junia.
thalia, Billings, 1860, Can. Nat. Geo., vol. 5, Trenton Gr. [Ety. mythological name.]
transversalis, see Leptæna transversalis.
trilobata, Owen, 1852, (Leptæna trilobata) Geo. Sur. Wis., Iowa & Minn., Trenton Gr. [Sig. three-lobed.]
tullia, Billings, 1874, Pal. Foss., vol. 2, Low. Devonian. [Ety. proper name.]
undulata, syn. for S. rhomboidalis.
undulosa, Conrad, 1841, Ann. Rep. N. Y., Low. Held. Gr. [Sig. full of undulations.]
unicostata, Meek & Worthen, 1868, Geo. Sur. Ill., vol. 3, Cin'ti Gr. [Sig. single-ribbed.]
varistriata, see Strophodonta varistriata.
ventricosa, Shaler, (Brachyprion ventricosum). The name was preoccupied.
woolworthana, see Streptorhynchus woolworthanus.

SYNTRIELASMA, Meek & Worthen, 1865, Proc. Acad. Nat. Sci. [Ety. syn, together; treis, three; elasma, plate.]
hemiplicatum, Hall, 1852, (Spirifera hemiplicata) Stan's. Ex. to Great Salt Lake, Coal Meas. [Sig. half-folded or plaited.]

SYRINGOTHYRIS, Winchell, 1863, Proc. Acad. Nat. Sci. [Ety. syrinx, a pipe or channel; thyris, a small window or orifice.]
halli, Winchell, 1863, Proc. Acad. Nat. Sci., Low. Carb. [Ety. proper name.]
typus, Winchell, 1863, Proc. Acad. Nat. Sci. Phil., Low. Carb. [Ety. type of the genus.]

TEREBRATULA, Llhwyd, 1696, Lith. Brit. Ichn. [Ety. diminutive of terebratus, perforated.]
affinis, syn. for Atrypa reticularis.
aprinis, see Rhynchonella aprinis.
arcuata, Swallow, 1863, Trans. St. Louis Acad. Sci., Low. Carb. [Sig. bent or arched.] The name was preoccupied by Roemer in 1840.
argentea, see Athyris argentea.
aspera, see Atrypa aspera.
bidentata, see Rhynchonella bidentata.
bovidens, Morton, 1836, Am. Jour. Sci., vol. 29, Coal Meas. [Ety. bos, an ox; dens, a tooth.]
brevirostris, see Rhynchonella brevirostris.
brevilobata, Swallow, 1863, Trans. St. Louis Acad. Sci., Low. Carb. [Sig. short-lobed.]
burlingtonensis, White, 1860, Bost. Jour. Nat. Hist., Kinderhook Gr. [Ety. proper name.]
concentrica, syn. for Athyris spiriferoides.
crenulata, Sowerby, 1840, (Atrypa crenulata) Geo. Trans., 2nd series, vol. 5, Devonian. [Sig. crenulated.]
cuneata, see Rhynchonella cuneata.
elia, Hall, 1867, Pal. N. Y., vol. 4, Up. Held. Gr. [Ety. proper name.]
elongata, Schlotheim, 1817, Akad. Munich, vol. 6, Permian Gr. [Sig. elongated.]
formosa, Hall, 1858, Trans. Alb. Inst., vol. 4, Warsaw Gr. [Sig. beautiful.]
geniculosa, syn. for Terebratula bovidens.
gracilis, Swallow, 1863, Trans. St. Louis Acad. Sci., Low. Carb. [Sig. slender.] The name was preoccupied by Von Buch in 1834.
harmonia, Hall, 1867, Pal. N. Y., vol. 4, Cornif. Gr. [Ety. proper name.]
inornata, McChesney, 1860, New Pal. Foss., Carb. [Sig. not ornamented.]
insperata, Phillips, 1841, Pal. Foss., Devonian. [Sig. unexpected.]
interplicata, see Rhynchonella interplicata.
jucunda, Hall, 1867, Pal. N. Y., vol. 4, Up. Held. Gr. [Sig. welcome.]
lens, Hall, 1860, 13th Reg. Rep., Cornif. Gr. [Sig. convex on both sides.]
laticosta, see Atrypa laticosta.
lincklæni, Hall, 1860, 13th Reg. Rep., Ham. Gr. [Ety. proper name.]
lynx, see Orthis lynx.
marcyi, see Retzia marcyi.
marginalis, see Atrypa marginalis.
michelini, see Orthis michelini.
millepunctata, syn. for T. bovidens.
mormoni, see Retzia mormoni.
navicella, Hall, 1867, Pal. N. Y., vol. 4, Ham. Gr. [Sig. a little boat.]
ontario, Hall, 1867, Pal. N. Y., vol. 4, Ham. Gr. [Ety. proper name.]

BRACHIOPODA. 139

ovoides, see Rensselæria ovoides.
parva, Swallow, 1863, Trans. St. Louis Acad. Sci., Low. Carb. [Sig. small.] The name was preoccupied by d'Archiac in 1846.
perinflata, Shumard, 1859,Trans. St. Louis Acad. Sci., Permian Gr. [Sig. very much inflated.]
planirostra, see Cryptonella planirostra.
rectirostra, see Cryptonella rectirostra.
reticularis, see Atrypa reticularis.
roemingeri, Hall, 1863, 16th Reg. Rep., Ham. Gr. [Ety. proper name.]
sacculus, Martin, 1809, Petrif. Derb., Low. Carb. [Sig. a little sack.]
schlotheimi, see Camerophoria schlotheimi.
simulator, Hall, 1867, Pal. N. Y., vol. 4, Ham. Gr. [Sig. an imitator.]
spiriferoides, see Athyris spiriferoides.
subtilita, see Athyris subtilita.
sullivanti, Hall, Pal. N. Y., vol. 4, Up. Held. Gr. [Ety. proper name.]
traversensis, Winchell, 1866, Rep. Low. Penin. Mich., Ham. Gr. [Ety. proper name.]
trinuclea, Hall, 1858, Trans. Alb. Inst., vol. 4, Warsaw Gr. [Sig. three-kerneled.]
turgida, Hall, 1858, Trans. Alb. Inst., vol. 4, Warsaw Gr. [Sig. swollen-out.]
uta, see Rhynchonella uta.
wilsoni, see Rhynchonella wilsoni.
Terebratulites, Schlotheim, syn. for Spirifera.
biforatus, see Orthis biforata.
TREMATIS, Sharpe, 1847, Quar. Jour. Geo. Soc., vol. 13. [Ety. *trema*, an opening.]
cælata, see Obolella cælata.
cancellata, Sowerby, 1825, (Orbicula cancellata) Zool. Jour., vol. 2, Trenton Gr. [Sig cancelled.]
crassa, see Obolella crassa.
dyeri, S. A. Miller, 1874, Cin. Quar. Jour. Sci., vol. 1, Cin'ti Gr. [Ety. proper name.]
filosa, see Schizocrania filosa.
huronensis, Billings, 1862, Pal. Foss., vol. 1, Black Riv. Gr. [Ety. proper name.]
montrealensis, Billings, 1862, Pal. Foss., vol. 1, Trenton Gr. [Ety. proper name.]
millepunctata, Hall, 1866, Pamphlet, Cin'ti Gr. [Sig. many dotted.]
ottawaensis, Billings, 1862, Pal. Foss., vol. 1, Trenton Gr. [Ety. proper name.]
pannulus, White, 1874, Rep. on Invert. Foss., Potsdam Gr. [Sig. a small piece of cloth.]
punctostriata, Hall, 1873, 23rd Reg. Rep., Trenton & Hud. Riv. Gr. [Sig. punctured and striated.]
(?) pustulosa, Hall, 1866, Pamphlet, Hud. Riv. Gr. [Sig. covered with pustules.]
rudis, Hall, 1873, 23rd Reg. Rep., Trenton Gr. [Sig. rough.]
terminalis, Emmons, 1842, (Orbicula terminalis) Geo. Rep. N. Y., Trenton Gr. [Sig. terminating.]

TREMATOSPIRA, Hall, 1859, 12th Reg. Rep. [Ety. *trema*, foramen; *spira*, a spire; in allusion to the perforation in the beak of the ventral valve.]
acadiæ, Hall, 1860, Can. Nat. & Geo., vol. 5, Up. Sil. [Ety. proper name.]
camura, Hall, 1852, (Atrypa camura) Pal. N. Y., vol. 2, Low. Held. Gr. [Sig. an arch.]
costata, Hall, 1859, Pal. N. Y., vol. 3, Low. Held. Gr. [Sig. ribbed.]
deweyi, see Rhynchospira deweyi.
formosa, see Rhynchospira formosa.
gibbosa, Hall, 1860, 13th Reg. Rep., Ham. Gr. [Sig. gibbous.]
globosa, Hall, 1856, 9th Reg. Rep., (Waldheimia globosa) Low. Held. Gr. [Sig. globular.]
granulifera, Meek, 1872, Proc. Acad. Nat. Sci. Phil., Cin'ti Gr. [Sig. bearing granules.]
hirsuta, Hall, 1857, (Atrypa hirsuta) 10th Reg. Rep., Cornif. and Ham. Gr. [Sig. rough, hairy.]
imbricata, Hall, 1857, 10th Reg. Rep., Low. Held. Gr. [Sig. like one upon another, as slates on a roof.]
mathewsoni, McChesney, 1861, New Pal. Foss., Niagara Gr. [Ety. proper name.]
multistriata, Hall, 1857, 10th Reg. Rep., Low. Held. Gr. [Sig. many-striated.]
liniuscula, Winchell, 1866, Rep. Low. Peninsula Mich., Ham. Gr. [Sig. fine-lined.]
(?) nobilis, Hall, 1860, (Rhynchospira nobilis) 13th Reg. Rep., Ham. Gr. [Sig. remarkable, noted.]
perforata, Hall, 1857, 10th Reg. Rep., Low. Held. Gr. [Sig. perforated.]
(?) quadriplicata, S. A. Miller, 1875, Cin. Quar. Jour. Sci., vol. 2, Cin'ti Gr. [Sig. four-plicated; in allusion to the four plications on the mesial fold.]
rectirostris, Hall, 1856, (Waldheimia rectirostra) 9th Reg. Rep., Low. Held. Gr. [Sig. straight-beaked.]
simplex, Hall, 1856, Pal. N. Y., vol. 3, Low. Held. Gr. [Sig. simple.]
TRIGONOTRETA, Konig, 1825, Icon. Foss. Sect. [Ety. *trigonos*, a triangle; *tretos* perforated.] Subgenus of Spirifera.
TRIMERELLA, Billings, 1862, Pal. Foss., vol. 1. [Ety. *treis*, three; *meros*, part; *ella*, diminutive.]
acuminata, Billings, 1862, Pal. Foss., vol. 1, Guelph Gr. [Sig. sharp-pointed.]
billingsi, Dall, 1871, Am. Jour. Conch., vol. 7, Guelph Gr. [Ety. proper name.]
dalli, Davidson & King, 1872, Brighton Meeting Brit. Assoc., Guelph Gr. [Ety. proper name.]
galtensis, Billings, 1862, (Obolus galtensis) Pal. Foss., vol. 1, Guelph Gr. [Ety. proper name.]
grandis, Billings, 1862, Pal. Foss., vol. 1, Guelph Gr. [Sig. large.]

minor, syn. for T. galtensis.
  ohioensis, Meek, 1871, Am. Jour. Sci., 2nd series, vol. 1, Niagara Gr. [Ety. proper name.]
Triplesia, Hall, Oct., 1858. Camarella has priority because it was published in August, 1858.
  congesta, see Camarella congesta.
  cuspidata, see Camarella cuspidata.
  extans, see Camarella extans.
  nucleata, see Camarella nucleus.
  ortoni, see Camarella ortoni.
TROPIDOLEPTUS, Hall, 1857, proposed in 10th Reg. Rep., but described in 1859 in 12th Reg. Rep. [Ety. *tropis*, the keel or bottom of a ship; *leptos*, slender.]
  carinatus, Conrad, 1839, (Strophomena carinata) Ann. Geo. Rep. N. Y., Ham. Gr. [Sig. keeled.]
  occidens, Hall, 1860, 13th Reg. Rep., Ham. Gr. [Sig. western.]
VITULINA, Hall, 1860, 13th Reg. Rep. [Ety. mythological name.]
  pustulosa, Hall, 1860, 13th Reg. Rep., Tully limestone. [Ety. covered with pustules; in allusion to the minute papillæ which cover the surface.]
WALDHEIMIA, King, 1849, Monograph of Permian Fossils. [Ety. proper name.]
  compacta, White & St. John, 1868, Trans. Chi. Acad. Sci., Up. Coal Meas. [Sig. compact, pressed together.]
  deweyi, see Rhynchospira deweyi.
  formosa, see Rhynchospira formosa.
  globosa, see Trematospira globosa.
  rectirostra, see Trematospira rectirostra.
ZYGOSPIRA, Hall, 1862, 15th Reg. Rep. N. Y. [Ety. *zygos*, a yoke; *spira*, a spire.]
  headi, Billings, 1862, (Athyris headi) Pal. Foss., vol. 1, Cin'ti & Hud. Riv. Gr. [Ety. proper name.]
  headi *var.* anticostiensis, Billings, 1862, (Athyris h e a d i *var.* anticostiensis) Pal. Foss., vol. 1, Hud. Riv. Gr. [Ety. proper name.]
  headi *var.* borealis, Billings, 1862, (Athyris headi *var.* borealis) Pal. Foss., vol. 1, Hud. Riv. Gr. [Sig. northern.]
  modesta, Say, 1847, (Atrypa modesta) Pal. N. Y., vol. 1, Trenton & Hud. Riv. Gr. [Sig. not large.]
  modesta *var.* cincinnatiensis, James, Pal. Ohio, vol. 1, Cin'ti Gr. [Ety. proper name.]
  paupera, Billings, 1866, Catal. Sil. Foss. Antic., Anticosti Gr. [Sig. impoverished, small.]
  subconcava, Meek & Worthen, 1868, Geo. Sur. Ill., vol. 3, Low. Held. Gr. [Sig. somewhat concave.]

CLASS PTEROPODA.

FAMILY *CONULARIIDÆ.*—Conularia, Coleoprion.
FAMILY *HYOLITHIDÆ.*—Hyolithellus, Hyolithes, Pterotheca, Stenotheca, Theca.
INCERTÆ SEDIS.—Aspidella, Scenella, Tentaculites.

ASPIDELLA, Billings, 1872, Am. Jour. Sci., 3rd series, vol. 3. [Ety. *aspidella*, a little shield.]
  terranovica, Billings, 1872, Am. Jour. Sci., 3rd series, vol. 3, Huronian Gr. [Ety. proper name.]
Clioderma, Hall, 1861, syn. for Pterotheca.
  attenuata, see Pterotheca attenuata.
  expansa, see Pterotheca expansa.
COLEOPRION, Sandberger, 1847, Jahrbuch. [Ety. *koleos*, a sheath; *prion*, a saw.]
  tenuicinctum, Hall, 1876, Illust. Devonian Foss., Cornif. & Ham. Gr. [Sig. fine-girded.]
CONULARIA, Miller, 1818, in Sowerby's Min. Conch., vol. 3. [Ety. *conulus*, a little cone.]
  asperata, Billings, 1866, Catal. Sil. Foss. Antic., Hud. Riv. Gr. [Sig. rough.]
  byblis, White, 1862, Proc. Bost. Soc. Nat. Hist., vol. 9, Chemung Gr. [Ety. mythological name.]
  cayuga, Hall, 1876, Illust. Devonian Foss., Ham. Gr. [Ety. proper name.]
  congregata, Hall, 1876, Illust. Devonian Foss., Portage Gr. [Sig. congregated together.]
  continens, Hall, 1876, Illust. Devonian Foss., Marcellus shale. [Sig. holding together.]
  crebristriata, Hall, 1876, Illust. Devonian Foss., Ham. Gr. [Sig. from the crowded striæ.]
  elegantula, Meek, 1871, Proc. Acad. Nat. Sci. Phil., Cornif. Gr. [Sig. quite elegant.]
  gattingeri, Safford, 1869, Geo. of Tenn., Trenton Gr. [Ety. proper name.]
  gracilis, Hall, 1847, Pal. N. Y., vol. 1, Trenton Gr. [Sig. slender.]
  granulata, Hall, 1847, Pal. N. Y., vol. 1, Trenton Gr. [Sig. granulated.]
  hudsoni, Emmons, 1856, Am. Geo., Utica Gr. [Ety. proper name.]
  huntana, Hall, 1859, Pal. N. Y., vol. 3, Low. Held. Gr. [Ety. proper name.]
  laqueata, Conrad, 1841, Ann. Rep. N. Y., Up. Held. Gr. [Sig. paneled.]
  lata, Hall, 1859, Pal. N. Y., vol. 3, Oriskany sandstone. [Sig. broad.]
  longa, Hall, 1852, Pal. N. Y., vol. 2, Niagara Gr. [Sig. long.]
  marionensis, Swallow, 1860, Trans. St. Louis Acad. Sci., Ham. Gr. [Ety. proper name.]
  micronema, Meek, 1871, Proc. Acad. Nat. Sci. Phil., Waverly Gr. [Sig. small-threaded or lined.]
  missouriensis, Swallow, 1860, Trans. St. Louis Acad. Sci., St. Louis Gr. [Ety. proper name.]
  molaris, White, 1876, Proc. Acad. Nat. Sci., Devonian. [Sig. a grinder.]
  newberryi, Winchell, 1865, Proc. Acad. Nat. Sci., Wav. Gr. [Ety. proper name.]
  niagarensis, Hall, 1852, Pal. N. Y., vol. 2, Niagara Gr. [Ety. proper name.]
  osagensis, Swallow, 1863, Trans. St. Louis Acad. Sci., Low. Carb. [Ety. proper name.]
  papillata, Hall, 1847, Pal. N. Y., vol. 1, Tren. Gr. [Sig. covered with papilli.]
  planocostata, Dawson, 1868, Acad. Geol., Carb. [Sig. plain-ribbed.]
  pyramidalis, Hall, 1859, Pal. N. Y., vol. 3, Low. Held. Gr. [Sig. pyramidal.]
  quadrata, Walcott, 1876, Desc. New Sp. Foss., Trenton Gr. [Sig. quadrate.]
  quadrisulcata, (?) Miller, 1826, Min. Conch., Niagara Gr. [Sig. having four depressions.]
  splendida, Billings, 1866, Catal. Sil. Foss. Antic., Hud. Riv. Gr. [Sig. splendid.]
  subcarbonaria, Meek & Worthen, 1873, Geo. Sur. Ill., vol. 5, Keokuk Gr. [Sig. in Lower Carboniferous rocks.]
  subulata, Hall, 1858, Trans. Alb. Inst., vol. 4, Warsaw Gr. [Sig. awl-shaped.]
  trentonensis, Hall, 1847, Pal. N. Y., vol. 1, Trenton & Hud. Riv. Gr. [Ety. proper name.]
  triplicata, Swallow, 1860, Trans. St. Louis Acad. Sci., Ham. Gr. [Sig. three-plicated.]
  undulata, Conrad, 1842, Jour. Acad. Nat. Sci., vol. 8, Ham. Gr. [Sig. wavy.]
  victa, White, 1862, Proc. Bost. Soc. Nat. Hist., vol. 9, Burlington Gr. [Sig. conquered, suppressed.]
HYOLITHELLUS, Billings, 1871, Can. Nat. & Geol., vol. 6. [Ety. diminutive of *Hyolithes*.]
  micans, Billings, 1871, Can. Nat. & Geol., vol. 6, Potsdam Gr. [Sig. shining.]
HYOLITHES, Eichwald.
  aclis, Hall, 1876, Illust. Devonian Foss., Ham. Gr. [Sig. a small javelin.]
  americanus, Billings, 1871, (Theca triangularis, Hall) Can. Nat. & Geol., vol. 6, Pots. Gr. [Ety. proper name.]

centennialis, Barrett, 1877, Ann. Lyc. Nat. Hist., vol. 11, Low. Held. Gr. [Ety. the one hundredth year, because the description was written in 1876.]
communis, Billings, 1871, Can. Nat. & Geol., vol. 6, Potsdam Gr. [Sig. common.]
emmonsi, Ford, 1873, Am. Jour. Sci., 3rd series, vol. 5, Potsdam Gr. [Ety. proper name.]
excellens, Billings, 1874, Pal. Foss., vol. 2, Potsdam Gr. [Sig. excellent.]
gibbosus, Hall, 1873, 23rd Reg. Rep. N. Y., Potsdam Gr. [Sig. gibbous.]
impar, Ford, 1872, Am. Jour. Sci., 3rd series, vol. 3, Potsdam Gr. [Sig. unequal.]
ligea, Hall, 1863, (Theca ligea) 15th Reg. Rep. N. Y., Up. Held. Gr. [Ety. mythological name.]
micans, see Hyolithellus micans.
primordialis, Hall, 1861, (Theca primordialis) Geo. Rep. Wis., Potsdam Gr. [Sig. first in order.]
princeps, Billings, 1871, Can. Nat. & Geol., vol. 6, Potsdam Gr. [Sig. primitive.]
principalis, Hall, 1876, Illust. Devonian Foss., Schoharie Grit. [Sig. principal, large.]
striatus, Hall, 1876, Illust. Devonian Foss., Ham. Gr. [Sig. striated.]
PTEROTHECA, Salter, 1852, Rep. Brit. Ass'n. [Ety. *Pterotheca*, a winged *Theca*.]
anatiformis, Hall, 1847, (Tellinomya anatiformis) Pal. N. Y., vol. 1, Trenton Gr. [Ety. like an *Anatifa*.]
attenuata, Hall, 1861, (Clioderma attenuata) 14th Reg. Rep., Trenton Gr. [Sig. drawn out, attenuated.]
canaliculata, Hall, 1861, 14th Reg. Rep., Trenton Gr. [Sig. channeled.]
expansa, Emmons, 1842, (Delthyris expansus) Geo. Rep. N. Y., Black Riv. & Trenton Gr. [Sig. spread out.]
saffordi, Hall, 1861, 14th Reg. Rep., Trenton Gr. [Ety. proper name.]
transversa, Salter, 1852, Rep. Brit. Ass'n, Hud. Riv. Gr. [Sig. crosswise.]
undulata, Hall, 1861, 14th Reg. Rep., Trenton Gr. [Sig. wavy.]
Pugiunculus aculeatus, see Theca aculeata.
SCENELLA, Billings, 1873, Can. Nat. & Geol. and Pal. Foss., vol. 2. [Ety. *scene*, a tent; *ella*, diminutive.]
reticulata, Billings, 1873, Pal. Foss., vol. 2, Huronian Gr. [Sig. like net-work.]
retusa, Ford, 1873, Am. Jour. Sci. & Arts, 3rd series, vol. 5, Low. Potsdam Gr. [Sig. blunted.]
STENOTHECA, Hicks, 1872, Quar. Jour. Geo. Soc. [Ety. *stenos*, narrow; *Theca*, a genus of Pteropods.]
pauper, Billings, 1874, Pal. Foss. vol. 2, Huronian Gr. [Sig. impoverished.]
TENTACULITES, Schlotheim, 1820, Petrefacten. [Ety. *tentaculum*, a feeler; *lithos*, stone.]
arenosus, Hall, 1876, Illust. Devon. Foss., Oriskany sandstone. [Sig. sandy.]

attenuatus, Hall, 1876, Illust. Devonian Foss., Ham. Gr. [Sig. attenuated.]
bellulus, Hall, 1876, Illust. Devonian Foss., Ham. Gr. [Sig. beautiful.]
distans, Hall, 1852, Pal. N. Y., vol. 2, Clinton Gr. [Sig. distant.]
elongatus, Hall, 1859, Pal. N. Y., vol. 3, Low. Held. Gr. [Sig. lengthened.]
fissurella, Hall, 1843, Geo. Rep. 4th Dist. N. Y., Marcellus shale. [Sig. a little cleft.]
flexuosa, see Conchicolites flexuosus.
hoyti, White, 1876, Proc. Acad. Nat. Sci., Devonian. [Ety. proper name.]
gyracanthus, Eaton, 1832, (Echinus gyracanthus) Geo. Text-book, Low. Held Gr. [Sig. round spine.]
incurvus, Shumard, 1856, Geo. Rep. Mo., Trenton Gr. [Sig. incurved.]
irregularis, Hall, 1859, Pal. N. Y., vol. 3, syn. for Tentaculites gyracanthus.
minutus, Hall, 1843, Geo. Rep. 4th Dist. N. Y., Clinton Gr. [Sig. small.]
niagarensis, Hall, 1852, Pal. N. Y., vol. 2, Niagara Gr. [Ety. proper name.]
ornatus, Sowerby, 1839, Murch. Sil. Syst., Water lime Gr. [Sig. adorned.]
oswegoensis, Meek & Worthen, 1865, Proc. Acad. Nat. Sci. Phil., Cin'ti Gr. [Ety. proper name.]
richmondensis, S. A. Miller, 1874, Cin. Quar. Jour. Sci., Cin'ti Gr. [Ety. proper name.]
scalariformis, Hall, 1876, Illust. Devonian Foss., Up. Held. Gr. [Sig. like *T. scalaris*.]
scalaris, Schlotheim, 1820, Petref. t. 29, Cornif. Gr. [Sig. like a ladder.] Not an American species.
sicula, Hall, 1876, Illust. Devonian Foss., Up. Held. Gr. [Sig. a little dagger.]
spicula, Hall, 1876, Illust. Devonian Foss., Chemung Gr. [Sig. a dart.]
sterlingensis, Meek & Worthen, 1865, Proc. Acad. Nat. Sci. Phil., Cin'ti Gr. [Ety. proper name.]
subtilis, Winchell, 1866, Rep. Low. Peninsula Mich., Ham. Gr. [Sig. very slender.]
tenuistriatus, Meek & Worthen, 1865, Proc. Acad. Nat. Sci. Phil., Cin'ti Gr. [Sig. finely striated.]
THECA, Sowerby, 1845, Morris' Memoir Strezelscki's N. S. Wales. [Ety. a sheath or case.]
aculeata, Hall, 1860, 13th Reg. Rep., Ham. Gr. [Sig. sharpened.]
gregaria, Meek & Hayden, 1861, Proc. Acad. Nat. Sci. Phil., Potsdam Gr. [Sig. occurring in masses.]
ligea, see Hyolithes ligea.
parviuscula, Hall, 1862, Geo. Rep. Wis., Hud. Riv. Gr. [Sig. very small.]
primordialis, see Hyolithes primordialis.
triangularis, Hall, 1847, Pal. N. Y., vol. 1, Hud. Riv. Gr. This name was preoccupied by Portlock in 1843. Billings described it as *Hyolithes americanus*.

CLASS GASTEROPODA.

FAMILY BELLEROPHONTIDÆ.— Bellerophon, Bucania, Carinaropsis, Cyrtolites, Ecculiomphalus, Microceras, Phragmostoma, Porcellia, Tremanotus.
FAMILY CALYPTRÆIDÆ.—Capulus,Conchopeltis,Metoptoma, Palæacmæa,Platyceras, Trochita.
FAMILY CHITONIDÆ.—Chiton.
FAMILY DENTALIIDÆ.—Dentalium.
FAMILY HELICIDÆ.—Dawsonella, Pupa, Streptaxis, Zonites.
FAMILY LITTORINIDÆ.—Xenophora.
FAMILY MACLURÆIDÆ.—Maclurea.
FAMILY MURICIDÆ.—Fusispira.
FAMILY NATICIDÆ.—Naticopsis, Trachydomia.
FAMILY PLEUROTOMARIIDÆ.—Helicotoma, Microdoma, Murchisonia, Pleurotomaria, Raphistoma, Scalites.
FAMILY PYRAMIDELLIDÆ.—Aclis, Chemnitzia, Loxonema, Macrocheilus, Polyphemopsis, Soleniscus, Subulites.
FAMILY ROTELLIDÆ.—Anomphalus.
FAMILY SCALARIIDÆ.—Holopella.
FAMILY SOLARIDÆ.—Euomphalus, Ophileta, Phanerotinus, Platyschisma, Platyostoma, Straparollus, Straparollina, Strophostylus.
FAMILY TURBINIDÆ.—Clisospira, Cyclonema, Cyclora, Eunema, Holopea, Isonema, Orthonema, Orthostoma, Trochonema, Turbo.
FAMILY TURRITELLIDÆ.—Turritella.

ACLIS, Loven. [Ety. *a*, without; *kleis*, a projection.]
   minuta, Stevens, 1858, Am. Jour. Sci., vol. 25, Coal Meas. [Sig. minute.]
   robusta, Stevens, 1858, Am. Jour. Sci., vol. 25, Coal Meas. [Sig. robust.]
   swallovana, Geinitz, 1866, (Turbonella swallovana) Carb. und Dyas in Neb., Coal Meas. [Ety. proper name.]
Acroculia, Phillips, 1841, syn. for Platyceras.
   angulata, see Platyceras angulata.
   erecta, see Platyceras erecta.
   ovalis, see Platyceras ovalis.
   niagarensis, see Platyceras niagarensis.
   trigonalis, see Platyceras trigonalis.
Ampullaria, Lamarck, 1801, Syst. An. sans Vert. [Ety. *ampulla*, a flask.]
   helicoides, see Platyschisma helicoides.
ANOMPHALUS, Meek & Worthen, 1866, Proc. Acad. Nat. Sci. [Ety. *Anomphalos*, without an umbilicus.]
   meeki, see Dawsonella meeki.
   rotulus, Meek & Worthen, 1866, Proc. Acad. Nat. Sci. Phil., Coal Meas. [Sig. a litttle wheel.]

BELLEROPHON, Montfort, 1808, Conch. Syst., vol. 1. [Ety. mythological name.]
   acutilira, Hall, 1862, 15th Reg. Rep., Ham. Gr. [Sig. sharp-ridged.]
   acutus, Sowerby, 1839, Murch. Sil. Syst., Low. Silurian. [Sig. sharp.]
   allegoricus, White, 1874, Rep. Invert. Foss., Quebec Gr. [Sig. allegorical.]
   angustata, see Bucania angustata.
   apertus, Sowerby, 1825, Min. Conch., vol. 5, Low. Carb. [Sig. an opening.]
   argo, Billings, 1860, Can. Nat. & Geol., vol. 5, Black Riv. & Trenton Gr. [Ety. mythological name.]
   auriculatus, Hall, 1852, Pal. N. Y., vol. 2, Coralline limestone. [Sig. having ear-like appendages.]
   barquensis, Winchell, 1862, Proc. Acad. Nat. Sci., Marshall Gr. [Ety. proper name.]
   bidorsatus, see Bucania bidorsata.
   bilobatus, Sowerby, 1839, Murch. Sil. Syst., Black Riv. to Mid. Sil. [Sig. two-lobed.]

bilobatus *var.* acutus, Hall, 1847, Pal. N. Y., vol. 1, Trenton Gr. [Sig. sharp.]
bilobatus *var.* corrugatus, Hall, 1847, Pal. N. Y., vol. 1, Trenton Gr. [Sig. wrinkled.]
blaneyanus, syn. for. B. carbonarius.
bowmani, White, 1876, Proc. Acad. Nat. Sci., Devonian. [Ety. proper name.]
brevilineatus, Conrad, 1842, Jour. Acad. Nat. Sci., vol. 8, Up. Sil. [Sig. shortlined.]
canadensis, Billings, 1866, Catal. Sil. Foss. Antic., Hud. Riv. Gr. [Ety. proper name.]
cancellatus, Hall, 1847, Pal. N. Y., vol. 1, Hud. Riv. Gr. [Sig. cancellated.]
cancellatus, Hall, 1858, Trans. Alb. Inst., vol. 4, Warsaw Gr. The name was preoccupied.
carbonarius, Cox, 1857, Geo. Rep. Ky., vol. 3, Coal Meas. [Sig. pertaining to the Coal Measures.]
carbonarius *var.* subpapillosus, White, 1876, Geo. Uinta Mountains, Up. Aubrey Gr. [Sig. somewhat papillose.]
carinatus, Sowerby, 1839, Murch. Sil. Syst., Devonian. [Sig. keeled.]
charon, Billings, 1860, Can. Nat. & Geol. vol. 5, Black Riv. & Trenton Gr. [Ety. mythological name.]
convolutus, Eaton, 1832, Geo. Text-book, Up. Sil. [Sig. spiral-whorled.]
crassus, Meek & Worthen, 1860, Proc. Acad. Nat. Sci., Coal Meas. [Sig. thick.]
crenistria, Hall, 1876. Illust, Devonian Foss., Ham. Gr. [Sig. having wrinkled lines.]
curvilineatus, Conrad, 1842, Jour. Acad. Nat. Sci., vol. 8, Onondaga, Schoharie and Up. Held. Gr. [Sig. having curved lines.]
cyrtolites, Hall, 1860, 13th Reg. Rep., Kinderhook Gr. [Sig. curved stone.]
declivis, Conrad, 1842, Jour. Acad. Nat. Sci., vol. 8, Trenton Gr. [Sig sloping.]
disculus, Billings, 1860, Can. Nat. & Geo., vol. 5, Black Riv. & Trenton Gr. [Sig. a little disc.]
ellipticus, McChesney, 1860, Desc. New Pal. Foss., Coal Meas. [Sig. elliptical.]
expansus, Hall, 1847. The name was preoccupied by Sowerby. See Bucania expansa.
fraternus, Billings, 1866, Catal. Sil. Foss. Antic., Hud. Riv. Gr. [Sig. fraternal, allied to *B. expansus*.]
galericulatus, Winchell, 1862, Proc. Acad. Nat. Sci., Marshall Gr. [Sig. small capped.]
globosus, Stevens, 1858, Am. Jour. Sci., vol. 25, Coal Meas. [Sig. globose.]
hiulcus, Sowerby, Min. Conch. Not American.
kansasensis, Shumard, 1858, Trans. St. Louis Acad. Sci., Coal Meas. [Ety. proper name.]

leda, Hall, 1862, 15th Reg. Rep. N. Y., Ham. Gr. [Ety. mythological name.]
lineolatus, Hall, 1860, 13th Reg. Rep., N. Y., Ham. Gr. [Sig. finely lined.]
lindsleyi, Safford, 1869, Geo. of Tenn., Nashville Gr. [Ety. proper name.]
lyra, Hall, 1862, 15th Reg. Rep., Ham. Gr. [Sig. a lyre.]
macer, Billings, 1865, Pal. Foss., vol. 1, Calciferous Gr. [Ety. proper name.]
maera, Hall, 1876, Illust. Devonian Foss., Chemung Gr. [Ety. mythological name.]
marcouanus, Geinitz, 1866, Carb. und Dyas. in Neb., Coal Meas. [Ety. proper name.]
meekanus, Swallow, 1858, Trans. St. Louis Acad. Sci., Coal Meas. [Ety. proper name.]
michiganensis, Winchell, 1862, Proc. Acad. Nat. Sci., Marshall Gr. [Ety. proper name.]
miser, Billings, 1866, Catal. Sil. Foss. Antic., Hud. Riv. Gr. [Sig. miserable.]
missouriensis, Swallow, 1863, Trans. St. Louis Acad. Sci., Chester Gr. [Ety. proper name.]
mohri, S. A. Miller, 1874, Cin. Quar. Jour. Sci., Cin'ti Gr. [Ety. proper name.]
montfortanus, Norwood & Pratten, 1855, Jour. Acad. Nat. Sci., Coal Meas. [Ety. proper name.]
nashvillensis, Troost, 1840, 5th Geo. Rep. Tenn., Low. Sil. [Ety. proper name.]
nautiloides, Winchell, 1862, Proc. Acad. Nat. Sci., Marshall Gr. [Sig. like a *nautilus*.]
neleus, Hall & Whitfield, 1876, Illust. Devonian Foss., Chemung Gr. [Ety. mythological name.]
newberryi, Meek, 1871, Proc. Acad. Nat. Sci., Corniferous Gr. [Ety. proper name.]
nodocarinatus, Hall, 1858, Geo. Rep. Iowa, Coal Meas. [Sig. knotty and keeled.]
obsoletus, Hall, 1876, Illust. Devonian Foss., Chemung Gr. [Sig. obsolete.]
otsego, Hall, 1862, 15th Reg. Rep., Ham. Gr. [Ety. proper name.]
palinurus, Billings, 1865, Pal. Foss., vol. 1, Quebec Gr. [Ety. mythological name.]
panneus, White, 1862, Proc. Bost. Soc. Nat. Hist., vol. 9, Chemung Gr. [Sig. ragged.]
patersoni, Hall, 1862, Geo. Rep. Wis., Hud. Riv. Gr. [Ety. proper name.]
patulus, Hall, 1843, Geo. Rep. 4th Dist. N. Y., Ham. Gr. [Sig. broad, spread out.]
pelops, Hall, 1862, 15th Reg. Rep., Schoharie & Up. Held. Gr. [Ety. mythological name.]

percarinatus, Conrad, 1842, Jour. Acad. Nat. Sci., vol. 8, Coal Meas. [Sig. very strongly keeled.]
perforatus, syn. for Tremanotus alpheus.
perlatus, Conrad, 1842, Jour. Acad. Nat. Sci., vol. 8, Coal Meas. [Sig. very wide.]
platystoma, Meek & Worthen, 1868, Geo. Sur. Ill., vol. 3, Galena Gr. [Sig. broad-mouthed.]
plenus, Billings, 1874, Pal. Foss., vol. 2, Devonian. [Sig. full, large.]
profundus, Emmons, Geo. Rep. N. Y., Trenton Gr. [Sig. deep.]
propinquus, Meek, 1871, Proc. Acad. Nat. Sci., Cornif. Gr. [Ety. from close resemblance to *B. newberryi*.]
punctifrons, Emmons, 1842, Geo. Rep. N. Y., Black Riv. & Trenton Gr. [Sig. dotted in front.]
rudis, Hall, 1862, 15th Reg. Rep., Ham. Gr. [Sig. rude, rough.]
rugosiusculus, Winchell, 1862, Proc.Acad. Nat. Sci., Marshall Gr. [Sig. covered with small wrinkles.]
scriptiferus, White, 1862, Proc. Bost. Soc. Nat. Hist., vol. 9, Chemung Gr. [Ety. *scriptum*, writing; *fero*, to carry.]
scissile, Conrad, 1846, Proc. Acad. Nat. Sci., vol. 2, Carb. [Sig. cleft, rent.]
solitarius, Billings, 1866, Catal. Sil. Foss. Antic., Hud. Riv. Gr. [Sig. alone, solitary.]
stamineus, Conrad, 1842, Jour. Acad. Nat. Sci., vol. 8, Low. Carb. [Sig. thready.]
stevensanus, McChesney, 1860, Desc. New Pal. Foss., Coal Meas. [Ety. proper name.]
striatus (?), Sowerby, 1839, Murch. Sil. Syst., Portage Gr. The name was preoccupied by D'Orbigny.
sublævis, Hall, 1856, Trans. Alb. Inst., vol. 4, Warsaw Gr. [Sig. somewhat smooth.]
sulcatinus, see Bucania sulcatina.
thalia, Hall, 1862, 15th Reg. Rep., Ham. Gr. [Ety. mythological name.]
tricarinatus, Shumard, 1858, Trans. St. Louis Acad. Sci., Coal Meas. [Sig. three-keeled.]
tricarinata, Hall, 1876, Illust. Devonian Foss., Chemung Gr. The name was preoccupied.
troosti, Safford, 1869, Geo. Rep. Tenn., Nashville Gr. [Ety. proper name.] This name was preoccupied by D'Orbigny in 1839.
tuber, Hall, 1876, 28th Reg. Rep., Niagara Gr. [Sig. a bump or knob.]
urei, Fleming, 1828, British Animals, Devonian. [Ety. proper name.] American species. (?)
vittatus, syn. for B. carbonarius.
volutus, Eaton, 1832, Geol. Text-book, Up. Sil. [Sig. whorled.]
whittleseyi, Winchell, 1865, Proc. Acad. Nat. Sci., Cuyahoga shale. [Ety. proper name.]

BUCANELLA nana, Meek, 1870, Proc. Am. Phil. Soc., vol. 11, Silurian. [Sig. dwarfish.]
BUCANIA, Hall, 1847, Pal. N. Y., vol. 1. [Ety. *bukane*, a trumpet.]
angustata, Hall, 1852, Pal. N. Y., vol. 2, Niagara & Guelph Gr. [Sig. narrowed or constricted.]
bellipuncta, Hall, 1852, Pal. N. Y., vol. 2, Clinton Gr. [Sig. beautifully dotted.]
bidorsata, Hall, 1847, (Bellerophon bidorsatus) Pal. N. Y., vol. 1, Trenton Gr. [Sig. having a double back.]
chicagoensis, McChesney, 1860, New Pal. Foss., Niagara Gr. [Ety. proper name.]
costata, James, 1872, (Cyrtolites costatus) Am. Jour. Sci., 3rd ser., vol. 3, Cin'ti Gr. [Sig. ribbed, having prominent ridges.]
crassolaris, McChesney, 1861, New Pal. Foss., Niagara Gr. [Ety. *crassus*, thick.]
devonica, Hall, 1872, 24th Reg. Rep., Low. Held. Gr. [Ety. proper name.]
expansa, Hall, 1847, (Bellerophon expansus) Pal. N. Y., vol. 1, Trenton Gr. [Sig. widely-spread.]
intexta, Hall, 1847, Pal. N. Y., vol. 1, Trenton Gr. [Sig. plaited, woven.]
lirata, Hall, 1862, Geo. Rep. Wis., Trenton Gr. [Sig. furrowed.]
pervoluta, McChesney, 1861, New Pal. Foss., Niagara Gr. [Sig. very much rolled.]
profunda, Conrad, 1841, Ann. Rep. N. Y., (Euomphalus profundus) Low. Held. Gr. [Sig. deep.]
profunda, Hall, 1859, Pal. N. Y., vol. 3, Low. Held. Gr. This name was preoccupied.
punctifrons, Hall, 1847, Pal. N. Y., vol. 1, Trenton Gr. [Sig. dotted in front.]
rotundata, Hall, 1847, Pal. N. Y., vol. 1, Chazy Gr. [Sig. rounded.]
rugosa, Emmons, 1856, Am. Geol., Utica Gr. [Sig. rugose.]
stigmosa, Hall, 1852, Pal. N. Y., vol. 2, Clinton Gr. [Sig. full of marks.]
sulcatina, Emmons, 1842, (Bellerophon sulcatinus) Pal. N. Y., vol. 1, Chazy, Black Riv. & Trenton Gr. [Sig. small-furrowed.]
trilobata, Conrad, 1838, (Planorbis trilobatus) Ann. Rep. N. Y., Medina sandstone & Clinton Gr. [Sig. three-lobed.]

BULIMELLA, Hall, 1858, Trans. Alb. Inst., vol. 4. [Ety. a small *Bulimus*.] This name was preoccupied by Pfeiffer in 1852.
bulimiformis, see Polyphemopsis bulimiformis.
canaliculata, see Polyphemopsis canaliculata.
elongata, see Polyphemopsis elongata.

GASTEROPODA.

CAPULUS, Montfort, 1810, Conch. Syst., vol. 2. [Ety. *capulus*, a coffin.]
acutirostris, Hall, 1856, Trans. Alb. Inst., vol. 4, Warsaw Gr. [Sig. sharp-beaked.]
auriformis, Hall, 1847, Pal. N. Y., vol. 1, Chazy Gr. [Sig. ear-shaped.]
parvus, Swallow, 1858, Trans. St. Louis Acad. Sci., Coal Meas. [Sig. small.]
triplicatus, Swallow, 1858, Trans. St. Louis Acad. Sci., Coal Meas. [Sig. three-plicated.]
CARINAROPSIS, Hall, 1847, Pal. N. Y., vol. 1, [Ety. from its resemblance to *Carinaria*.]
carinata, Hall, 1847, Pal. N. Y., vol. 1, Trenton Gr. [Sig. keeled or having ridges.]
orbiculata, Hall, 1847, Pal. N. Y., vol. 1, Hud. Riv. Gr. [Sig. orbicular.]
patelliformis, Hall, 1847, Pal. N. Y., vol. 1, Trenton & Hud. Riv. Gr. [Sig. patella or limpet-shaped.]
CHEMNITZIA, D'Orbigny, 1837, Mollusques, Echinodermes, Foraminiferes et Polypiers, etc. [Ety. proper name.] Prof. Meek was of the opinion that the species referred to this genus from the palæozoic rocks more properly belong to *Loxonema* and other genera.
attenuata, Stevens, 1858, Am. Jour. Sci., vol. 25, Coal Meas. [Sig. drawn out, attenuated.]
parva, Cox, 1857, Geo. Sur. Ky., vol. 3, Coal Meas. [Sig. small.]
swallovana, Shumard, 1859, Trans. St. Louis Acad. Sci., Permian Gr. [Ety. proper name.]
tenuilineata, Shumard, 1855, Geo. Rep. Mo., Chemung Gr. [Sig. fine lined.]
CHITON, Linnæus, 1758, Syst. Nat., ed. 10. [Ety. *chiton*, a coat of mail.]
canadensis, see Metoptoma canadense.
carbonarius, Stevens, 1859, Am. Jour. Sci., vol. 25, Coal Meas. [Sig. from the Coal Measures.]
parvus, Stevens, 1859, Am. Jour. Sci., vol. 25, Coal Meas. [Sig. small.]
CLISOSPIRA, Billings, 1865, Pal. Foss., vol. 1. [Ety. *kleio*, to lock; *spira*, a whorl.]
curiosa, Billings, 1865, Pal. Foss., vol. 1, Quebec Gr. [Sig. curious.]
CONCHOPELTIS, Walcott, 1876, Desc. New Sp. Foss. [Ety. *conche*, shell; *pelte*, shield.]
alternata, Walcott, 1876, Desc. New Sp. Foss., Trenton Gr. [Sig. alternate.]
minnesotensis, Walcott, 1876, Desc. New Sp. Foss., Trenton Gr. [Ety. proper name.]
CYCLONEMA, Hall, 1852, Pal. N. Y., vol. 2, [Ety. *kuklos*, a circle; *nema*, a thread.]
bellulum, Billings, 1866, Catal. Sil. Foss. Antic., Anticosti Gr. [Sig. beautiful.]
bilix, Conrad, 1842, (Pleurotomaria bilix) Jour. Acad. Nat. Sci., vol. 8, Trenton and Hud. Riv. Gr. [Sig. woven with a double thread.]
bilix *var.* conicum, S. A. Miller, 1874, Cin. Quar. Jour. Sci., vol. 1, Cin'ti Gr. [Sig. conical.]
bilix *var.* fluctuatum, James, 1874, (Cyclonema fluctuata) Cin. Quar. Jour. Sci., vol. 1, Cin'ti Gr. [Sig. wavy.]
cancellatum, Hall, 1843, (Littorina cancellata) Geo. Rep. 4th Dist. N. Y., Clinton Gr. [Sig. cancellated.]
commune, Billings, 1866, Catal. Sil. Foss. Antic., Anticosti Gr. [Sig. common.]
concinna, Hall, 1876, Illust. Devonian Foss., Chemung Gr. [Sig. beautiful.]
crenistriatum, Hall, 1876, Illust. Devonian Foss., Schoharie grit. [Sig. having wrinkled striæ.]
crenulatum, Meek, 1871, Proc. Acad. Nat. Sci., Cornif. Gr. [Sig. crenulated.]
decorum, Billings, 1866, Catal. Sil. Foss. Antic., Anticosti Gr. [Sig. seemly.]
elevatum, Hall, 1868, 20th Reg. Rep., Niagara Gr. [Sig. elevated.]
hageri, Billings, 1862, Pal. Foss., vol. 1, Trenton Gr. [Ety. proper name.]
hallanum, Salter, 1850, Can. Org. Rem., Decade 1, Black Riv. Gr. [Ety. proper name.]
hamiltoniæ, Hall, 1862, 15th Reg. Rep., Ham. Gr. [Ety. proper name.]
humile, Billings, 1866, Catal. Sil. Foss. Antic., Anticosti Gr. [Sig. poor, small.]
liratum, Hall, 1862, 15th Reg. Rep., Ham. Gr. [Sig. furrowed.]
mediocre, Billings, 1866, Catal. Sil. Foss. Antic., Anticosti Gr. [Sig. ordinary, middling.]
montrealense, Billings, 1862, Pal. Foss., vol. 1, Trenton Gr. [Ety. proper name.]
multiliratum, Hall, 1862, 15th Reg. Rep., Ham. Gr. [Sig. many-furrowed.]
obsoletum, Hall, 1852, Pal. N. Y., vol. 2, Clinton Gr. [Sig. obsolete, from the nearly obsolete spire.]
obsoleta, Hall, 1876, Illustr. Devonian Foss., Chemung Gr. The name was preoccupied.
percarinatum, Hall, 1847, (Pleurotomaria percarinata) Pal. N. Y., vol. 1, Trenton & Hud. Riv. Gr. [Sig. very much carinated.]
percingulatum, Billings, 1857, Rep. of Progr., Clinton & Niagara Gr. [Sig. encircled with many lines.]
phædra, Billings, 1865 Pal. Foss., vol. 1, Quebec Gr. [Ety. mythological name.]
pyramidatum, James, 1874, Cin. Quar. Jour. Sci., vol. 1, Cin'ti Gr. [Sig. pyramidal.]
rugælineatum, Hall, 1872, 24th Reg. Rep., Niagara Gr. [Sig. rugose-lined.]
semicarinatum, Salter, 1859, Can. Org. Rem., Decade 1, Black Riv. Gr. [Sig. half-carinated.]
sulcatum, Hall, 1852, Pal. N. Y., vol. 2, Guelph Gr. [Sig. furrowed.]

GASTEROPODA. 147

tennesseense, Roemer, 1860, (Turbo tennesseensis) Sil. Fauna. des West Tenn., Niagara Gr. [Ety. proper name.]
varians, Billings, 1857, Rep. of Progr., Mid. Sil. [Sig. variable.]
varicosum, Hall, 1870, 24th Reg. Rep. N. Y. (Published by mistake in 14th Reg. Rep., 1861, as *C. ventricosa*) Cin'ti Gr. [Sig. having the threads or lines enlarged.]
ventricosum, Hall, 1852, Pal. N.Y., vol. 2, Clinton Gr. [Sig. ventricose.]

CYCLORA, Hall, 1845, Am. Jour. Sci., vol. 48. [Ety. *kuklos*, a circle.]
hoffmanni, S. A. Miller, 1874, Cin. Quar. Jour. Sci., Cin'ti Gr. [Ety. proper name.]
minuta, Hall, 1845, Am. Jour. Sci.,vol. 48, Cin'ti Gr. [Sig. minute.]
nana, syn. for Cyclora minuta.
parvula, Hall, 1845, (Turbo parvula) Am. Jour. Sci., vol. 48, Cin'ti Gr. [Sig. very small.]

Cyclostoma, Lamarck, 1801, Syst. An. sans Vert. [Ety. *kuklos*, a circle; *stoma*, mouth.]
pervetusta, see Pleurotomaria pervetusta.

CYRTOLITES, Conrad, 1838, Ann. Rep. N. Y. [Ety. *kurtos*, curved; *lithos*, stone.]
carinatus, S. A. Miller, 1874, Cin. Quar. Jour. Sci., vol. 1, Cin'ti Gr. [Sig. keeled; from the ridges on the sides.]
compressus, Conrad, 1838, (Phragmolites compressus) Ann. Rep. N. Y., Black Riv. & Trenton Gr. [Sig. compressed.]
conradi, Hall,1862, Geo. Rep. Wis., Trenton Gr. [Ety. proper name.]
costatus, see Bucania costata.
cristatus, Safford, 1869, Geo. of Tenn., Nashville Gr. [Sig. crested.]
desideratus, Billings, 1866, Catal. Sil.Foss. Antic., Hud. Riv. Gr. [Sig. to be desired.]
dyeri, Hall, 1871, Pamphlet, Cin'ti Gr. [Ety. proper name.]
elegans, S. A. Miller, 1874, Cin. Quar. Jour. Sci., vol. 1, Cin'ti Gr. [Sig. elegant.]
expansus, Hall, 1859, Pal. N. Y., vol. 3, Oriskany sandstone. [Sig. spread out.]
filosus, Emmons, 1842, Geo. Rep. N. Y., Trenton Gr. [Sig. thready.]
gillanus, White & St. John, 1868, Trans. Chi. Acad. Sci., Coal Meas. [Ety. proper name.]
imbricatus, Meek & Worthen, 1868, Geo. Sur. Ill., vol. 3, Cin'ti Gr. [Sig. imbricated.]
mitella, Hall, 1862, 15th Reg. Rep., Ham. Gr. [Sig. a bandage.]
ornatus, Conrad, 1838, Ann. Rep. N. Y., Hud. Riv. Gr. [Sig. ornamented.]
pannosus, Billings, 1866, Catal. Sil. Foss. Antic., Hud. Riv. Gr. [Sig. ragged.]
pileolus, Hall, 1862,15th Reg. Rep., Ham. Gr. [Sig. a small cap.]
sinuosus, Hall, 1876, 28th Reg. Rep., Niagara Gr. [Sig. wavy.]
trentonensis, Conrad, 1842, Jour. Acad. Nat. Sci., vol. 8, Trenton Gr. [Ety. proper name.]

DAWSONELLA, Bradley, 1874, Am. Jour. Sci., 3rd series, vol. 7. [Ety. proper name.]
meeki, Bradley, 1872, (Anomphalus meeki) Am. Jour. Sci., 3rd series, vol. 4, Coal Meas. [Ety. proper name.]

DENTALIUM, Linnæus, 1740, Syst. Nat., 2nd Ed. [Ety. *dens*, a tooth.]
aciculatum, Hall, 1860, 13th Reg. Rep., Ham. Gr. [Sig. like a small needle.]
annulostriatum, Meek & Worthen, 1870, Proc. Acad. Nat. Sci., Coal Meas. [Sig. annular and striated.]
barquense, Winchell, 1862, Proc. Acad. Nat. Sci., Marshall Gr. [Ety. proper name.]
canna, White, 1874, Rep. Invert. Foss., Carb. [Sig. a reed.]
grandævum, Winchell, 1863, Proc. Acad. Nat. Sci., Chemung Gr. [Sig. very ancient.]
meekanum, Geinitz, 1866, Carb. und Dyas in Neb., Coal Meas. [Ety. proper name.]
missouriense, Swallow, 1863, Trans. St. Louis Acad. Sci., Chester Gr. [Ety. proper name.]
obsoletum, Hall, 1858, Geo. Rep. Iowa, Coal Meas. The name was preoccupied by Schlotheim in 1832.
primarium, Hall, 1858, Geo. Rep. Iowa, Warsaw Gr. [Sig. primary.]
venustum, Meek & Worthen, 1861, Proc. Acad. Nat. Sci., St. Louis Gr. [Sig. elegant.]

ECCULIOMPHALUS, Portlock, 1843, Geol. Rep. Lond. [Ety. *ecculiomphalus*, unrolled umbilicus.]
atlanticus, Billings, 1865, Pal. Foss., vol. 1, Quebec Gr. [Ety. proper name.]
canadensis, Billings, 1861, Can. Nat. & Geol., Quebec Gr. [Ety. proper name.]
distans, Billings, 1865, Pal. Foss., vol. 1, Quebec Gr. [Sig. distant, standing apart.]
intortus, Billings, 1861, Can. Nat. & Geol., Quebec Gr. [Sig. twisted, turned round.]
spiralis, Billings, 1861, Can. Nat. & Geol., Quebec Gr. [Sig. spiral.]
superbus, Billings, 1865, Pal. Foss., vol. 1, Quebec Gr. [Sig. magnificent.]
trentonensis, Conrad, 1842, Jour. Acad. Nat. Sci., vol. 8, Trenton Gr. [Ety. proper name.]
undulatus, Hall, 1861, Geo. Rep. Wis., Trenton Gr. [Sig. undulating.]

Eulima, Risso, 1826, Histoire Naturelle des principales, etc. [Sig. ravenous hunger.]
peracuta, see Polyphemopsis peracuta.

GASTEROPODA.

EUNEMA, Salter, 1859, Can.Org. Rem., Decade 1. [Ety. *eu*, beautiful; *nema*, a line.]
cerithloides, Salter, 1859, Can. Org. Rem., Decade 1, Black Riv. Gr. [Sig. like the genus *Cerithium*.]
erigone, Billings, 1862, Pal. Foss., vol. 1, Black Riv. Gr. [Ety. mythological name.]
pagoda, Salter, 1859, Can. Org. Rem., Decade 1, Black Riv. Gr. [Sig. an image of a supposed deity.]
priscum, Billings, 1859, Can. Nat. & Geo., vol. 4, Corniferous Gr. [Sig. old or ancient.]
salteri, see Orthonema salteri.
strigillatum, Salter, 1859, Can. Org. Rem., Decade 1, Black Riv. Gr. [Sig. furrowed, fluted.]
trilineatum, Hall, 1867, 20th Reg. Rep., Niagara Gr. [Sig. three-lined.]

EUOMPHALUS, Sowerby, 1814, Min. Conch., vol. 1. [Ety. *eu*, wide; *omphalos*, umbilicus.]
boonensis, Swallow, 1863, Trans. St. Louis Acad. Sci., Low. Carb. [Ety. proper name.]
catilloides, Conrad, 1842, (Inachus catilloides) Jour. Acad. Nat. Sci., vol. 8, Coal Meas. [Sig. like *E. catillus*.] The name was preoccupied by Koninck in 1841.
clymenioides, Hall, 1862, 15th Reg. Rep., Schoharie grit. [Sig. like a shell of the genus *Clymenia*.]
comes, Hall, 1876, Illust. Devonian Foss., Ham. Gr. [Sig. an associate.]
conradi, syn. for Euomphalus decewi.
cyclostomus, Hall, 1858, Geo. Rep. Iowa, vol. 1, pt. 2, Ham. Gr. [Sig. having a circular mouth.]
decewi, Billings, 1861, Can. Jour., Cornif. Gr. [Ety. proper name.]
depressus, Hall, 1843, Geo. Rep. 4th Dist. N. Y., Low. Carb. The name was preoccupied by Goldfuss in 1832.
disjunctus, Hall, 1859, Pal. N. Y., vol. 3, Low. Held. Gr. The name was preoccupied by Goldfuss.
eboracensis, Hall, 1862, 15th Reg. Rep., Ham. Gr. [Ety. proper name.]
exortivus, Dawson, 1868, Acad. Geol., Carb. [Sig. eastern.]
expansus, Conrad, 1842, Jour. Acad. Nat. Sci., vol. 8, Up. Sil. [Sig. expanded.]
hecale, Hall, 1876, Illust. Devonian Foss., Chemung Gr. [Ety. mythological name.]
hemispherica, see Platystoma hemisphericum.
inops, Hall, 1876, Illust. Devonian Foss., Up. Held. Gr. [Sig. meagre.]
latus, Hall, 1858, Geo. Rep. Iowa, Burlington Gr. [Sig. broad.]
laxus, Hall, 1862, 15th Reg. Rep. N. Y., Cornif. & Ham. Gr. [Sig. loose.]
lens, see Straparollus lens.
obtusus, Hall, 1858, Geo. Rep. Iowa, Kinderhook Gr. [Sig. obtuse.]

perspectivus, Swallow, 1863, Trans. St. Louis Acad. Sci., Kaskaskia Gr. [Sig. thoroughly viewed.]
pervetus, Conrad, 1843, (Inachus pervetus) Proc. Acad. Nat. Sci., vol. 1, Trenton Gr. [Sig. very ancient.]
pervetustus, Conrad, 1839, (Cyclostoma pervetusta) Ann. Rep. N. Y., Medina sandstone. [Sig. very ancient.]
planidorsatus, see Straparollus planidorsatus.
planispira, see Straparollus planispiratus.
planodiscus, Hall, 1860, 13th Reg. Rep., Ham. Gr. [Sig. having a plane disc.]
profundus, see Bucania profunda.
quadrivolvis, see Straparollus quadrivolvis.
roberti, White, 1862, Proc. Bost. Soc. Nat. Hist., vol. 9, Burlington Gr. [Ety. proper name.]
rotuliformis, Meek, 1870, Proc. Acad. Nat. Sci., Calciferous Gr. [Sig. wheel-shaped.]
rotundus, see Pleurotomaria rotunda.
rudis, Hall, 1876, Illust. Devonian Foss., Ham. Gr. [Sig. rough.]
rugælineatus, see Cyclonema rugælineatum.
rugosus, see Straparollus rugosus.
sinuatus, Hall, 1859, Pal. N. Y., vol. 3, Low. Held. Gr. [Sig. wavy.]
spergenensis, see Straparollus spergenensis.
spergenensis var. planorbiformis, see Straparollus spergenensis *var. planorbiformis*.
spirorbis, Hall, 1860, 13th Reg. Rep., Ham. Gr. [Sig. spiral-whorl.]
springvalensis, White, 1876, Proc. Acad. Nat. Sci., Kinderbook Gr. [Ety. proper name.]
subplanus, Hall, 1852, Stans. Ex. to Gt. Salt Lake, Coal Meas. [Sig. somewhat flat.]
sulcatus, Hall, 1843, Geo. Rep. 4th Dist. N. Y., Onondaga Gr. [Sig. furrowed.]
tioga, Hall, 1876, Illust. Devonian Foss., Chemung Gr. [Ety. proper name.]
triliratus, Conrad, 1843, Proc. Acad. Nat. Sci., Trenton Gr. [Sig. three-lined.]
trochiscus, Meek, 1870, Proc. Acad. Nat. Sci., Calciferous Gr. [Sig. wheel-shaped.]
umbilicatus, see Straparollus umbilicatus.
uniangulatus, see Ophileta uniangulata.
vaticinus, Hall, 1863, 16th Reg. Rep. N. Y., Potsdam Gr. [Sig. prophetical.]
whitneyi, see Straparollus whitneyi.

FUSISPIRA, Hall, 1871, Pamphlet. [Ety. *fusus*, a spindle; *spira*, a spire.]
elongata, Hall, 1871, Pamphlet, Trenton Gr. [Sig. lengthened, drawn out.]
subfusiformis, Hall, 1847, (Murchisonia subfusiforme) Pal. N. Y., vol. 1, Trenton and Hud. Riv. Gr. [Sig. somewhat spindle-shaped.]
terebriformis, Hall, 1871, Pamphlet, Cin'ti Gr. [Sig. like an auger or piercer.]
ventricosa, Hall, 1871, Pamphlet, Trenton Gr. [Sig. bulging out.]

GASTEROPODA.

vittata, Hall, 1847, (Murchisonia vittata) Pal. N. Y., vol. 1, Trenton Gr. [Sig. banded.]
Fusus, Bruguière, 1789, Encyc. Meth. [Ety. *fusus*, a spindle.] This genus is unknown in the palæozoic rocks.
inhabilis, see Macrocheilus primigenius.
HELICOTOMA, Salter, 1859, Can. Org. Rem., Decade 1. [Ety. *Helix*, a genus of shells; *tome*, a notch.]
declivis, Safford, 1869, Geo. of Tenn. Not defined.
eucharis, Billings, 1865, Pal. Foss., vol. 1, Quebec Gr. [Sig. beautiful.]
gorgonea, Billings, 1865, Pal. Foss., vol. 1, Quebec Gr. [Ety. mythological name.]
larvata, Salter, 1859, Can. Org. Rem., Decade 1, Black Riv. and Trenton Gr. [Sig. a ghost, masked.]
miser, Billings, 1865, Pal. Foss., vol. 1, Quebec Gr. [Sig. wretched.]
muricata, Salter, 1859, Can. Org. Rem., Decade 1, Black Riv. and Trenton Gr. [Sig. like the shell *Murex*.]
perstriata, Billings, 1859, Can. Nat.& Geo., vol. 4, Calciferous Gr. [Sig. very much striated.]
planulata, Salter, 1859, Decade 1, Black Riv. & Trenton Gr. [Sig. rather flat.]
proserpina, Billings, 1865, Pal. Foss., vol. 1, Quebec Gr. [Ety. mythological name.]
serotina, Nicholson, 1874, Rep. Pal. Ont. [Sig. backward.]
spinosa, Salter, 1859, Can. Org. Rem., Decade 1, Black Riv. Gr. [Sig. spiny, from the spines on the whorls.]
tennesseensis, Safford, 1869, Geo. of Tenn. Not defined.
tritonia, Billings, 1865, Pal. Foss., vol. 1, Quebec Gr. [Ety. mythological name.]
HOLOPEA, Hall, 1847, Pal. N. Y., vol. 1. [Ety. *holos*, entire; *ope*, an aperture.]
antiqua, Vanuxem, 1843, (Littorina antiqua) Geo. Rep. 3rd Dist. N. Y., Low. Held. Gr. [Sig. ancient.]
antiqua var. pervetusta, Hall, 1859, Low. Held. Gr. [Sig. very ancient.]
chicagoensis, Winchell & Marcy, 1865, Mem. Bost. Soc. Nat. Hist., Niagara Gr. [Ety. proper name.]
conica, Winchell, 1863, Proc. Acad. Nat. Sci., Chemung Gr. [Sig. conical.]
danai, Hall, 1859, Pal. N. Y., vol. 3, Low. Held. Gr. [Ety. proper name.]
dilucula, Hall, 1847, (Turbo dilucula) Pal. N. Y., vol. 1, Calciferous Gr. [Sig. very early, at break of day.]
(?) elongata, Hall, 1859, Pal. N. Y., vol. 3, Low. Held. Gr. [Sig. elongated.]
eriensis, Nicholson, 1874, Rep. Pal. Ont., Devonian. [Ety. proper name.]
gracia, Billings, 1862, Pal. Foss., vol. 1, Guelph Gr. [Sig. pleasant.]
guelphensis, Billings, 1862, Pal. Foss., vol. 1, Guelph Gr. [Ety. proper name.]
harmonia, Billings, 1862, Pal. Foss., vol. 1, Guelph Gr. [Ety. mythological name.]
lavinia, Billings, 1862, Pal. Foss., vol. 1, Trenton Gr. [Ety. mythological name.]
leiosoma, Billings, 1865, Pal. Foss., vol. 1, Quebec Gr. [Sig. smooth-bodied.]
nereis, Billings, 1862, Pal. Foss., vol. 1, Trenton & Black Riv. Gr. [Ety. mythological name.]
niagarensis, Winchell & Marcy, 1865, Mem. Bost. Soc. Nat. Hist., Niagara Gr. [Ety. proper name.]
obliqua, Hall, 1847, Pal. N. Y., vol. 1, Trenton & Hud. Riv. Gr. [Sig. oblique.]
obscura, Hall, 1847, (Turbo obscura) Pal. N. Y., vol. 1, Calciferous Gr. [Sig. obscure, doubtful.]
occidentalis, Nicholson, 1875, Quar. Jour. Geo. Soc. Lond., Guelph Gr. [Sig. western.]
ophelia, Billings, 1865, Pal. Foss., vol. 1, Quebec Gr. [Ety. proper name.]
ovalis, Billings, 1859, Can. Nat. & Geo., vol. 4, Calciferous Gr. [Sig. oval.]
paludiniformis, Hall, 1847, Pal. N. Y., vol. 1, Trenton Gr. [Sig. like a *Paludina*.]
proserpina, Billings, 1862, Pal. Foss., vol. 1, Calciferous & Chazy Gr. [Ety. mythological name.]
proutana, Hall, 1858, Trans. Alb. Inst., vol. 4, Warsaw Gr. [Ety. proper name.]
pyrene, Billings, 1862, Pal. Foss., vol. 1, Black Riv. Gr. [Ety. mythological name.]
reversa, Hall, 1860, Can. Nat. & Geo., vol. 5, Silurian. [Sig. reversed.]
subconica, Hall, 1859, Can. Nat. & Geo., vol. 3, Low. Held. Gr. [Sig. subconical.]
subconica, Winchell, 1863. This name was preoccupied.]
symmetrica, Hall, 1847, Pal. N. Y., vol. 1, Black Riv. Gr. [Sig. symmetrical.]
turgida, Hall, 1847, Pal. N. Y., vol. 1, Calciferous Gr. [Sig. turgid, swollen out.]
ventricosa, Hall, 1847, Pal. N. Y., vol. 1, Trenton Gr. [Sig. bulging out.]
HOLOPELLA, McCoy, 1855, Brit. Pal. Foss. [Ety. diminutive of *Holopea*.]
mira, Winchell, 1863, Proc. Acad. Nat. Sci., Chemung Gr. [Sig. wonderful.]
Inachus catilloides, see Euomphalus catilloides.
pervetustus, see Euomphalus pervetustus.
undatus, see Lituites undatus.
ISONEMA, Meek & Worthen, 1866, Proc. Acad. Nat. Sci. Phil. [Ety. *isos*, equal; *nema*, a thread.]
bellatulum, Hall, 1861, (Loxonema bellatula) 14th Reg. Rep. N. Y. [Sig. quite handsome.]
depressum, Meek & Worthen, 1865, Proc. Acad. Nat. Sci. Phil., Ham. Gr. [Sig. depressed.]
humilis, see Naticopsis humilis.

lichas, Hall, 1861, (Platyostoma lichas) 14th Reg. Rep. N. Y., Up. Held. Gr. [Ety. mythological name.]
Littorina, Ferussac, 1821, Tab. Syst. An. Mollusques, etc.
antiqua, see Holopea antiqua.
cancellata, see Cyclonema cancellatum.
wheeleri, see Naticopsis wheeleri.
LOXONEMA, Phillips, 1841, Pal. Foss. [Ety. *loxos*, oblique; *nema*, a thread.]
aculeatum, Billings, 1866, Catal. Sil. Foss. Antic., Anticosti Gr. [Sig. sharp-pointed.]
acutulum, Dawson, 1868, Acad. Geol., Carboniferous. [Sig. somewhat sharp-pointed.]
attenuatum, Hall, 1859, Pal. N. Y., vol. 3, Low. Held. Gr. [Sig. attenuated.]
bellatula, see Isonema bellatula.
bellona, Hall, 1876, Illust. Devonian Foss., Ham. Gr. [Ety. proper name.]
boydi, see Murchisonia boydi.
carinatum, Stevens, 1858, Am. Jour. Sci., 2d series, vol. 25, Coal Meas. [Sig. keeled.]
cerithiforme, Meek & Worthen, 1860, Proc. Acad. Nat. Sci., Coal Meas. [Sig. like a shell of the genus *Cerithium*.]
coaptum, Hall, 1876, Illust. Devonian Foss., Ham. Gr. [Sig. closely-joined.]
compactum, Hall, 1859, Pal. N. Y., vol. 3, Low. Held. Gr. [Sig. compact.]
danvillense, Stevens, 1858, Am. Jour. Sci., vol. 25, Coal Meas. [Ety. proper name.]
delphicola, Hall, 1862, 15th Reg. Rep., Ham. Gr. [Ety. mythological name.]
fasciatum, King, 1850, Permian Foss., Permian Gr. [Sig. banded.]
fitchi, Hall, 1859, Pal. N. Y., vol. 3, Low. Held. Gr. [Ety. proper name.]
halli, Norwood & Pratten, 1855, Jour. Acad. Nat. Sci., 2d series, vol. 3, Coal Meas. [Ety. proper name.]
hamiltoniæ, Hall, 1862, 15th Reg. Rep., Ham. Gr. [Ety. proper name.]
hydraulicum, Hall, 1872, 24th Reg. Rep., Ham. Gr. [Ety. from the hydraulic limestone.]
inornata, see Polyphemopsis inornata.
leda, Hall, 1868, 20th Reg. Rep. N. Y., Niagara Gr. [Ety. mythological name.]
minutum, Stevens, 1858, Am. Jour. Sci., 2d series, vol. 25, Coal Meas. [Sig. minute.]
multicostatum, Meek & Worthen, 1861, Proc. Acad. Nat. Sci., Coal Meas. [Sig. many-ribbed.]
murrayanum, Salter, 1859, Can. Org. Rem., Decade 1, Black Riv. Gr. [Ety. proper name.]
newberryi, see Macrocheilus newberryi.
nexile, Phillips, 1841, Pal. Foss., Ham. Gr. [Sig. interlaced.] American species. (?)
nitidula, see Polyphemopsis nitidula.

nodosum, Stevens, 1858, Am. Jour. Sci., 2d ser., vol. 25, Coal Meas. [Sig. knotted.]
obtusum, Hall, 1859, Pal. N. Y., vol. 3, Low. Held Gr. [Sig. obtuse.]
oligospiratum, Winchell, 1863, Proc. Acad. Nat. Sci., Chemung Gr. [Sig. having few whorls.]
pexatum, Hall, 1861, 14th Reg. Rep., Up. Held Gr. [Sig. clothed in a shell with a nap on it.]
pexatum *var*. obsoletum, Hall, 1876, Illust. Devonian Foss., Up. Held. Gr. [Sig. obsolete.]
planogyratum, Hall, 1839, Pal. N.Y., vol.3, Low. Held. Gr. [Sig. flattened whorl; from the flattening of the upper side of the last volution.]
politum, Stevens, 1858, Am. Jour. Sci., 2d series, vol. 25, Coal Meas. [Sig. smoothed.]
regulare, Cox, 1857, Geo. Sur. Ky., vol. 3, Coal Meas. [Sig. regular.]
robustum, Hall, 1862, 15th Reg. Rep., Schoharie grit. [Sig. robust.]
rugosum, Meek & Worthen, 1860, Proc. Acad. Nat. Sci., Coal Meas. [Sig. wrinkled.]
scitula, Meek & Worthen, 1860, Proc. Acad. Nat. Sci., Coal Meas. [Sig. neat, pretty.]
semicostatum, Meek, 1871, Proc. Aca. Nat. Sci., Coal Meas. [Sig. half-ribbed.]
solidum, Hall, 1862, 15th Reg. Rep., Schoharie grit. [Sig. solid.]
styliola, Hall, 1876, Illust. Devon. Foss., Chemung Gr. [Sig. a truncated column.]
subattenuatum, Hall, 1862, 15th Reg. Rep., Schoharie grit. [Sig. somewhat attenuated.]
subulata, see Murchisonia subulata.
tenuicarinatum, Stevens, 1858, Am. Jour. Sci., 2nd series, vol. 25, Coal Meas. [Sig. fine-lined.]
terebra, Hall, 1876, Illust. Devon. Foss., Chemung Gr. [Sig. an auger.]
teres, Hall, 1876, Illust. Devonian Foss., Corniferous Gr. [Sig. well-rounded.]
turritiformie, Hall, 1860, 13th Reg. Rep. N. Y., Ham. Gr. [Sig. tower-like.]
vincta, Hall, 1858, Trans. Alb. Inst., vol. 4, Warsaw Gr. [Sig. girded.]
yandellanum, Hall, 1858, Trans. Alb. Inst. vol. 4, Warsaw Gr. [Ety. proper name.]
MACLUREA, LeSueur, 1818, (Maclurites) Jour. Acad. Nat. Sci., vol. 1. [Ety. proper name.]
acuminata, Billings, 1865, Pal. Foss., vol. 1, Quebec Gr. [Sig. pointed.]
affinis, Billings, 1865, Pal. Foss., vol. 1, Quebec. Gr. [Sig. near to.]
atlantica, Billings, 1859, Can. Nat. & Geo., vol. 4, Chazy Gr. [Ety. proper name.]
bigsbyi, Hall, 1861, Geo. Rep. Wis., Trenton Gr. [Ety. proper name.]
crenulata, Billings, 1865, Pal. Foss., vol. 1, Quebec Gr. [Sig. crenulated.]

GASTEROPODA.

emmonsi, Billings, 1865, Pal. Foss., vol. 1, Quebec Gr. [Ety. proper name.]
labiata, see Raphistoma labiata.
logani, Salter, 1851, Rep. British Assoc., Black Riv. Gr. [Ety. proper name.]
magna, LeSueur, 1818, Jour. Acad. Nat. Sci., vol. 1, Chazy Gr. [Sig. large.]
matutina, Hall, 1847, Pal. N. Y., vol. 1, Calciferous Gr. [Sig. in the morning.]
oceana, Billings, 1865, Pal. Foss., vol. 1, Quebec Gr. [Ety. mythological name.]
ponderosa, Billings, 1865, Pal. Foss., vol. 1, Quebec Gr. [Sig. heavy.]
psyche, Billings, 1865, Pal. Foss., vol. 1, Quebec Gr. [Ety. mythological name.]
rotundata, Billings, 1865, Pal. Foss., vol. 1, Quebec Gr. [Sig. rounded.]
speciosa, Billings, 1865, Pal. Foss., vol. 1, Quebec Gr. [Sig. beautiful.]
sordida, see Ophileta sordida.
striata, Emmons, see Raphistoma striata.
striata, Troost, 1840. Not defined.
sylpha, Billings, 1865, Pal. Foss., vol. 1, Quebec Gr. [Ety. mythological name.]
transitionis, Billings, 1865, Pal. Foss., vol. 1, Quebec Gr. [Sig. a passing over, or transition between *M. affinis* and *M. emmonsi*.]

MACROCHEILUS, Phillips, 1841, Pal. Foss. [Ety. *macros*, long; *cheilos*, lip.]
altonensis, Worthen, 1873, Geo. Sur. Ill., vol. 5, Coal Meas. [Ety. proper name.]
angulifera, White, 1874, Rep. Invertebrate Foss., Carboniferous. [Sig. bearing angles.]
carinata, Stevens, 1858, Am. Jour. Sci., vol. 25, Coal Meas. [Sig. keeled.]
cooperensis, Swallow, 1863, Trans. St. Louis Acad. Sci., Low. Carb. [Ety. proper name.]
fusiformis, Hall, 1858, Geo. Rep. Iowa, Coal Meas. The name was preoccupied by Sowerby, see Morris' Catalogue.
gracilis, Cox, 1857, Geo. Sur. Ky., vol. 3, Coal Meas. [Sig. slender.]
hallanus, see Soleniscus hallanus.
hamiltoniæ, Hall, 1862, 15th Reg. Rep., Ham. Gr. [Ety. proper name.]
hebe, Hall, 1862, 15th Reg. Rep., Ham. Gr. [Ety. mythological name.]
hildrethi, Conrad, 1842, (Plectostylus hildrethi) Jour. Acad. Nat. Sci., vol. 8, Coal Meas. [Ety. proper name.]
inhabilis, see Macrocheilus primigenius.
intercalaris, Meek & Worthen, 1860, Proc. Acad. Nat. Sci., Coal Meas. [Sig. intercalated.]
kansasensis, Swallow, 1858, Trans. St. Louis Acad. Sci., Coal Meas. [Ety. proper name.]
klipparti, Meek, 1872, Proc. Acad. Nat. Sci., Coal Meas. [Ety. proper name.]
macrostomus, Hall, 1862, 15th Reg. Rep., Ham. Gr. [Sig. long-mouthed.]

medialis, Meek & Worthen, 1865, Proc. Acad. Nat. Sci., Coal Meas. [Sig. from its medium size.]
missouriensis, Swallow, 1858, Trans. St. Louis Acad. Sci., Coal Meas. [Ety. proper name.]
newberryi, Stevens, 1858, (Loxonema newberryi) Am. Jour. Sci., 2d series, vol. 25, Coal Meas. [Ety. proper name.]
paludinæformis, Hall, 1858, Geo. Rep. Iowa, Coal Meas. [Sig. like a shell of the genus *Paludina*.]
pinguis, Winchell, 1863, Proc. Acad. Nat. Sci., Chemung Gr. [Sig. fat, thick.]
ponderosus, Swallow, 1858, Trans. St. Louis Acad. Sci., Coal Meas. [Sig. heavy, bulky.]
primævus, Hall, 1876, Illust. Devonian Foss., Schoharie grit. [Sig. primeval.]
primigenius, Conrad, 1835, (Stylifer primigenia) Trans. Geo. Soc. Penn., vol. 1, Coal Meas. [Sig. first born.]
pulchellus, Meek & Worthen, 1860, Proc. Acad. Nat. Sci., Coal Meas. [Sig. beautiful.]
spiratus, McCoy, 1850, Brit. Pal. Foss., Coal Meas. [Sig. spiral.]
texanus, Shumard, 1859, Trans. St. Louis Acad. Sci., Coal Meas. [Ety. proper name.]
ventricosus, Hall, 1858, Geo. Rep. Iowa, Coal Meas. [Sig. ventricose.]

METOPTOMA, Phillips, 1836, Geo. of Yorkshire. [Ety. *metopon*, front; *tome*, incision.]
alceste, Billings, 1862, Pal. Foss., vol. 1, Hud. Riv. Gr. [Ety. mythological name.]
augusta, Billings, 1862, Pal. Foss., vol. 1, Quebec Gr. [Ety. proper name.]
anomala, Billings, 1862, Pal. Foss., vol. 1, Quebec Gr. [Sig. irregular.]
canadensis, Billings, 1865, Pal. Foss., vol. 1, (Chiton canadensis) Black Riv. Gr. [Ety. proper name.]
dubia, Hall, 1847, Pal. N. Y., vol. 1, Chazy Gr. [Sig. doubtful.]
erato, Billings, 1862, Pal. Foss., vol. 1, Black Riv. Gr. [Ety. mythological name.]
estella, Billings, 1862, Pal. Foss., vol. 1, Hud. Riv. Gr. [Ety. proper name.]
eubele, Billings, 1862, Pal. Foss., vol. 1, Calcif. & Black Riv. Gr. [Ety. proper name.]
hyrie, Billings, 1862, Pal. Foss., vol. 1, Quebec Gr. [Ety. proper name.]
instabilis, Billings, 1865, Pal. Foss., vol. 1, Quebec Gr. [Sig. not firm, changing.]
melissa, Billings, 1862, Pal. Foss., vol. 1, Quebec Gr. [Ety. mythological name.]
montrealensis, Billings, 1865, Pal. Foss., vol. 1, Chazy Gr. [Ety. proper name.]
niobe, Billings, 1862, Pal. Foss., vol. 1, Calcif. Gr. [Ety. mythological name.]
nycteis, Billings, 1862, Pal. Foss., vol. 1, Calcif. Gr. [Ety. mythological name.]

GASTEROPODA.

orithyia, Billings, 1862, Pal. Foss., vol. 1, Calcif. Gr. [Ety. mythological name.]
orphyne, Billings, 1862, Pal. Foss., vol. 1, Quebec Gr. [Ety. mythological name.]
quebecensis, Billings, 1865, Pal. Foss., vol. 1, Quebec Gr. [Ety. proper name.]
rugosa, Hall, 1847, Pal. N. Y., vol. 1, Hud. Riv. Gr. [Sig. wrinkled.]
simplex, Billings, 1865, Pal. Foss., vol. 1, Calcif. Gr. [Sig. simple.]
superba, Billings, 1865, Pal. Foss., vol. 1, Black Riv. Gr. [Sig. magnificent.]
trentonensis, Billings, 1862, Pal. Foss., vol. 1, Trenton Gr. [Ety. proper name.]
undata, Winchell, 1865, Proc. Acad. Nat. Sci. Phil., Kinderhook Gr. [Sig. wavy.]
umbella, Meek & Worthen, 1866, Proc. Acad. Nat. Sci. Phil., Burlington Gr. [Sig. umbrella-like.]
venilla, Billings, 1862, Pal. Foss., vol. 1, Quebec Gr. [Ety. mythological name.]
MICROCERAS, Hall, 1845, Am. Jour. Sci., vol. 48. [Ety. *mikros*, small; *keras*, horn.]
inornatus, Hall, 1845, Am. Jour. Sci., vol. 48, Cin'ti Gr. [Sig. not adorned, smooth.]
MICRODOMA, Meek & Worthen, 1866, Proc. Acad. Nat. Sci. [Ety. *mikros*, small; *domus*, house.]
conica, Meek & Worthen, 1866, Proc. Acad. Nat. Sci.Phil., Low. Coal Meas. [Sig. conical.]
MURCHISONIA, D'Archiac & Verneuil, 1841, Bull. Soc. Geo. Fr., vol. 12. [Ety. proper name.]
abbreviata, Hall, 1847, Pal. N. Y., vol. 1, Chazy Gr. The name was preoccupied by Koninck in 1841.
aciculata, Hall, 1860, Can. Nat. & Geo., vol. 5, Silurian. [Sig. needle-pointed.]
acrea, Billings, 1865, Pal. Foss., vol. 1, Quebec Gr. [Ety. proper name.]
ada, Billings, 1865, Pal. Foss., vol. 1, Calciferous Gr. [Ety. proper name.]
adelina, Billings, 1865, Pal. Foss., vol. 1, Quebec Gr. [Ety. proper name.]
agilis, Billings, 1865, Pal. Foss., vol. 1, Quebec Gr. [Sig. from the depths.]
alexandra, Billings, 1865, Pal. Foss., vol. 1, Black Riv. Gr. [Ety. proper name.]
angulata, Phillips, 1836, (Rostellaria angulata) Geol. of Yorkshire, Devonian. [Sig. angulated.] Very doubtfully identified in America.
angusta, Hall, 1847, Pal. N. Y., vol. 1, Birdseye Gr. [Sig. narrowed.]
anna, Billings, 1859, Can. Nat. & Geol., vol. 4, Calciferous Gr. [Ety. St. Annis in Canada.]
archimedea, McChesney, 1861, Desc. New Pal. Foss., Coal Meas. [Sig. pertaining to the machine invented by Archimedes.]
arenaria, Billings, 1859, Can. Nat. & Geo., vol. 4, Calciferous Gr. [Sig. sandy.]

arisaigensis, Hall, 1860, Can. Nat. & Geo., vol. 5, Silurian. [Ety. proper name.]
artemisia, Billings, 1865, Pal. Foss., vol. 1, Calciferous Gr. [Ety. proper name.]
aspera, Billings, 1859, Can. Nat. & Geo., vol. 4, Chazy Gr. [Sig. rough.]
attenuata, Hall, 1858, Trans. Alb. Inst., vol. 4, Warsaw Gr. [Sig. drawn out, attenuated.]
augustina, Billings, 1865, Pal. Foss., vol. 1, Quebec Gr. [Ety. proper name.]
bellicincta, Hall, 1847, Pal. N. Y., vol. 1, Trenton & Hud. Riv. Gr. [Sig. beautifully banded.]
bicincta, Hall, 1847, Pal. N. Y., vol. 1, Trenton & Hud. Riv. Gr. [Sig. doublebanded.] This name was preoccupied by McCoy in 1844.
bilirata, Hall, 1859, Pal. N. Y., vol. 3, Low. Held. Gr. [Sig. double-furrowed.]
bivittata, Hall, 1852, Pal. N. Y., vol. 2, Guelph Gr. [Sig. double-banded.]
bowdeni, Safford, 1869, Geo. of Tenn., Nashville Gr. [Ety. proper name.]
boydi, Hall, 1843, (Loxonema boydi) Geo. Rep. 4th Dist. N. Y., Guelph Gr. [Ety. proper name.]
boylii, Nicholson, 1875, Quar. Jour. Geo. Soc. Lond., vol. 31, Guelph Gr. [Ety. proper name.]
carinifera, Shumard, 1863, Trans. St. Louis Acad. Sci., Calciferous Gr. [Sig. keelbearing.]
cassandra, Billings, 1865, Pal. Foss., vol. 1, Quebec Gr. [Ety. mythological name.]
catherina, Billings, 1865, Pal. Foss., vol. 1, Quebec Gr. [Ety. proper name.]
cicelia, Billings, 1865, Pal. Foss., vol. 1, Quebec Gr. [Ety. proper name.]
conoidea, Hall, 1852, Pal. N. Y., vol. 2, Medina Gr. [Sig. somewhat conical.]
conradi, Hall, 1867, 20th Reg. Rep., Niagara Gr. [Ety. proper name.]
desiderata, Hall, 1862, 15th Reg. Rep., Up. Held. Gr. [Sig. desired.]
egregia, Billings, 1874, Pal. Foss., vol. 2, Up. Held. Gr. [Sig. excellent.]
elegantula, Hall, 1858, Trans. Alb. Inst., vol. 4, Warsaw Gr. [Sig. quite elegant.]
estella, Billings, 1862, Pal. Foss. vol. 1, Guelph Gr. [Ety. proper name.]
extenuata, Hall, 1859, Pal. N. Y., vol. 3, Low. Held. Gr. [Sig. thinned-out.]
funata, Billings, 1866, Catal. Sil. Foss. Antic., Anticosti Gr. [Sig. corded.]
gigantea, Billings, 1857, Rep. of Progr., Mid. Sil. [Sig. unusually large.]
gracilis, Hall, 1847, Pal. N. Y., vol. 1, Trenton & Hud. Riv. Gr. [Sig. slender.]
gypsea, Dawson, 1868, Acad. Geol., Carboniferous. [Sig. gypsum.]
hebe, Billings, 1874, Pal. Foss., vol. 2, Devonian. [Ety. mythological name.]

helicteres, Salter, 1859, Can. Org. Rem., Decade 1, Black Riv. and Trenton Gr. [Sig. a round, smooth spire.]
hercynia, Billings, 1862, Pal. Foss., vol. 1, Guelph Gr. [Ety. proper name.] The name was preoccupied by Roemer in 1843.
hermione, Billings, 1862, Pal. Foss., vol. 1, Chazy or Black Riv. Gr. [Ety. mythological name.]
hyale, Billings, 1862, Pal. Foss., vol. 1, Chazy or Black Riv. Gr. [Ety. proper name.]
infrequens, Billings, 1859, Can. Nat. & Geo., vol. 4, Chazy Gr. [Sig. rare.]
inornata, Meck & Worthen, 1866, Proc. Acad. Nat. Sci., Coal Meas. [Sig. not adorned.]
insculpta, Hall, 1858, Trans. Alb. Inst., vol. 4, Warsaw Gr. [Sig. engraved.]
jessica, Billings, 1865, Pal. Foss., vol. 1, Quebec Gr. [Ety. proper name.]
kansasensis, Swallow, 1858, Trans. St. Louis Acad. Sci., Coal Meas. [Ety. proper name.]
laphami, Hall, 1861, Rep. of Progr. Wis., Niagara Gr. [Ety. proper name.]
leda, Hall, 1861, 14th Reg. Rep. N. Y., Up. Held. Gr. [Ety. mythological name.]
limitaris, Hall, 1860, 13th Reg. Rep. N. Y., Ham. Gr. [Sig. limited.]
linearis, Billings, 1859, Can. Nat. & Geol., vol. 4, Calciferous Gr. [Sig. marked with lines.]
logani, Hall, 1852, Pal. N. Y., vol. 2, Guelph Gr. [Ety. proper name.]
longispira, Hall, 1852, Pal. N. Y., vol. 2, Guelph Gr. [Sig. long-spired.]
macrospira, Hall, 1852, Pal. N. Y., vol. 2, Guelph Gr. [Sig. large or long-spired.]
maia, Hall, 1861, 14th Reg. Rep., Up. Held. Gr. [Ety. mythological name.]
major, Hall, 1851, Geo. Lake Sup. Land Dist., vol. 2, Trenton Gr. [Sig. the greater.]
melaniiformis, Shumard, 1855, Geo. Rep. Mo., Calciferous Gr. [Ety. formed like the *Melania*, a genus of shells.]
minima, Swallow, 1858, Trans. St. Louis Acad. Sci., Middle Coal Meas. [Sig. the least.]
minuta, Hall, 1859, Pal. N. Y., vol. 3, Low. Held. Gr. [Sig. small.]
missisquoi, Billings, 1865, Pal. Foss., vol. 1, Quebec Gr. [Ety. proper name.]
modesta, Billings, 1857, Rep. of Progr., Hud. Riv. Gr. [Sig. not large.]
mucro, Winchell, 1866, Rep. Low. Peninsula Mich., Ham. Gr. [Sig. a sharp point.]
multivolvis, Billings, 1857, Rep. of Progr., Hud. Riv. Gr. [Sig. many rolled.]
mylitta, Billings, 1862, Pal. Foss., vol. 1, Guelph Gr. [Ety. proper name.]
nebrascensis, Geinitz, 1866, Carb. und Dyas in Neb., Coal Meas. [Ety. proper name.]

neglecta, Winchell, 1863, Proc. Acad. Nat. Sci., Chemung Gr. [Sig. neglected.]
obtusa, Hall, 1852, Pal. N. Y., vol. 2, Coralline limestone. [Sig. obtuse.]
ozarkensis, Shumard, 1863, Trans. St. Louis Acad. Sci., Calciferous Gr. [Ety. proper name.]
papillosa, Billings, 1857, Rep. of Progr., Mid. Sil. [Sig. covered with tubercles.]
perangulata, Hall, 1847, Pal. N.Y., vol. 1, Black Riv. and Trenton Gr. [Sig. very angular.]
perversa, Swallow, 1858, Trans. St. Louis Acad. Sci., Up. Coal Meas. [Sig. turned around.]
petilla, Hall, 1872, 24th Reg. Rep., Niagara Gr. [Sig. thin, slender.]
placida, Billings, 1865, Pal. Foss., vol. 1, Quebec Gr. [Sig. placid, smooth.]
procris, Billings, 1862, Pal. Foss., vol. 1, Black Riv. Gr. [Ety. mythological name.]
quadricincta, Winchell, 1863, Proc. Acad. Nat. Sci., Chemung Gr. [Sig. four-girded.]
rugosa, Billings, 1857, Rep. of Progr., Hud. Riv. Gr. [Sig. wrinkled.]
serrulata, Salter, 1859, Can. Org. Rem., Decade 1, Black Riv. and Trenton Gr. [Sig. minutely-serrated.]
shumardana, Winchell, 1863, Proc. Acad. Nat. Sci. Phil., Chemung Gr. [Ety. proper name.]
simulatrix, Billings, 1865, Pal. Foss., vol. 1, Quebec Gr. [Sig. an imitator.]
sororcula, Billings, 1865, Pal. Foss., vol. 1, Quebec Gr. [Sig. a little sister.]
subfusiformis, see Fusispira subfusiformis.
subtæniata, see Orthonema subtæniatum.
subulata, Conrad, 1842, (Loxonema subulata) Jour. Acad. Nat. Sci., vol. 8, Clinton Gr. [Sig. awl-shaped.]
sumnerensis, Safford, 1869, Geo. of Tenn., Nashville Gr. [Ety. proper name.]
sylvia, Billings, 1865, Pal. Foss., vol. 1, Quebec Gr. [Ety. mythological name.]
terebralis, Hall, 1852, Pal. N. Y., vol. 2, Coralline limestone. [Sig. like an auger.]
terebriformis, Hall, 1858, Trans. Alb. Inst., vol. 4, Warsaw Gr. [Sig. in the form of an auger or borer.]
teretiformis, Billings, 1857, Rep. of Progr. Hud. Riv. Gr. [Sig. of a long, round shape.]
texana, Shumard, 1860, Trans. St. Louis Acad. Sci., Coal Meas. [Ety. proper name.]
tricarinata, Hall, 1847, Pal. N. Y., vol. 1, Trenton Gr. [Sig. three-keeled.]
tricingulata, Dawson, 1868, Acad. Geol., Carboniferous. [Sig. three-banded.]
turricula, Billings, 1857, Rep. of Progr., Mid. Sil. [Sig. a little tower.]
turricula, Hall, 1862, 15th Reg. Rep. N. Y., Ham. Gr. This name was preoccupied.

turritella, Hall, 1858, Trans. Alb. Inst., vol. 4, Warsaw Gr. [Sig. a little tower.]
turritiformis, Hall, 1852, Pal. N. Y., vol. 2, Guelph Gr. [Sig. like a tower.]
uniangulata, Hall, 1847, Pal. N. Y., vol. 1, Trenton & Hud. Riv. Gr. [Sig. having one angular line.]
uniangulata *var.* abbreviata, Hall, 1847, Pal. N. Y., vol. 1, Hud. Riv. Gr. [Sig. comparatively shortened.]
varians, Billings, 1857, Rep. of Progr., Hud. Riv. Gr. [Sig. variable.]
varicosa, Hall, 1847, Pal. N. Y., vol. 1, Birdseye Gr. [Sig. varicose.]
ventricosa, Hall, 1847, Pal. N. Y., vol. 1, Black Riv. & Trenton Gr. [Sig. ventricose.]
vermicula, Hall, 1858, Trans. Alb. Inst., vol. 4, Warsaw Gr. [Sig. worm-shaped.]
vesta, Billings, 1862, Pal. Foss., vol. 1, Calciferous Gr. [Ety. mythological name.]
vitellia, Billings, 1862, Pal. Foss., vol. 1, Guelph Gr. [Ety. proper name.]
vittata, see Fusispira vittata.
xanthippe, Billings, 1862, Pal. Foss., vol. 1, Guelph Gr. [Ety. proper name.]

Natica, Adanson, 1757, Histoire Naturelle du Senegal, etc. [Ety. *nato*, to swim with a fluctuating motion.] This genus is unknown in Palæozoic rocks.
altonensis, see Naticopsis altonensis.
carleyana, see Naticopsis carleyana.
chesterensis, see Naticopsis chesterensis.
littonana, see Naticopsis littonana.
magister, syn. for Naticopsis ventricosa.
shumardi, see Naticopsis shumardi.
ventricosa, see Naticopsis ventricosa.

NATICOPSIS, McCoy, 1844, Synop.Carb. Foss., Ireland. [Ety. from resemblance to the genus *Natica*.]
æquistriata, Meek, 1873, Ohio Pal., vol. 1, Cornif. Gr. [Sig. having equal striæ.]
altonensis, McChesney, 1865, (Natica altonensis) Desc. New Pal. Foss., Coal Meas. [Ety. proper name.]
carleyana, Hall, 1858, (Natica carleyana) Trans. Alb. Inst., vol. 4, Warsaw Gr. [Ety. proper name.]
chesterensis, Swallow, 1863, Trans. St. Louis Acad. Sci., Kaskaskia Gr. [Ety. proper name.]
cretacea, Hall, 1873, 23rd Reg. Rep. N.Y., Cornif. Gr. [Sig. chalk-like.]
depressa, Winchell, 1863, Proc. Acad. Nat. Sci., Chemung Gr. [Sig. depressed.]
dispassa, Dawson, 1868, Acad. Geol., Carbonif. [Sig. much spread out.]
gigantea, Hall, 1873, 23rd Reg. Rep. N. Y., Chemung Gr. [Sig. unusually large.]
hollidayi, see Trachydomia hollidayi.
howi, Hartt, 1868, Acad. Geol., Carboniferous. [Ety. proper name.]
humilis, Meek, 1871, (Isonema hnmilis) Proc. Acad. Nat. Sci. Phil., Cornif. Gr. [Sig. small, dwarfish.]
lævis, Meek, 1871, Proc. Acad. Nat. Sci., Cornif. Gr. [Sig. smooth.]
littonana, Hall, 1858, (Natica littonana) Trans. Alb. Inst., vol. 4, Warsaw Gr. [Ety. proper name.]
magister, syn. for N. ventricosa.
nana, Meek & Worthen, 1860, (Platystoma nana) Proc. Acad. Nat. Sci., Coal Meas. [Sig. dwarfish.]
nodosa, see Trachydomia nodosa.
pricei, Shumard, 1858, Trans. St. Louis Acad. Sci., Up. Coal Meas. [Ety. proper name.]
remex, White, 1876, Geo. Uinta Mountains, Low. Aubrey Gr. [Sig. a rower.]
shumardi, McChesney, 1860, (Natica shumardi) Desc. New Pal. Foss., Coal Meas. [Ety. proper name.]
subovata, Worthen, 1873, Geo. Sur. Ill., vol. 5, Coal Meas. [Sig. somewhat ovate.]
ventricosa, Norwood & Pratten, 1855, (Natica ventricosa) Jour. Acad. Nat. Sci., Coal Meas. [Sig. ventricose.]
wheeleri, Swallow, 1860, (Littorina wheeleri) Trans. St. Louis Acad. Sci., Coal Meas. [Ety. proper name.]

Omphalotrochus, Meek, 1864, Geo. California. A name proposed as a subgenus of *Euomphalus* or *Straparollus*.

OPHILETA, Vanuxem, 1842, Geo. Rep. N. Y. [Ety. *ophis*, a snake.]
abdita, Billings, 1865, Pal. Foss., vol. 1, Quebec Gr. [Sig. concealed.]
(?) bella, Billings, 1865, Pal. Foss., vol. 1, Quebec Gr. [Sig. beautiful.]
compacta, Salter, 1859, Can. Org. Rem., Decade 1, Calciferous Gr. [Sig. compact.]
complanata, Vanuxem, 1842, Geo. Rep. N.Y., Calciferous Gr. [Sig. smoothed.]
disjuncta, Billings, 1865, Pal. Foss., vol. 1, Calciferous Gr. [Sig. disjoined.]
levata, Vanuxem, 1842, Geo. Rep. N. Y., Calciferous Gr. [Sig. polished, smoothed.]
nerine, Billings, 1865, Pal. Foss., vol. 1, Quebec Gr. [Ety. mythological name.]
ottawaensis, Billings, 1860, Can. Nat. & Geol., vol. 5, Trenton Gr. [Ety. mythological name.]
owenana, Meek & Worthen, 1868, Geo. Sur. Ill., vol. 3, Galena Gr. [Ety. proper name.]
profunda, Billings, 1865, Pal. Foss., vol. 1, Quebec Gr. [Sig. deep.]
sordida, Hall, 1847, (Maclurea sordida) Pal. N. Y., vol. 1, Calciferous Gr. [Sig. paltry.]
uniangulata, Hall, 1847, (Euomphalus uniangulatus) Pal. N. Y., vol. 1, Calcifer. Gr. [Sig. one angled; from the angular line on one whorl.]

ORTHONEMA, Meek & Worthen, 1861, Proc. Acad. Nat. Sci. Phil. [Ety. *orthos*, straight; *nema*, a thread.]
conicum, Meek & Worthen, 1866, Proc. Acad. Nat. Sci. Phil., Coal Meas. [Sig. a cone.]
newberryi, Meek, 1871, Proc. Acad. Nat. Sci. Phil., Corniferous Gr. [Ety. proper name.]
salteri, Meek & Worthen, 1860, (Eunema (?) salteri) Proc. Acad. Nat. Sci. Phil., Low. Coal Meas. [Ety. proper name.]
subtæniatum, Geinitz, 1866, (Murchisonia subtæniata) Carb. und Dyas in Neb., Coal Meas. [Sig. somewhat banded.]
Orthonychia, Hall, 1843, syn. for Platyceras.
ORTHOSTOMA, Conrad, 1838, Ann. Rep. N. Y. [Ety. *orthos*, straight; *stoma*, mouth.]
commune, Conrad, 1838, Ann. Rep. N. Y., figured in 1841, Birdseye Gr. [Sig. common.]
PALÆACMÆA, Hall, 1873, 23d Reg. Rep. N. Y. [Ety. *palaios*, ancient; *Acmæa*, an existing genus of shells.]
typica, Hall, 1873, 23d Reg. Rep. N. Y., Potsdam Gr. [Ety. type of the genus.]
PHANEROTINUS, Sowerby, 1842, Min. Conch. [Ety. *phaneros*, aperture; *teino*, extended.]
paradoxus, Winchell, 1863, Proc. Acad. Nat. Sci., Chemung Gr. [Sig. paradoxical.]
Phragmolites, syn. for Cyrtolites.
compressus, see Cyrtolites compressus.
PHRAGMOSTOMA, Hall, 1861, 14th Reg. Rep. [Ety. *phragmos*, a partition, *stoma*, the mouth; from the septum within the aperture.]
cumulus, Hall, 1861, 14th Reg. Rep., Hud. Riv. Gr. [Sig. a heap.]
cymbula, Hall, 1861, 14th Reg. Rep., Hud. Riv. Gr. [Sig. a small boat.]
natator, Hall, 1862, 15th Reg. Rep., Portage and Ham. Gr. [Sig. a swimmer.] The same species Hall identified with *Bellerophon expansus* of Sowerby, in 1843, Geo. Rep. 4th Dist. N. Y.
Pileopsis tubifer, syn. for Platyceras dumosum.
vetustus, Sowerby. Not American.
Planorbis, Guettard, 1756, Mem. Acad. Sc. Paris.
trilobatus, see Bucania trilobatus.
PLATYCERAS, Conrad, 1840, Ann. Rep. N. Y. [Ety. *platys*, broad; *keras*, horn.]
agreste, Hall, 1859, Pal. N. Y., vol. 3, Low. Held. Gr. [Sig. pertaining to the fields, coarse.]
ammon, Hall, 1862, 15th Reg. Rep., Up. Held. Gr. [Ety. mythological name.]
angulatum, Hall, 1852, (Acroculia angulata) Pal. N. Y., vol. 2, Clinton and Niagara Gr. [Sig. angulated.]
arcuatum, Hall, 1859, Pal. N. Y., vol. 3, Low. Held. Gr. [Sig. bent.]
argo, Hall, 1862, 15th Reg. Rep., Up. Held. Gr. [Ety. mythological name.]
attenuatum, Hall, 1862, 15th Reg. Rep., Ham. Gr. [Sig. attenuated.]
attenuatum, Meek, 1871, Proc. Acad. Nat. Sci. Phil., Cornif. Gr. This name was preoccupied.]
auriculatum, Hall, 1876, Illust. Devonian Foss., Ham. Gr. [Sig. eared.]
billingsi, Hall, 1859, Pal. N. Y., vol. 3, Low. Held. Gr. [Ety. proper name.]
biserialis, Hall, 1860, Supp. to Geo. Iowa, vol. 1, pt. 2, Burlington Gr. [Sig. having two rows.]
bisinuatum, Hall, 1859, Pal. N. Y., vol. 3, Low. Held. Gr. [Sig. double-sinuated.]
bisulcatum, Hall, 1859, Pal. N. Y., vol. 3, Low. Held. G r. [Sig. double-furrowed.]
bucculentum, Hall, 1862, 15th Reg. Rep., Ham. Gr. [Sig. large mouthed.]
calantica, Hall, 1859, Pal. N. Y., vol. 3, Low. Held. Gr. [Sig. like a covering for the head.]
callosum, Hall, 1859, Pal. N. Y., vol. 3, Oriskany sandstone. [Sig. having a thick skin.]
campanulatum, Winchell & Marcy, 1865, Mem. Bost. Soc. Nat. Hist., Niagara Gr. [Sig. bell-shaped.]
capulus, Hall, 1860, Supp. Geo. Iowa, Burlington Gr. [Sig. a coffin.]
carinatum, Hall, 1862, 15th Reg. Rep., Up. Held. Gr. [Sig. keeled.]
cirriformis, Conrad, 1841, Ann. Rep. N. Y. Not clearly defined.
clavatum, Hall, 1859, Pal. N. Y., vol. 3, Low. Held. Gr. [Sig. club-like.]
concavum, Hall, 1862, 15th Reg. Rep., Up. Held. Gr. [Sig. concave.]
conicum, Hall, 1862, 15th Reg. Rep., Ham. Gr. [Sig. conical.]
corniforme, Winchell, 1863, Proc. Acad. Nat. Sci., Chemung Gr. [Sig. in the form of a horn.]
crassum, Hall, 1862, 15th Reg. Rep., Up. Held. Gr. [Sig. thick.]
curvirostrum, Hall, 1859, Pal. N. Y., vol. 3, Low. Held. Gr. [Sig. bent-beaked.]
cymbium, Hall, 1862, 15th Reg. Rep., Up. Held. Gr. [Sig. a small drinking vessel.]
cyrtolites, McChesney, 1859, Pal. Foss., Coal Meas. [Sig. curved stone.]
dentalium, Hall, 1862, 15th Reg. Rep., Up. Held. Gr. [Sig. a plough share.]
dilatatum, Hall, 1859, Pal. N. Y., vol. 3, Low. Held. Gr. [Sig. dilated.]
dumosum, Conrad, 1840, Ann. Rep. N. Y., Up. Held. Gr. [Sig. bushy.]
dumosum *var.* rarispinum, Hall, 1862, 15th Reg. Rep., Up. Held. Gr. [Sig. few spined.]
echinatum, Hall, 1862, 15th Reg. Rep., Ham. Gr. [Sig. set with spines.]
elongatum, Hall, 1859, Pal. N. Y., vol. 3, Low. Held. Gr. [Sig. lengthened.]

equilateralis, Hall, 1860, Supp. to vol. 1, pt. 2, Iowa Rep., Keokuk Gr. [Sig. equal-sided.]
erectum, Hall, 1843, (Acroculia erecta) Geo. 4th Dist. N. Y., Cornif. & Ham. Gr. [Sig. erect, straight.]
expansus, see Strophostylus expansus.
fissurella, Hall, 1860, Supp. to Geo. Rep. Iowa, vol. 1, pt. 2, Keokuk Gr. [Sig. a little cleft.]
fornicatum, Hall, 1862, 15th Reg. Rep., Up. Held. Gr. [Sig. arched.]
fornicatum *var*. contractum, Hall, 1876, Illust. Devonian Foss., Up. Held. Gr. [Sig. contracted.]
gebhardi, Conrad, 1840, Ann. Rep. N. Y., Low. Held. and Oriskany Gr. [Ety. proper name.]
gibbosum, Hall, 1859, Pal. N. Y., vol. 3, Low. Held. Gr. [Sig. gibbous.]
haliotoides, Meek & Worthen, 1866, Proc. Acad. Nat. Sci. Phil., Kinderhook Gr. [Sig. like the *Haliotus* or ear-shell.]
incile, Hall, 1859, Pal. N. Y., vol. 3, Low. Held. Gr. [Sig. having gutters.]
infundibulum, Meek & Worthen, 1866, Proc. Acad. Nat. Sci. Phil., Keokuk Gr. [Sig. a funnel.]
intermedium, Hall, 1859, Pal. N. Y., vol. 3, Low. Held. Gr. [Sig. intermediate in size.]
lamellosum, Hall, 1859, Pal. N. Y., vol. 3, Low. Held. Gr. [Sig. with many plates.]
lodiense, Meek, 1871, Proc. Acad. Nat. Sci. Phil., Waverly Gr. [Ety. proper name.]
magnificum, Hall, 1859, Pal. N. Y., vol. 3, Oriskany sandstone. [Sig. magnificent.]
multisinuatum, Hall, 1859, Pal. N. Y., vol. 3, Low. Held. Gr. [Sig. having many depressions.]
multispinosum, Meek, 1871, Proc. Acad. Nat. Sci. Phil., Cornif. Gr. [Sig. having many spines.]
nebrascense, Meek, 1872, Pal. E. Neb., Coal Meas. [Ety. proper name.]
newberryi, Hall, 1859, Pal. N. Y., vol. 3, Low. Held. Gr. [Ety. proper name.]
niagarense, Hall, 1852, (Acroculia niagarensis) Pal. N. Y., vol. 2, Niagara Gr. [Ety. proper name.]
nodosum, Conrad, 1841, Ann. Rep. N. Y., Oriskany sandstone. [Sig. knobbed.]
obesum, Hall, 1859, Pal. N. Y., vol. 3, Low. Held. Gr. [Sig. plump in form.]
ovale, Stevens, 1858, (Acroculia ovalis) Am. Jour. Sci., vol. 25, Carboniferous. [Sig. oval.]
patulum, Hall, 1859, Pal. N. Y., vol. 3, Oriskany sandstone. [Sig. spread out.]
pentalobus, Hall, 1859, Pal. N. Y., vol. 3, Low. Held. Gr. [Sig. five-lobed.]
perlatum, Hall, 1859, Pal. N. Y., vol. 3, Low. Held. Gr. [Sig. very wide.]
perplexum, Hall, 1876, Illust. Devonian Foss., Up. Held. Gr. [Sig. obscure.]

perplicatum, Hall, 1859, Pal. N. Y., vol. 3, Low. Held. Gr. [Sig. very much plicated.]
pileiforme, Hall, 1859, Pal. N. Y., vol. 3, Low. Held. Gr. [Sig. cap-shaped.]
platystoma, Hall, 1859, Pal. N. Y., vol. 3, Low. Held. Gr. [Sig. broad-mouthed.]
platystoma *var*. alveatum, Hall, 1859, Pal. N. Y., vol. 3, Low. Held. Gr. [Sig. channeled.]
plicatile, Hall, 1859, Pal. N. Y., vol. 3, Low. Held. Gr. [Sig. in small folds.]
plicatum, Hall, 1859, Pal. N. Y., vol. 3, Low. Held. Gr. [Sig. folded.]
primævum, Billings, 1871, Can. Nat. & Geol., vol. 6, Quebec Gr. [Sig. firstborn.]
primordiale, Hall, 1863, 16th Reg. Rep., Potsdam Gr. [Sig. the first of all.]
pyramidatum, Hall, 1859, Pal. N. Y., vol. 3, Low. Held. Gr. [Sig. pyramidformed.]
quincyense, McChesney, 1861, New Pal. Foss., Burlington Gr. [Ety. proper name.]
reflexum, Hall, 1859, Pal. N. Y., vol. 3, Oriskany sandstone. [Sig. turned back.]
retrorsum, Hall, 1859, Pal. N. Y., vol. 3, Low. Held. Gr. [Sig. turned backward.]
retrorsum *var*. abnorme, Hall, 1859, Pal. N. Y., vol. 3, Low. Held. Gr. [Sig. out of the usual order or form.]
reversum, Hall, 1860, Supp. to Geo. Rep. Iowa, vol. 1, pt. 2, Burlington Gr. [Sig. turned or bent back.]
rictum, Hall, 1862, 15th Reg. Rep., Ham. & Up. Held. Gr. [Sig. open-mouthed.]
robustum, Hall, 1859, Pal. N. Y., vol. 3, Low. Held. Gr. [Sig. robust.]
senex, Winchell & Marcy, 1865, (Porcellia senex) Mem. Bost. Soc. Nat. Hist., Niagara Gr. [Sig. old, wrinkled.]
sinuatum, Hall, 1859, Pal. N. Y., vol. 3, Low. Held. Gr. [Sig. marked with depressions.]
spinigerum, Worthen, 1873, Geo. Sur. Ill., vol. 5, Coal Meas. [Sig. spinebearing.]
spirale, Hall, 1859, Pal. N. Y., vol. 3, Low. Held. Gr. [Sig. spiral.]
subnodosum, Hall, 1859, Pal. N. Y., vol. 3, Oriskany sandstone. [Sig. somewhat knotty, or marked with short projections.]
subplicatum, Meek & Worthen, 1866, Proc. Acad. Nat. Sci. Phil., Kinderhook Gr. [Sig. somewhat folded.]
subrectum, Hall, 1859, 12th Reg. Rep. N. Y., Up. Held. Gr. [Sig. somewhat straight.]
subrectum, Hall, 1860, Supp. to Iowa Rep. The name being preoccupied, Meek & Worthen proposed P. infundibulum.
subundatum, Conrad, 1841, Ann. Rep. N. Y., Up. Held. Gr. [Sig. somewhat waved.]

GASTEROPODA. 157

subundatum, Meek & Worthen, 1868, Geo. Sur. Ill., vol. 3, Low. Held. Gr. The name was preoccupied.
sulcatum, Conrad, 1841, Ann. Rep. N. Y., Oriskany sandstone. [Sig. furrowed.]
sulcoplicatum, Hall, 1859, Pal. N. Y., vol. 3, Low. Held. Gr. [Sig. thrown up in plications.]
symmetricum, Hall, 1862, 15th Reg. Rep., Ham. & Up. Held. Gr. [Sig. symmetrical.]
tenuiliratum, Hall, 1859, Pal. N. Y., vol. 3, Low. Held. Gr. [Sig. fine-lined.]
thetis, Hall, 1862, 15th Reg. Rep., Ham. Gr. [Ety. mythological name.]
thetis *var.* subspinosum, Hall, 1876, Illust. Devonian Foss., Ham. Gr. [Sig. somewhat spiny.]
tortum, Meek, 1871, Proc. Acad. Nat. Sci., Coal Meas. [Sig. twisted.]
tortuosum, Hall, 1859, Pal. N. Y., vol. 3, Oriskany sandstone. [Sig. very much twisted.]
trigonale, Stevens, 1858, (Acroculia trigonalis) Am. Jour. Sci. & Arts, vol. 25, Carboniferous. [Sig. triangular.]
trilobatum, Hall, 1859, Pal. N. Y., vol. 3, Low. Held. Gr. [Sig. three-lobed.]
tubæforme, Hall, 1859, Pal. N. Y., vol. 3, Low. Held. Gr. [Sig. trumpet-shaped.]
uncum, Meek & Worthen, 1866, Proc. Acad. Nat. Sci., Keokuk Gr. [Sig. crooked, hooked.]
undatum, Hall, 1876, Illust. Devonian Foss., Up. Held. Gr. [Sig. wavy.]
undulostriatum, Hall, 1859, Pal. N. Y., vol. 3, Low. Held. Gr. [Sig. having waved striæ.]
unguiforme, Hall, 1859, Pal. N. Y., vol. 3, Low. Held. Gr. [Sig. claw-shaped.]
uniseriale, Nicholson, 1874, Rep. Pal. Ont., Devonian. [Sig. having a single row or series.]
unisulcatum, Hall, 1859, Pal. N. Y., vol. 3, Low. Held. Gr. [Sig. one-furrowed.]
ventricosum, Conrad, 1840, Ann. Rep. N. Y., Low. Held. Gr. [Sig. bulging out.]
vomerium, Winchell, 1863, Proc. Acad. Nat. Sci., Chemung Gr. [Sig. a plow share.]

PLATYSTOMA, Conrad, 1842, Jour. Acad. Nat. Sci., vol. 8. [Ety. *platys*, broad; *stoma*, mouth.]
affine, Billings, 1874, Pal. Foss., vol. 2, Devonian. [Sig. related to.]
aplata, Hall, 1876, Illust. Devonian Foss., Schoharie grit. [Sig. without lines.]
arenosum, Conrad, 1842, Jour. Acad. Nat. Sci., vol. 8, Low. Held. Gr. [Sig. sandy.]
defiguratum, Hall, 1876. Illust. Devonian Foss., Ham. Gr. [Sig. disfigured.]
depressum, Hall, 1859, Pal. N. Y., vol. 3, Low. Held. Gr. [Sig. depressed.]
euomphaloides, Hall, 1876, Illust. Devonian Foss., Ham. Gr. [Sig. like a shell of the genus *Euomphalus*.]

hemisphericum, Hall, 1843, (Euomphalus hemispherica) Geo. Rep. 4th Dist. N. Y., Niagara Gr. [Sig. hemispherical.]
lichas, see Isonema lichas.
lineatum, Conrad, 1842, Jour. Acad. Nat. Sci., vol. 8, Cornif. Gr. [Sig. lined.]
lineatum, Hall, 1862, 15th Reg. Rep. The name was preoccupied.
lineatum *var.* amplum, Hall, 1876, Illust. Devonian Foss., Ham. Gr. [Sig. full, large.]
lineatum *var.* callosum, Hall, 1876, Illust. Devonian Foss., Ham. Gr. [Sig. thick, hard.]
lineatum *var.* sinuosum, Hall, 1876, Illust. Devonian Foss., Ham. Gr. [Sig. wavy, sinuous.]
nana, see Naticopsis nana.
niagarense, Hall, 1852, Pal. N. Y., vol. 2, Niagara Gr. [Ety. proper name.]
peoriense, McChesney, 1860, Desc. New Pal. Foss., Up. Coal Meas. [Ety. proper name.]
plebeium, Hall, 1876, 28th Reg. Rep. N. Y., Niagara Gr. [Sig. common.]
pleurotoma, Hall, 1876, Illust. Devonian Foss., Up. Held. Gr. [Sig. from the genus *Pleurotoma*.]
strophium, Hall, 1862, 15th Reg. Rep., Cornif. Gr. [Sig. twisted or turned.]
subangulatum, Hall, 1859, Pal. N. Y., vol. 3, Low. Held. Gr. [Sig. somewhat angular.]
trigonostoma, Meek, 1871, Proc. Acad. Nat. Sci., Niagara Gr. [Sig. triangular-mouthed.]
tumida, see Pleurotomaria tumida.
turbinatum, Hall, 1861, 14th Reg. Rep., Up. Held. Gr. [Sig. cone-shaped.]
turbinatum *var.* cochleatum, Hall, 1876, Illust. Devonian Foss., Up. Held. Gr. [Sig. spiral-formed.]
unisulcatum, Conrad, 1842, (Pleurotomaria unisulcata) Jour. Acad. Nat. Sci., vol. 8, Up. Held Gr. [Sig. having one depression.]
ventricosum, Conrad, 1842, Jour. Acad. Nat. Sci., vol. 8, Low. Held. Gr. [Sig. ventricose.]

PLATYSCHISMA, McCoy, 1844, Syn. Carb. Foss., Ireland. [Ety. *platys*, wide; *schisma*, a slit.]
dubium, Dawson, 1868, Acad. Geol., Carboniferous. [Sig. doubtful.]
helicoides, (?) Sowerby, 1829, (Ampularia helicoides) Min. Conch., vol. 8, Coal Meas. [Sig. resembling a shell of the genus *Helix*.]

Plectostylus, Conrad, 1842, Syn. for Macrocheilus.
hildrethi, see Macrocheilus hildrethi.

PLEUROTOMARIA, Defrance, 1826, Dict. Sci. Nat., 41. [Ety. *Pleura*, side; *tome*, cut or notch.]
abrupta, Billings, 1859, Can. Nat. & Geo., vol. 4, Calciferous Gr. [Sig. terminating suddenly.]

GASTEROPODA.

advena, Winchell, 1864, Am. Jour. Sci. & Arts, 2d series, vol. 37. Potsdam Gr. [Sig. a stranger.]

agarista, Billings, 1865, Pal. Foss., vol. 1, Quebec Gr. [Ety. proper name.]

agave, Billings, 1865, Pal. Foss., vol. 1, Trenton Gr. [Ety. mythological name.]

ambigua, Hall, 1847, Pal. N. Y., vol. 1, Trenton Gr. [Sig. doubtful.]

americana, Billings, 1860, Can. Nat. & Geo., vol. 5, Trenton Gr. [Ety. proper name.]

amphitrite, Billings, 1862, Pal. Foss., vol. 1, Chazy or Black Riv. Gr. [Ety. mythological name.]

ungulata, Conrad, 1843, Proc. Acad. Nat. Sci. Phil. This name was preoccupied by Sowerby.

antiquata, Hall, 1847, Pal. N. Y., vol. 1, Chazy Gr. [Sig. antiquated.]

aperta, see Raphistoma apertum.

apicalis, Hall, 1876, Illust. Devonian Foss., Chemung Gr. [Sig. apical.]

arabella, Billings, 1865, Pal. Foss., vol. 1, Calciferous Gr. [Ety. proper name.]

arachne, Billings, 1862, Pal. Foss., vol. 1, Black Riv. Gr. [Ety. mythological name.]

arata, Hall, 1862, 15th Reg. Rep., Schoharie grit. [Sig. furrowed.]

axion, Hall, 1867, 20th Reg. Rep., Niagara Gr. [Sig. furrowed.]

beckwithana, McChesney, 1860, Desc. New Pal. Foss., Coal Meas. [Ety. proper name.]

biangulata, Hall, 1847, Pal. N. Y., vol. 1, Chazy Gr. [Sig. double-angled, from the two angular elevations on each whorl.]

bicarinata, McChesney, 1860. This name was preoccupied and the species is now called P. turbiniformis.

bilix, see Cyclonema bilix.

bispiralis, Hall, 1852, Pal. N. Y., vol. 2, Guelph Gr. [Sig. two-whorled.]

bonharborensis, Cox, 1857, Geo. Sur. Ky., vol. 3, Coal Meas. [Ety. proper name.]

brazoensis, Shumard, 1860, Trans. St. Louis Acad. Sci., Coal Meas. [Ety. proper name.]

calcifera, Billings, 1859, Can. Nat. & Geo., Calciferous Gr. [Ety. from the Calciferous Group.]

calphurnia, Billings, 1865, Pal. Foss., vol. 1, Quebec Gr. [Ety. mythological name.]

calyx, Billings, 1859, Can. Nat. & Geo., vol. 4, Chazy Gr. [Sig. a cup.]

canadensis, Billings, 1865, Pal. Foss., vol. 1, Calciferous Gr. [Ety. proper name.]

capillaria, Conrad, 1842, Jour. Acad. Nat. Sci., vol. 8, Ham. Gr. [Sig. pertaining to hair.]

carbonaria, Norwood & Pratten, 1855, Jour. Acad. Nat. Sci., 2d series, vol. 3, Coal Meas. [Sig. pertaining to coal.]

casii, Meek & Worthen, 1868, Geo. Sur. Ill., vol. 3, Niagara Gr. [Ety. proper name.]

cavumbilicata, Winchell, 1866, Rep. Low. Peninsula Mich., Ham. Gr. [Sig. having a hollow umbilicus.]

chesterensis, Meek & Worthen, 1860, Proc. Acad. Nat. Sci., Chester Gr. [Ety. proper name.].

chesterensis, Swallow, 1863, Trans. St. Louis Acad. Sci., Chester Gr. The name was preoccupied.

circe, Billings, 1857, Rep. of Progr., Hud. Riv. Gr. [Ety. mythological name.]

concava, Hall, 1858, Trans. Alb. Inst., vol. 4, Warsaw Gr. The name was preoccupied by Deshayes in 1824-'36.

conoides, Meek & Worthen, 1866, Proc. Acad. Nat. Sci., Coal Meas. The name was preoccupied by Deshayes in 1831.

conulus, Hall, 1858, Trans. Alb. Inst., vol. 4, Warsaw Gr. [Sig. a little cone.]

coronula, syn. for P. sphærulata.]

coxana, Meek & Worthen, 1866, Proc. Acad. Nat. Sci., Coal Meas. [Ety. proper name.]

crevieri, Billings, 1859. Can. Nat. & Geol., vol. 4, Chazy Gr. [Ety. proper name.]

cryptata, Billings, 1866, Catal. Sil. Foss., Antic., Anticosti Gr. [Sig. concealed.]

cyclonemoides, Meek & Worthen, 1868, Geo. Sur. Ill., vol. 3, Niagara Gr. [Sig. like a *Cyclonema.*]

deiopea, Billings, 1862, Pal. Foss., vol. 1, Guelph Gr. [Ety. mythological name.]

delia, Billings, 1874, Pal. Foss., vol. 2, Devonian. [Ety. mythological name.]

delicatula, Hall, 1876, Illust. Devonian Foss., Up. Held. Gr. [Sig. very delicate.]

delphinuloides, Goldfuss, as identified by d'Archiac & Verneuil. Not American.

depauperata, Hall, 1862, Geo. Rep. Wis., Hud. Riv. Gr. [Sig. impoverished.]

depressa, Cox, 1857, Geo. Sur. Ky., vol. 3, Coal Meas. The name was preoccupied by Passy in 1832, by Phillips in 1836, and by Koninck in 1841.

disjuncta, Hall, 1876, Illust. Devonian Foss., Ham. Gr. [Sig. disjoined.]

dispersa, Dawson, 1868, Acad. Geol., Carboniferous. [Sig. dispersed.]

docens, Billings, 1859, Can. Nat. & Geol., vol. 4, Chazy Gr. [Sig. teaching.]

doris, Hall, 1862, 15th Reg. Rep., Schoharie & Cornif. Gr. [Ety. mythological name.]

dryope, Billings, 1865, Pal. Foss, vol. 1, Black Riv. Gr. [Ety. mythological name.]

ella, Hall, 1876, Illust. Devonian Foss., Ham. Gr. [Ety. proper name.]

elora, Billings, 1862, Pal. Foss., vol. 1, Guelph Gr. [Ety. proper name.]
emmetensis, Winchell, 1866, Rep. Low. Peninsula Mich., Ham. Gr. [Ety. proper name.]
estella, Hall, 1872, 24th Reg. Rep., Up. Held. Gr. [Ety. proper name.]
etna, Billings, 1865, Pal. Foss., vol. 1, Quebec Gr. [Ety. proper name.]
eugenia, Billings, 1862, Pal. Foss., vol. 1, Black Riv. Gr. [Ety. proper name.]
euomphaloides, Hall, 1862, 15th Reg.Rep., Ham. Gr. [Sig. like an *Euomphalus*.]
exigua, Winchell, 1862, Proc. Acad. Nat. Sci., Marshall Gr. [Sig. small, scanty.]
filitexta, Hall, 1876, Illust. Devonian Foss., Ham. Gr. [Sig. woven like threads.]
galtensis, Billings, 1862, Pal. Foss., vol. 1, Guelph Gr. [Ety. proper name.]
glandula, Shumard, 1860, Trans. St.Louis Acad. Sci., Coal Mens. [Sig. a little kernel.]
gonopleura, Winchell & Marcy, 1865, Bost. Soc. Nat. Hist., Niagara Gr. [Sig. having an angular rib.]
granulostriata, Meek & Worthen, 1860, Proc. Acad. Nat. Sci., Coal Meas. [Sig. granular and striated.]
grayvillensis, Norwood & Pratten, 1855, Jour. Acad. Nat. Sci., 2nd series, vol. 3, Coal Meas. [Ety. proper name.]
gregaria, Billings, 1859, Can. Nat. & Geo., vol. 4, Calciferous Gr. [Sig. occurring in masses.]
halei, Hall, 1861, Rep. of Progr. Wis. Sur., Niagara Gr. [Ety. proper name.]
hallana, Shumard, 1859, Trans. St. Louis Acad. Sci., Permian Gr. [Ety. proper name.]
halli, S. A. Miller, 1874, Cin. Quar. Jour. Sci., vol. 1, Cin'ti Gr. [Ety. proper name.]
harpya, Billings, 1865, Pal. Foss., vol. 1, Quebec Gr. [Ety. mythological name.]
haydenana, Geinitz, 1866, Carb. und Dyas in Neb., Coal Meas. [Ety. proper name.]
hebe, Hall, 1861, 14th Reg. Rep. N. Y., Up. Held. Gr. [Ety. mythological name.]
helena, Billings, 1860, Can. Nat. & Geo., vol. 5, Hud. Riv. Gr. [Ety. proper name.]
hickmanensis, Winchell, Geo. of Tenn., Low. Carb. [Ety. proper name.]
hortensia, Billings, 1865, Pal. Foss., vol. 1, Quebec Gr. [Ety. proper name.]
hoyi, Hall, 1861, Rep. of Progr. Wis. Sur., Niagara Gr. [Ety. proper name.]
humerosa, Meek & Hayden, 1858, Proc. Acad. Nat. Sci. Phil., Coal Meas. [Sig. humped.]
humilis, Hall, 1858, Trans. Alb. Inst., vol. 4, Warsaw Gr. [Sig. small.]
humilis, Winchell, 1862. This name was preoccupied.

huronensis, Winchell, 1862, Proc. Acad. Nat. Sci. Phil., Portage Gr. [Ety. proper name.]
hyale, Billings, 1865, Pal. Foss., vol. 1, Quebec Gr. [Ety. proper name.]
ida, Hall, 1861, Rep. of Progr. Wis. Sur., Niagara Gr. [Ety.mythological name.]
ignobilis, Dawson, 1868, Acad. Geol., Carboniferous. [Sig. ignoble.]
imitator, Hall, 1872, 24th Reg. Rep., Up. Held. Gr. [Sig. resembler.]
immatura, Billings, 1859, Can. Nat. & Geo., vol. 4, Chazy Gr. [Sig. immature.]
indenta, Hall, 1847, Pal. N. Y., vol. 1, Trenton Gr. [Sig. notched.]
inexspectans, Hall & Whitfield, 1875, Ohio Pal., vol. 1, Clinton Gr. [Sig. not expected.]
inornata, Meek, 1872, Pal. E. Neb., Coal Meas. [Sig. not adorned.]
insolita, Hall, 1876, Illust. Devonian Foss., Ham. Gr. [Sig. rare.]
isaacsi, Hall, 1873, 23rd Reg. Rep., Chemung Gr. [Ety. proper name.]
itys, Hall, 1876, Illust. Devonian Foss., Ham. Gr. [Ety. mythological name.]
kearneyi, Hall, 1861, 12th Reg. Rep., Up, Held. Gr. [Ety. proper name.]
labrosa, Hall, 1859, Pal. N. Y., vol. 3, Low. Held Gr. [Sig. having lips.]
lapicida, see Raphistoma lapicidum.
laurentina, Billings, 1859,Can. Nat. &Geo., vol. 4, Calcif. Gr. [Ety. proper name.]
leavenworthana, Hall, 1858, Trans. Alb. Inst., vol. 4, Warsaw Gr. [Ety. proper name.]
lenticularis, see Raphistoma lenticulare.
lineata, Hall, 1843, (Turbo lineatus) Geo. Rep. 4th Dist. N. Y. This name was preoccupied, and the fossil is now called P. itys.
litorea, Hall, 1852, Pal. N. Y., vol. 2, Medina sandstone. [Sig. on the shore.]
lucina, Hall, 1862, 15th Reg. Rep., Cornif. Gr. [Ety. proper name.]
lucina *var.* perfasciata, Hall, 1876, Illust. Devonian Foss., Ham. Gr. [Sig. much banded.]
lydia, Billings, 1874, Pal. Foss., vol. 2, Devonian. [Ety. proper name.]
marcouana, Geinitz, 1866, Carb. und Dyas in Neb., Coal Meas. [Ety. proper name.]
meekana, Hall, 1858, Trans. Alb. Inst., vol. 4, Warsaw Gr. [Ety. proper name.]
micula, Hall, 1862, Geo. Rep. Wis., Hud. Riv. Gr. [Sig. very small.]
miser, Billings, 1859, Can. Nat. & Geo., vol. 4, Calcif. Gr. [Sig. paltry.]
missisquoi, Billings, 1865, Pal. Foss., vol. 1, Quebec Gr. [Ety. proper name.]
missouriensis, Swallow, 1860, (Trochus missouriensis) Trans. St. Louis Acad. Sci., Coal Meas. [Ety. proper name.]
mitigata, Hall, 1860, 13th Reg. Rep., Ham. Gr. [Sig. lessened.]

muralis, Owen, 1852, Geo. Sur. Wis., Iowa & Minn., Calciferous Gr. [Sig. mural.]
nasoni, Hall, 1861, Geo. Rep. Wis., Trenton Gr. [Ety. proper name.]
niota, Hall, 1861, Geo. Rep. Wis., Trenton Gr. [Ety. proper name.]
nodomarginata, McChesney, 1860, Desc. New Pal. Foss., Ham. Gr. [Sig. having a nodose margin.]
nodulosa, Hall, 1847, Pal. N. Y., vol. 1. The name was preoccupied by Sandberger in 1842, and by King in 1844.
nodulostriata, Hall, 1848, Trans. Alb. Inst., vol. 4, Warsaw Gr. [Sig. nodulose and striated.]
normani, Billings, 1865, Pal. Foss., vol. 1, Quebec Gr. [Ety. proper name.]
nucleolata, Hall, 1847, Pal. N. Y., vol. 1, Birdseye Gr. [Sig. like a little nut.]
numeria, Billings, 1865, Pal. Foss., vol. 1, Quebec Gr. [Ety. mythological name.]
obsoleta, Hall, 1847, Pal. N. Y., vol. 1, Birdseye Gr. [Sig. obsolete.]
obtusispira, Shumard, 1859, Trans. St. Louis Acad. Sci., Coal Meas. [Sig. having a blunt spire.]
occidens, Hall, 1867, 20th Reg. Rep., Niagara Gr. [Sig. western.]
parvispira, Winchell, 1862, Rep. Low. Peninsula Mich., Ham. Gr. [Sig. having a small spire.]
pauper, Billings, 1859, Can. Nat. & Geol., vol. 4, Chazy Gr. [Sig. poor.]
pauper, Hall, 1865, 20th Reg. Rep. The name was preoccupied.
percarinata, see Cyclonema percarinatum.
perhumerosa, Meek, 1872, Pal. E. Neb., Coal Meas. [Sig. very humid.]
perlata, Hall, 1852, Pal. N. Y., vol. 2, Guelph Gr. [Sig. very wide.]
perornata, Shumard, 1859, Trans. St. Louis Acad. Sci., Coal Meas. [Sig. highly ornamented.]
pervetusta, Conrad, 1838, (Cyclostoma pervetusta) Ann. Rep. N. Y., Medina sandstone. [Sig. very ancient.]
piasaensis, Hall, 1858, Trans. Alb. Inst., vol. 4, Warsaw Gr. [Ety. proper name.]
planidorsalis, Hall, 1876, Illust. Devonian Foss., Ham. Gr. [Sig. plane-backed.]
plena, Hall, 1876, Illust. Devonian Foss., Ham. Gr. [Sig. full, large.]
postumia, Billings, 1862, Pal. Foss., vol. 1, Quebec Gr. [Ety. proper name.]
poulsoni, Conrad, 1842, Jour. Acad. Nat. Sci., vol. 8, Onondaga Gr. [Ety. proper name.]
pratteni, Meek & Worthen, 1860, Proc. Acad. Nat. Sci., Low. Coal Meas. [Ety. proper name.]
princessa, Billings, 1874, Pal. Foss., vol. 2, Up. Held. Gr. [Sig. a princess.]
progne, Billings, 1860, Can. Nat. & Geol., vol. 5, Black Riv. & Trenton Gr. [Ety. mythological name.]

proutana, Shumard, 1859, Trans. St. Louis Acad. Sci., Coal Meas. [Ety. proper name.]
quadricostata, Hall, 1847, Pal. N. Y., vol. 1, Birdseye Gr. [Sig. four-ribbed.]
quebecensis, Billings, 1865, Pal. Foss., vol. 1, Quebec Gr. [Ety. proper name.]
quinquesulcata, Winchell, 1865, Proc. Acad. Nat. Sci., Chemung Gr. [Sig. five-furrowed.]
ramsayi, Billings, 1859, Can. Nat. & Geol., vol. 4, Calciferous Gr. [Ety. proper name.]
regulata, Hall, 1860, 13th Reg. Rep., Ham. Gr. [Sig. regular.]
riddlei, Shumard, 1860, Trans. St. Louis Acad. Sci., Coal Meas. [Ety. proper name.]
rota, Winchell, 1863, Proc. Acad. Nat. Sci., Chemung Gr. [Sig. a wheel.]
rotalia, Hall, 1862, 15th Reg. Rep., Ham. Gr. [Sig. wheeled.]
rotuloides, Hall, 1847, Pal. N. Y., vol. 1, Black Riv. and Trenton Gr. [Sig. like a little wheel.]
rotunda, Hall, 1843, (Euomphalus (?) rotundus) Geo. Rep. 4th Dist. N. Y., Corniferous Gr. [Sig. rounded.]
rotundata, Hall, 1858, Trans. Alb. Inst., vol. 4, Warsaw Gr. This name was preoccupied by Munster.
rotundispira, Billings, 1865, Pal. Foss., vol. 1, Quebec Gr. [Sig. having a round spire.]
scitula, Meek & Worthen, 1860, Proc. Acad. Nat. Sci. Phil., Low. Coal Meas. [Sig. pretty, neat.]
selecta, Billings, 1865, Pal. Foss., vol. 1, Quebec Gr. [Sig. choice, select.]
semele, Hall, 1861, Geo. Rep. Wis., Hud. Riv. Gr. [Ety. mythological name.]
shumardi, Meek & Worthen, 1860, Proc. Acad. Nat. Sci. Phil., Keokuk Gr. [Ety. proper name.]
sigaretoides, Winchell & Marcy, 1865, Bost. Soc. Nat. Hist., vol. 1, Niagara Gr. [Sig. like a shell of the genus *Sigaretus*.]
sinistrorsa, Swallow, 1858, Trans. St. Louis Acad. Sci., Coal Meas. [Sig. turned to the left.]
solarioides, Hall, 1852, Pal. N. Y., vol. 2, Guelph Gr. [Sig. like a shell of the genus *Solarium*.]
speciosa, Meek & Worthen, 1860, Proc. Acad. Nat. Sci. Phil., Low. Coal Meas. [Sig. beautiful.]
sphærulata, Conrad, 1842, Jour. Acad. Nat. Sci., vol. 8, Coal Meas. [Sig. a widened sphere.]
spironema, Meek & Worthen, 1866, Proc. Acad. Nat. Sci. Phil., Coal Meas. [Sig. spiral-lined.]
sponsa, Billings, 1865, Pal. Foss., vol. 1, Quebec Gr. [Sig. betrothed.]
stella, Winchell, 1862, Proc. Acad. Nat. Sci., Marshall Gr. [Sig. a star.]

subangulata, Hall, 1858, Trans. Alb. Inst., vol. 4, Warsaw Gr. [Sig. somewhat angular.]
subconica, Hall, 1847, Pal. N. Y., vol. 1, Black Riv., Trenton & Hud. Riv. Gr. [Sig. somewhat conical.]
subconstricta, Meek & Worthen, 1860, Proc. Acad. Nat. Sci. Phil., Low. Coal Meas. [Sig. somewhat constricted.]
subdecussata, Geinitz, 1866, Carb. und Dyas in Neb., Coal Meas. [Sig. somewhat arranged in pairs that cross each other.]
subdepressa, Hall, 1852, Pal. N.Y., vol. 2, Coralline limestone. [Sig. somewhat depressed.]
subscalaris, Meek & Worthen, 1860, Proc. Acad. Nat. Sci. Phil., Low. Coal Meas. [Sig. somewhat ladder-shaped.]
subsinuata, Meek & Worthen, 1860, Proc. Acad. Nat. Sci. Phil., Coal Meas. [Sig. somewhat sinuated.]
subtilistriata, Hall, 1847, Pal. N. Y., vol. 1, Trenton Gr. [Sig. finely shaped.]
subturbinata, Meek & Hayden, 1872, Proc. Acad. Nat. Sci. Phil., Coal Meas. [Sig. somewhat top-shaped.]
sulcomarginata, Conrad, 1842, Jour.Acad. Nat. Sci., vol. 8, Ham. Gr. [Sig. having the margin deeply furrowed.]
supracingulata, Billings, 1857, Rep. of Progr., Trenton Gr. [Sig. encircled with lines on the upper part.]
swallovana, Hall, 1858, Trans. Alb. Inst., vol. 4, Warsaw Gr. [Ety. proper name.]
sybillina, Billings, 1866, Catal. Sil. Foss. Antic., Anticosti Gr. [Ety. proper name.]
tabulata, Conrad, 1835, (Turbo tabulata) Trans. Geo. Soc. Penn., vol. 1, Coal Meas. [Sig. tabulated.]
tectoria, Winchell, 1863, Proc. Acad. Nat. Sci., Chemung Gr. [Sig. rough cast.]
tenuicincta, Meek & Worthen, 1860, Proc. Acad. Nat. Sci., Coal Meas. [Sig. finely girded.]
tenuistriata, Shumard, 1860, Trans. St. Louis Acad. Sci., Coal Meas. [Sig. finely striated.]
textiligera, Meek, 1871, Proc. Acad. Nat. Sci., Waverly Gr. [Sig. web-bearing.]
thalia, Billings, 1857, Rep. of Progr., Mid. Sil. [Ety. mythological name.]
trilineata, Hall, 1858, Trans. Alb. Inst., vol. 4, Warsaw Gr. [Sig. three-lined.]
trilix, Hall, 1862, 15th Reg. Rep. N. Y., Ham. Gr. [Sig. three-lined.]
trochiformis, Swallow, 1863,Trans. St. Louis Acad. Sci., Chester Gr. [Sig. like the *Trochus*, or wheel-shell.] The name was preoccupied by Portlock in 1843.
tropidophora, Meek, 1872, Am. Jour. Sci. & Arts, 3rd series, vol. 4, Cin'ti Gr. [Sig. keel-bearing.]
tumida, Meek & Worthen, 1860, (Platystoma tumida) Proc. Acad. Nat. Sci., Coal Meas. [Sig. tumid.] The name was preoccupied by Phillips in 1836.

turbiniformis, Meek & Worthen, 1860, Proc. Acad. Nat. Sci., Coal Meas. [Sig. top-shaped.]
turgida, Hall, 1847, Pal. N. Y., vol. 1, Calciferous Gr. [Sig. swollen.]
umbilicata, see Trochonema umbilicatum.
unisulcata, see Platyceras unisulcatum.
vadosa, Hall, 1860, 13th Reg. Rep., N. Y., Ham. Gr. [Sig. shallow.]
vagans, Billings, 1862, Pal. Foss., vol. 1, Quebec Gr. [Sig. dispersed.]
valeria, Billings, 1865, Pal. Foss., vol. 1, Guelph Gr. [Ety. proper name.]
valvatiformis, Meek & Worthen, 1866, Proc. Acad. Nat. Sci. Phil., Coal Meas. [Sig. like the genus *Valvata*.]
viola, Billings, 1865, Pal. Foss., vol. 1, Guelph Gr. [Sig. a violet.]
virgo, Billings, 1865, Pal. Foss., vol. 1, Quebec Gr. [Sig. a virgin.]
virguncula, Billings, 1865, Pal. Foss., vol. 1, Quebec Gr. [Sig. a little maid.]
vitruvia, Billings, 1865, Pal. Foss., vol. 1, Black Riv. Gr. [Ety. proper name.]
voltumna, Billings, 1874, Pal. Foss., vol. 2, Devonian. [Ety. mythological name.]
whitii, Winchell, 1862, Proc. Acad. Nat. Sci., Marshall Gr. [Ety. proper name.]
wortheni, Hall, 1856, Trans. Alb. Inst., vol. 4, Warsaw Gr. [Ety. proper name.]

POLYPHEMOPSIS, Portlock, 1843, Geol. Londonderry. [Ety. *Polyphemus*, a genus of shells; *opsis*, appearance.]
bulimiformis, Hall, 1858, (Bulimella bulimiformis) Trans. Alb. Inst., vol. 4, Warsaw Gr. [Sig. like a *Bulimus*.]
canaliculata, Hall, 1858, (Bulimella canaliculata) Trans. Alb. Inst., vol. 4, Warsaw Gr. [Sig. channeled or grooved.]
chrysalis, Meek & Worthen, 1866, Proc. Acad. Nat. Sci. Phil., Coal Meas. [Sig. chrysalis.]
elongata, Hall, 1858, (Bulimella elongata) Trans. Alb. Inst., vol. 4, Warsaw Gr. [Sig. lengthened.] The name was preoccupied by Portlock in 1843.
inornata, Meek & Worthen, 1860, (Loxonema inornata) Proc. Acad. Nat. Sci. Phil., Up. Coal Meas. [Sig. not adorned.]
louisvillæ, Hall, 1872, 24th Reg. Rep., Up. Held. Gr. [Ety. proper name.]
nitidula, Meek & Worthen, 1860, (Loxonema nitidula) Proc. Acad. Nat. Sci. Phil., Up. Coal Meas. [Sig. quite neat.]
peracuta, Meek & Worthen, 1860, (Eulima (?) peracuta) Proc. Acad. Nat. Sci. Phil., Up. Coal Meas. [Sig. very sharp-pointed.]

PORCELLIA, Leveille, 1835, Mem. Soc. Geol. France. [Ety. *porcellus*, a little pig.]
hertzeri, Hall, 1876, Illust. Devonian Foss., Up. Held. Gr. [Ety. proper name.]

nais, Hall, 1862, (Gyroceras nais) 15th Reg. Rep. N. Y., Chemung Gr. [Ety. mythological name.]
nodosa, Hall, 1860, Supp. to vol. 1, pt. 2, Iowa Geo. Sur., Kinderhook Gr. [Sig. knotty.]
obliquinoda, White, 1862, Proc. Bost. Soc. Nat. Hist., vol. 9, Chemung Gr. [Sig. oblique-knotted.]
rectinoda, Winchell, 1863, Proc. Acad. Nat. Sci., Chemung Gr. [Sig. straight-noded.]
(?) rotatoria, Hall, 1876, Illust. Devonian Foss., Up. Held. Gr. [Sig. whorled.]
sciota, Hall, 1873, 23rd Reg. Rep. N. Y., Up. Held. Gr. [Ety. proper name.]
senex, see Platyceras senex.

PUPA, Humphrey, 1797, Museum Calonnianum. [Ety. *Pupa*, chrysalis shell.]
vermilionensis, Bradley, 1872, Am. Jour. Sci., 3rd series, vol. 4, Coal Meas. [Ety. proper name.]
vetusta, Dawson, 1860, Quar. Jour. Geo. Soc., vol. 16, Coal Meas. [Sig. ancient.]

RAPHISTOMA, Hall, 1847, Pal. N. Y., vol. 1. [Ety. *raphe*, seam or suture; *stoma*, mouth.]
apertum, Salter, 1859, Can. Org. Rem., Decade 1, Black Riv. & Trenton Gr. [Sig. open.]
labiatum, Emmons, 1842, (Maclurea labiata) Geo. Rep. N. Y., Calciferous Gr. [Sig. having lips.]
lapicida, Salter, 1859, Can. Org. Rem., Decade 1, Black Riv. & Trenton Gr. [Ety. from its resemblance to *Helix lapicida*.]
lenticulare, Emmons, 1842, (Pleurotomaria lenticularis) Geo. Rep. N. Y., Trenton & Hud. Riv. Gr. [Sig. lens-shaped.]
planistria, Hall, 1847, Pal. N. Y., vol. 1, Chazy Gr. [Sig. having plane striæ.]
planistria *var.* parvum, Hall, 1847, Pal. N. Y., vol. 1, Chazy Gr. [Sig. small.]
stamineum, Hall, 1847, Pal. N. Y., vol. 1, Chazy Gr. [Sig. thready; having reference to the thread like striæ on the surface.]
striatum, Emmons, 1842, (Macluria striata) Geo. Rep. N. Y., Chazy Gr. [Sig. striated.]
subplanum, Shumard, 1863, Trans. St. Louis Acad. Sci., Calciferous Gr. [Sig. somewhat flat.]

SCALITES, Conrad, 1842, Geo. Rep. N. Y. by Emmons. [Ety. *scala*, a stair case.]
angulatus, Conrad, 1842, Geo. Rep. N. Y. by Emmons, Chazy Gr. [Sig. angular.]

Solarium, Lamarck, 1801, Syst. An. sans Vert.
leai, one of Troost's catalogue names.

SOLENISCUS, Meek & Worthen, 1860, Proc. Acad. Nat. Sci. Phil. [Ety. *soleniskos*, a little channel or gutter.]
hallanus, Geinitz, 1866, (Macrocheilus hallanus) Carb. und Dyas in Neb., Coal Meas. [Ety. proper name.]
typicus, Meek & Worthen, 1860, Proc. Acad. Nat. Sci. Phil., Up. Coal Meas. [Ety. the type of the genus.]

STRAPAROLLINA, Billings, 1865, Pal. Foss., vol. 1. [Ety. from the resemblance to shells of the genus *Straparollus*.]
asperostriata, Billings, 1860, (Straparollus asperostriatus) Can. Nat. & Geol., vol. 5, Black Riv. Gr. [Sig. roughly striated.]
circe, Billings, 1860, (Straparollus circe) Can. Nat. & Geol., vol. 5, Black Riv. Gr. [Ety. mythological name.]
eurydice, Billings, 1860, (Straparollus eurydice) Can. Nat. & Geol., vol. 5, Black Riv. Gr. [Ety. mythological name.]
pelagica, Billings, 1865, Pal. Foss., vol. 1, Quebec Gr. [Sig. belonging to the deep sea.]
remota, Billings, 1874, Pal. Foss., vol. 2, Potsdam Gr. [Sig. remote, at a distance.]

STRAPAROLLUS, Montfort, 1810, Conch. Syst., vol. 2. [Ety. *strabos*, turned about.]
angulatus, Emmons, 1856, Am. Geo. Chazy Gr. [Sig. angular.]
asperostriatus, see Straparollina asperostriata.
barrisi, Winchell, 1863, Proc. Acad. Nat. Sci. Phil., Chemung Gr. [Ety. proper name.]
canadensis, Billings, 1861, Can. Jour., Devonian. [Ety. proper name.]
circe, see Straparollina circe.
clymenioides, see Euomphalus clymenioides.
cornudanus, Shumard, 1859, Trans. St. Louis Acad. Sci., Coal Meas. [Ety. proper name.]
daphne, Billings, 1862, Pal. Foss., vol. 1, Guelph Gr. [Ety. mythological name.]
eurydice, see Straparollina eurydice.
hippolyte, Billings, 1862, Pal. Foss., vol. 1, Guelph Gr. [Ety. mythological name.]
labiatus, Emmons, 1856, Am. Geo., Chazy Gr. [Sig. lipped.]
lens, Hall, 1860, (Euomphalus lens) 13th Reg. Rep., Kinderhook Gr. [Sig. lens-shaped.]
macromphalus, Winchell, 1863, Proc. Acad. Nat. Sci. Phil., Chemung Gr. [Sig. having a large umbilicus.]
magnificus, Shumard, 1863, Trans. St. Louis Acad. Sci., Carboniferous. [Sig. magnificent.]
minnesotensis, Owen, 1852, Geo. Sur. Wis., Iowa and Minn., Calcif. Gr. [Ety. proper name.]
mopsus, Hall, 1867, 20th Reg. Rep., Niagara Gr. [Ety. mythological name.]
niagarensis, Hall & Whitfield, 1875, Ohio Pal., vol. 2, Niagara Gr. [Ety. proper name.]

GASTEROPODA.

pernodosus, Meek & Worthen, 1870, Proc. Acad. Nat. Sci. Phil., Coal Meas. [Sig. very nodose.]
planidorsatus, Meek & Worthen, 1860, (Euomphalus planidorsatus) Proc. Acad. Nat. Sci. Phil., Chester Gr. [Sig. flat-backed.]
planispira, Hall, 1858, (Euomphalus planispira) Trans. Alb. Inst., vol. 4, Warsaw Gr. [Sig. having a flat spire.]
primordialis, Winchell, 1864, Am. Jour. Sci. & Arts, 2nd series, vol. 37, Potsdam Gr. [Sig. first in order.]
quadrivolvis, Hall, 1858, (Euomphalus quadrivolvis) Trans. Alb. Inst., vol. 4, Warsaw Gr. [Sig. having four volutions.]
similis, Meek & Worthen, 1861, Proc. Acad. Nat. Sci. Phil., St. Louis Gr. [Sig. similar.]
similis var. planus, Meek & Worthen, 1861, Proc. Acad. Nat. Sci. Phil., St. Louis Gr. [Sig. level, flat.]
spergenensis, Hall, 1858, (Euomphalus spergenensis) Trans. Alb. Inst., vol. 4, Warsaw Gr. [Ety. proper name.]
spergenensis var. planorbiformis, Hall, 1858, (Euomphalus spergenensis var. planorbiformis) Trans. Alb. Inst., vol. 4, Warsaw Gr. [Sig. like a shell of the genus *Planorbis*.
subquadratus, Meek & Worthen, 1870, Proc. Acad. Nat. Sci., Coal Meas. [Sig. somewhat squared.]
subrugosus, Meek & Worthen, 1873, Geo. Sur. Ill., vol. 5, Coal Meas. [Sig. somewhat rugose.] Proposed instead of *Euomphalus rugosus* of Hall, which was preoccupied.
umbilicatus, Meek & Worthen, 1860, (Euomphalus umbilicatus) Proc. Acad. Nat. Sci. Phil., Coal Meas. [Ety. from the large umbilicus.]
valvatiformis, Shumard, 1863, Trans. St. Louis Acad. Sci., Calciferous Gr. [Sig. like the genus *Valvata*.]
whitneyi, Meek, 1864, Pal. California, Coal Meas. [Ety. proper name.]

STREPTAXIS, Gray, 1837, in Mag. Nat. Hist., [Ety. *stroptes*, twisted; *axis*, axis.]
whitfieldi, Meek, 1871, Proc. Acad. Nat. Sci. Phil., Coal Meas. [Ety. proper name.]

STROPHOSTYLUS, Hall, 1859, Pal. N. Y., vol. 3. [Ety. *strophe*, turning round; *stylos*, column.]
andrewsi, Hall, 1859, Pal. N. Y., vol. 3, Oriskany sandstone. [Ety. proper name.]
cancellatus, Meek & Worthen, 1868, Geo. Sur. Ill., vol. 3, Oriskany sandstone. [Sig. latticed.]
cyclostomus, Hall, 1863, Trans. Alb. Inst., vol. 4, Niagara Gr. [Sig. circular-mouthed.]
depressus, Hall, 1859, Pal. N. Y., vol. 3, Low. Held. Gr. [Sig. depressed.]
elegans, Hall, 1859, Pal. N. Y., vol. 3, Low. Held. Gr. [Sig. elegant.]
expansus, Conrad, 1841, (Platyceras expansus) Ann. Rep. N. Y., Oriskany sandstone. [Sig. spread out.]
fitchi, Hall, 1859, Pal. N. Y., vol. 3, Low. Held. Gr. [Ety. proper name.]
globosus, Hall, 1859, Pal. N. Y., vol. 3, Low. Held. Gr. [Sig. globular.]
matheri, Hall, 1859, Pal. N. Y., vol. 3, Oriskany sandstone. [Ety. proper name.]
obliquus, Nicholson, 1874, Rep. Pal. Ont., Devonian. [Sig. oblique.]
obtusus, Hall, 1859, Pal. N. Y., vol. 3, Low. Held. Gr. [Sig. obtuse.]
ovatus, Nicholson, 1874, Rep. Pal. Ont., Devonian. [Sig. egg-shaped.]
rotundatus, Hall, 1859, Pal. N. Y., vol. 3, Low. Held. Gr. [Sig. rounded.]
subglobosus, Nicholson, 1874, Rep. Pal. Ont., Devonian. [Sig. somewhat globose.]
transversus, Hall, 1859, Pal. N. Y., vol. 3, Oriskany sandstone. [Sig. crosswise.]
unicus, Hall, 1862, 15th Reg. Rep., Scoharie grit. [Sig. single, alone.]
varians, Hall, 1876, Illust. Devonian Foss., Up. Held. Gr. [Sig. variable.]

STYLIFER, Broderip, 1829, in Sowerby, Gen. Shells.
primigenia, see Macrocheilus primigenium.

SUBULITES, Conrad, 1847, Pal. N. Y., vol. 1. [Ety. *subulites*, awl-shaped—subulate.]
abbreviatus, Hall, 1850, 3rd Reg. Rep., Trenton Gr. [Sig. shortened.]
brevis, Winchell & Marcy, 1865, Mem. Bost. Soc. Nat. Hist., Niagara Gr. [Sig. short.]
calciferus, Billings, 1859, Can. Nat. & Geo., vol. 4, Calciferous Gr. [Ety. from the Group.]
daphne, Billings, 1865, Pal. Foss., vol. 1, Quebec Gr. [Ety. mythological name.]
elongatus, Emmons, 1842, Geo. Rep. N. Y., Trenton Gr. [Sig. lengthened.]
inflatus, Meek & Worthen, 1869, Proc. Acad. Nat. Sci. Phil., Cin'ti Gr. [Sig. inflated, swollen.]
notatus, Billings, 1866, Catal. Sil. Foss. Antic., Anticosti Gr. [Sig. marked.]
parvulus, Billings, 1862, Pal. Foss., vol. 1, Black Riv. Gr. [Sig. very small.]
psyche, Billings, 1865, Pal. Foss., vol. 1, Quebec Gr. [Ety. mythological name.]
richardsoni, Billings, 1857, Rep. of Progr., Hud. Riv. Gr. [Ety. proper name.]
terebriformis, Hall & Whitfield, 1875, Ohio Pal., vol. 2, Niagara Gr. [Sig. auger-shaped.]
ventricosus, Hall, 1852, Pal. N. Y., vol. 2, Niagara and Guelph Gr. [Sig. bulging out.]

TRACHYDOMIA, Meek & Worthen, 1866, Geo. Sur. Ill., vol. 2. [Ety. *trachys*, rough; *doma*, a house.]

hollidayi, Meek & Worthen, 1860, (Naticopsis hollidayi) Proc. Acad. Nat. Sci. Phil., Low. Coal Meas. [Ety. proper name.]
nodosum, Meek & Worthen, 1860, (Naticopsis nodosa) Proc. Acad. Nat. Sci. Phil., Low. Coal Meas. [Ety. from the nodes on the surface.]

TREMANOTUS, Hall, 1868, 20th Reg. Rep. [Ety. *trema*, a hole; *notus*, the back.]
alpheus, Hall, 1864, 18th Reg. Rep., Niagara Gr. [Ety. mythological name.]
trigonostoma, Hall & Whitfield, 1875, Ohio Pal., vol. 2, Niagara Gr. [Sig. triangular-mouthed.]

TROCHITA, Schumacher, 1817, Essai N. Syst. [Ety. *trochus*, a wheel.]
antiqua, see Xenophora antiqua.
carbonaria, Meek, 1866, Proc. Acad. Nat. Sci., Coal Meas. [Sig. pertaining to coal.]

TROCHONEMA, Salter, 1859, Can. Org. Rem., Decade 1. [Ety. *trochos*, a wheel; *nenna*, a thread.]
emaceratum, Hall, 1872, 24th Reg. Rep., Up. Held. Gr. [Sig. thin, lean.]
fatua, Hall, 1867, 20th Reg. Rep., Niagara Gr. [Ety. mythological name.]
pauper, Hall, 1867, 20th Reg. Rep., Niagara Gr. [Sig. paltry, poor.]
pauper *var.* ohioense, Hall & Whitfield, 1875, Ohio Pal., vol. 2, Niagara Gr. [Ety. proper name.]
rectilatera, Hall, 1872, 24th Reg. Rep., Up. Held. Gr. [Sig. straight-sided.]
tricarinatum, Billings, 1859, Can. Nat. & Geo., vol. 4, Calciferous Gr. [Sig. three-keeled.]
tricarinata, Meek, 1871, Proc. Acad. Nat. Sci. Phil., Corniferous Gr. The name was preoccupied.
umbilicatum, Hall, 1847. (Pleurotomaria umbilicata) Pal. N. Y., vol. 1, Chazy to Hud. Riv. Gr. [Sig. navel-shaped.]
yandellanum, Hall, 1872, 24th Reg. Rep., Up. Held. Gr. [Ety. proper name.]

Trochus, Adanson, 1757, Voy. Senegal. [Ety. *trochus*, a hoop.]
missouriensis, see Pleurotomaria missouriensis.

TURBO, Klein, 1753, Tent. Meth. Ostr. [Ety. *turbo*, a top.]
bicarinatus, Troost, 1840. Not defined.
dilucula, see Holopea dilucula.
guadalupensis, Shumard, 1859, Trans. St. Louis Acad. Sci., Permian Gr. [Ety. proper name.]
lineatus, see Pleurotomaria lineata.
obesus, Shumard, 1858, Trans. St. Louis Acad. Sci., Up. Coal Meas. [Sig. plump in form.]
(?) *obscura*, see Holopea obscura.
tabulata, see Pleurotomaria tabulata.
tennesseensis, see Cyclonema tennesseense.
texanus, Shumard, 1859, Trans. St. Louis Acad. Sci., Coal Meas. [Ety. proper name.]

Turbonilla, Leach, 1826, Risso Eur. Merid. 4. [Ety. diminutive of *Turbo*, a genus of shells.]
swallovana, see Aclis swallovana.

TURRITELLA, Lamarck, 1801, Syst. An. sans Vert. [Ety. *turritella*, a little tower.]
stevenana, Meek & Worthen, 1866, Geo. Sur. Ill., vol. 2, Up. Coal Meas. [Ety. proper name.]

XENOPHORA, Fischer, 1806, Museum Demidovianum. [Ety. *xenos*, a log or body; *phoros*, bearing.]
antiqua, Meek, 1871, (Trochita antiqua) Proc. Acad. Nat. Sci. Phil., Corniferous Gr. [Sig. ancient.]

ZONITES, Montfort, 1810, Conch. Syst. [Ety. *zone*, a belt.]
priscus, Carpenter, 1867, Quar. Jour. Geo. Soc., vol. 23, Coal Meas. [Sig. ancient.]

CLASS CEPHALOPODA.

FAMILY ASCOCERATIDÆ.—Ascoceras, Conoceras, Glossoceras.
FAMILY CERIOLIDÆ.—Beatricea.
FAMILY CYRTOCERATIDÆ.—Cyrtoceras, Cyrtocerina, Oncoceras.
FAMILY GOMPHOCERATIDÆ.—Gomphoceras.
FAMILY GONIATITIDÆ.—Goniatites.
FAMILY GYROCERATIDÆ.—Gyroceras.
FAMILY NAUTILIDÆ.—Discites, Lituites, (?) Nautilus, Pteronautilus, Solenocheilus, Temnocheilus, Trematodiscus, Trocholites. (?)
FAMILY ORTHOCERATIDÆ.—Actinoceras, Cameroceras, Colpoceras, Diploceras, Endoceras, Gonioceras, Huronia, Ormoceras, Orthoceras, Piloceras.
FAMILY PHRAGMOCERATIDÆ.—Phragmoceras, Streptoceras.
FAMILY TROCHOCERATIDÆ.—Trochoceras.
INCERTÆ SEDIS.—Discosorus, Særichnites.

ACTINOCERAS, Bronn, 1837, Lethaea Geognostica. [Ety. *aktin*, a ray; *keras*, a horn.]
  bigsbyi, Stokes, 1840, Trans. Geo. Soc., 2nd series, vol. 5, Chazy Gr. [Ety. proper name.]
  inops, Dawson, 1868, Acad. Geol., Carb. [Ety. meagre.]
  lyoni, Stokes, 1840, Trans. Geol. Soc., vol. 5, Black Riv. Gr. [Ety. proper name.]
  richardsoni, Stokes, 1840, Trans. Geol. Soc., 2nd series, vol. 5, Silurian. [Ety. proper name.]
  simmsi, Stokes, 1840, Trans. Geo. Soc., 2nd series, vol. 5, Sil. [Ety. proper name.]
ASCOCERAS, Barrande, 1848, Haidinger's Berichte. [Ety. *askos*, a leather bottle; *keras*, horn.]
  anticostiense, Billings, 1866, Catal. Sil. Foss. Antic., Anticosti Gr. [Ety. proper name.]
  canadense, Billings, 1857, Rep. of Progr., Hud. Riv. Gr. [Ety. proper name.]
  newberryi, Billings, 1862, Pal. Foss., vol. 1, Hud. Riv. & Anticosti Gr. [Ety. proper name.]
BEATRICEA, Billings, 1857, Rep. of Progr. [Ety. proper name.] This genus is supposed, by Hyatt, (Am. Jour. Sci. & Arts, 1865) to belong to the class Cephalopoda, and he proposed a new order for the genus, to wit: *Ceriolites*, from *kerion*, a honey comb; *lithos*, a stone, and a new family *Ceriolidæ*.
  nodulosa, Billings, 1857, Rep. of Progr., Hud. Riv. & Mid. Sil. [Sig. having small knots.]

  undulata, Billings, 1857, Rep. of Progr., Mid. Sil. [Sig. wavy.]
CAMEROCERAS, Conrad, 1842, Jour. Acad. Nat. Sci. Phil., vol. 8. [Ety. *kamara*, a chamber; *keras*, horn.]
  trentonense, Conrad, 1842, Jour. Acad. Nat. Sci. Phil., vol. 8, Trenton Gr. [Ety. proper name.]
COLPOCERAS, Hall, 1850, 3rd Reg. Rep. N. Y. [Ety. *kolpos*, a furrow; *keras*, horn.]
  virgatum, Hall, 1850, 3rd Reg. Rep. N. Y., Birdseye and Black Riv. Gr. [Sig. wand-like.]
Clymenia, Munster, 1832. [Ety. mythological name.]
complanata, see Goniatites complanatus.
erato, see Goniatites erato.
Conilites, Schlotheim, 1820, Petrefactenkunde, etc. [Ety. *konos*, a cone; *lithos*, stone.]
capricornulus, Troost, 1840, 5th Geo. Rep. Tenn. Not satisfactorily defined.
CONOCERAS, Bronn, 1835, Leth. Geogn. [Ety. *konos*, a cone; *keras*, horn.]
  angulosum, Bronn, 1834, Leth. Geogn., Black Riv. Gr. [Sig. full of corners.]
Conotubularia, Troost, syn. for Orthoceras.
brongniarti, see Orthoceras brongniarti.
cuvieri, see Orthoceras cuvieri.
defrancii, see Orthoceras defrancii.
goldfussi, see Orthoceras goldfussi.
Conulites, Cozzens, 1848. Not satisfactorily defined.
  angulosum, Cozzens, 1848. Not satisfactorily defined. It may be a plant.

Cryptoceras, D'Orbigny, 1850. [Ety. *kryptos*, concealed; *keras*, horn.] This name was preoccupied by Latreille for a genus of insects, and had been previously used by Barrande for a genus of Cephalopods.

capax, see Solenocheilus capax.

CYRTOCERAS, Goldfuss, 1832, in De la Beche's Handbuch der Geognosie bearbeitet von v. Deschen. [Ety. *kurtos*, curved; *keras*, horn.]

absens, Hall, 1876, Illust. Devonian Foss., Schoharie grit. [Sig. absent, distant.]

alethes, Billings, 1865, Pal. Foss., Quebec Gr. [Ety. proper name.]

amplicorne, Hall, 1867, 20th Reg. Rep., Niagara Gr. [Sig. a large horn.]

annulatum, Hall, 1847, Pal. N. Y., vol. 1, Black Riv. & Trenton Gr. [Sig. ringed.] This name was preoccupied by Goldfuss in 1832.

arcticameratum, Hall, 1852, Pal. N.Y., vol. 2, Guelph Gr. [Sig. close-chambered.]

arcuatum, Hall, 1847, Pal. N. Y., vol. 1, Trenton Gr. [Sig. bent; in allusion to the arched chambers.] The name was preoccupied by Steininger in 1830, see D'Archiac & Verneuil, Memoir on Pal. Foss., 1842.

aristides, Billings, 1865, Pal. Foss., Quebec Gr. [Ety. mythological name.]

beta, Hall, 1862, (Gomphoceras beta) 15th Reg. Rep., Schoharie grit. [Sig. a Greek letter.]

billingsi, Salter, 1859, Can. Org. Rem., Decade 1, Chazy or Black Riv. Gr. [Ety. proper name.]

bondi, Safford, 1869, Geo. of Tenn., Nashville Gr. [Ety. proper name.]

brevicorne, Hall, 1867, 20th Reg. Rep., Niagara Gr. [Sig. short-horned.]

camurum, Hall, 1847, Pal. N. Y., vol. 1, Trenton Gr. [Sig. crooked.]

cancellatum, Hall, 1852, Pal. N. Y., vol. 2, Niagara Gr. [Sig. cancellated.] The name was preoccupied by Roemer in 1844.

carrollense, Worthen, 1875, Geo. Sur. Ill. vol. 6, Galena Gr. [Ety. proper name.]

clavatum, Hall, 1876, Illust. Devonian Foss., Schoharie grit. [Sig. club-shaped.]

clitus, Billings, 1866, Catal. Sil. Foss. Antic., Niagara Gr. [Ety. proper name.]

conicum, Owen, 1840, Rep. on Min. Lands, Up. Magnesian Gr. [Sig. conical, tapering to a point.]

conradi, Hall, 1860, (Gomphoceras conradi) 13th Reg. Rep., Ham. Gr. [Ety. proper name.]

constrictostriatum, Hall, 1847, Pal. N.Y., vol. 1, Trenton Gr. [Sig. constricted and striated.]

corniculum, Hall, 1862, Geo. Rep. Wis., Trenton Gr. [Sig. a little horn.] The name was preoccupied by Barrande in 1848, and again by Eichwald in 1860.

corydon, Billings, 1866, Catal. Sil. Foss. Antic., Niagara Gr. [Ety. proper name.]

curtum, Meek & Worthen, 1860, Proc. Acad. Nat. Sci. Phil., Up. Coal Meas. [Sig. short.] Was this name preoccupied by Eichwald?

dardanus, Hall, 1861, Rep. of Progr. Geo. Sur. of Wis., Niagara Gr. [Ety. mythological name.]

dictys, Billings, 1865, Pal. Foss., Quebec Gr. [Ety. mythological name.]

dictyum, White, 1870, Proc. Acad. Nat. Sci., Devonian. [Sig. a net.]

dilatatum, Meek & Worthen, 1860, Proc. Acad. Nat. Sci. Phil., Up. Coal Meas. [Sig. widened, spread out.]

dorsatum, Swallow, 1858, Trans. St. Louis Acad. Sci., Permian Gr. [Sig. high-backed.]

eugenium, Hall, 1862, 15th Reg. Rep., Schoharie grit. [Ety. proper name.]

eugium, Hall, 1861, Rep. of Progr. Wis., Chazy & Black Riv. Gr. [Sig. fertile.]

exiguum, Billings, 1860, Can. Nat. & Geo., vol. 5, Trenton Gr. [Sig. little, small.]

falx, Billings, 1857, Rep. of Progr., Black Riv. & Trenton Gr. [Sig. a sickle.]

filosum, Emmons, 1842, Nat. Hist. N. Y., vol. 4, Trenton Gr. [Sig. covered with thread-like markings.]

fosteri, Hall, 1861, Rep. of Progr. Geo. Sur. Wis., Niagara Gr. [Ety. proper name.]

fragile, Billings, 1866, Catal. Sil. Foss. Antic., Anticosti Gr. [Sig. frail, easily broken.]

gibbosum, Hall, 1876, Illust. Devonian Foss., Ham. Gr. [Sig. gibbous.]

giganteum, McChesney, Jan. 1860, New Pal. Foss., Niagara Gr. In 1861 McChesney referred this species to the genus *Lituites*, and proposed for it the name *Lituites cancellatus*. Prof. Hall, in the meantime, described it as *Lituites occidentalis*. It is now referred to the genus *Nautilus*, and as both the earlier names were preoccupied, McChesney's name *cancellatus* has precedence.

hercules, Winchell & Marcy, 1865, (Lituites hercules) Mem. Bost. Soc. Nat. Hist., Niagara Gr. [Ety. mythological name.]

hertzeri, Hall & Whitfield, 1875, Ohio Pal., vol. 2, Niagara Gr. [Ety. proper name.]

huronense, Billings, 1865, Pal. Foss., Black Riv. or Trenton Gr. [Ety. proper name.]

isidorus, Billings, 1865, Pal. Foss., Black Riv. or Trenton Gr. [Ety. proper name.]

janus, see Streptoceras jadus.
jason, Hall, 1862, 15th Reg. Rep., Schoharie grit. [Ety. mythological name.]
juvenalis, Billings, 1865, Pal. Foss., Trenton Gr. [Ety. proper name.]
lamellosum, Hall, 1847, Pal. N. Y., vol. 1, Trenton Gr. [Sig. having many thin plates.] The name was preoccupied by D'Archiac & Verneuil in 1842.
laterale, Hall, 1867, 20th Reg. Rep., Niagara Gr. [Sig. lateral.]
ligarius, Billings, 1875, Pal. Foss., Hud. Riv. Gr. [Ety. proper name.]
liratum, Hall, 1862, 15th Reg. Rep., Ham. Gr. [Sig. furrowed.]
loculosum, Hall, 1861, Rep. of Prog. Wis., Trenton Gr. [Sig. partitioned.]
lucillus, Hall, 1867, 20th Reg. Rep., Niagara Gr. [Ety. proper name.]
lysander, Billings, 1862, Pal. Foss., Hud. Riv. Gr. [Ety. proper name.]
macrostomum, Hall, 1847, Pal. N. Y., vol. 1, Black Riv. and Trenton Gr. [Sig. long-mouthed.]
magister, S. A. Miller, 1875, Cin. Quar. Jour. of Sci., vol. 2, Cin'ti Gr. [Sig. the chief.]
marginale, Conrad, 1843, Proc. Acad. Nat. Sci. [Sig. bordered.] The name was preoccupied by Phillips in 1841.
massiense, Safford, 1869, Geo. of Tenn., Nashville Gr. [Ety. proper name.]
matheri, see Gyroceras matheri.
maccoyi, Billings, 1859, Can. Nat. & Geo., vol. 4, Chazy Gr. [Ety. proper name.]
maximum, see Nautilus maximus.
mercurius, see Cyrtocerina mercurius.
metellus, Billings, 1865, Pal. Foss., Quebec Gr. [Ety. proper name.]
metula, Hall, 1862, 15th Reg. Rep., Up. Held. Gr. [Sig. a little obelisk.]
missisquoi, Billings, 1865, Pal. Foss., (Orthoceras missisquoi) Quebec Gr. [Ety. proper name.]
morsum, Hall, 1862, 15th Reg. Rep., Up. Held. Gr. [Sig. bitten off.]
multicameratum, Hall, 1847, Pal. N. Y., vol. 1, Black Riv. & Trenton Gr. [Sig. many-chambered.]
myrice, Hall & Whitfield, 1875, Ohio Pal., vol. 2, Niagara Gr. [Ety. mythological name.]
neleus, Hall, 1861, Rep. of Progr. Wis., Chazy & Black Riv. Gr. [Ety. mythological name.]
obscurum, S. A. Miller, changed to *magister* because *obscurum* was preoccupied.
ohioense, Meek, 1871, Proc. Acad. Nat. Sci. Phil., Corniferous Gr. [Ety. proper name.]
orcas, Hall, 1861, Rep. of Progr. Geo. Sur. Wis., Niagara Gr. Subsequently referred by Hall to the genus Oncoceras, see 20th Reg. Rep.
orestes, Billings, 1865, Pal. Foss., Niagara Gr. [Ety. mythological name.]

orion, Hall, 1876, Illust. Devonian Foss., Schoharie grit. [Ety. mythological name.] This name was preoccupied by Barrande.
orodes, Billings, 1862, Pal. Foss., Guelph Gr. [Ety. mythological name.]
postumius, Billings, 1865, Pal. Foss., Hud. Riv. Gr. [Ety. proper name.]
pusillum, Hall, 1867, 20th Reg. Rep., Niagara Gr. [Sig. very small.]
regulare, Billings, 1857, Rep. of Progr. Black Riv. and Trenton Gr. [Sig. regular.]
rigidum, Hall, 1867, 20th Reg. Rep., Niagara Gr. [Sig. rigid.]
rockfordense, Winchell, 1865, Proc. Acad. Nat. Sci., Kinderhook (?) Gr. [Ety. proper name.]
sacculus, Meek & Worthen, 1866, (Gomphoceras sacculum) Proc. Acad. Nat. Sci. Phil., Ham. Gr. [Sig. a little sack.]
septoris, Hall, see Gomphoceras septore.
simplex, Billings, 1857, Rep. of Progr., Black Riv. & Trenton Gr. [Sig. simple.]
sinuatum, Billings, 1857, Rep. of Progr., Black Riv. Gr. [Sig. marked with depressions.]
spinosum, see Gyroceras spinosum.
stonenese, Safford, 1869, Geo. of Tenn., Trenton Gr. [Ety. proper name.]
subrectum, Hall, 1859, Pal. N. Y., vol. 3, Low. Held. Gr. [Sig. somewhat straight.]
subturbinatum, Billings, 1857, Rep. of Progr., Chazy & Black Riv. Gr. [Sig. somewhat top-shaped.]
surgens, Barrande, 1869, Syst. Sil. de Boh., 4me serie, Quebec Gr. [Sig. rising.]
syphax, Billings, 1865, Pal. Foss., Quebec Gr. [Ety. mythological name.]
tesselatum, de Koninck. Not American.
transversum, see Gyroceras transversum.
trentonense, Emmons, 1842, (Orthoceras trentonensis) Geo. Rep. N.Y., Trenton Gr. [Ety. proper name.]
trivolvis, see Gyroceras trivolve.
typicum, see Cyrtocerina typica.
undulatum, Hall, 1876, Illust. Devonian Foss., Schoharie grit. [Sig. wavy.]
undulatum, Vanuxem, see Gyroceras undulatum.
unicorne, Winchell, 1863, Proc. Acad. Nat. Sci., Chemung Gr. [Sig. one-horned.]
vallandighami, S. A. Miller, 1874, Cin. Quar. Jour. Sci., Cin'ti Gr. [Ety. proper name.]
ventricosum, S. A. Miller, 1875, Cin. Quar. Jour. Sci., vol. 2, Cin'ti Gr. [Sig. bulging out; rapidly enlarging.]
whitneyi, Hall, 1861, Rep. of Progr. Wis., Hud. Riv. Gr. [Ety. proper name.]
CYRTOCERINA, Billings, 1865, Pal. Foss. [Ety. from the termination *inus*, signifying resemblace to *Cyrtoceras*.]
mercurius, Billings, 1865, Pal. Foss., Quebec Gr. [Ety. mythological name.]

typica, Billings, 1865, Pal. Foss., Black Riv. Gr. [Sig. type of the genus.]
DIPLOCERAS, Conrad, 1842, Jour. Acad. Nat. Sci., vol. 8. [Ety. *diploos*, double; *keras*, horn.]
vanuxemi, Conrad, 1842, Jour. Acad. Nat. Sci., vol. 8, Trenton Gr. [Ety. proper name.]
DISCITES, DeHaan, 1825, Mongr. Ammon., etc. [Ety. *diskos*, a quoit.]
disciformis, Meek & Worthen, 1865, Proc. Acad. Nat. Sci. Phil., Keokuk Gr. [Sig. in the shape of a quoit.]
highlandensis, Worthen, 1875, Geo. Sur. Ill., vol. 6, Coal Meas. [Ety. proper name.]
ornatus, syn. for Nautilus marcellensis.
tuberculatus, Owen, 1852, Geo. Sur. Wis., Iowa and Minn., Low. Carb. [Sig. covered with tubercles.]
DISCOSORUS, Hall, 1852, Pal. N. Y., vol. 2. [Ety. *diskos*, a quoit; *soros*, a heap or pile.]
conoideus, Hall, 1852, Pal. N. Y., vol. 2. Clinton & Niagara Gr. [Sig. conical.]
ENDOCERAS, Hall, 1847, Pal. N. Y., vol. 1. [Ety. *endos*, within; *keras*, horn.] This genus seems to be founded upon the same fossils that Conrad previously founded his genus Diploceras upon.
angusticameratum, Hall, 1847, Pal. N.Y., vol. 1, Trenton Gr. [Sig. narrow-chambered.]
annulatum, Hall, 1847, Pal. N. Y., vol. 1, Trenton Gr. [Sig. ringed.]
approximatum, Hall, 1847, Pal. N. Y., vol. 1, Trenton Gr. [Sig. approximating.]
arctiventrum, Hall, 1847, Pal. N. Y., vol. 1, Trenton Gr. [Ety. *arctus*, close; *venter*, a cavity.]
atlanticum, Barrande, 1869, Syst. Sil. de Boh., 2d ser., 4me, Quebec Gr. [Sig. Atlantic.]
distans, Hall, 1847, Pal. N. Y., vol. 1, Trenton Gr. [Sig. distant.]
duplicatum, Hall, 1847, Pal. N. Y., vol. 1, Trenton Gr. [Sig. doubled.]
gemelliparum, Hall, 1847, Pal. N. Y., vol. 1, Black Riv. Gr. [Sig. twins, in allusion to the two embryo-tubes enclosed within the siphuncle.]
insulare, Barrande, 1869, Syst. Sil. de Boh., 2d ser., 4me, Quebec Gr. [Sig. belonging to an island.]
lativentrum, Hall, 1850, 3rd Reg. Rep. N. Y., Trenton Gr. [Ety. *latus*, broad; *venter*, cavity.]
longissimum, Hall, 1847, Pal. N. Y., vol. 1, Black Riv. & Trenton Gr. [Sig. of unusual length.]
magniventrum, Hall, 1847, Pal. N. Y., vol. 1, Trenton Gr. [Ety. *magnus*, large; *venter*, cavity.]
marcoui, Barrande, 1869, Syst. Sil. de Boh., 2d ser., 4me, Quebec Gr. [Ety. proper name.]
multitubulatum, Hall, 1847, Pal. N. Y., vol. 1, Black Riv. & Trenton Gr. [Sig. having many pipes.]
proteiforme, Hall, 1847, Pal. N. Y., vol. 1, Hud. Riv. and Trenton Gr. [Sig. having many shapes.]
proteiforme *var*. elongatum, Hall, 1847, Pal. N. Y., vol. 1, Trenton Gr. [Sig. lengthened.]
proteiforme *var*. lineolatum, Hall, 1847, Pal. N. Y., vol. 1, Trenton Gr. [Sig. marked with fine lines.]
proteiforme *var*. strangulatum, Hall, 1847, Pal. N. Y., vol. 1, Trenton Gr. [Sig. constricted.]
proteiforme *var*. tenuistriatum, Hall, 1847, Pal. N. Y., vol. 1, Trenton Gr. [Sig. fine-lined.]
proteiforme *var*. tenuitextum, Hall, 1847, Pal. N. Y., vol. 1, Trenton Gr. [Sig. finely woven.]
rapax, Billings, 1860, (Orthoceras rapax) Can. Nat. & Geol., vol. 5, Black Riv. Gr. [Sig. ravenous.]
rottermundi, Barrande, 1866, (Orthoceras rottermundi) Syst. Sil. de Boh., 2d ser., 2me, Trenton Gr. [Ety. proper name.]
subcentrale, Hall, 1847, Pal. N. Y., vol. 1, Black Riv. Gr. [Sig. nearly central.]
Endolobus, Meek & Worthen, 1865, Proc. Acad. Nat. Sci. Phil. [Ety. *endos*, within; *lobos*, a lobe.] Prof. Meek said later that this genus is not distinct from *Temnocheilus*, and if distinct it would probably be a synonym for Montfort's genus *Bisiphites*.
perumplus, see Temnocheilus peramplum.
spectabilis, see Temnocheilus spectabile.
GLOSSOCERAS, Barrande, 1865, Cephalopods of Bohemia, vol. 2. [Ety. *glosse*, the tongue; *keras*, a horn.]
desideratum, Billings, 1866, Catal. Sil. Foss. Antic., Clinton Gr. [Sig. rare.]
GOMPHOCERAS, Sowerby, 1839, Murch. Sil. Syst. [Ety. *gomphos*, a club; *keras*, a horn.]
beta, see Cyrtoceras beta.
conradi, see Cyrtoceras conradi.
eos, Hall & Whitfield, 1875, Ohio Pal., vol. 2, Cin'ti Gr. [Sig. the dawn.]
eximium, Hall, 1861, 14th Reg. Rep., Up. Held. Gr. [Sig. choice, unparalleled.]
fischeri, Hall, 1860, 13th Reg. Rep., Ham. Gr. [Ety. proper name.]
marcyæ, Winchell & Marcy, 1865, Mem. Bost. Soc. Nat. Hist., Niagara Gr. Syn. for *G. scrinium*.
obesum, Billings, 1857, Rep. of Progr., Utica Gr. [Sig. plump in form.]
omicron, Winchell, 1866, Rep. Low. Peninsula Mich., Ham. Gr. [Ety. a letter of the Greek alphabet.]
oviforme, Hall, 1860, 13th Reg. Rep. N. Y., Ham. Gr. [Sig. egg-shaped.]
sacculum, see Cyrtoceras sacculum.
scrinium, Hall, 1864, 20th, Reg. Rep. N. Y., Niagara Gr. [Sig. a casket.]

CEPHALOPODA.

septore, Hall, 1867, 20th Reg. Rep., Niagara Gr. [Ety. *septem*, seven; *os-oris*, a mouth.]
subgracile, Billings, 1857, Rep. of Progr., Up. Sil. [Sig. somewhat slender.]
turbiniforme, Meek & Worthen, 1860, Proc. Acad. Nat. Sci. Phil., Ham. Gr. [Sig. formed like a top.]

GONIATITES, DeHaan, 1825, Monogr. Ammonites et Goniatites. [Ety. *gonia*, an angle: *lithos*, stone.] This name, it seems, should be spelled Gonialites.

allei, Winchell, 1862, Am. Jour. Sci., 2nd series, vol. 33, Marshall Gr. [Ety. proper name.]
bicostatus, Hall, 1843, Geo. Rep. 4th Dist. N. Y., Portage Gr. [Sig. double-ribbed.]
chemungensis, Vanuxem, 1843, Geo. Rep. 3rd Dist. N. Y., Chemung Gr. [Ety. proper name.]
chemungensis *var.* æquicostatus, Hall, 1875, 27th Reg. Rep. N. Y., Chemung Gr. [Sig. equal-ribbed.]
choctawensis, Shumard, 1863, Trans. St. Louis Acad. Sci., Coal Meas. [Ety. proper name.]
compactus, Meek & Worthen, 1875, Proc. Acad. Nat. Sci. Phil., Coal Meas. [Sig. compact.]
complanatus, Hall, 1843, (Clymenia (?) complanatus) Geo. Rep. 4th Dist. N. Y., Chemung Gr. [Sig. smoothed.]
complanatus *var.* perlatus, Hall, 1875, 27th Reg. Rep. N. Y., Chemung Gr. [Sig. very wide.]
discoideus, Hall, 1860, 13th Reg. Rep. N. Y., Ham. Gr. [Sig. disc-like.]
discoideus *var.* ohioensis, Hall, 1874, 27th Reg. Rep. N. Y., Up. Held. Gr. [Ety. proper name.]
entogonus, Gabb, 1861, Proc. Acad. Nat. Sci., Carboniferous. [Sig. angular within.]
erato, Hall, 1862, (Clymenia erato) 15th Reg. Rep., Ham. Gr. [Ety. mythological name.]
expansus, Vanuxem, 1842, Geo. Rep. 3rd Dist. N. Y., Marcellus shale. [Sig. expanded.]
globulosus, Meek & Worthen, 1860, Proc. Acad. Nat. Sci. Phil., Up. Coal Meas. [Sig. like a small ball.]
hathawayanus, McChesney, 1860, Desc. New Pal. Foss., Coal Meas. [Ety. proper name.]
holmesi, Swallow, 1860, Trans. St. Louis Acad. Sci., Chemung Gr. [Ety. proper name.]
houghtoni, Winchell, 1862, Am. Jour. Sci., 2d ser., vol. 33, Marshall Gr. [Ety. proper name.]
hyas, Hall, Dec., 1860, 13th Reg. Rep., Low. Carb. [Ety. mythological name.]
iowensis, Meek & Worthen, 1860, Proc. Acad. Nat. Sci. Phil., Coal Meas. [Ety. proper name.]

ixion, Hall, 1860, 13th Reg. Rep., Ham. Gr. This species is founded on the form which has been identified with the European species *G. rotatorius*. [Ety. mythological name.]
lyoni, Meek & Worthen, 1860, Proc. Acad. Nat. Sci. Phil., Kinderhook (?) Gr. [Ety. proper name.]
marcellensis, see Nautilus marcellensis.
marshallensis, Winchell, 1862, Am. Jour. Sci., 2d ser., vol. 33, Marshall Gr. [Ety. proper name.]
minimus, Shumard, 1858, Trans. St. Louis Acad. Sci., Coal Meas. [Sig. very small.]
mithrax, Hall, 1860, 13th Reg. Rep. N. Y., Up. Held. Gr. [Ety. *mithrax*, a precious stone.]
morganensis, Swallow, 1860, Trans. St. Louis Acad. Sci., Chouteau or Chemung Gr. [Ety. proper name.]
nolinensis, Cox, 1857, Geo. Sur. Ky., vol. 3, Coal Meas. [Ety. proper name.]
nundaia, Hall, 1875, 27th Reg. Rep. N. Y., Portage Gr. [Ety. proper name.]
orbicella, Hall, 1860, 13th Reg. Rep. N. Y., Ham. Gr. [Sig. a little circle.]
osagensis, Swallow, 1860, Trans. St. Louis Acad. Sci., Chemung Gr. [Ety. proper name.]
oweni, Hall, 1860, 13th Reg. Rep. N. Y., Ham. Gr. [Ety. proper name.]
oweni *var.* parallelus, Hall, 1860, 13th Reg. Rep., Ham. Gr. [Sig. parallel.]
parvus, Shumard, 1858, Trans. St. Louis Acad. Sci., Coal Meas. [Sig. small.]
patersoni, Hall, 1860, 13th Reg. Rep. N. Y., Portage Gr. [Ety. proper name.]
peracutus, Hall, 1876, Illust. Devonian Foss., Up. Held. Gr. [Sig. very acute.]
planorbiformis, Shumard, 1855, Geo. Sur. Mo., Coal Meas. [Sig. like a shell of the genus *Planorbis*.]
politus, Shumard, 1858, Trans. St. Louis Acad. Sci., Coal Meas. [Sig. smoothed.]
propinquus, Winchell, 1862, Am. Jour. Sci. & Arts, 2nd series, vol. 33, Marshall Gr. [Sig. related.]
punctatus, Conrad, 1838, Ann. Rep. N. Y., Ham. Gr. [Sig. dotted.]
pygmæus, Winchell, 1862, Am. Jour. Sci. and Arts, 2nd series, vol. 33, Marshall Gr. [Ety. mythological name.]
rotatorius, DeKoninck, 1843, Desc. des Anim. Foss. du Terr. Carb. See *G. ixion*.
shumardanus, Winchell, 1865, Am. Jour. Sci. & Arts, 2nd series, vol. 33, Marshall Gr. [Ety. proper name.]
simulator, Hall, 1875, 27th Reg. Rep., Chemung Gr. [Sig. a dissembler.]
sinuosus, Hall, 1843, Geo. Rep. 4th Dist. N. Y., Portage Gr. [Sig. full of furrows.]
sulciferus, Winchell. Not defined.
texanus, Shumard, 1863, Trans. St. Louis Acad. Sci., Coal Meas. [Ety. proper name.]

CEPHALOPODA.

uniangularis, Conrad, 1842, Jour. Acad. Nat. Sci., vol. 8, Ham. Gr. [Sig. one-angled.]
unilobatus, Hall, 1875, 27th Reg. Rep., N.Y., Ham. Gr. [Sig. one-lobed.]
whitii, Winchell, 1862, Proc. Acad. Nat. Sci., Portage Gr. [Ety. proper name.]
GONIOCERAS, Hall, 1847, Pal. N. Y., vol. 1. [Ety. *gonia*, an angle; *keras*, a horn.]
anceps, Hall, 1847, Pal. N. Y., vol. 1, Black Riv. Gr. [Sig. doubtful.]
occidentale, Hall, 1861, Rep. of Progr. Wis., Trenton Gr. [Sig. western.]
GYROCERAS, DeKoninck, 1841, Desc. An. Foss. Belg. [Ety. *gyros*, circle; *keras*, a horn.] Not Gyroceratites of Meyer, 1829.
americanum, Billings, 1857, Rep. of Progr. Can. Geo. Sur., Up. Sil. [Ety. proper name.]
bannisteri, Winchell & Marcy, 1865, Mem. Bost. Soc. Nat. Hist. See Trochoceras bannisteri.
burlingtonense, Owen, 1852, Geo. Sur. Wis., Iowa & Minn., Low. Carb. [Ety. proper name.]
constrictum, Meek & Worthen, 1868, Geo. Sur. Ill., vol. 3, Ham. Gr. [Sig. constricted.]
cornutum, Owen, 1840, Rep. on Min. Lands, Devonian. [Sig. horned.]
cyclops, Hall, 1862, 15th Reg. Rep., Up. Held. Gr. [Ety. mythological name.]
eryx, Hall, 1862, 15th Reg. Rep., Up. Held. Gr. [Ety. mythological name.]
erpansum, Saeman, Dunker & Von Meyer, 1853, Palæontographica, vol. 4, Marcellus shale. See Nautilus buccinum.
gracile, Hall, 1860, 13th Reg. Rep. N. Y., Ham. Gr. [Sig. slender.]
hartti, Dawson, 1868, Acad. Geol., Carboniferous. [Ety. proper name.]
inelegans, Meek, 1871, Proc. Acad. Nat. Sci. Phil., Corniferous Gr. [Sig. not elegant.]
jason, Hall, 1876, Illust. Devonian Foss., Schoharie grit. [Ety. proper name.]
liratum, Hall, 1860, 13th Reg. Rep., Marcellus shale. [Sig. furrowed.]
logani, Meek, 1868, Trans. Chi. Acad. Sci., Devonian. [Ety. proper name.]
magnificum, see Lituites magnificus.
matheri, Conrad, 1840, Ann. Rep. N. Y., (Cyrtoceras matheri) Low. Held. Gr. [Ety. proper name.]
nais, see Porcellia nais.
nereus, Hall, 1862, 15th Reg. Rep., Corniferous Gr. [Ety. mythological name.]
numa, Billings, 1875, Can. Nat. & Geol. Corniferous Gr. [Ety. mythological name.]
ohioense, Meek, 1871, Proc. Acad. Nat. Sci. Phil., Corniferous Gr. [Ety. proper name.]
paucinodum, Hall, 1876, Illust. Devonian Foss., Up. Held. Gr. [Sig. having few knots.]
(?) rockfordense, Meek & Worthen, 1866, (Nautilus (Cryptoceras) rockfordensis) Proc. Acad. Nat. Sci. Phil., Kinderhook Gr. [Ety. proper name.]
spinosum, Conrad, 1840, (Phragmoceras spinosum) Ann. Rep. N. Y., Schoharie grit. [Sig. spiny.]
subliratum, Hall, 1876, Illust. Devonian Foss., Ham Gr. [Sig. somewhat like *G. liratum*.]
transversum, Hall, 1860, (Cyrtoceras transversum) 13th Reg. Rep., Ham. Gr. [Sig. crosswise.]
trivolve, Conrad, 1840, (Cyrtoceras trivolvis) Ann. Rep. N. Y., Low. Held. Gr. [Sig. three-whorled.]
undulatum, Vanuxem, 1843, (Cyrtoceras undulatum) Geo. Rep. N. Y., Low. Held. Gr. [Sig. wavy.]
vagans, Billings, 1857, Rep. of Progr. Can. Geo. Sur., Black Riv. Gr. [Sig. wandering.]
validum, Hall, 1876, Illust. Devonian Foss., Schoharie grit. [Sig. strong.]
HURONIA, Stokes, 1823, Geo. Trans., n. s., vol. 1. [Ety. proper name.] When this genus was proposed, the author thought he was describing a coral. Prof. Billings says that the name was proposed for the siphuncle of an Orthoceras, and is, therefore, merely a synonym.
annulata, Hall, 1851, Lake Superior Land Dist. by Foster & Whitney, Niagara Gr. [Sig. ringed.]
bigsbyi, Stokes, 1823, Trans. Geo. Soc., vol. 1, Clinton Gr. [Ety. proper name.]
minuens, Barrande, 1869, Syst. Sil. de Boh., 2d series, vol. 4, Clinton Gr. [Sig. diminished.]
obliqua, Stokes, 1823, Trans. Geo. Soc., 2nd series, vol. 1, Clinton Gr. [Sig. oblique.]
portlocki, Stokes, 1840, Trans. Geo. Soc., 2nd series, vol. 5, Clinton Gr. [Ety. proper name.]
sphæroidalis, Stokes, 1823, Trans. Geo. Soc., 2nd series, vol. 1, Clinton Gr. [Sig. spherical.]
turbinata, Stokes, 1823, Trans. Geo. Soc., 2nd series, vol. 1, Clinton Gr. [Sig. top-shaped.]
vertebralis, Stokes, 1823, Trans. Geo. Soc., 2nd series, vol. 1, Niagara and Clinton Gr. [Sig. vertebra-like.] See Orthoceras canadense.
Hydnoceras, Conrad, 1842, Jour. Acad. Nat. Sci., vol. 8. [Ety. *hydnos*, a truffle; *keras*, a horn.] See *Dictyophyton*, a plant.
tuberosum, see Dictyophyton tuberosum.
LITUITES, Montfort, 1808, Conch. Syst. [Ety. *lituus*, a trumpet.]
apollo, Billings, 1862, Pal. Foss., Calcif. Gr. [Ety. mythological name.]

CEPHALOPODA.

cancellatus, McChesney, 1861, New Pal. Foss., Niagara Gr. See *L. occidentalis* and *Nautilus cancellatus* and *N. occidentalis*. If this species, as Prof. Hall suggests, is a true Nautilus, McChesney's name has precedence.
capax, Hall, 1860, Rep. of Progr. Geo. Sur. Wis. See Nautilus capax.
complanatus, Shumard, 1863, Trans. St. Louis Acad. Sci., Calciferous Gr. [Sig. smoothed.]
convolvans, (?) Schlotheim, 1813, in Jahrbuch, (Hall, Pal. N. Y., vol. 1) Black Riv. Gr. [Sig. wrapping together.]
farnsworthi, Billings, 1861, Pal. Foss., vol. 1, Calciferous Gr. [Ety. proper name.]
graftonensis, Meek & Worthen, 1869, Proc. Acad. Nat. Sci. Phil., Niagara Gr. [Ety. proper name.]
hercules, Winchell & Marcy, 1865, Mem. Bost. Soc. Nat. Hist., Niagara Gr. Syn. for Cyrtoceras amplicorne. See 20th Reg. Rep. N. Y.
imperator, Billings, 1861, Pal. Foss., Calciferous Gr. [Sig. the chief.]
magnificus, Billings, 1857, (Gyroceras magnificum) Rep. of Progr., Hud. Riv. Gr. [Sig. magnificent.]
marshi, Hall, 1867, 20th Reg. Rep., Niagara Gr. [Ety. proper name.]
occidentalis, Hall, 1861, Rep. of Progr. Geo. Sur. Wis., Niagara Gr. This species is now referred by Prof. Hall to the genus *Nautilus*, see 20th Reg. Rep. It was first described by McChesney, Jan. 1860, as *Cyrtoceras giganteum*, but that name being preoccupied, in 1861 he proposed *Lituites cancellatus*. If it is a *Nautilus*, the word *occidentalis* being preoccupied, McChesney's name *cancellatus* has precedence.
ortoni, Meek, 1873, Ohio Pal., vol. 1, Niagara Gr. [Ety. proper name.]
palinurus, Billings, 1862, Pal. Foss., Calciferous Gr. [Ety. mythological name.]
pluto, Billings, 1865, Pal. Foss., Quebec Gr. [Ety. mythological name.]
robertsoni, Hall, 1861, Rep. of Progr. Wis., Chazy & Black Riv. Gr. [Ety. proper name.]
undatus, Emmons, 1842, (Inachus undatus) Geo. Rep. N.Y., Black Riv. and Trenton Gr. [Sig. wavy.]
undatus var. occidentalis, Hall, 1861, Rep. of Progr. Wis., Black Riv. & Trenton Gr. [Sig. western.]
NAUTILUS, Breynius, 1732, Dissert. Polyth. [Ety. *Nautilos*, a sailor or navigator.]
avonensis, Dawson, 1868, Acad. Geol., Carboniferous. [Ety. proper name.]
avus, Barrande, 1869, Syst. Sil. de Boh., vol. 4, Quebec Gr. [Sig. an ancestor.]
barrandi, Hall, 1876, Illust. Devonian Foss., Ham. Gr. [Ety. proper name.]
biserialis, Hall, 1860, Supp. to vol. 1, pt. 2, Iowa Geo. Sur., Coal Meas. [Sig. in two series.] This is probably a syn. for N. occidentalis.

buccinum, Hall, 1876, Illust. Devonian Foss., Ham. Gr. [Sig. a sea-trumpet.] This species was referred to *Gyroceras expansum*.
calciferus, Billings, 1865, Pal. Foss.,Calciferous Gr. [Ety. from the Calciferous Group.]
cancellatus, McChesney, 1861, (Lituites cancellatus) New Pal. Foss., Niagara Gr. [Sig. cross-barred.]
canaliculatus, Cox, 1857, Geo. Sur. Ky., vol. 3, Coal Meas. [Sig. grooved.]
capax, Hall, 1860, (Lituites capax) Rep. of Progr. Geo. Sur. Wis., Niagara Gr. [Sig. large.]
capax, Meek & Worthen, 1865. This was preoccupied and must yield unless it can be retained in the subgenus *Solenocheilus*, which see.
chesterensis, Meek & Worthen, 1860, Proc. Acad. Nat. Sci. Phil., Chester Gr. [Ety. proper name.]
clarkanus, Hall, 1858, Trans. Alb. Inst., vol. 4, Low. Carb. [Ety. proper name.]
collectus, see Solenocheilus collectum.
cornulum, Hall, 1876, Illust. Devonian Foss., Ham. Gr. [Sig. a little horn.]
coxanus, see Temnocheilus coxanum.
decoratus, Cox, 1857, Geo. Sur. Ky., vol. 3, Coal Meas. [Sig. decorated.]
desertus, Billings, 1865, Pal. Foss., Quebec Gr. [Sig. deserted.]
digonus, Meek & Worthen, 1860, Proc. Acad. Nat. Sci. Phil., Kinderhook Gr. [Sig. two-cornered.]
disciformis, see Discites disciformis.
discoidalis, see Trematodiscus discoidalis.
divinus, White & St. John, 1868, Trans. Chi. Acad. Sci., Up. Coal Meas. The name was preoccupied by Meyer in 1831.
eccentricus, Meek & Hayden, 1858, Trans. Alb. Inst., vol. 4, Permian Gr. [Ety. *ek*, from ; *kentron*, center.]
ferox, Billings, 1865, Pal. Foss., Calciferous Gr. [Sig. fierce.]
ferratus, Cox, 1857, Geo. Sur. Ky., vol. 3, Coal Meas. [Sig. covered with iron.]
forbesanus, McChesney, 1860, Desc. New Pal. Foss., Coal Meas. [Ety. proper name.]
gilpini, Swallow, 1860, Trans. St. Louis Acad. Sci., Coal Meas. [Ety. proper name.]
globatus, Sowerby, 1825, Min. Conch., Chester Gr. [Sig. made round.]
hercules, Billings, 1857, Rep. of Progr. Can. Geo. Sur., Hud. Riv. Gr. [Ety. mythological name.]
highlandensis, see Discites highlandensis.
illinoisensis, McChesney, 1860, Desc. New Pal. Foss., Coal Meas. [Ety. proper name.]
ingentior, Winchell, 1862, Am. Jour. Sci., 2nd series, vol. 33, Marshall Gr. [Sig. larger, enormous.]
insolens, Billings, 1865, Pal. Foss., Quebec Gr. [Sig. rare, unusual.]

CEPHALOPODA.

jason, Billings, 1859, Can. Nat. & Geol., vol. 4, Chazy Gr. [Ety. mythological name.]
lasellensis, Meek & Worthen, 1866, Proc. Acad. Nat. Sci., Phil., Up. Coal Meas. [Ety. proper name.]
latus, see Temnocheilus latum.
lawsi, Swallow, 1860, Trans. St. Louis Acad. Sci., Ham. Gr. [Ety. proper name.]
leidyi, see Solenocheilus leidyi.
marcellensis, Vanuxem, 1842, (Goniatites marcellensis) Geo. Rep. N. Y., Ham. Gr. [Ety. proper name.]
maximus, Conrad, 1838, (Cyrtoceras maximus) Ann. Rep. N. Y., Ham. Gr. [Sig. the largest.]
meekanus, see Trematodiscus meekanus.
missouriensis, Swallow, 1858, Trans. St. Louis Acad. Sci., Coal Meas. [Ety. proper name.]
natator, Billings, 1859, Can. Nat. & Geo., vol. 4, Chazy Gr. [Sig. a swimmer.]
niotensis, see Temnocheilus niotense.
nodocarinatus, McChesney, syn. for N. occidentalis.
nodoso-dorsatus, Shumard, 1858, Trans. St. Louis Acad. Sci., Coal Meas. [Ety. *nodosus*, knotty; *dorsatus*, high-backed.]
occidentalis, Swallow, 1858, Trans. St. Louis Acad. Sci., Permian Gr. [Sig. western.]
occidentalis, Hall, 1860, 20th Reg. Rep., Niagara Gr. This name being preoccupied, McChesney's name, *cancellatus*, has precedence. See Lituites cancellatus.
oriens, Hall, 1876, Illust. Devonian Foss., Marcellus shale. [Sig. the east.]
ornatus, Hall, 1860, 13th Reg. Rep. See Discites ornatus.
permianus, Swallow, 1858, Trans. St. Louis Acad. Sci., Permian Gr. [Ety. proper name.]
planidorsalis, see Trematodiscus planidorsalis.
planorbiformis, Meek & Worthen, 1860, Proc. Acad. Nat. Sci. Phil., Coal Meas. [Sig. shaped like a shell of the genus *Planorbis*.]
planovolvis, Shumard, 1858, Trans. St. Louis Acad. Sci., Coal Meas. [Sig. plane-whorled.]
pomponius, Billings, 1862, Pal. Foss., Calciferous Gr. [Ety. mythological name.]
ponderosus, White, 1872, Pal. of E. Neb., Coal Meas. [Sig. heavy.]
quadrangularis, McChesney, 1860, Desc. New Pal. Foss., Coal Meas., syn. for N. occidentalis.
rockfordensis, Meek & Worthen, 1866, Proc. Acad. Nat. Sci. Phil., Kinderhook Gr. [Ety. proper name.] Probably a *Gyroceras*. See Ill. Geo. Sur., vol. 3.
sangamonensis, Meek & Worthen, 1860, Proc. Acad. Nat. Sci. Phil., Coal Meas. [Ety. proper name.]

seebachanus, Geinitz, 1866, Carb. und Dyas in Neb., Permian Gr. [Ety. proper name.] This species was made the type by Prof. Meek of the new genus Pteronautilus, which see.
spectabilis, see Temnocheilus spectabile.
springeri, White & St. John, 1868, Trans. Chi. Acad. Sci., Coal Meas. [Ety. proper name.]
striatulus, see Trematodiscus striatulus.
subglobosus, Meek & Worthen, 1860, Proc. Acad. Nat. Sci. Phil., Chester Gr. Syn. for N. globatus, see Geo. Sur. Ill., vol. 3.
subsulcatus, Phillips, 1836, Geol. York. Not clearly identified in this country.
trigonus, see Trematodiscus trigonus.
trisulcatus, see Trematodiscus trisulcatus.
tyrans, Billings, 1859, Can. Nat. & Geol., vol. 4, Chazy Gr. [Sig. a tyrant.]
versutus, Billings, 1865, Pal. Foss., Quebec Gr. [Sig. complicated.]
winslowi, see Temnocheilus winslowi.

ONCOCERAS, Hall, 1847, Pal. N. Y., vol. 1. [Ety. *onkos*, a swelling; *keras*, horn.]
abruptum, Hall, 1861, Rep. of Progr. Wis., Trenton Gr. [Sig. abrupt.]
alceus, Hall, 1861, Rep. of Progr. Wis., Chazy & Black Riv. Gr. [Ety. mythological name.]
amator, Billings, 1866, Catal. Sil. Foss. Antic., Clinton Gr. [Sig. a lover.]
clitus, Billings, 1866, Catal. Sil. Foss. Antic., Niagara Gr. [Ety. proper name.]
constrictum, Hall, 1847, Pal. N. Y., vol. 1, Black Riv. & Trenton Gr. [Sig. constricted.]
corydon, Billings, 1866, Catal. Sil. Foss. Antic., Niagara Gr. [Ety. proper name.]
dilatatum, Hall, 1860, 13th Reg. Rep., Ham. Gr. [Sig. spread out.]
expansum, Hall, 1852, Pal. N. Y., vol. 2, Coralline Gr. [Sig. expanded.]
futile, Billings, 1866, Catal. Sil. Foss. Antic., Clinton Gr. [Sig a water jar, fusiform.]
gibbosum, Hall, 1852, Pal. N. Y., vol. 2, Medina sandstone. [Sig. gibbous.]
lycus, Hall, 1861, Rep. of Progr. Wis., Chazy & Black Riv. Gr. [Ety. proper name.]
orcas, Hall, 1861, (Cyrtoceras orcas) Rep. of Progr. Geo. Sur. of Wis., Niagara Gr. [Ety. proper name.]
ovoides, Hall, 1859, Pal. N. Y., vol. 3, Low. Held. Gr. [Sig. egg-like.]
pandion, Hall, 1861, Rep. of Progr. Wis., Chazy & Black Riv. Gr. [Ety. mythological name.]
pettiti, Billings, 1866, Catal. Sil. Foss. Antic., Niagara Gr. [Ety. proper name.]
plebeium, Hall, 1861, Geo. Rep. Wis., Trenton Gr. [Sig. the common sort.]
subrectum, Hall, 1852, Pal. N. Y., vol. 2, Clinton Gr. [Sig. nearly straight.]

teucer, Billings, 1866, Catal. Sil. Foss. Antic., Niagara Gr. [Ety. mythological name.]
thales, Billings, 1866, Catal. Sil. Foss. Antic., Niagara Gr. [Ety. proper name.]
ORMOCERAS, Stokes, 1840, Trans. Geo. Soc., 2nd series, vol. 5. [Ety. *ormos*, a chain or necklace; *keras*, horn; from the appearance of the siphuncle.]
backi, Stokes, 1840, Trans. Geo. Soc., 2nd series, vol. 5, Clinton Gr. [Ety. proper name.]
bayfieldi, Stokes, 1840, Trans. Geo. Soc., 2nd series, vol. 5, Clinton Gr. [Ety. proper name.]
crebriseptum, Hall, 1847, Pal. N. Y., vol. 1, Hud. Riv. Gr. [Sig. having many septa.]
(?) gracile, Hall, 1847, Pal. N. Y., vol. 1, Black Riv. Gr. [Sig. slender.]
moniliforme, Hall, 1847, Pal. N. Y., vol. 1, Chazy Gr. [Sig. like a string of beads.]
remotiseptum, Hall, 1850, 3rd Reg. Rep., Trenton Gr. [Sig. having distant septa.]
tenuifilum, Hall, 1847, Pal. N. Y., vol. 1, Black Riv. & Trenton Gr. [Sig. fine-lined.]
tenuifilum var. distans, Hall, 1847, Pal. N. Y., vol. 1, Black Riv. Gr. [Sig. distant.]
vertebratum, Hall, 1852, Pal. N. Y., vol. 2, Clinton Gr. [Sig. like a back-bone.]
whitii, Stokes, 1840, Trans. Geo. Soc., 2nd series, vol. 5, Clinton Gr. [Ety. proper name.]

ORTHOCERAS, Breynius, 1732, Dissert. Polyth. [Ety. *orthos*, straight; *keras*, horn.]
abnorme, Hall, 1867, 20th Reg. Rep., Niagara Gr. [Sig. out of the usual order.]
abruptum, Hall, 1852, Pal. N. Y., vol. 2, Clinton Gr. [Sig. terminating suddenly.]
acicula, Hall, 1843, 4th Dist. Rep. N. Y., Portage Gr. [Sig. a little needle.]
aculeatum, Swallow, 1858, Trans. St. Louis Acad. Sci., Coal Meas. [Sig. armed with a sharp point.]
ægea, Hall, 1862, 15th Reg. Rep., Ham. Gr. [Ety. mythological name.]
æqualis, Emmons, 1842, Geo. Rep. N. Y., Trenton Gr. [Sig. equal.]
alienum, Hall, 1867, 20th Reg. Rep., Niagara Gr. [Sig. another sort.]
allumettense, Billings, 1857, Rep. of Progr., Chazy & Black Riv. Gr. [Ety. proper name.]
amplicameratum, Hall, 1847, Pal. N. Y., vol. 1, Black Riv. & Trenton Gr. [Sig. large-chambered.]
anax, Billings, 1875, Can. Nat. & Geol., Corniferous Gr. [Ety. mythological name.]
angulatum, (?) Wahlenberg, 1821, Nova. Acta., Niagara Gr. See remarks on this species by Prof. Hall in 20th Reg. Rep.
anellum, Conrad, 1843, Proc. Acad. Nat. Sci. Phil., Black Riv. & Trenton Gr. [Sig. a little ring.]
annulatum, Sowerby, 1818, Min. Conch., vol. 2, Clinton & Niagara Gr. [Sig. ringed.]
annulato-costatum, Meek & Worthen, 1861, Proc. Acad. Nat. Sci. Phil., Chester Gr. [Ety. *annulatus*, ringed; *costatus*, ribbed.] This name was preoccupied by Boll in 1857.
antenor, Billings, 1859, Can. Nat. & Geo., vol. 4, Chazy Gr. [Ety. mythological name.]
anticostiense, Billings, 1857, Rep. of Progr., Hud. Riv. Gr. [Ety. proper name.]
arcuatellum, Winchell, 1862, Am. Jour. Sci., 2nd series, vol. 33, Marshall Gr. [Sig. small and arched.]
arcuoliratum, Hall, 1847, Pal. N. Y., vol. 1, Black Riv. & Trenton Gr. [Sig. in arched ridges.]
arenosum, Hall, 1859, Pal. N. Y., vol. 3, Oriskany sandstone. [Sig. sandy.]
atticus, Billings, 1865, Pal. Foss., Quebec Gr. [Ety. proper name.]
autolycus, Billings, 1862, Pal. Foss., Quebec Gr. [Ety. mythological name.]
backi, Stokes, 1840, Geo. Trans., 2nd ser., vol. 4, Trenton Gr. See Geo. Sur. Ill., and see *Ormoceras backi*.
baculum, Meek, 1860, Proc. Acad. Nat. Sci., Low. Carb. [Sig. a staff or cudgel.]
baculum, Hall, 1862, 15th Reg. Rep., Schoharie grit. The name was preoccupied.
balteatum, Billings, 1857, Rep. of Progr., Hud. Riv. Gr. [Sig. belted.]
barquianum, Winchell, 1862, Am. Jour. Sci., 2nd series, vol. 33, Marshall Gr. [Ety. proper name.]
bebryx, Hall, 1876, Illust. Devonian Foss., Ham. Gr. [Ety. proper name.]
becki, Billings, 1859, Can. Nat. & Geol., vol. 4, Calciferous Gr. [Ety. proper name.]
bellatulum, Billings, 1866, Catal. Sil.Foss. Antic., Clinton Gr. [Sig. pretty, neat.]
bilineatum, Hall, 1847, Pal. N. Y., vol. 1, Chazy, Black Riv., Trenton and Hud. Riv. Gr. [Sig. double-lined.]
bilineatum var. a, Hall, 1847, Pal. N. Y., vol. 1, Trenton Gr.
brongniarti, Troost, 1838, (Conotubularia brongniarti) Mem. Soc. Geo. de France, 3, Low. Sil. [Ety. proper name.]
brontes, Billings, 1866, Catal. Sil. Foss. Antic., Niagara Gr. [Ety. mythological name.]
bucklandi, Billings, 1857, Rep. of Progr., Up. Sil. [Ety. proper name.]

bullatum, (?) Sowerby, 1839, Murch. Sil. Syst., Trenton Gr. [Sig. puffed up.]
byrnesi, S. A. Miller, 1875, Cin. Quar. Jour. Sci., vol. 2, Cin'ti Gr. [Ety. proper name.]
cadmus, Billings, 1866, Catal. Sil. Foss. Antic., Niagara Gr. [Ety. mythological name.]
cameolare, McChesney, 1861, New Pal. Foss., Niagara Gr. [Ety. (?).]
canadense, Billings, 1857, Rep. of Progr., Mid. Sil. [Ety. proper name.] Prof. Billings proposed this name as a substitute for *Huronia vertebralis* for the reason that *Huronia* is a syn. for *Orthoceras*, and there is one O. vertebralis.
cancellatum, Hall, 1852, Pal. N. Y., vol. 2, Niagara Gr. The name was preoccupied by Eichwald in 1842.
capitolinum, Safford, 1869, Geo. of Tenn., Trenton Gr. [Sig. a great tower.]
carleyi, Hall & Whitfield, 1875, Ohio Pal., vol. 2, Cin'ti Gr. [Ety. proper name.]
catilina, Billings, 1865, Pal. Foss., Quebec Gr. [Ety. proper name.]
cato, Billings, 1865, Pal. Foss., Quebec Gr. [Ety. proper name.]
catullus, Billings, 1865, Pal. Foss., Quebec Gr. [Ety. proper name.]
chemungense, Swallow, 1860, Trans. St. Louis Acad. Sci., Chemung Gr. [Ety. proper name.]
chesterense, Swallow, 1863, Trans. St. Louis Acad. Sci., St. Genevieve Gr. [Ety. proper name.]
chouteauense, Swallow, 1860, Trans. St. Louis Acad. Sci., Chouteau Gr. [Ety. proper name.]
cincinnatiense, S. A. Miller, 1875, Cin. Quar. Jour. Sci., vol. 2, Cin'ti Gr. [Ety. proper name.]
clathratum, Hall, 1847, Pal. N. Y., vol. 1, Trenton Gr. [Sig. cross-barred.]
clavatum, Hall, 1852, Pal. N. Y., vol. 2, Clinton Gr. [Sig. club-shaped.]
clavatum, Hall, 1859, Pal. N. Y., vol. 3, Low. Held. Gr. The name was appropriated.
clinocameratum, Winchell, 1862, Am. Jour. Sci., 2d series, vol. 33, Marshall Gr. [Ety. *clino*, bent; *cameratus*, chambered.]
clouei, Barrande, 1869, Syst. Sil. de Boh., 4me series, Quebec Gr. [Ety. proper name.]
colon, White, 1874, Rep. Invert. Foss., Quebec Gr. [Ety. *colon*, the great intestine.]
columnare, Hall, 1860, Rep. of Progr. Geo. Sur. Wis., Niagara Gr. [Sig. columnar.] The name was preoccupied by Mark. in 1857.
constrictum, Conrad, 1838, Ann. Rep. N. Y., Ham. Gr. [Sig. constricted.]
constrictum, Hall, see Oncoceras constrictum.

coralliferum, Hall, 1847, Pal. N. Y., vol. 1, Utica and Hud. Riv. Gr. [Sig. bearing a coral.]
cornuum, Billings, 1857, Rep. of Progr., Chazy Gr. [Sig. a horn.]
crebescens, Hall, 1867, 20th Reg. Rep., Niagara Gr. [Sig. frequent, increasing.]
crebristriatum, Meek & Worthen, 1865, Proc. Acad. Nat. Sci. Phil., Niagara Gr. [Sig. closely-striated.]
cribrosum, Geinitz, 1866, Carb. und Dyas in Neb., Coal Meas. [Sig. full of holes, like a seive.]
crotalum, Hall, 1862, 15th Reg. Rep.,Ham. Gr. [Sig. a rattle.]
crocus, Billings, 1866, Catal. Sil. Foss. Antic., Hud. Riv. Gr. [Ety. mythological name.]
cuvieri, Troost, 1838, (Conotubularia cuvieri) Mem. Soc. Geo. de France 3, Low. Sil. [Ety. proper name.]
darwini, Billings, 1868, Pal. Foss., Guelph Gr. [Ety. proper name.]
decrescens, Billings, 1857, Rep. of Progr., Black Riv. and Trenton Gr. [Sig. decreasing, growing less.]
defrancii, Troost, 1838, (Conotubularia defrancei) Mem. Soc. Geo. de France 3, Low. Sil. [Ety. proper name.]
deparcum, Billings, 1859, Can. Nat. & Geo., vol. 4, Calciferous Gr. [Sig. very scarce.]
diffidens, Billings, 1865, Pal. Foss., Chazy Gr. [Sig. diffident.]
dolatum, Dawson, 1868, Acad. Geol., Carb. [Sig. hewed.]
drummondi, Billings, 1865, Pal. Foss., Chazy Gr. [Ety. proper name.]
duseri, Hall & Whitfield, 1875, Ohio Pal., vol. 2, Cin'ti Gr. [Ety. proper name.]
dyeri, S. A. Miller, 1875, Cin. Quar. Jour. Sci., vol. 2, Cin'ti Gr. [Ety. proper name.]
edax, Billings, 1865, Pal. Foss., Calcif. Gr. [Sig. voracious.]
elegantulum, Dawson, 1860, Can. Nat. & Geo., vol. 4, Low. Sil. [Sig. elegant.]
emaceratum, Hall, 1862, 15th Reg. Rep. N. Y., Ham. Gr. [Sig. thin.]
epigrus, Hall, 1858, Trans. Alb. Inst., vol. 4, Low. Carb. [Sig. a wooden pin.]
eriense, Hall, 1877. The name is proposed here instead of *O. robustum*, Ham. Gr. [Ety. proper name.]
exile, Hall, 1862, 15th Reg. Rep., Ham. Gr. [Sig. small, slender.]
exornatum, Dawson, 1860, Can. Nat. & Geo., vol. 5, Up. Sil. [Sig. adorned.]
expansum, Meek & Worthen, 1860, Proc. Acad. Nat. Sci. Phil., St. Louis Gr. [Sig. spread out.]
explorator, Billings, 1865, Pal. Foss., Quebec Gr. [Sig. a scout, an examiner.]
ferum, Billings, 1866, Catal. Sil. Foss. Antic., Hud. Riv. & Anticosti Gr. [Sig. wild, cruel.]
flavius, Billings, 1865, Pal. Foss., Quebec Gr. [Ety. proper name.]

CEPHALOPODA.   175

foliatum, syn. for Cyrtoceras eugenium.
formosum, Billings, 1857, Rep. of Progr., Trenton, Hud. Riv. & Anticosti Gr. [Sig. beautiful.]
fosteri, S. A. Miller, 1875, Cin. Quar.Jour. Sci., vol. 2, Cin'ti Gr. [Ety. proper name.]
foxense, Safford, 1869, Geo. of Tenn. Not defined.
fulgur, Billings, 1866, Catal. Sil. Foss. Antic., Hud. Riv. Gr. [Sig. a thunder bolt.]
furtivum, Billings, 1865, Pal. Foss., Calcif. Gr. [Sig. hard to find.]
fusiforme, Hall, 1847, Pal. N. Y., vol. 1, Black Riv. and Trenton Gr. [Sig. tapering at both ends.]
glaucus, Billings, 1865, Pal. Foss., Calciferous Gr. [Ety. mythological name.]
goldfussi, Troost, 1838, (Conotubularia goldfussi) Mem. Soc. Geo. de France 3, Low. Sil. [Ety. proper name.]
gracilius, Winchell, 1862, Proc. Acad. Nat. Sci., Portage Gr. [Sig. more slender.]
gregarium, Hall, 1861, Rep. of Progr. Wis., Hud. Riv. Gr. [Sig. occuring in flocks or masses.] This name was preoccupied by Sowerby in 1839, Murch. Sil. Syst.
hæsitans, Billings, 1865, Pal. Foss., Quebec Gr. [Sig. doubtful.]
hageri, Hall, 1861, Geol. of Vermont, Calciferous Gr. [Ety. proper name.]
halli, S. A. Miller, 1875, Cin. Quar. Jour. Sci., vol. 2, Cin'ti Gr. [Ety. proper name.]
harperi, S. A. Miller, 1875, Cin. Quar. Jour. Sci., vol. 2, Cin'ti Gr. [Ety. proper name.]
hastatum, Billings, 1857, Rep. of Progr., Black Riv. and Trenton Gr. [Sig. formed like a spear.]
helderbergiæ, Hall, 1859, Pal. N. Y., vol. 3, Low. Held. Gr. [Ety. proper name.]
heterocinctum, Winchell, 1863, Proc. Acad. Nat. Sci., Kinderhook Gr. [Sig. irregularly girdled.]
hoyi, McChesney, 1861, New Pal. Foss., Niagara Gr. [Ety. proper name.]
huronense, Billings, 1857, Rep. of Progr., Trenton Gr. [Ety. proper name.]
hyas, Hall, 1862, 15th Reg. Rep., Schoharie grit. [Ety. mythological name.]
imbricatum, Sowerby, 1839, Murch. Sil. Syst., Niagara Gr. [Sig. lapping over.]
indagator, Billings, 1865, Pal. Foss., Calciferous Gr. [Sig. a diligent hunter.]
indianense, Hall, 1860, 13th Reg. Rep., Ham. Gr. [Ety. proper name.]
infelix, Billings, 1866, Catal. Sil. Foss. Antic., Clinton Gr. [Sig. miserable, useless.]
irregulare, McChesney, 1861, New Pal. Foss., Niagara Gr. The name was preoccupied, and is a syn. for O. woodworthi.

jamesi, Hall & Whitfield, 1875, Ohio Pal., vol. 2, Clinton Gr. [Ety. proper name.]
jolietense, Meek & Worthen, 1865, Proc. Acad. Nat. Sci. Phil., Niagara Gr. [Ety. proper name.]
junceum, Hall, 1847, Pal. N. Y., vol. 1, Trenton Gr. [Sig. rush-stem-like.]
kickapooense, Swallow, 1858, Trans. Acad. Sci. St. Louis, vol. 1, Up. Permian Gr. [Ety. proper name.]
knoxense, McChesney, 1860, New Pal. Foss., Coal Meas. [Ety. proper name.]
læve, Hall, 1843, Geo. Rep. 4th Dist. N. Y., Onondaga Gr. [Sig. smooth.] The name was preoccupied by Fleming in 1825.
lamarcki, Billings, 1859, Can. Nat. & Geo., vol. 4, Calciferous Gr. [Ety. proper name.]
lamellosum, Hall, 1847, Pal. N. Y., vol. 1, Hud. Riv. Gr. [Sig. in thin plates.]
laphami, McChesney, 1861, New Pal. Foss., Niagara Gr. [Ety. proper name.]
laqueatum, Hall, 1847, Pal. N. Y., vol. 1, Calciferous to Trenton Gr. [Sig. adorned.]
laqueatum var. a, Hall, 1847, Pal. N. Y., vol. 1, Trenton Gr.
laqueatum, Hartt, 1868, Acad. Geol. The name was preoccupied.
latiannulatum, Hall, 1847, Pal. N. Y., vol. 1, Trenton Gr. [Sig. widely annulated.]
lineolatum, McChesney, 1861, New Pal. Foss., Niagara Gr. The name was preoccupied by Phillips in 1841.
longicameratum, Hall, 1859, Pal. N. Y., vol. 3, Low. Held. Gr. [Sig. long-chambered.]
loxias, Hall, 1867, 20th Reg. Rep., Low. Sil. [Ety. mythological name.]
luxum, Hall, 1876, Illust. Devonian Foss., Schoharie grit. [Sig. dislocated.]
lyelli, Billings, 1857, Rep. of Progr., Hud. Riv. Gr. [Ety. proper name.]
magnisulcatum, Billings, 1857, Rep. of Progr., Hud. Riv. Gr. [Sig. deeply furrowed.]
marcellense, Vanuxem, 1842, Geo. Rep. N.Y., Ham. Gr. [Ety. proper name.]
marginale, Owen, 1840, Rep. on Min. Lands, Up. Magnesian Gr. [Sig. margined.]
maro, Billings, 1859, Can. Nat. & Geol., vol. 4, Chazy Gr. [Ety. proper name.]
marshallense, Winchell, 1862, Am. Jour. Sci., 2nd series, vol. 33, Marshall Gr. [Ety. proper name.]
medon, Billings, 1866, Catal. Sil. Foss. Antic., Clinton Gr. [Ety. proper name.]
medullare, Hall, 1860, Rep. of Progr. Geo. Sur. Wis., Niagara Gr. [Sig. like a pith.]

meeki, S. A. Miller, 1875, Cin. Quar. Jour. Sci., vol. 2, Cin'ti Gr. [Ety. proper name.]
menelaus, Billings, 1862, Pal. Foss., Black Riv. Gr. [Ety. mythological name.]
minganense, Billings, 1857, Rep. of Progr., Chazy & Black Riv. Gr. [Ety. proper name.]
missisquoi, Billings, 1865, Pal. Foss., Quebec Gr. [Ety. proper name.]
mohri, S. A. Miller, 1875, Cin. Quar. Jour. Sci., vol. 2, Cin'ti Gr. [Ety. proper name.]
molestum, Hall, 1876, Illust. Devonian Foss., Up. Held. Gr. [Sig. difficult.]
moniliforme, Hall, 1847, Pal. N. Y., vol. 1, Chazy Gr. [Sig. necklace-like.]
moniliforme, Swallow, 1858, Trans. St. Louis Acad. Sci., vol. 1, Coal Meas. The name was preoccupied.
montrealense, Billings, 1859, Can. Nat. & Geo., vol. 4, Calciferous Gr. [Ety. proper name.]
multicameratum, Emmons, 1842, Geo. Rep. N. Y., Birdseye Gr. [Sig. many-chambered.]
multicinctum, Hall, 1862, 15th Reg. Rep., Schoharie grit. [Sig. many-banded.]
multicinctum, Winchell, 1862, Proc. Acad. Nat. Sci., Marshall Gr. [Sig. many-banded.] It is not clear which of these authors is entitled to his name.
multilineatum, Emmons, 1842, Geo. Rep. N. Y., Trenton Gr. [Sig. many-lined.]
multiseptum, Hall, 1852, Pal. N. Y., vol. 2, Medina Gr. [Sig. having many divisions.]
murrayi, Billings, 1857, Rep. of Progr., Black Riv. and Trenton Gr. [Ety. proper name.]
niagarense, Hall, 1867, 20th Reg. Rep., Niagara Gr. [Ety. proper name.]
nodocostum, McChesney, 1861, New Pal. Foss., Niagara Gr. [Ety. *nodus*, a knot; *costa*, a rib.]
novamexicanum, Marcou, 1858, Geol. North America, Low. Carb. [Ety. proper name.]
nummularium, (?) 1839, Murch. Sil. Syst., Up. Sil. [Sig. like a coin.]
nuntium, Hall, 1862, 15th Reg. Rep., Ham. Gr. [Sig. news, a messenger.]
oberon, Billings, 1866, Catal. Sil. Foss. Antic., Niagara Gr. [Ety. the king of fairies in *Midsummer Night's Dream*.]
occidentale, Swallow, 1858, Trans. St. Louis Acad. Sci., Coal Meas., Permian Gr. [Sig. western.]
occidentale, Winchell, 1862, Am. Jour. Sci., Marshall Gr. This name was preoccupied.
ommanei, Salter, 1868, Bigsby, Thesaurus Siluricus, Niagara Gr. [Ety. proper name.]
ordinatum, Billings, 1865, Pal. Foss., Calciferous Gr. [Sig. well arranged.]
ortoni, Meek, 1872, Proc. Acad. Nat. Sci. Phil., Cin'ti Gr. [Ety. proper name.]

ottawaense, Billings, 1857, Rep. of Progr., Black Riv. & Trenton Gr. [Ety. proper name.]
ozarkense, Shumard, 1863, Trans. St. Louis Acad. Sci., Calciferous Gr. [Ety. proper name.]
pauciseptum, Hall, 1859, Pal. N. Y., vol. 3, Low. Held. Gr. [Sig. having few septa.]
pelops, Hall, 1862, 15th Reg. Rep., Schoharie grit. [Ety. mythological name.]
pelops var. ohioense, Hall, 1876, Illust. Devonian Foss., Up. Held. Gr. [Ety. proper name.]
perannulatum, Billings, 1857, Rep. of Progr., Hud. Riv. Gr. This name was preoccupied by Portlock in 1843. The species is now named *O. crocus*.
perelegans, Salter, 1848, Mem. Geo. Sur. Gr. Brit., vol. 2, Ham. Gr. [Sig. very elegant.]
perparvum, Billings, 1862, Pal. Foss., Black Riv. Gr. [Sig. very small.]
perseus, Billings, 1865, Pal. Foss., Quebec Gr. [Ety. mythological name.]
persiphonatum, Billings, 1857, Rep. of Progr., Mid. Sil. [Sig. having a very large siphuncle.] If the genus *Huronia* is valid, this species will belong to it.
perstriatum, Hall, 1859, Pal. N. Y., vol. 3, Low. Held. Gr. [Sig. many-lined.]
perstrictum, Dawson, 1868, Acad. Geol., Carb. [Sig. very much banded.] The name was preoccupied by Barrande.
pertinax, Billings, 1860, Can. Nat. & Geo., vol. 5, Black Riv. Gr. [Sig. persistent, constant.]
pileolum, Billings, 1866, Catal. Sil. Foss. Antic., Medina Gr. [Sig. a little cap.]
piscator, Billings, 1865, Pal. Foss., Quebec Gr. [Sig. a fisherman.]
piso, Billings, 1862, Pal. Foss., Hud. Riv. Gr. [Ety. proper name.]
planoconvexum, Hall, 1861, Rep. of Progr. Wis., Black Riv. & Trenton Gr. [Ety. *planus*, flat, plain; *convexus*, convex.]
pressum, Rogers, 1868, Bigsby, Thesaurus Siluricus, Trenton Gr. [Sig. pressed.]
priamus, Billings, 1865, Pal. Foss., Quebec Gr. [Ety. proper name.]
primigenium, Vanuxem, 1842, Geo. Rep. N. Y., Calciferous Gr. [Sig. original, first born.]
procerus, Hall, 1876, Illust. Devonian Foss., Schoharie grit. [Sig. long.]
profundum, Hall, 1862, 15th Reg. Rep. N. Y., Up. Held. Gr. [Sig. deep.]
propinquum, Billings, 1857, Rep. of Progr., Hud. Riv. Gr. [Sig. related to.]
punctostriatum, Hall, 1860, Can. Nat. & Geo., vol. 5, Low. Sil. [Sig. dotted and striated.]
pustulosum, Winchell, 1866, Rep. Low. Peninsula Mich., Ham. Gr. [Sig. covered with pustules.]

pylades, Billings, 1866, Catal. Sil. Foss. Antic., Niagara Gr. [Ety. mythological name.]
python, Billings, 1857, Rep. of Progr., Trenton Gr. [Ety. mythological name.]
rapax, see Endoceras rapax.
raptor, Billings, 1866, Catal. Sil. Foss. Antic., Medina Gr. [Sig. a robber.]
recedens, Barrande, 1869, Syst. Sil. de Boh., 4me ser., Quebec Gr. [Sig. receding.]
rectiannulatum, Hall, 1847, Pal. N. Y., vol. 1, Chazy & Birdseye Gr. [Ety. *rectus*, straight; *annulatus*, ringed.]
recticameratum, Hall, 1847, Pal. N. Y., vol. 1, Birdseye Gr. [Ety. *rectus*, straight; *cameratus*, chambered.]
rectum, Worthen, 1875, Geo. Sur. Ill., vol. 6, Niagara Gr. [Sig. straight.]
remus, Billings, 1866, Catal. Sil. Foss. Antic., Niagara Gr. [Ety. proper name.]
repens, Billings, 1865, Pal. Foss., Quebec Gr. [Sig. creeping.]
reticulatum, Phillips, 1836, Geol. York., Chemung Gr. Not clearly identified in this country.
richardsoni, Stokes, 1840, Trans. Geo. Soc., 2nd series, vol. 5, Black Riv. Gr. [Ety. proper name.]
rigidum, Hall, 1859, Pal. N. Y., vol. 3, Low. Held. Gr. [Sig. rigid.]
robustum, Winchell, 1862, Am. Jour. Sci., 2nd series, vol. 33, Marshall Gr. [Sig. robust.]
robustum, Hall, 1876, Illust. Devonian Foss., Ham. Gr. The name was preoccupied. See O. eriense.
rotulatum, Billings, 1857, Rep. of Progr., Niagara Gr. [Sig. rounded.]
rude, Hall, 1859, Pal. N. Y., vol. 3, Low. Held. Gr. [Sig. rude, not polished.]
rudicula, Hall, 1876, Illust. Devonian Foss., Up. Held. Gr. [Sig. a spatula.]
rushense, McChesney, 1860, New Pal. Foss., Coal Meas. [Ety. proper name.]
sayi, Billings, 1865, Pal. Foss., Quebec Gr. [Ety. proper name.]
scammoni, McChesney, 1861, New Pal. Foss., Niagara Gr. [Ety. proper name.]
sedgwicki, Billings, 1857, Rep. of Progr., Hud. Riv. Gr. [Ety. proper name.]
selwyni, Billings, 1862, Pal. Foss., Guelph Gr. [Ety. proper name.]
servile, Billings, 1865, Pal. Foss., Quebec Gr. [Sig. paltry.]
shumardi, Billings, 1859, Can. Nat. and Geo., vol. 4, Chazy Gr. [Ety. proper name.]
sieboldi, Billings, 1866, Catal. Sil. Foss. Antic., Hud. Riv. and Anticosti Gr. [Ety. proper name.]
simpsoni, Billings, 1859, Rep. of Progr. Assiniboine & Saskatchewan Ex. Exp., Silurian. [Ety. proper name.]
simulator, Hall, 1876, 28th Reg. Rep., Niagara Gr. [Sig. an imitator.]

sordidum, Billings, 1859, Can. Nat. and Geo., vol. 4, Calciferous Gr. [Sig. coarse.]
striatum, (?) Sowerby, 1812, Min. Conch., Devonian. [Sig. striated.]
striatolineatum, McChesney. 1861, New Pal. Foss., Niagara Gr. [Ety. *striatus*, striated; *lineatus*, lined.] Syn. (?) for O. medullare.
strigatum, Hall, 1847, Pal. N. Y., vol. 1, Trenton Gr. [Sig. furrowed.]
strix, Hall & Whitfield, 1875, Ohio Pal., vol. 2, Niagara Gr. [Sig. a channel.]
subarcuatum, Hall, 1847, Pal. N. Y., vol. 1, Chazy Gr. This name was preoccupied by Portlock in 1843.
subbaculum, Worthen, 1866, Geo. Sur. Ill., vol. 1, Niagara Gr. Not defined.
subtextile, Hall, 1859, Pal. N. Y., vol. 3, Low. Held. Gr. [Sig. somewhat like net-work.]
subulatum, Hall, 1843, 4th Dist. Geo. Rep. N. Y., Marcellus shale. [Sig. awl-shaped.]
tenerum, Billings, 1860, Can. Nat. & Geo., vol. 5, Black Riv. Gr. [Sig. tender.]
tenui-annulatum, Hall, 1850, Pal. N. Y., vol. 3, Low. Held. Gr. [Sig. having slight annulations.]
tenuiseptum, Hall, 1847, Pal. N. Y., vol. 1, Chazy Gr. [Sig. having slender partitions.]
teretiforme, Hall, 1847, Pal. N. Y., vol. 1, Trenton Gr. [Sig. long, round and smooth.]
tetricum, Hall, 1862, 15th Reg. Rep., Schoharie grit. [Sig. rude, rough.]
textile, Hall, 1847, Pal. N. Y., vol. 1, Trenton Gr. [Sig. woven, like a web.]
thoas, Hall, 1862,15th Reg.Rep., Schoharie grit. [Ety. mythological name.]
tityrus, Billings, 1865, Pal. Foss., Quebec Gr. [Ety. proper name.]
transversum, S. A. Miller, 1875, Cin. Quar. Jour. Sci., vol. 2, Cin'ti Gr. [Ety. from the transverse lines on the shell.]
trentonense, see Cyrtoceras trentonense.
turbidum, Hall & Whitfield, 1875, Ohio Pal., vol. 2, Cin'ti Gr. [Sig. disordered.]
typus, Saemann, as identified by Hall, 1876, Illust. Devonian Foss., Marcellus shale. [Sig type of the genus.]
undulatum, Owen, 1840, Rep. on Min. Lands, Niagara Gr. [Sig. wavy.] The name was preoccupied by Sowerby in 1812.
undulostriatum, Hall, 1847, Pal. N. Y., vol. 1, Trenton Gr. [Sig. wavy-lined.]
unionense, Worthen, 1875, Geo. Sur. Ill., vol. 6, Niagara Gr. [Ety. proper name.]
varro, Billings, 1866, Catal. Sil. Foss. Antic., Niagara Gr. [Ety. proper name.]
velox, Billings, 1865, Pal. Foss., Chazy Gr. [Sig. fitted for motion.]

CEPHALOPODA.

vertebrale, Hall, 1847, Pal. N. Y., vol. 1, Trenton Gr. This name was preoccupied by Schlotheim in 1820, and by Eichwald in 1840.
veterator, Billings, 1865, Pal. Foss., Calciferous Gr. [Sig. old.]
vindobonense, Dawson, 1868, Acad. Geol., Carboniferous. [Ety. proper name.]
virgatum, (?) Sowerby, 1839, Murch. Sil. Syst., (Hall, Pal. N. Y., vol. 2,) Niagara Gr. [Sig. rod-like.]
virgulatum, Hall, 1852, Pal. N. Y., vol. 2, Clinton & Niagara Gr. [Sig. like a small rod.]
vittatum, Sandberger. Not American.
vulgatum, Billings, 1857, Rep. of Progr., Trenton Gr. [Sig. common.]
whitii, Winchell, 1863, Proc. Acad. Nat. Sci., Kinderhook Gr. [Ety. proper name.]
winchelli, Meek & Worthen, 1866, Proc. Acad. Nat. Sci. Phil., Devonian. [Ety. proper name.]
woodworthi, McChesney, 1861, New Pal. Foss., Niagara Gr. [Ety. proper name.]
xerxes, Billings, 1865, Pal. Foss., Quebec Gr. [Ety. proper name.]
xiphias, Billings, 1857, Rep. of Progr., Trenton Gr. [Sig. having a point like a sword.]
PHRAGMOCERAS, Broderip, 1839, Murch. Sil. Syst. [Ety. *phragmos*, a partition; *keras*, a horn.]
byronense, Worthen, 1875, Geo. Sur. Ill., vol. 6, Niagara Gr. [Ety. proper name.]
ellipticum, Hall & Whitfield, 1875, Ohio Pal., vol. 2, Niagara Gr. [Sig. elliptical.]
expansum, Winchell, 1863, Proc. Acad. Nat. Sci., Kinderhook Gr. [Sig. expanded.]
hector, Billings, 1862, Pal. Foss., Guelph Gr. [Ety. proper name.]
nestor, Hall, 1867, 20th Reg. Rep., Niagara Gr. [Ety. mythological name.]
parvum, Hall & Whitfield, 1875, Ohio Pal., vol. 2, Niagara Gr. [Sig. small.]
præmaturum, Billings, 1866, Can. Nat. & Geo., vol. 5, Black Riv. & Trenton Gr. [Sig. premature.]
spinosum, see Gyroceras spinosum.
walshi, Meek & Worthen, 1866, Proc. Acad. Nat. Sci. Phil., Ham. Gr. [Ety. proper name.]
PILOCERAS, Salter, 1859, Quar. Jour. Geo. Soc., vol. 15. [Ety. *pilum*, a pounder or pestle; *keras*, a horn.]
canadense, Billings, 1860, Can. Nat. and Geol., vol. 5, Calciferous Gr. [Ety. proper name.]
gracile, Billings, 1865, Pal. Foss., Quebec Gr. [Sig. slender.]
triton, Billings, 1865, Pal. Foss., Quebec Gr. [Ety. mythological name.]
wortheni, Billings, 1865, Pal. Foss., Quebec Gr. [Ety. proper name.]

Polycronites haani, Troost, 1840, 5th Geo. Rep. Tenn., Devonian. Not clearly defined, but probably a Gyroceras.
PTERONAUTILUS, Meek, 1864, Pal. of Up. Mo. [Ety. *pteron*, a wing; *Nautilus*, a genus of shells.]
seebachanus, Geinitz, (Nautilus seebachanus) Carb. und Dyas, Permian Gr. [Ety. proper name.]
SÆRICHNITES, Billings, 1866, Catal. Sil. Foss. Antic. [Ety. *sairo*, to show the teeth; *ichnos*, a footstep.] The author supposed the tracks might have been made by a species of Cephalopoda.
abruptus, Billings, 1866, Catal. Sil. Foss. Antic., Hud. Riv. Gr. [Sig. from the more abrupt termination and deeper impression at one end than the other.]
SOLENOCHEILUS, Meek & Worthen, 1870, Proc. Acad. Nat. Sci. Phil., vol. 20. [Ety. *solen*, a channel; *cheilos*, a lip.]
capax, Meek & Worthen, 1865, (Cryptoceras capax) Proc. Acad. Nat. Sci. Phil., Coal Meas. [Sig. large.]
collectum, Meek & Worthen, 1870, Proc. Acad. Nat. Sci. Phil., St. Louis Gr. [Sig. gathered together.]
leidyi, Meek & Worthen, 1865, (Nautilus leidyi) Proc. Acad. Nat. Sci. Phil., Keokuk Gr. [Ety. proper name.]
Spirula, Lamarck, 1801, Syst. An. sans Vert.
mortoni, Troost, 1840, 5th Geo. Rep. Tenn., Niagara Gr. Not clearly defined.
STREPTOCERAS, Billings, 1866, Catal. Sil. Foss. Antic. [Ety. *streptos*, twisted; *keras*, horn.]
heros, Billings, 1866, Catal. Sil. Foss. Antic., Niagara Gr. [Ety. mythological name.]
janus, Billings, 1866, Catal. Sil. Foss. Antic., Niagara Gr. [Ety. proper name.]
TEMNOCHEILUS, McCoy, 1844, Synop. Carb. Foss. Ireland. [Ety. *temno*, I divide; *cheilos*, a lip.]
coxanum, Meek & Worthen, 1869, Proc. Acad. Nat. Sci. Phil., St. Louis Gr. [Ety. proper name.]
latum, Meek & Worthen, 1870, Proc. Acad. Nat. Sci. Phil., Coal Meas. [Sig. wide.]
niotense, Meek & Worthen, 1865, Proc. Acad. Nat. Sci. Phil., Keokuk Gr. [Ety. proper name.]
peramplum, Meek & Worthen, 1865, (endolobus peramplus) Proc. Acad. Nat. Sci. Phil., Chester Gr. [Sig. very large.]
spectabile, Meek & Worthen, 1860, (Nautilus spectabilis) Proc. Acad. Nat. Sci. Phil., Chester Gr. [Sig. showy, worth seeing.]
winslowi, Meek & Worthen, 1870, Proc. Acad. Nat. Sci. Phil., Coal Meas. [Ety. proper name.]

TREMATODISCUS, Meek & Worthen, 1861, Proc. Acad. Nat. Sci. Phil. [Ety. *trema*, a hole; *diskos*, a quoit.]
  altidorsalis, Winchell, 1862, Proc. Acad. Nat. Sci., Marshall Gr. [Sig. high-backed.]
  discoidalis, Winchell, 1862, Am. Jour. Sci., vol. 33, 2d series, Marshall Gr. [Sig. discoidal.]
  meekanus, Winchell, 1862, Am. Jour. Sci., 2nd series, vol. 33, Marshall Gr. [Ety. proper name.]
  planidorsalis, Winchell, 1862, Am. Jour. Sci., 2nd series, vol. 33, Marshall Gr. [Sig. smooth-backed.]
  striatulus, Winchell, 1862, Proc. Acad. Nat. Sci., 2nd series, vol. 33, Marshall Gr. [Sig. small-channeled.]
  strigatus, Winchell, 1862, Proc. Acad. Nat. Sci., Marshall Gr. [Sig. grooved.]
  trigonus, Winchell, 1862, Am. Jour. Sci., 2nd series, vol. 33, Marshall Gr. [Sig. three-cornered.]
  trisulcatus, Meek & Worthen, 1860, Proc. Acad. Nat. Sci. Phil., Kinderhook Gr. [Sig. three-furrowed.]
TROCHOCERAS, Hall, 1852, Pal. N. Y., vol. 2. [Ety. *trochos*, a hoop; *keras*, a horn.] This name was proposed by Barrande at about the same time.
  baeri, Meek & Worthen, 1865, Proc. Acad. Nat. Sci. Phil., Cin'ti Gr. [Ety. proper name.]
  bannisteri, Winchell & Marcy, 1865, (Gyroceras bannisteri) Mem. Bost. Soc. Nat. Hist., vol. 1, Niagara Gr. [Ety. proper name.]
  clio, Hall, 1861, 14th Reg. Rep., Schoharie grit. [Ety. mythological name.]
  costatum, Hall, 1861, Geo. Rep. of Wis., Niagara Gr. [Sig. ribbed.]
  desplainense, McChesney, 1860, New Pal. Foss., Niagara Gr. [Ety. proper name.]
  discoideum, Hall, 1862, 15th Reg. Rep., Schoharie grit. [Sig. disc-like.]
  eugenium, Hall, 1861, 14th. Reg. Rep., Schoharie grit. [Ety. proper name.]
  gebhardi, Hall, 1852, Pal. N. Y., vol. 2, Coralline Gr. [Ety. proper name.]
  incipiens, Barrande, 1869, Syst. Sil. de Boh., 4me ser., Quebec Gr. [Sig. beginning.]
  notum, Hall, 1867, 20th Reg. Rep. N. Y., Niagara Gr. [Sig. well known.]
  obliquatum, Hall, 1876, Illust. Devonian Foss., Up. Held. Gr. [Sig. bent.]
  pandion, Hall, 1876, Illust. Devonian Foss., Schoharie grit. [Ety. mythological name.]
  turbinatum, Hall, 1852, Pal. N. Y., vol. 2, Coralline Gr. [Sig. top-shaped.]
  waldronense, Hall, 1876, 28th Reg. Rep. N.Y., Niagara Gr. [Ety. proper name.]
TROCHOLITES, Conrad, 1838, Ann. Geo. Rep. N. Y. [Ety. *trochos*, a hoop; *lithos*, stone.]
  ammonius, Conrad, 1838, Ann. Geo. Rep. N. Y., Trenton, Utica & Hud. Riv. Gr. [Ety. mythological name.]
  planorbiformis, Conrad, 1842, Jour. Acad. Nat. Sci. Phil., Utica and Hud. Riv. Gr. [Sig. like a shell of the genus *Planorbis*.]

CLASS LAMELLIBRANCHIATA.

FAMILY AMBONYCHIIDÆ. — Ambonychia, Anomalodonta, Eopteria, Euchasma, Limoptera, Mytilarca.
FAMILY ANATINIDÆ.—Allorisma, Amphicœlia, Anatina, Chænomya, Chænocardia, Clinopistha, Cuneamya, Dexiobia, Ilionia, Leptodomus, Promacrus, Prothyris, Sedgwickia.
FAMILY ARCIDÆ.—Carbonarca, Dystactella, Macrodon, Megalomus, Megambonia, Ptychodesma, Solenomya, Tellinomya.
FAMILY AVICULIDÆ.—Actinodesma, Avicula, Aviculopinna, Bakevellia, Entolium, Euchondria, Gervillia, Inoceramus, Leiopteris, Monopteria, Monotis, Posidonia, Posidonomya, Pseudomonotis.
FAMILY CARDIIDÆ.—Cardiola, Cardiopsis, Cardium, Conocardium, Lunulacardium.
FAMILY CARDIOMORPHIDÆ.—Cardiomorpha, Edmondia.
FAMILY CYPRINIDÆ.—Astarte, Astartella, Cardinia, Cleidophorus, Cycloconcha, Cypricardella, Cypricardia, Cypricardinia, Cypricardites, Isocardia, Matheria, Pleurophorus.
FAMILY CYTHERODONTIDÆ.—Cytherodon, Schizodus.
FAMILY GRAMMYSIDÆ.—Grammysia.
FAMILY LUCINIDÆ.—Axinus, Lucina, Paracyclas.
FAMILY MODIOMORPHIDÆ.—Goniophora, Modiomorpha.
FAMILY MYACIDÆ.—Anthracomya.
FAMILY MYTILIDÆ.—Anodontopsis, Anthracoptera, Lithophaga, Modiola, Modiolopsis, Myalina, Mytilus.
FAMILY NUCULIDÆ.—Nucula, Nuculana, Nuculites, Pyrenomœus, Yoldia.
FAMILY NYASSIDÆ.—Nyassa.
FAMILY ORTHONOTIDÆ.—Orthodesma, Orthonota.
FAMILY OSTREIDÆ.—Lima, Ostrea, Placunopsis.
FAMILY PALANATINIDÆ.—Palanatina.
FAMILY PECTINIDÆ.—Aviculopecten, Lyriopecten, Pernopecten, Streblopteria.
FAMILY PHOLADELLIDÆ.—Cimitaria, Palæoneilo, Pholadella, Phthonia.
FAMILY PINNIIDÆ.—Pinna, Pinnopsis.
FAMILY PTERINIIDÆ.—Pterinea, Pteronites, Pteronitella.
FAMILY SANGUINOLITIDÆ.—Sanguinolites.
FAMILY SOLENIDÆ.—Solen, Solenopsis.
FAMILY TELLINIDÆ.—Sanguinolaria, Tellinopsis.
FAMILY TRIGONIIDÆ.—Dolabra, Ischyrinia, Lyrodesma.
FAMILY UNIONIDÆ.—Anthracosia.

ACTINODESMA, Hall, 1877, Pal. N. Y., vol. 5. [Ety. *aktin*, a beam; *desma*, a ligament.]
cruciforme, Conrad, 1841, (Avicula cruciformis) Ann. Rep. N. Y., Ham. Gr. [Sig. cross-shaped.]
erectum, Conrad, 1842, (Avicula erecta) Jour. Acad. Nat. Sci., vol. 8, Ham. Gr. [Sig. erect.]

ALLORISMA, King, 1844, Ann. Mag. Nat. Hist., vol. 14. [Ety. *allos*, variable; *ereisma*, support, expressive of the variable nature of the cartilage support or fulcrum.]
altirostrata, see Sedgwickia altirostrata.
antiqua, Swallow, 1863, Trans. St. Louis Acad. Sci., Low. Carb. [Sig. ancient.]

LAMELLIBRANCHIATA.

capax, Newberry, 1861, Ives' Col. Ex. Exped., Coal Meas. [Sig. capacious, large.]
clavata, McChesney, 1860, New Pal. Foss., Chester Gr. [Sig. club-shaped.]
cooperi, see Chænomya cooperi.
costata, Meek & Worthen, 1869, Proc. Acad. Nat. Sci. Phil., Coal Meas. [Sig. ribbed.]
cuneata, Swallow, 1858, Trans. St. Louis Acad. Sci., Mid. Coal Meas. [Sig. wedge-shaped.]
curta, Swallow, 1858, Trans. St. Louis Acad. Sci., Permian Gr. [Sig. short.]
elegans, King, as identified by Geinitz. See *A. geinitzi*.
ensiformis, Swallow, 1860, Trans. St. Louis Acad. Sci., Coal Meas. [Sig. in the form of a sword.]
geinitzi, Meek, 1867, Am. Jour. Sci., vol. 44, Coal Meas. [Ety. proper name.]
granosa, Shumard, 1858, (Leptodomus granosus) Trans. St. Louis Acad. Sci., Coal Meas. [Sig. full of grains.]
hannibalensis, see Grammysia hannibalensis.
hybrida, Meek & Worthen, 1865, (Chænomya hybrida) Proc. Acad. Nat. Sci. Phil., Keokuk Gr. [Sig. a hybrid.]
lanceolata, Swallow, 1858, Trans. St. Louis Acad. Sci., Permian Gr. [Sig. spear-shaped.]
lata, Swallow, 1858, Trans. St. Louis Acad. Sci., Mid. Coal Meas. [Sig. broad.]
leavenworthensis, see Chænomya leavenworthensis.
marionensis, White, 1876, Proc. Acad. Nat. Sci., St. Louis Gr. [Ety. proper name.]
minnehaha, see Chænomya minnehaha.
pleuropistha, Meek, 1871, Proc. Acad. Nat. Sci. Phil., Waverly Gr. [Ety. *pleuron*, the side; *opisthe*, behind.]
reflexa, Meek, 1872, Pal. E. Neb., Coal Meas. [Sig. turned back.]
sinuata, McChesney, 1860, New Pal. Foss., Chester Gr. [Sig. wavy.]
subcuneata, Meek & Hayden, 1858, Proc. Acad. Nat. Sci. Phil., Coal Meas. [Sig. somewhat wedge-shaped.]
subelegans, Meek, 1872, Pal. E. Neb., Coal Meas. [Sig. somewhat elegant.]
terminalis, Hall, 1852, Stans. Ex. to Gt. Salt Lake, Coal Meas. [Sig. terminal.]
ventricosa, Meek, 1871, Proc. Acad. Nat. Sci. Phil., Waverly Gr. [Sig. bulging out.]
winchelli, Meek, 1871, Proc. Acad. Nat. Sci. Phil., Waverly Gr. [Ety. proper name.]

AMBONYCHIA, Hall, 1847, Pal. N. Y., vol. 1. [Ety. *ambon*, the boss of a shield; *onyx*, a claw or talon.]
acutirostra, Hall, 1867, 20th Reg. Rep., Niagara Gr. [Sig. acute-beaked.]
alata, see Anomalodonta alata.
amygdalina, see Cypricardites amygdalinus.
aphæa, Hall, 1867, 20th Reg. Rep., Niagara Gr. [Ety. mythological name.]
attenuata, Hall, 1861, Geo. Rep. Wis., Trenton Gr. [Sig. diminished, drawn out.]
bellistriata, Hall, 1847, Pal. N. Y., vol. 1, Trenton Gr. [Sig. beautifully striated.]
cancellosa, Hall, 1861, Geo. Rep. Wis., Trenton Gr. [Sig. made crosswise, like net work.]
carinata, Goldfuss, 1826, (Pterinea carinata) Germ. Petref., Trenton & Hud. Riv. Gr. [Sig. keeled.]
casii, Meek & Worthen, 1866, Proc. Acad. Nat. Sci. Phil., Cin'ti Gr. [Ety. proper name.]
costata, James, 1873, Ohio Pal., vol. 1, Cin'ti Gr. [Sig. ribbed.]
erecta, Hall, 1861, Geo. Rep. Wis., Trenton Gr. [Sig. erect, straight.]
illinoisensis, Worthen, 1875, Geo. Rep. Ill., vol. 6, Cin'ti Gr. [Ety. proper name.]
intermedia, Meek & Worthen, 1868, Geo. Sur. Ill., vol. 3, Galena Gr. [Sig. intermediate, between *A. bellistriata* and *A. radiata*.
lamellosa, Hall, 1861, Geo. Rep. Wis., Trenton Gr. [Sig. made of many thin plates.]
maxima, Safford, 1869, Geo. of Tenn. Not defined.
mytiloides, Hall, 1847, Pal. N. Y., vol. 1, Chazy Gr. [Sig. resembling a *Mytilus*.]
neglecta, see Amphicœlia neglecta.
nitida, Billings, 1866, Catal. Sil. Foss. Antic., Anticosti Gr. [Sig. neat, pretty.]
obtusa, see Cypricardites obtusus.
orbicularis, Emmons, 1842, (Pterinea orbicularis) Geo. Rep. N. Y., Trenton Gr. [Sig. orb-shaped.]
planistriata, Hall, 1861, Geo. Rep. Wis., Trenton Gr. [Sig. plane-striated.]
radiata, Hall, 1847, Pal. N. Y., vol. 1, Trenton, Hud. Riv. Gr. & Mid. Sil. [Sig. radiated.]
striæcostata, see Pterinea striæcostata.
superba, Billings, 1866, Catal. Sil. Foss. Antic., Anticosti Gr. [Sig. superb, large and fine.]
swanana, Safford, 1869, Geo. of Tenn. Not defined.
undata, Emmons, 1842, (Pterinea undata) Geo. Rep. N. Y., Black Riv. & Trenton Gr. [Sig. wavy.]

AMPHICŒLIA, Hall, 1868, 20th Reg. Rep. N. Y. [Ety. *amphi*, both; *koilos*, hollow.]
costata, Hall & Whitfield, 1875, Ohio Pal., vol. 2, Niagara Gr. [Sig. ribbed.]
leidyi, Hall, 1867, 20th Reg. Rep. N. Y., Niagara Gr. [Ety. proper name.]
neglecta, McChesney, 1861, (Ambonychia neglecta) Pal. Foss., Niagara Gr. [Sig. neglected, overlooked.]

ANATINA, Lamarck, 1809, Phil. Zool. [Ety. pertaining to the duck, or like the duck's bill.]

leda, Hall, 1860, 13th Reg. Rep. N. Y., Ham. Gr. [Ety. mythological name.]
sinuata, see Ilionia sinuata.
ANODONTOPSIS, McCoy, 1851, Ann. & Mag. Nat. Hist., 2nd series, vol. 7. [Ety. from the resemblance to the shells of the genus *Anodonta*.]
(?) milleri, Meek, 1871, Am. Jour. Sci., 3rd series, vol. 2, Cin'ti Gr. [Ety. proper name.]
(?) unionoides, Meek, 1871, Am. Jour. Sci., 3rd series, vol. 2, Cin'ti Gr. [Sig. like a *Unio*.] This species is probably a *Modiolopsis*.
ventricosa, Billings, 1874, Pal. Foss., vol. 2, Devonian. [Sig. bulging out.]
ANOMALODONTA, S. A. Miller, 1874, Cin. Quar. Jour. Sci., vol. 1. [Sig. anomalous-toothed.]
alata, Meek, 1872, (Ambonychia alata) Proc. Acad. Nat. Sci. Phil., Cin'ti Gr. [Sig. winged.]
gigantea, S. A. Miller, 1874, Cin. Quar. Jour. Sci., vol. 1, Cin'ti Gr. [Sig. very large.]
ANTHRACOMYA, Salter, 1861, Mem. Geo. Sur. Gr. Brit. [Ety. *anthrax*, coal; *Mya*, a genus of shells.]
angulata, Dawson, 1860, (Naiadites angulata) Acadian Geology, Coal Meas. [Ety. from the angular outline of the posterior extremity.]
arenacea, Dawson, 1860, (Naiadites arenacea) Acadian Geology, Coal Meas. [Ety. from the arenaceous shale.]
carbonaria, Dawson, 1860, (Naiadites carbonaria) Acadian Geology, Coal Meas. [Ety. from the Coal Measures.]
elongata, Dawson, 1860, (Naiadites elongata) Acadian Geology, Coal Meas. [Ety. from the elongation laterally.]
lævis, Dawson, 1860, (Naiadites lævis) Acadian Geology, Coal Meas. [Sig. smooth.]
obtusa, Dawson, 1860, (Naiadites obtusa) Acadian Geology, Coal Meas. [Ety. from the obtuse anterior end.]
ovalis, Dawson, 1860, (Naiadites ovalis) Acadian Geology, Coal Meas. [Sig. oval.]
ANTHRACOPTERA, Salter, 1862, Mem. Geo. Sur. Gr. Brit. [Ety. *anthrax*, coal; *pteron*, a wing.]
carbonaria, see Anthracomya carbonaria.
fragilis, Meek & Worthen, 1866, Proc. Chi. Acad. Sci., Keokuk Gr. [Sig. easily broken.]
lævis, see Anthracomya lævis.
Arca, Linne, 1758. This genus is unknown in the palæozoic rocks.
carbonaria, Cox, 1857, Geo. Sur. Ky., vol. 3. See Macrodon carbonarius.
cuspidata, Swallow, 1858, Trans. St. Louis Acad. Sci., Up. Coal Meas. [Sig. pointed.]
modesta, Winchell, 1863, Proc. Acad. Nat. Sci., Chemung Gr. [Sig. modest, small.]
striata, Schlotheim, as identified by Geinitz, is *Macrodon tenuistriatus*.
punctifera, Dawson, 1868, Acad. Geol., Carb. [Sig. bearing dots.] The name was preoccupied by Deshayes in his work, 1824-1836.
ANTHRACOSIA, King, 1844, in Mag. Nat. Hist. [Ety. *anthrax*, coal.]
bradorica, Dawson, 1868. Acad. Geol., Carb. [Ety. proper name.]
ASTARTE, Sowerby, 1818, Min. Conch., vol. 2. [Ety. mythological name.]
mortonensis, Geinitz, 1866, Carb. und Dyas in Neb., Coal Meas. [Ety. proper name.]
nebrascensis, Geinitz, 1866, Carb. und Dyas in Neb., Coal Meas. [Ety. proper name.]
subtextilis, see Cardiomorpha subtextilis.
ASTARTELLA, Hall, 1858, Geo. Rep. Iowa. [Ety. diminutive of *Astarte*.]
concentrica, McChesney, 1860,(Edmondia concentrica) Descr. New Pal. Foss., Coal Meas. [Sig. from the concentric wrinkles or folds.]
newberryi, Meek, 1875, Ohio Pal., vol. 2, Coal Meas. [Ety. proper name.]
varica, McChesney, 1860, Descr. New Pal. Foss., Coal Meas. [Sig. straddling.]
vera, Hall, 1858, Geo. Rep. Iowa, Coal Meas. [Sig. true, genuine.]
AVICULA, Klein, 1753, Ostrac. [Ety. *avicula*, a little bird.]
acanthoptera, Hall, 1843, Geo. Rep. 4th Dist. N. Y., Chemung Gr. [Ety. *akantha*, a spine; *pteron*, a wing.]
acosta, Cox, 1857, Geo. Sur. Ky., vol. 3, Coal Meas. [Sig. without ribs.] The correct etymology would make this word *incosta*.
æquilatera, see Aviculopecten æquilaterus.
æquiradiata, Hall, 1859, Pal. N. Y., vol. 3, Low. Held. Gr. [Sig. having equal radiating striæ.]
æsopus, Conrad, 1842, Jour. Acad. Nat. Sci., vol. 8, Up. Sil. [Ety. proper name.]
angustirostra, Conrad, 1842, Jour. Acad. Nat. Sci., Up. Sil. [Sig. narrow-beaked.]
antiqua, see Bakevellia antiqua.
arenaria. Not American.
aviformis, Conrad, 1842, Jour. Acad. Nat. Sci., vol. 8, Low. Sil. [Sig. bird-like.]
bella, Conrad, 1841, Ann. Rep. N. Y., Devonian. [Sig. beautiful.]
bellula, Hall, 1859, Pal. N. Y., vol. 3, Low. Held. Gr. [Sig. quite beautiful.]
boydi, see Pterinea boydi.
chemungensis, see Pterinea chemungensis.
circulus, Shumard, 1855, Geo. Rep. Mo., Kinderhook Gr. [Sig. a circle.]
communis, Hall, 1859, Pal. N. Y., vol. 3, Low. Held. Gr. [Sig. a common form.]
cooperensis, see Entolium cooperense.
corrugata, James, 1874, Cin. Quar. Jour. Sci., vol. 1, Cin'ti Gr. [Sig. wrinkled.]
cruciformis, see Actinodesma cruciforme.

LAMELLIBRANCHIATA. 183

custa, (a misprint) see Eumicrotis curta.
damnoniensis, Sowerby, as identified in the early N. Y. Reports, see Pteronites chemungensis.
decussata, Hall, 1843, see Pteronites decussatus. Not Münster, 1834.
demissa, Conrad, 1842, Jour. Acad. Nat. Sci., vol. 8, Hud. Riv. Gr. [Sig. hanging down.]
desquamata, see Obolella desquamata.
elliptica, Hall, 1847, Pal. N. Y., vol. 1, Trenton Gr. [Sig. elliptical.]
emacerata, Conrad, 1842, Jour. Acad. Nat. Sci., Clinton & Niagara Gr. [Sig. made thin.]
erecta, see Actinodesma erecta.
ferruginea, Conrad, 1848, Proc. Acad. Nat. Sci., vol. 3, Up. Sil. [Sig. rusty.]
flabella, see Pterinea flabellum.
fragilis, see Aviculopecten fragilis. The name was preoccupied by DeFrance.
gebhardi, Conrad, 1841, Ann. Rep. N. Y., Oriskany sandstone. [Ety. proper name.]
hermione, Billings, 1862, Pal. Foss., vol. 1, Trenton Gr. [Ety. mythological name.]
honeymani, Hall, 1860, Can. Nat. & Geo., vol. 5, Sil. [Ety. proper name.]
insueta, Emmons, 1842, Geo. Rep. N. Y., Hud. Riv. and Cin'ti Gr. [Sig. unusual.]
lævis, see Pteronites lævis.
leptonota, Hall, 1843, Geo. Rep. 4th Dist. N. Y., Clinton Gr. [Ety. *leptos*, slender; *notos*, back.]
limiformis, Hall, 1852, Pal. N. Y., vol. 2, Coralline limestone. [Sig. resembling the genus *Lima*.]
longa, Geinitz, 1866, (Gervillia longa) Carb. und Dyas in Neb., Coal Meas. [Sig. long.]
longispina, see Pterinea longispina.
magna, Swallow, 1863, Trans. St. Louis Acad. Sci., Low. Carb. [Sig. large.]
manticula, Conrad, 1842, Jour. Acad. Nat. Sci., vol. 8, Low. Held. Gr. [Sig. a little wallet.]
morganensis, Meek & Worthen, 1866, (Pterinea morganensis) Proc. Acad. Nat. Sci., Coal Meas. [Ety. proper name.]
multilineata, Conrad, 1842, Jour. Acad. Nat. Sci., vol. 8, Chemung Gr. [Sig. many-lined.]
muricata, see Pteronites muricatus.
naviformis, Conrad, 1842, Jour. Acad. Nat. Sci., Low. Held. Gr. [Sig. boat-shaped.]
obliquata, Hall, 1859, Pal. N. Y., vol. 3, Low. Held. Gr. [Sig. oblique, slanting.]
oblonga, see Aviculopecten oblongus.
obscura, Hall, 1859, Pal. N. Y., vol. 3, Low. Held. Gr. [Sig. obscure.]
orbicularis, Stevens, 1858, Am. Jour. Sci., vol. 25, 2d ser., Coal Meas. The name was preoccupied by Sowerby in 1839.
orbiculata, Hall, 1843, see Lyriopecten orbiculatus.
orbiculata, Hall, 1852, Pal. N. Y., vol. 2, Niagara Gr. [Sig. orbicular.]
parilis, see Aviculopecten parilis.
pauciradiata, Hall, 1859, Pal. N. Y., vol. 3, Low. Held. Gr. [Sig. few-rayed.]
pectiniformis, see Aviculopecten pectiniformis.
perobliqua, Conrad, 1842, Jour. Acad. Nat. Sci., vol. 8, Ham. Gr. [Sig. very oblique.]
pleuroptera, Conrad, 1842, Jour. Acad. Nat. Sci., vol. 8, Ham. Gr. [Sig. side-winged.]
protexta, see Pterinea protexta.
quadrula, see Pterinea quadrula.
rectalaterarea, see Aviculopecten rectilaterarius.
recticosta, Hall, 1859, Pal. N. Y., vol. 3, Oriskany sandstone. [Sig. having straight costæ.]
rhomboidea, Hall, 1852, Pal. N. Y., vol. 2, Clinton Gr. [Sig. lozenge-shaped.]
rugosa, Conrad, 1841, Ann. Geo. Rep. N. Y., Water Lime Gr. The name was preoccupied by Munster in 1826.
scohariæ, Hall, 1859, Pal. N. Y., vol. 3, Low. Held. Gr. [Ety. proper name.]
securiformis, Hall, 1852, Pal. N. Y., vol. 2, Coralline limestone. [Sig. axe-shaped.]
securiformis, Hall, 1859, Pal. N. Y., vol. 3. This name was preoccupied.
semielliptica, Shumard, 1858, Trans. St. Louis Acad. Sci., Up. Coal Meas. [Sig. half-elliptical.]
shawneensis, Shumard, 1858, Trans. St. Louis Acad. Sci., Up. Coal Meas. [Ety. proper name.]
shumardi, see Entolium shumardi.
signata, see Aviculopecten signatus.
speciosa, Hall, 1843, Geo. Rep. 4th Dist. N. Y., Portage Gr. [Sig. beautiful.]
spinigera, see Pteronites spinigerus.
spinulifera, Hall, 1859, Pal. N. Y., vol. 3, Low. Held. Gr. [Sig. bearing spines.]
subæquilatera, Hall, 1859, Pal. N. Y., vol. 3, Low. Held. Gr. [Sig. somewhat equal-sided.]
subfalcata, Conrad, 1842, Jour. Acad. Nat. Sci., vol. 8, Ham. Gr. [Sig. somewhat sickle-shaped.]
subplana, Hall, 1852, Pal. N. Y., vol. 2, Niagara Gr. [Sig. somewhat flat.]
subquadrans, Conrad, 1842, Jour. Acad. Nat. Sci., vol. 8, Up. Sil. [Sig. somewhat squared.]
subrecta, Conrad, 1841, Ann. Rep. N. Y., Corralline limestone. See Pterinea subrecta.
subrecta, Hall, 1852. See Aviculopecten subrecta.
sulcata, Geinitz, 1866, (Gervillia sulcata) Carb. und Dyas in Neb., Coal Meas. If this form is properly referred to the genus Avicula, then the name was preoccupied by Reuss in 1843.

tenuilamellata, Hall, 1859, Pal. N. Y., vol. 3, Low. Held. Gr. [Sig. very thin-plated.]
textilis, Hall, 1859, Pal. N. Y., vol. 3, Low. Held. Gr. [Sig. woven.]
textilis var. arenaria, Hall, 1859, Pal. N. Y., vol. 3, Low. Held. Gr. [Sig. sandy.]
trentonensis, Conrad, 1842, Jour. Acad. Nat. Sci., vol. 8, Trenton Gr. [Ety. proper name.]
tricostata, Vanuxem, 1842, Geo. Rep. N. Y., Chemung Gr. [Sig. three-ribbed.]
trilobata, Conrad, 1842, Jour. Acad. Nat. Sci., vol. 8, Up. Sil. [Sig. three-lobed.]
triplistriata, Stevens, 1858, Am. Jour. Sci., vol. 25, Coal Meas. [Sig. three-lined.]
triquetra, Hall, 1843, Geo. Rep. N. Y., Onondaga Gr. [Sig. a triangle or three-cornered figure.]
triradiata, Conrad, 1842, Geo. Rep. 3rd Dist. N. Y., Portage Gr. [Sig. three-plicated.]
tuberculata, Conrad, 1838, Ann. Rep. N. Y., Corniferous Gr. [Sig. tuberculated.]
umbonata, Hall, 1859, Pal. N. Y., vol. 3, Low. Held. Gr. [Sig. protuberant.]
undata, Hall, 1852, Pal. N. Y., vol. 2, Niagara Gr. [Sig. wavy.]
welchi, James, 1874, Cin. Quar. Jour. Sci., vol. 1, Cin'ti Gr. [Ety. proper name.]
whitii, Winchell, 1863, Proc. Acad. Nat. Sci., Chemung Gr. [Ety. proper name.]

AVICULOPECTEN, McCoy, 1851, Ann. Mag. Nat. Hist., 2nd series, vol. 7. [Ety. *Aviculopecten*, from the resemblance to the shells of the genera *Avicula* and *Pecten*.]
acadicus, Hartt, 1868, Acad. Geol., Carb. [Ety. proper name.]
acutialatus, Swallow, 1858, (Avicula acutialata) Trans. St. Louis Acad. Sci., Permian Gr. [Sig. sharp-winged.]
æquilaterus, Hall, 1843, (Avicula æquilatera) Geo. Rep. 4th Dist. N. Y., Up. Held. Gr. and Marcellus shale. [Sig. equal-sided.]
amplus, Meek & Worthen, 1860, Proc. Acad. Nat. Sci. Phil., Keokuk Gr. [Sig. full, large.]
armigerus, Harlan, 1835, (Pecten armigerus) Trans. Geo. Soc. Penn., Carb. [Sig. armed.]
aviculatus, Swallow, 1858, (Pecten aviculatus) Trans. St. Louis Acad. Sci., Coal Meas. [Sig. spread out like an *Avicula*.]
burlingtonensis, Meek & Worthen, 1860, Proc. Acad. Nat. Sci., Burlington Gr. [Ety. proper name.]
cancellatus, Hall, 1843, (Pecten cancellatus) Geo. Rep. 4th Dist. N. Y., Chemung Gr. [Sig. cancelled.]
carboniferus, Stevens, 1858, (Pecten carboniferus) Am. Jour. Sci. & Arts, Coal Meas. [Sig. from the coal formation.]
caroli, Winchell, 1863, Proc. Acad. Nat. Sci., Chemung Gr. [Ety. proper name.]
clevelandicus, Swallow, 1858, (Pecten clevelandicus) Trans. St. Louis Acad. Sci., Permian Gr. [Ety. proper name.]
coloradoensis, Newberry, 1861, Ives' Col. Ex. Exped., Coal Meas. [Ety. proper name.]
convexus, Hall, 1843, (Pecten convexus) Geo. Rep. 4th Dist. N. Y., Chemung Gr. [Sig. convex.]
cora, Dawson, 1868, Acad. Geol., Carb. [Ety. mythological name.]
corcyanus, White, 1874, Rep. Invert. Foss. [Ety. proper name.]
coxanus, Meek & Worthen, 1860, Proc. Acad. Nat. Sci. Phil., Coal Meas. [Ety. proper name.]
crassicostatus, Hall, 1872, 24th Reg. Rep., Up. Held. Gr. [Sig. thick-ribbed.]
crenistriatus, Meek, 1871, Proc. Acad. Nat. Sci. Phil., Waverly Gr. [Sig. having crenulated striæ.]
debertanus, Dawson, 1868, Acad. Geol., Carboniferous. [Ety. proper name.]
dolabriformis, Hall, 1843, (Pecten (?) dolabriformis) Geo. Rep. 4th Dist. N. Y., Chemung Gr. [Sig. resembling a mattock or pick axe.]
duplicatus, Hall, 1843, (Pecten duplicatus) Geo. Rep. 4th Dist. N. Y., Chemung Gr. [Sig. doubled, folded in two.]
fragilis, Hall, 1843, (Avicula fragilis) Geo. Rep. 4th Dist. N. Y., Genessee slate. [Sig. frail, easily broken.]
glaber, see Pernopecten glaber.
gradicostus, White, 1862, Proc. Bost. Soc. Nat. Hist., vol. 9, Chemung Gr. [Sig. ribbed in steps.]
halli, Swallow, 1860, (Avicula halli) Trans. St. Louis Acad. Sci., Coal Meas. [Ety. proper name.]
hertzeri, Meek, 1875, Ohio Pal., vol. 2, Coal Meas. [Ety. proper name.]
indianensis, Meek & Worthen, 1866, Proc. Acad. Nat. Sci. Phil., Keokuk Gr. [Ety. proper name.]
intercostalis, Winchell, 1866, Rep. Low. Peninsula Mich., Ham. Gr. [Sig. lined between costæ.]
interlineatus, Meek & Worthen, 1860, Proc. Acad. Nat. Sci. Phil., Low. Coal Meas. [Sig. interlined.]
konincki, Meek & Worthen, 1860, Proc. Acad. Nat. Sci. Phil., Coal Meas. [Ety. proper name.]
limaformis, see Pernopecten limiformis.
lyelli, Dawson, 1868, Acad. Geol., Carb. [Ety. proper name.]
maccoyi, Meek & Hayden, 1865, Pal., Up. Mo., Permian Gr. [Ety. proper name.]
missouriensis, Shumard, 1855, (Pecten missouriensis) Geo. Rep. Mo., St. Louis Gr. [Ety. proper name.]
neglectus, Geinitz, 1860, (Pecten neglectus) Carb. und Dyas in Neb., Coal Meas. [Sig. overlooked.]

oblongus, Meek & Worthen, 1860, (Avicula oblonga) Proc. Acad. Nat. Sci. Phil., Keokuk Gr. [Sig. oblong.]
occidentalis, Shumard, 1855, (Pecten occidentalis) Geo. Rep. Mo., Carboniferous & Permian. [Sig. western.]
occidentalis, Winchell, 1863, Proc. Acad. Nat. Sci., Chemung Gr. This name was preoccupied.
orbiculatus, see Lyriopecten orbiculatus.
oweni, Meek & Worthen, 1860, Proc. Acad. Nat. Sci. Phil., Keokuk Gr. [Ety. proper name.]
parilis, Conrad, 1842, (Avicula parilis) Jour. Acad. Nat. Sci. Phil., vol. 8, Cornif. Gr. [Sig. equal, like.]
pecteniformis, Conrad, 1842, (Avicula pecteniformis) Jour. Acad. Nat. Sci., vol. 8, Up. Held. Gr. and Marcellus shale. [Sig. like a *Pecten*.]
pellucidus, Meek & Worthen, 1860, Proc. Acad. Nat. Sci. Phil., Coal Meas. [Sig. clear, transparent.]
princeps, Conrad, Ham. Gr. [Sig. principal.]
providencensis, Cox, 1857, (Pecten providencensis) Geo. Sur. Ky., vol. 3, Coal Meas. [Ety. proper name.]
rectilaterarius, Cox, 1857, (Avicula rectalaterarea) Geo. Sur. Ky., vol. 3, Coal Meas. [Sig. straight-sided.]
reticulatus, Dawson, 1868, Acad. Geol., Carboniferous. [Sig. reticulated.]
ringens, Swallow, 1858, (Pecten ringens) Trans. St. Louis Acad. Sci., Permian Gr. [Sig. gaping.]
rugæstriatus, Hall, 1843, (Lima rugæstriata) Geo. Rep. 4th Dist. N. Y., Chemung Gr. [Sig. having wrinkled striæ.]
sanduskyensis, syn. for Aviculopecten parilis.
signatus, Hall, 1843, (Avicula signata) Geo. Rep. 4th Dist. N. Y., Chemung Gr. [Sig. marked.]
simplex, Dawson, 1868, Acad. Geol., Carboniferous. [Sig. simple.]
striatus, Hall, 1843, (Pecten striatus) Geo. Rep. 4th Dist. N. Y., Chemung Gr. [Sig. striated.]
suborbicularis, Hall, 1843, (Pterinea suborbicularis) Geo. Rep. 4th Dist. N. Y., Chemung Gr. [Sig. somewhat orbicular.]
subrectus, Conrad, 1841, (Avicula subrecta) Ann. Rep. N. Y., Coralline limestone. [Sig. somewhat-straight.]
tenuicostus, Winchell, 1863, Proc. Acad. Nat. Sci., Chemung Gr. [Sig. slender-ribbed.]
undulatus, Hall, 1877, Pal. N. Y., vol. 5, Ham. Gr. [Sig. wavy.]
unionensis, Worthen, 1875, Geo. Sur. Ill., vol. 6, Corniferous Gr. [Ety. proper name.]

utahensis, Meek, 1860, (Pecten utahensis) Proc. Acad. Nat. Sci., Coal Meas. [Ety. proper name.]
whitei, Meek, 1872, Pal. E. Neb., Coal Meas. [Ety. proper name.]
williamsi, Meek, 1871, Proc. Acad. Nat. Sci., Chouteau limestone. [Ety. proper name.]
winchelli, Meek, 1875, Ohio Pal., vol. 2, Waverly Gr. [Ety. proper name.]

AVICULOPINNA, Meek, 1867, Am. Jour. Sci., vol. 44. [Ety. from the resemblance to the genera *Avicula* and *Pinna*.]
americana, Meek, 1867, Am. Jour. Sci., vol. 44, Coal Meas. [Ety. proper name.]

AXINUS, Sowerby, 1821, Min. Conch., vol. 3. [Ety. *axine*, battle-axe.] This genus is unknown in palæozoic rocks.
ovatus, see Schizodus ovatus.
securis, Shumard, 1859, Trans. St. Louis Acad. Sci., Permian Gr. [Sig. axe or hatchet-shaped.]

BAKEVELLIA, King, 1849, Perm. Foss. [Ety. proper name.]
antiqua, Munster, 1826, (Avicula antiqua) Goldfuss Germ. Petref. Not American.
parva, Meek & Hayden, 1858, Trans. Alb. Inst., vol. 4, Permian Gr. [Sig. small.]
(?) pulchra, Swallow, 1858, Trans. St. Louis Acad. Sci., Permian Gr. [Sig. beautiful.]

CARBONARCA, Meek & Worthen, 1870, Proc. Acad. Nat. Sci. Phil. [Ety. *carbo*, coal; *Arca*, a genus of shells; from the minute teeth on the cardinal margin, as in *Arca*.]
gibbosa, Meek & Worthen, 1870, Proc. Acad. Nat. Sci. Phil., Coal Meas. [Sig. gibbous.]

CARDINIA, Agassiz, 1838, in Societ. Basil. [Ety. *cardo*, the hinge of a door.]
æquimarginalis, Winchell, 1862, Proc. Acad. Nat. Sci., Marshall Gr. [Sig. equal-margined.]
antigonesensis, Dawson, 1868, Acad. Geo., Carb. [Ety. proper name.]
complanata, Winchell, 1862, Proc. Acad. Nat. Sci., Portage Gr. [Sig. smoothed.]
concentrica, Winchell, 1862, Proc. Acad. Nat. Sci., Marshall Gr. [Sig. concentric.]
cordata, Swallow, 1858, Trans. St. Louis Acad. Sci., Permian Gr. [Sig. heart-like.]
(?) fragilis, Cox, 1857, Geo. Sur. Ky., vol. 3, Coal Meas. [Sig. frail, easily broken.]
occidentalis, Swallow, 1860, Trans. St. Louis Acad. Sci., Chemung Gr. [Sig. western.]
subangulata, Swallow, 1858, Trans. St. Louis Acad. Sci., Permian Gr. [Sig. somewhat angular.]
subangulata, Dawson, 1868, Acad. Geol This name was preoccupied.

CARDIOLA, Broderip, 1834, Trans. Geo. Soc. [Ety. *kardia*, the heart.]
lincklaeni, Hall, 1877, Pal. N. Y., vol. 5, Ham. Gr. [Ety. proper name.]
radians, Hall, 1877, Pal. N. Y., vol. 5, Ham. Gr. [Sig. radiated.]
speciosa, Hall, 1877, Pal. N. Y., vol. 5, Genessee slate. [Sig. beautiful.]
ventricosa, Hall, 1870, Prelim. Notice Lam. shells, Goniatite limestone. [Sig. ventricose.]
vetusta, Hall, 1843, (Cardium vetustum) Geo. Rep. 4th Dist. N. Y., Portage Gr. [Sig. ancient.]
CARDIOMORPHA, DeKoninck, 1844, Anim. Foss. Carb. Belg. [Ety. *kardia*, heart; *morphe*, form.]
archiacana, Koninck, 1843, Desc. An. Foss. Belg., Carboniferous. [Ety. proper name.]
bellatula, Hall, 1870, Prelim. Notice Lam. shells, Ham. Gr. [Sig. small and beautiful.]
capuloides, Winchell, 1862, Proc. Acad. Nat. Sci., Marshall Gr. [Sig. resembling a shell of the genus *Capulus*.]
eriopia, Hall, 1870, Prelim. Notice. Lam. shells, Ham. Gr. [Ety. mytho. name.]
julia, Winchell, 1862, Proc. Acad. Nat. Sci., Marshall Gr. [Ety. proper name.]
kansasensis, Swallow, 1858, Trans. St. Louis Acad. Sci., Permian Gr. [Ety. proper name.]
missouriensis, Shumard, 1858, Trans. St. Louis Acad. Sci., Coal Meas. [Ety. proper name.]
modiolaris, Winchell, 1862, Proc. Acad. Nat. Sci., Marshall Gr. [Sig. like a shell of the genus *Modiola*.]
(?) obliquata, Meek, 1872, Proc. Acad. Nat. Sci. Phil., Cin'ti Gr. [Sig. oblique, slanting.]
ovata, see Dexiobia ovata.
parvirostris, White, 1862, Proc. Bost. Soc. Nat. Hist., vol. 8, Kinderhook Gr. [Sig. small-beaked.]
radiata, see Cardiopsis radiata.
(?) rhomboidea, Hall, 1858, Geo. Rep. Iowa, vol. 1, pt. 2, Kinderhook Gr. See Cardiomorpha subrhomboidea.
(?) rhomboidea, Swallow, 1858, Trans. St. Louis Acad. Sci., Permian Gr. [Sig. rhomboidal.]
subglobosa, Meek, 1875, Ohio Pal., vol. 2, Waverly Gr. [Sig. somewhat globose.]
suborbicularis, Hall, 1840, (Ungulina suborbicularis) Geo. Rep. 4th Dist. N. Y., Portage Gr. [Sig. somewhat orbicular.]
subrhomboidea, Hall, 1877, (Proposed instead of *Cypricardites rhomboidea*, in Geo. Rep. Iowa, which was preoccupied) Kinderhook Gr. [Sig. somewhat rhomb-like.]
subtextilis, Hall, 1843, (Astarte subtextilis) Geo. Rep. 4th Dist. N. Y., Portage Gr. [Sig. somewhat woven.]
textilis, Hall, 1877, Pal. N. Y., vol. 5, Chemung Gr. [Sig. woven.]

triangulata, Swallow, 1860, Trans. St. Louis Acad. Sci., Chemung Gr. [Sig. three-cornered, triangular.]
trigonalis, Winchell, 1863, Proc. Acad. Nat. Sci., Chemung Gr. [Sig. triangular.]
(?) *vetusta*, see Cypricardites vetustus.
vindobonensis, Hartt, 1868, Acad. Geol., Carboniferous. [Ety. proper name.]
CARDIOPSIS, Meek & Worthen, 1861, Proc. Acad. Nat. Sci. Phil. [Ety. *kardia*, the heart; *opsis*, appearance.]
. crassicostata, Hall, 1873, 24th Reg. Rep. N. Y., Schoharie grit and Corniferous limestone. [Sig. thick-ribbed.]
crenistriata, Winchell, 1862, Proc. Acad. Nat. Sci., Marshall Gr. [Sig. having wrinkled striæ.]
jejuna, Winchell, 1862, Proc. Acad. Nat. Sci., Marshall Gr. [Sig. poor, contemptible.]
megambonata, Winchell, 1862, Proc. Acad. Nat. Sci., Marshall Gr. [Sig. having a great umbone.]
parvirostris, White, 1862, Proc. Bost. Soc. Nat. Hist., vol. 9, Chemung Gr. [Sig. small-beaked.]
radiata, Meek & Worthen, 1860, (Cardiomorpha radiata) Proc. Acad. Nat. Sci. Phil., Kinderhook Gr. [Ety. from radiating striæ.]
robusta, Hall, 1877, Pal. N. Y., vol. 5, Schoharie grit and Portage Gr. [Sig. robust.]
CARDIUM, Linnæus, 1758, Syst. Nat., 10th Ed. [Ety. *kardia*, the heart.]
iowensis, see Cypricardites iowensis.
lexingtonense, Swallow, 1858, Trans. St. Louis Acad. Sci., Mid. Coal Meas. [Ety. proper name.]
napoleonense, Winchell, 1862, Proc. Acad. Nat. Sci., Marshall Gr. [Ety. proper name.]
(?) *vetustum*, see Cardiola vetusta.
CHÆNOCARDIA, Meek & Worthen, 1869, Proc. Acad. Nat. Sci. [Ety. *chaino*, to gape; *kardia*, the heart.]
ovata, Meek & Worthen, 1869, Proc. Acad. Nat. Sci. Phil., Coal Meas. [Sig. ovate.]
CHÆNOMYA, Meek, 1864, Pal. of Up. Mo. [Ety. *chaino*, to open or gape; *Mya*, a genus of shells.]
cooperi, Meek & Hayden, 1858, (Panopæa cooperi) Trans. Alb. Inst., vol. 4, Coal Meas. [Ety. proper name.]
hybrida, see Allorisma hybrida.
leavenworthensis, Meek & Hayden, 1858, (Allorisma leavenworthensis) Proc. Acad. Nat. Sci. Phil., Coal Meas. [Ety. proper name.]
minnehaha, Swallow, 1858, (Allorisma (?) minnehaha) Trans. St. Louis Acad. Sci., Coal Meas. [Ety. proper name.]
rhomboidea, Meek & Worthen, 1865, Proc. Acad. Nat. Sci. Phil., St. Louis Gr. [Sig. rhomboidal — lozenge-shaped.]

LAMELLIBRANCHIATA. 187

CIMITARIA, Hall, 1870, Prelim. Notice Lam. shells. [Ety. from resemblance to a Cimitar.]
corrugata, Conrad, 1842, (Cypricardites corrugata) Jour. Acad. Nat. Sci., vol. 8, Ham. Gr. [Sig. corrugated.]
elongata, Conrad, 1841, (Cypricardites elongatus) Ann. Rep. N. Y., Ham. Gr. [Sig. elongated.]
recurva, Conrad, 1842, (Cypricardites recurva) Jour. Acad. Nat. Sci., vol. 8, Ham. Gr. [Sig. recurved.]
CLEIDOPHORUS, Hall, 1847, Pal. N. Y., vol. 1. [Ety. *kleidos*, a clavicle; *phoros*, bearing; in allusion to the depression on the side of the cast, anterior to the beak, indicating an interior ridge in each valve.]
concentricus, Hall, 1860, Can. Nat. & Geo., vol. 5, Low. Sil. [Sig. marked with concentric lines.]
concentricus, Dawson, 1868. The name was preoccupied.
cuneatus, Hall, 1860, Can. Nat. & Geo., vol. 5, Low. Sil. [Sig. wedge-shaped.]
elongatus, Hall, 1860, Can. Nat. & Geo., vol. 5, Low. Sil. [Sig. elongated.]
erectus, Hall, 1860, Can. Nat. & Geo., vol. 5, Up. Sil. [Sig. erect.]
erectus, Dawson, 1868. The name was preoccupied.
fabula, Hall, 1845, Am. Jour. Sci. & Arts, vol. 43, (Nucula fabula) Cin'ti Gr. [Sig. a little bean.]
*macch*œ*neyanus*, syn. for Modiolopsis recta.
neglectus, Hall, 1862, Geo. Rep. Wis., Hud. Riv. Gr. [Sig. overlooked.]
nuculiformis, Hall, 1860,Can. Nat. & Geo., vol. 5, Up. Sil. [Sig. like a shell of the genus *Nucula*.]
oblongus, Hall, 1843, (Nucula oblonga) Geo. Rep. 4th Dist. N. Y., Ham. Gr. [Sig. oblong.]
planulatus, Conrad, 1841, (Nuculites planulatus) Ann. Rep. N. Y., Hud. Riv. Gr. [Sig. flattened.]
semiradiatus, Hall, 1860, Can. Nat. & Geo., vol. 5, Low. Sil. [Sig. half-rayed.]
solenoides, see Solenopsis solenoides.
subovatus, Hall, 1860, Can. Nat. & Geo., vol. 5, Low. Sil. [Sig. somewhat ovate.]
CLINOPISTHA, Meek & Worthen, 1870, Proc. Acad. Nat. Sci. Phil. [Ety. *klino*, I lean; *opisthe*, backwards.]
antiqua, Meek, 1871, Proc. Acad. Nat. Sci. Phil., Corniferous Gr. [Sig. ancient.]
lævis, Meek & Worthen, 1870, Proc. Acad. Nat. Sci. Phil., Coal Meas. [Sig. smooth.]
radiata, Hall, 1858, (Edmondia radiata) Geo. Rep. Iowa, Coal Meas. [Sig. radiated.]
CONOCARDIUM, Bronn, 1835, Læth. Geo. [Ety. *konos*, a cone; *kardia*, the heart.]
acadianum, Hartt, 1868, Acad. Geol., Carb. [Ety. proper name.]
æquilaterale, Hall, 1858, Trans. Alb. Inst. vol. 4, Warsaw Gr. [Sig. equal-sided.]

attenuatum, Conrad, 1842, (Pleurorhynchus attenuatus) Jour. Acad. Nat.Sci., vol. 8, Up. Sil. [Sig. drawn out.]
bifarium, Winchell, 1856, Rep. Low. Peninsula Mich., Ham. Gr. [Sig. double.]
blumenbachium, see Euchasma blumenbachi.
carinatum, Hall, 1858, Trans. Alb. Inst., vol. 4, Warsaw Gr. [Sig. keeled.]
catastomum, Hall, 1858, Trans.Alb. Inst., vol. 4, Warsaw Gr. [Sig. gaping at the lower end.]
crassifrons, Conrad, 1842, (Pleurorhynchus crassifrons) Jour. Acad. Nat.Sci., vol. 8, Ham. Gr. [Sig. thick front.]
cuneatum, Hall, 1858, Trans. Alb. Inst., vol. 4, Warsaw Gr. [Sig. wedge-shaped.]
cuneus, Conrad, 1840, (Pleurorhynchus cuneus) Ann. Rep. N. Y., Up. Held. Gr. [Sig. a wedge.]
eboraceum, Hall, 1860, 13th Reg. Rep., Ham. Gr. [Ety. *Eboracum*, York.]
elegantulum, Billings, 1866, Catal. Sil. Foss. Antic., Anticosti Gr. [Sig. quite handsome.]
emmetense, Winchell, 1866, Rep. Low. Peninsula Mich., Ham. Gr. [Ety. proper name.]
immaturum, Billings, 1862, Pal. Foss. vol. 1, Black Riv. Gr. [Sig. immature.]
inceptum, Hall, 1859, Pal. N. Y., vol. 3, Low. Held. Gr. [Sig. a beginning.]
meekanum, Hall, 1858, Trans. Alb. Inst., vol. 4, Warsaw Gr. [Ety. proper name.]
niagarense, Winchell & Marcy, 1865, Mem. Bost. Soc. Nat. Hist., Niagara Gr. [Ety. proper name.]
obliquum, Meek & Worthen, 1865, Proc. Acad. Nat. Sci. Phil., Coal Meas. [Sig. oblique, slanting.]
ohioense, Meek, 1871, Proc. Acad. Nat. Sci. Phil., Corniferous Gr. [Ety. proper name.]
ornatum, Winchell & Marcy, 1865, Mem. Bost. Soc. Nat. Hist., Niagara Gr. [Sig. adorned.]
prattenanum, Hall, 1858, Trans. Alb. Inst., vol. 4, Warsaw Gr. [Ety. proper name.]
trigonale, Hall, 1843,(Pleurorhynchus trigonalis) Geo. Rep. 4th Dist. N. Y., Corniferous Gr. [Sig. triangular.]
ventricosum, Hall, 1860, 13th Reg. Rep. N. Y., Ham. Gr. [Sig. ventricose.]
vomer, Conrad, 1842, (Pleurorhynchus vomer) Jour. Acad. Nat. Sci., vol. 8, Devonian. [Sig. a plough-share.]
Ctenodonta, Salter, 1851. Syn. for Tellinomya.
abrupta, see Tellinomya abrupta.
angela, see Tellinomya angela.
astartiformis, see Tellinomya astartiformis.
contracta, see Tellinomya contracta.
gibberula, see Tellinomya gibberula.
hartsvillensis, see Tellinomya hartsvillensis.
hubbardi, syn. for Nuculites sulcatinus.
iphigenia, see Tellinomya iphigenia.
logani, see Tellinomya logani.

LAMELLIBRANCHIATA.

Cucullæa, Lamarck, 1801, Syst. An. [Ety. *Cucullus*, a hood.]
 opima, syn. for Nucula lirata.
CUNEAMYA, Hall & Whitfield, 1875, Ohio Pal., vol. 2. [Ety. *cuneus*, a wedge; *Mya*, a genus of shells.]
 miamiensis, Hall & Whitfield, 1875, Ohio Pal., vol. 2, Cin'ti Gr. [Ety. proper name.]
 scapha, Hall & Whitfield, 1875, Ohio Pal., vol. 2, Cin'ti Gr. [Sig. a skiff or boat.]
CYCLOCONCHA, S. A. Miller, 1874, Cin. Quar. Jour. Sci., vol. 1. [Ety. in allusion to the nearly circular form of the shell.]
 mediocardinalis, S. A. Miller, 1874, Cin. Quar. Jour. Sci., vol. 1, Cin'ti Gr. [Ety. in allusion to the position of the cardinal teeth near the middle of the hinge line.]
CYPRICARDIA, Lamarck, 1801, Syst. An. sans Vert. [Ety. from the two genera *Cyprina* and *Cardium*.]
 alata, see Modiomorpha alata.
 angusta, see Cypricardites angustus.
 angustata, see Cypricardites angustatus.
 chouteauensis, Swallow, 1863, Trans. St. Louis Acad. Sci., Chemung Gr. [Ety. proper name.]
 contracta, see Cypricardites contractus.
 indianensis, Hall, 1858, Trans. Alb. Inst., vol. 4, Warsaw Gr. [Ety. proper name.]
 insecta, Dawson, 1868, Acad. Geol., Carboniferous. [Sig. cut into.]
 leidyi, Lea, see Isaia leidyi.
 obsoleta, see Cypricardites obsoletus.
 occidentalis, Hall, 1852, Stans. Ex. to Great Salt Lake, Coal Meas. [Sig. western.]
 occidentalis, Swallow, 1863, Trans. St. Louis Acad. Sci., Carb. This name was preoccupied.
 pikensis, Swallow, 1863, Trans. St. Louis Acad. Sci., Coal Meas. [Ety. proper name.]
 plicatula, Swallow, 1858, Trans. St. Louis Acad. Sci., Mid. Coal Meas. [Sig. having small plications.]
 rhombea, see Cypricardites rhombeus.
 rigidi, see Sanguinolites rigidi.
 shumardana, Swallow, 1863, Trans. St. Louis Acad. Sci., Low. Carb. [Ety. proper name.]
 subplana, Hall, 1858, Trans. Alb. Inst., vol. 4, Warsaw Gr. [Sig. somewhat flat.]
 ventricosa, Hall, 1860, 13th Reg. Rep. N. Y., Ham. Gr. [Sig. ventricose.]
 wheeleri, see Schizodus wheeleri.
CYPRICARDELLA, Hall, 1858, Trans. Alb. Inst., vol. 4. [Ety. diminutive of *Cypricardia*.]
 nucleata, Hall, 1858, Trans. Alb. Inst., vol. 4, Warsaw Gr. [Sig. like a kernel.]
 oblonga, Hall, 1858, Trans. Alb. Inst., vol. 4, Warsaw Gr. [Sig. oblong.]
 plicata, Hall, 1858, Trans. Alb. Inst., vol. 4, Warsaw Gr. [Sig. folded, plicated.]
 subelliptica, Hall, 1858, Trans. Alb. Inst., vol. 4, Warsaw Gr. [Sig. somewhat elliptical.]
CYPRICARDINIA, Hall, 1859, Pal. N. Y., vol. 3. [Ety. *Cypricardinia*, from its resemblance to *Cypricardia*.]
 arata, Hall, 1867, 20th Reg. Rep. N. Y., Niagara Gr. [Sig. plowed.]
 carbonaria, Meek, 1871, Proc. Acad. Nat. Sci. Phil., Coal Meas. [Sig. pertaining to coal.]
 concentrica, Hall, 1859, Pal. N. Y., vol. 3, Low. Held. Gr. [Sig. from the concentric ridges that mark the surface.]
 crassa, Hall, 1859, Pal. N. Y., vol. 3, Low. Held. Gr. [Sig. thick, from the thick lamellæ.]
 (?) cylindrica, Hall, 1872, 24th Reg. Rep. N. Y., Corniferous Gr. [Sig. cylindrical.]
 distincta, Billings, 1874, Pal. Foss., vol. 2, Devonian. [Sig. distinct.]
 dorsata, Hall, 1859, Pal. N. Y., vol. 3, Low. Held. Gr. [Sig. high-backed.]
 indenta, Conrad, 1842, (Cypricardites indenta) Jour. Acad. Nat. Sci., vol. 8, Up. Held. Gr. [Sig. indented.]
 inflata, Emmons, 1842, (Nuculites inflata) Emmons' Geo. Rep. N. Y., Trenton Gr. [Sig. swollen out, inflated.]
 inflata var. subæquivalvis, Hall, 1872, 24th Reg. Rep. N. Y., Up. Held. Gr. [Sig. valves somewhat equal.]
 lamellosa, Hall, 1859, Pal. N. Y., vol. 3, Up. Held. Gr. [Sig. in allusion to the distant lamellæ on the surface.]
 planulata, Conrad, 1842, (Pterinea planulata) Jour. Acad. Nat. Sci., vol. 8, Low. Held. Gr. [Sig. flattened.]
 pygmæa, Conrad, 1842, (Pterinea pygmæa) Jour. Acad. Nat. Sci., vol. 8, Up. Held. Gr. [Sig. a pigmy, very small.]
 sublamellosa, Hall, 1859, Pal. N. Y., vol. 3, Low. Held. Gr. [Sig. somewhat lamellose; from the concentric striæ on the surface.]
CYPRICARDITES, Conrad, 1841, Ann. Geo. Rep. N. Y. [Ety. from resemblance to the genus *Cypricardia*.]
 acutumbonus, Billings, 1866, (Cyrtodonta acutumbona) Catal. Sil. Foss. Antic., Anticosti Gr. [Ety. from the strongly carinated, or sharp-keeled umbones.]
 alta, see Modiomorpha alta.
 alveatus, Conrad, 1843, Geo. Rep. 3rd Dist. N. Y., Ham. Gr. [Sig. hollowed out.]
 amygdalinus, Hall, 1847, (Ambonychia amygdalina) Pal. N. Y., vol. 1, Black Riv. and Trenton Gr. [Sig. like an almond.]
 angustus, Hall, 1843, (Cypricardia angusta) Geo. Rep. 4th Dist. N. Y., Clinton Gr. [Sig. narrow, slender.]
 angustatus, see Modiomorpha angustata.
 angustifrons, syn. for Modiolopsis modiolaris.

anodontoides, see Modiolopsis anodontoides.
anticostiensis, Billings, 1866, (Cyrtodonta (?) anticostiensis) Catal. Sil. Foss. Antic., Hud. Riv. Gr. [Ety. proper name.]
bayfieldi, Billings, 1858, (Vanuxemia bayfieldi) Can. Nat. & Geo. vol. 4, Hud. Riv. Gr. [Ety. proper name.]
bisulcata, see Grammysia bisulcata.
breviusculus, Billings, 1859, (Cyrtodonta breviuscula) Can. Nat. & Geo., vol. 4, Chazy Gr. [Sig. very short.]
canadensis, Billings, 1858, (Cyrtodonta canadensis) Can. Nat. & Geo., vol. 3, Black Riv. & Trenton Gr. [Ety. proper name.]
carinatus, see Sanguinolites carinatus.
carinatus, Meek, 1872, (Dolabra carinata) Proc. Acad. Nat. Sci. Phil., Cin'ti Gr. This name was preoccupied.
cariniferus, Conrad, 1842, Jour. Acad. Nat. Sci., vol. 8, Chemung Gr. [Sig. bearing a ridge; from the strongly carinated umbonal slope.]
catskillensis, see Modiomorpha catskillensis.
chemungensis, see Sanguinolites chemungensis.
concentrica, see Modiomorpha concentrica.
contractus, Hall, 1843, (Cypricardia contracta) Geo. Rep. 4th Dist. N. Y., Low. Carb. [Sig. contracted.]
cordiformis, Billings, 1858, (Cyrtodonta cordiformis) Can. Nat. & Geo., vol. 3, Black Riv. & Trenton Gr. [Sig. heart-shaped.]
corrugatus, Conrad, 1842, Jour. Acad. Nat. Sci., vol. 8, Ham. Gr. [Sig. wrinkled.]
curtus, Conrad, 1841, Ann. Rep. N. Y., Cin'ti Gr. [Sig. short.]
dixonensis, Meek & Worthen, 1866, Proc. Chi. Acad. Sci., vol. 1, Trenton Gr. [Ety. proper name.]
elongatus, Conrad, 1841, Ann. Rep. N. Y., Up. Held. Gr. [Sig. elongated.]
emma, Billings, 1862, (Cyrtodonta emma) Pal. Foss., vol. 1, Hud. Riv. Gr. [Ety. proper name.]
ferrugineus, Hall & Whitfield, 1875, Ohio Pal., vol. 2, Clinton Gr. [Sig. of the color of rusty iron; from iron ore beds.]
ganti, Safford, 1869, (Cyrtodonta ganti) Geo. of Tenn., Trenton & Hud. Riv. Gr. [Ety. proper name.]
hainesi, S. A. Miller, 1874, Cin. Quar. Jour.Sci.,vol. 1, Cin'ti Gr. [Ety. proper name.]
harrietta, Billings, 1862, (Cyrtodonta harrietta) Pal. Foss., vol. 1, Hud. Riv. Gr. [Ety. proper name.]
haynanus, Safford, 1869, (Cyrtodonta hayniana) Geo. of Tenn., Trenton & Hud. Riv. Gr. [Ety. proper name.]
hindi, Billings, 1862, (Cyrtodonta hindi) Pal. Foss., vol. 1, Hud. Riv. Gr. [Ety. proper name.]

huronensis, Billings, 1858, (Cyrtodonta huronensis) Can. Nat. & Geo., vol. 3, Black Riv. and Trenton Gr. [Ety. proper name.]
inconstans, Billings, 1858, (Vanuxemia inconstans) Can. Nat. & Geo., vol. 3, Chazy Gr. [Sig. changing.]
indenta, see Cypricardinia indenta.
inflatus, Conrad, 1842, Jour. Acad. Nat. Sci., vol. 8, Onondaga Gr. [Sig. inflated.]
insularis, Billings, 1866, (Cyrtodonta insularis) Catal. Sil. Foss. Antic., Hud. Riv. Gr. [Sig. belonging to an island.]
islandicus, Hall, 1877, (Proposed here instead of *Cypricardites ventricosus,* Hall, 1859) Low. Held. Gr. [Ety. from the type locality, St. Joseph's Island.]
iowensis, Owen, 1840, Rep. on Mineral lands, Calciferous Gr. [Ety. proper name.]
latus, Hall, 1847, (Modiolopsis latus) Pal. N. Y., vol. 1, Trenton Gr. [Sig. wide.]
leucothea, Billings, 1862, (Cyrtodonta leucothea) Pal. Foss., vol. 1, Black Riv. Gr. [Ety. mythological name.]
marcellensis, see Lunulacardium marcellense.
modiolaris, Emmons, syn. for Modiolopsis nasuta.
montrealensis, Billings, 1859, (Vanuxemia montrealensis) Can. Nat. & Geo., vol. 4, Chazy Gr. [Ety. proper name.]
mytiloides, Conrad, 1841, Ann. Rep. N. Y., Ham. Gr. [Ety. from the resemblance to *Mytilus.*]
nasutus, see Modiolopsis nasuta.
niota, Hall, 1861, Geo. Rep. Wis., Trenton Gr. [Ety. proper name.]
obliquus, Meek & Worthen, 1868, Geo. Sur. Ill., vol. 3, Galena Gr. [Sig. oblique, slanting.]
oblongus, syn. for Modiomorpha concentrica.
obsoletus, Hall, 1843, (Cypricardia obsoleta) Geo. Rep. 4th Dist. N. Y., Clinton Gr. [Sig. obsolete.]
obtusus, Hall, 1847, (Ambonychia obtusa) Pal. N. Y., vol. 1, Black Riv. and Trenton Gr. [Sig. obtuse.]
ovata, syn. for Modiolopsis modiolaris.
plebeius, Billings, 1866, (Cyrtodonta plebeia) Catal. Sil. Foss. Antic., Hud. Riv. Gr. [Sig. common.]
ponderosus, Billings, 1862, (Cyrtodonta ponderosa) Pal. Foss., vol. 1, Hud. Riv. Gr. [Sig. thick, heavy.]
quadrilateralis, Hall, 1867, 20th Reg. Rep. N. Y., Niagara Gr. [Sig. four-sided.]
radiatus, Conrad, 1841, Ann. Rep. N. Y., Ham. Gr. [Ety. from the radiating striæ.]
rectus, Conrad, 1841, Ann. Rep. N. Y., Up. Held. Gr. [Sig. straight.]
rectirostris, Hall, 1861, Geo. Rep. Wis., Trenton Gr. [Sig. straight-beaked.]
recurvus, Conrad, 1842, Jour. Acad. Nat. Sci., vol. 8, Ham. Gr. [Sig. recurved.]

rhombeus, Hall, 1843, (Cypricardia rhombea) Geo. Rep. 4th Dist. N. Y., Low. Carb. [Sig. rhomboidal.]
rotundatus, Hall, 1861, Geo. Rep. Wis., Trenton Gr. [Sig. rounded.]
rugosus, Billings, 1858, (Cyrtodonta rugosa) Can. Nat. & Geo., vol. 3, Black Riv. Gr. [Sig. wrinkled.]
rugosa, see Goniophora rugosa.
saffordi, Hall, 1852, (Palæarca saffordi) 12th Reg. Rep., Low. Held. Gr. [Ety. proper name.]
sectifrons, see Phthonia sectifrons.
sigmoideus, Billings, 1859, (Cyrtodonta sigmoidea) Can. Nat. & Geo., vol. 3, Black Riv. Gr. [Sig. like the Greek letter Sigma.] [toides.
sinuata, syn. for Modiolopsis anodonspiniferus, Billings, 1858, (Cyrtodonta spinifera) Can. Nat. & Geo., vol. 3, Black Riv. Gr. [Sig. bearing spines.]
sterlingensis, Meek & Worthen, 1866, (Dolabra sterlingensis) Proc. Acad. Nat. Sci. Phil., Cin'ti Gr. [Ety. proper name.]
subalatus, see Modiomorpha subalata.
subangulatus, Hall, 1847, (Edmondia subangulata) Pal. N. Y., vol. 1, Black Riv. and Trenton Gr. [Sig. somewhat angular.]
subcarinatus, Billings, 1858, (Cyrtodonta subcarinata) Can. Nat. & Geo., vol. 3, Black Riv. Gr. [Sig. somewhat carinated.]
subspatulatus, Hall, 1847, (Modiolopsis subspatulata) Pal. N. Y., vol. 1, Black Riv. & Trenton Gr. [Sig. somewhat spatula-shaped.]
subtentus, Conrad, 1843, Geo. Rep. 3rd Dist. N. Y., Ham. Gr. [Sig. somewhat curved.]
subtruncatus, Hall, 1847, (Edmondia subtruncata) Pal. N. Y., vol. 1., Black Riv. & Trenton Gr. [Sig. somewhat truncated.]
truncatus, see Sanguinolites truncatus.
ungulatus, Billings, 1866, (Cyrtodonta ungulata) Catal. Sil. Foss. Antic., Hud. Riv. Gr. [Sig. having claws.]
ventricosus, Hall, 1847, (Edmondia ventricosa) Pal. N. Y., vol. 1, Trenton Gr. [Sig. bulging out.]
ventricosus, Hall, 1859, (Palæarca ventricosa) Pal. N.Y., vol. 3, Low. Held. Gr. This name was preoccupied. See Cypricardites islandicus.
vetustus, Hall, 1847, (Cardiomorpha vetusta) Pal. N. Y., vol. 1, Trenton Gr. [Sig. ancient.]
winchelli, Safford, 1869, (Cyrtodonta winchelli) Geo. Tenn., Trenton and Hud. Riv. Gr. [Ety. proper name.]
Cyrtodonta, syn. for Cypricardites.
acutumbona, see Cypricardites acutumbonus.
anticostiensis, see C. anticostiensis.
breviuscula, see C. breviusculus.
canadensis, see C. canadensis.
cordiformis, see C. cordiformis.
emma, see C. emma.
ganti, see C. ganti.
harrietta, see C. harrietta.
hayniana, see C. haynanus.
hindi, see C. hindi.
huronensis, see C. huronensis.
insularis, see C. insularis.
leucothea, see C. leucothea.
normanensis, Safford. Not defined.
plebeia, see Cypricardites plebeius.
ponderosa, see C. ponderosus.
rugosa, see C. rugosus.
saffordi, see C. saffordi.
sigmoidea, see C. sigmoideus.
spinifera, see C. spiniferus.
subcarinata, see C. subcarinatus.
ungulata, see C. ungulatus.
winchelli, see C. winchelli.

Cytherodon, Hall, 1873, in 23rd Reg. Rep. N. Y. [Ety. *Cythere*, a genus of shells; *odous*, a tooth.]
appressus, Conrad, (Nuculites appressa) 1842, Jour. Acad. Nat. Sci., vol. 8, Ham. Gr. [Sig. pressed together.]
(?) placidus, Billings, 1874, Pal. Foss., vol. 2, Up. Sil. [Sig. smooth.]
socialis, Billings, 1874, Pal. Foss., vol. 2, Up. Sil. [Sig. social, clustered together.]
tumidus, Hall, 1870, (Schizodus tumidus) Prelim. Notice Lam. shells, etc., Up. Held. Gr. [Sig. swollen, tumid.]

Dexionia, Winchell, 1863, Proc. Acad. Nat. Sci. [Ety. *dexios*, on the right side; *bia*, strength.]
halli, Winchell, 1863, Proc. Acad. Nat. Sci., Chemung Gr. [Ety. proper name.]
ovata, Hall, 1858, (Cardiomorpha ovata) Geo. Rep. Iowa, Kinderhook Gr. [Sig. ovate.]
whitei, Winchell, 1863, Proc. Acad. Nat. Sci., Chemung Gr. [Ety. proper name.]

Dolabra, McCoy, 1844, Syn. Carb. Foss. Ireland. [Ety. *dolabra*, a mattock or pick axe.]
alpina, Hall, 1858, Geo. Rep. Iowa, Coal Meas. [Sig. alpine.]
carinata, see Cypricardites carinatus.
sterlingensis, see Cypricardites sterlingensis.

Dystactella, Hall, 1877, Pal. N. Y., vol. 5. [Ety. *dystaktos*, hard to arrange.]
subnasuta, Hall, 1877, Pal. N. Y., vol. 5, Up. Held. Gr. [Sig. somewhat nasute.]

Edmondia, DeKoninck, 1844, Desc. Anim. Foss., Carb. Belg. [Ety. proper name.]
anomala, Dawson, 1868, Acad. Geo., Carb. [Sig. irregular, out of order.]
aspenwallensis, Meek, 1871, Hayden's Rep. Sur. Wyoming, Coal Meas. [Ety. proper name.]
bicarinata, Winchell, 1863, Proc. Acad. Nat. Sci., Chemung Gr. [Sig. doublekeeled.] Prof. Hall regards this as a syn. for *Sanguinolites rigidus*.

binumbonata, Winchell, 1862, Proc. Acad. Nat. Sci., Marshall Gr. [Sig. having double umbones.]
burlingtonensis, White & Whitfield, 1862, Proc. Bost. Soc. Nat. Hist., vol. 8, Kinderhook Gr. [Ety. proper name.]
calhouni, see Pleurophorus calhouni.
concentrica, see Astartella concentrica.
depressa, Hall, 1870, Prelim. Notice Lam. shells, Waverly Gr. [Sig. depressed.]
elliptica, Winchell, 1863, Proc. Acad. Nat. Sci., Chemung Gr. [Sig. elliptical.]
gibbosa, Swallow, 1858, Trans. St. Louis Acad. Sci., Permian Gr. [Sig. gibbous, tumid.]
glabra, Meek, 1872, Pal. E. Neb., Coal Meas. [Sig. smooth.]
hartti, Dawson, 1868, Acad. Geol., Carb. [Ety. proper name.]
hawni, Swallow, 1858, Trans. St. Louis Acad. Sci., Coal Meas. [Ety. proper name.]
inlesi, Winchell & Marcy, 1865, Proc. Bost. Soc. Nat. Hist., Niagara Gr. [Ety. proper name.]
ledoides, Winchell, 1866, Rep. Low. Peninsula Mich., Ham. Gr. [Sig. resembling a shell of the genus *Leda*.]
mactroides, Winchell, 1866, Rep. Low. Peninsula Mich., Ham. Gr. [Sig. resembling a shell of the genus *Mactra*.]
marionensis, Swallow, 1860, Trans. St. Louis Acad. Sci., Chemung Gr. [Ety. proper name.]
nebrascensis, Geinitz, 1866, Carb. und Dyas in Neb., Coal Meas. [Ety. proper name.]
nilesi, Winchell & Marcy, 1865, Proc. Bost. Soc. Nat. Hist., Niagara Gr. [Ety. proper name.]
nitida, Winchell, 1863, Proc. Acad. Nat. Sci., Chemung Gr. [Sig. smooth, neat.]
nuptialis, Winchell, 1863, Proc. Acad. Nat. Sci., Chemung Gr. [Sig. nuptial.]
otoensis, Swallow, 1858, Trans. St. Louis Acad. Sci., Permian Gr. [Ety. proper name.]
peroblonga, Meek & Worthen, 1866, Proc. Acad. Nat. Sci. Phil., Coal Meas. [Sig. somewhat oblong.]
philipi, Hall, 1870, Prelim. Notice Lam. shells, Chemung Gr. [Ety. proper name.]
radiata, see Clinopistha radiata.
reflexa, Meek, 1872, Pal. E. Neb., Coal Meas. [Sig. turned back, reversed.]
rhomboidea, Hall, 1877, Pal. N.Y., vol. 5, Chemung Gr. [Sig. rhomboidal.]
semiorbiculata, Swallow, 1858, Trans. St. Louis Acad. Sci., Permian Gr. [Sig. half orb-shaped.]
strigillata, Winchell, 1863, Proc. Acad. Nat. Sci., Chemung Gr. [Sig. furrowed.]
subangulata, see Cypricardites subangulatus.
subtruncata, Hall, 1847, see Cypricardites subtruncatus.
subtruncata, Meek, 1872, Pal. E. Neb., Coal Meas. [Sig. somewhat truncated.]
tapetiformis, Meek, 1875, (E. tapesiformis) Ohio Pal., vol. 2, Waverly Gr. [Ety. *tapet*, a carpet; *formis*, like.]
undulata, Hall, 1870, Prelim. Notice Lam. shells, Chemung Gr. [Sig. undulated.]
unioniformis, Phillips, 1836, (Isocardia unioniformis) Geol. Yorkshire, vol. 2, Carboniferous. [Sig. resembling the shell of the genus *Unio*.]
ventricosa, Hall, see Cypricardites ventricosus.

ENTOLIUM, Meek, 1864, Cal. Geo. Sur., vol. 2. [Ety. *entos*, inside; *leion*, smooth.]
avicula, Swallow, 1858, (Pecten aviculus) Trans. St. Louis Acad. Sci., Coal Meas. [Sig. like a shell of the genus *Avicula*.]
cooperensis, Shumard, 1855, (Avicula cooperensis) Low. Carb. [Ety. proper name.]

EOPTERIA, Billings, 1865, Pal. Foss., vol. 1. [Ety. *eos*, dawn; *pteron*, a wing.] Prof. Billings says if *Euchasma* is the same as *Eopteria*, then he desires *Eopteria* to be withdrawn from science.
(?) ornata, Billings, 1865, Pal. Foss., vol. 1, Quebec Gr. [Sig. adorned.]
richardsoni, Billings, 1865, Pal. Foss., vol. 1, Quebec Gr. [Ety. proper name.]
typica, Billings, 1865, Pal. Foss., vol. 1, Quebec Gr. [Ety. type of the genus.]

EUCHASMA, Billings, 1865, Pal. Foss., vol. 1. [Ety. *eu*, well; *chasma*, a hollow.]
blumenbachi, Billings, 1859, (Conocardium blumenbachi) Can. Nat. & Geo., vol. 4, Quebec Gr. [Ety. proper name.]

EUCHONDRIA, Meek, 1874, Am. Jour. Sci., 3rd series, vol. 7, a subgeneric name proposed for *Aviculopecten neglectus*, on account of its peculiar hinge line.

Eumicrotis, syn. for Pseudomonotis.
curta, see Pseudomonotis curta.
hawni, see Pseudomonotis hawni.
hawni var. ovata, see Pseudomonotis hawni var. ovata.
hawni var. sinuata, see Pseudomonotis hawni var. sinuata.

Exochorhynchus, Meek, 1864, Pal. Up. Mo. [Ety. *exochos*, prominent; *rhynchos*, beak.] This name was suggested as a probable genus or subgenus to include *Sedgwickia altirostrata*.

GERVILLIA, DeFrance, 1820, Dict. Sci. Nat., xviii. [Ety. proper name.] This genus is probably unknown in the palæozoic rocks.
auricula, Stevens, 1858, Am. Jour. Sci., vol. 25, Coal Meas. [Sig. the ear lap.]
longa, see Avicula longa.
longispina, see Monopteria longispina.

strigosa, White, 1862, Proc. Bost. Soc. Nat. Hist., vol. 9, Chemung Gr. [Sig. lean, thin.] This species may belong to the genus Bakevellia.
sulcata, see Avicula sulcata.
GONIOPHORA, Phillips, 1848. [Ety. *gonia*, an angle; *phoros*, bearing.] *Goniophorus* was used by Agassiz for a genus of Echinoderms in 1840.
acuta, Hall, 1877, Pal. N. Y., vol. 5, Ham. Gr. [Sig. acute.]
bellula, Billings, 1874, Pal. Foss., vol. 2, Up. Sil. [Sig. small and beautiful.]
consimilis, Billings, 1874, Pal. Foss., vol. 2, Up. Sil. [Sig. very like, just such another.]
glabra, Hall, 1877, Pal. N. Y., vol. 5, Chemung Gr. [Sig. smooth.]
hamiltonensis, Hall, 1870, (Sanguinolites hamiltonensis) Prelim. Notice Lam. Shells, Ham. Gr. [Ety. proper name.]
mediocris, Billings, 1874, Pal. Foss., vol. 2, Up. Sil. [Sig. middling.]
perangulata, Hall, 1877, Pal. N. Y., vol. 5, Up. Held. Gr. [Sig. very angular.]
rugosa, Conrad, 1841, (Cypricardites rugosa) Ann. Rep. N. Y., Ham. Gr. [Sig. wrinkled.]
transiens, Billings, 1874, Pal. Foss., vol. 2, Up. Sil. [Sig. transient, passing.]
GRAMMYSIA, DeVerneuil, 1847, Bull. Soc. Geo. France. [Ety. *gramme*, a line of writing; *Mys*, a mussel shell, in allusion to the transverse furrows which cross the valves from the umbones to the middle of the ventral margin.]
acadica, Billings, 1874, Pal. Foss., vol. 2, Up. Sil. [Ety. proper name.]
alveata, Conrad, 1841,(Posidonia alveata) Ann. Rep. N. Y., Ham. Gr. [Sig. hollowed out.]
arcuata, Conrad, 1841, (Posidonia arcuata) Ann. Rep. N. Y., Ham. Gr. [Sig. arched.]
bisulcata, Conrad, 1838, (Pterinea bisulcata, 1841,) (Cypricardites bisulcata) Ann. Rep. N. Y., Ham. Gr. [Sig. two-furrowed.]
canadensis, Billings, 1874, Pal. Foss., vol. 2, Up. Sil. [Ety. proper name.]
chemungensis, Pitt, 1874, Buff. Soc. Nat. Hist., Chemung Gr. [Ety. proper name.]
circularis, Hall, 1870, Prelim. Notice Lam. shells, Ham. and Chemung Gr. [Sig. circular.]
constricta, Hall, 1870, Prelim. Notice Lam. shells, Ham. Gr. [Sig. constricted.]
elliptica, Hall, 1870, Prelim. Notice Lam. shells, Chemung Gr. [Sig. elliptical.]
erecta, Hall, 1870, Prelim. Notice Lam. shells, Ham. Gr. [Sig. erect.]
globosa, Hall, 1870, Prelim. Notice Lam. shells, Ham. Gr. [Sig. globose.]
hamiltonensis, syn. for G. bisulcata.
hannibalensis, Shumard, 1855, (Allorisma hannibalensis) Geo. Sur. Mo., Kinderhook Gr. [Ety. proper name.]

lirata, Hall, 1870, Prelim. Notice Lam. shells, Ham. Gr. [Sig. furrowed.]
magna, Hall, 1870, Prelim. Notice Lam. shells, Ham. Gr. [Sig. large.]
nodocostata, Hall, 1870, Prelim. Notice Lam. shells, Ham. Gr. [Sig. having nodes on the ribs.]
obsoleta, Hall, 1870, Prelim. Notice Lam. shells, Ham. Gr. [Sig. obsolete.]
parallela, Hall, 1870, Prelim. Notice Lam. shells, Ham. Gr. [Sig. parallel.]
præcursor, Hall, 1870, Prelim. Notice Lam. shells, Schoharie grit. [Sig. a forerunner.]
remota, Billings, 1874, Pal. Foss., vol. 2, Up. Sil. [Sig. remote.]
rhomboidalis, Meek & Worthen, 1865, Proc. Acad. Nat. Sci. Phil., Ham. Gr. [Sig. rhomboidal.]
rhomboides, Meek, 1871, Proc. Acad. Nat. Sci. Phil., Waverly Gr. [Sig. rhomb-like.]
rustica, Billings, 1874, Pal. Foss., vol. 2, Up. Sil. [Sig. rough, rustic.]
secunda, Hall, 1870, Prelim. Notice Lam. shells, Up. Held. Gr. [Sig. following.]
subarcuata, Hall, 1870, Prelim. Notice Lam. shells, Chemung Gr. [Sig. somewhat like G. *arcuata*.]
ventricosa, Meek, 1871, Proc. Acad. Nat. Sci. Phil., Waverly Gr. [Sig. ventricose.]
Gryphorhynchus, Meek, 1864, Am. Jour. Sci. Not defined.
ILIONIA, Billings, 1875, Can. Nat. & Geol. [Ety. proper name.]
canadensis, Billings, 1875, Can. Nat. & Geol., Corniferous Gr. [Ety. proper name.]
sinuata, Hall, 1859, (Anatina sinuata) Pal. N. Y., vol. 3, Low. Held. Gr. [Sig. wavy.]
Inoceramus, Sowerby, 1818, Min. Conch., vol. 2. This genus is unknown in American palæozoic rocks.
chemungensis, see Mytilarca chemungensis.
mytilimeris, see Mytilarca mytilimeris.
oviformis, see Mytilarca oviformis.
ISCHYRINIA, Billings, 1866, Catal. Sil. Foss. Antic. [Ety. *ischyros*, strong.]
plicata, Billings, 1866, Catal. Sil. Foss. Antic., Anticosti Gr. [Sig. folded.]
winchelli, Billings, 1866, Catal. Sil. Foss. Antic., Hud. Riv. Gr. [Ety. proper name.]
ISOCARDIA, Klein, 1753, Tent. Meth. Ostr. [Ety. *isos*, like; *kardia*, the heart.] This is an existing littoral genus that burrows in the sand. It is not known in the palæozoic rocks.
(?) curta, Shumard, 1858, Trans. St. Louis Acad. Nat. Sci., Chemung Gr. [Sig. short.]
jennæ, Winchell, 1863, Proc. Acad. Nat. Sci., Chemung Gr. [Ety. proper name.]
unioniformis, see Edmondia unioniformis.

LAMELLIBRANCHIATA.

Leda, Schumacher, 1817, syn. for Nuculana.
   barrisi, see Palæoneilo barrisi.
   bellistriata, see Nuculana bellistriata.
   brevirostris, see Nuculana brevirostris.
   curta, see Nuculana curta.
   dens-mamillata, see Nuculana dens-mamillata.
   gibbosa, see Yoldia gibbosa.
   knoxensis, see Yoldia knoxensis.
   levistriata, see Yoldia levistriata.
   nuculiformis, see Nuculana nuculiformis.
   oweni, see Yoldia oweni.
   pandoriformis, see Nuculana pandoriformis.
   polita, see Nuculana polita.
   rushensis, see Yoldia rushensis.
   saccata, see Nuculana saccata.
   subscitula, see Yoldia subscitula.
LEIOPTERIS, Hall, 1877, Pal. N. Y., vol. 5. [Ety. *leios*, smooth; *pteron*, a wing.] A proposed genus in the family *Aviculidæ*.
LEPTODOMUS, McCoy, 1844, Synopsis Carb. Foss. Ireland. [Ety. *leptos*, slender; *domus*, house.]
   arata, Hall, 1860, Can. Nat. & Geo., vol. 5, Silurian. [Sig. furrowed.]
   canadensis, Billings, 1874, Pal. Foss., vol. 2, Devonian. [Ety. proper name.]
   clavata, Winchell, 1862, Proc. Acad. Nat. Sci., Portage Gr. [Sig. club-shaped.]
   granosus, see Allorisma granosum.
   topekaensis, see Sedgwickia topekaensis.
LIMA, Brugueire, 1791, Encycl. Meth. and Deshayes, 1824, Descrip. de Coquilles fossiles des environs de Paris. [Ety. *lima*, a file.]
   glabra, see Pernopecten glaber.
   macroptera, see Limoptera macroptera.
   obsoleta, see Pernopecten obsoletus.
   retifera, Shumard, 1858, Trans. St. Louis Acad. Sci., Coal Meas. [Sig. net-bearing.]
   rugæstriata, see Aviculopecten rugæstriatus.
LIMOPTERA, Hall, 1870, Prelim. Notice Lam. shells, Up. Held. Gr. [Ety. *Lima*, a genus of shells; *pteron*, a wing.]
   cancellata, Hall, 1870, Prelim. Notice Lam. shells, Up. Held. Gr. [Sig. cancellated.]
   curvata, Hall, 1870, Prelim. Notice Lam. shells, Ham. Gr. [Sig. curved.]
   macroptera, Conrad, 1838, (Lima macroptera) Ann. Rep. N. Y., Ham. Gr. [Sig. long-winged.]
   obsoleta, Hall, 1870, Prelim. Notice Lam. shells, Ham. Gr. [Sig. obsolete.]
   pauperata, Hall, 1870, Prelim. Notice Lam. shells, Up. Held. Gr. [Sig. poor.]
   triquetra, Conrad, Ham. Gr. [Sig. a triangle.]
LITHOPHAGA, Lamarck, 1812, Hist. An. sans Vert. [Ety. *lithos*, stone; *phago*, I eat.]
   lingualis, Phillips, 1836, (Modiola lingualis) Geol. Yorkshire, vol. 2, Keokuk Gr. [Sig. tongue-like.]
   pertenuis, Meek & Worthen, 1865, Proc. Acad. Nat. Sci. Phil., St. Louis Gr. [Sig. very slender.]

LUCINA, Bruguiere, 1792, Encyclop. Meth. [Ety. mythological name.]
   elliptica, see Paracyclas elliptica.
   elliptica var. occidentalis, see Paracyclas elliptica var. occidentalis.
   hamiltonensis, Winchell, 1866, Rep. Low. Peninsula Mich., Ham. Gr. [Ety. proper name.]
   lirata, see Paracyclas lirata.
   occidentalis, Billings, 1859, Assiniboine & Saskatchewan Ex. Exped., Devonian. This name was preoccupied by Conrad for an Eocene species.
   ohioensis, see Paracyclas ohioensis.
   retusa, Hall, 1843, Geo. Rep. 4th Dist. N. Y., Portage Gr. [Sig. blunt.]
LUNULACARDIUM, Munster, 1840, Beitrage zur Petrefaktenkunde. [Ety. *lunula*, a little moon; *Cardium*, a genus of shells.]
   acutirostrum, Hall, 1843, (Pinnopsis acutirostra) Geo. Rep. 4th Dist. N. Y., Portage Gr. [Sig. acute-beaked.]
   curtum, Hall, 1870, Prelim. Notice Lam. shells, Up. Held. Gr. [Sig. short.]
   fragilis, Hall, 1877, Pal. N. Y., vol. 5, Ham. Gr. [Sig. frail.]
   marcellense, Vanuxem, 1843, (Cypricardites marcellensis) Geo. Rep. 3rd Dist. N. Y., Marcellus shale. [Ety. proper name.]
   ornatum, Hall, 1843, (Pinnopsis ornata) Geo. Rep. 4th Dist. N. Y., Portage Gr. [Sig. adorned.]
Lyonsia, Turton, 1822. Not found in palæozoic rocks.
   concava, see Sedgwickia concava.
LYRIOPECTEN, Hall, 1877, Pal. N. Y., vol. 5. [Ety. *lyrion*, a lyre; *Pecten*, a genus of shells.]
   anomiæformis, Hall, 1877, Pal. N. Y., vol. 5, Up. Held. Gr. [Sig. like a shell of the genus *Anomia*.]
   macrodonta, Hall, 1877, Pal. N. Y., vol. 5, Up. Held. Gr. [Sig. from the large tooth-like ridges.]
   orbiculatus, Hall, 1843, (Avicula orbiculata) Geo. Rep. 4th Dist. N. Y., Ham. Gr. [Sig. orbicular.]
   parallelodonta, Hall, 1877, Pal. N. Y., vol. 5, Up. Held. Gr. [Sig. from the broad straight hinge line.]
LYRODESMA, Conrad, 1841, Ann. Geo. Rep. N. Y. [Ety. *lyra*, a harp; *desma*, a ligament.]
   cincinnatiense, Hall, 1871, Pamphlet, Cin'ti Gr. [Ety. proper name.]
   planum, Conrad, 1841, Ann. Geo. Rep., Hud. Riv. Gr. [Ety. *planus*, flat.]
   poststriatum, Emmons, 1842, (Nuculana poststriata) Geo. Rep. N. Y., Black Riv. Gr. [Ety. striated posteriorly.]
   pulchellum, Hall, 1847, Pal. N. Y., vol. 1, Hud. Riv. Gr. [Ety. *pulchellus*, beautiful.]

MACRODON, Lycett, 1845, Murch. Geo. Chelt. [Ety. *macros*, long; *odous*, a tooth.]
carbonarius, Cox, 1857, (Arca carbonarius) Geo. Sur. Ky., vol. 3, Coal Meas. [Sig. pertaining to coal.]
chemungensis, Hall, 1870, Prelim. Notice Lam. Shells, Chemung Gr. [Ety. proper name.]
cochlearis, Winchell, 1863, Proc. Acad. Nat. Sci., Chemung Gr. [Sig. spoonshaped.] Prof. Hall suggests that it is a syn. for *M. parvus*.
curtus, Dawson, 1868, Acad. Geol., Carb. [Sig. short.]
delicatus, Meek & Worthen, 1870, Proc. Acad. Nat. Sci., Coal Meas. [Sig. delicate.]
hamiltoniæ, Hall, 1870, Prelim. Notice Lam. shells, Ham. Gr. [Ety. proper name.]
hardingi, Dawson, 1868, Acad. Geol., Carb. [Ety. proper name.]
micronema, Meek & Worthen, 1866, Proc. Acad. Nat. Sci., Chester Gr. [Sig. a small thread.]
obsoletus, Meek, 1871, Reg. Rep. University W. Va., Coal Meas. [Sig. obsolete.]
ovatus, Hall, 1870, Prelim. Notice Lam. shells, Waverly Gr. [Sig. ovate.]
parvus, White & Whitfield, 1862, Proc. Bost. Soc. Nat. Hist., vol. 8, Kinderhook Gr. [Sig. small.]
shubenacadiensis, Dawson, 1868, Acad. Geo., Carb. [Ety. proper name.]
tenuistriata, Meek & Worthen, 1867, Proc. Acad. Nat. Sci., Coal Meas. [Sig. finelined.]

MATHERIA, Billings, 1858, Can. Nat. & Geo., vol. 3. [Ety. proper name.]
tenera, Billings, 1858, Can. Nat. & Geo., vol. 3, Trenton Gr. [Sig. tender, delicate.]

MEGALOMUS, Hall, 1852, Pal. N. Y., vol. 2. [Ety. *megale*, great; *omos*, shoulder.]
canadensis, Hall, 1852, Pal. N. Y., vol. 2, Guelph Gr. [Ety. proper name.]

MEGAMBONIA, Hall, 1859, Pal. N. Y., vol. 3. [Ety. *mega*, great; *ambon*, the boss of a shield.]
aviculoidea, Hall, 1859, Pal. N. Y., vol. 3, Low. Held. Gr. [Sig. like an *Avicula*.]
bellistriata, Hall, 1859, Pal. N. Y., vol. 3, Oriskany sandstone. [Sig. beautifully striated.]
cancellata, Hall, 1860, Can. Nat. & Geo., vol. 5, Silurian. [Sig. cancellated.]
cardiiformis, Hall, 1843, (Pterinea cardiiformis) Geo. Rep. 4th Dist. N. Y., Cornif. Gr. [Sig. like a *Cardium*.]
cordiformis, see Mytilarca cordiformis.
jamesi, Meek, 1872, Proc. Acad. Nat. Sci. Phil., Cin'ti Gr. [Ety. proper name.] This shell probably belongs to the genus Ambonychia.
lamellosa, Hall, 1859, Pal. N. Y., vol. 3, Oriskany sandstone. [Sig. made of thin plates.]
lata, Hall, 1859, Pal. N. Y., vol. 3, Low. Held. Gr. [Sig. wide, expanded on the posterior slope.]
lyoni, syn. for Cardiopsis radiata.
mytiloidea, Hall, 1859, Pal. N. Y., vol. 3, Low. Held. Gr. [Sig. like the *Mytilus*.]
oblonga, Hall, 1859, Pal. N. Y., vol. 3, Low. Held. Gr. [Sig. oblong.]
obscura, Hall, 1859, Pal. N. Y., vol. 3, Low. Held. Gr. [Sig. characters obscure.]
orata, syn. for Mytilarca mytilimeris.
ovoidea, Hall, 1859, Pal. N. Y., vol. 3, Low. Held. Gr. [Sig. ovoidal.]
rhomboidea, Hall, 1859, Pal. N. Y., vol. 3, Low. Held. Gr. [Sig. rhomboidal.]
spinneri, Hall, 1859, Pal. N. Y., vol. 3, Low. Held. Gr. [Ety. proper name.]
striata, Hall, 1860, Can. Nat. & Geo., vol. 5, Silurian. [Sig. striated.]
subcordiformis, Hall, 1877, Pal. N. Y., vol. 5, Up. Held. Gr. [Sig. somewhat like *M. cordiformis*.]
suborbicularis, Hall, 1859, Pal. N. Y., vol. 3, Low. Held. Gr. [Sig. somewhat orbicular.]

Megaptera, Meek & Worthen, 1866. The name was preoccupied.

Microdon, Conrad, 1842, Jour. Acad. Nat. Sci., vol. 8. This name was applied by Agassiz to a genus of fish in 1833, and was also preoccupied for a genus of insects.
bellistriata, Conrad, 1842, Jour. Acad. Nat. Sci., vol. 8, Ham. Gr. [Sig. beautifully lined.]
complanatus, Hall, 1870, Prelim. Notice Lam. shells, Ham. Gr. [Sig. smoothed.]
gregaria, Hall, 1870, Prelim. Notice Lam. shells, Ham. Gr. [Sig. gregarious.]
reservatus, Hall, 1870, Prelim. Notice Lam. shells, Waverly Gr. [Sig. preserved.]
tenuistriata, Hall, 1870, Prelim. Notice Lam. shells, Ham. Gr. [Sig. finely lined.]

MODIOLA, Lamarck, 1801, Syst. An. sans Vert. [Ety. *modiolus*, a small measure or drinking vessel.]
avonia, Dawson, 1868, Acad. Geol., Low. Carb. [Ety. proper name.]
concentrica, see Modiomorpha concentrica.
lingualis, see Lithophaga lingualis.
metella, Hall, 1870, Prelim. Notice Lam. shells, Chemung Gr. [Ety. proper name.]
minor, Lea, 1852, Jour. Acad. Nat. Sci., 2nd series, vol. 2, Coal Meas. [Sig. less.]
obtusa, see Modiolopsis obtusa.
pooli, Dawson, 1868, Acad. Geol., Low. Carb. [Ety. proper name.]
præcedens, Hall, 1870, Prelim. Notice Lam. shells, Chemung Gr. [Sig. surpassing, going before.]
wyomingensis, Lea, 1852, Jour. Acad. Nat. Sci., 2d series, vol. 2, Coal Meas. [Ety. proper name.]

LAMELLIBRANCHIATA. 195

MODIOLOPSIS, Hall, 1847, Pal. N. Y., vol. 1. [Ety. *Modiola*, a genus of shells; *opsis*, appearance; from its resemblance to Modiola.]
adrastia, Billings, 1862, Pal. Foss., vol. 1, Black Riv. Gr. [Ety. mythological name.]
anodontoides, Conrad, 1847, (Cypricardites anodontoides) Pal. N. Y., vol. 1, Hud. Riv. Gr. [Ety. from its resemblance to the genus *Anodonta*.]
arcuata, Hall, 1847, Pal. N. Y., vol. 1, Trenton Gr. [Sig. bent, bow-shaped.]
aviculoides, Hall, 1847, Pal. N. Y., vol. 1, Trenton Gr. [Sig. resembling an *Avicula*.]
carinata, Hall, 1847, Pal. N. Y., vol. 1, Trenton Gr. [Sig. keeled.]
cincinnatiensis, Hall & Whitfield, 1875, Ohio Pal., vol. 2, Cin'ti Gr. [Ety. proper name.]
concentrica, Hall & Whitfield, 1872, Ohio Pal., vol. 2, Cin'ti Gr. [Sig. concentric.]
curta, Hall, 1847, Pal. N. Y., vol. 1, Hud. Riv. Gr. [Sig. short.]
dicteus, Hall, 1867, 20th Reg. Rep. N. Y., Niagara Gr. [Ety. mythological name.]
(?) dubia, Hall, 1859, Pal. N. Y., vol. 3, Low. Held. Gr. [Sig. doubtful.]
exilis, Billings, 1874, Pal. Foss., vol. 2, Up. Sil. [Sig. slender.]
faba, Conrad, 1842, in Emmons' Geo. Rep. N. Y., Black Riv., Trenton and Hud. Riv. Gr. [Sig. a bean.]
gesneri, Billings, 1862, Pal. Foss., vol. 1, Trenton and Black Riv. Gr. [Ety. proper name.]
latus, see Cypricardites latus.
maia, Billings, 1862, Pal. Foss., vol. 1, Trenton Gr. [Ety. mythological name.]
meyeri, Billings, 1862, Pal. Foss., vol. 1, Trenton Gr. [Ety. proper name.]
modiolaris, Conrad, 1838, (Pterinea modiolaris) Ann. Geo. Rep. N. Y., Hud. Riv. Gr. [Sig. like a *Modiola*.]
modioliformis, Meek & Worthen, 1868, Geo. Sur. Ill., vol. 3, Trenton Gr. [Sig. like a *Modiola*.]
mytiloides, Hall, 1847, Pal. N. Y., vol. 1, Black Riv. & Trenton Gr. [Sig. like a shell of the genus *Mytilus*.]
nais, Billings, 1862, Pal. Foss., vol. 1, Black Riv. Gr. [Ety. mythological name.]
nasuta, C o n r a d , 1841, (Cypricardites nasutus) Ann. Rep. N. Y., Trenton & Hud. Riv. Gr. [Sig. nasute.]
nuculiformis, see Tellinomya nuculiformis.
obtusa, Hall, 1847, Pal. N. Y., vol. 1, Birdseye Gr. [Sig. obtuse.]
orthonota, Conrad, 1839, (Unio orthonotus) Ann. Rep. N. Y., Medina sandstone. [Sig. straight-backed.]

orthonota, Meek & Worthen, 1868, Geo. Sur. Ill., vol. 3, Trenton Gr. This name was preoccupied.
ovata, Hall, 1852, Pal. N. Y., vol. 2, Clinton Gr. [Sig. ovate.]
parallela, Conrad, 1847, Pal. N. Y., vol. 1, Trenton Gr. [Sig. parallel.]
parviuscula, Billings, 1859, Can. Nat. & Geo., vol. 4, Chazy Gr. [Sig. very small.]
perlata, Hall, 1876, 28th Reg. Rep. N. Y., Niagara Gr. [Sig. very wide.]
perovata, see Modiomorpha perovata.
pholadiformis, Hall, 1851, Lake Sup. Land Dist., vol. 2, Hud. Riv. Gr. [Sig. like the *Pholas*.]
plana, Hall, 1861, Geo. Rep. Wis., Trenton Gr. [Sig. smooth.]
primigenia, Conrad, 1838, (Unio primigenius) Ann. Rep. N. Y., Medina sandstone. [Sig. first-born.]
recta, Hall, 1867, 20th Reg. Rep. N. Y., Niagara Gr. [Sig. straight.]
rhomboidea, Hall, 1860, Can. Nat. & Geo., vol. 5, Up. Sil. [Sig. rhomboidal.]
rudis, Billings, 1874, Pal. Foss., vol. 2, Up. Sil. [Sig. rude.]
striata, Billings, 1866, Catal. Sil. Foss. Antic., Anticosti Gr. [Sig. striated.]
subalata, Hall, 1852, Pal. N. Y., vol. 2, Clinton & Niagara Gr. [Sig. somewhat winged.]
subcarinata, Hall, 1852, Pal. N. Y., vol. 2, Clinton Gr. [Sig. somewhat carinated.]
subnasuta, Hall, 1860, Can. Nat. & Geo., vol. 5, Up. Sil. [Sig. somewhat nasute.]
subnasuta, Meek & Worthen, 1870. This name was preoccupied.
subspatulata, Hall, 1847. See Cypricardites subspatulatus.
superba, Hall, 1861, Geo. Rep. Wis., Trenton Gr. [Sig. grand.]
terminalis, Hall, 1847, Pal. N. Y., vol. 1, Cin'ti Gr. [Sig. terminating.]
trentonensis, Hall, 1847, Pal. N. Y., vol. 1, Trenton Gr. [Ety. proper name.]
truncata, Hall, 1847, Pal. N. Y., vol. 1, Hud. Riv. Gr. [Sig. cut short.]
undulostriata, Hall, 1852, Pal. N. Y., vol. 2, Niagara Gr. [Sig. having waved striæ.]
varia, Billings, 1874, Pal. Foss., vol. 2, Low. Held. Gr. [Sig. different.]
versaillesensis, S. A. Miller, 1874, Cin. Quar. Jour. Sci., Cin'ti Gr. [Ety. proper name.]
MODIOMORPHA, Hall, 1870, Prelim. Notice Lam. shells. [Ety. contracted from *Modiola*, a genus of shells; *morphe*, form.]
alta, Conrad, 1841, (Cypricardites alta) Ann. Rep. N. Y., Ham. Gr. [Ety. from the elevation of the hinge margin.]
alata, Hall, 1843, (Cypricardites alata) Geo. Rep. N. Y., Chemung Gr. [Sig. winged.]

angustata, Vanuxem, 1842, (Cypricardites angustata) Geo. Rep. N. Y., Catskill Gr. [Sig. narrowed.]

catskillensis, Vanuxem, 1842, (Cypricardites catskillensis) Geo. Rep. N. Y., Catskill Gr. [Ety. proper name.]

complanata, Hall, 1870, Prelim. Notice Lam. shells, Up. Held. Gr. [Sig. smooth.]

concentrica, Conrad, 1838, (Cypricardites concentrica) Geo. Rep. 4th Dist. N. Y., Ham. Gr. [Sig. marked with concentric lines.]

cymbula, Hall, 1870, Prelim. Notice Lam. shells, Ham. Gr. [Sig. a small boat.]

hyalea, Hall, 1870, Prelim. Notice Lam. shells, Waverly Gr. [Ety. proper name.]

inornata, Billings, 1874, Pal. Foss., vol. 2, Devonian. [Sig. unadorned.]

macilenta, Hall, 1870, Prelim. Notice Lam. shells, Ham. Gr. [Sig. lean, meager.]

perovata, Meek & Worthen, 1865, (Modiolopsis perovata) Proc. Acad. Nat. Sci. Phil., Ham. Gr. [Sig. very ovate.]

planulata, Hall, 1870, Prelim. Notice Lam. shells, Ham. Gr. [Sig. flattened.]

ponderosa, Hall, 1877, Pal. N. Y., vol. 5, Up. Held. Gr. [Sig. large.]

quadrula, Hall, 1870, Prelim. Notice Lam. shells, Chemung Gr. [Sig. a small square.]

subalata, Conrad, 1841, (Cypricardites subalata) Ann. Rep. N. Y., Ham. Gr. [Sig. somewhat winged.]

MONOPTERIA, Meek & Worthen, 1866, Proc. Chi. Acad. Nat. Sci., vol. 1. [Ety. *monos*, single; *pteron*, a wing.]

gibbosa, Meek & Worthen, 1866, Proc. Chi. Acad. Sci., Coal Meas. [Sig. gibbous, tumid.]

longispina, Cox, 1857, (Gervillia longispina) Geo. Sur. Ky., vol. 3, Coal Meas. [Sig. long-spined.]

marian, White, 1874, Rep. Invert. Foss., Carboniferous. [Ety. proper name.]

MONOTIS, Bronn, 1824, System Urweltlicher Konchylien. [Ety. *monos*, one; *ousotos*, ear.]

elevata, Conrad, 1848, Proc. Acad. Nat. Sci., vol. 3, Chemung Gr. [Sig. elevated.]

gregaria, Meek & Worthen, 1870, Proc. Acad. Nat. Sci., Phil., Coal Meas., [Sig. clustered together, found in flocks.]

halli, Swallow, 1858, Trans. St. Louis Acad. Sci., Permian Gr. [Ety. proper name.]

hawni, see Eumicrotis hawni.

poulsoni, Conrad, 1848, Proc. Acad. Nat. Sci., vol. 3, Chemung Gr. [Ety. proper name.]

princeps, Conrad, 1838, Ann. Rep. N. Y., Ham. Gr. [Sig. primitive, original.]

radialis, Phillips, 1834, (Pecten radialis) Permian Gr. See Pseudomonotis radialis.

radians, Conrad, 1842, (Pterinea radians) Jour. Acad. Nat. Sci. vol. 8, Ham. Gr. [Sig. radiating.]

speluncaria, Schlotheim, 1816, (Gryphites speluncarius) Permian Gr. [Sig. like a cave, den, or hole in a rock.]

variabilis, Swallow, 1858, Trans. St. Louis Acad. Sci., Permian Gr. [Sig. variable, changing.]

MYALINA, Koninck, 1844, Desc. Anim. Foss. Carb. Belg. [Ety. *Mya*, a genus of shells; *inus*, like.]

angulata, Meek & Worthen, 1860, Proc. Acad. Nat. Sci. Phil., Chester Gr. [Sig. angulated.]

apachei, Marcou, 1858, Geol. North America, Carboniferous. [Ety. proper name.]

aviculoides, Meek & Hayden, 1860, Proc. Acad. Nat. Sci. Phil., Permian Gr. [Sig. like an *Avicula*.]

aviculoides, Winchell, 1862. This name was preoccupied.

concentrica, Meek & Worthen, 1860, Proc. Acad. Nat. Sci. Phil., St. Louis Gr. [Sig. having concentric lines.]

deltoidea, Galb, 1859, Proc. Acad. Nat. Sci. Phil., Low. Carb. [Sig. like the Greek letter *Delta*.]

imbricaria, Winchell, 1862, Proc. Acad. Nat. Sci., Marshall Gr. [Sig. imbricated.]

iowensis, Winchell, 1865, Proc. Acad. Nat. Sci., Chemung Gr. [Ety. proper name.]

kansasensis, Shumard, 1858, Trans. St. Louis Acad. Sci., Coal Meas. [Ety. proper name.]

keokuk, Worthen, 1875, Geo. Sur. Ill., vol. 6, Keokuk Gr. [Ety. proper name.]

meliniformis, Meek & Worthen, 1866, Proc. Chi. Acad. Sci., Coal Meas. [Sig. in the form of a purse.]

michiganensis, Winchell, 1862, Proc. Acad. Nat. Sci., Marshall Gr. [Ety. proper name.]

mytiliformis, Hall, 1852, Pal. N. Y., vol. 2, Clinton Gr. [Sig. like a *Mytilus*.]

perattenuata, Meek & Hayden, 1862, Trans. Alb. Inst., vol. 4. Coal Meas. [Sig. very much drawn out, attenuated.]

permiana, Swallow, 1858, (Mytilus permianus) Trans. St. Louis Acad. Sci., Permian Gr. [Ety. proper name.]

perniformis, Cox, 1857, Geo. Sur. Ky., vol. 3, Coal Meas. [Sig. in the form of a *Perna*.]

pterineæformis, Winchell, 1862, Proc. Acad. Nat. Sci., Marshall Gr. [Sig. in the form of a *Pterinea*.]

recta, Shumard, 1858, Trans. St. Louis Acad. Sci., Permian Gr. [Sig. straight.]

recurvirostris, Meek & Worthen, 1860, Proc. Acad. Nat. Sci. Phil., Up. Coal Meas. [Sig. beaked, recurved.]
squamosa, Sowerby, 1827, Trans. Geo. Soc. Lond., 2d ser., vol. 3, Permian Gr. [Sig. rough, scaly.]
subquadrata, Shumard, 1855, Geo. Rep. Mo., Coal Meas. [Sig. somewhat quadrate.]
st. ludovici, Worthen, 1873, Geo. Sur. Ill., vol. 5, St. Louis Gr. [Ety. proper name; *Sanctus Ludovicus*, St. Louis.]
swallovi, McChesney, 1860, New Pal. Foss., Coal Meas. [Ety. proper name.]

MYTILARCA, Hall, 1870, Prelim. Notice Lam. shells. [Ety. from the two genera *Mytilus* and *Arca*.]
arenacea, Hall, 1870, Prelim. Notice Lam. shells, Schoharie grit. [Sig. sandy.]
attenuata, Hall, 1870, Prelim. Notice Lam. shells, Chemung Gr. [Sig. elongated.]
canadensis, Billings, 1874, Pal. Foss., vol. 2, Devonian. [Ety. proper name.]
carinata, Hall, 1877, Pal. N. Y., vol. 5, Chemung Gr. [Sig. carinated.]
chemungensis, Conrad, 1842, (Inoceramus chemungensis) Jour. Acad. Nat. Sci. Phil., vol. 8, Chemung Gr. [Ety. proper name.]
cordiformis, Hall, 1859, (Megambonia cordiformis) Pal. N. Y., vol. 3, Low. Held. Gr. [Sig. heart-shaped.]
fibristriata, White & Whitfield, 1862, (Mytilus fibristriatus) Proc. Bost. Soc. Nat. Hist., Kinderhook Gr. [Sig. fibre-lined.]
mytilimeris, Conrad, 1842, (Inoceramus mytilimeris) Jour. Acad. Nat. Sci., vol. 8, Low. Held. Gr. [Ety. *Mytilus* a genus of shells; *meros*, a part.]
nitida, Billings, 1874, Pal. Foss., vol. 2, Devonian. [Sig. neat, pretty.]
occidentalis, White & Whitfield, 1862, (Mytilus occidentalis) Proc. Bost. Soc. Nat. Hist., vol. 8, Kinderhook Gr. [Sig. western.]
oviformis, Conrad, 1842, (Inoceramus oviformis) Jour. Acad. Nat. Sci., vol. 8, Ham. Gr. [Sig. egg-shaped.]
ponderosa, Hall, 1870, Prelim. Notice Lam. shells, Up. Held. Gr. [Sig. large, heavy.]
radiata, Hall, 1877, Pal. N. Y., vol. 5, Chemung Gr. [Sig. radiated.]
sigillum, Hall, 1876, 28th Reg. Rep., Niagara Gr. [Sig. a seal.]
triquetra, Conrad, 1838, (Pterinea triquetra) Ann. Rep. N. Y., Ham. Gr. [Sig. a triangle.]

MYTILUS, Linnæus, 1758, Syst. Nat., 10th ed. [Ety. *Mytilus*, the fish mussel.] This genus does not, so far as known, exist in palæozoic rocks. Most of the species referred to it belong to the genus *Myalina*.

concavus, Swallow, 1858, Trans. St. Louis Acad. Sci., Permian Gr. [Sig. concave.]
fibristratus, see Mytilarca fibristriata.
occidentalis, see Mytilarca occidentalis.
ottawensis, Shumard, 1858, Trans. St. Louis Acad. Sci., Up. Coal Meas. [Ety. proper name.]
permianus, see Myalina permiana.
squamosus, Sowerby, 1839, Trans. Geol. Soc. Lond., vol. 4, Permian Gr. [Sig. rough, scaly.]
tenuiradiatus, Shumard, 1858, Trans. St. Louis Acad. Sci., Up. Coal Meas. [Sig. slender-rayed.]
whitfieldanus, Winchell, 1862, Proc. Acad. Nat. Sci., Marshall Gr. [Ety. proper name.] Prof. Hall suggests that this is a syn. for *Mytilarca fibristriata*.

Naiadites, Dawson, 1860, Acad. Geol., but not defined. The name was used for a genus of plants by Buckman in 1843. The fossils were defined by Salter in 1861, under the name of *Anthracomya*.
angulata, see Anthracomya angulata.
arenaceus, see A. arenacea.
carbonarius, see A. carbonaria.
elongata, see A. elongata.
lævis, see A. lævis.
obtusa, see A. obtusa.
ovalis, see A. ovalis.

NUCULA, Lamarck, 1815, Hist. Nat. des An. sans Vert. [Ety. *nucula*, a little nut.]
anodontoides, Meek, 1871, Reg. Rep. University W. Va., Coal Meas. [Sig. like an *Anodonta*.]
arata, Hall, 1852, Stansbury's Ex. to Gr. Salt Lake, Coal Meas. [Sig. furrowed.]
bellatula, syn. for N. bellistriata.
bellistriata, Conrad, 1841, (Nuculites bellistriatus) Ann. Rep. N. Y., Ham. Gr. [Sig. beautifully striated.]
beyrichia, Schlotheim, as identified by Geinitz. See Nucula parva.
corbuliformis, Hall, 1870, Prelim. Notice Lam. shells, Ham. & Chemung Gr. [Sig. resembling a *Corbula*.]
cylindricus, syn. for Cardiomorpha missouriensis.
donaciformis see Tellinomya donaciformis.
fabula, see Cleidophorus fabula.
hians, Hall, 1860, 13th Reg. Rep., Ham. Gr. [Sig. gaping.]
houghtoni, Stevens, 1858, Am. Jour. Sci., vol. 25, Coal Meas. [Ety. proper name.]
hubbardi, Winchell, 1862, Proc. Acad. Nat. Sci., Marshall Gr. Syn. for Nuculites sulcatinus.
iowensis, White & Whitfield, 1862, Proc. Acad. Nat. Sci., Marshall Gr. [Ety. proper name.]
kazanensis, as identified by Geinitz is Nuculana bellistriata.
levata, see Tellinomya levata.
lineata, see Tellinomya lineata.

lineolata, Hall, 1843, Geo. Rep. 4th Dist. N. Y., Portage Gr. [Sig. marked with small lines.]
lirata, Conrad, 1842, (Nuculites liratus) Jour. Acad. Nat. Sci., vol. 8, Ham. Gr. [Sig. furrowed.]
machæræformis, see Tellinomya machæræformis.
mactræformis, see Tellinomya mactræformis.
mercerensis, syn. for Cardiomorpha missouriensis.
microdonta, Winchell, 1863, Proc. Acad. Nat. Sci., Chemung Gr. [Sig. small-toothed.]
minuta, Owen, 1840, Rep. on Min. Lands, Devonian. The name was preoccupied by De France in 1825.
nasuta, Hall, 1858, Trans. Alb. Inst., vol. 4, Warsaw Gr. [Sig. nasute.]
neda, Hall, 1872, 24th Reg. Rep. N. Y., Up. Held. Gr. [Ety. proper name.]
niotica, Hall, 1872, 24th Reg. Rep. N. Y., Up. Held. Gr. [Ety. proper name.]
obliqua, see Tellinomya obliqua.
oblonga, see Cleidophorus oblongus.
parva, McChesney, 1860, New Pal. Foss., Coal Meas. [Sig. small.]
poststriata, see Lyrodesma poststriatum.
randalli, Hall, 1870, Prelim. Notice Lam. shells, Ham. and Chemung Gr. [Ety. proper name.]
rectangula, McChesney, 1860, Desc. New Pal. Foss., Ham. Gr. [Sig. rectangular.]
sectoralis, Winchell, 1862, Proc. Acad. Nat. Sci., Marshall Gr. [Sig. like a sector.]
shumardana, Hall, 1858, Trans. Alb. Inst., vol. 4, Warsaw Gr. [Ety. proper name.]
stella, Winchell, 1862, Proc. Acad. Nat. Sci., Marshall Gr. [Sig. a star.]
subnasuta, Hall, 1872, 24th Reg. Rep. N. Y., Up. Held. Gr. [Sig. somewhat nasute.]
varicosa, Hall, 1870, Prelim. Notice Lam. shells, Ham. Gr. [Sig. varicose.]
ventricosa, Hall, 1858, Geo. Sur. Iowa, Coal Meas. [Sig. bulging out.]
NUCULANA, Link, 1807, Rost. Samml., vol. 3. [Ety. like a shell of the genus Nucula.]
bellistriata, Stevens, 1858, (Leda bellistriata) Am. Jour. Sci., vol. 25, Coal Meas. [Sig. beautifully striated.]
bellistriata var. attenuata, Meek, 1872, Pal. E. Neb., Coal Meas. [Sig. drawn out, attenuated.]
brevirostris, Hall, 1870, (Leda (?) brevirostris) Prelim. Notice Lam. shells, Ham. Gr. [Sig. short-beaked.]
curta, Meek, 1861, (Leda curta) Proc. Acad. Nat. Sci. Phil., St. Louis Gr. [Sig. short.]
dens-mamillata, Stevens, 1858, Am. Jour. Sci., vol. 25, Coal Meas. [Sig. having mammillated teeth.]
nuculiformis, Stevens, 1858, Am. Jour. Sci., vol. 25, Coal Meas. [Ety. from the resemblance to *Nucula*.]
pandoriformis, Stevens, 1858, Am. Jour. Sci., vol. 25, Coal Meas. [Sig. like a shell of the genus *Pandora*.]
saccata, Winchell, 1863, (Leda saccata) Proc. Acad. Nat. Sci., Chemung Gr. [Sig. like a little bag.]
vaseyana, McChesney, 1860, (Nuculites vaseyana) Desc. New Pal. Foss., Ham. Gr. [Ety. proper name.]

NUCULITES, Conrad, 1841, Ann. Geo. Rep. N. Y. [Ety. *Nucula*, a genus of shells.]
altus, Conrad, 1842, Jour. Acad. Nat. Sci., vol. 8, Devonian. [Sig. high.]
appressus, see Cytherodon appressus.
bellistriatus, see Nucula bellistriata.
carinatus, Hall, 1860, Can. Nat. & Geol., vol. 5, Up. Sil. [Sig. keeled.]
chemungensis, see Schizodus chemungensis.
concentricus, Conrad, 1842, Jour. Acad. Nat. Sci., vol. 8, Coal Meas. [Sig. concentric.]
constrictus, see Palæoneilo constricta.
cuneiformis, Conrad, 1841, Ann. Rep. N. Y., Ham Gr. [Sig. wedge-shaped.]
emarginatus, see Palæoneilo emarginata.
faba, see Modiolopsis faba.
filosus, see Palæoneilo filosa.
inflatus, see Cypricardinia inflata.
lamellosus, Conrad, 1841, Ann. Geo. Rep. N. Y., Up. Sil. [Sig. having lamellæ.]
liratus, see Nucula lirata.
mactroides, Conrad, 1842, Jour. Acad. Nat. Sci., vol. 8, Low. Carb. [Sig. like a shell of the genus *Mactra*.]
maxima, see Palæoneilo maxima.
multilineatus, Conrad, 1842, Jour. Acad. Nat. Sci., vol. 8, Ham. Gr. [Sig. many-lined.]
nyssa, Hall, 1870, Prelim. Notice Lam. shells, Ham. Gr. [Ety. proper name.]
oblongus, see Cleidophorus oblongus.
oblongatus, Conrad, 1841, Ann. Geo. Rep. N. Y., Ham. Gr. [Sig. oblong.]
planulatus, see Cleidophorus planulatus.
poststriata, see Lyrodesma poststriatum.
radiatus, see Pholadella radiata.
rostellatus, Conrad, 1841, Ann. Rep. N. Y., Ham. Gr. [Sig. little-beaked.]
scitula, syn. for Cleidophorus planulatus.
subemarginata, see Tellinopsis subemarginata.
sulcatinus, Conrad, 1842, Jour. Acad. Nat. Sci., vol. 8, Low. Carb. [Sig. furrowed.]
triqueter, Conrad, 1841, Ann. Rep. N. Y., Ham. Gr. [Sig. triangular.]
vaseyana, see Nuculana vaseyana.

NYASSA, Hall, 1870, Prelim. Notice Lam. shells. [Ety. proper name.]
arguta, Hall, 1870, Prelim. Notice Lam. shells, Ham. Gr. [Sig. clearly marked.]
elliptica, Hall, 1870, Prelim. Notice Lam. shells, Up. Held. Gr. [Sig. elliptical.]

recta, Hall, 1870, Prelim. Notice Lam. shells, Ham. Gr. [Sig. from the straight umbonal ridge.]
subalata, Hall, 1870, Prelim. Notice Lam. shells, Ham. Gr. [Sig. somewhat winged.]

Opisthoptera, Meek. Not defined.

ORTHODESMA, Hall & Whitfield, 1875, Ohio Pal., vol. 2. [Ety. *orthos*, straight; *desma*, a ligament.]
contractum, Hall, 1847, (Orthonota contracta) Pal. N. Y., vol. 1, Hud. Riv. Gr. [Sig. contracted.]
curvatum, Hall & Whitfield, 1875, Ohio Pal., vol. 2, Cin'ti Gr. [Sig. curved.]
parallelum, Hall, 1847, (Orthonota parallela) Pal. N. Y., vol. 1, Hud. Riv. Gr. [Sig. parallel.]
rectum, Hall & Whitfield, 1875, Ohio Pal., vol. 2, Cin'ti Gr. [Sig. straight.]

ORTHONOTA Conrad, 1841, Ann. Rep. N. Y. [Ety. *orthos*, straight; *notos*, the back.]
angulifera, (?) McCoy, 1850, Brit. Pal. Rocks, Up. Sil. [Sig. having angles.]
carinata, Conrad, 1841, Ann. Rep. N. Y., Ham. Gr. [Sig. keeled.]
contracta, see Orthodesma contractum.
curta, Hall, 1843, Geo. Rep. 4th Dist. N. Y., Clinton and Niagara Gr. [Sig. short.]
ensiformis, Hall, 1870, Prelim. Notice Lam. shells, Ham. Gr. [Sig. ensiform.]
incerta, Billings, 1874, Pal. Foss., vol. 2, Up. Sil. [Sig. doubtful.]
parallela, see Orthodesma parallelum.
parvula, Hall, 1870, Prelim. Notice Lam. shells, Ham. Gr. [Sig. very small.]
phaselia, Winchell, 1863, Proc. Acad. Nat. Sci., Chemung Gr. [Sig. a kidney bean.]
pholadis, Conrad, 1838, (Pterinea pholadis) Ann. Geo. Rep. N. Y., Hud. Riv. Gr. [Sig. like a shell of the genus *Pholas*.]
rectidorsalis, Winchell, 1862, Proc. Acad. Nat. Sci., Marshall Gr. [Sig. straight-backed.]
siliquoidea, Hall, 1870, Prelim. Notice Lam. shells, Ham. Gr. [Sig. like a pod.]
simulans, Billings, 1874, Pal. Foss., vol. 2, Up. Sil. [Sig. resembling.]
(?) speciosa, Billings, 1874, Pal. Foss., vol. 2, Up. Sil. [Sig. beautiful.]
undulata, Conrad, 1841, Ann. Rep. N. Y., Ham. Gr. [Sig. wavy.]
venusta, Billings, 1874, Pal. Foss., vol. 2, Up. Sil. [Sig. elegant.]

OSTREA, Linnæus, 1758, Syst. Nat., 10th ed. [Ety. *ostrea*, an oyster.]
patercula, Winchell, 1865, Proc. Acad. Nat. Sci., Burlington Gr. [Ety. proper name.]

Palæarca, syn. for Cypricardites.
saffordi, see Cypricardites saffordi.
ventricosa, see Cypricardites ventricosus

PALÆOCARDIA, Hall, 1867, 20th Reg. Rep. N. Y. [Ety. *palaois*, ancient; *kardia*, a heart.]
cordiformis, Hall, 1867, 20th Reg. Rep. N. Y., Niagara Gr. [Sig. heart-shaped.]

PALÆONEILO, Hall, 1870, Prelim. Notice Lam. shells. [Ety. *palaios* ancient; *Neilo*, a genus of shells.]
attenuata, Hall, 1870, Prelim. Notice Lam. shells, Waverly Gr. [Sig. attenuated.]
barrisi, White & Whitfield, 1862, (Leda barrisi) Proc. Bost. Soc. Nat. Hist., vol. 8, Kinderhook Gr. [Ety. proper name.]
bedfordensis, Meek, 1875, Pal. Ohio, vol. 2, Waverly Gr. [Ety. proper name.]
bisulcata, Hall, 1870, Prelim. Notice Lam. shells, Ham. Gr. [Sig. double-sulcated.]
brevis, Hall, 1870, Prelim. Notice Lam. shells, Chemung Gr. [Sig. short.]
carbonaria, see Yoldia carbonaria.
constricta, Conrad, 1842, (Nuculites constricta) Jour. Acad. Nat. Sci., vol. 8, Chemung Gr. [Sig. constricted.]
emarginata, Conrad, 1841, (Nuculites emarginata) Ann. Rep. N. Y., Ham. Gr. [Sig. emarginated.]
filosa, Conrad, 1842, (Nuculites filosa) Jour. Acad. Nat. Sci., vol. 8, Chemung Gr. [Sig. thready.]
fœcunda, Hall, 1870, Prelim. Notice Lam. shells, Ham. Gr. [Sig. abundant.]
maxima, Conrad, 1841, (Nuculites maxima) Ann. Rep. N. Y., Ham. Gr. [Sig. the largest.]
muta, Hall, 1870, Prelim. Notice Lam. shells, Ham. Gr. [Sig. inconspicuous.]
parallela, Hall, 1870, 23rd Reg. Rep. N. Y., Waverly Gr. [Sig. parallel.]
perplana, Hall, 1870, Prelim. Notice Lam. shells, Ham. Gr. [Sig. very flat.]
plana, Hall, 1870, Prelim. Notice Lam. shells, Ham. Gr. [Sig. flat.]
tenuistriata, Hall, 1870, Prelim. Notice Lam. shells, Ham. Gr. [Sig. fine-lined.]

PALÆANATINA, Hall, 1870, Prelim. Notice Lam. shells. [Ety. *palaios*, ancient; *Anatina*, a genus of shells.] The family name at the head of this class should be spelled *Palæanatinidæ* instead of *Palanatinidæ*.
quadrata, Hall, 1877, Pal. N. Y., vol. 5, Chemung Gr. [Sig. quadrate.]
typus, Hall, 1870, Prelim. Notice Lam. shells, Chemung Gr. [Ety. the type of the genus.]

Panopæa, Menard de la Groye, 1807, Ann. du Mus. 9. [Ety. mythological name.]
cooperi, see Chænomya cooperi.

PARACYCLAS, Hall, 1843, Geo. Rep. 4th Dist. N. Y. [Ety. *para*, allied to; *Cyclas*, a genus of shells.]
elliptica, Hall, 1843, Geo. Rep. 4th Dist. N. Y., Cornif. Gr. [Sig. elliptical.]

elliptica *var.* occidentalis, Hall, 1872, 24th Reg. Rep., Up. Held. Gr. [Sig. western.]
lirata, Conrad, 1838, (Posidonia lirata) Ann. Rep. N. Y., Corniferous Gr. [Sig. furrowed.]
ohioensis, Meek, 1871, Proc. Acad. Nat. Sci. Phil., Cornif. Gr. [Ety. proper name.]
sabini, White, 1876, Proc. Acad. Nat. Sci., Devonian. [Ety. proper name.]
Pecten, Mueller, 1776. This genus is unknown in the Palæozoic rocks.
acutialatus, see Aviculopecten acutialatus.
armigerus, see A. armigerus.
aviculatus, see A. aviculatus.
broadheadi, syn. for Aviculopecten carboniferus.
cancellatus, see Aviculopecten cancellatus.
carboniferus, see A. carboniferus.
clevelandicus, see A. clevelandicus.
coloradoensis, see A. coloradoensis.
convexus, see A. convexus.
crenulatus, see Pernopecten crenulatus.
dolabriformis, see Aviculopecten dolabriformis.
duplicatus, see A. duplicatus.
halli, see A. halli.
hawni, syn. for A. carboniferus.
missouriensis, see A. missouriensis.
neglectus, see A. neglectus.
occidentalis, see A. occidentalis.
providencensis, see A. providencensis.
radialis, see Pseudomonotis radialis.
ringens, see Aviculopecten ringens.
striatus, see A. striatus. [ata.
tenuilineatus, see Streblopteria tenuilineutahensis, see Aviculopecten utahensis.
PERNOPECTEN, Winchell, 1865, Proc. Acad. Nat. Sci. Phil. [Ety. from the shells *Perna* and *Pecten*.]
crenulatus, Hall, 1843, (Pecten crenulatus) Geo. Rep. 4th Dist. N. Y., Chemung Gr. [Sig. crenulated.]
fasciculatus, Hall, 1877, Pal. N. Y., vol. 5, Chemung Gr. [Sig. fasciculated.]
glaber, Hall, 1843, (Lima glaber) Geo. Rep. 4th Dist. N. Y., Chemung Gr. [Sig. smooth.]
limiformis, White & Whitfield, 1862, (Aviculopecten limaformis) Proc. Bost. Soc. Nat. Hist., vol. 8, St. Louis Gr. [Sig. like a shell of the genus *Lima*.]
limatus, Winchell, 1865, Proc. Acad. Nat. Sci., Chemung Gr. [Sig. polished.]
obsoletus, Hall, 1843, (Lima obsoleta) Geo. Rep. 4th Dist. N. Y., Chemung Gr. [Sig. obsolete.]
shumardanus, Winchell, 1865, Proc. Acad. Nat. Sci. Phil., Kinderhook Gr. [Ety. proper name.]
PHOLADELLA, Hall, 1870, Prelim. Notice Lam. shells. [Ety. diminutive of the recent genus *Pholas*.]
cuneata, Hall, 1870, Prelim. Notice Lam. shells, Kinderhook Gr. [Sig. wedge-shaped.]

newberryi, Hall, 1870, Prelim. Notice Lam. shells, Waverly Gr. [Ety. proper name.]
ornata, Hall, 1870, Prelim. Notice Lam. shells, Chemung Gr. [Sig. ornamented.]
radiata, Conrad, 1842, (Nuculites radiata) Jour. Acad. Nat. Sci., vol. 8, Ham. Gr. [Sig. radiated.]
truncata, Hall, 1870, Prelim. Notice Lam. shells, Ham. Gr. [Sig. truncated.]
PHTHONIA, Hall, 1870, Prelim. Notice Lam. shells.
nodicostata, Hall, 1870, Prelim. Notice Lam. shells, Ham. Gr. [Sig. from the nodes on the costæ.]
sectifrons, Conrad, 1842, (Cypricardites sectifrons) Jour. Acad. Nat. Sci., Ham. Gr. [Sig. having a divided front.]
PINNA, Linnæus, 1758, Syst. Nat., 10th Ed. [Ety. *pinna*, a feather.]
adamsi, syn. for Pinna peracuta.
hinrichsana, White & St. John, 1868, Trans. Chi. Acad. Sci., St. Louis Gr. [Ety. proper name.]
marshallensis, Winchell, 1865, Proc.Acad. Nat. Sci., Marshall Gr. [Ety. proper name.]
missouriensis, Swallow, 1863, Trans. St. Louis Acad. Sci., Low. Carb. [Ety. proper name.]
peracuta, Shumard, 1858, Trans. St.Louis Acad. Sci., Coal Meas. [Sig. very acute.]
subspatulata, Worthen, 1875, Geo. Sur. Ill., vol. 6, Warsaw Gr. [Sig. somewhat spatulate or blade-shaped.]
Pinnopsis, syn. for Lunulacardium.
acutirostra, see Lunulacardium acutirostrum.
ornatus, see Lunulacardium ornatum.
PLACUNOPSIS, Morris & Lycett, 1853, Monogr. Foss. Great Oolite. [Ety. *Placuna*, a genus of shells; *opsis*, resemblance.]
carbonaria, Meek & Worthen, 1866, Proc. Chi. Acad. Sci., vol. 1, Up. Coal Meas. [Sig. pertaining to coal.]
recticardinalis, Meek, 1875, Ohio Pal., vol. 2, Coal Meas. [Sig. straight on the cardinal line.]
PLEUROPHORUS, King, 1844, Ann. Mag. Nat. Hist., vol. 14. [Ety. *pleuron*, a rib; *phoros*, bearing.]
angulatus, Meek & Worthen, 1865, Proc. Acad. Nat. Sci. Phil., Coal Meas. [Sig. angulated.]
calhouni, Meek & Hayden, 1858, (Edmonia calhouni) Trans. Alb. Inst., vol. 4, Permian Gr. [Ety. proper name.]
costatiformis, Meek & Worthen, 1865, Proc. Acad. Nat. Sci. Phil., Keokuk Gr. [Sig. like *P. costatus*.]
oblongus, Meek, 1872, Pal. E. Neb., Coal Meas. [Sig. oblong.]
occidentalis, Meek & Hayden, 1862, Trans. Alb. Inst., vol. 4, Coal Meas. [Sig. western.]

LAMELLIBRANCHIATA. 201

pallasi, as identified by Geinitz, is P. oblongus.
permianus, Swallow, 1858, Trans. St. Louis Acad. Sci., Permian Gr. [Ety. proper name.]
quadricostatus, Dawson, 1868, Acad. Geo., Carboniferous. [Sig. four-ribbed.]
simplus, as identified by Geinitz, is P. subcuneatus.
subcostatus, Meek & Worthen, 1865, Proc. Acad. Nat. Sci. Phil., Up. Coal Meas. [Sig. somewhat costated.]
subcuneatus, Meek & Hayden, 1858, Trans. Alb. Inst., vol. 4, Permian Gr. [Sig. somewhat wedge-shaped.]
(?) subellipticus, Meek, 1867, Am. Jour. Sci., vol. 44, Coal Meas. [Sig. somewhat elliptical.]
tropidophorus, Meek, 1875, Ohio Pal., vol. 2, Coal Meas. [Sig. keel-bearing.]
Pleurorhynchus, Phillips, syn. for Conocardium.
attenuatus, see Conocardium attenuatum.
crassifrons, see Conocardium crassifrons.
cuneus, see Conocardium cuneus.
trigonalis, see Conocardium trigonale.
vomer, see Conocardium vomer.
POSIDONIA, Bronn, 1824, Syst. Urweltlicher Konchylien. [Ety. proper name.]
alata, Hall, 1843, Geo. Rep. 4th Dist. N. Y., Clinton Gr. [Sig. winged.]
alveata, see Grammysia alveata.
arcuata, see Grammysia arcuata.
clathrata, Lea, 1852, Jour. Acad. Nat. Sci., 2d ser., vol. 2, Coal Meas. [Sig. latticed.]
distans, Lea, 1852, Jour. Acad. Nat. Sci., 2d ser., vol. 2, Coal Meas. [Sig. distant.]
lirata, see Paracyclas lirata.
moorei, Gabb, 1859, Proc. Acad. Nat. Sci., Coal Meas. [Ety. proper name.]
perstriata, Lea, 1852, Jour. Acad. Nat. Sci., Coal Meas. [Sig. closely lined.]
POSIDONOMYA, Bronn, 1837, Leth. Geogn. [Ety. *Poseidon*, a proper name; *Mya*, a genus of shells.]
ambigua, Winchell, 1863, Proc. Acad. Nat. Sci., Chemung Gr. [Sig. doubtful.]
fracta, Meek, 1875, Ohio Pal., vol. 2, Coal Meas. [Sig. frail, easily broken.]
mesambonata, Winchell, 1862, Proc. Acad. Nat. Sci., Marshall Gr. [Ety. *mesos*, middle; *ambon*, umbone.]
rhomboidea, Hall, 1852, Pal. N. Y., vol. 2, Niagara Gr. [Sig. rhomboidal.]
romingeri, Winchell, 1862, Proc. Acad. Nat. Sci., Marshall Gr. [Ety. proper name.]
striata, Stevens, 1858, Am. Jour. Sci., vol. 25, Coal Meas. [Sig. striated.]
whiteana, Winchell, 1862, Proc. Acad. Nat. Sci., Marshall Gr. [Ety. proper name.]
PROMACRUS, Meek, 1871. [Ety. *pro*, forward; *makros*, long.]
andrewsi, Meek, 1871, Am. Jour. Conch., vol. 7, Waverly Gr. [Ety. proper name.]

missouriensis, see Sanguinolites missouriensis.
nasutus, see Sanguinolites nasutus.
PROTHYRIS, Meek, 1869, Proc. Acad. Nat. Sci. Phil. [Ety. *pro*, forward; *thyris*, an orifice.]
elegans, Meek, 1871, Am. Jour. Conch., vol. 7, Coal Meas. [Sig. elegant.]
meeki, Winchell, 1875, Ohio Pal., vol. 2, Waverly Gr. [Ety. proper name.]
PSEUDOMONOTIS, Beyrich, 1862, Zeit. der Deutsch., Geol. Gesselsch., vol. 14. [Ety. *pseudos*, false; *Monotis*, a genus of shells.]
curta, Hall, 1852, (Avicula custa, a typographical error) Stansbury's Ex. Ex. to Gt. Salt Lake, Coal Meas. [Sig. short.]
hawni, Meek & Hayden, 1858, (Monotis hawni) Trans. Alb. Inst., vol. 4, Up. Coal Meas. [Ety. proper name.]
hawni *var.* ovata, Meek & Hayden, 1865, (Eumicrotis hawni *var.* ovata) Pal. Up. Mo., Permian Gr. [Sig. ovate.]
hawni *var.* sinuata, Meek & Worthen, 1866, (Eumicrotis hawni *var.* sinuata) Geo. Sur. Ill., vol. 2, Up. Coal Meas. [Sig. sinuated.]
radialis, (?) Phillips, 1834, (Pecten radialis) Encyc. Meth., vol. 4, Coal Meas. [Sig. radiated.]
PTERINEA, Goldfuss, 1826, Germ. Petref. [Ety. *pteron*, a wing.]
appressa, Conrad, 1838, Ann. Rep. N. Y., Ham. Gr. [Sig. pressed together.]
arenacea, Hall, 1877, Pal. N. Y., vol. 5, Chemung Gr. [Sig. sandy.]
bellilineata, Billings, 1866, Catal. Sil. Foss. Antic. Hud. Riv. Gr. [Sig. beautifully lined.]
bisulcata, see Grammysia bisulcata.
brisa, Hall, 1867, 20th Reg. Rep. N. Y., Niagara Gr. [Sig. pressed.]
boydi, Conrad, 1842, (Avicula boydi) Jour. Acad. Nat. Sci. Phil., vol. 8, Ham. Gr. [Ety. proper name.]
cardiiformis, see Mcgambonia cardiiformis.
cardinata, Winchell, 1862, Proc. Acad. Nat. Sci., Marshall Gr. [Sig. hinged.]
carinata, Goldfuss, see Ambonychia carinata.
chemungensis, Conrad, 1842, (Avicula chemungensis) Jour. Acad. Nat. Sci., vol. 8, Chemung Gr. [Ety. proper name.]
concentrica, Conrad, 1838, Ann. Rep. N. Y., Ham. Gr. [Sig. concentrically lined.]
cuneata, see Sanguinolites cuneatus.
curiosa, Billings, 1866, Catal. Sil. Foss. Antic., Anticosti Gr. [Sig. curious.]
cyrtodontoides, Winchell & Marcy, 1865, Mem. Bost. Soc. Nat. Hist., Niagara Gr. [Sig. like a shell of the genus *Cyrtodonta*.]
flabellum, Conrad, 1842, (Avicula flabella) Jour. Acad. Nat. Sci., vol. 8, Up. Held. and Ham. Gr. [Sig. a little fan.]

longispina, Hall, 1843, (Avicula longispina) Geo. Rep. 4th Dist. N. Y., Ham. and Chemung Gr. [Sig. long-spined.]
modiolaris, see Modiolopsis modiolaris.
morganensis, see Avicula morganensis.
orbicularis, see Ambonychia orbicularis.
pholadis, see Orthonota pholadis.
planulata, see Cypricardinia planulata.
prolifica, Billings, 1866, Catal. Sil. Foss. Antic., Hud. Riv. Gr. [Sig. abundant.]
protexta, Conrad, 1842, (Avicula protexta) Jour. Acad. Nat. Sci., vol. 8, Chemung Gr. [Sig. closely-woven.]
punctulata, Conrad, 1838, Ann. Rep. N. Y., Ham. Gr. [Sig. marked with small dots.]
pygmæa, see Cypricardinia pygmæa.
quadrula, Conrad, 1842, (Avicula quadrula) Jour. Acad. Nat. Sci., vol. 8, Ham. Gr. [Sig. a small square.]
radians, see Monotis radians.
revoluta, Winchell & Marcy, 1865, Mem. Bost. Soc. Nat. Hist., Niagara Gr. [Sig. curled back.]
striato-costata, McChesney, 1861, (Ambonychia striæcosta) New Pal. Foss., Niagara Gr. [Sig. striated and ribbed.]
suborbicularis, see Aviculopecten suborbicularis.
subpapyracea, Meek & Worthen, 1866, Proc. Chi. Acad. Sci., Ham. Gr. [Sig. somewhat like *papyrus*.]
thebesensis, Meek & Worthen, 1868, Geo. Sur. Ill., vol. 3, Niagara Gr. [Ety. proper name.]
thisbe, Billings, 1866, Catal. Sil. Foss. Antic., Anticosti Gr. [Ety. proper name.]
triquetra, see Mytilarca triquetra.
undata, see Ambonychia undata.
undulata, Meek & Worthen, 1868, Geo. Sur. Ill., vol. 3, Kinderhook Gr. [Sig. wavy.]
variostriata, Billings, 1866, Catal. Sil. Foss. Antic., Anticosti Gr. [Sig. variably-lined.]
volans, Winchell & Marcy, 1865, Mem. Bost. Soc. Nat. Hist., Niagara Gr. [Sig. winged.]

PTERONITELLA, Billings, 1874, Pal. Foss., vol. 2. [Ety. diminutive of *Pteronites*.]
curta, Billings, 1874, Pal. Foss., vol. 2, Low. Held. Gr. [Sig. short.]
oblonga, Billings, 1874, Pal. Foss., vol. 2, Low. Held. Gr. [Sig. oblong.]
venusta, Billings, 1874, Pal. Foss., vol. 2, Low. Held. Gr. [Sig. beautiful, lovely.]

PTERONITES, McCoy, 1844, Syn. Carb. Foss. Ireland. [Ety. *Pteron*, a wing.]
(?) chemungensis, Hall, 1843, Geo. Rep. 4th Dist. N. Y., Chemung Gr. [Ety. proper name.]
decussata, Hall, 1843, (Avicula decussata) Geo. Rep. 4th Dist. N. Y., Ham. Gr. [Sig. arranged in pairs that cross each other.]
gayensis, Dawson, 1868, Acad. Geo., Low. Carb. [Ety. proper name.]
lævis, Hall, 1843, (Avicula lævis) Geo. Rep. 4th Dist. N., Y., Ham. Gr. [Sig. smooth.]
muricatus, Hall, 1843,(Avicula muricata) Geo. Rep. 4th Dist. N. Y., Ham. Gr. [Sig. spiny like a *Murex*.]
spinigerus, Conrad, 1842, (Avicula spinigera) Jour. Acad. Nat. Sci., vol. 8, chemung Gr. [Sig. bearing spines.]
subdecussata, Hall, 1877, Pal. N. Y., vol. 5, Ham. Gr. [Sig. somewhat like *P. decussata*.]

PTYCHODESMA, Hall, 1872, 24th Reg. Rep. [Ety. *ptychos*, a folding; *desma*, a ligament or band.]
knappanum, Hall, 1872, 24th Reg. Rep., Corniferous Gr. [Ety. proper name.]

PYRENOMŒUS, Hall, 1852, Pal. N. Y., vol. 2. [Ety. *pyrenos*, Nucula; *omoios*, similar; from its resemblance in general form to the shells of the genus *Nucula*.]
cuneatus, Hall, 1852, Pal. N. Y., vol. 2, Clinton Gr. [Sig. wedge-shaped.]

SANGUINOLARIA, Lamarck, 1801, Syst. An. sans Vert. [Ety. *Sanguinolarius*, having blood.]
leptogaster, Winchell, 1863, Proc. Acad. Nat. Sci., Chemung Gr. [Ety. *leptos*, thin ; *gaster*, belly.]
rostrata, Winchell, 1865, Proc. Acad. Nat. Sci., Marshall Gr. [Sig. beaked.]
sectoralis, Winchell, 1862, Proc. Acad. Nat. Sci., Marshall Gr. [Sig. like a sector.]
septentrionalis, Winchell, 1862, Proc. Acad. Nat. Sci., Marshall Gr. [Sig. northern.]
similis, Winchell, 1862, Proc. Acad. Nat. Sci., Marshall Gr. [Sig. similar.]

SANGUINOLITES, McCoy, 1844, Synop. Carb. Foss. Ireland. [Ety. *Sanguinolaria*, a genus of shells ; *lithos*, stone.]
acutus, Hall, 1870, Prelim. Notice Lam. shells, Ham. Gr. [Sig. acute.]
æolus, Hall, 1870, Prelim. Notice Lam. shells, Wavorly Gr. [Ety. mythological name.]
amygdalinus, Winchell, 1863, Proc. Acad. Nat. Sci., Chemung Gr. [Sig. like an almond.]
arcæformis, Hall, 1870, Prelim. Notice Lam. shells, Ham. Gr. [Sig. like a shell of the genus *Arca*.]
borealis, Winchell, 1862, Proc. Acad. Nat. Sci., Marshall Gr. [Sig. northern.]
carinatus, Conrad, 1841, (Cypricardites carinata) Ann. Rep. N. Y., Ham. Gr. [Sig. carinated.]
chemungensis, Vanuxem, 1842, (Cypricardites chemungensis) Geo. Rep. N. Y., Chemung Gr. [Ety. proper name.]
clavulus, Hall, 1870, Prelim. Notice Lam. shells, Chemung Gr. [Sig. a small nail.]

LAMELLIBRANCHIATA.

cuneatus, Conrad, 1838, (Pterinea cuneata) Ann. Rep. N. Y., Ham. Gr. [Sig. wedge-shaped.]
cylindricus, Winchell, 1863, Proc. Acad. Nat. Sci., Marshall Gr. [Sig. cylindrical.]
flavius, Hall, 1870, Prelim. Notice Lam. shells, Waverly Gr. [Ety. proper name.]
glaucus, Hall, 1870, Prelim. Notice Lam. shells, Ham. Gr. [Ety. mythological name.]
hamiltonensis, see Goniophora hamiltonensis.
ida, Hall, 1870, Prelim. Notice Lam. shells, Ham. Gr. [Ety. mythological name.]
iowensis, Winchell, 1863, Proc. Acad. Nat. Sci., Chemung Gr. [Ety. proper name.]
jejunus, Winchell, 1863, Proc. Acad. Nat. Sci., Chemung Gr. [Sig. scanty, poor.]
marshallensis, Winchell, 1862, Proc. Acad. Nat. Sci., Marshall Gr. [Ety. proper name.]
missouriensis, Swallow, 1860, (Solen (?) missouriensis) Trans. St. Louis Acad. Sci., Low. Carb. [Ety. proper name.]
nasutus, Meek, 1871, Am. Jour. Conch., vol. 7, Low. Carb. [Sig. having a prominent nose.]
obliquus, Meek, 1871, Proc. Acad. Nat. Sci., Waverly Gr. [Sig. oblique.]
perangulatus, Hall, 1870, Prelim. Notice Lam. shells, Schoharie grit. [Sig. very angular.]
ponderosus, Hall, 1870, Prelim. Notice Lam. shells, Up. Held. Gr. [Sig. bulky, heavy.]
rigidus, White & Whitfield, 1862, (Cypricardia rigidi) Proc. Bost. Soc. Nat. Hist., Kinderhook Gr. [Sig. rigid.]
sanduskyensis, Meek, 1871, Proc. Acad. Nat. Sci., Corniferous Gr. [Ety. proper name.]
solenoides, Hall, 1870, Prelim. Notice Lam. shells, Ham. Gr. [Sig. like a shell of the genus *Solen*.]
strigatus, Winchell, 1865, Proc. Acad. Nat. Sci., Chemung Gr. [Sig. furrowed.]
subtortuosus, Hall, 1870, Prelim. Notice Lam. shells, Ham. Gr. [Sig. somewhat twisted.]
sulcifer, Winchell, 1866, Rep. Low Peninsula Mich., Ham. Gr. [Sig. bearing furrows.]
tethys, Billings, 1874, Pal. Foss., vol. 2, Devonian. [Ety. mythological name.]
truncatus, Conrad, 1842, (Cypricardites truncatus) Jour. Acad. Nat. Sci., vol. 8, Ham. Gr. [Sig. truncated.]
undatus, Hall, 1870, Prelim. Notice Lam. shells, Chemung Gr. [Sig. wavy.]
unioniformis, Winchell, 1862, Proc. Acad. Nat. Sci., Marshall Gr. [Sig. like a shell of the genus *Unio*.]
valvulus, Hall, 1870, Prelim. Notice Lam. shells, Waverly Gr. [Sig. having small valves, a pod.]
ventricosus, White & Whitfield, 1862, Proc. Bost. Soc. Nat. Hist., Chemung and Kinderhook Gr. [Sig. ventricose.]

SCHIZODUS, King, 1844, Ann. Mag. Nat. Hist., vol. 14. [Ety. *schizo*, I split; *odous*, a tooth.]
amplus, Meek & Worthen, 1870, Proc. Acad. Nat. Sci. Phil., Coal Meas. [Sig. full, large.]
appressus, see Cytherodon appressus.
cayuga, Hall, 1870, Prelim. Notice Lam. shells, Ham. Gr. [Ety. proper name.]
chemungensis, Conrad, 1842, (Nuculites chemungensis) Jour. Acad. Nat. Sci., vol. 8, Chemung Gr. [Ety. proper name.]
chesterensis, Meek & Worthen, 1865, Proc. Acad. Nat. Sci. Phil., Chester Gr. [Ety. proper name.]
cuneatus, Meek, 1875, Ohio Pal., vol. 2, Coal Meas. [Sig. wedge-shaped.]
curtus, Meek & Worthen, 1866, Proc. Chi. Acad. Sci., Coal Meas. [Sig. short.]
ellipticus, Hall, 1870, Prelim. Notice Lam. shells, Ham. Gr. [Sig. elliptical.]
gregarius, Hall, 1870, Prelim. Notice Lam. shells, Ham. Gr. [Sig. gregarious.]
medinaensis, Meek, 1871, Proc. Acad. Nat. Sci. Phil., vol. 23, Waverly Gr. [Ety. proper name.]
oblatus, Hall, 1870, Prelim. Notice Lam. shells, Chemung Gr. [Sig. oblate.]
obscurus, see Schizodus wheeleri.
ovatus, Meek & Hayden, 1858, (Axinus ovatus) Proc. Acad. Nat. Sci. Phil., Permian Gr. [Sig. ovate.]
perelegans, Meek & Worthen, 1870, Proc. Acad. Nat. Sci. Phil., Coal Meas. [Sig. very elegant.]
quadrangularis, Hall, 1870, Prelim. Notice Lam. shells, Chemung Gr. [Sig. quadrangular.]
rhombeus, Hall, 1870, Prelim. Notice Lam. shells, Ham. Gr. [Sig. rhombshaped.]
rossicus, Verneuil, 1845, Geo. Russ., vol. 2, Permian Gr. [Ety. proper name.]
triangularis, Swallow, 1858, Trans. St. Louis Acad. Sci., Permian Gr. [Sig. triangular.]
tumidus, see Cytherodon tumidus.
wheeleri, Swallow, 1862, (Cypricardia (?) wheeleri) Trans. St. Louis. Acad. Sci., Coal Meas. [Ety. proper name.]

SEDGWICKIA, McCoy, 1844, Synop. Carb. Foss. Ireland. [Ety. proper name.]
altirostrata, Meek & Hayden, 1858, (Allorisma (?) altirostrata) Proc. Acad. Nat. Sci. Phil., Coal Meas. [Sig. highbeaked.]
(?) compressa, Meek, 1872, Proc. Acad. Nat. Sci. Phil., Cin'ti Gr. [Sig. compressed.]

concava, Meek & Hayden, 1858, (Lyonsia concava) Trans. Alb. Inst., vol. 4, Coal Meas. [Sig. concave.]
divaricata, Hall & Whitfield, 1875, Ohio Pal., vol. 2, Cin'ti Gr. [Sig. from the diverging plications.]
(?) fragilis, Meek, 1872, Proc. Acad. Nat. Sci., Phil., Cin'ti Gr. [Sig. frail, easily broken.]
(?) neglecta, Meek, 1872, Proc. Acad. Nat. Sci. Phil., Cin'ti Gr. [Sig. neglected, overlooked.]
subarcuata, Meek & Worthen, 1865, Proc. Acad. Nat. Sci. Phil., Keokuk Gr. [Sig. somewhat arched.]
topekaensis, Shumard, 1858, (Leptodomus topekaensis) Trans. St. Louis Acad. Sci., Coal Meas. [Ety. proper name.]

Solemya, Lamarck, 1818, Hist. Nat. An. sans Vert., vol. 5. See Solenomya—the correct orthography, first used by Menke, 1828, Syn. Meth. Edit.

SOLEN, Linnæus, 1758, Syst. Nat., 10th ed. [Ety. *Solen*, a tube or pipe.]
scalpriformis, Winchell, 1862, Proc. Acad. Nat. Sci., Marshall Gr. [Sig. lancet-shaped.]
missouriensis, see Sanguinolites missouriensis.
permianus, Swallow, 1858, Trans. St. Louis Acad. Sci., Permian Gr. [Ety. proper name.]
priscus, Winchell, 1862, Proc. Acad. Nat. Sci., Portage Gr. [Sig. ancient.]
quadrangularis, Winchell, 1862, Proc. Acad. Nat. Sci., Marshall Gr. [Sig. quadrangular.]

SOLENOMYA, Lamarck, 1818, (Solemya) Hist. Nat. Anim. sans Vert., vol. 5. [Ety. from the resemblance to the two genera *Solen* and *Mya*.]
anodontoides, Meek, 1875, Ohio Pal., vol. 2, Coal Meas. [Sig. like a shell of the genus *Anodonta*.]
biarmica, Verneuil, 1845, Geo. Russ. and Ural Mountains, Permian Gr. This species has not been satisfactorily identified in this country.
radiata, Meek & Worthen, 1860, (Solemya radiata) Proc. Acad. Nat. Sci. Phil., Coal Meas. [Sig. radiated.]
recurvata, Swallow, 1858, Trans. St. Louis Acad. Sci., Up. Coal Meas. [Sig. curved backward.]
soleniformis, Cox, 1857, Geo. Sur. Ky., vol. 3, Coal Meas. [Sig. like a shell of the genus *Solen*.]
vetusta, Meek, 1871, Proc. Acad. Nat. Sci. Phil., Corniferous Gr. [Sig. ancient.]

SOLENOPSIS, McCoy, 1844, Carb. Foss. Ireland. [Ety. *Solenopsis*, resembling a shell of the genus *Solen*.]
solenoides, Geinitz, 1866, (Clidophorus solenoides) Carb. und Dyas in Neb., Coal Meas. [Sig. like a shell of the genus *Solen*.]

STREBLOPTERIA, McCoy, 1851, Ann. Mag. Nat. Hist., 2d series, vol. 7. [Ety. *streblos*, turned the wrong way; *pteron*, a wing.]
tenuilineata, Meek & Worthen, 1860, (Pecten tenuilineatus) Proc. Acad. Nat. Sci. Phil., Coal Meas. [Sig. fine-lined.]

Tellina, Linnæus, 1758, Syst. Nat., 10th ed. [Ety. *telline*, a sort of mussel.] This genus is unknown in the palæozoic rocks.
(?) ovata, Hall, 1843, Geo. Rep. 4th Dist. N. Y., Ham. Gr. Syn. for Palæoneilo maxima.

TELLINOMYA, Hall, 1847, Pal. N. Y., vol. 1. [Ety. from a resemblance to the genera *Tellina* and *Mya*.]
abrupta, Billings, 1862, (Ctenodonta abrupta) Pal. Foss., vol. 1, Black River Gr. [Sig. abrupt.]
æquilatera, Hall, 1852, Pal. N. Y., vol. 2, Coralline limestone. [Sig. equal-sided.]
alta, Hall, 1861, Geo. Rep. Wis., Trenton Gr. [Sig. high.]
anatiniformis, see Pterotheca anatiniformis.
angela, Billings, 1865, (Ctenodonta angela) Pal. Foss., vol. 1, Quebec Gr. [Ety. proper name.]
angustata, Hall, 1860, Can. Nat. & Geo., vol. 5, Low. Sil. [Sig. narrowed.]
astartiformis, Salter, 1859, (Ctenodonta astarteformis) Can. Org. Rem., Decade 1, Black Riv. Gr. [Ety. like an *Astarte*.]
attenuata, Hall, 1860, Can. Nat. & Geo., vol. 5, Silurian. [Sig. attenuated.]
contracta, Salter, 1859, (Ctenodonta contracta) Can. Org. Rem., Decade 1, Black Riv. & Trenton Gr. [Sig. contracted.]
curta, Hall, 1852, Pal. N. Y., vol. 2, Clinton Gr. [Sig. short.]
donaciformis, Hall, 1847, Pal. N. Y., vol. 1, Trenton Gr. [Sig. like a shell of the genus *Donax*.]
dubia, Hall, 1847, Pal. N. Y., vol. 1, Black Riv. & Trenton Gr. [Sig. doubtful.]
elliptica, Hall, 1852, Pal. N. Y., vol. 2, Clinton Gr. [Sig. elliptical.]
gibberula, Salter, 1859, (Ctenodonta gibberula) Can. Org. Rem., Decade 1, Black Riv. & Trenton Gr. [Sig. a little hunch-backed.]
gibbosa, Hall, 1847, Pal. N. Y., vol. 1, Black Riv. & Trenton Gr. [Sig. gibbous.]
hartsvillensis, Safford, 1859, (Ctenodonta hartsvillensis) Geo. of Tenn., Nashville Gr. [Ety. proper name.]
hilli, S. A. Miller, 1874, Cin. Quar. Jour. Sci., Cin'ti Gr. [Ety. proper name.]
inflata, Hall, 1861, Geo. Rep. Wis., Trenton Gr. [Sig. inflated.]

LAMELLIBRANCHIATA. 205

iphigenia, Billings, 1862, (Ctenodonta iphigenia) Pal. Foss., vol. 1, Hud. Riv. Gr. [Ety. proper name.]
lata, Hall, 1852, Pal. N. Y., vol. 2, Clinton Gr. [Sig. wide.]
levata, Hall, 1847, (Nucula levata) Pal. N. Y., vol. 1, Black Riv., Trenton and Hud. Riv. Gr. [Sig. smoothed.]
lineata, Phillips, 1836, (Nucula lineata) Pal. Foss., Ham. Gr. [Sig. lined.]
logani, Salter, 1851, (Ctenodonta logani) Rep. Brit. Assoc., Hud. Riv. Gr. [Ety. proper name.]
machæriformis, Hall, 1843,(Nucula machariformis) Geo. Rep. 4th Dist. N. Y., Clinton Gr. [Sig. sword-shaped.]
mactræformis, Hall, 1843, (Nucula mactræformis) Geo. Rep. 4th Dist. N. Y., Clinton Gr. [Sig. trough-like.]
nasuta, Hall, 1847, Pal. N. Y., vol. 1, Black Riv. & Trenton Gr. [Sig. having a prominent nose.]
nucleiformis, Hall, 1859, Pal. N. Y., vol. 3, Low. Held. Gr. [Sig. nut-shaped.]
nuculiformis, Hall, 1847, (Modiolopsis nuculiformis) Pal. N. Y., vol. 1, Hud. Riv. Gr. [Sig. like a *Nucula*.]
obliqua, Hall, 1845, (Nucula obliqua) Am. Jour. Sci., vol. 43, Cin'ti Gr. [Sig. oblique.]
ovata, Hall, 1861, Geo. Rep. Wis., Trenton Gr. [Sig. ovate.]
pectunculoides, Hall, 1871, Pamphlet, Cin'ti Gr. [Sig. like a shell of the genus *Pectunculus*.]
protensa, Hall, 1852, Stans. Ex. to Gt. Salt Lake, Coal Meas. [Sig. stretched out.]
sanguinolarioidea, Hall, 1847, Pal. N. Y., vol. 1, Trenton Gr. [Sig. like *Sanguinolaria*.]
subnasuta, Hall, 1872, 24th Reg. Rep. N. Y., Up. Held. Gr. [Sig. somewhat nasute.]
ventricosa, Hall, 1861, Geo. Rep. Wis., Trenton Gr. [Sig. ventricose.]

TELLINOPSIS, Hall, 1870, Prelim. Notice Lam. shells. [Ety. *tellinopsis*, resembling a shell of the genus *Tellina*.]
submarginata, Conrad, 1842, (Nuculites submarginatus) Jour. Acad. Nat. Sci., vol. 8, Ham. Gr. [Sig. slightly emarginated.]
Ungulina, Daudin, 1802, Bosc. Hist. Nat. Coq. 3. [Ety. *ungulina*, like a hoof.]
suborbicularis, see Cardiomorpha suborbicularis.
Vanuxemia, syn. for Cypricardites.
bayfieldi, see Cypricardites bayfieldi.
inconstans, see Cypricardites inconstans.
montrealensis, see Cypricardites montrealensis.
dixonensis, see Cypricardites dixonensis.
YOLDIA, Muller, 1842, Kroyer's Nat. Tid. [Ety. proper name.]
carbonaria, Meek, 1871, Rep. Reg. University W. Va., Coal Meas. [Sig. pertaining to coal.]
gibbosa, McChesney, 1859, (Leda gibbosa) Pal. Foss., Coal Meas. [Sig. gibbous.]
knoxensis, McChesney, 1865, (Leda knoxensis) Expl. Pal. Foss., Coal Meas. [Ety. proper name.]
levistriata, Meek & Worthen, 1860, (Leda levistriata) Proc. Acad. Nat. Sci. Phil., St. Louis Gr. [Sig. small-lined.]
oweni, McChesney, 1860, (Leda oweni) Desc. New Pal. Foss., Coal Meas. [Ety. proper name.]
polita, McChesney, 1859, (Leda polita) Pal. Foss., Coal Meas. [Sig. smoothed.]
rushensis, McChesney, 1865, (Leda rushensis) Expl. Pal. Foss., Coal Meas. [Ety. proper name.]
stevensoni, Meek, 1871, Rep. Reg. University W. Va., Coal Meas. [Ety. proper name.]
subscitula, Meek & Hayden, 1858, (Leda subscitula) Trans. Alb. Inst., vol. 4, Permian Gr. [Sig. somewhat pretty.]
valvulus, Hall, 1872, 24th Reg. Rep., Corniferous Gr. [Sig. like the shell of a bean.]

SUB-KINGDOM ARTICULATA.

First Class,	- - - -	ANNELIDA.
Second Class,	- - - -	CRUSTACEA.
Third Class,	- - - -	ARACHNIDA.
Fourth Class,	- - - -	MYRIAPODA.
Fifth Class,	- - - -	INSECTA.

CLASS ANNELIDA.

GENERA.—Arenicolites, Conchicolites, Cornulites, Salterella, Serpulites, Spirorbis.

ARENICOLITES, Salter, 1856, Quar. Jour. Geo. Soc. [Ety. *arena*, sand; *colo*, I inhabit; *lithos*, stone; circular holes which appear in twos on the surface of sandstones, and having the appearance of worm-burrows like those of the *Arenicola*.]
  spiralis, as identified by Billings, and others, Pal. Foss., vol. 2, Huronian Gr. [Sig. spiral.]

CONCHICOLITES, Nicholson, 1872, Am. Jour. Sci. [Ety. *concha*, a shell; *colo*, I dwell; *lithos*, a stone.]
  corrugatus, Nicholson, 1872, Lond. Geo. Mag., vol. 9, Cin'ti Gr. [Sig. corrugated.]
  flexuosus, Hall, 1847, (Tentaculites flexuosus) Pal. N. Y., vol. 1, Trenton & Hud. Riv. Gr. [Sig. flexuous.]
  intermedius, Nicholson, 1874, (Ortonia intermedia) Geo. Mag., n. s., vol. 1, Ham. Gr. [Sig. intermediate.]
  minor, Nicholson, 1873, (Ortonia minor) Lond. Geo. Mag., vol. 10, Cin'ti Gr. [Sig. less.]

CORNULITES, Schlotheim, 1820, Petrefactenkunde. [Ety. *cornu*, horn; *lithos*, stone.]
  arcuatus, Conrad, 1848, Jour. Acad. Nat. Sci., vol. 8, Niagara Gr. [Sig. bent, bow-shaped.]
  flexuosus, Hall, 1852, Pal. N. Y., vol. 2, Clinton Gr. [Sig. wavy.]
  flexuosus var. gracilis, Hall, 1860, Can. Nat. & Geo., vol. 5, Silurian. [Sig. slender.]
  proprius, Hall, 1876, 28th Reg. Rep. N. Y., Niagara Gr. [Sig. peculiar, lasting.]

Ortonia, Nicholson, 1872, Lond. Geo. Mag., Vol. 9. This is a synonym for Conchicolites, if indeed both are not synonyms for Cornulites.
  conica, syn. for Conchicolites flexuosus.
  intermedia, see Conchicolites intermedius.
  minor, see Conchicolites minor.

SALTERELLA, Billings, 1861, Pal. Foss., vol. 1. [Ety. proper name.]
  billingsi, Safford, 1869, Geo. of Tenn., Trenton Gr. [Ety. proper name.]
  obtusa, Billings, 1861, Pal. Foss., vol. 1, Potsdam Gr. [Sig. obtuse.]
  pulchella, Billings, 1861, Pal. Foss., vol. 1, Potsdam Gr. [Sig. very beautiful.]
  rugosa, Billings, 1861, Pal. Foss., vol. 1, Potsdam Gr. [Sig. wrinkled.]

Serpula, Linnæus, 1758, Syst. Nat., 10th ed. [Ety. *serpo*, to creep.]
  omphalodes, see Spirorbis omphalodes.
  valvata, see Spirorbis valvatus.

SERPULITES, McLeay, 1839, Murch. Sil. Syst. [Ety. *Serpula*, a genus of annelids.]
  annulatus, Dawson, 1868, Acad. Geol., Carboniferous. [Sig. ringed.]
  dissolutus, Billings, 1862, Pal. Foss., vol. 1, Trenton Gr. [Sig. weak, broken.]
  hortonensis, Dawson, 1868, Acad. Geol., Carboniferous. [Ety. proper name.]
  inelegans, Dawson, 1868, Acad. Geol., Carboniferous. [Sig. not elegant.]
  murchisoni, Hall, 1861, Geo. Rep. Wis., Potsdam Gr. [Ety. proper name.]
  splendens, Billings, 1859, Can. Nat. and Geo., vol. 4, Chazy Gr. [Sig. splendid.]

SPIRORBIS, Lamarck, 1801, Syst. An. sans Vert. [Sig. spiral-whorl.]
ammon, Winchell, 1866, Rep. Low. Peninsula Mich., Ham. Gr. [Ety. mythological name.]
angulatus, Hall, 1862, 15th Reg. Rep. N.Y., Ham. Gr. [Sig. angular.]
angulatus, Dawson, 1868. The name was preoccupied.
annulatus, Hall, 1858, Trans. Alb. Inst., vol. 4, Warsaw Gr. [Sig. ringed.]
annulatus *var.* nodulosus, Hall, 1858, Trans. Alb. Inst., vol. 4, Warsaw Gr. [Sig. full of knots.]
arietinus, Dawson, 1869, Rep. of Progr., Coal Meas. [Sig. resembling a ram's horn.]
arkonensis, Nicholson, 1874, Geo. Mag., vol. 1, Ham. Gr. [Ety. proper name.]

carbonarius, Dawson, 1845, Quar. Jour. Geo. Soc., vol. 1, Coal Meas. [Sig. pertaining to coal.]
flexuosus, Hall, 1863, Trans. Alb. Inst., vol. 4, Niagara Gr. [Sig. bent.]
inornatus, Hall, 1863, Trans. Alb. Inst., vol. 4, Niagara Gr. [Sig. not ornamented.]
laxus, Hall, 1859, Pal. N. Y., vol. 3, Low. Held. Gr. [Sig. loose.]
obesus, Winchell, 1866, Rep. Low. Peninsula Mich., Ham. Gr. [Sig. plump in form.]
orbiculostoma, Swallow, 1858, Trans. St. Louis Acad. Sci., Permian Gr. [Sig. orbicular-mouthed.]
valvatus, Goldfuss, 1826, (Serpula valvata). Not American.

CLASS CRUSTACEA.

FAMILY ACIDASPIDÆ.—Acidaspis, Terataspis.
FAMILY AGLASPIDÆ.—Aglaspis.
FAMILY AGNOSTIDÆ.—Agnostus, Shumardia.
FAMILY ASAPHIDÆ.—Asaphus, Bathyurellus, Bathyurus, Dolichometopus, Illænus, Illænurus, Isotelus, Megalaspis, Nileus, Ogygia.
FAMILY BRONTEIDÆ.—Bronteus.
FAMILY CALYMENIDÆ.—Arges, Calymene, Homalonotus.
FAMILY CERAURIDÆ.—Amphion, Ceraurus, Encrinurus, Sphærexochus.
FAMILY CONOCEPHALIDÆ.—Conocephalites, Conocoryphe, Crepicephalus, Dicellocephalus, Ptychaspis, Solenopleura.
FAMILY CYPHASPIDÆ.—Cyphas.
FAMILY CYTHERIDÆ.—Cythere, Cytherina, Cytheropsis.
FAMILY EURYPTERIDÆ.—Anthraconectes, Dolichopteris, Eurypterus, Pterygotus.
FAMILY HARPEDIDÆ.—Harpes, Harpides.
FAMILY LEPERDITIIDÆ.—Beyrichia, Isochilinia, Leperditia, Primitia.
FAMILY LICHASIDÆ.—Lichas.
FAMILY PARADOXIDÆ.—Agraulos, Anapolenus, Arionellus, Bathynotus, Chariocephalus, Loganellus, Menocephalus, Olenellus, Olenus, Paradoxides, Remopleurides, Telephus, Triarthrella, Triarthrus.
FAMILY PHACOPIDÆ.—Dalmanites, Phacops, Thaleops.
FAMILY PROETIDÆ.—Phillipsia, Prœtus.   [cleus.
FAMILY TRINUCLEIDÆ.—Ampyx, Endymionia, Holometopus, Microdiscus, Trinu-
ORDER AMPHIPODA.—Diplostylus.
ORDER CIRRHOPODA.—Plumulites.
ORDER ISOPODA.—Acanthotelson.
ORDER MACRURA.—Anthracopalæmon, Archæocaris, Palæocaris.
ORDER PHYLLOPODA.—Ceratiocaris, Dithyrocaris, Leaia, Solenocaris.
ORDER STOMAPODA.—Amphipeltis.
ORDER XYPHOSURA.—Euproops.
INCERTÆ SEDIS.—Climachtichnites, Diplichnites, Helminthoidichnites, Protichnites, Rusichnites.

Acantholoma, syn. for Acidaspis.
spinosa, syn. for Acidaspis tuberculata.
ACANTHOTELSON, Meek & Worthen, 1860, Proc. Acad. Nat. Sci. [Ety. *akantha*, a spine; *telson*, the end.]
  eveni, Meek & Worthen, 1868, Am. Jour. Sci., vol. 46, Coal Meas. [Ety. proper name.]
inæqualis, syn. for Palæocaris typus.
  stimpsoni, Meek & Worthen, 1865, Proc. Acad. Nat. Sci., Coal Meas. [Ety. proper name.]

ACIDASPIS, Murchison, 1839, Sil. Syst. [Ety. *akis*, a spear-point; *aspis*, a shield.]
  anchoralis, S. A. Miller, 1875, Cin. Quar. Jour. Sci., Cin. Gr. [Sig. anchor-like.]
  ceralepta, Anthony, 1838, (Ceratocephala ceralepta) Am. Jour. Sci., Cin'ti Gr. Not clearly defined.
  cincinnatiensis, Meek, 1873, Ohio Pal., vol. 1, Cin'ti Gr. [Ety. proper name.]
  crosotus, Locke, 1843, Am. Jour. Sci., vol. 44, Cin'ti Gr. [Ety. *crossotos*, fringed.] The word is misspelled—it should be *crossota*.

CRUSTACEA. 209

danai, Hall, 1862, Geo. Sur. Wis., Niagara Gr. [Ety. proper name.]
eriopis, see Terataspis eriopis.
grandis, see Terataspis grandis.
halli, Shumard, 1855, Geo. Sur. Mo., Trenton Gr. [Ety. propor name.]
hamata, Conrad, 1841, (Dicranurus hamatus) Ann. Rep. N. Y., Low. Held. Gr. [Sig. hooked.]
horani, Billings, 1859, Rep. of Progr., Trenton Gr. [Ety. proper name.]
ida, syn. for Acidaspis danai.
oncalli, S. A. Miller, 1875, Cin. Quar. Jour. Sci., vol. 2, Cin'ti Gr. [Ety. proper name.]
parvula, Walcott, 1877, 29th Reg. Rep. N. Y., Trenton Gr. [Sig. very small.]
spiniger, see Bathyurus spiniger.
trentonensis, Hall, 1847, Pal. N. Y., vol. 1, Trenton Gr. [Ety. proper name.]
tuberculata, Conrad, 1840, Ann. Rep. N. Y., Low. Held. Gr. [Sig. tuberculated.]

AGLASPIS, Hall, 1863, 16th Reg. Rep. & 1862, Can. Nat. & Geo., vol. 6. [Ety. *Aglaos*, bright; *aspis*, shield.]
barrandi, Hall, 1863, 16th Reg. Rep. N. Y., Potsdam Gr. [Ety. proper name.]

AGNOSTUS, Brongniart, 1822, Hist. Nat. Crust. Foss. [Ety. *agnostos*, obscure.]
acadicus, Hartt, 1868, Acad. Geol., St. John's Gr. [Ety. proper name.]
americanus, Billings, 1860, Can. Nat. & Geol., vol. 5, Quebec Gr. [Ety. proper name.]
bidens, Meek, 1872, Hayden's Geo. Rep., Potsdam Gr. [Sig. two-pronged.]
canadensis, Billings, 1860, Can. Nat. & Geol., vol. 5, Quebec Gr. [Ety. proper name.]
coloradoensis, Shumard, 1861, Am. Jour. Sci. & Arts, Potsdam Gr. [Ety. proper name.]
disparilis, Hall, 1863, 16th Reg. Rep. N. Y., Potsdam Gr. [Sig. unequal.]
fabius, Billings, 1865, Pal. Foss., vol. 1, Quebec Gr. [Ety. proper name.]
galba, Billings, 1865, Pal. Foss., vol. 1, Quebec Gr. [Ety. proper name.]
interstrictus, White, 1874, Rep. Invert. Foss., Potsdam Gr. [Sig. drawn together.]
josepha, Hall, 1863, 16th Reg. Rep. N. Y., Potsdam Gr. [Ety. proper name.]
latus, see Beyrichia lata.
lobatus, Hall, 1847, Pal. N. Y., vol. 1, Hud. Riv. Gr. [Sig. lobate.]
maladensis, Meek, 1872, Hayden's Geo. Rep., Calciferous Gr. [Ety. proper name.]
nobilis, Ford, 1872, Am. Jour. Sci., 3rd series, vol. 3, Potsdam Gr. [Sig. excellent.]
orion, Billings, 1860, Can. Nat. & Geol., vol. 5, Quebec Gr. [Ety. mythological name.]
parilis, Hall, 1863, 16th Reg. Rep. N. Y., Potsdam Gr. [Sig. like, equal.]

similis, Hartt, 1868, Acad. Geol., St. John's Gr. [Sig. similar, from resemblance to *A. acadicus*.]

AGRAULOS, Corda, 1847, Bohemian Trilobites. [Ety. proper name.]
affinis, Billings, 1874, Pal. Foss., vol. 2, Potsdam Gr. [Sig. near to; closely allied to *A. socialis*.]
oweni, Meek & Hayden, 1861, (Arionellus oweni) Proc. Acad. Nat. Sci. Phil., Potsdam Gr. [Ety. proper name.]
socialis, Billings, 1874, Pal. Foss., vol. 2, Potsdam Gr. [Sig. living in groups.]
strenuus, Billings, 1874, Pal. Foss., vol. 2, Potsdam Gr. [Sig. vigorous.]

AMPHION, Pander, 1830, Beitrage zur Geognosie des Russischen Reiches. [Ety. mythological name.]
barrandi, Billings, 1865, Pal. Foss., vol. 1, Quebec Gr. [Ety. proper name.]
canadensis, Billings, 1859, Can. Nat. and Geol., vol. 4, Chazy Gr. [Ety. proper name.]
convexus, Billings, 1865, Pal. Foss., vol. 1, Quebec Gr. [Sig. convex.]
insularis, Billings, 1865, Pal. Foss., vol. 1, Quebec Gr. [Sig. on an island.]
julius, Billings, 1865, Pal. Foss., vol. 1, Quebec Gr. [Ety. proper name.]
matutinus, Hall, 1863, 16th Reg. Rep. N. Y., Potsdam Gr. [Sig. in the morning.]
multisegmentatus, see Encrinurus multisegmentatus.
salteri, Billings, 1861, Can. Nat. & Geol., vol. 6, Calciferous Gr. [Ety. proper name.]
westoni, Billings, 1865, Pal. Foss., vol. 1, Quebec Gr. [Ety. proper name.]

AMPHIPELTIS, Salter, 1863, Quar. Jour. Geo. Soc., vol. 19. [Ety. *amphi*, doubtful; *peltis*, provided with a shield or buckler.]
paradoxus, Salter, 1863, Quar. Jour. Geo. Soc., vol. 19, Devonian. [Sig. extraordinary.]

AMPYX, Dalman, 1827, Monograph of Trilobites. [Ety. mythological name.]
halli, Billings, 1861, Pal. Foss., vol. 1, Chazy Gr. [Ety. proper name.]
laeviusculus, Billings, 1865, Pal. Foss., Quebec Gr. [Sig. quite smooth.]
normalis, Billings, 1865, Pal. Foss., vol. 1, Quebec Gr. [Sig. according to the square.]
rutilius, Billings, 1865, Pal. Foss., vol. 1, Quebec Gr. [Ety. proper name.]
semicostatus, Billings, 1865, Pal. Foss., vol. 1, Quebec Gr. [Sig. half-ribbed.]

ANOPLOLENUS, Salter, 1864, Geo. Mag., vol. 1. [Ety. *a*, without; *ops*, an eye; *olena*, convexity.]
venustus, Billings, 1874, Pal. Foss., vol. 2, Low. Potsdam Gr. [Sig. beautiful.]

ANTHRACONECTES, Meek & Worthen, 1868, Am. Jour. Sci., vol. 46. [Ety. *anthrax*, coal; *nectos*, swimming.] A subgenus of Eurypterus, founded upon *E. mazonensis*, as the type.

210   CRUSTACEA.

ANTHRACOPALÆMON, Salter, 1861, Quar. Jour. Geo. Soc. Lond., vol. 17. [Ety. *anthrax*, coal; *palæmon*, ancient prawn or shrimp.]
- gracilis, Meek & Worthen, 1865, Proc. Acad. Nat. Sci. Phil., Coal Meas. [Sig. slender.]
- hillanus, Dawson, 1877, Geo. Mag., Coal Meas. [Ety. proper name.]

ARCHÆOCARIS, Meek, 1872, Proc. Acad. Nat. Sci. Phil. [Ety. *archaios*, ancient; *karis*, a shrimp.]
- vermiformis, Meek, 1872, Proc. Acad. Nat. Sci. Phil., Low. Carb. [Sig. worm-shaped.]

ARGES, Goldfuss, 1839, Nova Acta Phys. Acad. Caes. Leop. Nat. Cur. [Ety. mythological name.]
- phlyctainodes, Green, 1837, (Calymene phlyctainodes) Am. Jour. Sci., vol. 32, Niagara Gr. [Sig. pimply, pustulous.]

ARIONELLUS, Barrande, 1852, Syst. Sil. Boh. [Ety. diminutive of Arion, a generic name.]
- bipunctatus, Shumard, 1863, Trans. St. Louis Acad. Sci., Potsdam Gr. [Sig. double-dotted.]
- cylindricus, Billings, 1860, Can. Nat. & Geo., vol. 5. Quebec Gr. [Sig. cylindrical.]
- oweni, see Agraulos oweni.
- planus, Shumard, 1861, Am. Jour. Sci., Potsdam Gr. [Sig. flat.]
- pustulatus, Walcott, 1877, 29th Reg. Rep. N. Y., Chazy Gr. [Sig. pustulated.]
- subclavatus, Billings, 1860, Can. Nat. & Geo., vol. 5, Quebec Gr. [Sig. somewhat club-shaped.]
- texanus, Shumard, 1861, Am. Jour. Sci., Potsdam Gr. [Ety. proper name.]
- tripunctatus, Whitfield, 1876, Rep. Recon. Up. Mo. to Yel. Nat. Park, Potsdam Gr. [Sig. three-dotted.]

Asaphiscus, Meek. Syn. for Bathyurellus.
wheeleri, see Bathyurellus wheeleri.

ASAPHUS, Brongniart, 1822, Hist. Nat. Crust. Foss. [Ety. *asaphus*, uncertain, obscure.]
- *acantholeurus*, see Dalmanites acantholeurus.
- alacer, Billings, 1866, Catal. Sil. Foss. Antic., Hud. Riv. Gr. [Sig. lively, quick.]
- *aspectans*, Conrad, see Dalmanites aspectans.
- astragalotes, Green, 1834, Am. Jour. Sci., vol. 25, (?) Gr. [Sig. having vertebræ.]
- barrandi, Hall, 1851, Lake Sup. Land. Dist., Birdseye Gr. [Ety. proper name.]
- canadensis, Chapman, 1856, Can. Jour., vol. 2, Trenton Gr. [Ety. proper name.]
- *canalis*, see Isotelus canalis.
- *corycœus*, see Proetus corycœus.
- crypturus, Green, 1827, Am. Jour. Sci., vol. 13, Up. Sil. [Sig. concealed tail.]
- curiosus, Billings, 1865, Pal. Foss., vol. 1, Quebec Gr. [Sig. curious.]
- *denticulatus*, see Dalmanites denticulatus.
- diurus, Green, 1840, Am. Jour. Sci., vol. 37, Niagara Gr. Probably the fragment of a Dalmanite.
- *extans*, see Bathyurus extans.
- *gigas*, see Isotelus gigas.
- goniocercus, Meek, 1872, Hayden's Geo. Rep., Quebec Gr. [Sig. angular-tailed.]
- goniurus, Billings, 1868, Can. Nat., vol. 5, Quebec Gr. [Sig. angular-tailed.]
- *halli*, Conrad, 1840, syn. for Dalmanites boothi.
- halli, Chapman, 1858, Ann. & Mag. Nat. Hist., 3rd ser., vol. 2, Trenton Gr. [Ety. proper name.]
- *hausmani*, Brongniart, as identified by D'Archiac and Verneuil. Not American.
- hincksi, Salter, 1859, Ann. & Mag. Nat. Hist., 3rd series, vol. 4, Trenton Gr. [Ety. proper name.]
- homalonotoides, Walcott, 1877, 29th Reg. Rep. N. Y., Trenton Gr. [Sig. like *Homalonotus*.]
- huttoni, Billings, 1865, Pal. Foss., Quebec Gr. [Ety. proper name.]
- illænoides, Billings, 1860, Can. Nat., vol. 5, Quebec Gr. [Sig. like an Illænus.]
- iowensis, Owen, 1852, Geo. Wis., Iowa and Minn., Trenton Gr. [Ety. proper name.]
- *laticostatus*, syn. for Dalmanites anchiops.
- latimarginatus, Hall, 1847, Pal. N. Y., vol. 1, Trenton and Hud. Riv. Gr. [Sig. broad-margined.]
- *limulurus*, see Dalmanites limulurus.
- marginalis, Hall, 1847, Pal. N. Y., vol. 1, Chazy Gr. [Sig. having a margin.]
- *megalopthalmus*, Troost, 1840, 5th Geo. Tenn., Niagara Gr. Not clearly defined, but probably a Dalmanite.
- *megistos*, see Isotelus megistos.
- *micrurus*, see Dalmanites micrurus.
- morrisi, Billings, 1865, Pal. Foss., Quebec Gr. [Ety. proper name.]
- *myrmecophorus*, see Dalmanites myrmecophorus.
- *nasutus*, see Dalmanites nasutus.
- notans, Billings, 1866, Catal. Sil. Foss. Antic., Hud. Riv. Gr. [Sig. marked.]
- obtusus, Hall, 1847, Pal. N. Y., vol. 1, Chazy Gr. [Sig. obtuse.]
- pelops, Billings, 1865, Pal. Foss., Quebec Gr. [Ety. mythological name.]
- platycephalus, Stokes, 1823, Geo. Trnns., vol. 1, Trenton Gr. [Sig. flat-headed.]
- platypleurus, Green, 1837, Am. Jour. Sci., vol. 32, Low. Sil. [Sig. flat-sided.]
- *pleuropteryx*, see Dalmanites pleuropteryx.
- polypleurus, Green, 1838, Am. Jour. Sci., vol. 34, Keokuk Gr. [Ety. many-ribbed.] Probably a *Phillipsia*.
- quadraticaudatus, Billings, 1865, Pal. Foss., Quebec Gr. [Sig. square-tailed.]

romingeri, Walcott, 1876, Desc. New Pal. Foss., Black Riv. & Trenton Gr. [Ety. proper name.]
selenurus, see Dalmanites selenurus.
stokesi, see Proetus stokesi.
tetragonocephalus,Green, 1834,Am. Jour. Sci., vol. 25. [Sig. square-headed.]
trentonensis, see Lichas trentonensis.
trimblii, Green, 1837, Jour. Acad. Nat. Sci. Phil., vol. 7, Niagara Gr. [Ety. proper name.]
vetustus, Hall, 1847, (Ogygia vetustus) Pal. N. Y., vol. 1, Trenton & Hud. Riv. Gr. [Sig. ancient.]
wisconsinensis, Walcott, 1876, Desc. New Sp. Foss., Trenton Gr. [Ety. proper name.]

Atops, Emmons. Syn. for Triarthrus.
trilineatus, see Triarthrus trilineatus.

Barrandia, Hall, 1860. The name was preoccupied by McCoy 1849, and Olenellus was afterwards substituted.
thompsoni, see Olenellus thompsoni.
vermontana, see Olenellus vermontanus.

BATHYXOTUS, Hall, 1860, 18th Reg. Rep. N. Y. [Ety. *bathys*, ample; *notos*, the back; in allusion to the ample central lobe or axis of the typical species.]
holopyga, Hall, 1859, (Olenellus holopyga) 12th Reg. Rep. N. Y., Potsdam Gr. [Ety. *holos*, entire, all; *pygos*, rump.]

BATHYURELLUS, Billings, 1865, Pal. Foss. [Ety. diminutive of *Bathyurus*,of which it is a subgenus.]
abruptus, Billings, 1865, Pal. Foss., Quebec Gr. [Sig. abrupt.]
bradleyi, Meek, 1872, Hayden's Geo. Rep., Quebec Gr. [Ety. proper name.]
expansus, Billings, 1865, Pal. Foss., Quebec Gr. [Sig. spread out.]
formosus, Billings, 1865, Pal. Foss., Quebec Gr. [Sig. beautiful.]
fraternus, Billings, 1865, Pal. Foss., Quebec Gr. [Sig. fraternal.]
litoreus, Billings, 1865, Pal. Foss., Quebec Gr. [Sig. occurring on the beach.]
marginatus, Billings, 1865, Pal. Foss., Quebec Gr. [Sig. bordered.]
nitidus, Billings, 1865, Pal. Foss., Quebec Gr. [Sig. neat, pretty.]
rarus, Billings, 1865, Pal. Foss., Quebec Gr. [Sig. rare.]
truncatus, Meek, 1872, Hayden's Geo. Rep., Potsdam Gr. [Sig. truncated.]
validus, Billings, 1865, Pal. Foss., Quebec Gr. [Sig. strong.]
wheeleri, Meek, 1872, Hayden's Geo. Rep., Quebec Gr. [Ety. proper name.]

BATHYURUS, Billings, 1859, Can. Nat. & Geo., vol. 4. [Ety. *bathys*, deep; *oura*, the tail.]
amplimarginatus, Billings, 1859, Can. Nat. & Geo., vol. 4, Calciferous Gr. [Sig. full-bordered.]
angelini, Billings, 1859, Can. Nat. & Geol., vol. 4, Chazy Gr. [Ety. proper name.]
arcuatus, Billings, 1865, Pal. Foss., vol. 1, Quebec Gr. [Sig. bent.]
armatus, Billings, 1860, Can. Nat. & Geo., vol. 5, Quebec Gr. [Sig. armed.]
bituberculatus, Billings, 1860, Can. Nat. & Geol., vol. 5, Quebec Gr. [Sig. double-tuberculated.]
breviceps, Billings, 1865, Pal. Foss., vol. 1, Quebec Gr. [Sig. short-headed.]
capax, Billings, 1860, Can. Nat. & Geol., vol. 5, Quebec Gr. [Sig. large.]
caudatus, Billings, 1865, Pal. Foss., vol. 1, Quebec Gr. [Sig. having a tail.]
conicus, Billings, 1859, Can. Nat. & Geo. vol. 4, Calciferous Gr. [Sig. conical.]
cordai, Billings, 1860, Can. Nat. & Geol., vol. 5, Calciferous Gr. [Ety. proper name.]
cybele, Billings, 1859, Can. Nat. & Geol., vol. 4, Calciferous Gr. [Ety. mythological name.]
dubius, Billings, 1860, Can. Nat. & Geol., vol. 5, Quebec Gr. [Sig. doubtful.]
extans, Hall, 1847, (Asaphus extans) Pal. N. Y., vol. 1, Trenton Gr. [Sig. standing up.]
gregarius, Billings, 1865, Pal. Foss., vol. 1, Potsdam Gr. [Sig. occurring in flocks.]
haydeni, Meek, 1872, Hayden's Geo. Rep., Potsdam Gr. [Ety. proper name.]
longispinus, Walcott, 1876, Desc. New Sp. Foss., Black Riv. & Trenton Gr. [Sig. long-spined.]
minganensis, Billings, 1865, Pal. Foss., vol. 1, Calciferous Gr. [Ety. proper name.]
nero, Billings. 1865, Pal. Foss., vol. 1, Quebec Gr. [Ety. proper name.]
oblongus, Billings, 1860, Can. Nat. & Geol., vol.5, Quebec Gr. [Sig. oblong.]
parvulus, Billings, 1861, Pal. Foss., vol. 1, Potsdam Gr. [Sig. very small.]
perplexus, Billings, 1865, Pal. Foss., vol. 1, Potsdam Gr. [Sig. intricate, perplexing.]
perspicator, Billings, 1865, Pal. Foss., vol. 1, Quebec Gr. [Sig. sharp-sighted; from the large eyes.]
quadratus, Billings, 1860, Can. Nat. & Geo., vol. 5, Quebec Gr. [Sig. square-shaped.]
saffordi, Billings, 1860, Can. Nat. & Geo., vol. 5, Quebec Gr. [Ety. proper name.]
senectus, Billings, 1861, Pal. Foss., vol. 1, Potsdam Gr. [Sig. extreme age.]
serratus, Meek, 1872, Hayden's Geo. Rep., Potsdam Gr. [Sig. serrated.]
smithi, Billings, 1862, Pal. Foss., vol. 1, Black Riv. Gr. [Ety. proper name.]
solitarius, Billings, 1865, Pal. Foss., vol. 1, Quebec Gr. [Sig. solitary.]
spiniger, Hall, 1847, (Acidaspis spiniger) Pal. N. Y., vol. 1, Black Riv. & Trenton Gr. [Sig. spiny.]
strenuus, Billings, 1865, Pal. Foss., vol. 1, Quebec Gr. [Sig. vigorous.]

timon, Billings, 1865, Pal. Foss. vol. 1, Quebec Gr. [Ety. proper name.]
vetulus, Billings, 1865, Pal. Foss., vol. 1, Potsdam Gr. [Sig. old.]

Belinurus, Konig, 1825, Icones Fossilium Sectiles. [Ety. *belos*, a dart; *oura*, the tail.]

danæ, see Euproops danæ.

BEYRICHIA, McCoy, 1850, Syn. Sil. Foss. Ireland. [Ety. proper name.]
æquilatera, Hall, 1860, Can. Nat. & Geo., vol. 5, Silurian. [Sig. equal sided.]
atlantica, Billings, 1865, Pal. Foss., vol. 1, Quebec Gr. [Ety. mythological name.]
chambersi, S. A. Miller, 1874, Cin. Quar. Jour. Sci., vol. 1, Cin'ti Gr. [Ety. proper name.]
ciliata, Emmons, 1855, American Geo., Cin'ti Gr. [Sig. haired on the margin.]
cincinnatiensis, S. A. Miller, 1875, Cin. Quar. Jour. Sci.,vol. 2, Cin'ti Gr. [Ety. proper name.]
clathrata, Jones, 1858, Ann. & Mag. Nat. Hist., 3rd series, vol. 1, Niagara Gr. [Sig. latticed.]
decora, Billings, 1866, Catal. Sil. Foss. Antic., Anticosti Gr. [Sig. suitable.]
duryi, S. A. Miller, 1874, Cin. Quar. Jour. Sci., vol. 1, Cin'ti Gr. [Ety. proper name.]
foetoidea, White & St. John, 1868, Trans. Chi. Acad. Sci., Up. Coal Meas. [Sig. like a tumor.]
granulosa, Hall, 1876, 28th Reg. Rep. N. Y., Niagara Gr. [Sig. granular.]
granulata, Hall, 1859, Pal. N. Y., vol. 3, Low. Held. Gr. [Sig. granulated.]
jonesi, Dawson, 1868, Acad. Geol., Carboniferous. [Ety. proper name.]
lata, Vanuxem, 1842, (Agnostus latus) Geo. Rep. N. Y., Clinton Gr. [Sig. wide.]
lithofactor, White & St. John, 1868, Prel. Notice of New Foss.,Coal Meas. [Sig. made of stone.]
logani, Jones, 1858, Ann. Nat. Hist., 3rd ser., vol. 1, Chazy Gr. [Ety. proper name.]
logani *var.* leperditoides, Jones, 1858, Can. Org. Rem., Decade 3, Chazy Gr. [Sig. like the genus Leperditia.]
reniformis, Jones, 1858, Can. Org. Rem., Decade 3, Chazy Gr. [Sig. kidney-shaped.]
maccoyana, Jones, 1855, Ann. & Mag. Nat. Hist., 2d ser., vol. 16, Onondaga Gr. [Ety. proper name.]
notata, Hall, 1859, Pal. N. Y., vol. 3, Low. Held. Gr. [Sig. marked.]
notata *var.* ventricosa, Hall, 1859, Pal. N. Y., vol. 3, Low. Held. Gr. [Sig. bulged out.]
oculifera, Hall, 1871, Pamphlet,Cin'ti Gr. [Sig. bearing eyes.]
oculina, Hall, 1859, Pal. N. Y., vol. 3, Low. Held. Gr. [Sig. having eyes.]

pennsylvanica, Jones, 1858, Ann. & Mag. Nat. Hist., 3rd ser., vol. 1, Onondaga Gr. [Ety. proper name.]
petrifactor, White & St. John,1868, Trans. Chi. Acad. Sci., St. Louis Gr. [Sig. made of stone.]
petrifactor *var.* velata, White & St. John, 1868, Trans. Chi. Acad. Sci., St. Louis Gr. [Sig. covered.]
plagosa, Jones, 1858, Ann. & Mag. Nat. Hist., 3rd ser., vol. 1, Niagara Gr. [Sig. full of stripes.]
punctulifera, Hall, 1862, 15th Reg. Rep. N. Y., Ham. Gr. [Sig. bearing punctures.]
pustulosa, Hall, 1860, Can. Nat. & Geo., vol. 5, Silurian. [Sig. covered with pustules.]
quadrilirata, syn. for Beyrichia regularis.
regularis, Emmons, 1855, Am. Geol., Cin'ti Gr. [Sig. formed in bars.]
richardsoni, S. A. Miller, 1874, Cin. Quar. Jour. Sci., vol. 1, Cin'ti Gr. [Ety. proper name.]
rugulifera, Jones, 1858, Ann. & Mag. Nat. Hist., 3rd series, vol. 1, Niagara Gr. [Sig. bearing wrinkles.]
sigillata, Jones, 1858, Ann. & Mag. Nat. Hist., 3rd series, vol. 1, Niagara Gr. [Sig. adorned with figures.]
spinosa, Hall, 1852, (Cytherina spinosa) Pal. N. Y., vol. 2, Niagara Gr. [Sig. covered with spines.]
striato-marginata, S. A. Miller, 1874, Cin. Quar. Jour. Sci., vol. 1, Cin'ti Gr. [Sig. having a striated margin.]
symmetrica, Hall, 1852, Pal. N. Y., vol. 2, Niagara Gr. [Sig. symmetrical.]
trisulcata, Hall, 1859, Pal. N. Y., vol. 3, Low. Held. Gr. [Sig. three-furrowed.]
tumifrons, syn. for Beyrichia ciliata.
venusta, Billings, 1866, Catal. Sil. Foss. Antic., Anticosti Gr. [Sig. beautiful.]

Brongniartia, Eaton, 1832, Geo. Text Book. Syn. for Asaphus.

BRONTEUS, Goldfuss, 1843, in Burmeister, Mon. Tri. [Ety. mythological name.]
acamas, Hall, 1865, 20th Reg. Rep. N. Y., Niagara Gr. [Ety. mythological name.]
barrandi, Hall, 1859, Pal. N. Y., vol. 3, Low. Held. Gr. [Ety. proper name.]
insularis, Billings, 1866, Catal. Sil. Foss. Antic., Anticosti Gr. [Sig. upon an island.]
lunatus, Billings, 1857, Rep. of Progr., Trenton Gr. [Sig. crescent-formed.]
niagarensis, Hall, 1852, Pal. N. Y., vol. 2, Niagara Gr. [Ety. proper name]
occanus, syn. for Bronteus acamas.

Bumastus, Murchison, 1839, Sil. Syst. See Illænus.
barriensis, see Illænus barriensis.
trentonensis, see Illænus trentonensis.

CALYMENE, Brongniart, 1822, Hist. Nat. Crust. Foss. [Ety. *kekalymenos*, concealed.]
anchiops, see Dalmanites, anchiops.
becki, see Triarthrus becki.
blumenbachi, Brongniart, 1822, Hist. Nat. Crust. Foss., Niagara Gr. [Ety. proper name.]
bucklandi, syn. for Ceraurus pleurexanthemus.
bufo, see Phacops bufo.
callicephala, Green, 1832, Monograph Trilobites, Cin'ti Gr. [Sig. beautiful head.]
camerata, Conrad, 1842, Jour. Acad. Nat. Sci., vol. 8, Coralline limestone. [Sig. arched.]
christyi, Hall, 1860, 13th Reg. Rep. N.Y., Cin'ti Gr. [Ety. proper name.]
clintoni, Vanuxem, (Hemicrypturus clintoni) Geo. Rep. 3rd Dist. N. Y., Clinton Gr. [Ety. proper name.]
crassimarginata, see Proetus crassimarginatus.
mammillata, Hall, 1861, Geo. Rep. Wis., Trenton Gr. [Sig. covered with nipples.]
marginalis, see Proetus marginalis.
multicosta, Hall, 1847, Pal. N. Y., vol. 1, Trenton Gr. [Sig. having many ribs.]
niagarensis, Hall, 1843, Geo. Rep. 4th Dist. N. Y., Niagara Gr. [Ety. proper name.] This is the American variety of *C. blumenbachi*.
nupera, see Phacops nuperus.
odontocephala, syn. for Dalmanites selenurus.
phlyctainodes, see Arges phlyctainodes.
platys, Green, 1832, Monograph of Trilobites, Schoharie grit. [Sig. broad.]
rowii, see Proetus rowii.
rugosa, Shumard, 1855, Geo. Rep. Mo., Low. Held. Gr. [Sig. wrinkled.]
senaria, Conrad, 1841, Ann. Rep. N. Y., Trenton & Hud. Riv. Gr. [Sig. is said to be in allusion to six tubercles on the buckler, but the application is not evident.] — It is a syn. for *C. callicephala.*
spinifera. Not defined.
trisulcata, Hall, 1843, Geo. Rep. 4th Dist. N. Y., Clinton Gr. [Sig. three-furrowed.]

CERATIOCARIS, McCoy, 1849, Ann. & Mag. Nat. Hist., 2d ser., vol. 4. [Ety. *keration*, a pod; *karis*, shrimp.]
aculeata, Hall, 1859, Pal. N. Y., vol. 3, Waterline Gr. [Sig. armed with sharp points.]
acuminata, Hall, 1859, Pal. N. Y., vol. 3, Waterline Gr. [Sig. sharp-pointed.]
armata, Hall, 1863, 16th Reg. Rep. N. Y., Ham. Gr. [Sig. armed.]
bradleyi, Meek, 1872, Proc. Acad. Nat. Sci. Phil., Low. Carb. [Ety. proper name.]
deweyi, Hall, 1859, (Onchus deweyi) Pal. N. Y., vol. 2, Niagara Gr. [Ety. proper name.]
elytroides, Meek, 1872, Proc. Acad. Nat. Sci. Phil., Low. Carb. [Sig. like the elytra of Beetles.]
longicauda, Hall, 1863, 16th Reg. Rep. N. Y., Genesee slate. [Sig. long-tailed.]
maccoyana, Hall, 1859, Pal. N. Y., vol. 3, Waterlime Gr. [Ety. proper name.]
punctata, Hall, 1863, 16th Reg. Rep. N. Y., Ham. Gr. [Sig. punctured.]
sinuata, Meek & Worthen, 1868, Am. Jour. Sci., vol. 46, Coal Meas. [Sig. marked with depressions.]
strigata, Meek, 1872, Proc. Acad. Nat. Sci. Phil., Low. Carb. [Sig. furrowed, grooved.]

Ceratocephala, Warder. Not defined so as to be recognized.
ceralepta, Anthony, a fragment of the tail of a *Ceraurus pleurexanthemus* or of an *Acidaspis*.
goniata, Warder, a fragment of a *Dalmanites* or an *Acidaspis*.

CERAURUS, Green, 1832, Monograph Trilobites. [Ety. *kerus*, a horn; *oura*, the tail.]
apollo, Billings, 1860, Can. Nat. & Geol., vol. 5, Quebec Gr. [Ety. mythological name.]
bimucronatus, see Ceraurus niagarensis.
crosotus, see Acidaspis crosotus.
eryx, Billings, 1860, Can. Nat. & Geo., vol. 5, Quebec Gr. [Ety. mythological name.]
glaucus, Billings, 1865, Pal. Foss., Quebec Gr. [Ety. mythological name.]
icarus, Billings, 1850, Can. Nat. & Geol., vol. 5, Hud. Riv. Gr. [Ety. mythological name.]
insignis, see Ceraurus niagarensis.
mercurius, Billings, 1865, Pal. Foss., Quebec Gr. [Ety. mythological name.]
niagarensis, Hall, 1867, 20th Reg. Rep., Niagara Gr. [Ety. proper name.]
numitor, Billings, 1866, Catal. Sil. Foss. Antic., Hud. Riv. Gr. [Ety. mythological name.]
nuperus, Billings, 1866, Catal. Sil. Foss. Antic., Anticosti Gr. [Sig. late, recent.]
perforator, Billings, 1865, Pal. Foss., Quebec Gr. [Sig. a borer through.]
pleurexanthemus, Green, 1832, Monog. Trilobites, Trenton and Hud. Riv. Gr. [Ety. *pleura*, a side; *exanthema*, breaking out.]
polydorus, Billings, 1865, Pal. Foss., Quebec Gr. [Ety. mythological name.]
pompilius, Billings, 1865, Pal. Foss., Chazy & Black Riv. Gr. [Ety. proper name.]
prolificus, Billings, 1865, Pal. Foss., Quebec Gr. [Sig. prolific, common.]
pustulosus, syn. for Ceraurus pleurexanthemus.

rarus, Walcott, 1877, 29th Reg. Rep. N.Y., Trenton Gr. [Sig. rare.]
satyrus, Billings, 1865, Pal. Foss., Chazy Gr. [Ety. mythological name.]
sol, Billings, 1865, Pal. Foss., Quebec Gr. [Ety. mythological name.]
solitarius, Billings, 1865, Pal. Foss., Quebec Gr. [Sig. solitary, alone.]
vigilans, see Encrinurus vigilans.
vulcanus, Billings, 1865, Pal. Foss., Quebec Gr. [Ety. mythological name.]

CHARIOCEPHALUS, Hall, 1863, 16th Reg. Rep. [Ety. *charis*, charming or graceful; *kephale*, head.]
whitfieldi, Hall, 1863, 16th Reg. Rep., Potsdam Gr. [Ety. proper name.]

Cheirurus, Beyrich, syn. for Ceraurus.

CLIMACTICHNITES, Logan, 1860, Can. Nat. & Geo., vol. 5. [Ety. *klimax*, ladder; *ichnos*, a foot step.]
wilsoni, Logan, 1860, Can. Nat. & Geo., vol. 5, Potsdam Gr. [Ety. proper name.]

Colpocaris, Meek, 1872, a subgenus of Ceratiocaris.

Conocephalus, Zenker, 1833, Beitr. Naturg. Preoccupied for a genus of Orthoptera.

CONOCEPHALITES, Adams, 1848, Am. Jour. Sci., 2d series, vol. 5. [Ety. *konos*, a cone; *kephale*, head.]
adamsi, Billings, 1861, Pal. Foss., vol. 1, Potsdam Gr. [Ety. proper name.]
anatinus, Hall, 1863, 16th Reg. Rep. N. Y., Potsdam Gr. [Sig. duck-like.]
antiquatus, Salter, 1859, Jour. Geo. Soc., vol. 15, Potsdam Gr. [Sig. very ancient.]
arenosus, Billings, 1861, Pal. Foss., Potsdam Gr. [Sig. sandy.]
aurora, Hartt, 1868, Acad. Geol., St. John's Gr. [Sig. morning.]
baileyi, Hartt, 1868, Acad. Geol., St. John's Gr. [Ety. proper name.]
billingsi, Shumard, 1861, Am. Jour. Sci., vol. 32, Potsdam Gr. [Ety. proper name.]
binodus, Hall, 1863, 16th Reg. Rep. N. Y., Potsdam Gr. [Sig. two-knotted.]
chippewensis, Owen, 1852, (Lonchocephalus chippewaensis) Geo. Iowa, Wis. & Minn., Potsdam Gr. [Ety. proper name.]
depressus, Shumard, 1861, Am. Jour. Sci., vol. 32, Potsdam Gr. [Sig. depressed.]
diadematus, Hall, 1863, 16th Reg. Rep. N. Y., Potsdam Gr. [Sig. adorned with a diadem.]
dorsalis, Hall, 1863, 16th Reg. Rep. N. Y., Potsdam Gr. [Sig. from the back.]
elegans, Hartt, 1868, Acad. Geol., St. John's Gr. [Sig. elegant.]
eos, Hall, 1863, 16th Reg. Rep. N. Y., Potsdam Gr. [Sig. the dawn.]
eryon, Hall, 1863, 16th Reg. Rep. N. Y., Potsdam Gr. [Sig. from the genus *Eryon*.]
formosus, Hartt, 1868, Acad. Geol., St. John's Gr. [Sig. beautiful.]
gemini-spinosus, Hartt, 1868, Acad. Geol., St. John's Gr. [Sig. covered with twin-spines.]
halli, Hartt, 1868, Acad. Geol., St. John's Gr. [Ety. proper name.]
hamulus, Owen, 1852, (Lonchocephalus hamulus) Geo. Iowa, Wis. & Minn., Potsdam Gr. [Sig. a little hook.]
iowensis, Owen, 1852, (Dikelocephalus iowensis) Geo. Iowa, Wis. & Minn., Potsdam Gr. [Ety. proper name.]
mathewi. Hartt, 1868, Acad. Geol., St. John's Gr. [Ety. proper name.]
minor, Shumard, 1863, Trans. St. Louis Acad. Sci., Potsdam Gr. [Sig. less.]
minutus, Bradley, 1860, Am. Jour. Sci., vol. 30, Potsdam Gr. [Sig. small.]
miser, Billings, 1861, Pal. Foss., vol. 1, Potsdam Gr. [Sig. wretched, poor.]
nasutus, Hall, 1863, 16th Reg. Rep. N. Y., Potsdam Gr. [Sig. having a prominent nose.]
neglectus, Hartt, 1868, Acad. Geol., St. John's Gr. [Sig. overlooked.]
optatus, Hall, 1863, 16th Reg. Rep. N. Y., Potsdam Gr. [Sig. desired.]
orestes, Hartt, 1868, Acad. Geol., St. John's Gr. [Ety. mythological name.]
ouangondianus, Hartt, 1868, Acad. Geol., St. John's Gr. [Ety. proper name.]
oweni, Hall, 1863, 16th Reg. Rep. N. Y., Potsdam Gr. [Ety. proper name.]
pattersoni, Hall, 1863, 16th Reg. Rep. N. Y., Potsdam Gr. [Ety. proper name.]
perseus, Hall, 1863, 16th Reg. Rep. N. Y., Potsdam Gr. [Ety. mythological name.]
quadratus, Hartt, 1868, Acad. Geol., St. John's Gr. [Sig. squared.]
robbi, Hartt, 1868, Acad. Geol., St. John's Gr. [Ety. proper name.]
shumardi, Hall, 1863, 16th Reg. Rep. N. Y., Potsdam Gr. [Ety. proper name.]
tener, Hartt, 1868, Acad. Geol., St. John's Gr. [Sig. delicate.]
teucer, Billings, 1861, Pal. Foss., vol. 1, Potsdam Gr. [Ety. mythological name.]
thersites, Hartt, 1868, Acad. Geol., St. John's Gr. [Ety. mythological name.]
vulcanus, Billings, 1861, Pal. Foss., vol. 1, Potsdam Gr. [Ety. mythological name.]
winona, Hall, 1863, 16th Reg. Rep. N. Y., Potsdam Gr. [Ety. proper name.]
wisconsinensis, Owen, 1852, (Crepicephalus wisconsinensis) Geo. Iowa, Wis. & Minn., Potsdam Gr. [Ety. proper name.]
zenkeri, Billings, 1860, Can. Nat. & Geo., vol. 5, Quebec Gr. [Ety. proper name.]

CONOCORYPHE, Corda, 1847, Prodr. [Ety. *konos*, a cone; *koryphe*, the top of the head.]
kingi, Meek, 1870, Proc. Acad. Nat. Sci., Potsdam Gr. [Ety. proper name.]
gallatinensis, Meek, 1872, Hayden's Geo. Rep., Pots. Gr. [Ety. proper name.]

CRUSTACEA. 215

CREPICEPHALUS, Owen, 1852, Geo. Sur. Wis., Iowa & Minn. [Ety. *krepis*, a horseshoe; *kephale*, head.]
miniscaensis, Owen, 1852, Geo. Sur. Wis., Iowa & Minn., Potsdam Gr. [Ety. proper name.]
montanensis, Whitfield, 1876, Rep. Recon. Up. Mo. to Yel. Nat. Park, Potsdam Gr. [Ety. proper name.]
wisconsinensis, see Conocephalites wisconsinensis.
Cryphæus, Green, 1837, Jour. Acad. Nat. Sci., vol. 7, syn. for Dalmanites. This name has priority over *Dalmanites*, but the definition has not given satisfaction.
boothi, Green, 1837, Jour. Acad. Nat. Sci., vol. 7. See Dalmanites boothi.
callitelus, see Dalmanites calliteles.
greeni, syn. for Dalmanites boothi.
Cryptolithus, syn. for Trinucleus.
tesselatus, see Trinucleus concentricus.
Cybele, Loven, 1845, in Ofversigt of Vetensk. Acad. Handl.
punctata, Hall, 1852. This species belongs to the genus *Encrinurus*, and the specific name being preoccupied, the name is changed to *E. ornatus*.
CYPHASPIS, Burmeister, 1843, Monograph of Trilobites. [Ety. *cyphos*, convex; *aspis*, a shield.]
christyi, Hall, 1863, Trans. Alb. Inst., vol. 4, Niagara Gr. [Ety. proper name.]
girardeauensis, Shumard, 1855, Geo. Rep. Mo., Up. Sil. [Ety. proper name.]
CYTHERE, Muller, 1785, Entomostraca sue Insecta, etc. [Ety. proper name.]
americana, Shumard, 1858, Trans. St. Louis Acad. Sci., Up. Coal Meas. [Ety. proper name.]
carbonaria, Hall, 1858, Trans. Alb. Inst., vol. 4, Warsaw Gr. [Sig. from the Carboniferous Group.]
cincinnatiensis, Meek, 1872, Proc. Acad. Nat. Sci., Cin'ti Gr. [Ety. proper name.]
crassimarginata, Winchell, 1862, Proc. Acad. Nat. Sci., Marshall Gr. [Sig. thick-margined.]
nebrascensis, Geinitz, 1866, Carb. und Dyas in Neb., Coal Meas. [Ety. proper name.]
okeni, see Leperditia okeni.
simplex, White & St. John, 1868, Trans. Chi. Acad. Sci., St. Louis Gr. [Sig. simple.]
sublævis, see Leperditia sublævis.
subrecta, see Leperditia subrecta.
CYTHERINA, Lamarck, 1818, Anim. sans Vert. [Ety. diminutive of Cythere.]
alta, see Leperditia alta.
crenulata, Emmons, 1856, Am. Geol., Trenton Gr. [Sig. crenulated.]
cylindrica, see Isochilina cylindrica.
fabulites, see Leperditia fabulites.
spinosa, Hall, 1852, see Beyrichia spinosa. Not Reuss in 1844.
subcylindrica, Emmons, 1856, Am. Geol., Trenton Gr. [Sig. somewhat cylindrical.] This name was preoccupied by Munster in 1830.
subelliptica, Emmons, 1856, Am. Geol., Black Riv. Gr. [Sig. somewhat elliptical.]
CYTHEROPSIS, McCoy, 1849, Ann. & Mag. Nat. Hist., 2d series, vol. 4. [Ety. *Cytheropsis*, resembling Cythere.]
concinna, Jones, 1858, Ann. Nat. Hist., 3rd series, vol. 1, Black Riv. Gr. [Sig. neat, pretty.] In 1865 Jones established the genus *Primitia*, in which he included this species.
rugosa, Jones, 1858, Ann. Nat. Hist., 3rd series, vol. 1, Black Riv. Gr. [Sig. wrinkled.]
siliqua, Jones, 1858, Ann. Nat. Hist., 3d series, vol. 1, Black Riv. Gr. [Sig. a pod.]
Dalmania, Emmrich, 1845. This name having been preoccupied, Dalmanites has been substituted, though many authors prefer to use *Odontochile*, a name proposed by Corda.
DALMANITES, Emmrich, 1845, (Dalmania) Barrande, 1852, Sil. Syst. Boh. [Ety. proper name.]
acantholeurus, Conrad, 1841, (Asaphus acantholeurus) Ann. Rep. N. Y., Onondaga limestone. [Sig. smooth-spined.]
achates Billings, 1860, Can. Nat. & Geo., vol. 5, Trenton Gr. [Ety. mythological name.]
ægeria, Hall, 1861, 15th Reg. Rep. N. Y., Up. Held. Gr. [Ety. mythological name.]
anchiops, Green, 1832, (Calymene anchiops) Monograph, Schoharie grit. [Sig. from the closeness of the eyes.]
anchiops var. armatus, Hall, 1861, 15th Reg. Rep. N. Y., Schoharie grit. [Sig. armed.]
aspectans, Conrad, 1841, (Asaphus aspectans) Ann. Rep. N. Y., Up. Held. Gr. [Sig. seeing at a glance.]
bebryx, Billings, 1860, Can. Nat. & Geol., vol. 5, Trenton Gr. [Ety. mythological name.]
bifidus, Hall, 1862, 15th Reg. Rep. N. Y., Up. Held. Gr. [Sig. divided.]
boothi, Green, 1837, (Cryphæus boothi) Jour. Acad. Nat. Sci., vol. 7, Ham. Gr. [Ety. proper name.]
bicornis, Hall, 1876, 28th Reg. Rep. N.Y., Niagara Gr. [Sig. two-horned.]
breviceps, Hall, 1866, Pamplet, Cin'ti Gr. [Sig. short-headed.]
callicephalus, Hall, 1847, (Phacops callicephalus) Pal. N. Y., vol. 1, Trenton Gr. [Sig. beautiful head.]
calliteles, Green, 1837, (Cryphæus callitelus) Am. Jour. Sci. & Arts, Ham. Gr. [Ety. proper name.]

calypso, Hall, 1861, 15th Reg. Rep. N. Y., Up. Held. Gr. [Ety. mythological name.]
carleyi, Meek, 1872, Am. Jour. Sci., 3rd ser., vol. 3, Cin'ti Gr. [Ety. proper name.]
caudatus, see Dalmanites limulurus.
concinnus, Hall, 1876, Illust. Devonian Foss., Schoharie grit. [Sig. handsome.]
coronatus, Hall, 1861, 15th Reg. Rep. N. Y., Ham. Gr. [Sig. crowned.]
danæ, Meek & Worthen, 1865, Proc. Acad. Nat. Sci. Phil., Niagara Gr. [Ety. proper name.]
dentatus, Barrett, 1876, Am. Jour. Sci. & Arts, vol. 11, Low. Held. Gr. [Ety. from the dentate margin of the cephalic shield.]
denticulatus, Conrad, 1841, (Asaphus denticulatus) Ann. Rep. N. Y., Up. Held. Gr. [Ety. from the denticulate termination of the ribs.]
emarginatus, Hall, 1876, Illust. Devonian Foss., Up. Held. Gr. [Sig. emarginated.]
erina, Hall, 1861, 15th Reg. Rep. N. Y., Up. Held. Gr. [Ety. proper name.]
helena, Hall, 1861, 15th Reg. Rep. N. Y., Corniferous Gr. [Ety. mythological name.]
intermedius, Walcott, 1877, 29th Reg. Rep. N.Y., Trenton Gr. [Sig. intermediate.]
laticaudatus, Hall, 1847. Prof. Hall says this name may be erased from the list of fossils.
limulurus, Green, 1832, (Asaphus limulurus) Monograph Trilobites, Niagara Gr. [Ety. having a pointed tail like *Limulus* or king crab.]
logani, Hall, 1860, Can. Nat. & Geol., vol. 5, Silurian. [Ety. proper name.]
macrops, Hall, 1862, 15th Reg. Rep. N. Y., Up. Held. Gr. [Ety. *makros*, large; *ops*, the eye.]
micrurus, Green, 1832, (Asaphus micrurus) Monograph Trilobites, Low. Held. Gr. [Sig. small-tailed.]
myrmecophorus, Green, 1835, (Asaphus myrmecophorus) Supp. to Monograph of Trilobites, Up. Held. Gr. [Sig. wart-bearing.]
nasutus, Conrad, 1841, (Asaphus nasutus) Ann. Rep. N. Y., Low. Held. Gr. [Sig. having a prominent nose.]
ohioensis, Meek & Worthen, 1871, Proc. Acad. Nat. Sci. Phil., Corniferous Gr. [Ety. proper name.]
pleione, Hall, 1861, 15th Reg. Rep. N. Y., Up. Held. Gr. [Ety. mythological name.]
pleuropteryx, Green, 1832, (Asaphus pleuropteryx) Monograph Trilobites, Low. Held. Gr. [Sig. side-winged.]
regalis, Hall, 1876, Illust. Devonian Foss., Schoharie grit. [Sig. regal.]
selenurus, Eaton, 1832, (Asaphus selenurus) Geo. Text Book, Corniferous Gr. [Sig. from the crescent-shaped tail.]
tridens, Hall, 1859, Pal. N. Y., vol. 3, Low. Held. Gr. [Sig. three-toothed.]
tridentiferus, Shumard, 1855, Geo. Rep. Mo., Low. Held. Gr. [Sig. bearing three teeth.]
troosti, Safford. Not defined.
verrucosus, Hall, 1863, Trans. Alb. Inst., vol. 4, Niagara Gr. [Sig. warty.]
vigilans, Hall, 1861, Rep. Progr. Wis.Sur., Niagara Gr. [Sig. watchful, eyes opened.]

DICELLOCEPHALUS, Owen, 1852, Geo. Sur. Wis., Iowa & Min., (written by Owen, Dikelocephalus). [Ety. *dikella*, a mattock; *kephale*, head.]
affinis, Billings, 1865, Pal. Foss., Quebec Gr. [Sig. related to.]
belli, Billings, 1860, Can. Nat. & Geol., vol. 5, Quebec Gr. [Ety. proper name.]
(?) corax, Billings, 1865, Pal. Foss., Quebec Gr. [Ety. proper name.]
cristatus, Billings, 1860, Can. Nat. & Geo. vol. 5, Quebec Gr. [Sig. crested.]
devinci, Billings, 1865, Pal. Foss., Quebec Gr. [Ety. proper name.]
flagricaudus, White, 1874, Rep. Invert. Foss., Quebec Gr. [Sig. whip-tailed.]
granulosus, see Ptychaspis granulosa.
hisingeri, Billings, 1865, Pal. Foss., Quebec Gr. [Ety. proper name.]
iowensis, see Conocephalites iowensis.
latifrons, Shumard, 1863, Trans. St. Louis Acad. Sci., vol. 2, Potsdam Gr. [Sig. broad-fronted.]
limbatus, Hall, 1863, 16th Reg. Rep. N. Y., Potsdam Gr. [Sig. bordered.]
magnificus, Billings, 1860, Can. Nat. & Geo., vol. 5, Quebec Gr. [Sig. magnificent.]
megalops, Billings, 1860, Can. Nat. & Geo., vol. 5, Quebec Gr. [Sig. large-eyed.]
miniscaensis, see Ptychaspis miniscaensis.
minnesotensis, Owen, 1852, Rep. Wis., Iowa & Minn., Potsdam Gr. [Ety. proper name.]
minnesotensis, var. limbatus, Hall, 1863, 16th Reg. Rep. N. Y., Potsdam Gr. [Sig. bordered.]
misa, Hall, 1873, 16th Reg. Rep. N. Y., Potsdam Gr. [Ety. proper name.]
missisquoi, Billings, 1865, Pal. Foss., Quebec Gr. [Ety. proper name.]
osceola, Hall, 1863, 16th Reg. Rep. N. Y., Potsdam Gr. [Ety. proper name.]
oweni, Billings, 1860, Can. Nat. & Geo., vol. 5, Quebec Gr. [Ety. proper name.]
pauper, Billings, 1865, Pal. Foss., Quebec Gr. [Sig. poor, small.]
pepinensis, Owen, 1852, Rep. Wis., Iowa and Minn., Potsdam Gr. [Ety. proper name.]
planifrons, Billings, 1860, Can. Nat. & Geo., vol. 5, Quebec Gr. [Sig. having a plane front.]

roemeri, Shumard, 1861, Am. Jour. Sci., vol. 32, Potsdam Gr. [Ety. proper name.]
selectus, Billings, 1865, Pal. Foss., Quebec Gr. [Sig. select.]
sesostris, Billings, 1865, Pal. Foss., Quebec Gr. [Ety. proper name.]
spiniger, Hall, 1863, 16th Reg. Rep. N. Y., Potsdam Gr. [Sig. bearing spines.]
Dicranurus, syn. for Acidaspis.
hamatus, see Acidaspis hamata.
DIONIDE, Barrande, 1847, in Lith. Proc. [Ety. from the mythological name *Dione*.]
(?) perplexa, Billings, 1866, Catal. Sil. Foss. Antic., Anticosti Gr. [Sig. difficult, perplexing.]
Dipleura, Green, syn. for Homalonotus.
dekayi, see Homalonotus dekayi.
DIPLICHNITES, Dawson, 1863, Am. Jour. Sci. and Arts, 3rd series, vol. 5. [Ety. *diplos*, double; *ichnos*, foot-print or track.]
ænigma, Dawson, 1863, Am. Jour. Sci. & Arts, 3rd series, vol. 5, Coal Meas. [Sig. obscure, perplexing.]
DIPLOSTYLUS, Salter, 1863, Quar. Jour. Geo. Soc., vol. 19. [Ety. *Diplostylus*, double-tail; in allusion to the two pairs of appendages to the last segment or telson.]
dawsoni, Salter, 1863, Quar. Jour. Geo. Soc., vol. 19, Coal Meas. [Ety. proper name.]
DITHYROCARIS, Scouler, 1855, British Pal. Rocks. [Ety. *dithros*, having two valves; *caris*, shrimp.]
belli, H. Woodward, 1870, Geo. Mag., vol. 8, Mid. Devonian. [Ety. proper name.]
carbonaria, Meek & Worthen, 1869, Proc. Acad. Nat. Sci. Phil., Coal Meas. [Sig. from the Coal Measures.]
neptuni, Hall, 1863, 16th Reg. Rep. N. Y., Chemung Gr. [Ety. mythological name.]
DOLICHOMETOPUS, Angelin, 1852, Palæontologia Scandinavica. [Ety. *dolichos*, long; *metope*, panel, or space between two hollows.]
(?) convexus, Billings, 1865, Pal. Foss., Quebec Gr. [Sig. convex.]
gibberulus, Billings, 1865, Pal. Foss., Quebec Gr. [Sig. somewhat humped or convex.]
rarus, Billings, 1865, Pal. Foss., vol. 1, Calciferous Gr. [Sig. rare.]
DOLICHOPTERUS, Hall, 1859, Pal. N. Y., vol. 3. [Ety. *dolichos*, long; *pteron*, a wing.]
macrocheirus, Hall, 1859, Pal. N. Y., vol. 3, Waterlime Gr. [Sig. long-handed.]
Elliptocephala, Emmons, 1844, syn. for Olenus.
asaphoides, see Olenellus asaphoides.
ENCRINURUS, Emmrich, 1844, Zur Naturgeschichte der Trilobiten, im Pruefungs-Prog. der Real-Schule, etc. [Ety. (?).]
deltoideus, Shumard, 1855, Geo. Sur. Mo., Up. Sil. [Sig. shaped like the Greek letter *Delta*.]

elegantulus, Billings, 1866, Catal. Sil. Foss. Antic., Anticosti Gr. [Sig. quite elegant.]
excedrensis, Safford. Not defined.
lævis, Angelin, 1852, Palæontologia Scandinavica, Up. Sil. [Sig. smooth.]
mirus, Billings, 1865, Pal. Foss., vol. 1, Quebec Gr. [Sig. extraordinary.]
multisegmentatus, Portlock, 1843, (Amphion multisegmentatus) Rep. Geo. of Londonderry, etc., Anticosti Gr. [Sig. having many segments.]
nereus, Hall, 1867, 20th Reg. Rep. N. Y., Niagara Gr. [Ety. mythological name.]
ornatus, Hall & Whitfield, 1875, Ohio Pal., vol. 2, Niagara Gr. [Sig. ornamented.]
punctatus, Wahlenberg, 1821, Nova Acta Soc. Upsal., Anticosti Gr. [Sig. punctated.]
raricostatus, Walcott, 1877, 29th Reg. Rep. N. Y., Trenton Gr. [Sig. few-ribbed.]
trentonensis, Walcott, 1877, 29th Reg. Rep. N. Y., Trenton Gr. [Ety. proper name.]
vigilans, Hall, 1847, (Ceraurus vigilans) Pal. N. Y., vol. 1, Black Riv. & Trenton Gr. [Sig. vigilant.]
Endymion, Billings, 1862. The name being preoccupied for a genus of plants the author proposed *Endymionia*.
meeki, see Endymionia meeki.
ENDYMIONIA, Billings, 1865, Pal. Foss., vol. 1. [Ety. proper name.]
meeki, Billings, 1862, (Endymion meeki) Pal. Foss., vol. 1, Quebec Gr. [Ety. proper name.]
EUPROOPS, Meek, 1867, Am. Jour. Sci., vol. 43. [Ety. *eu*, very; *pro*, forward; *ops*, an eye.]
danæ, Meek & Worthen, 1865, (Bellinurus danæ) Proc. Acad. Nat. Sci. Phil., Coal Meas. [Ety. proper name.]
EURYPTERUS, DeKay, 1825, Ann. Lyc. Nat. Hist. N.Y., vol. 1. [Ety. *euros*, breadth; *pteron*, wing or fin.]
dekayi, Hall, 1859, Pal. N. Y., vol. 3, Waterlime Gr. [Ety. proper name.]
lacustris, Harlan, 1834, Trans. Geo. Soc. Penn., vol. 1, Waterlime Gr. [Sig. pertaining to a lake or swamp.]
lacustris var. robustus, Hall, 1859, Pal. N. Y., vol. 3, Waterlime Gr. [Sig. robust.]
mazonensis, Meek & Worthen, 1868, Am. Jour. Sci., vol. 46, Coal Meas. [Ety. proper name.]
micropthalmus, Hall, 1855, Pal. N. Y., vol. 3, Waterlime Gr. [Sig. having small eyes.]
pachycheirus, Hall, 1859, Pal. N. Y., vol. 3, Waterlime Gr. [Sig. thick-handed.]
pulicaris, Salter, 1863, Quar. Jour. Geo. Soc., vol. 19, Coal Meas. [Ety. *pulis*, a little gate; *karis*, a shrimp.]
pustulosus, Hall, 1859, Pal. N. Y., vol. 3, Waterlime Gr. [Sig. covered with pustules.]

CRUSTACEA.

remipes, DeKay, 1825, Ann. Lyc. Nat. Hist. N. Y., Waterlime Gr. [Sig. oar-footed.]
tetragonopthalmus, Fischer, 1839, Bull. Soc. Imper. Nat. Moscou., Waterlime Gr. [Sig. having square eyes.]
HARPES, Goldfuss, 1839, Nova Acta Physico-medicaAcademiæ Cæsareæ Leopoldino Carolinæ Naturæ Curiosorum, vol. 19: [Ety. *harpe*, a hook or sickle.]
antiquatus, Billings, 1859,Can.Nat.& Geo., vol. 4, Chazy Gr. [Sig. ancient, old.]
consuetus, Billings, 1866, Catal. Sil. Foss. Antic., Anticosti Gr. [Sig. related to, from its close alliance to H. ottawænsis.]
dentoni, Billings, 1863, Can. Nat. & Geo., Hud. Riv. Gr. [Ety. proper name.]
escanabiæ, Hall, 1851, Geo. Lake Sup. Land Dist., vol. 2, Trenton Gr. [Ety. proper name.]
granti, Billings, 1865, Pal. Foss., Quebec Gr. [Ety. proper name.]
ottawensis, Billings, 1865, Pal. Foss., Trenton Gr. [Ety. proper name.]
HARPIDES, Beyrich, 1846, Untersuchungen Trilobiten als Fort. [Ety. from resemblance to the genus *Harpes*.]
atlanticus, Billings, 1865, Pal. Foss., Quebec Gr. [Ety. mythological name.]
concentricus, Billings, 1865, Pal. Foss., Quebec Gr. [Sig. concentrical.]
desertus, Billings, 1865, Pal. Foss., Quebec Gr. [Sig. deserted, solitary.]
HELMINTHOIDICHNITES, Fitch, 1848, Geo. Sur. Wash. Co. [Sig. like little worm tracks.] See *Gordia marina*.
marinus, Fitch, 1848, Geo. Sur.Wash. Co., Potsdam Gr. [Sig. marine.]
tenuis, Fitch, 1848, Geo. Sur. Wash. Co., Potsdam Gr. [Sig. slender.]
Hemicrypturus, Green, syn. for Asaphus.
clintoni, Vanuxem, 1843, Geo. Rep. 3rd Dist. N. Y., Clinton Gr.
rasoumowski, syn. for Asaphus expansus.
HOLOMETOPUS, Angelin, 1852, Palæontologia Scandinavica. [Ety. *holos*, entire; *metopon*, space between the eyes.]
angelini, Billings, 1862, Pal. Foss., Quebec Gr. [Ety. proper name.]
HOMALONOTUS, Konig, 1825, Icones. Foss. Sectiles. [Ety. *homalos*, on the same level; *notos*, the back.]
dawsoni, Hall, 1860, Can. Nat. & Geo., vol. 5, Silurian. [Ety. proper name.]
dekayi, Green, 1832, (Dipleura dekayi) Monograph of Trilobites, Ham. Gr. [Ety. proper name.]
delphinocephalus, Green, 1832, (Trimerus delphinocephalus) Monograph of Trilobites, Clinton & Niagara Gr. [Sig. dolphin-headed.]
jacksoni, Green, 1837, (Trimerus jacksoni) Am. Jour. Sci., vol. 32, Up. Sil. [Ety. proper name.]
knighti, Koninck, 1825, Icones. Foss. Sectiles, Low. Held. Gr. [Ety. proper name.]

vanuxemi, Hall, 1859, Pal. N. Y., vol. 3, Low. Held. Gr. [Ety. proper name.]
ILLÆNURUS, Hall, 1863, 16th Reg. Rep. N. Y. [Ety. from the genus *Illænus*; *oura* the tail.]
quadratus, Hall, 1863, 16th Reg. Rep. N. Y., Potsdam Gr. [Sig. square-shaped.]
ILLÆNUS, Dalman, 1828, ueber die Palæaden oder die sogenannten Trilobiten. [Ety. *illaino*, to look awry, to squint.]
americanus, Billings, 1859, Can. Nat. & Geo., vol. 4, Trenton Gr. [Ety. proper name.]
angusticollis, Billings, 1859, Can. Nat. & Geo., vol. 4, Black Riv. Gr. [Sig. having a narrow neck or column.]
arcturus, Hall, 1847, Pal. N. Y., vol. 1, Chazy & Black. Riv. Gr. [Ety. proper name.]
arcuatus, Billings, 1865, Pal. Foss., Quebec Gr. [Sig. arched.]
armatus, Hall, 1867, 20th Reg. Rep., Niagara Gr. [Sig. armed.]
barriensis, Murch, 1839, Sil. Syst. The species formerly identified with this one is described by Prof. Hall as *Illænus ioxus*.
bayfieldi, Billings, 1859, Can. Nat. & Geo., vol. 4, Chazy Gr. [Ety. proper name.]
clavifrons, Billings, 1859, Can. Nat. & Geo., Chazy & Black Riv. Gr. [Sig. club-fronted.]
conifrons, Billings, 1859, Can. Nat. & Geo., Black Riv. Gr. [Sig. having a conical front.]
conradi, Billings, 1859, Can. Nat. & Geo., vol. 4, Black Riv. Gr. [Ety. proper name.]
consimilis, Billings, 1865, Pal. Foss., Quebec Gr. [Sig. similar in all parts.]
consobrinus, Billings, 1865, Pal. Foss., Quebec Gr. [Sig. nearly allied.]
cornigerus, Hall, 1872, 24th Reg. Rep. N. Y., Niagara Gr. [Sig. horned.]
crassicauda, Wahlenberg, 1821,(Entomostracites crassicauda) Nov. Act. Soc. Upsal., vol. 8, Trenton & Galena Gr. [Sig. thick-tailed.]
cuniculus, Hall, 1867, 20th Reg. Rep. N. Y., Niagara Gr. [Sig. a cradle.]
daytonensis, Hall & Whitfield, 1875, Ohio Pal., vol. 2, Clinton Gr. [Ety. proper name.]
fraternus, Billings, 1865, Pal. Foss., Quebec Gr. [Sig. fraternal.]
globosus, Billings, 1859, Can. Nat. & Geo., vol. 4, Chazy Gr. [Sig. globose.]
graftonensis, Meek & Worthen, 1869, Proc. Acad. Nat. Sci. Phil., Niagara Gr. [Ety. proper name.]
grandis, Billings, 1859, Can. Nat. & Geo., vol. 4, Hud. Riv. Gr. & Mid. Sil. [Sig. grand.]
imperator, Hall, 1861, Rep. of Progr.Wis., Niagara Gr. [Sig. commander-in-chief.]
incertus, Billings, 1865, Pal. Foss., Quebec Gr. [Sig. uncertain.]

CRUSTACEA.

indeterminatus, Walcott, 1877, 29th Reg. Rep. N. Y., Black Riv. Gr. [Sig. not determined.]
insignis, Hall, 1867, 20th Reg. Rep. N.Y., Niagara Gr. [Sig. distinguished by a mark.]
ioxus, Hall, 1847, 20th Reg. Rep. N. Y., Niagara Gr. [Ety. mythological name.]
latidorsatus, Hall, 1847, Pal. N.Y., vol. 1, Trenton Gr. [Sig. broad-backed.]
milleri, Billings, 1859, Can. Nat. & Geo., vol. 4, Black River and Trenton Gr. [Ety. proper name.]
orbicaudatus, Billings, 1859, Can. Nat. & Geo., vol. 4, Hud. Riv. Gr. & Mid. Sil. [Sig. having a circular tail.]
ovatus, Conrad, 1843, Proc. Acad. Nat. Sci. Phil., vol. 1, Black Riv. Gr. [Sig. egg-shaped.]
simulator, Billings, 1865, Pal. Foss., Quebec Gr. [Sig. imitator.]
taurus, Hall, 1861, Rep. of Progr. Wis. Sur., Trenton & Galena Gr. [Ety. mythological name.]
trentonensis, Emmons, 1842, (Bumastus trentonensis) Geo. Rep. N. Y., Trenton Gr. [Ety. proper name.]
tumidifrons, Billings, 1865, Pal. Foss., Quebec Gr. [Sig. swollen out in front.]
vindex, Billings, 1865, Pal. Foss., Chazy Gr. [Sig. a defender, a revenger.]
worthenanus, syn. for Illænus insignis.
Isochilina, Jones, 1858, Can. Org. Rem., Decade 3. [Ety. *isos*, equal; *cheilos*, lip.]
cylindrica, Hall, 1852, (Cytherina cylindrica) Pal. N. Y., vol. 2, Medina Gr. [Sig. cylindrical.]
gracilis, Jones, 1858, Can. Org. Rem., Decade 3, Black Riv. Gr. [Sig. slender.]
ottawa, Jones, 1858, Can. Org. Rem., Decade 3, Black Riv. Gr. [Ety. proper name.]
Isotelus, DeKay, 1825, Annals Lyceum Nat. Hist. N. Y., vol. 1. [Ety. *isos*, equal; *telos*, end.] A subgenus of Asaphus.
canalis, Conrad, 1847, Pal. N. Y., Trenton & Hud. Riv. Gr. [Sig. a groove or channel.]
gigas, DeKay, 1825, Ann. Lyc. Nat. Hist. N. Y., vol. 1, Trenton & Hud. Riv. Gr. [Sig. a giant.]
maximus, see Isotelus megistos.
megistos, Locke, 1841, Proc. Am. Assoc., Trenton & Hud. Riv. Gr. [Sig. very large.]
vigilans, Meek & Worthen, 1870, Proc. Acad. Nat. Sci. Phil., Cin'ti. Gr. [Sig. watching.]
Leaia, Jones, 1862, App. to Mon. Foss., Estheriæ. [Ety. proper name.]
leidyi, Lea, 1856, (Cypricardia leidyi) Proc. Acad. Nat. Sci., vol. 7, Coal Meas. [Ety. proper name.]
tricarinata, Meek & Worthen, 1868, Geo. Sur. Ill., vol. 3, Coal Meas. [Sig. three-keeled.]

Leperditia, Rouault, 1851, Bull. Soc. Geo. France, 2nd series, t. 8. [Ety. *lepis*, a scale; *dittos*, double.]
alta, Conrad, 1843, (Cytherina alta) Geo. Rep. 3rd Dist. N. Y., Low. Held. Gr. [Sig. high.]
amygdalina, Jones, 1858, Can. Org. Rem., Decade 3, Black Riv. Gr. [Sig. like an almond.]
anna, Jones, 1858, Can. Org. Rem., Decade 3, Calciferous Gr. [Ety. proper name.]
anticostiana, Jones, 1858, Can. Org. Rem., Decade 3, Hud. Riv. Gr. [Ety. proper name.]
arctica, Jones, 1856, Ann. & Mag. Nat. Hist., 2nd series, vol. 17, Up. Sil. [Sig. northern, arctic.]
bivia, White, 1874, Rep. Invert. Foss., Quebec Gr. [Sig. having two ways or passages.]
byrnesi, S. A. Miller, 1874, Cin. Quar. Jour. Sci., vol. 1, Cin'ti Gr. [Ety. proper name.]
canadensis, Jones, 1858, Ann. Nat. Hist., 3rd series, vol. 1, Chazy to Trenton Gr. [Ety. proper name.]
capax, Safford. Not defined.
cayuga, Hall, 1862, 15th Reg. Rep. N. Y., Cornif. Gr. [Ety. proper name.]
concinnula, Billings, 1865, Pal. Foss., vol. 1, Quebec Gr. [Sig. small and beautiful.]
cylindrica, Hall, 1871, Pamphlet, Cin'ti Gr. [Sig. cylindrical.]
faba, Hall, 1876, 28th Reg. Rep. N. Y., Niagara Gr. [Sig. a bean.]
fabulites, Conrad, 1843, (Cytherina fabulites) Proc. Acad. Nat. Sci. Phil., Trenton Gr. [Sig. a small bean.]
fonticola, Hall, 1867, 20th Reg. Rep. N. Y., Niagara Gr. [Sig. fountain-dwelling.]
gibbera, Jones, 1856, Ann. & Mag. Nat. Hist., 2nd series, vol. 17, Niagara Gr. [Sig. bumpbacked.]
gracilis, see Isochilina gracilis.
hudsonica, Hall, 1859, Pal. N. Y., vol. 3, Low. Held. Gr. [Ety. proper name.]
jonesi, Hall, 1859, Pal. N. Y., vol. 3, Low. Held. Gr. [Ety. proper name.]
josephina, Jones, 1858, Can. Org. Rem., Decade 3, Black Riv. to Trenton Gr. [Ety. proper name.]
labrosa, Jones, 1858, Can. Org. Rem., Decade 3, Chazy Gr. [Sig. having a rim or border.]
louckana, Jones, 1858, Can. Org. Rem., Decade 3, Black Riv. Gr. [Ety. proper name.]
marginata, Keyserling, 1846, Wissenschaftliche Beobachtungen, etc., Niagara Gr. [Sig. bordered.]
minutissima, Hall, 1871, Pamphlet, Cin'ti Gr. [Sig. very small.]
morgani, Safford. Not defined.
nana, Jones, 1858, Can. Org. Rem., Decade 3, Calciferous Gr. [Sig. dwarfish.]

CRUSTACEA.

okeni, Munster, 1830, (Cythere okeni) Jahrbuch fur Min., Geo. und Petrif., Carboniferous. [Ety. proper name.]
ottawa, see Isochilina ottawa.
ovata, Jones, 1858, Ann. & Mag. Nat. Hist., 3rd ser., vol. 1, Black Riv. Gr. [Sig. egg-shaped.]
paquettana, Jones, 1858, Can. Org. Rem., Decade 3, Black Riv. Gr. [Ety. proper name.]
parasitica, Hall, 1859, Pal. N. Y., vol. 3, Low. Held. Gr. [Sig. parasitic.]
parvula, Hall, 1859, Pal. N. Y., vol. 3, Low. Held. Gr. [Sig. very small.]
pennsylvanica, Jones, 1858, Ann. & Mag. Nat. Hist., 3rd ser., vol. 1, Clinton Gr. [Ety. proper name.]
punctulifera, Hall, 1860, 13th Reg. Rep. N. Y., Ham. Gr. [Sig. bearing dots.]
scalaris, Jones, 1858, Ann. & Mag. Nat. Hist., 3rd ser., vol. 1, Waterlime Gr. [Sig. like a ladder.]
seneca, Hall, 1862, 15th Reg. Rep. N. Y., Ham. Gr. [Ety. proper name.]
sinuata, Hall, 1860, Can. Nat. & Geo., vol. 5, Silurian. [Sig. marked with depressions.]
troyensis, Ford, 1863, Am. Jour. Sci. & Arts, 3rd ser., vol. 6, Low. Potsdam Gr. [Ety. proper name.]
turgida, Billings, 1865, Pal. Foss., vol. 1, Quebec Gr. [Sig. swollen out.]
ventralis, Billings, 1865, Pal. Foss., vol. 1, Quebec Gr. [Sig. ventral.]
spinulifera, Hall, 1862, 15th Reg. Rep. N. Y., Cornif. Gr. [Sig. bearing spines.]
sublævis, Shumard, 1855, (Cythere sublævis) Geo. Rep. Mo., Low. Magnesian Gr. [Sig. somewhat smooth.]

LICHAS, Dalman, 1826, Monograph of Trilobites. [Ety. mythological name.]
armatus, Hall, 1862. Being preoccupied, the name is changed to L. eriopia.
bigsbyi, Hall, 1859, Pal. N. Y., vol. 3, Low. Held. Gr. [Ety. proper name.]
boltoni, Bigsby, 1825, (Paradoxides boltoni) Jour. Acad. Nat. Sci., vol. 4, Niagara Gr. [Ety. proper name.]
boltoni var. occidentalis, Hall, 1863, Trans. Alb. Inst., vol. 4, Niagara Gr. [Sig. western.]
breviceps, Hall, 1863, Trans. Alb. Inst., vol. 4, Niagara Gr. [Sig. short-headed.]
canadensis, Billings, 1866, Catal. Sil. Foss. Antic., Antic. Gr. [Ety. proper name.]
cucullus, Meek & Worthen, 1865, Proc. Acad. Nat. Sci. Phil., Trenton Gr. [Sig. a cap.]
decipiens, Winchell & Marcy, 1865, Mem. Bost. Soc. Nat. Hist., Niagara Gr. [Sig. doubtful.]
eriopia, see Terataspis eriopia.
grandis, see Terataspis grandis.
jukesi, Billings, 1865, Pal. Foss., vol. 1, Quebec Gr. [Ety. proper name.]
minganensis, Billings, 1865, Pal. Foss., vol. 1, Chazy and Black Riv. Gr. [Ety. proper name.]

nereus, Hall, 1863, 16th Reg. Rep. N. Y., Niagara Gr. [Ety. mythological name.]
pugnax, Winchell & Marcy, 1865, Mem. Bost. Soc. Nat. Hist., Niagara Gr. [Sig. war-like.]
pustulosus, Hall, 1859, Pal. N. Y., vol. 3, Low. Held. Gr. [Sig. covered with pustules.]
superbus, Billings, 1875, Can. Nat. & Geol., Corniferous Gr. [Sig. superb.]
trentonensis, Conrad, 1842, (Asaphus trentonensis) Jour. Acad. Nat. Sci., vol. 8, Black Riv., Trenton and Hud. Riv. Gr. [Ety. proper name.]
Loganellus, Devine, 1863, Can. Nat. & Geo., vol. 8. [Ety. proper name.]
quebecensis, see Olenus (?) logani.
Lonchocephalus, syn. for Conocephalites.
chippewensis, see Conocephalites chippewensis.
hamulus, see Conocephalites hamulus.

MEGALASPIS, Angelin, 1852, Palæontologia Scandinavica. [Ety. megale, great; aspis, shield.]
belemnurus, White, 1874, Rep. Invert. Foss., Quebec Gr. [Ety. belemnon, a dart; oura, a tail.]

MENOCEPHALUS, Owen, 1852, Geo. Sur. Wis., Iowa & Minn. [Ety. menos, strength; kephale, head.]
globosus, Billings, 1860, Can. Nat. & Geo., vol. 5, Quebec Gr. [Sig. globose.]
minnesotensis, Owen, 1852, Rep. Iowa, Wis. and Minn., Potsdam Gr. [Ety. proper name.]
salteri, Devine, 1863, Can. Nat. & Geo., vol. 8, Quebec Gr. [Ety. proper name.]
sedgwicki, Billings, 1860, Can. Nat. & Geo., vol. 5, Quebec Gr. [Ety. proper name.]

MICRODISCUS, Emmons, 1856, Am. Geol. [Ety. mikros, small; diskos, a quoit.]
dawsoni, Hartt, 1868, Acad. Geol., St. John's Gr. [Ety. proper name.]
lobatus, Hall, 1847, (Agnostus lobatus) Pal. N. Y., vol. 1, Hud. Riv. Gr. [Sig. divided into segments.]
meeki, Ford, 1876, Am. Jour. Sci. & Arts, Low. Potsdam Gr. [Ety. proper name.]
quadricostatus, Emmons, 1856, Am. Geo., Quebec Gr. [Sig. four-ribbed.]
speciosus, Ford, 1863, Am. Jour. Sci. & Arts, 3rd series, vol. 6, Low. Potsdam Gr. [Sig. beautiful.]

NILEUS, Dalman, 1826, Monograph of Trilobites. [Ety. mythological name.]
affinis, Billings, 1865, Pal. Foss., vol. 1, Quebec Gr. [Sig. near to, related.]
macrops, Billings, 1865, Pal. Foss., vol. 1, Quebec Gr. [Sig. having large eyes.]
scrutator, Billings, 1865, Pal. Foss., vol. 1, Quebec Gr. [Sig. a searcher.]

Nuttainia, syn. for Trinucleus.
concentrica, see Trinucleus concentricus.
sparsa, syn. for Homalonotus dekayi.

Odontocephalus, Conrad, 1840, Ann. Geo. Rep. N. Y.
selenurus, see Dalmanites selenurus.
Odontochile, syn. for Dalmanites.
Ogygia, Brongniart, 1822, Hist. Nat. Crust. Foss. [Ety. mythological name.]
vetusta, see Asaphus vetustus.
OLENELLUS, Hall, 1861, 14th Reg. Rep. N. Y. [Ety. diminutive of *Olenus*.]
asaphoides, Emmons, 1844, (Ellipto-cephalus asaphoides) Geo. Rep. N.Y., Potsdam Gr. [Sig. like an *Asaphus*.]
gilberti, Meek, 1874, Rep. Invert. Foss., Potsdam Gr. [Ety. proper name.]
howelli, Meek, 1874, Rep. Invert. Foss., Potsdam Gr. [Ety. proper name.]
thompsoni, Hall, 1859, (Olenus thompsoni) 12th Reg. Rep. N. Y., (Barrandia thompsoni) 13th Reg. Rep., (Paradoxides thompsoni) Emmons, Man. Geo., Potsdam Gr. [Ety. proper name.]
vermontanus, Hall, 1859, (Olenus vermontana) 13th Reg. Rep. N. Y., Potsdam Gr. [Ety. proper name.]
OLENUS, Dalman, 1826, Monograph of Trilobites. [Ety. mythological name.]
asaphoides, see Olenellus asaphoides.
(?) logani, Devine, 1863, Can. Nat. & Geo., vol. 8, Quebec Gr. [Ety. proper name.]
thompsoni, see Olenellus thompsoni.
undulostriatus, Hall, 1847, Pal. N. Y., vol. 1, Hud. Riv. Gr. [Sig. having waved striæ.]
vermontana, see Olenellus vermontanus.
PALÆOCARIS, Meek & Worthen, 1865, Proc. Acad. Nat. Sci. Phil. [Ety. *palaios*, ancient; *karis*, a shrimp.]
typus, Meek & Worthen, 1865, Proc. Acad. Nat. Sci. Phil., Coal Meas. [Ety. type of the genus.]
PARADOXIDES, Brongniart, 1822, Hist. Nat. Crust. Foss. [Ety. *paradoxos*, marvelous, paradoxical.]
arcuatus, Harlan, 1835, Trans. Geo. Soc. Penn., Potsdam Gr. [Sig. arched.]
becki, see Triarthrus becki.
bennetti, Salter, 1859, Quar. Jour. Geo. Soc., vol. 15, Potsdam Gr. [Ety. proper name.]
boltoni, see Lichas boltoni.
decorus, Billings, 1865, Pal. Foss., vol. 1, Potsdam Gr. [Sig. beautiful.]
eatoni, syn. for Triarthrus becki.
harlani, Green, 1834, Am. Jour. Sci., vol. 25, Potsdam Gr. [Ety. proper name.]
lamellatus, Hartt, 1868, Acad. Geol., St. John's Gr. [Sig. from the laminæ on the anterior lobe of the glabella.]
micmac, Hartt, 1868, Acad. Geol., St. John's Gr. [Ety. proper name.]
nevadensis, Meek, 1870, Proc. Acad. Nat. Sci., Potsdam Gr. [Ety. proper name.]
tenellus, Billings, 1874, Pal. Foss., vol. 2, Potsdam Gr. [Sig. very delicate.]
thompsoni, see Olenellus thompsoni.
triarthrus, Harlan, 1835, Trans. Geo. Soc. Penn., Potsd. Gr. [Sig. three-jointed.]
vermontana, see Olenellus vermontanus.

Peltura, M. Edwards, 1840, Crustaces, etc.
holopyga, Hall, 1859, see Bathynotus holopyga.
PEMPHIGASPIS, Hall, 1863, 16th Reg. Rep. N. Y. [Ety. *pemphix*, a pustule; *aspis*, a shield.]
bullata, Hall, 1863, 16th Reg. Rep. N. Y., Potsdam. Gr. [Sig. studded, bossed.]
PHACOPS, Emmrich, 1839, de Trilobites. [Ety. *phakos*, a lens; *ops*, the eye.]
bufo, Green, 1832, (Calymene bufo) Monograph of Trilobites, Ham. Gr. [Sig. a toad.]
bombifrons, Hall, 1861, 15th Reg. Rep. N. Y., Up. Held. Gr. [Sig. a hollow-sounding front.]
cacapona, Hall, 1861, 15th Reg. Rep. N. Y., Up. Held. Gr. [Ety. proper name.]
callicephala, see Dalmanites callicephalus.
cristata, Hall, 1861, 15th Reg. Rep. N.Y., Up. Held. Gr. [Sig. tufted.]
hudsonica, Hall, 1859, Pal. N. Y., vol. 3, Low. Held. Gr. [Ety. proper name.]
laticaudus, see Dalmanites laticaudus.
logani, Hall, 1859, Pal. N. Y., vol. 3, Low. Held. Gr. [Ety. proper name.]
nupera, Hall, 1843, (Calymene nupera) Geo. Rep. 4th Dist. N. Y., Chemung Gr. [Sig. late, recent.]
orestes, Billings, 1860, Can. Nat. & Geo., vol. 5, Mid. Sil. [Ety. mythological name.]
rana, Green, 1832, (Calymene bufo *var.* rana) Monograph of Trilobites, Ham. Gr. [Sig. a frog.]
trisulcata, Hall, 1843, (Calymene (?) trisulcatus) Geo. Rep. 4th Dist. N. Y., Clinton Gr. [Sig. three-furrowed.]
PHILLIPSIA, Portlock, 1843, Rep. Geol. Londonderry. [Ety. proper name.]
bufo, Meek & Worthen, 1870, Proc. Acad. Nat. Sci., Keokuk Gr. [Sig. a toad.]
cliftonensis, Shumard, 1858, Trans. St. Louis Acad. Sci., Coal Meas. [Ety. proper name.]
coronata, Hall, 1877, Ham. Gr. Proposed instead of *P. ornata*, which was preoccupied. [Sig. crowned.]
howi, Billings, 1863, Can. Nat. & Geol., vol. 8, Carboniferous. [Ety. proper name.]
insignis, Winchell, 1863, Proc. Acad. Nat. Sci., Burlington Gr. [Sig. marked.]
lævis, Hall, 1876, Illust. Devonian Foss., Tully limestone. [Sig. smooth.]
lodiensis, Meek, 1875, Ohio Pal., vol. 2, Waverly Gr. [Ety. proper name.]
major, Shumard, 1858, Trans. St. Louis Acad. Sci., vol. 1, Coal Meas. [Sig. larger.]
meramecensis, Shumard, 1855, Geo. Rep. Mo., Archimedes limestone. [Ety. proper name.]
minuscula, Hall, 1876, Illust. Devonian Foss., Up. Held. Gr. [Sig. rather small.]

missouriensis, Shumard, 1858, Trans. St. Louis Acad. Sci., Coal Meas. [Ety. proper name.]
ornata, Hall, 1876, Illust. Devonian Foss., Ham. Gr. The name was preoccupied by Portlock in 1843. See P. coronata.
perannulata, Shumard, 1858, Trans. St. Louis Acad. Sci., Permian Gr. [Sig. very annulated.]
portlocki, Meek & Worthen, 1865, Proc. Acad. Nat. Sci., Keokuk Gr. [Ety. proper name.]
rockfordensis, Winchell, 1865, Proc. Acad. Nat. Sci., Kinderhook (?) Gr. [Ety. proper name.]
sangamonensis, Meek & Worthen, 1865, Proc. Acad. Nat. Sci., Coal Meas. [Ety. proper name.]
scitula, Meek & Worthen, 1865, Proc. Acad. Nat. Sci., Coal Meas. [Sig. pretty.]
stevensoni, Meek, 1871, Reg. Rep. University W. Va., Chester Gr. [Ety. proper name.]
vindobonensis, Hartt, 1868, Acad. Geol., Carboniferous. [Ety. proper name.]
Piliolites, Cozzens, 1848. Not sufficiently defined to be identified.
ohioensis, Cozzens, 1848. Not sufficiently defined to be identified, but probably the fragment of a Dalmanites.
Platynotus, syn. for Lichas.
boltoni, see Lichas boltoni.
trentonensis, see Lichas trentonensis.
PLUMULITES, Barrande. [Ety. *plumula*, a feather; *lithos*, stone.]
jamesi, Hall & Whitfield, 1875, Pal. Ohio, vol. 2. [Ety. proper name.]
PRIMITIA, Jones, 1865, Ann. & Mag. Nat. Hist., 3rd series, vol. 16. A subgenus of Beyrichia, founded upon *B. logani* and *Cytheropsis concinna*.
PROETUS, Steininger, 1830, Bemerkungen uber die Versteinerungen welche im Uebergangs-Gebirge der Eifel. [Ety. mythological name.]
alaricus, Billings, 1860, Can. Nat. & Geo., vol. 5, Hud. Riv. Gr. [Sig. winged.]
angustifrons, Hall, 1861, 15th Reg. Rep. N. Y., Schoharie grit. [Sig. having a front.]
auriculatus, Hall, 1861, 15th Reg. Rep. N. Y., Chemung Gr. [Sig. eared.]
canaliculatus, Hall, 1861, 15th Reg. Rep. N. Y., Up. Held. Gr. [Sig. channeled.]
clarus, Hall, 1861, 15th Reg. Rep. N. Y., Up. Held. Gr. [Sig. clear, manifest.]
conradi, Hall, 1861, 15th Reg. Rep. N. Y., Schoharie grit. [Ety. proper name.]
corycœus, Conrad, 1842, (Asaphus corycœus) Jour. Acad. Nat. Sci., vol. 8, Niagara Gr. [Ety. proper name.]
crassimarginatus, Hall, 1843, (Calymene crassimarginata) Geo. Rep. 4th Dist. N. Y., Corniferous Gr. [Sig. thick-margined.]
doris, Hall, 1860, 13th Reg. Rep. N. Y., Ham. Gr. [Ety. mythological name.]
ellipticus, Meek & Worthen, 1865, Proc. Acad. Nat. Sci., Kinderhook Gr. [Sig. elliptical.]
haldemani, Hall, 1861, 15th Reg. Rep. N. Y., Ham. Gr. [Ety. proper name.]
hesione, Hall, 1861, 15th Reg. Rep. N. Y., Schoharie grit. [Ety. mythological name.]
longicaudus, Hall, 1861, 15th Reg. Rep. N. Y., Ham. Gr. [Sig. long-tailed.]
macrocephalus, Hall, 1861, 15th Reg. Rep. N. Y., Ham. Gr. [Sig. long-headed.]
marginalis, Conrad, 1839, (Calymene marginalis) Ann. Geo. Rep. N. Y., Tully limestone. [Sig. margined.]
missouriensis, Shumard, 1855, Geo. Rep. Mo., Low. Carb. [Ety. proper name.]
occidens, Hall, 1861, 15th Reg. Rep. N. Y., Ham. Gr. [Sig. western.]
parviusculus, Hall, 1866, Pamphlet, Cin'ti Gr. [Sig. very small.]
phocion, Billings, 1874, Pal. Foss., vol. 2, Devonian. [Ety. proper name.]
planimarginatus, Meek, 1871, Proc. Acad. Nat. Sci. Phil., Corniferous Gr. [Sig. flat-margined.]
protuberans, Hall, 1859, Pal. N. Y., vol. 3, Low. Held. Gr. [Sig. protuberant.]
prouti, Shumard, 1863, Trans. St. Louis Acad. Sci., Ham. Gr. [Ety. proper name.]
rowii, Green, 1838, (Calymene rowii) Am. Jour. Sci., vol. 38, Ham Gr. [Ety. proper name.]
spurlocki, Meek, 1872, Am. Jour. Sci., 3rd series, vol. 3, Cin'ti Gr. [Ety. proper name.]
stokesi, Murchison, 1839, (Asaphus stokesii) Sil. Syst., Niagara Gr. [Ety. proper name.]
swallovi, Shumard, 1855, Geo. Rep. Mo., Chemung Gr. [Ety. proper name.]
verneuili, Hall, 1861, 15th Reg. Rep. N. Y., Up. Held. Gr. [Ety. proper name.]
PROTICHNITES, Owen, 1852, Jour. Geo. Soc., vol. 8. [Ety. *protos*, first; *ichnos*, footprint or track; *lithos*, stone.]
acadicus, Dawson, 1863, Am. Jour. Sci. & Arts, 3rd ser., vol. 5, Coal Meas. [Ety. proper name.]
alternans, Owen, 1852, Jour. Geo. Soc., vol. 8, Quebec Gr. [Sig. alternating.]
carbonarius, Dawson, 1863, Am. Jour. Sci. & Arts, 3rd ser., vol. 5, Coal Meas. [Sig. pertaining to coal.]
latus, Owen, 1852, Jour. Geo. Soc., vol. 8, Quebec Gr. [Sig. wide.]
lineatus, Owen, 1852, Jour. Geo. Soc., vol. 8, Quebec Gr. [Sig. lined.]
logananus, Marsh, 1869, Am. Jour. Sci. & Arts, 2d ser., vol. 48, Potsdam Gr. [Ety. proper name.]
multinotatus, Owen, 1852, Jour. Geo. Soc., vol. 8, Quebec Gr. [Sig. many marks or tracks.]
octo-notatus, Owen, 1852, Jour. Geo. Soc., vol. 8, Quebec Gr. [Sig. having eight marks or tracks.]

CRUSTACEA. 223

septem-notatus, Owen, 1852, Jour. Geo. Soc., vol. 8, Quebec Gr. [Sig. having seven marks.]
PTERYGOTUS, Agassiz, 1839, in Murch. Sil. Syst. [Ety. *pteron*, wing; *ous*, ear.]
 cobbi, Hall, 1859, Pal. N. Y., vol. 8, Waterlime Gr. [Ety. proper.]
 macrophthalmus, Hall, 1859, Pal. N. Y., vol. 3, Waterlime Gr. [Sig. long-eyed.]
 osborni, Hall, 1859, Pal. N. Y., vol. 3, Waterlime Gr. [Ety. proper name.]
PTYCHASPIS, Hall, 1863, 16th Reg. Rep. N. Y. [Ety. *ptyche*, folding or doubling; *aspis*, a shield.]
 barabuensis, Winchell, 1864, Am. Jour. Sci. & Arts, 2d series, vol. 37, Potsdam Gr. [Ety. proper name.]
 granulosa, Owen, 1852, (Dikelocephalus granulosus) Rep. Iowa, Wis. & Minn., Potsdam Gr. [Sig. covered with granules.]
 miniscaensis, Owen, 1852, (Dikelocephalus miniscaensis) Rep. Wis., Iowa & Minn., Potsdam Gr. [Ety. proper name.]
 sesostris, Billings, 1865, (Dikelocephalus sesostris) Pal. Foss., vol. 1, Quebec Gr. [Ety. proper name.]
REMOPLEURIDES, Portlock, 1843, Rep. Geol. Lond. [Ety. *remus*, an oar; *pleura*, a rib.]
 affinis, Billings, 1865, Pal. Foss., vol. 1, Quebec Gr. [Sig. related to.]
 canadensis, Billings, 1865, Pal. Foss., vol. 1, Chazy Gr. [Ety. proper name.]
 panderi, Billings, 1865, Pal. Foss., vol. 1, Quebec Gr. [Ety. proper name.]
 schlotheimi, Billings, 1865, Pal. Foss., vol. 1, Quebec Gr. [Ety. proper name.]
 striatulus, Walcott, 1875, Cin. Quar. Jour. Sci., vol. 2, Trenton Gr. [Sig. having little striæ.]
Rhabdichnites, Dawson, 1875, Can. Nat. & Geo. Name proposed.
RUSICHNITES, Dawson, 1861, Can. Nat. & Geo. [Ety. *rusos*, wrinkled; *ichnos*, a track.] Supposed track of a small *Limulus*.
 acadicus, Dawson, 1861, Can. Nat. & Geo., Coal Meas. [Ety. proper name.]
SHUMARDIA, Billings, 1862, Pal. Foss., vol. 1. [Ety. proper name.]
 glacialis, Billings, 1865, Pal. Foss., vol. 1, Quebec Gr. [Sig. icy, frozen.]
 granulosa, Billings, 1862, Pal. Foss., vol. 1, Quebec Gr. [Sig. covered with granules.]
Solenocaris, Meek, 1872, subgenus of Ceratiocaris.
SOLENOPLEURA, as used by Billings, 1874, Pal. Foss., vol. 2. [Ety. *Solen*, the razor fish; *pleura*, a rib.]
 communis, Billings, 1874, Pal. Foss., vol. 2, Potsdam Gr. [Sig. common, in large numbers.]
SPHÆREXOCHUS, Beyrich, 1845, Monograph of Trilobites. [Ety. *sphaira*, a ball; *exochos*, prominent.]
 canadensis, Billings, 1866, Catal. Sil. Foss. Antic., Anticosti Gr. [Ety. proper name.]
 mirus, Beyrich, see Sphærexochus romingeri.
 parvus, Billings, 1865, Pal. Foss., Chazy & Black Riv. Gr. [Sig. small.]
 romingeri, Hall, 1867, 20th Reg. Rep. N. Y., Niagara Gr. [Ety. proper name.]
SPHÆROCORYPHE, Angelin, 1852, Palæontologia Scandinavica. [Ety. *sphaira*, a ball; *koryphe*, the top of the head.]
 robustus, Walcott, 1875, Cin. Quar. Jour. Sci., vol. 2, Trenton Gr. [Sig. robust.]
 salteri, Billings, 1866, Catal. Sil. Foss. Antic., Anticosti Gr. [Ety. proper name.]
TELEPHUS, Barrande. [Ety. mythological name.]
 americanus, Billings, 1865, Pal. Foss., vol. 1, Quebec Gr. [Ety. proper name.]
TERATASPIS, Hall, 1863, 16th Reg. Rep. N. Y. [Ety. *teras*, a marvel; *aspis*, shield.] A subgenus of Acidaspis.
 grandis, Hall, 1862, 15th Reg. Rep. N. Y., (Lichas grandis) Schoharie grit. [Sig. grand.]
 eriopia, Hall, 1863, (Lichas eriopis) 16th Reg. Rep. N. Y., Up. Held. Gr. [Ety. mythological name.]
THALEOPS, Conrad, 1843, Proc. Acad. Nat. Sci., vol. 1. [Ety. *thalos*, a stalk; *ops*, an eye.]
 ovata, Conrad, 1843, Proc. Acad. Nat. Sci., vol. 1, Trenton Gr. [Sig. ovate.]
TRIARTHRELLA, Hall, 1863, 16th Reg. Rep. N. Y. [Ety. diminutive of *Triarthrus*.]
 auroralis, Hall, 1863, 16th Reg. Rep. N. Y., Potsdam Gr. [Sig. in the morning.]
TRIARTHRUS, Green, 1832, Monograph of Trilobites. [Ety. *triarthrus*, three-jointed.]
 becki, Green, 1832, Monograph of Trilobites, Trenton & Hud. Riv. Gr. [Ety. proper name.]
 spinosus, Billings, 1857, Rep. of Progr., Trenton Gr. [Sig. full of spines.]
 trilineatus, Emmons, 1844, (Atops trilineatus) Taconic Syst., Potsdam Gr. [Sig. three-lined.]
Trimerus, syn. for Homalonotus.
 delphinocephalus, see Homalonotus delphinocephalus.
 jacksoni, see Homalonotus jacksoni.
TRINUCLEUS, L h w y d, (or as he spelt it Lhwydd) 1698, Phil. Trans., vol. 20. [Ety. *trinucleus*, three-kerneled.]
 concentricus, Eaton, 1832, (Nuttainia concentrica) Geo. Text Book, Trenton & Hud. Riv. Gr. [Sig. from the concentrical arrangement of the punctures.]
 tessellatus, Green, 1832, (Cryptolithus tessellatus) syn. for Trinucleus concentricus.

CLASS ARACHNIDA.

GENERA.—Architarbus, Arthrolycosa, Eoscorpius, Mazonia.

ARCHITARBUS, Scudder, 1868, Geo. Sur. Ill., vol. 3. [Ety. *arche*, beginning; *tarbos*, object of alarm.]
    rotundatus, Scudder, 1868, Geo. Sur. Ill., vol. 3, Coal Meas. [Sig. rounded.]

ARTHROLYCOSA, Harger, 1874, Am. Jour. Sci., 3rd series, vol. 7. [Ety. *arthron*, a joint; *lykos*, a spider.]
    antiqua, Harger, 1874, Am. Jour. Sci., 3d ser., vol. 7, Coal Meas. [Sig. ancient.]

EOSCORPIUS, Meek & Worthen, 1868, Am. Jour. Sci., vol. 45. [Ety. *eos*, dawn; *scorpius*, a scorpion.]
    carbonarius, Meek & Worthen, 1868, Am. Jour. Sci., vol. 45, Coal Meas. [Sig. pertaining to coal.]

MAZONIA, Meek & Worthen, 1868, Geo. Sur. Ill., vol. 3. [Ety. proper name.]
    woodana, Meek & Worthen, 1868, Geo. Sur. Ill., vol. 3, Coal Meas. [Ety. proper name.]

CLASS MYRIAPODA.

GENERA.—Anthracerpes, Archiulus, Euphoberia, Xylobius.

ANTHRACERPES, Meek & Worthen, 1865, Proc. Acad. Nat. Sci. Phil. [Ety. *anthrax*, coal; *erpo*, to creep; in allusion to its Carboniferous age and probable habits.]
    typus, Meek & Worthen, 1865, Proc. Acad. Nat. Sci., Coal Meas. [Ety. type of the genus.] This is probably a worm.

ARCHIULUS, Scudder, 1868, Mem. Bost. Soc. Nat. Hist., vol. 2. [Ety. *arche*, beginning; *ioulos*, a wood louse.]
    xylobioides, Scudder, 1868, Mem. Bost. Soc. Nat. Hist., vol. 2, Coal Meas. [Ety. from its resemblance to the genus Xylobius.]

EUPHOBERIA, Meek & Worthen, 1868, Am. Jour. Sci., vol. 45. [Ety. *eu*, very; *phoberos*, formidable.]
    armigera, Meek & Worthen, 1868, Am. Jour. Sci., vol. 45, Coal Meas. [Sig. armed.]
    major, Meek & Worthen, 1868, Am. Jour. Sci., vol. 45, Coal Meas. [Sig. large.]

XYLOBIUS, Dawson, 1860, Quar. Jour. Geo. Soc., vol. 16. [Ety. *xylobius*, living in wood.]
    dawsoni, Scudder, 1868, Mem. Bost. Soc. Nat. Hist., vol. 1, Coal Meas. [Ety. proper name.]
    fractus, Scudder, 1868, Mem. Bost. Soc. Nat. Hist., vol. 2, Coal Meas. [Sig. frail.]
    sigillariæ, Dawson, 1860, Quar. Jour. Sci., vol. 16, Coal Meas. [Sig. from the *Sigillaria* in which it was found.]
    similis, Scudder, 1868, Mem. Bost. Soc. Nat. Hist., vol. 2, Coal Meas. [Sig. similar.]

CLASS INSECTA.

ORDER NEUROPTERA.—Chrestotes, Dyscritus, Ephemerites, Gerephemera, Haplophlebium, Hemeristia, Homothetus, Libellula, Lithentomum, Megathentomum, Miamia, Paolia, Platephemera, Xenoneura.
ORDER ORTHOPTERA.—Archegogryllus, Archimylacris, Blattina, Mylacris.
INCERTÆ SEDIS—Palæocampa.

ARCHEGOGRYLLUS, Scudder, 1868, Proc. Bost. Soc. Nat. Hist. [Ety. *archegos*, first in time; *gryllus*, a cricket.]
  priscus, Scudder, 1868, Proc. Bost. Soc. Nat. Hist., Coal Meas. [Sig. ancient.]
ARCHIMYLACRIS, Scudder, 1868, Acad. Geol. [Ety. *arche*, beginning; *Mylacris*, cockroach.]
  acadica, Scudder, 1868, Acad. Geol., Low. Carb. [Ety. proper name.]
  mantis, Scudder, 1868, Geo. Sur. Ill., vol. 3, Coal Meas. [Ety. *Mantis*, a genus of insects.]
BLATTINA, Burmeister, 1838, Handbuch der Entomologie. [Ety. *Blatta*, a cockroach.]
  bretonensis, Scudder, 1875, Can. Nat., n. s., vol. 7, Carboniferous. [Ety. proper name.]
  heeri, Scudder, 1868, Geo. Sur. Ill., vol. 3, Coal Meas. [Ety. proper name.]
  sepulta, Scudder, 1876, Can. Nat. & Geo., vol. 8, Carboniferous. [Sig. lulled to sleep, buried.]
  venusta, Lesquereux, 1860, Geo. Ark., vol. 2, Coal Meas. [Sig. handsome.]
CHRESTOTES, Scudder, 1868, Geo. Sur. Ill., vol. 3. [Ety. *chrestotes*, good of its kind.]
  lapidea, Scudder, 1868, Geo. Sur. Ill., vol. 3, Coal Meas. [Sig. stony.]
DYSCRITUS, Scudder, 1868, Geo. Mag., vol. 5. [Ety. *dyscritos*, hard to determine, doubtful.]
  vetustus, Scudder, 1868, Geo. Mag., vol. 5, Devonian. [Sig. ancient.]
EPHEMERITES, Scudder, 1868, Geo. Sur. Ill., vol. 3. [Ety. from resemblance to the *Ephemera*.]
  affinis, Scudder, 1868, Geo. Sur. Ill., vol. 3, Coal Meas. [Sig. related to.]
  gigas, Scudder, 1868, Geo. Sur. Ill., vol. 3, Coal Meas. [Sig. large.]
  simplex, Scudder, 1868, Geo. Sur. Ill., vol. 3, Coal Meas. [Sig. simple.]

GEREPHEMERA, Scudder, 1868, Geo. Mag., vol. 5. [Ety. *geros*, old; *Ephemera*, a genus of insects.]
  simplex, Scudder, 1868, Geo. Mag., vol. 5, Devonian. [Sig. simple.]
HAPLOPHLEBIUM, Scudder, 1867, Can. Nat. & Geo., 2d ser., vol. 3. [Ety. *haplos*, simple; *phlebion*, a vein.]
  barnesi, Scudder, 1867, Can. Nat. & Geo., 2d ser., vol. 3, Coal Meas. [Ety. proper name.]
HEMERISTIA, Dana, 1864, Am. Jour. Sci. & Arts, 2d ser., vol. 37. [Ety. *hemera*, day.]
  occidentalis, Dana, 1864, Am. Jour. Sci. & Arts, 2d ser., vol. 37, Coal Meas. [Sig. western.]
HOMOTHETUS, Scudder, 1867, Can. Nat. & Geo., 2d series, vol. 3. [Ety. *homos*, similar; *thesis*, placed.]
  fossilis, Scudder, 1867, Can. Nat. & Geo., 2nd series, vol. 3, Ham. or Chemung Gr. [Sig. fossil, extinct.]
LIBELLULA, Fabricius, 1776, as identified by Scudder.
  carbonaria, Scudder, 1876, Can. Nat. & Geo., vol. 8, Coal Meas. [Sig. pertaining to coal.]
LITHENTOMUM, Scudder, 1867, Can. Nat. & Geo. 2d series, vol. 3. [Ety. *lithos*, stone; *entomon*, an insect.]
  harti, Scudder, 1867, Can. Nat. & Geo., 2nd series, vol. 3, Ham. or Chemung Gr. [Ety. proper name.]
MEGATHENTOMUM, Scudder, 1868, Geo. Sur. Ill., vol. 3. [Ety. *megathos*, largeness; *entomon*, an insect.]
  pustulatum, Scudder, 1868, Geo. Sur. Ill., vol. 3, Coal Meas. [Sig. covered with pustules.]
MIAMIA, Dana, 1864, Am. Jour. Sci. & Arts, 2d ser., vol. 37. [Ety. proper name.]
  bronsoni, Dana, 1864, Am. Jour. Sci. & Arts, 2d ser., vol. 37, Coal Meas. [Ety. proper name.]

danæ, Scudder, 1868, Geo. Sur. Ill., vol. 3, Coal Meas. [Ety. proper name.]

MYLACRIS, Scudder, 1868, Geo. Sur. Ill., vol. 3. [Ety. *mulakris*, a kind of cockroach.]

anthracophila, Scudder, 1868, Geo. Sur. Ill., vol. 3, Coal Meas. [Ety. *anthrax*, coal; *philo*, I like.]

PALÆOCAMPA, Meek & Worthen, 1865, Proc. Acad. Nat. Sci. [Ety. *palaios*, ancient; *kampe*, a caterpillar.]

anthrax, Meek & Worthen, 1865, Proc. Acad. Nat. Sci., Coal Meas. [Ety. *anthrax*, coal; in allusion to the Carboniferous age of the species.]

PAOLIA, Smith, 1871, Am. Jour. Sci. & Arts, 3rd series, vol. 1. [Ety. proper name.]

vetusta, Smith, 1871, Am. Jour. Sci., 3rd series, vol. 1, Coal Meas. [Sig. ancient.]

PLATEPHEMERA, Scudder, 1867, Can. Nat. & Geo., 2nd series, vol. 3. [Ety. *platys*, flat; *ephemera*, an insect.]

antiqua, Scudder, 1867, Can. Nat. & Geo., 2d series, vol. 3, Ham. or Chemung Gr. [Sig. ancient.]

XENONEURA, Scudder, 1867, Can. Nat. & Geo., 2nd series, vol. 3. [Ety. *xenos*, new, strange; *neura*, a nerve.]

antiquorum, Scudder, 1867, Can. Nat. & Geo., 2nd series, vol. 3, Ham. or Chemung Gr. [Sig. ancient.]

SUB-KINGDOM VERTEBRATA.

First Class, - - - - - PISCES.
Second Class, - - - - - REPTILIA.

CLASS PISCES.

ORDER GANOIDEI.—Acanthaspis, Acantholepis, Acrolepis, Amblypterus, Anaclitacanthus, Aspidichthys, Asterosteus, Catopterus, Cephalaspis, Coccosteus, Cœlacanthus, Conchodus, Ctenodus, Cyrtacanthus, Dinichthys, Dipterus, Eurylepis, Heliodus, Holoptychius, Liognathus, Macropetalichthys, Mecolepis, Onychodus, Palæoniscus, Peplorhina, Platysomus, Pterichthys, Pygopterus, Rhizodus, Rhynchodus.

ORDER SELACHII.—Acondylacanthus, Agassizodus, Amacanthus, Antliodus, Apedodus, Aspidodus, Asteroptychius, Batacanthus, Bathycheilodus, Bythiacanthus, Calopodus, Carcharopsis, Cheirodus, Cholodus, Chomatodus, Cladodus, Climaxodus, Cochliodus, Compsacanthus, Ctenacanthus, Ctenopetalus, Ctenoptychius, Cymatodus, Dactylodus, Deltodus, Desmiodus, Diplodus, Drepanacanthus, Edestes, Erismacanthus, Fissodus, Gampsacanthus, Geisacanthus, Glymmatacanthus, Gyracanthus, Harpacodus, Helodus, Hybocladodus, Lambdodus, Lecracanthus, Leiodus, Leptophractus, Lisgodus, Listracanthus, Machæracanthus, Marracanthus, Mesodmodus, Orucanthus, Orodus, Orthacanthus, Peltodus, Periplectrodus, Peripristis, Petalodus, Petalorhynchus, Petrodus, Phœbodus, Physonemus, Platyodus, Pleuracanthus, Pnigeacanthus, Pœcilodus, Polyrhizodus, Pristicladodus, Pristodus, Psammodus, Psephodus, Ptyctodus, Sandalodus, Stemmatodus, Stenacanthus, Tanaodus, Thrinacodus, Trigonodus, Venustodus, Xystracanthus, Xystrodus.

ACANTHASPIS, Newberry, 1875, Ohio Pal., vol. 2. [Ety. *akantha*, spine; *aspis*, a shield.] armata, Newberry, 1875, Ohio Pal., vol. 2, Corniferous Gr. [Sig. armed.]
ACANTHOLEPIS, Newberry, 1875, Ohio Pal., vol. 2. [Ety. *akantha*, a spine; *lepis*, a scale.]
pustulosa, Newberry, 1875, Ohio Pal., vol. 2, Corniferous Gr. [Sig. covered with pustules.]
ACONDYLACANTHUS, St. John & Worthen, 1875, Geo. Sur. Ill., vol. 6. [Ety. *akondylos*, without bony knobs; *akantha*, a spine.]
æquicostatus, St. John & Worthen, 1875, Geo. Sur. Ill., vol. 6, Keokuk Gr. [Sig. equal ribbed.]

gracilis, St. John & Worthen, 1875, Geo. Sur. Ill., vol. 6, Kinderhook Gr. [Sig. slender.]
occidentalis, Newberry & Worthen, 1866, (Leptacanthus occidentalis) Geo. Sur. Ill., vol. 2, St. Louis Gr. [Sig. western.]
ACROLEPIS, Agassiz, 1843, Recherches Poiss. Foss. [Ety. *akros*, sharp; *lepis*, a scale.]
hortonensis, Dawson, 1868, Acad. Geol., Carboniferous. [Ety. proper name.]
Agassichthys, syn. for Macropetalichthys.
manni, see Macropetalichthys manni.
sullivanti, see Macropetalichthys sullivanti.

AGASSIZODUS, St. John & Worthen, 1875, Geo. Sur. Ill., vol. 6. [Ety. proper name; *odous*, tooth.]
corrugatus, Newberry & Worthen, 1870, (Orodus corrugatus) Geo. Sur. Ill., vol. 4, Coal Meas. [Sig. wrinkled.]
scitulus, St. John & Worthen, 1875, Geo. Sur. Ill., vol. 6, Coal Meas. [Sig. pretty, neat.]
variabilis, Newberry & Worthen, 1870, (Lophodus variabilis) Geo. Sur. Ill., vol. 4, Coal Meas. [Sig. variable.]
virginianus, St. John & Worthen, 1875, Geo. Sur. Ill., vol. 6, Coal Meas. [Ety. proper name.]
AMACANTHUS, St. John & Worthen, 1875, Geo. Sur. Ill., vol. 6. [Ety. *anakanthos*, destitute of prickles.]
gibbosus, Newberry & Worthen, 1866, (Homacanthus gibbosus) Geo. Sur. Ill., vol. 2, St. Louis Gr. [Sig. gibbous.]
AMBLYPTERUS, Agassiz, 1833, Recherches sur les Poissons Fossiles. [Ety. *amblys*, blunt; *pteron*, fin.]
macropterus, Agassiz, 1835, Recherch. Poiss. Foss., vol. 2, Coal Meas. [Sig. large-finned.]
ANACLITACANTHUS, St. John & Worthen, 1875, Geo. Sur. Ill., vol. 6. [Ety. *anaklitos*, leant upon; *akantha*, a spine.]
semicostatus, St. John & Worthen, 1875, Geo. Sur. Ill., vol. 6, Burlington Gr. [Sig. half-ribbed.]
ANTLIODUS, Newberry & Worthen, 1866, Geo. Sur. Ill., vol. 2. [Ety. *antlia*, a depression; *odous*, a tooth.]
cucullus, Newberry & Worthen, 1866, Geo. Sur. Ill., vol. 2, Keokuk Gr. [Sig. a covering.]
gracilis, St. John & Worthen, 1875, Geo. Sur. Ill., vol. 6, Warsaw Gr. [Sig. slender.]
minutus, Newberry & Worthen, 1866, Geo. Sur. Ill., vol. 2, Keokuk Gr. [Sig. small.]
mucronatus, Newberry & Worthen, 1866, Geo. Sur. Ill., vol. 2, St. Louis Gr. [Sig. sharp-pointed.]
parvulus, Newberry & Worthen, 1866, Geo. Sur. Ill., vol. 2, Burlington Gr. [Sig. very small.]
perovalis, St. John & Worthen, 1875, Geo. Sur. Ill., vol. 6, Warsaw Gr. [Sig. very ovate.]
robustus, Newberry & Worthen, 1866, Geo. Sur. Ill., vol. 2, Chester Gr. [Sig. robust.]
politus, Newberry & Worthen, 1866, Geo. Sur. Ill., vol. 2, Keokuk Gr. [Sig. smoothed.]
sarcululus, Newberry & Worthen, 1870, Geo. Sur. Ill., vol. 4, Burlington Gr. [Sig. a little hoe.]
similis, Newberry & Worthen, 1866, Geo. Sur. Ill., vol. 2, Keokuk Gr. [Sig. similar.]
simplex, Newberry & Worthen, 1866, Geo. Sur. Ill., vol. 2, Burlington Gr. [Sig. simple.]
sulcatus, Newberry & Worthen, 1866, Geo. Sur. Ill., vol. 2, Keokuk Gr. [Sig. furrowed.]
APEDODUS, Leidy, 1856, Jour. Acad. Nat. Sci., 2d ser., vol. 3. [Ety. *apedos*, level, smooth; *odous*, tooth.]
priscus, Leidy, 1856, Jour. Acad. Nat. Sci., 2d ser., vol. 3, Chemung Gr. [Sig. ancient.]
ASPIDICHTHYS, Newberry, 1873, Ohio Pal., vol. 1. [Ety. *aspis*, a shield; *ichthys*, a fish.]
clavatus, Newberry, 1873, Ohio Pal., vol. 1, Portage Gr. [Sig. club-shaped.]
ASPIDODUS, Newberry & Worthen, 1866, Geo. Sur. Ill., vol. 2. [Ety. *aspis*, a shield; *odous*, tooth.]
convolutus, Newberry & Worthen, 1866, Geo. Sur. Ill., vol. 2, Chester Gr. [Sig. rolled as it were together.]
crenulatus, Newberry & Worthen, 1866, Geo. Sur. Ill., vol. 2, Chester Gr. [Sig. crenulated, zigzag.]
ASTERACANTHUS, Agassiz, 1837, Recherches sur les Poissons Fossiles. [Ety. *aster*, star; *akantha*, spine.]
siderius, Leidy, see Bythiacanthus siderius.
ASTEROPTYCHIUS, Agassiz, 1843, Recherches sur les Poissons Fossiles. [Ety. *aster*, a star; *ptyche*, a wrinkle.]
bellulus, St. John & Worthen, 1875, Geo. Sur. Ill., vol. 6, Coal Meas. [Sig. pretty, neat.]
keokuk, St. John & Worthen, 1875, Geo. Sur. Ill., vol. 6, Keokuk Gr. [Ety. proper name.]
st. ludovici, St. John & Worthen, 1875, Geo. Sur. Ill., vol. 6, St. Louis Gr. [Ety. proper name.]
tenuis, St. John & Worthen, 1875, Geo. Sur. Ill., vol. 6, Chester Gr. [Sig. slender.]
triangularis, Newberry & Worthen, 1870, Geo. Sur. Ill., vol. 4, Burlington Gr. [Sig. triangular.]
vetustus, St. John & Worthen, 1875, Geo. Sur. Ill., vol. 6, Kinderhook Gr. [Sig. ancient.]
ASTEROSTEUS, Newberry, 1875, Ohio Pal., vol. 2. [Ety. *aster*, a star; *osteon*, a bone.]
stenocephalus, Newberry, 1875, Ohio Pal., vol. 2, Corniferous Gr. [Sig. narrow-headed.]
BATACANTHUS, St. John & Worthen, 1875, Geo. Sur. Ill., vol. 6. [Ety. *batos*, a prickly bush; *akantha*, a spine.]
baculiformis, St. John & Worthen, 1875, Geo. Sur. Ill., vol. 6, Keokuk Gr. [Sig. staff-shaped.]
stellatus, Newberry & Worthen, 1866, (Drepanacanthus (?) stellatus) Geo. Sur. Ill., vol. 2, Keokuk Gr. [Sig. starred.]

PISCES. 229

BATHYCHEILODUS, St. John & Worthen, 1875, Geo. Sur. Ill., vol. 6. [Ety. *bathys*, deep; *cheilos*, lip; *odous*, tooth.]
macisaacsi, St. John & Worthen, 1875, Geo. Sur. Ill., vol. 6, Devonian. [Ety. proper name.]
BYTHIACANTHUS, St. John & Worthen, 1875, Geo. Sur. Ill., vol. 6. [Ety. *bythios*, deep; *akantha*, a spine.]
siderius, Leidy, 1873, (Asteracanthus siderius) Ext. Vert. Fauna, Low. Carb. [Sig. starred.]
vanhornei, St. John & Worthen, 1875, Geo. Sur. Ill., vol. 6, St. Louis Gr. [Ety. proper name.]
CALOPODUS, St. John & Worthen, 1875, Geo. Sur. Ill., vol. 6. [Sig. a shoemaker's last.]
apicalis, St. John & Worthen, 1875, Geo. Sur. Ill., vol. 6, Coal Meas. [Sig. sharp-pointed.]
CARCHAROPSIS, Agassiz, 1843, Recherches sur les Poissons Fossiles. [Ety. *carcharopsis*, shark-like.]
wortheni, Newberry, 1866, Geo. Sur. Ill., vol. 2, Low. Carb. [Ety. proper name.]
CATOPTERUS, Redfield, 1841, Annals N. Y. Lyceum, vol. 4. [Ety. *kata*, on the lower side; *pteron*, a fin.]
macrurus, Redfield, 1841, Am. Jour. Sci., vol. 41, Carboniferous. [Sig. large-finned.]
CEPHALASPIS, Agassiz, 1835, Recherch. Pois. Foss. [Ety. *kephale*, head; *aspis*, a shield.]
dawsoni, Lankester, 1870, London Geo. Mag., Devonian. [Ety. proper name.]
CHEIRODUS, McCoy, 1848, Ann. & Mag. Nat. Hist. [Ety. *cheir*, the hand; *odous*, tooth.]
acutus, Newberry, 1857, Proc. Acad. Nat. Sci., vol. 8, Coal Meas. [Sig. sharp-pointed.]
CHOLODUS, St. John & Worthen, 1875, Geo. Sur. Ill., vol. 6. [Ety. *cholos*, defective; *odous*, tooth.]
inæqualis, St. John & Worthen, 1875, Geo. Sur. Ill., vol. 6, Coal Meas. [Sig. unequal.]
CHOMATODUS, Agassiz, 1838, Recherches sur les Poissons Fossiles. [Ety. *choma*, a pile or heap; *odous*, tooth.]
affinis, Newberry & Worthen, 1866, Geo. Sur. Ill., vol. 2, Keokuk Gr. [Sig. near to.]
angularis, see Tanaodus angularis.
arcuatus, St. John, 1870, Proc. Am. Phil. Soc., vol. 2, Coal Meas. [Sig. arched.]
chesterensis, St. John & Worthen, 1875, Geo. Sur. Ill., vol. 6, Chester Gr. [Ety. proper name.]
comptus, St. John & Worthen, 1875, Geo. Sur. Ill., vol. 6, Burlington Gr. [Sig. elegant.]
costatus, Newberry & Worthen, 1866, Geo. Sur. Ill., vol. 2, Chester Gr. [Sig. ribbed.]

cultellus, Newberry & Worthen, 1866, Geo. Sur. Ill., vol. 2, Chester Gr. [Sig. a small knife.]
elegans, Newberry & Worthen, 1866, Geo. Sur. Ill., vol. 2, Keokuk Gr. [Sig. elegant.]
gracillimus, see Tanaodus gracillimus.
inconstans, St. John & Worthen, 1875, Geo. Sur. Ill., vol. 6, St. Louis Gr. [Sig. not constant in form or size.]
incrassatus, St. John & Worthen, 1875, Geo. Sur. Ill., vol. 6, St. Louis Gr. [Sig. thickened.]
insignis, Leidy, 1856, Trans. Am. Phil. Soc., vol. 11, St. Louis Gr. [Sig. marked.]
linearis, Agassiz, 1838, (Psammodus linearis) Recherches Pois. Foss., Low. Carb. [Sig. lined.]
loriformis, Newberry & Worthen, 1866, Geo. Sur. Ill., vol. 2, Keokuk Gr. [Sig. like a thong or whip.]
molaris, Newberry & Worthen, 1866, Geo. Sur. Ill., vol. 2, Keokuk Gr. [Sig. like a molar or grinder.]
multiplicatus, see Tanaodus multiplicatus.
obscurus, see Tanaodus obscurus.
parallelus, St. John & Worthen, 1875, Geo. Sur. Ill., vol. 6, Warsaw Gr. [Sig. parallel.]
pusillus, Newberry & Worthen, 1866, Geo. Sur. Ill., vol. 2, Keokuk Gr. [Sig. very small.]
varsoviensis, St. John & Worthen, 1875, Geo. Sur. Ill., vol. 6, Warsaw Gr. [Ety. proper name.]
venustus, Leidy, see Venustodus leidyi, where the specific name is made to designate the genus, and the author the specific name, contrary to the rules of nomenclature; also see *Venustodus venustus*.
CLADODUS, Agassiz, 1843, Recherches sur les Poissons Fossiles. [Ety. *klados*, a twig; *odous*, tooth.]
acuminatus, Newberry, 1857, Proc. Acad. Nat. Sci. Phil., vol. 8, Low. Carb. [Sig. sharp-pointed; from the extreme sharpness of the central and lateral denticles.]
alternatus, St. John & Worthen, 1875, Geo. Sur. Ill., vol. 6, Kinderhook Gr. [Sig. alternating.]
angulatus, Newberry & Worthen, 1866, Geo. Sur. Ill., vol. 2, Keokuk Gr. [Sig. angular.]
bellifer, St. John & Worthen, 1875, Geo. Sur. Ill., vol. 6, Burlington Gr. [Sig. war-like, martial.]
carinatus, St. John & Worthen, 1875, Geo. Sur. Ill., vol. 6, Coal Meas. [Sig. keeled.]
concinnus, Newberry, 1875, Ohio Pal., vol. 2, Portage Gr. [Sig. handsome.]
costatus, Newberry & Worthen, 1866, Geo. Sur. Ill., vol. 2, Chester Gr. [Sig. ribbed, having prominent ridges.]

deflexus, Newberry & Worthen, 1870, Geo. Sur. Ill., vol. 4, Burlington Gr. [Sig. downward, turned aside.]
eccentricus, St. John & Worthen, 1875, Geo. Sur. Ill., vol. 6, St. Louis Gr. [Sig. eccentric.]
elegans, Newberry & Worthen, 1870, Geo. Sur. Ill., vol. 6, St. Louis Gr. [Sig. elegant.]
euglypheus, St. John & Worthen, 1875, Geo. Sur. Ill., vol. 6, St. Louis Gr. [Sig. well carved, distinctly marked.]
exiguus, St. John & Worthen, 1875, Geo. Sur. Ill., vol. 6, Kinderhook Gr. [Sig. small.]
exilis, St. John & Worthen, 1875, Geo. Sur. Ill., vol. 6, Kinderhook Gr. [Sig. small, slender.]
ferox, Newberry & Worthen, 1866, Geo. Sur. Ill., vol. 2, St. Louis Gr. [Sig. fierce, sharp-pointed.]
fulleri, St. John & Worthen, 1875, Geo. Sur. Ill., vol. 6, Coal Meas. [Ety. proper name.]
gomphoides, St. John & Worthen, 1875, Geo. Sur. Ill., vol. 6, Burlington Gr. [Sig. like a club.]
gracilis, Newberry & Worthen, 1866, Geo. Sur. Ill., vol. 2, Coal Meas. [Sig. slender.]
grandis, Newberry & Worthen, 1866, Geo. Sur. Ill., vol. 2, Chester Gr. [Sig. great.]
hertzeri, Newberry, 1875, Ohio Pal., vol. 2, Portage Gr. [Ety. proper name.]
intercostatus, St. John & Worthen, 1875, Geo. Sur. Ill., vol. 6, Burlington Gr. [Sig. intercostated, ridge within ridge.]
ischypus, Newberry & Worthen, 1870, Geo. Sur. Ill., vol. 4, St. Louis Gr. [Sig. strong-footed.]
lamnoides, Newberry & Worthen, 1866, Geo. Sur. Ill., vol. 2, Kinderhook Gr. [Ety. from resemblance to the teeth of the living *Lamna*.]
magnificus, Tuomey, 1858, 2d Rep. Geo. Ala., Low. Carb. [Sig. magnificent.]
micropus, Newberry & Worthen, 1866, Geo. Sur. Ill., vol. 2, Keokuk Gr. [Sig. small footed.]
mortifer, Newberry & Worthen, 1866, Geo. Sur. Ill., vol. 2, Coal Meas. [Sig. death dealing, deadly.]
newmani, Tuomey, 1856, Geo. Alabama, Low. Carb. [Ety. proper name.]
pandatus, St. John & Worthen, 1875, Geo. Sur. Ill., vol. 6, Coal Meas. [Sig. bent downward.]
parvulus, Newberry, 1875, Ohio Pal., vol. 2, Portage Gr. [Sig. very small.]
pattersoni, Newberry, 1875, Ohio Pal., vol. 2, Waverly Gr. [Ety. proper name.]
politus, Newberry & Worthen, 1875, Geo. Sur. Ill., vol. 2, Chester Gr. [Sig. smoothed.]
prænuntius, St. John & Worthen, 1875, Geo. Sur. Ill., vol. 6, Burlington Gr. [Sig. a foreteller, harbinger.]

raricostatus, St. John & Worthen, 1875, Geo. Sur. Ill., vol. 6, Keokuk Gr. [Sig. having few ridges; from a few diverging thread-like striæ on the outer face of the principal cone.]
robustus, Newberry & Worthen, 1866, Geo. Sur. Ill., vol. 2, Keokuk Gr. [Sig. robust.]
romingeri, Newberry, 1875, Ohio Pal., vol. 2, Waverly Gr. [Ety. proper name.]
spinosus, Newberry & Worthen, 1866, Geo. Sur. Ill., vol. 2, St. Louis Gr. [Sig. full of spines.]
springeri, St. John & Worthen, 1875, Geo. Sur. Ill., vol. 6, Kinderhook Gr. [Ety. proper name.]
stenopus, Newberry & Worthen, 1866, Geo. Sur. Ill., vol. 2, St. Louis Gr. [Sig. narrow-footed.]
subulatus, Newberry, 1875, Ohio Pal., vol. 2, St. Louis Gr. [Sig. awl-shaped, subulate.]
succinctus, St. John & Worthen, 1875, Geo. Sur. Ill., vol. 6, Kinderhook Gr. [Sig. girded, contracted, short.]
turritus, Newberry & Worthen, 1866, Geo. Sur. Ill., vol. 2, Keokuk Gr. [Sig. fortified, turreted.]
vanhornei, St. John & Worthen, 1875, Geo. Sur. Ill., vol. 6, St. Louis Gr. [Ety. proper name.]
wachsmuthi, St. John & Worthen, 1875, Geo. Sur. Ill., vol. 6, Kinderhook Gr. [Ety. proper name.]
zygopus, Newberry & Worthen, 1866, Geo. Sur. Ill., vol. 2, Chester Gr. [Sig. with joined feet.]

CLIMAXODUS, McCoy, 1848, Ann. & Mag. Nat. Hist. [Ety. *klimax*, a ladder; *odous*, a tooth.]
brevis, Newberry, 1857, Proc. Acad. Nat. Sci., vol. 8, Coal Meas. [Sig. short.]

COCCOSTEUS, Agassiz, 1843, Recherch. Pois. Foss., vol. 2. [Ety. *kokkos*, a berry; *osteon*, a bone.]
occidentalis, Newberry, 1874, Ohio Pal., vol. 2, Corniferous Gr. [Sig. western.]

COCHLIODUS, Agassiz, 1838, Recherches sur les Poissons Fossiles. [Ety. *kochlias*, a cockle; *odous*, a tooth.]
costatus, Newberry & Worthen, 1870, Geo. Sur. Ill., vol. 4, Burlington Gr. [Sig. ribbed.]
crassus, Newberry & Worthen, 1866, Geo. Sur. Ill., vol. 2, Keokuk Gr. [Sig. thick.]
nobilis, Newberry & Worthen, 1866, Geo. Sur. Ill., vol. 2, Keokuk Gr. [Sig. noble.]

CŒLACANTHUS, Agassiz, 1843, Recherches sur les Poissons Fossiles. [Ety. *koilos*, hollow; *akantha*, spine.]
elegans, Newberry, 1857, Proc. Acad. Nat. Sci. Phil., vol. 8, Coal Meas. [Sig. elegant.]
ornatus, Newberry, 1857, Proc. Acad. Nat. Sci. Phil., vol. 8, Coal Meas. [Sig. adorned.]

robustus, Newberry, 1857, Proc. Acad. Nat. Sci. Phil., vol. 8, Coal Meas. [Sig. robust.]
COMPSACANTHUS, Newberry, 1857, Proc. Acad. Nat. Sci., vol. 8. [Ety. *compsos*, elegant; *akantha*, a spine.]
lævis, Newberry, 1857, Proc. Acad. Nat. Sci. Phil., vol. 8, Coal Meas. [Sig. smooth.]
Conchiopsis, syn. for Coelacanthus.
anguliferus, syn. for Coelacanthus elegans.
exanthematicus, syn. for Peplorhina anthracina.
filiferus, syn. for Coelacanthus elegans.
CONCHODUS, McCoy, 1848, Ann. & Mag. Nat. Hist., 2d ser., vol. 2. [Ety. *conchos*, a shell; *odous*, a tooth.]
plicatus, Dawson, 1868, Acad. Geol., Coal Meas. [Sig. plicated.]
CTENACANTHUS, Agassiz, 1837, Recherches sur les Poissons Fossiles. [Ety. *ktenos*, a comb; *akantha*, a spine.]
angulatus, Newberry & Worthen, 1866, Geo. Sur. Ill., vol. 2, Chester Gr. [Sig. angular.]
burlingtonensis, St. John & Worthen, 1875, Geo. Sur. Ill., vol. 6, Burlington Gr. [Ety. proper name.]
costatus, Newberry & Worthen, 1866, Geo. Sur. Ill., vol. 2, St. Louis, Gr. [Sig. ribbed.]
elegans, Tuomey, 1858, Geo. Ala., Low. Carb. [Sig. elegant.]
excavatus, St. John & Worthen, 1866, Geo. Sur. Ill., vol. 6, Keokuk Gr. [Sig. excavated.]
formosus, Newberry, 1873, Ohio Pal., vol. 1, Waverly Gr. [Sig. beautiful.]
furcicarinatus, Newberry, 1875, Ohio Pal., vol. 2, Waverly Gr. [Sig. forked and keeled.]
gemmatus, St. John & Worthen, 1875, Geo. Sur. Ill., vol. 6, St. Louis Gr. [Sig. having buds.]
gracillimus, Newberry & Worthen, 1866, Geo. Sur. Ill., vol. 2, St. Louis Gr. [Sig. very slender.]
grado-costatus, St. John & Worthen, 1875, Geo. Sur. Ill., vol. 6, Burlington Gr. [Sig. having steps and ribs.]
keokuk, St. John & Worthen, 1875, Geo. Sur. Ill., vol. 6, Keokuk Gr. [Ety. proper name.]
marshi, Newberry, 1873, Ohio Pal., vol. 1, Coal Meas. [Ety. proper name.]
mayi, Newberry & Worthen, 1866, Ohio Pal., vol. 1, Burlington Gr. [Ety. proper name.]
parvulus, Newberry, 1875, Ohio Pal., vol. 2, Cleveland shale. [Sig. very small.]
pugiunculus, St. John & Worthen, 1875, Geo. Sur. Ill., vol. 6, St. Louis Gr. [Sig. a small dagger.]
sculptus, St. John & Worthen, 1875, Geo. Sur. Ill., vol. 6, Kinderhook Gr. [Sig. engraved.]

similis, St. John & Worthen, 1875, Geo. Sur. Ill., vol. 6, Chester Gr. [Sig. similar.]
speciosus, St. John & Worthen, 1875, Geo. Sur. Ill., vol. 6, Kinderhook Gr. [Sig. beautiful.]
spectabilis, St. John & Worthen, 1875, Geo. Sur. Ill., vol. 6, Kinderhook Gr. [Sig. remarkable.]
triangularis, Newberry, 1873, Ohio Pal., vol. 1, Waverly Gr. [Sig. triangular.]
varians, St. John & Worthen, 1875, Geo. Sur. Ill., vol. 6, Kinderhook Gr. [Sig. variable.]
vetustus, Newberry, 1873, Ohio Pal., vol. 1, Portage Gr. [Sig. ancient.]
CTENODUS, Agassiz, 1838, Recherches sur les Poissons Fossiles. [Ety. *ktenos*, a comb; *odous*, a tooth.]
ohioensis, Cope, 1874, Proc. Acad. Nat. Sci. Phil., Coal Meas. [Ety. proper name.]
reticulatus, Newberry, 1875, Ohio Pal., vol. 2, Coal Meas. [Sig. like network.]
serratus, Newberry, 1875, Ohio Pal., vol. 2, Coal Meas. [Sig. saw-edged, serrated.]
CTENOPETALUS, Agassiz, 1869, Catal. Foss. Fish, Collection of Earl of Enniskillen. [Ety. *ktenos*, a comb; *petalos*, broad, full grown.]
bellulus, St. John & Worthen, 1875, Geo. Sur. Ill., vol. 6, St. Louis Gr. [Sig. very beautiful.]
limatulus, St. John & Worthen, 1875, Geo. Sur. Ill., vol. 6, Chester Gr. [Sig. like a little file.]
medius, St. John & Worthen, 1875, Geo. Sur. Ill., vol. 6, Chester Gr. [Sig. intermediate.]
occidentalis, St. John & Worthen, 1875, Geo. Sur. Ill., vol. 6, Coal Meas. [Sig. western.]
vinosus, St. John & Worthen, 1875, Geo. Sur. Ill., vol. 6, Keokuk Gr. [Sig. full of wine.]
CTENOPTYCHIUS, Agassiz, 1838, Recherches sur les Poissons Fossiles. [Ety. *ktenos*, a comb; *ptyche*, a wrinkle.]
cristatus, Dawson, 1868, Acad. Geol., Coal Meas. [Sig. tufted.]
pertenuis, St. John & Worthen, 1875, Geo. Sur. Ill., vol. 6, Chester Gr. [Sig. very slender.]
semicircularis, Newberry & Worthen, 1866, Geo. Sur. Ill., vol. 2, Coal Meas. [Sig. half-circular.]
stevensoni, St. John & Worthen, 1875, Geo. Sur. Ill., vol. 6, Coal Meas. [Ety. proper name.]
CYRTACANTHUS, Newberry, 1873, Ohio Pal., vol. 1. [Ety. *kurtos*, curved; *akantha*, a spine.]
dentatus, Newberry, 1873, Ohio Pal., vol. 1, Corniferous Gr. [Sig. toothed.]

CYMATODUS, Newberry & Worthen, 1870, Geo. Sur. Ill., vol. 4. [Ety. *cymatos*, wavy; *odous*, tooth.]
  oblongus, Newberry & Worthen, 1870, Geo. Sur. Ill., vol. 4, Up. Coal Meas. [Sig. oblong.]
DACTYLODUS, Newberry & Worthen, 1866, Geo. Sur. Ill., vol. 2. [Ety. *daktylos*, a finger; *odous*, a tooth.]
  concavus, St. John & Worthen, 1875, Geo. Sur. Ill., vol. 6, St. Louis Gr. [Sig. concave.]
  excavatus, St. John & Worthen, 1875, Geo. Sur. Ill., vol. 6, Chester Gr. [Sig. hollowed out.]
  inflexus, Newberry & Worthen, 1866, Geo. Sur. Ill., vol. 2, Chester Gr. [Sig. bent, curved.]
  lobatus, Newberry & Worthen, 1866, Geo. Sur. Ill., vol. 2, St. Louis Gr. [Sig. lobate.]
  minimus, St. John & Worthen, 1875, Geo. Sur. Ill., vol. 6, St. Louis Gr. [Sig. very small.]
  princeps, Newberry & Worthen, 1866, Geo. Sur. Ill., vol. 2, St. Louis Gr. [Sig. chief.]
DELTODUS, Newberry & Worthen, 1866, Geo. Sur. Ill., vol. 2. [Ety. *delta*, triangular; *odous*, tooth.]
  alatus, Newberry & Worthen, 1870, Geo. Sur. Ill., vol. 4, Keokuk Gr. [Sig. winged.]
  angularis, Newberry & Worthen, 1866, Geo. Sur. Ill., vol. 2, Coal Meas. [Sig. angular.]
  angustus, Newberry & Worthen, 1870, Geo. Sur. Ill., vol. 4, Chester Gr. [Sig. narrow.]
  cingulatus, Newberry & Worthen, 1866, Geo. Sur. Ill., vol. 2, Chester Gr. [Sig. encircled with lines.]
  complanatus, Newberry & Worthen, 1866, Geo. Sur. Ill., vol. 2, Burlington Gr. [Sig. smoothed.]
  fasciatus, Newberry & Worthen, 1870, Geo. Sur. Ill., vol. 4, Keokuk Gr. [Sig. banded, girded.]
  grandis, Newberry & Worthen, 1866, Geo. Sur. Ill., vol. 2, Keokuk Gr. [Sig. grand.]
  littoni, Newberry & Worthen, 1870, Geo. Sur. Ill., vol. 4, Low. Carb. [Ety. proper name.]
  rhomboideus, Newberry & Worthen, 1866, Geo. Sur. Ill., vol. 2, St. Louis Gr. [Sig. lozenge or diamond-shaped.]
  spatulatus, Newberry & Worthen, 1866, Geo. Sur. Ill., vol. 2, Burlington Gr. [Sig. spatulate, blade-shaped.]
  stellatus, Newberry & Worthen, 1866, Geo. Sur. Ill., vol. 2, Keokuk Gr. [Sig. starred.]
  undulatus, Newberry & Worthen, 1866, Geo. Sur. Ill., vol. 2, Keokuk Gr. [Sig. undulated, waved.]

DESMIODUS, St. John & Worthen, 1875, Geo. Sur. Ill., vol. 6. [Ety. *desmos*, a ligament; *odous*, tooth.] This name was applied to a genus of bats, in 1826, by Prinz, Neu. Wied. in Beitrage zur Naturg. Brasilieus.
  costelliferus, St. John & Worthen, 1875, Geo. Sur. Ill., vol. 6, St. Louis Gr. [Sig. bearing faint ribs.]
  flabellum, St. John & Worthen, 1875, Geo. Sur. Ill., vol. 6, Keokuk Gr. [Sig. a little fan.]
  ligoniformis, St. John & Worthen, 1875, Geo. Sur. Ill., vol. 6, Keokuk Gr. [Sig. like a mattock.]
  minusculus, Newberry & Worthen, 1866, (Orodus minusculus) Geo. Rep. Ill., vol. 2, Keokuk Gr. [Sig. rather small.]
  tumidus, St. John & Worthen, 1875, Geo. Sur. Ill., vol. 6, St. Louis Gr. [Sig. swollen out.]
DINICHTHYS, Newberry, 1873, Ohio Pal., vol. 1. [Ety. *deinos*, terrible; *ichthys*, a fish.]
  hertzeri, Newberry, 1873, Ohio Pal., vol. 1, Corniferous Gr. [Ety. proper name.]
  terrelli, Newberry, 1873-'75, Ohio Pal., vol. 1 and 2, Corniferous Gr. [Ety. proper name.]
DIPLODUS, Agassiz, 1843, Recherches sur les Poissons Fossiles. [Ety. *diploos*, double; *odous*, a tooth.] This name was used by Rafinesque for a genus of *Sparidæ* in 1810, Indice d'Littologia Siciliana.
  acinaces, Dawson, 1860, Acad. Geol., Coal Meas. [Sig. a short saber.]
  compressus, Newberry, 1857, Proc. Acad. Nat. Sci. Phil., vol. 8, Coal Meas. [Sig. compressed.]
  duplicatus, see Thrinacodus duplicatus.
  gracilis, Newberry, 1857, Proc. Acad. Nat. Sci. Phil., vol. 8, Coal Meas. [Sig. slender.]
  incurvus, see Thrinacodus incurvus.
  latus, Newberry, 1857, Proc. Acad. Nat. Sci. Phil., vol. 8, Coal Meas. [Sig. broad.]
  penetrans, Dawson, 1860, Acad. Geol., Coal Meas. [Sig. piercing, penetrating.]
DIPTERUS, Sedgwick & Murchison, 1828, Geo. Trans., 2d series, vol. 3. [Ety. *dipteros*, two winged.] *Diptera* is an order of insects established by Linnæus.
  sherwoodi, Newberry, 1875, Ohio Pal., vol. 2, Catskill Gr. [Ety. proper name.]
DREPANACANTHUS, Newberry & Worthen, 1866, Geo. Sur. Ill., vol. 2. [Ety. *drepane*, a sickle; *akantha*, a spine.]
  anceps, see Xystracanthus anceps.
  geminatus, Newberry & Worthen, 1866, Geo. Sur. Ill., vol. 2, Keokuk Gr. [Sig. budded or having knots.]
  reversus, St. John & Worthen, 1875, Geo. Sur. Ill., vol. 6, St. Louis Gr. [Sig. reversed.]
  stellatus, see Batacanthus stellatus.

EDESTES, Leidy, 1856, Jour. Acad. Nat. Sci., 2nd series, vol. 3. [Ety. *edestes*, a devourer.]
heinrichsi, Newberry & Worthen, 1870, Geo. Sur. Ill., vol. 4, Coal Meas. [Ety. proper name.]
minor, Newberry, 1866, Geo. Sur. Ill., vol. 2, Coal Meas. [Sig. less.]
vorax, Leidy, 1856, Jour. Acad. Nat. Sci. Phil., vol. 3, 2d series, Coal Meas. [Sig. ravenous, voracious.]
Elonichthys peltigerus, see Palæoniscus peltigerus.
ERISMACANTHUS, McCoy, 1848, Ann. & Mag. Nat. Hist., 2d series, vol. 2. [Ety. *ereisma*, a prop or stay; *akantha*, a spine.]
maccoyanus, St. John & Worthen, 1875, Geo. Sur. Ill., vol. 6, St. Louis Gr. [Ety. proper name.]
EURYLEPIS, Newberry, 1856, Proc. Acad. Nat. Sci. Phil. [Ety. *eurys*, broad; *lepis*, scale.]
corrugata, Newberry, 1856, Proc. Acad. Nat. Sci. Phil., Coal Meas. [Sig. corrugated.]
granulata, Newberry, 1856, Proc. Acad. Nat. Sci. Phil., Coal Meas. [Sig. granular.]
insculpta, Newberry, 1856, Proc. Acad. Nat. Sci. Phil., Coal Meas. [Sig. carved, engraved.]
lineata, Newberry, 1856, Proc. Acad. Nat. Sci. Phil., Coal Meas. [Sig. lined.]
minima, Newberry, 1873, Ohio Pal., vol. 1, Coal Meas. [Sig. very small.]
ornatissima, Newberry, 1856, Proc. Acad. Nat. Sci. Phil., Coal Meas. [Sig. highly ornamented.]
ovoidea, Newberry, 1856, Proc. Acad. Nat. Sci. Phil., Coal Meas. [Sig. egg-shaped.]
striolata, Newberry, 1873, Ohio Pal., vol. 1, Coal Meas. [Sig. very minutely striated.]
tuberculata, Newberry, 1856, Proc. Acad. Nat. Sci., Coal Meas. [Sig. tuberculated.]
FISSODUS, St. John & Worthen, 1875, Geo. Sur. Ill., vol. 6. [Ety. *fissus*, split; *odous*, tooth.]
bifidus, St. John & Worthen, 1875, Geo. Sur. Ill., vol. 6, Chester Gr. [Sig. divided into two parts.]
tricuspidatus, St. John & Worthen, 1875, Geo. Sur. Ill., vol. 6, Chester Gr. [Sig. three-pointed.]
GAMPSACANTHUS, St. John & Worthen, 1875, Geo. Sur. Ill., vol. 6. [Ety. *gampsos*, curved; *akantha*, a spine.]
(?) latus, St. John & Worthen, 1875, Geo. Sur. Ill., vol. 6, Keokuk Gr. [Sig. wide.]
squamosus, St. John & Worthen, 1875, Geo. Sur. Ill., vol. 6, St. Louis Gr. [Sig. scaly.]

typus, St. John & Worthen, 1875, Geo. Sur. Ill., vol. 6, St. Louis Gr. [Ety. the type of the genus.]
GEISACANTHUS, St. John & Worthen, 1875, Geo. Sur. Ill., vol. 6. [Ety. *geison*, a border; *akantha*, a spine.]
bullatus, St. John & Worthen, 1875, Geo. Sur. Ill., vol. 6, Chester Gr. [Sig. bossed, studded.]
stellatus, St. John & Worthen, 1875, Geo. Sur. Ill., vol. 6, St. Louis Gr. [Sig. starred.]
GLYMMATACANTHUS, St. John & Worthen, 1875, Geo. Sur. Ill., vol. 6. [Ety. *glymmatos*, engraved; *akantha*, a spine.]
irishi, St. John & Worthen, 1875, Geo. Sur. Ill., vol. 6, Kinderhook Gr. [Ety. proper name.]
GYRACANTHUS, Agassiz, 1837, Recherches sur les Poissons Fossiles. [Ety. *gyros*, a circle; *akantha*, a spine.]
alleni, Newberry, 1873, Ohio Pal., vol. 1, Cuyahoga shale. [Ety. proper name.]
compressus, Newberry, 1873, Ohio Pal., vol. 1, Cuyahoga shale. [Sig. compressed.]
duplicatus, Dawson, 1868, Acad. Geol., Coal Meas. [Sig. doubled.]
magnificus, Dawson, 1868, Acad. Geol., Low. Carb. [Sig. magnificent, splendid.]
HARPACODUS, Agassiz, 1869, Catal. Foss. Fish, Collection of Earl of Enniskillen. [Ety. *harpe*, a hook; *odous*, a tooth.]
compactus, St. John & Worthen, 1875, Geo. Sur. Ill., vol. 6, Chester Gr. [Sig. compact.]
occidentalis, St. John & Worthen, 1875, Geo. Sur. Ill., vol. 6, St. Louis Gr. [Sig. western.]
HELIODUS, Newberry, 1875, Ohio Pal., vol. 2. [Ety. *helios*, the sun; *odous*, a tooth.]
lesleyi, Newberry, 1875, Ohio Pal., vol. 2, Chemung Gr. [Ety. proper name.]
HELODUS, Agassiz, 1838, Recherches sur les Poissons Fossiles. [Ety. *helos*, a nail or rudder; *odous*, a tooth.]
angulatus, Newberry & Worthen, 1866, Geo. Sur. Ill., vol. 6, Burlington Gr. [Sig. angular.]
biformis, Newberry & Worthen, 1866, Geo. Sur. Ill., vol. 2, Kinderhook Gr. [Sig. double-formed.]
carbonarius, Newberry & Worthen, 1866, Geo. Sur. Ill., vol. 2, Coal Meas. [Sig. pertaining to coal.]
compressus, Newberry & Worthen, 1870, Geo. Sur. Ill., vol. 4, Burlington Gr. [Sig. compressed.]
compressus, see Hybocladodus compressus.
coniculus, Newberry & Worthen, 1866, Geo. Sur. Ill., vol. 2, Burlington Gr. [Sig. a little cone.]
consolidatus, Newberry & Worthen, 1866, Geo. Sur. Ill., vol. 2, Keokuk Gr. [Sig. consolidated.]

crenulatus, Newberry & Worthen, 1866, Geo. Sur. Ill., vol. 2, Keokuk Gr. [Sig. crenulated.]
denshumani, Newberry & Worthen, 1866, Geo. Sur. Ill., vol. 2, Keokuk Gr. [Ety. proper name.]
denticulatus, Newberry & Worthen, 1866, Geo. Sur. Ill., vol. 2, Keokuk Gr. [Sig. denticulated.]
elytra, Newberry & Worthen, 1866, Geo. Sur. Ill., vol. 2, Keokuk Gr. [Ety. *elytron*, the wing covering as the *elytra* of the beetle.]
gibbosus, Newberry & Worthen, 1866, Geo. Sur. Ill., vol. 2, Keokuk Gr. [Sig. gibbous.]
limax, Newberry & Worthen, 1866, Geo. Sur. Ill., vol. 2, Burlington Gr. [Sig. a slug, dew-snail.]
nobilis, Newberry & Worthen, 1866, Geo. Sur. Ill., vol. 2, Keokuk Gr. [Sig. noble, excellent.]
placenta, Newberry & Worthen, 1866, Geo. Sur. Ill., vol. 2, Kinderhook Gr. [Sig. a cake.]
politus, Newberry & Worthen, 1866, Geo. Sur. Ill., vol. 2, Keokuk Gr. [Sig. smooth.]
rugosus, Newberry & Worthen, 1866, Geo. Sur. Ill., vol. 4, Coal Meas. [Sig. wrinkled.]
sulcatus, Newberry & Worthen, 1866, Geo. Sur. Ill., vol. 2, Keokuk Gr. [Sig. furrowed.]
undulatus, Newberry & Worthen, 1866, Geo. Sur. Ill., vol. 2, Keokuk Gr. [Sig. waved.]
HOLOPTYCHIUS, Agassiz, 1843, Recherches sur les Poissons Fossiles. [Ety. *holos*, entire; *ptyche*, a wrinkle.]
americanus, Leidy, 1856, Jour. Acad. Nat. Sci., 2nd series, vol. 3, Catskill Gr. [Ety. proper name.]
nobilissimus, Agassiz, as identified by Hall, 1843, Geo. Rep. 4th Dist. N. Y., is described as *H. americanus*.
taylori, Hall, 1843, (Sauripteris taylori) Geo. Rep. 4th Dist. N. Y., Catskill Gr. [Ety. proper name.]
Homacanthus, Agassiz, 1845, Mon. Pois. Foss. Syst., Devonian.
gibbosus, see Amacanthus gibbosus.
rectus, see Marracanthus rectus.
HYBOCLADODUS, St. John & Worthen, 1875, Geo. Sur. Ill., vol. 6. [Ety. *hybos*, a hump; *Cladodus*, a genus of fossil fish.]
compressus, Newberry & Worthen, 1866, (Helodus compressus) Geo. Sur. Ill., vol. 2, Burlington Gr. [Sig. compressed.]
intermedius, St. John & Worthen, 1875, Geo. Sur. Ill., vol. 6, Keokuk Gr. [Sig. intermediate.]
nitidus, St. John & Worthen, 1875, Geo. Sur. Ill., vol. 6, Chester Gr. [Sig. neat, pretty.]

plicatilus, St. John & Worthen, 1875, Geo. Sur. Ill., vol. 6, Burlington Gr. [Sig. in small folds.]
tenuicostatus, St. John & Worthen, 1875, Geo. Sur. Ill., vol. 6, Keokuk Gr. [Sig. finely lined.]
LAMBDODUS, St. John & Worthen, 1875, Geo. Sur. Ill., vol. 6. [Ety. *Lambda*, a Greek letter; *odous*, a tooth.]
calceolus, St. John & Worthen, 1875, Geo. Sur. Ill., vol. 6, Burlington Gr. [Sig. a sandal, a slipper.]
calceolus var. robustus, St. John & Worthen, 1866, Geo. Sur. Ill., vol. 6, Keokuk Gr. [Sig. robust.]
costatus, St. John & Worthen, 1875, Geo. Sur. Ill., vol. 6, Burlington Gr. [Sig. ribbed.]
hamulus, St. John & Worthen, 1875, Geo. Sur. Ill., vol. 6, Chester Gr. [Sig. a small hook.]
reflexus, St. John & Worthen, 1875, Geo. Sur. Ill., vol. 6, Chester Gr. [Sig. turned back.]
transversus, St. John & Worthen, 1875, Geo. Sur. Ill., vol. 6, St. Louis Gr. [Sig. crosswise.]
LECRACANTHUS, St. John & Worthen, 1875, Geo. Sur. Ill., vol. 6. [Ety. *lekroi*, the antlers of a stag; *akantha*, a spine.]
unguiculus, St. John & Worthen, 1875, Geo. Sur. Ill., vol. 6, St. Louis Gr. [Sig. furnished with claw-like processes.]
LEIODUS, St. John & Worthen, 1875, Geo. Sur. Ill., vol. 6. [Ety. *leios*, smooth; *odous*, tooth.]
calcaratus, St. John & Worthen, 1875, Geo. Sur. Ill., vol. 6, Burlington Gr. [Sig. spurred, spur-shaped.]
calcaratus var. grossipunctatus, St. John & Worthen, 1875, Geo. Sur. Ill., vol. 6, Keokuk Gr. [Sig. thick-punctured.]
Leptacanthus, Agassiz, 1837, Poiss. Foss., vol. 3. [Ety. *leptos*, slender; *akantha*, spine.]
occidentalis, see Acondylacanthus occidentalis.
LEPTOPHRACTUS, Cope, 1873, Proc. Acad. Nat. Sci. Phil. [Ety. *leptos*, thin; *phraktos*, walled.]
obsoletus, Cope, 1873, Proc. Acad. Nat. Sci. Phil., Coal Meas. [Sig. obsolete.]
LIOGNATHUS, Newberry, 1873, Ohio Pal., vol. 1. [Ety. *lis*, smooth; *gnathos*, the jaw.]
spatulatus, Newberry, 1873, Ohio Pal., vol. 1, Corniferous Gr. [Sig. spatulate, blade-shaped.]
LISGODUS, St. John & Worthen, 1875, Geo. Sur. Ill., vol. 6. [Ety. *lisgos*, a spade; *odous*, a tooth.]
curtus, St. John & Worthen, 1875, Geo. Sur. Ill., vol. 6, Burlington Gr. [Sig. short.]
selluliformis, St. John & Worthen, 1875, Geo. Sur. Ill., vol. 6, St. Louis Gr. [Sig. like a little seat or stool.]
serratus, St. John & Worthen, 1875, Geo. Sur. Ill., vol. 6, Burlington Gr. [Sig. saw-edged.]

LISTRACANTHUS, Newberry & Worthen, 1870, Geo. Sur. Ill., vol. 4. [Ety. *listron*, a shovel; *akantha*, a spine.]
hildrethi, Newberry, 1875, Ohio Pal., vol. 2, Coal Meas. [Ety. proper name.]
hystrix, Newberry & Worthen, 1870, Geo. Sur. Ill., vol. 4, Burlington Gr. [Sig. covered or beset with spines.]
Lophodus, Newberry & Worthen, 1870, Geo. Sur. Ill., vol. 4. This name was preoccupied by Romanowsky in 1864.
variabilis, see Agassizodus variabilis.
MACHÆRACANTHUS, Newberry, 1853, Annals of Science, vol. 1, and 1857, Bull. Nat. Inst. [Ety. *machaira*, a sabre; *akantha*, a spine.]
major, Newberry, 1857, Bull. Nat. Inst., Corniferous Gr. [Sig. greater.]
peracutus, Newberry, 1857, Bull. Nat. Inst., Corniferous Gr. [Sig. very sharp-pointed.]
sulcatus, Newberry, 1857, Bull. Nat. Inst., Corniferous Gr. [Sig. furrowed.]
MACROPETALICHTHYS, Norwood & Owen, 1846, Am. Jour. Sci., 2d ser., vol. 1. [Ety. *makros*, large; *petalos*, expanded or spread out; *ichthys*, a fish.]
manni, Newberry, 1853, (Agassichthys manni) Annals of Science, vol. 1, and 1857, Bull. Nat. Inst., Corniferous Gr. [Ety. proper name.]
rapheidolabis, Norwood & Owen, 1846, Am. Jour. Sci., 2d ser., vol. 1, Up. Held. Gr. [Ety. *raphis*, a needle; *eidos*, form; *labis*, forceps.]
sullivanti, Newberry, 1857, (Agassichthys sullivanti) Bull. Nat. Inst., Corniferous Gr. [Ety. proper name.]
MARRACANTHUS, St. John & Worthen, 1875, Geo. Sur. Ill., vol. 6. [Ety. *marron*, a spade; *akantha*, a spine.]
rectus, Newberry & Worthen, 1866, (Homacanthus (?) rectus) Geo. Sur. Ill., vol. 2, St. Louis Gr. [Sig. straight.]
MECOLEPIS, Newberry, 1857, Proc. Acad. Nat. Sci., vol. 8. [Ety. *mekos*, large; *lepis*, a scale.]
corrugata, Newberry, 1857, Proc. Acad. Nat. Sci., vol. 8, Coal Meas. [Sig. wrinkled.]
granulata, Newberry, 1857, Proc. Acad. Nat. Sci., vol. 8, Coal Meas. [Sig. granulated.]
insculpta, Newberry, 1857, Proc. Acad. Nat. Sci., vol. 8, Coal Meas. [Sig. engraved.]
lineata, Newberry, 1857, Proc. Acad. Nat. Sci., vol. 8, Coal Meas. [Sig. lined.]
ornatissima, Newberry, 1857, Proc. Acad. Nat. Sci., vol. 8, Coal Meas. [Sig. highly ornamented.]
ovoidea, Newberry, 1857, Proc. Acad. Nat. Sci., vol. 8, Coal Meas. [Sig. ovoidal.]
serrata, Newberry, 1857, Proc. Acad. Nat. Sci., vol. 8, Coal Meas. [Sig. serrated.]

tuberculata, Newberry, 1857, Proc. Acad. Nat. Sci., vol. 8, Coal Meas. [Sig. tuberculated.]
MESODMODUS, St. John & Worthen, 1875, Geo. Sur. Ill., vol. 6. [Ety. *mesodme*, something between; *odous*, a tooth.]
explanatus, St. John & Worthen, 1875, Geo. Sur. Ill., vol. 6, Kinderhook Gr. [Sig. spread out.]
exculptus, St. John & Worthen, 1875, Geo. Sur. Ill., vol. 6, Kinderhook Gr. [Sig. deeply sculptured.]
ornatus, St. John & Worthen, 1875, Geo. Sur. Ill., vol. 6, Burlington Gr. [Sig. ornamented.]
Onchus, Agassiz, 1837, Recherches sur les Poissons Fossiles. [Ety. *onchos*, bent or hooked like a talon or arrow-barb.]
deweyi, see Ceratiocaris deweyi.
ONYCHODUS, Newberry, 1853, Annals of Science, vol. 1, and 1857, Bull. Nat. Inst. [Ety. *onyx*, a claw; *odous*, tooth.]
hopkinsi, Newberry, 1853, Annals of Science, vol. 1, and 1857, Bull. Nat. Inst., Corniferous Gr. [Ety. proper name.]
sigmoides, Newberry, 1853, Annals of Science, vol. 1, and 1857, Bull. Nat. Inst., Corniferous Gr. [Sig. like the letter *Sigma*.]
ORACANTHUS, Agassiz, 1837, Recherches sur les Poissons Fossiles. [Ety. *oraios*, beautiful; *akantha*, spine.]
abbreviatus, Newberry, 1853, Annals of Science, vol. 1, Corniferous Gr. [Sig. abbreviated.]
consimilis, St. John & Worthen, 1875, Geo. Sur. Ill., vol. 6, St. Louis Gr. [Sig. wholly similar.]
(?) obliquus, St. John & Worthen, 1875, Geo. Sur. Ill., vol. 6, Keokuk Gr. [Sig. oblique.]
pnigeus, Newberry & Worthen, 1866, Geo. Sur. Ill., vol. 2, Keokuk Gr. This species is made the type of the genus Pnigeacanthus, see P. deltoides.
vetustus, Leidy, 1856, Jour. Acad. Nat. Sci. Phil., 2d ser., vol. 3, Low. Carb. [Sig. ancient.]
fragilis, Newberry, 1853, Annals of Science, vol. 1, Corniferous Gr. [Sig. frail, easily broken.]
granulatus, Newberry, 1853, Annals of Science, vol. 1, Corniferous Gr. [Sig. granulated.]
multiseriatus, Newberry, 1853, Annals of Sci., vol. 1, Corniferous Gr. [Sig. in many series.]
ORODUS, Agassiz, 1838, Recherches sur les Poissons Fossiles. [Ety. *oraios*, beautiful; *odous*, tooth.]
alleni, St. John & Worthen, 1875, Geo. Sur. Ill., vol. 6, Coal Meas. [Ety. proper name.]
carinatus, St. John & Worthen, 1875, Geo. Sur. Ill., vol. 6, Keokuk Gr. [Sig. keeled.]
corrugatus, see Agassizodus corrugatus.

dædaleus, St. John & Worthen, 1875, Geo. Sur. Ill., vol. 6, Kinderhook Gr. [Ety. mythological name.]
decussatus, St. John & Worthen, 1875, Geo. Sur. Ill., vol. 6, Kinderhook Gr. [Sig. arranged in pairs that alternately cross each other.]
elegantulus, Newberry & Worthen, 1866, Geo. Sur. Ill., vol. 2, Burlington Gr. [Sig. very elegant.]
fastigiatus, St. John & Worthen, 1875, Geo. Sur. Ill., vol. 6, Burlington Gr. [Sig. pointed.]
major, St. John & Worthen, 1875, Geo. Sur. Ill., vol. 6, Burlington Gr. [Sig. large.]
mammillaris, Newberry & Worthen, 1866, Geo. Sur. Ill., vol. 2, Keokuk Gr. [Sig. pap-shaped.]
minusculus, see Desmiodus minusculus.
minutus, Newberry & Worthen, 1866, Geo. Sur. Ill., vol. 2, Keokuk Gr. [Sig. very small.]
multicarinatus, Newberry & Worthen, 1866, Geo. Sur. Ill., vol. 2, Kinderhook Gr. [Sig. many-keeled.]
neglectus, St. John & Worthen, 1875, Geo. Sur. Ill., vol. 6, St. Louis Gr. [Sig. overlooked.]
ornatus, Newberry & Worthen, 1866, Geo. Sur. Ill., vol. 2, Keokuk Gr. [Sig. ornamented.]
parallelus, St. John & Worthen, 1875, Geo. Sur. Ill., vol. 6, Kinderhook Gr. [Sig. parallel.]
parvulus, St. John & Worthen, 1875, Geo. Sur. Ill., vol. 6, St. Louis Gr. [Sig. very small.]
plicatus, Newberry & Worthen, 1866, Geo. Sur. Ill., vol. 2, St. Louis Gr. [Sig folded.]
tuberculatus, Newberry & Worthen, 1866, Geo. Sur. Ill., vol. 2, Burlington Gr. [Sig. tuberculated.]
turgidus, St. John & Worthen, 1875, Geo. Sur. Ill., vol. 6, Chester Gr. [Sig. turgid, swollen out.]
variabilis, Newberry, 1875, Ohio Pal., vol. 2, Waverly Gr. [Sig. variable.]
variocostatus, St. John & Worthen, 1875, Geo. Sur. Ill., vol. 6, Burlington Gr. [Sig. variably ribbed.]
whitei, St. John & Worthen, 1875, Geo. Sur. Ill., vol. 6, Kinderhook Gr. [Ety. proper name.]

ORTHACANTHUS, Agassiz, 1843, Poiss. Foss., vol. 3. [Ety. *orthos*, straight; *akantha*, spine.]
arcuatus, Newberry, 1857, Proc. Acad. Nat. Sci. Phil., Coal Meas. [Sig. bent, bow-shaped.]
gracilis, Newberry, 1875, Ohio Pal., vol. 2, Coal Meas. [Sig. slender.]

PALÆONISCUS, Agassiz, 1833, Recherches sur les Poissons Fossiles. [Ety. *palaios*, ancient; *oniscus*, a wood louse.]
alberti, Jackson, 1851, Rep. on the Albert Coal Mine, New Brunswick, Coal Meas. [Ety. proper name.]
brainardi, Thomas, 1853, Bost. Soc. Nat. Hist., vol. 4, Ohio Pal., vol. 1, Berea grit. [Ety. proper name.]
browni, Jackson, 1851, Rep. on Albert Coal Mine, Coal Meas. [Ety. proper name.]
cairnesi, Jackson, 1851, Rep. on Albert Coal Mine, New Brunswick, Coal Meas. [Ety. proper name.]
gracilis, Newberry & Worthen, 1870, Geo. Sur. Ill., vol. 4, Coal Meas. [Sig. slender.]
leidyanus, Lea, 1852, Jour. Acad. Nat. Sci., 2d series, vol. 2, Coal Meas. [Ety. proper name.]
peltigerus, Newberry, 1857, (Elonichthys peltigerus) Proc. Acad. Nat. Sci., vol. 8, Coal Meas. [Sig. bearing a small shield.]
scutigerus, Newberry, 1857, Proc. Acad. Nat. Sci. Phil., Coal Meas. [Sig. shield-bearing.]

PELTODUS, Newberry & Worthen, 1870, Geo. Sur. Ill., vol. 4. [Ety. *pelte*, a half-moon shield; *odous*, a tooth.]
plicomphalus, St. John & Worthen, 1875, Geo. Sur. Ill., vol. 6, Chester Gr. [Sig. folded in the middle.]
quadratus, St. John & Worthen, 1875, Geo. Sur. Ill., vol. 6, St. Louis Gr. [Sig. square-shaped.]
transversus, St. John & Worthen, 1875, Geo. Sur. Ill., vol. 6, Coal Meas. [Sig. crosswise.]
unguiformis, Newberry & Worthen, 1870, Geo. Sur. Ill., vol. 4, Coal Meas. [Sig. claw-shaped.]

PEPLORHINA, Cope, 1873, Proc. Acad. Nat. Sci. Phil. [Ety. *peplos*, a robe; *Rhine*, a kind of dog-fish.]
anthracina, Cope, 1873, Proc. Acad. Nat. Sci. Phil., Coal Meas. [Ety. *anthrax*, coal.] Prof. Newberry says this species is an amphibian and not a fish.

PERIPLECTRODUS, St. John & Worthen, 1875, Geo. Sur. Ill., vol. 6. [Ety. *peri*, near by; *Plectrodus*, a genus of fish.]
compressus, St. John & Worthen, 1875, Geo. Sur. Ill., vol. 6, St. Louis Gr. [Sig. compressed.]
expansus, St. John & Worthen, 1875, Geo. Sur. Ill., vol. 6, Chester Gr. [Sig. expanded.]
warreni, St. John & Worthen, 1875, Geo. Sur. Ill., vol. 6, Burlington Gr. [Ety. proper name.]

PETALODUS, Owen, 1840, Odontography. [Ety. *petalos*, spread out; *odous*, a tooth.]
alleghaniensis, Leidy, 1856, Jour. Acad. Nat. Sci., 2nd series, vol. 3, Coal Meas. [Ety. proper name.]
curtus, Newberry & Worthen, 1866, Geo. Sur. Ill., vol. 6, Keokuk Gr. [Sig. short.]
destructor, Newberry & Worthen, 1866, Geo. Sur. Ill., vol. 2, Coal Meas. [Sig. a destroyer.]

hybridus, St. John & Worthen, 1875, Geo. Sur. Ill., vol. 6, St. Louis Gr. [Sig. hybrid; intermediate between two species.]
linguifer, Newberry & Worthen, 1856, Geo. Sur. Ill., vol. 2, Chester Gr. [Sig. tongue-bearing.]
parvulus, Newberry & Worthen, Carboniferous. [Sig. very small.]
politus, Newberry & Worthen, Carboniferous. [Sig. smooth.]
sulcatus, Newberry & Worthen, Carboniferous. [Sig. furrowed.]
proximus, St. John & Worthen, 1875, Geo. Sur. Ill., vol. 6, Coal Meas. [Sig. approaching, near to.]
PETALORHYNCHUS, Agassiz, 1855, in British Pal. Rocks. [Ety. *petalos*, spread out; *rhynchos*, a beak.]
distortus, St. John & Worthen, 1875, Geo. Sur. Ill., vol. 6, St. Louis Gr. [Sig. distorted, twisted.]
pseudosagittatus, St. John & Worthen, 1875, Geo. Sur. Ill., vol. 6, St. Louis Gr. [Sig. false *sagittatus*, a species of *Petalorhynchus*.]
spatulatus, St. John & Worthen, 1875, Geo. Sur. Ill., vol. 6, St. Louis Gr. [Sig. spatulate, blade-shaped.]
striatus, Newberry & Worthen, 1866, Geo. Sur. Ill., vol. 2, Burlington Gr. [Sig. striated.]
PETRODUS, McCoy, 1848, Ann. & Mag. Nat. Hist., 2d series, vol. 2. [Ety. *petros*, a rock; *odous*, a tooth.]
acutus, Newberry & Worthen, 1866, Geo. Sur. Ill., vol. 2, Coal Meas. [Sig. sharp-pointed.]
occidentalis, Newberry & Worthen, 1866, Geo. Sur. Ill., vol. 2, Coal Meas. [Sig. western.]
pustulosus, Newberry & Worthen, 1870, Geo. Sur. Ill., vol. 4, Burlington Gr. [Sig. covered with pustules.]
PHŒBODUS, St. John & Worthen, 1875, Geo. Sur. Ill., vol. 6. [Ety. mythological name; *odous*, a tooth.]
sophiæ, St. John & Worthen, 1875, Geo. Sur. Ill., vol. 6, Devonian. [Ety. proper name.]
PHYSONEMUS, Agassiz, 1843, Recherches sur les Poissons Fossiles. [Ety. *physa*, a bladder; *nema*, a thread.]
altonensis, St. John & Worthen, 1875, Geo. Sur. Ill., vol. 6, St. Louis Gr. [Ety. proper name.]
carinatus, St. John & Worthen, 1875, Geo. Sur. Ill., vol. 6, Kinderhook Gr. [Sig. keeled.]
chesterensis, St. John & Worthen, 1875, Geo. Sur. Ill., vol. 6, Chester Gr. [Ety. proper name.]
depressus, St. John & Worthen, 1875, Geo. Sur. Ill., vol. 6, Kinderhook Gr. [Sig. depressed.]
gigas, Newberry & Worthen, 1870, Geo. Sur. Ill., vol. 4, Burlington Gr. [Sig. large.]

parvulus, St. John & Worthen, 1875, Geo. Sur. Ill., vol. 6, Keokuk Gr. [Sig. very small.]
proclivis, St. John & Worthen, 1875, Geo. Sur. Ill., vol. 6, Kinderhook Gr. [Sig. sloping, going downward.]
PLATYODUS, Newberry, 1875, Ohio Pal., vol. 2. [Ety. *platys*, broad; *odous*, a tooth.]
lineatus, Newberry, 1875, Ohio Pal., vol. 2, Waverly Gr. [Sig. lined.]
PLATYSOMUS, Agassiz, 1833, Recherches sur les Poissons Fossiles. [Ety. *platys*, broad; *soma*, body.]
circularis, Newberry & Worthen, 1870, Geo. Sur. Ill., vol. 4, Coal Meas. [Sig. circular.]
PLEURACANTHUS, Agassiz, 1837, Poiss. Foss., vol. 3. [Ety. *pleura*, side; *akantha*, spine.]
arcuatus, Newberry, 1857, Proc. Acad. Nat. Sci., vol. 8, Coal Meas. [Sig. bent.]
biserialis, Newberry, 1857, Proc. Acad. Nat. Sci., vol. 8, Coal Meas. [Sig. having two series.]
dilatatus, Newberry, 1857, Proc. Acad. Nat. Sci., vol. 8, Coal Meas. [Sig. dilated.]
PNIGEACANTHUS, St. John & Worthen, 1875, Geo. Sur. Ill., vol. 6. [Ety. from the specific name in *Oracanthus pnigeus*; *akantha*, a spine.]
deltoides, St. John & Worthen, 1875, Geo. Sur. Ill., vol. 6, Keokuk Gr. [Sig. like the Greek letter *Delta*.] But why should this species not be *Pnigeacanthus pnigeus?*
PŒCILODUS, Agassiz, 1843, Recherches Poiss. Foss., vol. 3. [Ety. *poikilos*, variegated; *odous*, tooth.]
convolutus, Newberry & Worthen, 1870, Geo. Sur. Ill., vol. 4, Keokuk Gr. [Sig. rolled together.]
ornatus, Newberry & Worthen, 1866, Geo. Sur. Ill., vol. 2, Keokuk Gr. [Sig. ornamented.]
rugosus, Newberry & Worthen, 1866, Geo. Sur. Ill., vol. 2, Keokuk Gr. [Sig. wrinkled.]
POLYRHIZODUS, McCoy, 1848, Ann. & Mag. Nat. Hist., 2nd series, vol. 2. [Ety. *polys*, many; *rhiza*, root; *odous*, tooth.]
amplus, St. John & Worthen, 1875, Geo. Sur. Ill., vol. 6, St. Louis Gr. [Sig. full-sized.]
carbonarius, St. John & Worthen, 1875, Geo. Sur. Ill., vol. 6, Coal Meas. [Sig. pertaining to coal.]
dentatus, Newberry & Worthen, 1866, Geo. Sur. Ill., vol. 2, Chester Gr. [Sig. toothed.]
littoni, Newberry & Worthen, 1870, Geo. Sur. Ill., vol. 4, St. Louis Gr. [Ety. proper name.]
modestus, Newberry, 1875, Ohio Pal., vol. 2, Cleveland shale. [Sig. moderate.]
nanus, St. John & Worthen, 1875, Geo. Sur. Ill., vol. 6, Keokuk Gr. [Sig. dwarfish.]

piasaensis, St. John & Worthen, 1875, Geo. Sur. Ill., vol. 6, Chester Gr. [Ety. proper name.]
ponticulus, Newberry & Worthen, 1866, Geo. Sur. Ill., vol. 2, Chester Gr. [Sig. a little bridge.]
porosus, Newberry & Worthen, 1866, Geo. Sur. Ill., vol. 2, Burlington Gr. [Sig. full of pores, porous.]
truncatus, Newberry & Worthen, 1870, Geo. Sur. Ill., vol. 4, Burlington Gr. [Sig. truncated, cut off.]
williamsi, St. John & Worthen, 1875, Geo. Sur. Ill., vol. 6, Keokuk Gr. [Ety. proper name.]

PRISTICLADODUS, McCoy, 1855, British Pal. Rocks. [Ety. from the two genera *Pristis* and *Cladodus*.]
springeri, St. John & Worthen, 1875, Geo. Sur. Ill., vol. 6, Kinderhook Gr. [Ety. proper name.]

PRISTODUS, Agassiz. [Ety. *pristis*, a saw; *odous*, a tooth.]
acuminatus, St. John & Worthen, 1875, Geo. Sur. Ill., vol. 6, Kinderhook Gr. [Sig. sharp-pointed.]

PSAMMODUS, Agassiz, 1838, Recherches sur les Poissons Fossiles. [Ety. *psammos*, sand; *odous*, tooth.]
angularis, Newberry & Worthen, 1866, Geo. Sur. Ill., vol. 2, Chester Gr. [Sig. angular.]
antiquus, Newberry, 1853, Annals of Science, vol. 1, and 1857, Bull. Nat. Inst., Corniferous Gr. [Sig. ancient.]
porosus, Agassiz, 1838, Recherch. Poiss. Foss., Chester Gr. [Sig. full of pores.]
reticulatus, Newberry & Worthen, 1866, Geo. Sur. Ill., vol. 2, Chester Gr. [Sig. reticulated, net work.]
rhomboideus, Newberry & Worthen, 1866, Geo. Sur. Ill., vol. 2, Keokuk Gr. [Sig. like a rhomb, lozenge-shaped.]
rugosus, Agassiz, 1838, Recherch. Poiss. Foss., St. Louis Gr. [Sig. wrinkled.]
semicylindricus, Newberry & Worthen, 1866, Geo. Sur. Ill., vol. 2, Keokuk Gr. [Sig. half cylindrical.]

PSEPHODUS, Agassiz. [Ety. *psephos*, a pebble; *odous*, a tooth.]
reticulatus, St. John & Worthen, 1875, Geo. Sur. Ill., vol. 6, Kinderhook Gr. [Sig. reticulated, like net work.]

PTERICHTHYS, Miller, 1840, in British Rep. [Ety. *pteron*, a fin; *ichthys*, a fish.]
norwoodensis, Owen, 1846, Am. Jour. Sci., 2nd series, vol. 1, Up. Held. Gr. [Ety. proper name.]

PTYCTODUS, Pander, Uber die Ctenodipterinen des Devonischens Systems. [Ety. *ptyktos*, folded; *odous*, a tooth.]
calceolus, Newberry & Worthen, 1866, (Rinodus calceolus) Geo. Sur. Ill., vol. 2, Ham. Gr. [Sig. a sandal or slipper.]

PYGOPTERUS, Agassiz, 1833, Poiss. Foss., vol. 2. [Ety. *pyge*, rump; *pteron*, a fin.]
scutellatus, Newberry, 1857, Proc. Acad. Nat. Sci., vol. 8, Coal Meas. [Sig. shielded.]

RHIZODUS, Owen, 1840, Odontography. [Ety. *rhiza*, a root; *odous*, a tooth.]
angustus, Newberry, 1857, Proc. Acad. Nat. Sci. Phil., vol. 8, Coal Meas. [Sig. narrow.]
hardingi, Dawson, 1868, Acad. Geol., Carboniferous. [Ety. proper name.]
incurvus, Newberry, 1857, Proc. Acad. Nat. Sci., vol. 8, Coal Meas. [Sig. incurved.]
lancifer, Newberry, 1857, Proc. Acad. Nat. Sci. Phil., vol. 8, Coal Meas. [Sig. bearing a lance.]
occidentalis, Newberry & Worthen, 1866, Geo. Sur. Ill., vol. 2, Coal Meas. [Sig. western.]
quadratus, Newberry, 1873, Ohio Pal., vol. 1, Coal Meas. [Sig. square-shaped.]
reticulatus, Newberry & Worthen, 1870, Geo. Sur. Ill., vol. 4, Coal Meas. [Sig. reticulated, net-work.]

RHYNCHODUS, Newberry, 1873, Ohio Pal., vol. 1. [Ety. *rhynchos*, a beak; *odous*, a tooth.]
crassus, Newberry, 1873, Ohio Pal., vol. 1, Corniferous Gr. [Sig. thick.]
frangens, Newberry, 1873, Ohio Pal., vol. 1, Corniferous Gr. [Sig. broken.]
secans, Newberry, 1873, Ohio Pal., vol. 1, Corniferous Gr. [Sig. divided up.]

Rinodus, syn. for Ptyctodus.
calceolus, see Ptyctodus calceolus.

SANDALODUS, Newberry & Worthen, 1866, Geo. Sur. Ill., vol. 2. [Ety. *sandalon*, a sandal; *odous*, a tooth.]
angustus, Newberry & Worthen, 1866, Geo. Sur. Ill., vol. 2, Keokuk Gr. [Sig. narrow.]
carbonarius, Newberry & Worthen, 1866, Geo. Sur. Ill., vol. 2, Coal Meas. [Sig pertaining to coal.]
crassus, Newberry & Worthen, 1870, Geo. Sur. Ill., vol. 4, St. Louis Gr. [Sig. thick.]
grandis, Newberry & Worthen, 1866, Geo. Sur. Ill., vol. 2, Keokuk Gr. [Sig. large.]
laevissimus, Newberry & Worthen, 1866, Geo. Sur. Ill., vol. 2, Keokuk Gr. [Sig. very smooth.]
parvulus, Newberry & Worthen, 1866, Geo. Sur. Ill., vol. 2, St. Louis Gr. [Sig. very small.]
spatulatus, Newberry & Worthen, 1866, Geo. Sur. Ill., vol. 2, St. Louis Gr. [Sig. spatulate, blade-shaped.]

Sauripteris, Hall, 1843, Geo. Rep. 4th Dist. N. Y. [Ety. *sauros*, a lizard; *pteron*, a wing.]
taylori, see Holoptychius taylori.

STEMMATODUS, St. John & Worthen, 1875, Geo. Sur. Ill., vol. 6. [Ety. *stemmatos*, a wreath; *odous*, a tooth.]
    bicristatus, St. John & Worthen, 1875, Geo. Sur. Ill., vol. 6, Burlington Gr. [Sig. double-peaked.]
    bifurcatus, St. John & Worthen, 1875, Geo. Sur. Ill., vol. 6, Burlington Gr. [Sig. forked.]
    cheiriformis, St. John & Worthen, 1875, Geo. Sur. Ill., vol. 6. Burlington Gr. [Sig. like a hand.]
    compactus, St. John & Worthen, 1875, Geo. Sur. Ill., vol. 6, Chester Gr. [Sig. compact.]
    keokuk, St. John & Worthen, 1875, Geo. Sur. Ill., vol. 6, Keokuk Gr. [Ety. proper name.]
    simplex, St. John & Worthen, 1875, Geo. Sur. Ill., vol. 6, Burlington Gr. [Sig. simple.]
    symmetricus, St. John & Worthen, 1875, Geo. Sur. Ill., vol. 6, Burlington Gr. [Sig. symmetrical.]
STENACANTHUS, Leidy, 1857, Proc. Acad. Nat. Sci., vol. 8. [Ety. *stenos*, narrow; *akantha*, a spine.]
    nitidus, Leidy, 1857, Proc. Acad. Nat. Sci., vol. 8, Devonian. [Sig. neat, smooth.]
TANAODUS, St. John & Worthen, 1875, Geo. Sur. Ill., vol. 6. [Ety. *tanaos*, long; *odous*, a tooth.]
    angularis, Newberry & Worthen, 1866, (Chomatodus angularis) Geo. Sur. Ill., vol. 2, Coal Meas. [Sig. angulated.]
    bellicinctus, St. John & Worthen, 1875, Geo. Sur. Ill., vol. 6, Chester Gr. [Sig. beautifully banded.]
    depressus, St. John & Worthen, 1875, Geo. Sur. Ill., vol. 6, Chester Gr. [Sig. depressed.]
    gracillimus, Newberry & Worthen, 1866, (Chomatodus gracillimus) Geo. Sur. Ill., vol. 6, Burlington Gr. [Sig. very slender.]
    grossiplicatus, St. John & Worthen, 1875, Geo. Sur. Ill., vol. 6, Chester Gr. [Sig. thick-folded.]
    multiplicatus, Newberry & Worthen, 1866, (Chomatodus multiplicatus) Geo. Sur. Ill., vol. 2, Burlington Gr. [Sig. many folded.]
    obscurus, Leidy, 1856, (Chomatodus obscurus) Trans. Am. Phil. Soc., vol. 11, Keokuk Gr. [Sig. obscure.]
    polymorphus, St. John & Worthen, 1875, Geo. Sur. Ill., vol. 6, Chester Gr. [Sig. many formed.]
    praenuntius, St. John & Worthen, 1875, Geo. Sur. Ill., vol. 6, St. Louis Gr. [Sig. forboding.]
    pumilus, St. John & Worthen, 1875, Geo. Sur. Ill., vol. 6, St. Louis Gr. [Sig. a dwarf.]
    sculptus, St. John & Worthen, 1875, Geo. Sur. Ill., vol. 6, St. Louis Gr. [Sig. engraved, sculptured.]
    sublunatus, St. John & Worthen, 1875, Geo. Sur. Ill., vol. 6, St. Louis Gr. [Sig. somewhat lunate.]
THRINACODUS, St. John & Worthen, 1875, Geo. Sur. Ill., vol. 6. [Ety. *thrinakos*, three-pronged; *odous*, a tooth.]
    nanus, St. John & Worthen, 1875, Geo. Sur. Ill., vol. 6, Kinderhook Gr. [Sig. dwarfish.]
    duplicatus, Newberry & Worthen, 1866, (Diplodus duplicatus) Geo. Sur. Ill., vol. 2, Keokuk Gr. [Sig. folded in two.]
    incurvus, Newberry & Worthen, 1866, (Diplodus incurvus) Geo. Sur. Ill., vol. 2, Keokuk Gr. [Sig. incurved.]
TRIGONODUS, Newberry & Worthen, 1866, Geo. Sur. Ill., vol. 2. [Ety. *trigonos*, three-cornered; *odous*, tooth.]
    major, Newberry & Worthen, 1866, Geo. Sur. Ill., vol. 2, Burlington Gr. [Sig. large.]
    minor, Newberry & Worthen, 1866, Geo. Sur. Ill., vol. 2, Keokuk Gr. [Sig. small.]
VENUSTODUS, St. John & Worthen, 1875, Geo. Sur. Ill., vol. 6. [Ety. *venustus*, beautiful; *odous*, tooth.]
    argutus, St. John & Worthen, 1875, Geo. Sur. Ill., vol. 6, Chester Gr. [Sig. pretty.]
    leidyi, St. John & Worthen, 1875, Geo. Sur. Ill., vol. 6, St. Louis Gr. This name is a syn. for V. venustus.
    robustus, St. John & Worthen, 1875, Geo. Sur. Ill., vol. 6, Burlington Gr. [Sig. robust.]
    tenuicristatus, St. John & Worthen, 1875, Geo. Sur. Ill., vol. 6, Keokuk Gr. [Sig. slender-peaked.]
    variabilis, St. John & Worthen, 1875, Geo. Sur. Ill., vol. 6, Burlington Gr. [Sig. variable.]
    venustus, Leidy, 1856, (Chomatodus venustus) Trans. Am. Phil. Soc. Phil., vol. 11, St. Louis Gr. [Sig. beautiful.]
XYSTRACANTHUS, Leidy, 1859, Proc. Acad. Nat. Sci. Phil. [Ety. *xystra*, a tool for scraping; *akantha*, a spine.]
    acinaciformis, St. John & Worthen, 1875, Geo. Sur. Ill., vol. 6, Coal Meas. [Sig. scimitar-shaped.]
    anceps, Newberry & Worthen, 1866, (Drepanacanthus anceps) Geo. Sur. Ill., vol. 2, Coal Meas. [Sig. doubtful.]
    arcuatus, Leidy, 1859, Proc. Acad. Nat. Sci. Phil., Up. Coal Meas. [Sig. bent, bow-shaped.]
    mirabilis, St. John & Worthen, 1875, Geo. Sur. Ill., vol. 6, Coal Meas. [Sig. wonderful.]
XYSTRODUS, Agassiz. [Ety. *xystra*, an instrument for scraping; *odous*, tooth.]
    occidentalis, St. John, Coal Meas. [Sig. western.]

CLASS REPTILIA.

SUB-CLASS BATRACHIA.

FAMILY COLOSTEIDÆ.—Amphibamus, Colosteus, Sauropleura.
FAMILY MOLGOPHIDÆ.—Molgophis, Pleuroptyx.
FAMILY PHLEGETHONTIIDÆ.—Phlegethontia.
FAMILY PTYONIIDÆ.—Hyphasma, Oestocephalus, Ptyonius.
FAMILY TUDITANIDÆ.— Brachydectes, Ceraterpeton, Dendrerpeton, Eurythorax, Hylerpeton, Hylonomus, Leptophractus, Pelion, Tuditanus.
ORDER PROTEIDA.—Cocytinus, Thyrsidium.
INCERTÆ SEDIS.—Baphetes, Cheirotherium, Clepsysaurus, Collettosaurus, Eosaurus, Ornithichnites, Sauropus, Spheropezium, Thenaropus.

AMPHIBAMUS, Cope, 1865, Proc. Acad. Nat. Sci. Phil. [Ety. *amphi*, both; *bama*, a step; from its two modes of progressing—swimming and walking.]
  grandiceps, Cope, 1865, Proc. Acad. Nat. Sci. Phil., Coal Meas. [Sig. big-headed.]
BAPHETES, Owen, 1853, Jour. Geo. Soc. London, vol. 9. [Ety. *bapto*, I dip or dive—a diving animal.]
  minor, Dawson, 1870, Can. Nat. & Geol., Coal Meas. [Sig. less.]
  planiceps, Owen, 1853, Jour. Geo. Soc. London, vol. 9, Coal Meas. [Ety. flat-headed.]
BRACHYDECTES, Cope, 1868, Proc. Acad. Nat. Sci. Phil. [Ety. *brachys*, short; *dektes*, a biter.]
  newberryi, Cope, 1868, Proc. Acad. Nat. Sci., Coal Meas. [Ety. proper name.]
CERATERPETON, Huxley. [Ety. *keras*, horn; *erpeton*, reptile.]
  punctolineatum, Cope, 1875, Ohio Pal., vol. 2, Coal Meas. [Sig. puncture-lined.]
  tenuicorne, Cope, 1875, Ohio Pal., vol. 2, Coal Meas. [Sig. slender-horned.]
CHEIROTHERIUM, Kaup, 1835, in Leonhard und Bronn, Neues Jahrbuch fur Mineralogie. [Ety. *cheir*, the hand; *therion*, beast.]
  reiteri, Moore, 1863, Am. Jour. Sci. & Arts, 3rd ser., vol. 5, Coal Meas. [Ety. proper name.]

CLEPSYSAURUS, Lea, 1852, Jour. Acad. Nat. Sci., vol. 2. [Ety. *klepsydra*, a sand-glass; *saurus*, a lizard; from the compression laterally of the vertebræ towards the center.]
  pennsylvanicus, Lea, 1852, Jour. Acad. Nat. Sci., vol. 2, Coal Meas. [Ety. proper name.]
COCYTINUS, Cope, 1871, Proc. Am. Phil. Soc. [Ety. *kokytos*, mythological name.]
  gyrinoides, Cope, 1874, Trans. Am. Phil. Soc., Coal Meas. [Sig. like a tadpole.]
COLLETTOSAURUS, Cox, 1873, Geo. Sur. Ind. [Ety. proper name; *saurus*, a lizard.]
  indianensis, Cox, 1873, Geo. Sur. Ind., Coal Meas. [Ety. proper name.]
COLOSTEUS, Cope, 1869, Trans. Am. Phil. Soc. [Ety. *kolos*, short; *osteon*, a bone.]
  crassiculatus, syn. for C. scutellatus.
  foveatus, 1869, Trans. Am. Phil. Soc., Coal Meas. [Sig. pitted.]
  marshi, see Ptyonius marshi.
  pauciradiatus, Cope, 1874, Trans. Am. Phil. Soc., Coal Meas. [Sig. few-lined.]
  scutellatus, Newberry, 1856, (Pygopterus scutellatus) Proc. Acad. Nat. Sci. Phil., Coal Meas. [Sig. having a shield.]
DENDRERPETON, Owen, 1853, Quar. Jour. Geo. Soc., vol. 9. [Ety. *dendron*, a tree; *erpeton*, a lizard; from the peculiar circumstances under which the fossil reptile was found.]

acadianum, Owen, 1853, Quar. Jour. Geo. Soc., vol. 9, Coal Meas. [Ety. from *Acadia*, the ancient name of Nova Scotia.]
oweni, Dawson, 1863, Quar. Jour. Geo. Soc., vol. 19, Coal Meas. [Ety. proper name.]
EOSAURUS, Marsh, 1862, Can. Nat. & Geo., vol. 7. [Ety. *eos*, the dawn; *sauros*, a lizard.]
acadianus, Marsh, 1862, Can. Nat. & Geo., vol. 7, Coal Meas. [Ety. proper name.]
EURYTHORAX, Cope, 1875, Ohio Pal., vol. 2. [Ety. *curys*, broad; *thorax*, the breast.]
sublævis, Cope, 1871, Proc. Am. Phil. Soc., Coal Meas. [Sig. somewhat smooth.]
HYLERPETON, Owen, 1862, Quar. Jour. Geo. Soc., vol. 18. [Ety. *hyle*, wood; *erpeton*, reptile.]
curtidentatum, Dawson, 1876, Am. Jour. Sci. & Arts, vol. 12, Coal Meas. [Sig. short-toothed.]
dawsoni, Owen, 1862, Quar. Jour. Geo. Soc., vol. 18, Coal Meas. [Ety. proper name.]
longidentatum, Dawson, 1876, Am. Jour. Sci. & Arts, vol. 12, Coal Meas. [Sig. long-toothed.]
HYLONOMUS, Dawson, 1860, Quar. Jour. Geo. Soc., vol. 16. [Ety. *hyle*, wood; *nomos*, an abode; forest dweller.]
aciedentatus, Dawson, 1860, Quar. Jour. Geo. Soc., vol. 16, Coal Meas. [Sig. needle-toothed, sharp-toothed.]
lyelli, Dawson, 1860, Quar. Jour. Geo. Soc., vol. 16, Coal Meas. [Ety. proper name.]
hymani, Dawson, 1860, Quar. Jour. Geo. Soc., vol. 16, Coal Meas. [Ety. proper name.]
HYPHASMA, Cope, 1875, Proc. Acad. Nat. Sci. Phil. [Ety. *hyphasma*, a web.]
lævis, Cope, 1875, Ohio Pal., vol. 2, Coal Meas. [Sig. smooth.]
LEPTOPHRACTUS, Cope, 1873, Proc. Acad. Nat. Sci. Phil. [Ety. *leptos*, stripped; *phraktos*, defence.]
obsoletus, Cope, 1873, Proc. Acad. Nat. Sci. Phil., Coal Meas. [Sig. obsolete.]
MOLGOPHIS, Cope, 1868, Proc. Acad. Nat. Sci. [Ety. *molgos*, a skin; *ophis*, a serpent.]
brevicostatus, Cope, 1875, Ohio Pal., vol. 2, Coal Meas. [Sig. short-ribbed.]
macrurus, Cope, 1868, Proc. Acad. Nat. Sci., Coal Meas. [Sig. long-tailed.]
wheatleyi, Cope, 1874, Trans. Am. Phil. Soc., Coal Meas. [Ety. proper name.]
OESTOCEPHALUS, Cope, 1868, Proc. Acad. Nat. Sci. Phil. [Ety. *oistos*, an arrow; *kephale*, the head.]
amphiuminus, Cope, 1868, Proc. Acad. Nat. Sci., Coal Meas. [Ety. from resemblance to *Amphiuma*.]
pectinatus, see Ptyonius pectinatus.
rectidens, Cope, 1874, Trans. Am. Phil. Soc., Coal Meas. [Sig. straight-toothed.]
remex, Cope, 1868, (Sauropleura remex) Proc. Acad. Nat. Sci., Coal Meas. [Sig. an oarsman, a rower.]
vinchellanus, see Ptyonius vinchellanus.
ORNITHICHNITES, Hitchcock, 1832, Am. Jour. Sci. & Arts. [Ety. *ornithos*, a bird; *ichnos*, a foot step.]
culbertsoni, King, 1845, Am. Jour. Sci., vol. 48, Coal Meas. [Ety. proper name.]
gallinuloides, King, 1845, Am. Jour. Sci., vol. 48, Coal Meas. [Sig. similar to a pullet.]
PELION, Wyman, 1868, Proc. Acad. Nat. Sci. Phil. [Ety. proper name.]
lyelli, Wyman, 1858, (Raniceps lyelli) Am. Jour. Sci. & Arts, Coal Meas. [Ety. proper name.]
PHLEGETHONTIA, Cope, 1871, Proc. Am. Phil. Soc. [Ety. *phlegetho*, to scorch, to burn.]
linearis, Cope, 1874, Trans. Am. Phil. Soc., Coal Meas. [Sig. lined.]
serpens, Cope, 1874, Trans. Am. Phil. Soc., Coal Meas. [Sig. a serpent.]
PLEUROPTYX, Cope, 1875, Ohio Pal., vol. 2. [Ety. *pleura*, a rib; *ptyx*, a fold, a wing.]
clavatus, Cope, 1875, Ohio Pal., vol. 2, Coal Meas. [Sig. club-shaped.]
PTYONIUS, Cope, 1874, Trans. Am. Phil. Soc. [Ety. *ptyon*, a fan.]
marshi, Cope, 1875, (Colosteus marshii) Ohio Pal., vol. 2, Coal Meas. [Ety. proper name.]
nummifer, Cope, 1875, Ohio Pal., vol. 2, Coal Meas. [Sig. coin-bearing.]
pectinatus, Cope, 1868, (Sauropleura pectinata) Proc. Acad. Nat. Sci., Coal Meas. [Sig. pectinated, toothed like a comb.]
serrula, Cope, 1874, Trans. Am. Phil. Soc., Coal Meas. [Sig. a saw.]
vinchellanus, Cope, 1871, (*Oestocephalus* vinchellanus) Proc. Am. Phil. Soc., Coal Meas. [Ety. proper name.]
Pygopterus, Agassiz, 1833, Recherch. Poiss. Foss.
scutellatus, see Colosteus scutellatus.
RANICEPS, Wyman, 1858, Am. Jour. Sci. & Arts. [Sig. frog-headed.] Name was preoccupied.
lyellii, see Pelion lyelli.
SAUROPLEURA, Cope, 1868, Proc. Acad. Nat. Sci. Phil. [Ety. *sauros*, a lizard; *pleura*, a rib.]
digitata, Cope, 1868, Proc. Acad. Nat. Sci. Phil., Coal Meas. [Sig. having fingers or toes.]
newberryi, Cope, 1875, Ohio Pal., vol. 2, Coal Meas. [Ety. proper name.]
pectinata, see Ptyonius pectinatus.
remex, Cope, 1868, Proc. Acad. Nat. Sci., Coal Meas. [Sig. a rower.]
SAUROPUS, Lea, 1849, Trans. Am. Phil. Soc., vol. 10. [Ety. *sauros*, a lizard; *pous*, a foot.]
primævus, Lea, 1849, Trans. Am. Phil. Soc., vol. 10, Coal Meas. [Sig. first.]

sydenensis, Dawson, 1868, Acad. Geol., Coal Meas. [Ety. proper name.]

unguifer, Dawson, 1872, Geo. Mag., vol. 9, Coal Meas. [Sig. claw-bearing; from the peculiar appendages to the hind foot.]

SPHEROPEZIUM, King, 1845, Am. Jour. Sci., vol. 48. [Ety. *sphaira*, sphere; *pezia*, sole of the foot.]

leptodactylum, King, 1845, Am. Jour. Sci., vol. 48, Coal Meas. [Sig. slender-fingered or slender-toed.]

ovoidactylum, King, 1845, Am. Jour. Sci., vol. 48, Coal Meas. [Sig. having egg-shaped toes.]

pachydactylum, King, 1845, Am. Jour. Sci., vol. 48, Coal Meas. [Sig. thick-toed.]

thærodactylum, King, 1845, Am. Jour. Sci., vol. 48, Coal Meas. [Sig. hinge-toed.]

THENAROPUS, King, 1846, Am. Jour. Sci., vol. 48. [Ety. *thenaros*, palm of the hand; *pous*, foot.]

heterodactylus, King, 1845, Am. Jour. Sci., vol. 48, Coal Meas. [Sig. irregular-toed.]

leptodactylus, King, 1846, Proc. Acad. Nat. Sci., vol. 2, Coal Meas. [Sig. slender-toed.]

ovoidactylus, King, 1846, Proc. Acad. Nat. Sci., vol. 48, Coal Meas. [Sig. ovoid-toed.]

pachydactylus, King, 1846, Proc. Acad. Nat. Sci., vol. 48, Coal Meas. [Sig. thick-toed.]

sphærodactylus, King, 1846, Proc. Acad. Nat. Sci., vol. 48, Coal Meas. [Sig. spherical-toed.]

THYRSIDIUM, Cope, 1875, Ohio Pal., vol. 2. [Ety. *thyrsos*, light shaft.]

fasciculare, Cope, 1875, Ohio Pal., vol. 2, Coal Meas. [Sig. a small bundle.]

TUDITANUS, Cope, 1871, Proc. Am. Phil. Soc. [Ety. proper name.]

brevirostris, Cope, 1874, Trans. Am. Phil. Soc., Coal Meas. [Sig. short-beaked.]

huxleyi, Cope, 1874, Trans. Am. Phil. Soc., Coal Meas. [Ety. proper name.]

longipes, Cope, 1874, (Sauropleura longipes) Trans. Am. Phil. Soc., Coal Meas. [Sig. long-footed.]

mordax, Cope, 1875, Ohio Pal., vol. 2, Coal Meas. [Sig. given to biting.]

obtusus, Cope, 1868, Proc. Acad. Nat. Sci. Phil., Coal Meas. [Sig. obtuse.]

punctulatus, Cope, 1874, Trans. Am. Phil. Soc., Coal Meas. [Sig. well punctured.]

radiatus, Cope, 1874, Trans. Am. Phil. Soc., Coal Meas. [Sig. radiating from a point.]

ADDENDA.

Prof. James Hall, of New York, has proposed new names for his species, that were found preoccupied in the course of this work. Part of these new names were received in time to fall in their proper places, others are added here.

AMPLEXUS hamiltoniæ, Hall, 1876, Illust. Devonian Foss., Ham. Gr. [Ety. proper name.]
intermittens, Hall, 1876, Illust. Devonian Foss., Ham. Gr. [Sig. interrupted, left at intervals.]
BELLEROPHON textilis, Hall, 1877, Warsaw Gr. [Sig. woven.] Proposed instead of B. cancellatus, Hall, 1858, which was preoccupied.
triliratus, Hall, 1877, Chemung Gr. [Sig. three-lined.] Proposed instead of B. tricarinatus, Hall, 1876, which was preoccupied.
BUCANIA euomphaloides, Owen, 1862, Geo. Sur. Ind. [Ety. from its resemblance to an Euomphalus.] Not very satisfactorily defined.
profunda, Hall, 1859, is the same as B. (Euomphalus) profunda, Conrad, 1841.
CHÆTETES fruticosus, Hall, 1876, Illust. Devonian Foss., Ham. Gr. [Sig. bushy.]
furcatus, Hall, 1876, Illust. Devonian Foss., Ham. Gr. [Sig. forked.]
humilis, Hall, 1876, Illust. Devonian Foss., Up. Held. Gr. [Sig. small.]
tabulatus, Hall, 1876, Illust. Devonian Foss., Low. Held. Gr. [Sig. tabulated.]
tenuis, Hall, 1876, Illust. Devonian Foss., Up. Held. Gr. [Sig. slender.]
CONULARIA crawfordsvillensis, Owen, 1862, Geo. Sur. Ind., Keokuk Gr. [Ety. proper name.]
CYATHOPHYLLUM conatum, Hall, 1876, Illust. Devonian Foss., Ham. Gr. [Sig. an effort.]
galerum, Hall, 1876, Illust. Devonian Foss., Ham. Gr. [Sig. a cap.]
gradatum, Hall, 1876, Illust. Devonian Foss., Ham. Gr. [Sig. made with steps.]
nanum, Hall, 1876, Illust. Devonian Foss., Ham. Gr. [Sig. a dwarf.]
nepos, Hall, 1876, Illust. Devonian Foss., Ham. Gr. [Ety. proper name.]
palum, Hall, 1876, Illust. Devonian Foss., Ham. Gr. [Sig. a stake.]

perlamellosum, Hall, 1876, Illust. Devonian Foss., Up. Held. Gr. [Sig. having many lamellæ.]
robustum, Hall, 1876, Illust. Devonian Foss., Ham. Gr. [Sig. robust.]
subcaespitosum, Meek, 1872, Hayden's Geo. Rep., Low. Carb. [Ety. from a resemblance to C. caespitosum.]
validum, Hall, 1876, Illust. Devonian Foss., Up. Held. Gr. [Sig. stout.]
CYCLONEMA obsolescens, Hall, 1877, Chemung Gr. [Sig. grown old.] Proposed instead of C. obsoleta, Hall, 1876, which was preoccupied.
CYRTOCERAS hallanum, D'Orbigny, 1850, Prodrome de Pal., Trenton Gr. [Ety. proper name.] Proposed instead of C. lamellosum, Hall, 1847, which was preoccupied.
olenus, Hall, 1877, Schoharie grit. [Ety. mythological name.] Proposed instead of C. orion, Hall, 1876, which was preoccupied.
subannulatum, D'Orbigny, 1850, Prodr. de Pal., Black Riv. & Trenton Gr. [Sig. somewhat annulated.] Proposed instead of C. annulatum, Hall, 1847, which was preoccupied.
subarcuatum, D'Orbigny, 1850, Prodr. de Pal., Trenton Gr. [Sig. somewhat arched.] Proposed instead of C. arcuatum, Hall, 1847, which was preoccupied.
subcancellatum, Hall, 1877, Niagara Gr. [Sig. somewhat cancellated.] Proposed instead of C. cancellatum, Hall, 1852, which was preoccupied.
tenuistriatum, Hall, 1877, Trenton Gr. [Sig. finely lined.] Proposed instead of C. corniculum, Hall, 1862, which was preoccupied.
CYSTIPHYLLUM conifollis, Hall, 1876, Illust. Devonian Foss., Ham. Gr. [Sig. an inflated cone.]
corrugatum, Hall, 1876, Illust. Devonian Foss., Ham. Gr. [Sig. corrugated.]

varians, Hall, 1876, Illust. Devonian Foss., Ham. Gr. [Sig. variable.]
DENTALIUM sublœve, Hall, 1877, Coal Meas. [Sig. somewhat smooth.] Proposed instead of *D. obsoletum*, Hall, 1858, which was preoccupied.
DISCINA saffordi, Winchell, 1869, Geo. Tenn., Low. Carb. [Ety. proper name.]
EODON, Hall, 1877. Proposed instead of Microdon, Conrad, which was preoccupied. [Ety. *eos*, beginning; *odous*, a tooth; from the rudimentary teeth.]
bellistriatus, Conrad, 1842, (Microdon bellistriata) Jour. Acad. Nat. Sci., vol. 8, Ham. Gr. [Sig. beautifully lined.]
complanatus, Hall, 1870, (Microdon complanatus) Prelim. Notice Lam. shells, Ham. Gr. [Sig. smoothed.]
gregarius, Hall, 1870, (Microdon gregaria) Prelim. Notice Lam. shells, Ham. Gr. [Sig. gregarious.]
reservatus, Hall, 1870, (Microdon reservatus) Prelim. Notice Lam. shells, Waverly Gr. [Sig. reserved.]
tenuistriatus, Hall, 1870, (Microdon tenuistriata) Prelim. Notice Lam. shells, Ham. Gr. [Sig. finely lined.]
EUOMPHALUS decollatus, Hall, 1877, Low. Held. Gr. [Sig. decollated, beheaded.] Proposed instead of *E. disjunctus*, Hall, 1859, which was preoccupied.
hecale, Hall, 1876, is the same species described by Hall, 1843, as *E. depressus*, which was preoccupied.
EUSARCUS, Grote & Pitt, 1876, Buff. Acad. Nat. Sci., vol. 3. [Ety. *eusarkos*, fleshy.] This belongs to the family *Eurypteridæ*.
scorpionis, Grote & Pitt, 1876, Buff. Acad. Nat. Sci., Waterlime Gr. [Sig. a scorpion.]
FAVOSITES arbuscula, Hall, 1876, Illust. Devonian Foss., Ham. Gr. [Ety. proper name.]
argus, Hall, 1876, Illust. Devonian Foss., Ham. Gr. [Ety. mythological name.]
emmonsi, Hall, 1876, Illust. Devonian Foss., Up. Held. Gr. [Ety. proper name.]
epidermatus var. biloculi, Hall, 1876, Illust. Devonian Foss., Up. Held. Gr. [Sig. having a double receptacle.]
epidermatus var. corticosus, Hall, 1876, Illust. Devonian Foss., Up. Held. Gr. [Sig. covered with a crust.]
explanatus, Hall, 1876, Illust. Devonian Foss., Ham. Gr. [Sig. spread out.]
hamiltoniae, Hall, 1876, Illust. Devonian Foss., Ham. Gr. [Ety. proper name.]
hemisphericus var. distortus, Hall, 1876, Illust. Devonian Foss., Up. Held. Gr. [Sig. distorted.]
hemisphericus var. rectus, Hall, 1876, Illust. Devonian Foss., Up. Held. Gr. [Sig. straight.]
GRAPTOLITHUS subtenuis, Hall, 1877, Hud. Riv. Gr. [Sig. somewhat slender.] Proposed instead of *G. tenuis*, Hall, which was preoccupied.

GYROCERAS rhombolineare, Owen, 1862, Geo. Sur. Indiana. [Sig. from the rhomb-like markings.]
HALYSITES sexto-attenuatus, Owen, 1862, Geo. Sur. Indiana, Niagara Gr. [Sig. having six contracted places.]
HELIOPHYLLUM arachne, Hall, 1876, Illust. Devonian Foss., Ham. Gr. [Ety. mythological name.]
confluens, Hall, 1876, Illust. Devonian Foss., Ham. Gr. [Sig. confluent, running together.]
degener, Hall, 1876, Illust. Devonian Foss., Ham. Gr. [Sig. degenerated.]
halli var. reflexum, Hall, 1876, Illust. Devonian Foss., Ham. Gr. [Sig. reflexed.]
halli var. obconicum, Hall, 1876, Illust. Devonian Foss., Ham. Gr. [Sig. inversely conical.]
irregulare, Hall, 1876, Illust. Devonian Foss., Ham. Gr. [Sig. irregular.]
proliferum, Hall, 1876, Illust. Devonian Foss., Ham. Gr. [Sig. prolific.]
LINGULA perplexa, Hall, 1877, Clinton Gr. [Sig. perplexing.] Proposed instead of *L. elliptica*, Hall, 1843, which was preoccupied.
LOXONEMA emaceratum, Hall, 1877, Low. Held. Gr. [Sig. thin, emacerated.] Proposed instead of *L. attenuatum*, Hall, 1859, which seems to have been preoccupied by Stevens, 1858, under the name *Chemnitzia attenuata*.
semicostata, Meek, is a syn. for *L.* (Chemnitzia) attenuata of Stevens.
MACROCHEILUS attenuatum, Hall, 1877, Chazy Gr. [Sig. attenuated.] Proposed instead of *M. fusiforme*, Hall, 1858, which was preoccupied.
MICHELINIA dividua, Hall, 1876, Illust. Devonian Foss., Ham. Gr. [Sig. divisible.]
stylopora, Eaton, 1832, (Astrea stylopora) Geo. Text Book, Ham. Gr. [Sig. having pointed cells.]
MURCHISONIA decurta, Hall, 1877, Chazy Gr. [Sig. cut off.] Proposed instead of *M. abbreviata*, Hall, 1847, which was preoccupied.
micula, Hall, 1877, Ham. Gr. [Sig. a little crumb.] Proposed instead of *M. turricula*, Hall, 1862, which was preoccupied.
milleri, Hall, 1877, Trenton & Cin'ti Gr. [Ety. proper name.] Proposed instead of *M. bicincta*, Hall, 1847, which was preoccupied.
ORTHOCERAS clintoni, Hall, 1877, Chazy Gr. [Ety. proper name.] Proposed instead of *O. subarcuatum*, Hall, 1847, which was preoccupied.
desideratum, Hall, 1877, Low. Held. Gr. [Sig. to be desired.] Proposed instead of *O. clavatum*, Hall, 1859, which was preoccupied.

hallanum, S. A. Miller, 1877, Cin'ti Gr. [Ety. proper name.] Proposed instead of *O. halli*, in Cin. Quar. Jour. Sci., which was preoccupied by Barrande.
olorus, Hall, 1877, Trenton Gr. [Ety. mythological name.] Proposed instead of *O. vertebrale*, Hall, 1847, which was preoccupied.
orus, Hall, 1877, Niagara Gr. [Ety. mythological name.] Proposed instead of *O. columnare*, Hall, 1860, which was preoccupied.
socialis, Hall, 1877, Hud. Riv. Gr. [Sig. social.] Proposed instead of *O. gre-'garium*, Hall, 1861, which was preoccupied.
stylus, Hall, 1877, Schoharie grit. [Sig. an upright pointed body.] Proposed instead of *O. baculum*, Hall, 1862, which was preoccupied.
subcancellatum, Hall, 1877, Niagara Gr. [Sig. somewhat cancellated.] Proposed instead of *O. cancellatum*, Hall, 1852, which was preoccupied.
sublæve, D'Orbigny, 1850, Prodrome de Pal., Onondaga Gr. [Sig. somewhat smooth.] Proposed instead of *O. læve*, Hall, 1843, which was preoccupied.

PHILLIPSIA tennesseensis, Winchell, 1869, Geo. of Tenn., Low. Carb. [Ety. proper name.

PLATYCERAS pabulocrinus, Owen, 1862, (Pileopsis pabulocrinus) Geo. Sur. Indiana, Keokuk Gr. [Ety. from an erroneous idea that the species subsisted on the crinoid.]

Platystoma lineatum, Hall, is identical with *P. lineatum* Conrad.

PLEUROTOMARIA subglobosa, Hall, 1877, Warsaw Gr. [Sig. somewhat globose.] Proposed instead of *P. rotundata*, Hall, 1858, which was preoccupied.
nodulosa, Hall, 1847, is probably a poor specimen of Trochonema umbilicata.

tenuimarginata, Hall, 1877, Warsaw Gr. [Sig. slender-margined.] Proposed instead of *P. concava*, Hall, 1858, which was preoccupied.

POLYPHEMOPSIS teretiformis, Hall, 1877, Warsaw Gr. [Sig. of a long round-shape.] Proposed instead of *P. elongata*, Hall, 1858, which was preoccupied.

PRODUCTUS incurvatus, Shepard, 1838, Am. Jour. Sci., vol. 34, Coal Meas. [Sig. from the incurved basal margin.]
pectenoideus, Shepard, 1838, Am. Jour. Sci., vol. 34, Coal Meas. [Sig. like a shell of the genus *Pecten*.]
semipunctatus, Shepard, 1858, Am. Jour. Sci., vol. 34, Coal Meas. [Sig. halfpunctated.]

Quenstedtia, a genus, on page 60, was preoccupied in the Class Lamellibranchiata, by Morris & Lycett, in 1853.

STREPTELASMA ungula, Hall, 1876, Illust. Devonian Foss., Ham. Gr. [Sig. a hoof.]
profundum, Hall, is probably identical with S. (Cyathophyllum) profundum, Conrad.

STRIATOPORA limbata, Eaton, 1832, (Madrepora limbata) Geo. Text Book, Ham. Gr. [Sig. bordered.]

Terebratula argentea, see Athyris argentea.

TREMATOPORA spiculata, Hall, 1877, Niagara Gr. [Sig. having little points.] Proposed instead of *T. spinulosa*, Hall, 1876, which was preoccupied.

ZAPHRENTIS ampla, Hall, 1876, Illust. Devonian Foss., Ham. Gr. [Sig. full, large.]
cannonensis, Winchell, 1869, Geo. Tenn., Low. Carb. [Ety. proper name.]
eccentrica, Meek, 1871, Hayden's Geo. Rep., Low. Carb. [Sig eccentric, from the center.]
halli, Edwards & Haime, 1851, Polyp. Foss. Terr. Palæoz., Ham. Gr. [Ety. proper name.]

CORRIGENDA.

In addition to the facilities offered in the Preface, in the article of Prof. Claypole, and in the Index of Genera for correcting specific names, the following corrections are deemed worthy of notice:

Read horny fracture for corny fracture, on page 13.
Read opusculum for opusculus, on page 67.
Read pistillum for pistillus, on page 72.
Read rhombiferus for rhombiferous, on page 85.
Read remus for remos, on page 85.
Read varsoviensis for varsouviensis, on page 86.
Read pumilus for pumilis, on page 87.
Read differens for differentis, on page 104.

THE

AMERICAN
PALÆOZOIC FOSSILS:

A CATALOGUE

OF THE

GENERA AND SPECIES,

WITH

NAMES OF AUTHORS, DATES, PLACES OF PUBLICATION, GROUPS OF ROCKS IN WHICH FOUND AND THE ETYMOLOGY AND SIGNIFICATION OF THE WORDS,

AND

AN INTRODUCTION

DEVOTED TO THE STRATIGRAPHICAL GEOLOGY OF THE PALÆOZOIC ROCKS.

SECOND EDITION—JANUARY, 1883.

BY S. A. MILLER.

CINCINNATI, OHIO:
PUBLISHED BY THE AUTHOR, No. 8 W. THIRD ST.

1883.

Entered according to act of Congress, in the year 1882, by
S. A. MILLER,
In the Office of the Librarian of Congress, at Washington.

PREFACE TO THE SECOND EDITION.

The first edition of this work was published by the author in 1877. No work of the kind had ever appeared, and no encouragement for the undertaking had been received. The author had confidence in its value to science, and published it himself. This condition of affairs is well expressed by the distinguished Prof. Leo. Lesquereux, in a letter addressed to the author, while the work was in press, on the 28th of March, 1877. He said:

"Your kind favor of the 25th, and the page of prospectus of your catalogue are received. You will please allow me to say freely, that when I received your first letter advising me of the preparation or publication of a catalogue of the American Palæozoic Fossils, I did not think much of a work of the kind, for the reason that I knew by experience what amount of time and research would be necessary for the preparation of such a work as the one you were about to publish, and could not suppose that in this country, where science is so generally considered under a remunerative point of view, there was a naturalist disposed to make the researches you have so well done for mere love of science. But after the perusal of your first proof sheets, my opinion was fully reversed, and I recognized immediately the great value of your catalogue, the assistance which it will afford to all the palæontologists, and the accuracy with which you have pursued the immense amount of researches necessitated for its preparation. For what concerns myself personally, I thank you sincerely for this production, and I am certain that every palæontologist will do the same."

After he had received a copy of the work, he again wrote as follows:

"I am sincerely glad to have the whole catalogue, though I have also good use of the separate copy of the botanical part, which, bound with intermixed blank leaves, serves me as a very commodious referendum, especially for new American species of coal plants. Your excellent, most useful catalogue, in its fullness receives a degree of interest from the Introductory remarks, those on the construction of systematic names, and the Geological introduction. The work is, indeed, one of devotion to science. It may not bring you pecuniary reward, but you will have the satisfaction to know that it has helped many, and been of great use, or will be hereafter, for the advancement of palæontological science. For, indeed, it is a true dictionary, and should be in the hands of every American geologist."

No testimonials of the value of the work were ever solicited, but they came, nevertheless, from all parts of this country and from abroad. Only a few more will be selected. Prof. James Hall, of New York, wrote:

"You have done a valuable work, for which the thanks of every describer of fossils are due; indeed, I do not recall the publication of a single book within the past twenty years so necessary and useful to the geologist and palæontologist."

Prof. John H. Klippart, of Columbus, Ohio:

"I like the work very much. It is just such a work as I have long since very much needed. It will prove to the student of fossils what Webster's unabridged proves to the scholar—that it is simply indispensable. There is no work to take its place, or that can be substituted for it."

Prof. G. C. Broadhead, of Missouri :

"I find it a useful book for a student of palæontology, and it supplies a necessity which experience has demanded."

Prof. Henry G. Hanks, of California :

"I am very much pleased with the work. It is just what I want."

Prof. R. H. Holbrook, of New Jersey :

"I consider it a valuable work worth many times the price."

Rev. J. Allen, President Alfred University, Pa. :

"We find your book a very valuable production—a great aid."

Prof. J. W. Dawson, of Montreal:

"I have found the book very useful, and hope you will follow it up with supplements as new species are described."

Prof. G. Lindstrom, of Stockholm, Sweden :

"I bought your excellent catalogue of the North American Palæozoic Fossils, for use in the Museum, and I dare say that scarcely a day passes by without my consulting it."

The notices this book has received in scientific periodicals have been no less flattering, as will be observed in the *Annals and Magazine of Natural History*, 1878, 5th ser., vol. i., p. 99, and the *Geological Magazine* of London, for October 1877, p. 472.

Under these circumstances, I have been induced to prepare a supplement, for the unbound copies remaining on hand, and to publish it in this form for a second edition. The supplement contains the genera and species described since 1877, those that were overlooked in the first edition, corrected references of species to genera, corrections of the more serious mistakes and typographical errors, and a new index of the genera. The Vegetable Kingdom is arranged in orders and classes, the Class Echinodermata is re-subdivided, some new orders and new families are briefly defined, and a few specific names have been proposed by the author for those preoccupied, which constitute the new features of the supplement. In addition to the corrections, the fossils described in the 23d and 24th Reports of the Reg. of the Univ. of N. Y., on the condition of the St. Cab. of Nat. Hist., should be accredited to Hall and Whitfield, instead of to Hall alone, and the Cincinnati Group should, in all cases, be stricken out, and, in nearly every instance, the Hud. Riv. Gr. inserted, a few species only being referable to the Trenton and Utica Slate.

S. A. MILLER.

CINCINNATI, OHIO, *December, 1882.*

SUPPLEMENT.

VEGETABLE KINGDOM.

In the first edition of this work, no attempt was made to arrange the fossils of the vegetable kingdom, represented in the palæozoic rocks, into classes, orders, or families, for reasons then expressed; since that time, however, the work of Leo Lesquereux on the coal flora of Pennsylvania has rendered possible a classification which will be of some utility, though only approximately correct.

CELLULAR CRYPTOGAMOUS PLANTS.

CLASS FUNGI.—Rhizomorpha.
CLASS THALASSOPHYTES.—Arthraria, Arthrophycus, Asterophycus, Blastophycus, Buthotrephis, Calamophycus, Chondrites, Conostichus, Cruziana, Dactylophycus, Discophycus, Dystactophycus, Eophyton, Heliophycus, Hippodophycus, Ichnophycus, Licrophycus, Palæochorda, Palæophycus, Phytopsis, Protostigma, Rusophycus, Sphenothallus, Taonurus, Trichophycus.

VASCULAR CRYPTOGAMOUS PLANTS.

CLASS EQUISETACEÆ.—Anarthrocanna, Annularia, Arthrostigma, Asterophyllites, Bechera, Bornia, Calamites, Calamodendron, Calamocladus, Calamostachys, Equisetites, Macrostachya, Nematophyllum, Solenoula, Sphenophyllum.
CLASS FILICACEÆ (ferns).
ORDER NEUROPTERIDS. — Danæites, Dictyopteris, Idiophyllum, Lesleya, Lonchopteris, Megalopteris, Neriopteris, Neuropteris, Odontopteris, Orthogoniopteris, Tæniopteris.
ORDER ALETHOPTERIDS. — Alethopteris, Callipteris, Callipteridium, Lescuropteris, Protoblechnum.
ORDER PSEUDOPECOPTERIDS.—Pseudopecopteris.
ORDER PECOPTERIDS.—Asterocarpus, Cymoglossa, Goulopteris, Oligocarpia, Pecopteris.
ORDER SPHENOPTERIDS.—Asplenites, Eremopteris, Sphenopteris.
ORDER ADIANTITES.—Aneimites, Archæopteris, Cyclopteris, Triphyllopteris.
FERNS OF UNCERTAIN AFFINITY.—Crematopteris, Hymenophyllites, Pachypteris, Rhacophyllum.
RACHIS OF FERNS.—Rhachiopteris.
RHIZOMAS OF FERNS.—Rhizomopteris, Stigmarioides.
SEPARATE FRUCTIFICATIONS OF FERNS.—Sorocladus.
STEMS OR TRUNKS OF FERNS.—Caulopteris, Megaphytum, Psaronius, Stemmatopteris.
CLASS LYCOPODIACEÆ.—Arthrostigma, Cyclostigma, Decheula, Diplostegium, Glyptodendron, Halonia, Knorria, Lepidocystis, Lepidodendron, Lepidophloios, Lepidophyllum, Lepidostrobus, Leptophloeum, Lycopodites, Psilophyton, Polyporites, Sporocystis, Sternbergia, Ulodendron.
ORDER TÆNIOPHYLLEÆ.—Tæniophyllum.
ORDER SIGILLARIEÆ.—Didymophyllum, Pinnularia, Rhizolithes, Sigillaria, Sigillarioides, Stigmaria, Syringodendron,
ORDER NŒGGERATHIEÆ.—Nœggerathia, Whittleseya.
CLASS CORDAITEÆ.—Antholithes, Cardiocarpon, Carpolithes, Cordaianthus, Cordaicarpus, Cordaites, Cordaistrobus, Desmiophyllum, Dicranophyllum, Guilleluites, Lepidoxylon, Rhabdocarpus, Trigonocarpum.
CONIFERÆ (?)—Araucarites, Baiera, Dadoxylon, Nematoxylon, Ormoxylon, Prototaxites, Saportæa, Syringoxylon, Walchia.
PLANTS OF UNCERTAIN AFFINITY. — Acanthophyton, Asteropteris, Celluloxylon, Palæoxyris, Spirangium, Sporangites.

ALETHOPTERIS *acuta* is Pecopteris acuta.
ambigua, Lesquereux, 1880, Coal Flora of Pa., Coal Meas. [Sig. uncertain.]
aquilina was first called *Filicites aquilinus* by Schlotheim in his Flor. d. Vorw.
crenulata, as identified by Lesquereux, in the Geo. Sur., Ill., vol. 2, is now described as Pseudopecopteris subcrenulata.
cristata is Pecopteris cristata.
distans is a variety of A. lonchitica.
emarginata is Pecopteris emarginata.
erosa is Pecopteris erosa.
gibsoni, Lesquereux, 1880, Coal Flora of Pa., Coal Meas. [Ety. proper name.]
halli is Pecopteris halli.
helenæ, Lesquereux, 1880, Coal Flora of Pa., Coal Meas. [Ety. proper name.]
holdeni is Protoblechnum holdeni.
hymenophylloides is Pseudopecopteris hymenophylloides.
inflata is Callipteridium inflatum.
lanceolata is Pecopteris lanceolata.
lonchitica instead of A. lonchitidis.
longifolia is Pecopteris longifolia.
macrophylla is Danæites macrophyllus.
massillonis is Callipteridium massillonicum.
mazonana is Pseudopecopteris mazonana.
muricata is Pseudopecopteris muricata.
nervosa is Pseudopecopteris nervosa.
obscura, syn. for Callipteridium rugosum.
oweni is Callipteridium oweni.
pluckeneti is Pseudopecopteris pluckeneti.
pteroides is Pecopteris pteroides.
rugosa is Callipteridium rugosum.
serrula is Pecopteris serrula.
solida is Pecopteris solida.
stellata is Pecopteris stellata.
tæniopteroides is Pecopteris tæniopteroides.
virginiana, Fontaine and White, 1880, Perm. or Up. Carb. Flora, Coal Meas. or Permian. [Ety. proper name.]
ANNULARIA was described by Sternberg, in 1820, Essai d'un expose Geognostico-botanique de la flore du monde primitif, 2d cahier.
acuminata is Sporangites acuminatus.
dawsoni is Asterophyllites dawsoni.
emersoni, Lesquereux, 1880, Coal Flora of Pa., Coal Meas. [Ety. proper name.]
romingeri, Lesquereux, 1877, Trans. Am. Phil. Soc., Low. Held. Gr. [Ety. proper name.]
ANTHOLITHES priscus was described in 1873, in Ohio Pal., vol. 1.
APHLEBIA *adnascens*, see Rhacophyllum adnascens.
flabellata, see Rhacophyllum flabellatum.
irregularis, see Rhacophyllum irregulare.
ARCHÆOPTERIS *bockschiana*, Lesquereux, 1858 (Nœggerathia bockschiana), Geo. of Penn., vol. 2, Coal Meas. [Ety. proper name.]
gaspensis, Dawson, 1881, Can. Nat. & Geo., vol. 10, Devonian. [Ety. proper name.]
ARCHÆOPTERIS *hallana* instead of Cyclopteris hallana.
hybernica, Forbes, 1852 (Cyclopteris hybernica), Proc. Brit. Assoc., Chemung Gr. [Ety. proper name.]
jacksoni instead of Cyclopteris jacksoni.
minor instead of Nœggerathia minor.
obliqua instead of Nœggerathia obliqua.
rogersi instead of Cyclopteris rogersi.
Aristophycus, Miller & Dyer, 1878, Cont. to Pal., No. 2. [Ety. *aristos*, excellent; *phukos*, a sea plant.] Probably inorganic.
ramosum, Miller & Dyer, 1878, Cont. to Pal., No. 2, Hud. Riv. Gr. [Sig. branchy.] Probably inorganic.
ramosum var. *germanum*, Miller & Dyer, 1878, Cont. to Pal., No. 2, Hud. Riv. Gr. [Sig. near of kin.] Probably inorganic.
Asolanus, Wood, 1860, Proc. Acad. Nat. Sci. This genus is not recognized by botanists, and is generally regarded as a synonym for *Sigillaria*. A. *camptotænia*, is classed by Lesquereux as a synonym for *Sigillaria monostigma*, and the other two names mentioned by Wood, A. *manephlæus*, and A. *ornithicnoides*, are referred to *Sigillaria*.
ASTEROPHYCUS simplex, Lesquereux, 1880, Coal Flora of Pa., Coal Meas. [Sig. simple.]
ASTEROPHYLLITES anthracinus, Heer, 1877, Fl. Foss. Helv., vol. 4, Coal Meas. [Sig. coal black.]
aperta, syn. for Macrostachya infundibuliformis.
equisetiformis was called by Schlotheim, Casuarinites equisetiformis, in 1804.
fasciculatus, Lesquereux, 1880, Coal Flora of Pa., Coal Meas. [Sig. bundled.]
lanceolatus syn. for A. foliosus.
latifolius should be restored.
longifollus was called Bruckmannia longifolia, by Sternberg, in Tent. flor. prim.
radiatus, Brongniart, 1822, Class. d. Veg. Foss. Coal Meas. [Sig. rayed.]
rigidus was called Bruckmannia rigida, by Sternberg, in Tent. flor. prim.
ASTEROPTERIS, Dawson, 1881, Lond. Quar. Jour. Geo. Soc., vol. 37. [Ety. *aster*, a star; *pteris*, a fern.]
novoboracensis, Dawson, 1881, Lond. Quar. Jour. Geo. Soc., vol. 37, Portage Gr. [Ety. proper name.]
BAIERA, Fr. Braun, 1840, Die Petrefakten d. Naturalien Samml. [Ety. proper name.]
virginiana, Fontaine and White, 1880, Perm. or Up. Carb. Flora, Coal Meas. or Permian. [Ety. proper name.]
Bergeria rhombica, see Lepidodendron rhombicum.

VEGETABLE KINGDOM.

BLASTOPHYCUS, Miller & Dyer, 1878, Jour. Cin. Soc. Nat. Hist., vol. 1. [Ety. *blastos*, a bud; *phukos*, a sea weed.]
diadematum, Miller & Dyer, 1878, Jour. Cin. Soc. Nat. Hist., vol. 1, Utica Slate Gr. [Sig. diademed.]
BORNIA, F. A. Rœmer, 1854, Palæontographica, vol. 3. [Ety. proper name.]
radiata, Brongniart, 1828 (Calamites radiatus), Hist. d. Veg. Foss., Sub-carboniferous. [Sig. radiated.]
Brachyphyllum obtusum is Lepidocystis obtusus.
Bruckmannia longifolia, see Asterophyllites longifolius.
rigida, see Asterophyllites rigidus.
BUTHOTREPHIS asteroides, Fitch, 1849, Trans. Ag. Soc., Quebec Gr. [Sig. star like.]
rigida, Emmons, 1844 (Fucoides rigidus) Taconic Syst., Quebec Gr. [Sig. rigid.]
CALAMITES *arenaceus* is a Triassic species.
decoratus may be erased.
dubius, was published in 1825.
major, Weiss, 1872, Fossil Flora d. jungsten steinkohlen formation, Coal Meas. [Sig. large.]
radiatus, see Bornia radiata.
ramifer, Stur, 1875, Culm. Flora d. Mährisch-Schlesischen Dachschiefers, Coal Meas. [Sig. bearing branches.]
ramosus, was published in 1825.
sulcatus may be erased.
transitionis is Gœppert's species, as identified by Dawson.
undulatus was described by Brongniart in 1828, in Hist. d. Veg. Foss., vol. 1.
CALAMODENDRON is Brongniart's Genus.
CALAMOPHYCUS, Lesquereux, 1877, Proc. Am. Phil. Soc. [Ety. *calamus*, a reed; *phukos*, a sea plant.]
septus, Lesquereux, 1877, Proc. Am. Phil. Soc., Devon. [Sig. enveloped.]
CALAMOSTACHYS, Schimper, 1869, Traite de Paleontologie Vegetale, vol. 1, p. 328. · [Ety. *Calamus*, a reed; *Stachys*, a plant.]
prælongus, Lesquereux, 1880, Coal Flora of Pa., Coal Meas. [Sig. very long.]
CALLIPTERIDIUM, Weiss, 1872, Foss. Flora d. jungsten steinkohlen formation. [Ety. from the genus *Callipteris*.]
aldrichi, Lesquereux, 1880, Coal Flora of Pa., Coal Meas. [Ety. proper name.]
dawsonanum, Fontaine & White, 1880, Perm. or Up. Carb. Flora, Coal Meas. or Permian. [Ety. proper name.]
grandifolium, Fontaine & White, 1880, Perm. or Up. Carb. Flora, Coal Meas. or Permian. [Sig. large leaved.]
inflatum, Lesquereux, 1870 (Alethopteris inflata), Geo. Sur. Ill., vol. 4. Coal Meas. [Sig. inflated.]
inæquale, Lesquereux, 1880, Coal Flora of Pa., Coal Meas. [Sig. unequal.]
mansfieldii, Lesquereux, 1880, Coal Flora of Pa., Coal Meas. [Ety. proper name.]

CALLIPTERIDIUM massilloneum instead of Alethopteris massillonis.
membranaceum, Lesquereux, 1880, Coal Flora of Pa., Coal Meas. [Sig. membranaceous].
neuropteroides, Lesquereux, 1880, Coal Flora of Pa., Coal Meas. [Sig. like *Neuropteris*.]
oblongifolium, Fontaine & White, Perm. or Up. Carb. Flora, Coal Meas. or Perm. [Sig. oblong leaved.]
odontopteroides, Fontaine & White, 1880, Perm. or Up. Carb. Flora, Coal Meas. or Perm. [Sig. like *Odontopteris*.]
oweni instead of Alethopteris oweni.
pardeei, Lesquereux, 1880, Coal Flora of Pa., Coal Meas. [Ety. proper name.]
rugosum instead of Alethopteris rugosa.
sullivanti instead of Callipteris sullivanti.
unitum, Fontaine & White, 1880, Perm. or Up. Carb. Flora, Coal Meas. or Permian. [Sig. united.]
CALLIPTERIS conferta, Sternberg (Sphenopteris conferta), Tent. flor. prim., Coal Meas. or Permian. [Sig. close together.]
pilosa instead of Sphenopteris pilosa.
CARDIOCARPON annulatum, C. elongatum, C. latum, C. minus, C. orbiculare, and C. samaræforme, were described in 1873, in Ohio Pal., vol. 1.
apiculatum, Gœppert & Berger, 1848, De fructibus et seminibus, Coal Meas. [Sig. having a pointed termination.]
bicornutum instead of Ptilocarpus bicornutus.
congruens, Grand'Eury, 1877, Flore Carbonifere, Coal Meas. [Sig. running together.]
diminutivum, Lesquereux, 1880, Coal Flora of Pa., Coal Meas. [Sig. diminutive.]
fasciculatum, Lesquereux, 1880, Coal Flora of Pa., Coal Meas. [Sig. bundled.]
late-alatum, Lesquereux, 1880, Coal Flora of Pa., Coal Meas. [Sig. broad winged.]
mamillatum instead of Rhabdocarpus mamillatus.
marginatum was described in 1825.
ovatum, Grand'Eury, 1877, Flore Carbonifere, Coal Meas. [Sig. ovate.]
pachytesta, Lesquereux, 1880, Coal Flora of Pa., Coal Meas. [Sig. having thick testa.]
regulare, Sternberg, 1821-38, Flor. d. Vorwelt., Coal Meas. [Sig. regular.]
simplex, Lesquereux, 1880, Coal Flora of Pa., Coal Meas. [Sig. simple.]
zonulatum, Lesquereux, 1880, Coal Flora of Pa., Coal Meas. [Sig. small girded.]
CARDIOPTERIS, Schimper, 1869, Traite de Paleontologie Vegetale, vol. 1, p. 451. [Ety. *kardia*, heart; *pteris*, a fern.]

CARDIOPTERIS eriana, Dawson, 1881, Quar. Jour. Geo. Soc. Lond., vol. 37, Devonian. [Ety. proper name.]
CARPOLITHES acuminatus, Sternberg, 1821-38, Flor d. Vorw., Coal Meas. [Sig. pointed.]
  bicarpus, Fontaine & White, 1880, Perm. or Up. Carb. Flora, Coal Meas. or Permian. [Sig. two fruited.]
  bullatus is Lepidocystis bullatus.
  disjunctus is a syn. for Trigonacarpon dawesi.
  jacksonensis is Rhabdocarpus jacksonensis.
  limatus, read C. lunatus.
  marginatus, Fontaine & White, 1880, Perm. or Up. Carb. Flora, Coal Meas. or Permian. [Sig. margined. This name was preoccupied by Artis in 1825.
  vesicularis is Lepidocystis vesicularis.
CASUARINITES *equisetiformis* is Asterophyllites equisetiformis.
CAULERPITES was described by Brongniart in 1828, Prodr. d. Hist. d. Veg. Foss.
  marginatus is Taonurus marginatus.
CAULOPTERIS *acanthophora* is, probably, Ulodendron punctatum.
  elliptica, Fontaine & White, 1880, Perm. or Up. Carb. Flora, Coal Meas. or Perm. [Sig. elliptical.]
  giffordi, Lesquereux, 1880, Coal Flora of Pa., Coal Meas. [Ety. proper name.]
  gigantea, Lesquereux, is Stemmatopteris gigantea.
  gigantea, Fontaine & White, 1880, Perm. or Up. Carb. Flora, Coal Meas. or Permian. [Sig. large.]
  insignis is Stemmatopteris insignis.
  lacoei, Lesquereux, 1880, Coal Flora of Pa., Coal Meas. [Ety. proper name.]
  mansfieldi, Lesquereux, 1880, Coal Flora of Pa., Coal Meas. [Ety. proper name.]
  punctata is Stemmatopteris punctata.
  wortheni is Stemmatopteris wortheni.
CELLULOXYLON, Dawson, 1881; Lond. Quar. Jour. Geo. Soc., vol. 37. [Ety. *cellula*, a small apartment; *xylon*, wood.]
  primævum, Dawson, 1880, Lond. Quar. Jour. Geo. Soc., vol. 37, Ham. Gr. [Sig. in the first period of life.]
Chlæphycus, Miller & Dyer, 1878, Cont. to Pal. No. 2. [Ety. *chlæ*, young grass; *phukos*, a sea plant.] This is probably inorganic.
  plumosum, Miller & Dyer, 1878, Cont. to Pal., No. 2, Hud. Riv. Gr. [Sig. feathered.] This is probably inorganic.
CHONDRITES *colletti* is Taonurus colletti.
CONOSTICHUS broadheadi, Lesquereux, 1880, Coal Flora of Pa., Coal Meas. [Ety. proper name.]
  prolifer, Lesquereux, 1880, Coal Flora of Pa., Coal Meas. [Sig. bearing offspring.]

CORDAIANTHUS, Grand'Eury, 1877, Flore Carbonifere. [Ety. *Cordaites*, a genus; *anthos*, a flower.]
  dichotomus, Lesquereux, 1880, Coal Flora of Pa., Coal Meas. [Sig. divided.]
  ovatus, Lesquereux, 1880, Coal Flora of Pa., Coal Meas. [Sig. ovate.]
  simplex, Lesquereux, 1880, Coal Flora of Pa., Coal Meas. [Sig. simple.]
CORDAICARPUS, Grand'Eury, 1877, Flore Carbonifere. [Ety. *Cordaites*, a genus; *karpos*, fruit.]
  apiculatus, Lesquereux, 1880, Coal Flora of Pa., Coal Meas. [Sig. pointed.]
  costatus, see Cordaites costatus.
  gutbieri, Grand'Eury, 1877, Flore Carbonifere, Coal Meas. [Ety. proper name.]
  ovatus, Grand'Eury, 1877, Flore Carbonifere, Coal Meas. [Sig. ovate.]
CORDAISTROBUS, Lesquereux, 1880, Coal Flora of Pa. [Ety. *Cordaites*, a genus; *strobus*, a cone.]
  grandeuryi, Lesquereux, 1880, Coal Flora of Pa., Coal Meas. [Ety. prop. name.]
CORDAITES communis, Lesquereux, 1878, Proc. Am. Phil. Soc., Coal Meas. [Sig. common.]
  costatus, Lesquereux, 1879, Proc. Am. Phil. Soc., Coal Meas. [Sig. costate.]
  crassinervis, Fontaine & White, 1880, Perm. or Up. Carb. Flora, Coal Meas. or Permian. [Sig. thick nerved.]
  crassus, Lesquereux, 1880, Coal Flora of Pa., Coal Meas. [Sig. thick.]
  diversifolius, Lesquereux, 1880, Coal Flora of Pa., Coal Meas. [Sig. diverse leaved.] Proposed instead of L. angustifolius, Lesquereux.
  gracilis, Lesquereux, 1878, Proc. Am. Phil. Soc., Coal Meas. [Sig. slender.]
  grandifolius, Lesquereux, 1878, Proc. Am. Phil. Soc., Coal Meas. [Sig. grand leaved.]
  lacoei, Lesquereux, 1880, Coal Flora of Pa., Coal Meas. [Ety. proper name.]
  lingulatus, Grand'Eury, 1877, Flore Carbonifere, Coal Meas. [Sig. tongue shaped.]
  mansfieldi, Lesquereux, 1880, Coal Flora of Pa., Coal Meas. [Ety. proper name.]
  radiatus, Lesquereux, 1880, Coal Flora of Pa., Coal Meas. [Sig. radiated.]
  serpens, Lesquereux, 1878, Proc. Am. Phil. Soc., Coal Meas. [Sig. creeping.]
  validus, Lesquereux, 1878, Proc. Am. Phil. Soc., Coal Meas. [Sig. strong.]
CYCLOPTERIS *alleghaniensis*, syn. for Archæopteris rogersi.
  browni is Rhacophyllum browni.
  elongata, Lesquereux, 1880, Coal Flora of Pa., Coal Meas. [Sig. elongated.]
  germari is Neuropteris germari.
  hallana is Archæopteris hallana.
  jacksoni is Archæopteris jacksoni.

CYCLOPTERIS *laciniata* is Neuropteris laciniata.
  lescuriana is Triphyllopteris lescuriana.
  problematica, Dawson, 1871, Foss. Plants, Dev. & Up. Sil., Middle Devonian. [Sig. problematical.]
  rogersi is Archæopteris rogersi.
  trichomanoides is Neuropteris trichomanoides.
  virginiana is Pseudopecopteris virginiana.
CYCLOSTIGMA affine, Dawson, 1881, Quar. Jour. Geo. Soc., Lond., vol. 37, Devonian. [Sig. near to.]
  kiltorkense, Haughton, 1860, Ann. & Mag. Nat. Hist., 3d ser., vol. 5, Subcarboniferous. [Ety. proper name.]
CYMOGLOSSA, Schimper, 1869, Traite de Paleontologie Vegetale, vol. 1, p. 553. [Ety. *kumo*, wavy; *glossa*, a tongue.]
  breviloba, Fontaine & White, 1880, Perm. & Up. Carb. Flora, Coal Meas. or Permian. [Sig. short lobed.]
  formosa, Fontaine & White, 1880, Perm. & Up. Carb. Flora, Coal Meas. or Permian. [Sig. beautiful.]
  lobata, Fontaine & White, 1880, Perm. & Up. Carb. Flora, Coal Meas. or Permian. [Sig. lobed.]
  obtusifolia, Fontaine & White, 1880, Perm. & Up. Carb. Flora, Coal Meas. or Permian. [Sig. obtuse leaved.]
DACTYLOPHYCUS, Miller & Dyer, 1878, Cont. to Pal., No. 2. [Ety. *dactylos*, a finger; *phukos*, a sea plant.]
  quadripartitum, Miller & Dyer, 1878, Cont. to Pal., No. 2, Utica Slate Gr. [Sig. four parted.]
  tridigitatum, Miller & Dyer, 1878, Cont. to Pal., No. 2, Utica Slate Gr. [Sig. three fingered.]
DANÆITES emersoni, Lesquereux, 1880, Coal Flora of Pa., Coal Meas. [Ety. proper name.]
  macrophyllus instead of Alethopteris macrophylla.
DECHENIA, Goeppert, 1841-48, Die Gattungen der fossilen Pflanzen. [Ety. proper name.]
  striata, Lesquereux, 1880, Coal Flora of Pa., Coal Meas. [Sig. striated.]
DESMIOPHYLLUM, Lesquereux, 1880, Coal Flora of Pa. [Ety. *desmos*, a band; *phyllon*, a leaf.]
  gracile, Lesquereux, 1880, Coal Flora of Pa., Coal Meas. [Sig. slender.]
DICRANOPHYLLUM, Grand'Eury, 1877, Flore Carbonifere. [Ety. *dikranos*, two pointed; *phyllon*, a leaf.]
  dichotomum, Lesquereux, 1880, Coal Flora of Pa., Coal Meas. [Sig. divided.]
  dimorphum, Lesquereux, 1878, Proc. Am. Phil. Soc., Coal Meas. [Sig. double formed.]
Dictyophyton is not a plant. It is supposed to be a sponge.
DIDYMOPHYLLUM oweni, Lesquereux, 1880, Coal Flora of Pa., Coal Meas. [Ety. proper name.]

DISCOPHYCUS, Walcott, 1879, Trans. Alb. Inst., vol. 10. [Ety. *diskos*, a disk; *phukos*, a sea weed.]
  typicale, Walcott, 1879, Trans. Alb. Inst. vol. 10, Utica Slate. [Sig. the type.]
DYSTACTOPHYCUS, Miller & Dyer, 1878, Cont. to Pal., No. 2. [Ety. *dystaktos*, hard to arrange; *phukos*, a sea plant.]
  mamillanum, Miller & Dyer, 1878, Cont. to Pal., No. 2, Hud. Riv. Gr. [Sig. protuberant.]
EQUISETITES macrodontus, Wood, 1860, Proc. Acad. Nat. Sci., Coal Meas. Not satisfactorily defined.
  wrightanus, Dawson, 1880, Lond. Geo. Mag., n. s. vol. 7, Up. Devonian. [Ety. proper name.]
EREMOPTERIS crenulata, Lesquereux, 1876, Geo. Rep. of Ala., Coal Meas. [Sig. crenulated.]
  dissecta, Lesquereux, 1876, Geo. Rep. of Ala., Coal Meas. [Sig. cut up.]
  elegans, Ettingshausen, 1852 (Asplenites elegans), Die Steinkohlen flora, v. Stradonitz in Böhmen, Coal Meas. [Sig. elegant.]
  flexuosa, Lesquereux, 1876, Geo. Rep. of Ala., Coal Meas. [Sig. flexuous.]
  microphylla, Lesquereux, 1880, Coal Flora of Pa., Coal Meas. [Sig. small leaved.]
  missouriensis, Lesquereux, 1880, Coal Flora of Pa., Coal Meas. [Ety. proper name.]
Ficoidites scabrosus, Hildreth, 1837, Am. Jour. Sci. & Arts, vol. 31, Lower Coal Meas. Not recognized.
Filicites arborescens, see Pecopteris arborescens.
  miltoni, see Pecopteris miltoni.
  plumosus, see Pecopteris plumosa.
GLYPTODENDRON, Claypole, 1878, Am. Jour. Sci. & Arts, 3d ser., vol. 15. [Ety. *glyptos*, sculptured; *dendron*, a tree.]
  eatonense, Claypole, 1878, Am. Jour. Sci. & Arts, 3d ser., vol. 15, Niagara Gr. [Ety. proper name.]
Goniopteris newberryana, see Pecopteris newberryana.
  oblonga, see Pecopteris oblonga.
GUILIELMITES, Geinitz, 1858, Leitpflanzen d. Rothleig. u. d. Zechstein; Sachsen. [Ety. from the genus *Gulielma*.]
  orbicularis, Fontaine & White, 1880, Perm. & Up. Carb. Flora, Coal Meas. or Permian. [Sig. orbicular.]
HALONIA flexuosa, Goldenburg, 1855 (Ulodendron flexuosum), Flora Sarrepontana fossilis, Coal Meas. [Sig. flexuous.]
  mansfieldi, Lesquereux, 1880, Coal Flora of Pa., Coal Meas. [Ety. proper name.]
  secreta, Lesquereux, 1880, Coal Flora of Pa., Coal Meas. [Sig. concealed.]
  tortuosa, Lindley & Hutton, 1835, Foss. Flora, vol. 2, p. 11, Coal Meas. [Sig. tortuous.]

HELIOPHYCUS, Miller & Dyer, 1878, Cont. to Pal., No. 2. [Ety. *helios*, the sun; *phukos*, a sea plant.]
stelliforme, Miller & Dyer, 1878, Cont. to Pal., No. 2, Hud. Riv. Gr. [Sig. star shaped.]
HYMENOPHYLLITES *adnascens*, refer to Rhacophyllum adnascens.
arborescens, refer to Rhacophyllum arborescens.
ballantini, refer to Sphenopteris ballantini.
clarki, refer to Rhacophyllum clarki.
flexicaulis, refer to Sphenopteris flexicaulis.
furcatus, refer to Sphenopteris furcata.
hildrethi, refer to Sphenopteris hildrethi.
inflatus, see Rhacophyllum inflatum.
lactuca, see Rhacophyllum lactuca.
mollis, see Rhacophyllum mollis.
spinosus, refer to Sphenopteris spinosa.
strongi, see Rhacophyllum strongi.
thalliformis, see Rhacophyllum thalliforme.
trichomanoides, refer to Sphenopteris trichomanoides.
tridactylites, refer to Sphenopteris tridactylites.
IDIOPHYLLUM, Lesquereux, 1880, Coal Flora of Pa. [Ety. *idios*, peculiar; *phyllon*, a leaf.]
rotundifolium, Lesquereux, 1880, Coal Flora of Pa., Coal Meas. [Sig. round leaved.]
LEPIDOCYSTIS, Lesquereux, 1880, Coal Flora of Pa. [Ety. *lepis*, a scale; *kustis*, a bladder.]
angularis, Lesquereux, 1880, Coal Flora of Pa., Coal Meas. [Sig. angular.]
bullatus instead of Carpolithes bullatus.
fraxiniformis, Gœppert & Berger, 1848 (Carpolithes fraxiniformis), De fruct. et sem., Coal Meas. [Sig. like Fraxinus.]
lineatus, Lesquereux, 1880, Coal Flora of Pa., Coal Meas. [Sig. lined.]
obtusus instead of Brachyphyllum obtusum.
pectinatus, Lesquereux, 1880, Coal Flora of Pa., Coal Meas. [Sig. pectinated.]
quadrangularis, Lesquereux, 1880, Coal Flora of Pa., Coal Meas. [Sig. quadrangular.]
vesicularis instead of Carpolithes vesicularis.
LEPIDODENDRON was described by Sternberg in 1820, in Essai d'un expose Geognostico-botanique de la flore du monde primitif, 1st Cahier. Also L. aculeatum, L. crenatum, L. dichotomum, L. obovatum, L. rimosum, L. selaginoides and L. undulatum.
andrewsi, Lesquereux, 1880, Coal Flora of Pa., Coal Meas. [Ety. proper name.]
alveolare, see Sigillaria alveolaris.
brittsi, Lesquereux, 1880, Coal Flora of Pa., Coal Meas. [Ety. proper name.]

LEPIDODENDRON *chilallœum*, syn. for L. distans.
cuspidatum, Lesquereux, 1880, Coal Flora of Pa., Coal Meas. [Sig. pointed.]
cyclostigma, Lesquereux, 1880, Coal Flora of Pa., Coal Meas. [Sig. round dotted.]
drepanaspis, Wood, 1860, Proc. Acad. Nat. Sci., Coal Meas. [Sig. sickle shield.]
dubium, syn. for L. rimosum.
elegans, was Lycopodiolithes elegans of Sternberg in Tent. Flor. Prim.
icthyolepis, Wood, 1860, Proc. Acad. Nat. Sci., Coal Meas. [Sig. fish scale.]
ingens, syn. for L. aculeatum.
lanceolatum, Lesquereux, 1880, Coal Flora of Pa., Coal Meas. [Sig. lanceolated.]
latifolium, Lesquereux, 1880, Coal Flora of Pa., Coal Meas. [Sig. wide leaved.]
lesquereuxi, Wood, syn. for L. aculeatum.
longifolium, Brongniart, 1828, Prodr. Hist. Veg. Foss., Coal Meas. [Sig. long leaved.]
magnum, Wood, 1860, Proc. Acad. Nat. Sci., Coal Meas. [Sig. large.]
mammillatum syn. for L. veltheimanum.
mekistom syn. for L. modulatum.
oculatum syn. for L. distans.
oweni syn. for L. vestitum.
politum syn. for L. modulatum.
quadrangulatum, Schlotheim, 1821-23, Petrefactenkunde, Coal Meas. [Sig. quadrangular.]
quadrilaterale, Lesquereux, 1880, Coal Flora of Pa., Coal Meas. [Sig. quadrilateral.]
rhombicum, Presl., 1833 (Bergeria rhombica), in Sternberg, Flor. d. Vorw., vol. 2, Coal Meas. [Sig. rhombic.]
rugosum syn. for L. dichotomum.
scutatum, Lesquereux, 1880, Coal Flora of Pa., Coal Meas. [Sig. bearing shields.]
squamiferum, Lesquereux, 1880, Coal Flora of Pa., Coal Meas. [Sig. scale bearing.]
tetragonum, Sternberg, 1821, Essai d'un expose Geognostico-botanique de la flore du monde primitif, 2 Cahier, Coal Meas. [Sig. quadrangular.]
venustum, syn. for L. obtusum.
LEPIDOPHLOIOS ichthyoderma, Lesquereux, 1880, Coal Flora of Pa., Coal Meas. [Sig. fish skinned.]
ichthyolepis, see Lepidodendron ichthyolepis.
sigillarioides, Lesquereux, 1880, Coal Flora of Pa., Coal Meas. [Sig. like Sigillaria.]
LEPIDOPHYLLUM *foliaceum* is Lepidostrobus foliaceus.
lanceolatum is Lepidostrobus lanceolatus, and is Brongniart's species 1828, in Prodr. d. Hist. Veg. Foss.

LEPIDOPHYLLUM linearifolium, Lesquereux, 1880, Coal Flora of Pa., Coal Meas. [Sig. having lined leaves.]
mansfieldi, Lesquereux, 1880, Coal Flora of Pa., Coal Meas. [Ety. proper name.]
morrisanum, Lesquereux, 1880, Coal Flora of Pa., Coal Meas. [Ety. proper name.]
trinerve is Brongniart's species, 1828, in Prodr. d. Hist. d. Veg. Foss.
LEPIDOSTROBUS aldrichi, Lesquereux, 1880, Coal Flora of Pa., Coal Meas. [Ety. proper name.]
foliaceus instead of Lepidophyllum foliaceum.
goldenbergi, Schimper, 1872, Traite de Palæontologie Vegetale, vol. 2, p. 61. Coal Meas. [Ety. proper name.]
hastifolius should be L. hastatus. The mistake is in Geo. Sur., Ill., vol. 2, p. 456.
incertus, Lesquereux, 1880, Coal Flora of Pa., Coal Meas. [Sig. uncertain.]
lacoei, Lesquereux, 1880, Coal Flora of Pa., Coal Meas. [Ety proper name.]
lanceolatus instead of Lepidophyllum lanceolatum.
mansfieldi, Lesquereux, 1880, Coal Flora of Pa., Coal Meas. [Ety. proper name.]
mirabilis, Newberry, 1873 (Polysporia mirabilis), Ohio Pal., vol. 1, Lower Coal Meas. [Sig. extraordinary.]
prælongus, Lesquereux, 1880, Coal Flora of Pa., Coal Meas. [Sig. very long.]
quadratus, Lesquereux, 1880, Coal Flora of Pa., Coal Meas. [Sig. quadrate.]
salisburyi, Lesquereux, 1880, Coal Flora of Pa., Coal Meas. [Ety. proper name.]
spectabilis, Lesquereux, 1880, Coal Flora of Pa., Coal Meas. [Sig. remarkable.]
stachyoides, Wood, 1860, Proc. Acad. Nat. Sci., Coal Meas. [Sig. like the plant *Stachys*.]
LEPIDOXYLON, Lesquereux, 1880, Coal Flora of Pa. [Ety. *lepis*, a scale; *xylon*, wood.]
anomalum, Lesquereux, 1880, Coal Flora of Pa., Coal Meas. [Sig. anomalous.]
LESCUROPTERIS adiantites instead of Neuropteris adiantites.
LESLEYA, Lesquereux, 1880, Coal Flora of Pa., Coal Meas. [Ety. proper name.]
grandis, Lesquereux, 1880, Coal Flora of Pa., Coal Meas. [Sig. grand.]
LICROPHYCUS flabellum, Miller & Dyer, 1878, Jour. Cin. Soc. Nat. Hist., vol. 1, Hud. Riv. Gr. [Sig. a fan.]
Lycopodiolithes elegans, see Lepidodendron elegans.
LYCOPODITES cavifolius instead of Selaginites cavifolius.
ortoni, Lesquereux, 1880, Coal Flora of Pa., Coal Meas. [Ety. proper name.]
pendulus, Lesquereux, 1880, Coal Flora of Pa., Coal Meas. [Sig. hanging down.]
LYCOPODITES strictus, Lesquereux, 1880, Coal Flora of Pa., Coal Meas. [Sig. pressed together.]
uncinnatus instead of Selaginites uncinnatus.
vanuxemi, syn. for Plumalina plumaria, which is a Graptolite.
MACROSTACHYA, Shimper, 1869, Traite de Palæontologie Vegetale, vol. 1, p. 332. [Ety. *makros*, long; *Stachys*, a plant.]
infundibuliformis, Brongniart, 1828 (Equisetum infundibuliforme). Hist. Veg. Foss., Coal Meas. [Sig. funnel shaped.]
MEGALOPTERIS abbreviata, Lesquereux, 1880, Coal Flora of Pa., Coal Meas. [Sig. abbreviated.]
fasciculata, Lesquereux, 1880, Coal Flora of Pa., Coal Meas. [Sig. bundled.]
marginata, Lesquereux, 1880, Coal Flora of Pa., Coal Meas. [Sig. margined.]
southwelli, Lesquereux, 1880, Coal Flora of Pa., Coal Meas. [Ety. proper name.]
MEGAPHYTON goldenbergi, Weiss, 1860, Zeitsch d. deutsch geol. Geselish XII., Coal Meas. [Ety. proper name.]
grandeuryi, Lesquereux, 1880, Coal Flora of Pa., Coal Meas. [Ety. proper name.]
Myrianites is a trail.
Nemapodia is a trail.
NEMATOPHYLLUM, Fontaine & White, 1880, Perm. or Up. Carb. Flora. [Ety. *nema*, thread; *phyllon*, a leaf.]
angustum, Fontaine & White, 1880, Perm. or Up. Carb. Flora, Coal Meas. or Permian. [Sig. narrow.]
Nereites is a trail.
NEUROPTERIS acuminata, Schlotheim, 1820 (Filicites acuminatus), Petref., Coal Meas. [Sig. acuminated.]
adiantites is Lescuropteris adiantites.
agassizi, Lesquereux, 1880, Coal Flora of Pa., Coal Meas. [Ety. proper name.]
anomala, Lesquereux, 1880, Coal Flora of Pa., Coal Meas. [Sig. anomalous.]
aspera, Lesquereux, 1880, Coal Flora of Pa., Coal Meas. [Sig. rough.]
biformis, Lesquereux, 1880, Coal Flora of Pa., Coal Meas. [Sig. two formed.]
callosa, Lesquereux, 1880, Coal Flora of Pa., Coal Meas. [Sig. thick skinned.]
cordato-ovata, Weiss, 1877, Fossile Flora, Coal Meas. [Sig. cordate-ovate.]
decipiens, Lesquereux, 1880, Coal Flora of Pa., Coal Meas. [Sig deceiving.]
dictyopteroides, Fontaine & White, 1880, Perm. or Up. Carb. Flora, Coal Meas. or Permian. [Sig. like *Dictyopteris*.]
dilatata, Lindley & Hutton, 1835 (Cyclopteris dilatata), Foss. Flora. vol. 2, Coal Meas. [Sig. dilated.]
elrodi, Lesquereux, 1880, Coal Flora of Pa., Coal Meas. [Ety. proper name.]
germari, Gœppert, 1836 (Adiantites

germari), Systema Filicum fossilium, Coal Meas. [Ety. proper name.]
NEUROPTERIS laciniata instead of Cyclopteris laciniata.
linnæifolia is a Triassic species.
missouriensis, Lesquereux, 1880, Coal Flora of Pa., Coal Meas. [Ety. proper name.]
obscura, Lesquereux, 1880, Coal Flora of Pa., Coal Meas. [Sig. obscure.]
odontopteroides, Fontaine & White, 1880, Perm. or Up. Carb. Flora, Coal Meas. or Permian. [Sig. like Odontopteris.]
platynervis, Fontaine & White, 1880, Perm. or Up. Carb. Flora, Coal Meas. [Sig. flat nerved.]
smithsi, Lesquereux, 1880, Coal Flora of Pa., Coal Meas. [Ety. proper name.]
subfalcata, Lesquereux, 1880, Coal Flora of Pa., Coal Meas. [Sig. subfalcate.]
tenuinervis is Odontopteris tenuinervis.
NŒGGERATHIA was described by Sternberg, in 1820, in Essai d'un expose Geognostico-botanique de la Flore du monde primitif, 2 Cahier.
bockschiana is Archæopteris bockschiana.
minor is Archæopteris minor.
obliqua, Lesquereux, is Archæopteris obliqua.
obliqua, Gœppert is not an American species.
obtusa is Archæopteris obtusa.
ODONTOPTERIS abbreviata, Lesquereux, 1880, Coal Flora of Pa., Coal Meas. [Sig. shortened.]
alpina, Sternberg, 1821-38 (Neuropteris alpina), Versuch einer Geognost.-botan. Darstellung der Flora der Vorwelt., vol. 2, Coal Meas. [Ety. proper name.]
brardii is Ad., Brongniart's species, 1822, in Class d. Veg. Foss.
cornuta, Lesquereux, 1880, Coal Flora of Pa., Coal Meas. [Sig. horned.]
crenulata, as first identified by Lesquereux, see O subcrenulata
deformata, Lesquereux, 1880, Coal Flora of Pa., Coal Meas. [Sig. deformed.]
densifolia, Fontaine & White, 1880, Perm. or Up. Carb. Flora, Coal Meas. or Perm. [Sig. dense leaved.]
lescurei, Wood, 1866, Trans. Am. Phil. Soc., vol. 13, Coal Meas. [Ety. proper name]
nervosa, Fontaine & White, 1880, Perm. or Up Carb. Flora, Coal Meas. or Perm. [Sig. full of veins.]
newberryi, Lesquereux, 1880, Coal Flora of Pa., Coal Meas. [Ety. proper name.]
neuropteroides, being preoccupied by Roemer, Lesquereux proposed O. newberryi, 1880, Coal Flora of Pa.
pachyderma, Fontaine & White, 1880, Perm. or Up. Carb. Flora, Coal Meas. or Permian. [Sig. thick skinned.]
rotundifolia, Wood, 1866, Trans. Am. Phil. Soc., vol. 13, Coal Meas. [Sig. round leaved.]
ODONTOPTERIS sphenopteroides, Lesquereux, 1880, Coal Flora of Pa., Coal Meas. [Sig. like *Sphenopteris*]
squamosa, Dawson, 1881, Quar. Jour. Geo. Soc., Lond., vol. 37, Devonian. [Sig. scaly.]
subcrenulata, Lesquereux, 1880, Coal Flora of Pa., Coal Meas. [Sig. somewhat crenulated.]
tenuinervis, instead of Neuropteris tenuinervis.
OLIGOCARPIA, Gœppert, 1841-48, Die Gattungen der fossilen Pflanzen. [Ety. *oligos*, few; *carpus*, fruit.]
alabamensis, Lesquereux, 1875, Geol. Rep. Ala., Coal Meas. [Ety. proper name.]
flagellaris instead of Sphenopteris flagellaris.
gutbieri, Gœppert, 1841-48, Die Gattungen der fossilen Pflanzen, Coal Meas. [Ety. proper name.]
PALÆOCHORDA prima, Whitfield, 1877, Rep. on the Pal. of the Black Hills, Potsdam Gr. [Sig. first.]
PALÆOPHYCUS plumosum, Whitfield, 1878, Ann. Rep. Geo. Sur. Wis., Potsdam Gr. [Sig. feathered.]
occidentale, Whitfield, 1877, Rep. on the Pal. of the Black Hills, Potsdam Gr. [Sig. western.]
PALÆOXYRIS *appendiculata* is Spirangium appendiculatum.
corrugata is Spirangium corrugatum.
prendeli is Spirangium prendeli.
Palmacites oculatus, see Sigillaria oculata.
nœggerathi, see Trigonocarpum nœggerathi.
PECOPTERIS angustipinna, Fontaine & White, 1880, Perm. or Up. Carb. Flora, Coal Meas. or Permian. [Sig. having narrow pinnæ.]
arborescens is Schlotheim's species, 1820 (Filicites arborescens) Petref.
arguta is Sternberg's species in Tent. Flor. Primord.
aspera, Brongniart, 1828, Prodr. d. Veg. Foss., Coal Meas. [Sig. rough.]
asplenioides, Fontaine & White, 1880, Perm. or Up. Carb. Flora, Coal Meas. or Permian. [Sig. like *Asplenium*.]
bullata is a Triassic species.
clarki, Lesquereux, 1880, Coal Flora of Pa., Coal Meas. [Ety. proper name.]
clintoni, Lesquereux, 1880, Coal Flora of Pa., Coal Meas. [Ety. proper name.]
elliptica, Fontaine & White (Goniopteris elliptica). The name was preoccupied.
germari, Weiss, 1869-72 (Cyatheites germari), Foss. Flora d. Jungsten Steink. Form., Up. Coal Meas. or Permian. [Ety. proper name.]
germari var. crassinervis, Fontaine & White, 1880, Perm. or Up. Carb. Flora, Coal Meas. or Permian. [Sig. thick nerved.]
germari var. cuspidata, Fontaine &

White, 1880, Perm. or Up. Carb. Flora, Coal Meas. or Permian. [Sig. toothed.]

PECOPTERIS goniopteroides, Fontaine & White, 1880, Perm. or Up. Carb. Flora, Coal Mens. or Permian. [Sig. like *Goniopteris*.]

halli instead of Alethopteris, halli.

heerana, Fontaine & White, 1880, Perm. or Up. Carb. Flora, Coal Meas. or Permian. [Ety. proper name.]

imbricata, Fontaine & White, Perm. or Up. Carb. Flora, Coal Meas. [Sig. imbricated.]

inclinata, Fontaine & White, 1880, Perm. or Up. Carb. Flora, Coal Meas. [Sig. inclined.]

lanceolata, Lesquereux, instead of Alethopteris lanceolata.

lanceolata, Fontaine & White. The name was preoccupied.

latifolia, Fontaine & White, 1880, Perm. or Up. Carb. Flora, Coal Meas. [Sig. wide leaved.]

lyratifolia, Gœppert, 1841-48 (Sphenopteris lyratifolia), Die Gattungen d. Foss. Pflanzen, Coal Meas. [Sig. having lyre-shaped leaves.]

marginata, see Alethopteris marginata.

merianopteroides, Fontaine & White, 1880, Perm. or Up. Carb. Flora, Coal Meas. or Permian. [Sig. like *Merianopteris*.]

microphylla, Brongniart, 1828, Prodr. d. Hist. Veg. Foss., Coal Meas. [Sig. small leaved.]

miltoni, Artis, 1825 (Filicites miltoni), Anted. Phytol., Coal Meas. [Ety. proper name.]

murrayana is identified by Lesquereux, Geo. Sur., Ill., vol. 2, see Sphenopteris pseudo-murrayana.

newberryana, Fontaine & White, 1880 (Goniopteris newberryana), Perm. or Up. Carb. Flora, Coal Meas. or Permian. [Ety. proper name.]

newberryi, refer to Pseudopecopteris newberryi.

nodosa, Gœppert, 1836 (Aspidites nodosus), Systema Filicum Fossilium, Upper Coal Meas. [Sig. nodose.]

oblonga, Fontaine & White, 1880 (Goniopteris oblonga), Perm. or Up. Carb. Flora, Coal Meas. or Perm. [Sig. oblong.]

ovoides, Fontaine & White, 1880, Perm. or Up. Carb. Flora, Coal Meas. or Permian. [Sig. somewhat ovoid.]

pachypteroides, Fontaine & White, 1880, Perm. or Up. Carb. Flora, Coal Meas. or Permian. [Sig. like *Pachypteris*.]

platynervis, Fontaine & White, 1880, Perm. or Up. Carb. Flora, Coal Meas. or Permian. [Sig. flat nerved.]

platyrachis, Brongniart, 1828, Prodr. d. Hist. d. Veg. Foss., Coal Meas. [Sig. having a flat rachis.]

PECOPTERIS plumosa is Artis' species (Filicites plumosa), 1825, Anted. Phytol.

pteroides instead of Alethopteris pteroides.

pusilla is Pseudopecopteris pusilla.

quadratifolia, Lesquereux, 1880, Coal Flora of Pa., Coal Meas. [Sig. quadrate leaved.]

rarinervis, Fontaine & White, 1880, Permian or Up. Carb. Flora, Coal Meas. or Permian. [Sig. rare veined.]

robusta, Lesquereux, 1880, Coal Flora of Pa., Coal Meas. [Sig. robust.]

rotundifolia, Fontaine & White, 1880, Perm. or Up. Carb. Flora, Coal Meas. or Permian. [Sig. round leaved.]

rotundiloba, Fontaine & White, 1880, Perm. or Up. Carb. Flora, Coal Meas. or Permian. [Sig. round lobed.]

schimperana, Fontaine & White. 1880, Perm. or Up. Carb. Flora, Coal Meas. or Permian. [Ety. proper name.]

serpillifolia, Lesquereux, 1880, Coal Flora of Pa., Coal Meas. [Sig. thyme leaved.]

serrula, Lesquereux, instead of Alethopteris serrula.

shœfferi is Pseudopecopteris shœfferi.

sillimani is Pseudopecopteris sillimani.

solida instead of Alethopteris solida.

stellata instead of Alethopteris stellata.

subfalcata, Fontaine & White, 1880, Perm. or Up. Carb. Flora, Coal Meas. or Permian. [Sig. somewhat falcate.]

tenuinervis, Fontaine & White, 1880, Perm. or Up. Carb. Flora, Coal Meas. or Permian. [Sig. fine veined.]

venulosa, Lesquereux, 1880, Coal Flora of Pa., Coal Meas. [Sig. full of small veins.]

vestita, Lesquereux, 1880, Coal Flora of Pa., Coal Meas. [Sig. clothed.]

Physophycus marginatus is Taonurus marginatus.

Phytolithus cancellatus is Lepidodendron veltheimianum.

Polyporites mirabilis is a mistake and may be erased.

polysporus is a mistake and may be erased.

Polyspora is a syn. for Lepidostrobus.

mirabilis is Lepidostrobus mirabilis.

PROTOBLECHNUM, Lesquereux, 1880, Coal Flora of Pa. [Ety. *protos*, first; *Blechnum*, a genus.]

holdeni instead of Alethopteris holdeni.

PROTOSTIGMA, Lesquereux, 1877, Proc. Am. Phil. Soc. [Sig. *protos*, first; *stigma*, a brand or dot.]

sigillarioides, Lesquereux, 1877, Proc. Am. Phil. Soc., Hud. Riv. Gr. [Sig. like *Sigillaria*.]

PSEUDOPECOPTERIS, Lesquereux, 1880, Coal Flora of Pa. [Ety. *pseudos*, false; *Pecopteris*, a genus.]

abbreviata instead of Sphenopteris. abbreviata.

acuta instead of Sphenopteris acuta,

PSEUDOPECOPTERIS anceps, Lesquereux, 1880, Coal Flora of Pa., Coal Meas. [Sig. double.]
callosa instead of Pecopteris callosa.
cordato-ovata, Weiss, 1869-22 (Neuropteris cordata-ovata), Foss. Flor. d. jungst. steink. form., Coal Meas. [Sig. cordate ovate.]
decipiens instead of Sphenopteris decipiens.
decurrens instead of Pecopteris decurrens, Lesquereux.
denudata, Lesquereux, 1880, Coal Flora of Pa., Coal Meas. [Sig. denuded.]
dimorpha, Lesquereux, 1880, Coal Flora of Pa., Coal Meas. [Sig. two-formed.]
glandulosa instead of Sphenopteris glandulosa.
hymenophylloides instead of Alethopteris hymenophylloides.
irregularis instead of Sphenopteris irregularis.
latifolia instead of Sphenopteris latifolia.
macilenta instead of Sphenopteris macilenta.
mazonana instead of Alethopteris mazonana.
muricata instead of Alethopteris muricata.
nervosa instead of Alethopteris nervosa.
newberryi instead of Sphenopteris newberryi.
pluckeneti instead of Alethopteris pluckeneti.
polyphylla instead of Sphenopteris polyphylla.
pusilla instead of Pecopteris pusilla.
shaefferi instead of Pecopteris shaefferi.
sillimani instead of Pecopteris sillimani.
speciosa, Lesquereux, 1880, Coal Flora of Pa., Coal Meas. [Sig. beautiful.]
spinulosa instead of Alethopteris spinulosa.
subcrenulata, Lesquereux, 1880, Coal Flora of Pa., Coal Meas. [Sig. somewhat crenulated.]
subnervosa Rœmer, F. A., 1860, Palæontographica, vol. 9, Coal Meas. [Sig. somewhat veiny.]
trifoliata instead of Sphenopteris trifoliata, and credit to Artis, 1825 (Filicites trifoliatus) Anted. Phytol.
virginiana instead of Cyclopteris virginiana.

PSILOPHYTON cornutum, Lesquereux, 1877, Proc. Am. Phil. Soc., Low. Held. Gr. [Sig. horned.]
gracillimum, see Dendrograptus gracillimus.
princeps, var. ornatum, Dawson, 1871, Foss. Plants, Devonian. [Sig. ornate.]

Ptilocarpus bicornutus, see Cardiocarpon bicornutum.

Ptilophyton, Dawson, 1878, Scottish Devonian Plants. [Ety. ptilon, a wing;
• phyton, a plant.] This name is proposed for Lycopodites vanuxemi, which is Plumalina plumaria, and L. plumula. If the types are not Graptolites, the genus may stand.

RHABDOCARPUS apiculatus, syn. for R. curinatus.
costatus, syn. for R. acuminatus.
cornutus, Lesquereux, 1880, Coal Flora of Pa., Coal Meas. [Sig. horned.]
howardi, Lesquereux, 1880, Coal Flora of Pa., Coal Meas. [Ety. proper name.]
insignis, Lesquereux, 1880. The name having been preoccupied by Dawson, see R. lescuriana.
jacksonensis instead of Carpolithes jacksonensis.
lescurianus, n. sp. Coal Meas. Proposed instead of R. insignis, Lesquereux, 1880, Coal Flora of Pa., p. 575, pl. 85, fig. 26.
mammillatus is Cardiocarpon mammillatum.
multistriatus instead of Carpolithes multistriatus.
oblongus, Fontaine & White, 1880, Perm. or Up. Carb. Flora, Coal Meas. or Perm. [Sig. oblong.]
platimarginatus instead of Carpolithes platimarginatus.

RHACHIOPTERIS affinis instead of Stigmarioides affinis.
selago instead of Stigmarioides selago.

RHACOPHYLLUM adnascens instead of Hymenophyllites adnascens.
arborescens instead of Hymenophyllites arborescens.
clarki instead of Hymenophyllites clarki.
cornutum, Lesquereux, 1880, Coal Flora of Pa., Coal Meas. [Sig. horned.]
corallum, Lesquereux, 1880, Coal Flora of Pa., Coal Meas. [Sig. a coral.]
expansum, Lesquereux, 1880, Coal Flora of Pa., Coal Meas. [Sig. expanded.]
filiciforme, Gutbier, 1842 (Fucoides filiciformis). Abdr. u. Verst. d. zwickaur schwarzk, u. sein. Ungeb., Coal Meas. [Sig fern like.]
flabellatum, Sternberg, 1821-38 (Aphlebia flabellata), Flora der Vorwelt, vol. 2, Coal Meas. [Sig. fan like.]
fucoideum, Lesquereux, 1880, Coal Flora of Pa., Coal Meas. [Sig. fucus like.]
hamulosum, Lesquereux, 1880, Coal Flora of Pa., Coal Meas. [Sig. full of hooks]
inflatum instead of Hymenophyllites inflatus.
irregulare, Germar, 1844 (Aphlebia irregularis), Verst d. Steink. v. Wettin u Löbejün, Coal Meas. [Sig. irregular.]
laciniatum, Fontaine & White, 1880, Perm. or Up. Carb. Flora, Coal Meas. or Permian. [Sig. jagged.]
membranaceum, Lesquereux, 1880, Coal Flora of Pa., Coal Meas. [Sig. membranaceous.]
molle, instead of Hymenophyllites mollis.

RHACOPHYLUM scolopendrites instead of Scolopendrites dentatus.
  spinosum, Lesquereux, 1880, Coal Flora of Pa., Coal Meas. [Sig. full of spines.]
  strongi instead of Hymenophyllites strongi.
  thalliforme instead of Hymenophyllites thalliformis.
  trichoideum, Lesquereux, 1880, Coal Flora of Pa., Coal Meas. [Sig. hair-like.]
  truncatum, Lesquereux, 1880, Coal Flora of Pa., Coal Meas. [Sig. truncated.]
RHIZOMORPHA SIGILLARIÆ, Lesquereux, 1877, Proc. Am. Phil. Soc., Coal Meas. [Sig. of Sigillaria.]
RUSOPHYCUS ASPERUM, Miller & Dyer, 1878, Jour. Cin. Soc. Nat. Hist., vol. 1, Utica Slate. [Sig. rough.]
SAPORTÆA, Fontaine & White, 1880, Perm. or Up. Carb. Flora. [Ety. proper name.]
  grandifolia, Fontaine & White, 1880, Perm. or Up. Carb. Flora, Coal Meas. or Permian. [Sig. large leaved.]
  salisburioides, Fontaine & White, 1880, Perm. or Up. Carb. Flora, Coal Meas. or Permian. [Sig. like Salisburia.]
SCHUTZIA bracteata refer to Cordaianthus bracteatus.
Scolithus has no place among the plants. It represents the work of a borer, and is referred therefore to the Annelida.
Scolopendrites dentatus is Rhacophyllum scolopendrites.
Selaginites cavifolius is Lycopodites cavifolius.
  crassus is a syn. for Lycopodites cavifolius.
  formosus is not a plant. It was founded upon fragments of a crustacean.
  uncinnatus is Lycopodites uncinnatus.
SIGILLARIA acuminata, Newberry, 1874, Proc. Cleveland Acad. Sci., Coal Meas. [Sig. pointed.]
  alveolaris is Sternberg's species (Lepidodendron alveolare), 1820 in essai d'un expose Geognostico-botanique de la Flore du monde primitif. 1st cahier.
  approximata, Fontaine & White, 1880, Perm. or Up. Carb. Flora, Coal Meas. or Permian. [Sig near to.]
  biercei, syn. for S. ichthyolepis.
  brardi was described in 1822. Class d. Veg. Foss.
  brongniarti, Geinitz, 1855. Die Verst. d. Steink. form. Sachsen , Coal Meas. [Ety. proper name]
  cancellata, syn. for Lepidodendron velthelmianum.
  chemungensis is Lepidodendron chemungense.
  cortei, Brongniart, 1828, Prodr. d. Hist. d. Veg. Foss., Coal Meas. [Ety. proper name.]
  cuspidata, Brongniart, 1828, Prodr. d.

Hist. d. Veg. Foss., Coal Meas. [Sig. pointed.]
SIGILLARIA dentata, Newberry, 1874, Proc. Cleveland Acad. Sci., Coal Meas. [Sig. toothed.]
  elegans is Sternberg's species (Favularia elegans) in Tent. flor. primord.
  hexagona, Schlotheim, 1820 (Palmacites hexagonatus), Petref., Coal Meas. [Sig. hexagonal.]
  ichthyolepis, Sternberg, 1821-38, Flor. d. Vorw., Coal Meas. [Sig. fish-scaled.]
  lacoei, Lesquereux, 1880, Coal Flora of Pa., Coal Meas. [Ety. proper name.]
  leioderma, Brongniart, 1828-44, Hist. d. Veg. Foss., Coal Meas. [Sig. smooth skinned.]
  leptoderma, Lesquereux, 1880, Coal Flora of Pa., Coal Meas. [Sig. thin skinned.]
  lorenzi, Lesquereux, 1880, Coal Flora of Pa., Coal Meas. [Ety. proper name.]
  mammillaris, Brongniart, 1828-44, Hist. d. Veg. Foss. Coal Meas. [Sig. mammillated.]
  marginata, Lesquereux, 1880, Coal Flora of Pa., Coal Meas. [Sig. margined.]
  marineria, Hildreth, 1837, Am Jour. Sci. & Arts, vol. 31, Low. Coal Meas. [Ety. proper name.]
  martini may be erased.
  oculata is Schlotheim's species (Palmacites oculatus), 1820, Petref.
  orbicularis, Brongniart, 1828, Prodr. d. Hist. d. Veg. Foss., Coal Meas. [Sig. orbicular.]
  ornithicnoides, Wood, 1866, Trans. Am. Phil. Soc., vol. 13, Coal Meas. [Sig. like bird tracks.]
  ovalis, Lesquereux, 1880, Coal Flora of Pa., Coal Meas. [Sig. oval.]
  oweni, see Didymophyllum oweni.
  perplexa, Wood, 1866, Trans. Am. Phil. Soc., vol. 13, Coal Meas. [Sig. perplexing.]
  pittstonana, Lesquereux, 1880, Coal Flora of Pa., Coal Meas. [Ety. proper name.]
  pulchra, Newberry, 1874, Proc. Cleveland Acad. Sci., Coal Meas. [Sig. beautiful.]
  reticulata, Steinhaur, may be erased and *reticulata*, Lesquereux, restored.
  solanus, Wood, 1860, Proc. Acad. Nat. Sci., Coal Meas [*solanus*, in text; *solenotus*, on plate; *solena*, in Trans. Am. Phil. Soc., vol. 13.]
  tessellata, Steinhaur, is correct, but tessellata. Brongniart, may be struck out as he did not claim the name.
  voltzi, Brongniart, 1828, Prodr. d. Hist. d. Veg. Foss., Coal Meas. [Ety. proper name.]
  williamsi, Lesquereux, 1880, Coal Flora of Pa., Coal Meas. [Ety. proper name]
  yardleyi was described by Lesquereux, in 1858, in Catal. Potts. Foss.

SIGILLARIOIDES *stellatus*, refer to Stigmaria stellata.
SOLENOULA, Wood, 1860, Proc. Acad. Nat. Sci., vol. 2. [Ety. *solen*, a channel; *oulos*, entire.]
  psilophlœus, Wood, 1860, Proc. Acad. Nat. Sci., Coal Meas. [Sig. smooth barked.]
SOROCLADUS, Lesquereux, 1880, Coal Flora of Pa. [Ety. *soros*, a heap, one of the fruit dots on the back of the frond; *klado*, to break in pieces.]
  asteroides instead of Staphylopteris asteroides.
  ophioglossoides, Lesquereux, 1880, Coal Flora of Pa., Coal Meas. [Sig. like *Ophioglossus*.]
  sagittatus instead of Staphylopteris sagittata.
  stellatus instead of Staphylopteris stellata.
  wortheni instead of Staphylopteris wortheni.
SPHENOPHYLLUM densifoliatum, Fontaine & White, 1880, Perm. or Up. Carb. Flora, Coal Meas. or Permian. [Sig. dense leaved.]
  fontainianum, n. sp., Upper Coal Meas. or Permian. Proposed instead of S. latifolium of Fontaine & White, in Perm. or Up. Carb. Flora, p. 36, pl. 1, figs. 10 & 11.
  latifolium, Wood, 1866, Trans Am. Phil. Soc., vol. 13, Coal Meas. [Sig. wide leaved.]
  latifolium, Fontaine & White, 1880, Perm. or Up. Carb. Flora. The name was preoccupied. See S. fontainianum.
  primævum, Lesquereux. 1877. Proc. Am. Phil. Soc., Hud. Riv. Gr. I think there is no probability that this is a plant.
  tenuifolium, Fontaine & White. 1880, Perm. or Up. Carb. Flora. Coal Meas. or Permian. [Sig. slender leaved.]
SPHENOPTERIS *abbreviata*, refer to Pseudopecopteris abbreviata.
  acrocarpa, Fontaine & White, 1880, Perm. or Up. Carb. Flora, Coal Meas. or Permian. [Sig. pointed fruit.]
  acuta, refer to Pseudopecopteris acuta.
  alabamensis, refer to Oligocarpia alabamensis.
  auriculata, Fontaine & White, 1880, Perm. or Up. Carb. Flora. Coal Meas. or Permian. [Sig. auriculated.]
  ballantini instead of Hymenophyllites ballantini.
  britsi, Lesquereux, 1880, Coal Flora of Pa., Coal Meas. [Ety. proper name.]
  chærophylloides instead of Pecopteris chærophylloides.
  coriacea, Fontaine & White. 1880, Perm. or Up. Carb. Flora, Coal Meas. or Permian. [Sig. coriaceous.]
  cristata, Sternberg, 1821-38, Flor. d. Vorw., vol. 2, Coal Meas. [Sig. tufted.]
  decipiens, refer to Pseudopecopteris decipiens.

SPHENOPTERIS dentata, Fontaine & White, 1880, Perm. or Up. Carb. Flora, Coal Meas. or Permian. [Sig. toothed.]
  flaccida, Crepin, 1874, Bull. Acad. Roy. of Belgium, Sub-carboniferous. [Sig. flaccid.]
  flagellaris refer to Oligocarpia flagellaris.
  flexicaulis instead of Hymenophyllites flexicaulis.
  follosa, Fontaine & White, 1880, Perm. or Up. Carb. Flora, Coal Meas. or Permian. [Sig. leafy.]
  furcata instead of Hymenophyllites furcatus
  glandulosa, refer to Pseudopecopteris glandulosa.
  goniopteroides, Lesquereux, 1880, Coal Flora of Pa., Coal Meas. [Sig. like *Goniopteris*.]
  hastata, Fontaine & White, 1880, Perm. or Up. Carb. Flora, Coal Meas. or Permian. [Sig. armed with a spear.]
  hildrethi instead of Hymenophyllites hildrethi.
  intermedia, Lesquereux, The name was preoccupied in 1852 by Ettingshausen. It is now described as S. mediana.
  irregularis, refer to Pseudopecopteris irregularis.
  larischi, Stur, 1877 (Calymmotheca larischi), Culm. Flor. d. Ostr. u. Waldenburger Schichten, Coal Meas. [Ety. proper name.]
  latifolia, refer to Pseudopecopteris latifolia.
  laxa, Hall, is Archæopteris hallana.
  lescuriana, Fontaine & White, 1880, Perm. or Up. Carb. Flora, Coal Meas. or Permian. [Ety. proper name.]
  linearis, Sternberg, Tent. flor. prim. p. 15, tab. 42, fig. 4, Lower Coal Meas. [Sig. lined.]
  macilenta, refer to Pseudopecopteris macilenta.
  mediana, Lesquereux, 1880, Coal Flora of Pa., Coal Meas. [Sig middle.] Proposed instead of S. intermedia.
  microcarpa, Lesquereux, 1880, Coal Flora of Pa., Coal Meas. [Sig. small fruited.]
  minutisecta, Fontaine & White, 1880, Perm or Up. Carb. Flora, Coal Meas. or Permian. [Sig. finely marked.]
  newberryi refer to Pseudopecopteris newberryi.
  pachynervis, Fontaine & White, 1880, Perm. or Up. Carb. Flora, Coal Meas. or Permian. [Sig. thick veined.]
  pilosa, refer to Callipteris pilosa.
  polyphylla, refer to Pseudopecopteris polyphylla.
  pseudomurrayana, Lesquereux, 1880, Coal Flora of Pa., Coal Meas. [Sig. from its resemblance to *Pecopteris murrayana*.]
  pterota, Wood, 1866, Trans. Am. Phil.

Soc., vol. 13, Coal Meas. [Sig. feathered.]
SPHENOPTERIS quercifolia, Gœppert, 1836, Systema Filicum fossilium, Coal Meas. [Sig. oak leaved.]
 spinosa instead of Hymenophyllites spinosus.
 squamosa is Pseudopecopteris anceps.
 subalata, Weiss, 1869-72, Foss. Flor. d. jungst. Steink. form., Coal Meas. [Sig. somewhat alate.]
 trichomanoides instead of Hymenophyllites trichomanoides.
 tridactylites instead of Hymenophyllites tridactylites.
 trifoliata, refer to Pseudopecopteris trifoliata and to Artis (Filicites trifoliatus), 1825, Antediluvian Phytology.
SPIRANGIUM, Schimper, 1874, Traite de Palæontologie Vegetale. [Ety. from the coiled and twisted marking around the pod or vegetable substance.]
 appendiculatum instead of Palæoxyris appendiculata.
 corrugatum instead of Palæoxyris corrugata.
 intermedium, Lesquereux, 1880, Coal Flora of Pa., Coal Meas. [Sig. intermediate.]
 multiplicatum, Lesquereux, 1880, Coal Flora of Pa., Coal Meas. [Sig. many plicated.]
 prendeli instead of Palæoxyris prendeli.
Spirophyton, Hall, is classed by Prof. Lesquereux as a synonym for Taonurus.
SPORANGITES acuminatus instead of Annularia acuminata.
 huronensis, Dawson, 1871, Am. Jour. Sci. & Arts, Ham. Gr. [Ety. proper name.]
SPOROCYSTIS, Lesquereux, 1880, Coal Flora of Pa. [Ety. *sporos*, seed; *kustis*, bladder.]
 planus, Lesquereux, 1880, Coal Flora of Pa., Coal Meas. [Sig. even.]
STAPHYLOPTERIS *asteroides*, *S. sagittata*, *S. stellata*, and *S. wortheni*, are referred to the genus Sorocladus.
STEMMATOPTERIS, Corda, 1845, Beitrage zur Flora der Vorwelt. [Ety. *stemmatos*, a wreath; *pteris*, a fern.]
 angustata, Lesquereux, 1880, Coal Flora of Pa., Coal Meas. [Sig. narrowed.]
 cyclostigma, Lesquereux, 1880, Coal Flora of Pa, Coal Meas. [Sig. round dotted.]
 emarginata, Lesquereux, 1880, Coal Flora of Pa., Coal Meas. [Sig. emarginated.]
 gigantea, instead of Caulopteris gigantea.
 hirsuta, Lesquereux, 1880, Coal Flora of Pa., Coal Meas. [Sig. hairy.]
 insignis, instead of Caulopteris insignis.
 mimica, Lesquereux, 1880, Coal Flora of Pa., Coal Meas. [Sig. mimic.]

STEMMATOPTERIS polita, Lesquereux, 1880, Coal Flora of Pa., Coal Meas. [Sig. polished.]
 punctata, instead of Caulopteris punctata.
 schimperi, Lesquereux, 1880, Coal Flora of Pa., Coal Meas. [Ety. proper name.]
 squamosa, Lesquereux, 1880, Coal Flora of Pa., Coal Meas. [Sig. scaly.]
 wortheni, instead of Caulopteris wortheni.
STIGMARIA amœna, Lesquereux, 1880, Coal Flora of Pa., Coal Meas. [Sig. charming.]
 stellaris instead of Sigillarioides stellaris.
STIGMARIOIDES *affinis*, refer to Rhachiopteris affinis.
 selago, refer to Rhachiopteris selago.
Strobilus caryophyllus, Hildreth, 1837, Am. Jour. Sci. & Arts, vol. 31, Coal Meas.
SYRINGODENDRON was described by Sternberg in 1820 in Essai d'un exposé Geognostico-botanique de la Flore du monde primitif. 1st cahier, and S. pes-capreoli was also described.
 brongniarti, Geinitz, 1855 (Sigillaria brongniarti), Verst. d. Steink. form. in Sachsen, Coal Meas. [Ety. proper name.]
 gracile is from the Waverly Group.
 kirtlandium, Hildreth, 1837, Am. Jour. Sci. & Arts, vol. 31, Coal Meas. [Ety. proper name.]
 magnificum, Wood, 1866, Trans. Am. Phil. Soc., vol. 13, Coal Meas. [Sig. magnificent.]
 pachyderma instead of Sigillaria pachyderma.
TÆNIOPHYLLUM, Lesquereux, 1878, Proc. Am. Phil. Soc. [Ety. *tainia*, a ribbon; *phyllon*, a leaf.]
 contextum, Lesquereux, 1878, Proc. Am. Phil. Soc., Coal Meas. [Sig. entwined.]
 decurrens, Lesquereux, 1878, Proc. Am. Phil. Soc., Coal Meas. [Sig. extending downward.]
 deflexum, Lesquereux, 1878, Proc. Am. Phil. Soc., Coal Meas. [Sig. bent downward.]
TÆNIOPTERIS lescuriana, Fontaine & White, 1880, Perm. or Up. Carb. Flora, Coal Meas. or Permian. [Ety. proper name.]
 magnifolia is a Jurassic species.
 newberryana, Fontaine & White, 1880, Perm. or Up. Carb. Flora, Coal Meas. or Permian. [Ety. proper name.]
 smithi, Lesquereux, 1875, Geol. Rep. of Ala., Coal Meas. [Ety. proper name.]
TAONURUS, Fisher-Ooster, 1858, Foss. Fucoiden d. Schweizer Alpen. [Ety. *taon*, a peacock; *oura*, tail.]
 cauda-galli instead of Spirophyton cauda-galli, according to Lesquereux.
 colletti instead of Chondrites colletti.

TAONURUS crassus instead of Spirophyton crassum.
marginatus instead of Caulerpites marginatus.
typus instead of Spirophyton typus.
velum instead of Spirophyton velum.
TRICHOPHYCUS, Miller & Dyer, 1878, Jour. Cin. Soc. Nat Hist., vol. 1. [Ety. *trichos*, hair; *phukos*, sea weed.]
lanosum, Miller & Dyer, 1878, Jour. Cin. Soc. Nat. Hist., vol. 1, Hud. Riv. Gr. [Sig. woolly.]
sulcatum, Miller & Dyer, 1878. Cont. to Pal., No. 2, Hud. Riv. Gr. [Sig. furrowed.]
venosum, S. A. Miller, 1879, Jour. Cin. Soc. Nat. Hist., vol. 2, Hud. Riv. Gr. [Sig. veiny.]
TRIGONOCARPUM giffordi, Lesquereux, 1880, Coal Flora of Pa., Coal Meas. [Ety. proper name.]
nœggerathi is Sternberg's species (Palmacites nœggerathi), Tent. flor. primord.
saffordi, Lesquereux, 1880, Coal Flora of Pa., Coal Meas. [Ety. proper name.]
TRIPHYLLOPTERIS, Schimper, 1874, Traite de Pal. Veg. [Ety. *tria*, three; *phyllon*, a leaf; *pteris*, a fern.]
lescuriana instead of Cyclopteris lescuriana.

Trochophyllum was used by Edwards & Haime for a genus of Corals in 1851, and hence was preoccupied before Wood suggested its use instead of Annularia.
clavatum, Lesquereux, 1880, Coal Flora of Pa., Coal Meas. [Sig. knotted.]
lineare, Lesquereux, 1880, Coal Flora of Pa., Coal Meas. [Sig. linear.]
ULODENDRON commutatum, Schimper, 1874, Traite de Pal. Veg., Coal Meas. [Sig. changed.]
flexuosum, see Halonia flexuosa.
Uphantænia is referred to the Class Porifera.
WALCHIA gracilis was published in 1863, Can. Nat., vol. 8, and is the same as Araucarites gracilis. W robusta in 1871, in Rep. on Prince Edward's Island.
WHITTLESEYA was described as well as W. elegans, in 1874, in Proc. Cleveland Acad. Nat. Sci.
integrifolia, Lesquereux, 1880, Coal Flora of Pa., Coal. Meas. [Sig. whole leaved.]
undulata, Lesquereux, 1880, Coal Flora of Pa., Coal Meas. [Sig. undulated.]
ZAMITES *gramineus* and *Z. obtusifolius* are Triassic species.

ANIMAL KINGDOM.

SUBKINGDOM PROTISTA.

ARCHÆOCYATHELLUS, Ford, 1873, Am. Jour. Sci. & Arts, 3d ser., vol. 6. A proposed subgenus of Archæocyathus, founded upon A. rensselæricus.
ASTROCONIA, Sollas 1881, Lond. Quar. Jour. Geo. Soc., vol. 37. [Ety. *aster*, a star; *konia*, dust.]
granti, Sollas, 1881, Lond. Quar. Jour. Geo. Soc., vol. 37, Niagara Gr. [Ety. proper name.]
ASTYLOSPONGIA, præmorsa, var. nuxmoschata, Hall, 1876, 28th Reg. Rep., Niagara Gr. [Sig. nutmeg.]
BEATRICEA (p. 165) is a sponge, and should have been placed in this subkingdom.
CALCARINA, D'Orbigny, 1826, Tableau Methodique de la classe des Cephalopodes in Annales des Sciences Naturelles, Tome 7. [Ety. *calcis*, limestone.]
ambigua, Brady, 1876, Monograph of Carboniferous and Permian foraminifera, Carboniferous. [Sig. ambiguous.]
CALCISPHÆRA, Williamson, 1880, Mem. Org. of the Plants of the Coal Meas., pt. 10. [Ety. *calcis*, limestone; *sphæra*, a sphere.]
robusta, Williamson, 1880, Mem. Org. Plants, Coal Meas., Corniferous Gr. [Sig. robust.]
CYATHOPHYCUS, Walcott, 1879, Trans. Alb. Inst., vol. 10. [Ety. *kuathos*, a cup; *phukos*, a sea weed.] This genus is now supposed to represent sponges.
reticulatum, Walcott, 1879, Trans. Alb. Inst., vol. 10, Utica Slate Gr. [Sig. reticulated.]
subsphericum, Walcott, 1879, Trans. Alb Inst., vol. 10, Utica Slate Gr. [Sig. somewhat spherical.]
CYATHOSPONGIA, Hall, 1882, Foss. Corals, Niagara and U.p. Held. Gr. [Ety. *kuathos*, a cup; *spongia*, a sponge.]
excrescens, Hall, 1882, Foss. Corals, Niagara and Up. Held. Gr., Niagara Gr. [Sig. growing up preternaturally.]
DICTYOPHYTON catilliforme, Whitfield, 1881, Bull. No. 1, Am. Mus. Nat. Hist., Keokuk Gr. [Sig. dish like.]
cylindricum, Whitfield, 1881, Bull No. 1, Am. Mus. Nat. Hist., Keokuk Gr. [Sig. cyclindrical.]
DYSTACTOSPONGIA, S. A. Miller, 1882, Jour. Cin. Soc. Nat. Hist., vol. 5. [Ety. *dystaktos*, hard to arrange; *spongia*, a sponge.]

DYSTACTOSPONGIA insolens, S. A. Miller, 1882, Jour. Cin. Soc. Nat. Hist., vol. 5, Hud. Riv. Gr. [Sig. unusual.]
ENDOTHYRA, Phillips, 1845, Proc. Geol. & Polytech, Soc. W. Riding Yorks, vol. 2. [Ety. *endos*, within; *thura*, a door.]
baileyi instead of Rotalia baileyi.
FUSILINA hyperborea, Salter, 1855, Belcher's last Arctic Voyage, vol. 2, Carboniferous. [Sig. very far north.]
HINDSIA, Duncan, 1879, Ann. & Mag. Nat. Hist., 5th ser., vol. 4. [Ety. proper name]
sphæroidalis, Duncan, 1879, Ann. & Mag. Nat. Hist., 5th ser., vol. 4, Low. Held. Gr. [Sig. sphæroidal.]
Ischadites tessellatus see Receptaculites tessellatus.
LEPIDOLITES, Ulrich, 1879, Jour. Cin. Soc. Nat. Hist., vol. 2. [Ety. *lepis*, scale; *lithos*, stone.]
dickhauti. Ulrich, 1879, Jour. Cin. Soc. Nat. Hist., vol. 2, Hud. Riv. Gr. [Ety. proper name.] Another specimen he called *L. elongatus*, without characters to distinguish it. The Specimens are very poor, and unless they are spongoid and related to Cyathophycus, the name may as well be erased for want of definition.
LOFTUSIA, Carpenter & Brady, 1869, Trans Royal Soc. [Ety. proper name.]
columbiana, Dawson, 1879, Quar. Jour. Geo. Soc., vol. 35, Up. Carb. [Ety. proper name.]
MICROSPONGIA, Miller & Dyer, 1878, Jour. Cin. Soc. Nat. Hist., vol. 1. [Ety. *mikros*, small; *spongia*, a sponge.]
gregaria, Miller & Dyer, 1878, Jour. Cin. Soc. Nat. Hist., vol. 1, Hud. Riv. Gr. [Sig. belonging to a flock.]
NODOSINELLA, Brady, 1876, Monograph, Carb. & Perm. foraminifera. [Ety. *nodus*, a knot.]
priscilla instead of Dentalina priscilla.
PALÆACIS cuneiformis, Milne-Edwards, 1860, Hist. Nat. d. Corollaires, vol. 3, Warsaw Gr. [Sig. wedge formed.]
cuneatus is a syn. for P. cuneiformis, and P. compressus and P. enormis, should be dated 1860.
PALÆOMANON cratera should be referred to the Niagara Gr.
PATTERSONIA, S. A. Miller, 1882, Jour. Cin. Soc. Nat. Hist., vol. 5 [Ety. proper name.]
difficilis, S. A. Miller, 1882, Jour. Cin. Soc. Nat. Hist., vol. 5, Hud. Riv. Gr. [Sig. difficult.]
PROTOCYATHUS, Ford, 1878, Am. Jour. Sci. & Arts, 3d ser., vol. 15. [Ety. *protos*, first; *kuathos*, a cup.]
rarus, Ford, 1878, Am. Jour. Sci. & Arts, 3d ser., vol. 15, Low. Potsdam Gr. [Sig. rare.]

RECEPTACULITES arcticus, Etheridge, 1878, Quar. Jour. Geo. Soc., vol. 34, Silurian. [Sig. arctic.]
circularis, Emmons, 1856, Am. Geol., Hud. Riv. Gr. [Sig. circular.]
devonicus, Whitfield, 1882, Desc. new species Foss. from Ohio, Up. Held. Gr. [Sig Devonian.]
iowensis, should be referred to the Trenton Gr.
sacculus, Hall, 1879, Desc. new species Foss. from Waldron, Ind., Niagara Gr. [Sig. a little bag.]
tessellatus, Winchell & Marcy, 1865, Mem. Bost. Soc. Nat. Hist., vol. 1, Niagara Gr. [Sig. checkered.]
Rotalia baileyi refer to Endothyra baileyi.
SACCAMMINA, Sars, 1868, Vidensk-Selsk. Förhandl. [Ety. diminutive of *sakkos*, a bag.]
eriana, Dawson, 1881, Can. Nat., vol. 10, syn. for Calcisphæra robusta.
STROMATOCERIUM richmondense, S. A. Miller, 1882, Jour. Cin. Soc. Nat. Hist., vol. 5, Hud. Riv. Gr. [Ety. proper name.]
STROMATOPORA should be referred to this subkingdom. In 1867, Proc. Am. Ass. Ad. Sci., Winchell proposed the subgenus *Cœnostroma* to include *S. monticulifera*, *S. polymorpha*, *S. pustulosa*, *S. radiosa*, and *S. ramosa*; and the subgenus *Idiostroma* to include *S. cæspitosa*, and a proposed species *S. gordiaceum*.
hindi is from the Niagara Gr.
nulliporoides, Nicholson, 1875, Pal. Prov. Ont., Ham. Gr. [Sig. like *Nullipora*.]
Textilaria palæotrochus see Valvulina palæotrochus.
UPHANTÆNIA, dawsoni, Whitfield, 1881, Am. Jour. Sci. & Arts; also, Bull. No. 1, Am. Mus. Nat. Hist., Keokuk Gr. [Ety. proper name.]
VALVULINA, D'Orbigny, 1826, Tabl. Method, d. 1. Classe d. Cephalopodes. [Ety. *valva*, a door.]
bulloides, Brady, 1876, Monog. Carb. & Perm. foraminifera, Carboniferous. [Sig. like a bubble.]
decurrens, Brady, 1878, Mem. Geo. Sur. Scotland, Carboniferous. [Sig. extending down.]
palæotrochus, Ehrenberg, 1857 (Textilaria palæotrochus), Mikrogeologie, Carboniferous. [Sig. ancient *Trochus*.]
plicata, Brady, 1873, Mem. Geo. Sur. Scotland, Carboniferous. [Sig. plicated.]
rudis, Brady, 1876, Monog. Carb. & Perm. foraminifera, Carboniferous. [Sig. rude]

SUBKINGDOM RADIATA.

CLASS POLYPI.

ACANTHOGRAPTUS, Spencer, 1878, Can. Nat. vol. 8. [Ety. *akantha*, spine; *grapho*, to write.]
  granti, Spencer, 1878, Can. Nat., vol. 8, Niagara Gr. [Ety. proper name.]
ACERVULARIA adjunctive, White, 1880, Proc. U. S. Nat. Mus., vol. 2, Carboniferous. [Sig. joined.]
  pentagona, Goldfuss, 1826 (Cyathophyllum pentagonum), Petref. Germ., Devonian. [Sig. pentagonal.]
ACROPHYLLUM clarki, Davis (In press), Foss. Corals of Ky., in vol. 1, Pal. Ky. St. Geo. Sur., Upper Devonian. [Ety. proper name.]
  ellipticum, Davis (In press), Foss. Corals of Ky., in vol. 1, Pal. Ky. St. Geo. Sur., Lower Devonian. [Sig. elliptical].
ALVEOLITES arctica, Woodward, 1879, Lond. Geo. Mag. n. s. vol. 5, Devonian. [Sig. arctic.]
  constans, Davis (In press), Foss. Corals of Ky., in vol. 1, Pal. Ky. St. Geo. Sur., Lower Devonian. [Sig. regular, unchangeable.]
  exsul, refer to Callopora exsul.
  fibrosus, Davis (In press), Foss. Corals of Ky., in vol. 1, Pal. Ky. St. Geo. Sur., Niagara Gr.[Sig. fibrous.]
  fischeri (which is Pachypora fischeri), A. goldfussi, and A. rœmeri, should be dated 1860, Can. Jour. vol. 5, and are from Ham. Gr.
  frondosus, refer to Pachypora frondosa.
  irregularis, Whitfield, 1878, Ann. Rep. Geo. Sur. Wis., Hud. Riv. Gr. [Sig. irregular.]
  louisvillensis, Davis (In press) Foss. Corals of Ky , in vol. 1, Pal. Ky. St. Geo. Sur., Niagara Gr. [Ety. proper name.]
  minimus, Davis (In press), Foss. Corals of Ky., in vol. 1, Pal. Ky. St. Geo. Sur., Lower Devonian. [Sig. smallest.]
  mordax, Davis (In press), Foss. Corals of Ky., in vol. 1, Pal. Ky. St. Geo. Sur., Lower Devonian. [Sig. rough as a rasp.]
  multilamella, Meek, 1877, U. S. Geo. Sur., 40th Parallel, vol. 4, Devonian. [Sig. having many lamellæ.]
  niagarensis, Nicholson & Hinde, 1874, Can. Jour., Niagara Gr. [Ety. proper name.]
  niagarensis, Rominger, see Alveolites undosus.
  scandularis, Davis (In press), Foss. Corals of Ky., in vol. 1 Pal. Ky. St. Geo. Sur., Upper Devonian. [Sig. covered as with shingles.]
  ALVEOLITES undosus, n. sp., Niag. Gr. [Sig. wavy.] This name is proposed for the species described by Rominger in 1876, in his Foss. Corals, p. 40, pl. 16, figs. 1 and 2, under the name of A. niagarensis, as his name was preoccupied by Nicholson & Hinde.
AMPLEXUS annulatus, Whitfield, 1878, Ann. Rep. Geo. Sur. Wis., Niagara Gr. [Sig. annulated.]
  feildeni, Etheridge, 1878, Quar. Jour. Geo. Soc. vol. 34, Upper Silurian. [Ety. proper name.]
  fenestratus, Whitfield, 1878, Ann. Rep. Geo. Sur. Wis., Niagara Gr. [Sig. having open windows.]
  junctus, Hall, 1882, Foss. Corals, Niagara and Up. Held. Gr., Niagara Gr. [Sig. joined.]
  laxatus, Billings, Can. Nat. Corniferous (?) limestone. [Sig. spread out.]
  uniformis, Hall, 1882, Foss. Corals, Niagara and Up. Held. Gr., Niagara Gr. [Sig. uniform.]
ANISOPHYLLUM, Edwards and Haime, 1851, Mon. d. Pol. foss. d. Terr. Pal. [Ety. *anisos*, unequal; *phyllon*, a leaf.]
  agassizi, Edwards & Haime, 1851, Mon. d. Pol. foss. d. Terr. Pal., Low. Held. Gr. [Ety. proper name.]
  bilamellatum, Hall, 1882, Foss. Corals, Niagara and Up. Held. Gr., Niagara Gr. [Sig. having two lamellæ.]
  trifurcatum, Hall, 1882, Foss. Corals, Niagara and Up. Held. Gr., Niagara Gr. [Sig. three forked.]
  unilarguin, Hall, 1882, Foss. Corals, Niagara and Up. Held. Gr., Niagara Gr. [Sig. one large.]
ARACHNOPHYLLUM, Dana, 1848, Zoophytes, U. S. Expl. Exped., vol. 8. [Ety. *arachne*, a spider; *phyllon*, a leaf.]
  richardsoni, Salter, 1852, Sutherland's Jour., vol. 2, Upper Sil. [Ety. proper name.]
Astraea hennahi, see Smithia hennahi.
ASTRÆOPHYLLUM, Nicholson & Hinde, 1874, Can. Jour., vol. 14. [Ety. *aster*, a star; *phyllon*, a leaf.]
  gracile, Nicholson & Hinde, 1874, Can. Jour., vol. 14, Niagara Gr. [Sig. slender.]
AULACOPHYLLUM, Edwards & Haime, 1854, British Fossil Corals. [Ety. *aulas*, a furrow; *phyllon*, a leaf.]
  bilaterale, Hall, 1882, Foss. Corals

Niagara and Up. Held. Gr., Corniferous limestone. [Sig. bilateral.]
AULACOPHYLLUM conigerum, Davis (In press), Foss. Corals of Ky., in vol. 1, Pal. Ky. St. Geo. Sur., Upper Devonian. [Sig. bearing a cone.]
convergens, Hall, 1882. Foss. Corals, Niagara and Up. Held. Gr., Corniferous limestone. [Sig. converging.]
cruciforme, Hall, 1882, Foss. Corals, Niagara and Up. Held. Gr., Corniferous limestone. [Sig. cruciform.]
mutabile, Davis (In press), Foss. Corals of Ky., in vol. 1, Pal. Ky. St. Geo. Sur., Middle Devonian. [Sig. changeful.]
parvum, Davis (In press), Foss. Corals of Ky., in vol. 1, Pal. Ky. St. Geo. Sur., Middle Devonian. [Sig. small.]
pinnatum, Hall, 1882, Foss. Corals, Niagara and Up. Held. Gr., Corniferous limestone. [Sig. plumed.]
poculum, Hall, 1882, Foss. Corals, Niagara and Up. Held. Gr., Corniferous limestone. [Sig. a cup.]
praeciptum, Hall, 1882, Foss. Corals, Niagara and Up. Held. Gr., Corniferous limestone [Sig. anticipated.]
prateriforme. Hall, 1882, Foss. Corals, Niagara and Up. Held. Gr., Corniferous limestone. [Sig. prateriform?]
princeps, Hall, 1882, Foss. Corals, Niagara and Up. Held. Gr., Corniferous limestone. [Sig. first.]
reflexum, Hall, 1882, Foss. Corals, Niagara and Up. Held. Gr., Corniferous limestone. [Sig. turned back.]
sulcatum. D'Orbigny, 1850. Prodr. d. Pal., vol. 1, p. 105, Corniferous limestone. [Sig. furrowed.]
tripinnatum. Hall, 1882, Foss. Corals. Niagara and Up. Held. Gr., Corniferous limestone. [Sig. three pinnated.]
trisulcatum, Hall, 1882, Foss. Corals, Niagara and Up. Held Gr., Corniferous limestone. [Sig. three furrowed.]
unguloideum. Davis (In press), Foss. Corals of Ky., in vol. 1, Pal. Ky. St. Geo. Sur., Lower Devonian. [Sig. like a hoof.]
AULOPORA canadensis, 1875 (Alecto canadensis). Can. Nat. and Geo., vol. 7, Corniferous limestone. [Ety. proper name.]
culmula, Davis (In press), Foss. Corals of Ky., in vol. 1, Pal. Ky. St Geo. Sur., Devonian. [Sig. a little stalk.]
procumbens, Davis (In press), Foss. Corals of Ky., in vol. 1, Pal. Ky. St. Geo. Sur., Middle Devonian. [Sig. prostrate.]
BARYPHYLLUM fungulus, White, 1878, Proc. Acad. Nat. Sci., Niagara Gr. [Sig. a small mushroom.]
verneuilanum. Edwards & Haime, 1851, Mon. d. Pol. foss. d. Terr. Pal., Niagara Gr. [Ety. proper name.]

BLOTHROPHYLLUM cinctutum, Davis (In press), Foss. Corals of Ky., vol. 1, Pal. Ky. St. Geo. Sur., Middle Devonian. [Sig. girt.]
corium, Davis (In press), Foss. Corals of Ky., vol. 1, Pal. Ky. St. Geo. Sur., Middle Devonian. [Sig. bark, skin.]
louisvillense, Davis (In press), Foss. Corals of Ky., vol. 1, Pal. Ky. St. Geo. Sur., Middle and Upper Devonian. [Ety. proper name]
multicalicatum, Hall, 1882, Foss. Corals, Niagara and Up. Held. Gr., Corniferous limestone. [Sig. many plastered.]
niagarense, Davis (In press), Foss. Corals of Ky., vol. 1, Pal. Ky. St. Geo. Sur., Niagara Gr. [Ety. proper name.]
papulosum, Hall, 1882, Foss. Corals, Niagara and Up. Held Gr., Corniferous limestone. [Sig. pustuled.]
parvulum, Davis (In press), Foss. Corals of Ky., vol. 1, Pal. Ky. St. Geo. Sur., Middle Devonian. [Sig. very small.]
promissum, Hall., 1882, Foss. Corals, Niagara and Up. Held. Gr., Corniferous limestone. [Sig. hanging down.]
sessile, Davis (In press), Foss. Corals of Ky., vol. 1, Pal. Ky. St. Geo. Sur., Middle Devonian. [Sig. dwarfish, seeming to sit.]
sinuosum, Hall, 1882, Foss. Corals, Niagara and Up. Held. Gr., Corniferous limestone. [Sig. sinuous.]
Calamopora cellulata, see Favosites cellulata.
goldfussi, see Favosites goldfussi.
minuta, see Favosites minuta.
minutissima, see Favosites minutissima.
radians, see Favosites radians.
verneuili, see Favosites verneuili.
CALCEOLA should be removed from the Brachiopoda to this class.
attenuata, Lyon, 1879, Proc. Acad. Nat. Sci., Niagara Gr. [Sig. attenuated.]
corniculum, Lyon, 1879, Proc. Acad. Nat. Sci., Niagara Gr. [Sig. a little horn.] Is it a syn. for C. tennesseensis?
coxi, Lyon, 1879, Proc. Acad. Nat. Sci., Niagara Gr. [Ety. proper name.] Is it a syn. for C. tennesseensis?
proteus, Davis (In press), Foss. Corals of Ky., in vol. 1, Pal. Ky. St. Geo. Sur., Niagara Gr. [Ety. mythological name.]
pusilla, Hall, 1882, Foss. Corals, Niagara and Up. Held. Gr., Niagara Gr. [Sig. very small.]
CALLOGRAPTUS niagarensis, Spencer, 1878, Can. Nat., vol. 8, Niagara Gr. [Ety. proper name.]
CALOPHYLLUM, Dana, 1848, Zoophytes U.S. Expl. Exped., vol. 8. [Ety. *kalos*, beautiful; *phyllon*, a leaf.]
phragmoceras, Salter, 1852, Sutherland's Jour., vol. 2, Up. Sil. [Sig. a partitioned horn.]

CALYPTOGRAPTUS, Spencer, 1878, Can. Nat., vol. 8. [Ety. *kalyptos*, covered; *grapho*, to write.]
cyathiformis, Spencer, 1878, Can. Nat., vol. 8, Niagara Gr. [Sig. cup shaped.]
subretiformis, Spencer, 1878, Can. Nat., vol. 8, Niagara Gr. [Sig. somewhat net formed.]
CANNAPORA annulata, Nicholson & Hinde, 1874, Can. Jour., Niagara Gr. [Sig. ringed.]
Caryophyllia cornicula, see Zaphrentis cornicula.
gigantea, see Zaphrentis gigantea.
pulmonea, see Zaphrentis pulmonea.
CATENIPORA *michelini*, Castlenau, syn. for Halysites catenulatus.
CHETETES æquidistans, Hall, 1881. Bryozoans of the Up. Held. Gr. [Sig. equidistant.]
arcticus, Haughton, 1857, Jour. Roy. Dub. Soc., vol. 1. [Sig. arctic.]
compressus, Ulrich, 1879. Jour. Cin. Soc. Nat. Hist., vol. 2. Hud. Riv. Gr. [Sig. compressed.] This is a bryozoan.
crebrirama, Hall, 1881, Bryozoans of the Up. Held. Gr. [Sig. having dense branches.]
egenus, Hall, 1881, Bryozoans of the Up. Held. Gr. [Sig. destitute of.]
elegans, Ulrich, see Monticulipora elegans.
internascens, Hall, 1881. Bryozoans of the Up. Held. Gr. [Sig. growing between.]
irregularis, see Monticulipora irregularis.
filiasa, see Monticulipora filiasa.
fusiformis, Whitfield, 1878, Ann. Rep. Geo. Sur. Wis., Hud. Riv Gr. [Sig. fusiform.]
granuliferus, see Monticulipora granulifera.
milleporaceus, Edwards & Haime, 1851, Mon. d. Pol. foss. d. Terr. Pal. Carboniferous. [Sig. having innumerable pores.]
moniliformis, refer to Monticulipora moniliformis.
ramosus, see Monticulipora ramosa.
subglobosus, Ulrich, see Monticulipora subglobosa.
undulatus, see Monticulipora undulata.
venustus, see Monticulipora venusta.
CHONOPHYLLUM capax, Hall, 1882, Foss. Corals, Niagara and Up. Held. Gr., Niagara Gr. [Sig. capacious.]
magnificum, was published in Can. Jour., vol. 5, in 1860.
nanum, Davis (In press), Foss. Corals of Ky., in vol. 1, Pal. Ky. St. Geo. Sur., Upper Devonian. [Sig. a dwarf.]
sedaliense, White, 1880, 12th Rep. U S. Geo. Sur. Terr., Choteau Gr. [Ety. proper name.]
vadum, Hall, 1882, Foss. Corals, Niagara and Up. Held. Gr., Niagara Gr. [Sig. shallow.]

CHONOSTEGITES, Edwards & Haime, 1851, Pol. Foss. des Terr. Pal., p. 299. [Ety. *konos*, a cone; *stege*, a covering.]
clappi, Edwards & Haime, 1851, Pol. Foss. des Terr. Pal. p. 299, Up. Held. Gr. [Ety. proper name.]
CLADOPORA alcicornis, Davis (In press), Foss. Corals of Ky., in vol. 1, Pal. Ky. St. Geo. Sur., Up. Devonian. [Sig. elk's horn.]
aculeata, Davis (In press), Foss Corals of Ky., in vol. 1, Pal. Ky. St. Geo. Sur., Niagara Gr. [Sig. having prickles or spines.]
acupicta, Davis (In press), Foss. Corals of Ky., in vol. 1, Pal. Ky. St. Geo. Sur., Low. Devonian [Sig. punctured as with a needle.]
bifurca, Davis (In press), Foss. Corals of Ky., in vol. 1, Pal. Ky. St. Geo. Sur. Low. Devonian. [Sig. two pronged, forked.]
canadensis, Rominger, is a syn. for Pachypora frondosa.
crassa, Davis (In press), Foss. Corals of Ky., in vol. 1, Pal Ky. St. Geo. Sur., Low. Devonian. [Sig. thick.]
dentata, Davis (In press), Foss. Corals of Ky., in vol. 1, Pal. Ky. St. Geo. Sur., Low. Devonian. [Sig. having teeth.]
desquamata, Davis (In press), Foss. Corals of Ky., in vol. 1, Pal. Ky. St. Geo. Sur., Low. Devonian. [Sig. scaled, peeled.]
dispansa, Davis (In press), Foss. Corals of Ky., in vol. 1, Pal. Ky. St. Geo. Sur., Low. Devonian. [Sig. spread out flat.]
equisetalis, Davis (In press), Foss. Corals of Ky., in vol. 1, Pal. Ky. St. Geo. Sur., Niagara Gr. [Sig. from the plant Equisetum.]
fibrata, Davis (In press), Foss. Corals of Ky., in vol. 1, Pal. Ky. St. Geo. Sur., Low. Devonian. [Sig. fibrous.]
foliata, Davis (In press), Foss Corals of Ky., in vol. 1, Pal. Ky. St. Geo. Sur., Low Devonian. [Sig. arranged in leaves.]
gracilis, Davis (In press), Foss. Corals of Ky., in vol. 1, Pal. Ky. St. Geo. Sur., Mid. Devonian. [Sig. slender.]
gulielmi, Davis (In press), Foss. Corals of Ky., in vol. 1, Pal. Ky. St. Geo. Sur., Up. Devonian. [Ety. proper name.]
knappi, Davis (In press), Foss. Corals of Ky., in vol. 1, Pal. Ky. St Geo. Sur., Low Devonian. [Ety. proper name.]
lichenoides, Rominger, 1876. This name was preoccupied by Winchell & Marcy, in 1865. See C. winchellana.
menis, Davis (In press), Foss. Corals of Ky., in vol 1, Pal. Ky. St. Geo. Sur., Niagara Gr. [Sig. a crescent.]
ordinata, Davis (In press), Foss. Corals

of Ky., in vol. 1, Pal. Ky. St. Geo. Sur., Niagara Gr. [Sig. from the arrangement in rows.]
CLADOPORA proboscidalis, Davis (In press), Foss. Corals of Ky., in vol. 1, Pal. Ky. St. Geo. Sur., Niagara Gr. [Sig. having a proboscis.]
ricta, Davis (In press), Foss. Corals of Ky., in vol. 1, Pal. Ky. St. Geo. Sur., Low. Devonian. [Sig. open mouth.]
robusta *var*. tela, Davis (In press), Foss. Corals of Ky., in vol. 1, Pal. Ky. St. Geo. Sur., Low. Devonian. [Sig a web.]
sarmentosa, Hall, 1876, Desc. new species of fossils from Waldron, Niagara Gr. [Sig full of branches.]
striata, Davis (In press), Foss. Corals of Ky., in vol. 1, Pal. Ky. St. Geo. Sur., Niagara Gr. [Sig. striped.]
undosa, Davis (In press), Foss. Corals of Ky., in vol. 1, Pal. Ky. St. Geo. Sur., Low. Devonian. [Sig. wavy.]
winchellana, n. sp., Up. Held. Gr. This name is proposed for the species described by Rominger, under the preoccupied name of C. lichenoides, in his Fossil Corals, p. 47, pl. 17, figs. 1 and 4. Named in honor of Alexander Winchell.
CLISIOPHYLLUM austini, Salter, 1852 (Strephodes austini), Sutherland's Jour., vol. 2, Devonian [Ety. proper name.]
dannanum, Edwards & Haime, 1851, Mon. d. Pol. foss. d. Terr. Pal., Low. Held. Gr. [Ety. proper name.]
pluriradiale, Nicholson, 1874, Pal. Prov. Ontario, Corniferous limestone. [Sig. many radiated.]
tumulus, Salter, 1855, Belcher's last of the Arctic Voyages, vol. 2, Carboniferous. [Sig. a mound.]
Cœnites lunata, see Limaria lunata.
COLUMNARIA halli, Nicholson. 1879, Tabulate Corals, syn for C. alveolata.
mamillaris, and C. multiradiata, Castelnau, 1843, are not recognized by palæontologists.
sutherlandi, Salter, 1852, Sutherland's Jour., vol 2. Devonian. [Ety. proper name.]
COLUMNOPORA rayi, Davis (In press), Foss. Corals of Ky., in vol. 1, Pal. Ky. St. Geo. Sur., Hud. Riv. Gr. [Ety. proper name.]
CRASPEDOPHYLLUM, Dybowski, 1873, Beschreibung neuen aus Nordamerika, Stammenden Devonischen, art der Zoantharia rugosa. [Ety. *kraspedos*, an edge; *phyllon*, a leaf.]
americanum, Dybowski, 1873, Beschr. n a. Norda. s. Dev. a. d. Zoanth. rugosa., Up. Held. Gr., at Columbus, O. [Ety. proper name.]
Crepidophyllum, Nicholson & Thompson, 1877, Proc. Royal Soc. Edinburgh, vol.

9. A proposed name to include Diphyphyllum archiaci.
CYATHAXONIA columellata, Hall, 1882, Foss. Corals, Niagara and Up. Held. Gr., Niagara Gr. [Sig. pillared.]
cynodon, Rafinesque & Clifford, 1820, Monographie d. Turbinolides, in Ann. d. Phys. d. Brux., t. 5, Waverly Gr. [Sig. dog tooth.]
gainesi, Davis (In press), Foss. Corals of Ky., in vol. 1, Pal. Ky. St. Geo. Sur., Hud. Riv. Gr. [Ety. proper name.]
herzeri, Hall, 1882, Foss. Corals, Niagara and Up. Held. Gr., Niagara Gr. [Ety. proper name.]
profunda, Edwards & Haime, 1851, Mon. d. Pol. foss. d. Terr. Pal., Carboniferous. [Sig. profound.]
wisconsinensis, Whitfield, 1878, Ann. Rep. Geo. Sur. Wis., Niagara Gr. [Ety. proper name.]
CYATHOPHYLLUM *agglomeratum, C. ammonis, C. arborescens, C. atlas, C. conicum, C. distinctum, C. d'orbignyi, C. goldfussi, C. goliath, C. michelini, C. plicatulum, C. rollini, C. striatulum,* and *C. vicinum*, are Castlenau's species imperfectly described in 1843, in his Systeme silur. de l'Amerique septentr. They are not recognized by American authors.
articulatum, Wahlenberg (Madreporites articulatus), Nov. Act. Upsal., vol. 8, Up. Sil. This, though identified by some from America, is probably not an American species.
arctifossa, Hall, 1882, Foss. Corals Niagara and Up. Held. Gr., Corniferous limestone. [Sig. close wrinkled.]
brevicorne, Davis (In press), Foss. Corals of Ky., in vol. 1, Pal. Ky. St. Geo. Sur., Devonian. [Sig. short horn.]
bullatum, Hall, 1882, Foss. Corals, Niagara and Up. Held. Gr., Corniferous limestone. [Sig. vesicled.]
bullulatum, Hall, 1882, Foss. Corals, Niagara and Up. Held. Gr., Niagara Gr. [Sig. little vesicled.]
canaliculatum, Hall 1882, Foss. Corals, Niagara and Up. Held. Gr., Corniferous limestone. [Sig. canaliculated.]
cohærens, Hall, 1882, Foss. Corals, Niagara and Up. Held. Gr., Corniferous limestone. [Sig. coherent.]
concentricum, Hall, 1882, Foss. Corals, Niagara and Up. Held. Gr., Corniferous limestone. [Sig. concentric.]
depressum. Hall, 1882. Foss. Corals, Niagara and Up. Held. Gr., Corniferous limestone. [Sig. depressed.]
exfoliatum, Hall, 1882, Foss. Corals, Niagara and Up. Held. Gr., Corniferous limestone. [Sig. exfoliated.]
fimbriatum. Davis (In press), Foss. Corals of Ky., in vol. 1, Pal. Ky. St. Geo. Sur., Mid. Devonian. [Sig. fringed.]
flos, Davis (In press), Foss. Corals of

Ky., in vol. 1, Pal. Ky. St. Geo. Sur., Niagara Gr. and Devonian. [Sig. a flower.]
CYATHOPHYLLUM gemmiferum, Davis (In press), Foss Corals of Ky., in vol. 1, Pal. Ky. St. Geo. Sur., Devonian. [Sig. bearing buds.]
gigas is a syn. for Zaphrentis gigantea.
impositum, Hall, 1882, Foss. Corals, Niagara and Up. Held. Gr., Corniferous limestone. [Sig. laying over.]
infoveatum, Davis (In press), Foss. Corals of Ky., in vol. 1, Pal. Ky. St. Geo. Sur., Devonian. [Sig. without a fovea.]
intertrium, Hall, 1882, Foss. Corals, Niagara and Up. Held. Gr., Niagara Gr. [Ety. from the three smaller lamellæ between the larger ones.]
intervesiculum, Hall, 1882, Foss. Corals, Niagara and Up. Held. Gr., Corniferous limestone. [Sig. having vesicles between.]
juvenis, is without letter j in part of the 1st edition of this work.
lesueuri, should be accredited to Edwards & Haime, 1851, Mon. d. Pol. Foss. d. Terr. Pal.
multicrena, Davis (In press), Foss. Corals of Ky., in vol. 1, Pal. Ky. St. Geo. Sur., Devonian. [Sig. many notched.]
nevadense, Meek, 1877, U. S. Geo. Sur., 40th parallel, Carboniferous. [Ety. proper name.]
œdipus, Davis (In press), Foss. Corals of Ky., in vol. 1, Pal. Ky. St. Geo. Sur., Low. and Mid. Devonian. [Sig. having a club foot.]
palmeri, Meek, 1877, U. S. Geo. Sur., 40th parallel, Devonian. [Ety. proper name.]
perfossulatum, Hall, 1882, Foss. Corals, Niagara and Up. Held. Gr., Corniferous limestone. [Sig. having many little ditches.]
perplicatum, Hall, 1882. Foss. Corals, Niagara and Up. Held. Gr., Corniferous limestone. [Sig. many plicated.]
pickthorni, Salter, 1852 (Strephodes pickthorni). Sutherland's Jour., vol. 2, Devonian. [Ety. proper name.]
plicatum, Goldfuss, 1826, Germ. Petref. This specific name is condemned as American, because Goldfuss applied it to two distinct species at the same time—one from America and the other from Sweden.
pocillum, Davis (In press), Foss. Corals of Ky., in vol. 1, Pal. Ky. St. Geo. Sur., Mid. and Up. Devonian. [Sig. a little cup.]
pumilus, Davis (In press), Foss. Corals of Ky., in vol. 1, Pal. Ky. St. Geo. Sur., Mid. Devonian. [Sig. a dwarf.]
pustulatum, Conrad, 1848. Not clearly defined.
robustum, Hall, 1882, Foss. Corals, Niagara and Up Held. Gr., Corniferous limestone. [Sig. robust.]
CYATHOPHYLLUM scalenum, Hall, 1882, Foss. Corals, Niagara and Up. Held. Gr., Corniferous limestone. [Sig. having unequal sides.]
septatum, Hall, 1882, Foss. Corals, Niagara and Up. Held. Gr., Corniferous limestone. [Sig. partitioned.]
tornatum. Davis (In press) Foss. Corals of Ky., in vol. 1, Pal. Ky. St. Geo. Sur., Up. Devonian. [Sig. rounded as in a lathe.]
vesiculatum, Hall, 1882, Foss. Corals, Niagara and Up. Held. Gr., Corniferous limestone. [Sig. vesicled.]
zenkeri, Billings, 1860, Can. Jour., vol. 5, Corniferous limestone. [Ety. proper name.]
CYSTIPHYLLUM bifurcatum, Hall, 1882, Foss. Corals, Niagara and Up. Held. Gr., Corniferous limestone. [Sig. bifurcated.]
bipartitum, Hall, 1882, Foss. Corals, Niagara and Up. Held. Gr., Corniferous limestone. [Sig. two parted.]
cicatriciferum, Davis (In press), Foss. Corals of Ky., in vol. 1, Pal. Ky. St. Geo. Sur., Up. Devonian. [Sig. bearing a scar.]
crateriforme, Hall, 1882, Foss. Corals, Niagara and Up. Held. Gr., Corniferous limestone. [Sig. crateriform.]
granilineatum, Hall, 1882, Foss. Corals, Niagara and Up. Held. Gr., Niagara Gr. [Sig. granule lined.]
hispidum, Davis (In press), Foss. Corals of Ky., in vol. 1, Pal. Ky. St. Geo. Sur., Low. Devonian. [Sig. prickly, thorny.]
incurvum, Davis (In press), Foss. Corals of Ky., in vol. 1, Pal. Ky. St. Geo. Sur., Niagara Gr. [Sig. crooked.]
infundibulum, Hall, 1882, Foss. Corals, Niagara and Up. Held. Gr., Corniferous limestone. [Sig. a funnel.]
latiradius, Hall, 1882, Foss. Corals, Niagara and Up. Held. Gr., Corniferous limestone. [Sig. wide rayed.]
limbatum, Davis (In press), Foss. Corals of Ky., in vol. 1, Pal. Ky. St. Geo. Sur., Low. Devonian. [Sig. bordered, like a garment with flounces.]
lineatum, Davis (In press), Foss. Corals of Ky., in vol. 1, Pal. Ky. St. Geo. Sur., Niagara Gr. and Low. Devonian. [Sig. lined.]
muricatum, Hall, 1882, Foss. Corals, Niagara and Up. Held. Gr., Corniferous limestone. [Sig. pointed.]
nanum, Hall, 1882, Foss. Corals, Niagara and Up. Held. Gr., Corniferous limestone. [Sig. a dwarf.]
nettlerothi, Davis (In press), Foss. Corals of Ky., in vol, 1. Pal. Ky. St. Geo. Sur., Low. Devonian. [Ety. proper name.]

POLYPI.   267

CYSTIPHYLLUM obliquum, Hall, 1882, Foss.
  Corals, Niagara and Up. Held. Gr.,
  Corniferous limestone. [Sig. oblique.]
  os, Davis (In press), Foss. Corals of Ky.,
    in vol. 1, Pal. Ky. St. Geo. Sur.,
    Low. Devonian. [Sig. a bone.]
  plicatum, Davis (In press), Foss. Corals
    of Ky., in vol. 1, Pal. Ky. St. Geo.
    Sur., Low. and Mid. Devonian. [Sig.
    folded in plaits.]
  pustulatum, Hall, 1882, Foss. Corals,
    Niagara and Up. Held. Gr., Cor-
    niferous limestone. [Sig. pustulated.]
  quadrangulare, Hall, 1882, Foss. Corals,
    Niagara and Up. Held. Gr., Cor-
    niferous limestone. [Sig. quad-
    rangular.]
  scalatum, Hall, 1882, Foss. Corals,
    Niagara and Up. Held. Gr., Cor-
    niferous limestone. [Sig. having
    stairs.]
  striatum, Hall, 1882, Foss. Corals, Ni-
    agara and Up. Held. Gr., Corniferous
    limestone. [Sig. channeled.]
  sulcatum was described in 1858, in
    Can. Nat. & Geol., vol. 3.
  supraplanum, Hall, 1882, Foss. Corals,
    Niagara and Up. Held. Gr., Cor-
    niferous limestone. [Sig. very level.]
  tenuiradius, Hall, 1882, Foss. Corals,
    Niagara and Up. Held. Gr., Cor-
    niferous limestone. [Sig. fine rayed.]
  theissi, Davis (In press), Foss. Corals in
    Ky., in vol. 1, Pal. Ky. St. Geo. Sur.,
    Mid. Devonian. [Ety. proper name.]
  zaphrentiforme, Davis (In press), Foss.
    Corals of Ky., in vol. 1, Pal. Ky.
    St. Geo. Sur., Low Devonian. [Sig.
    formed like Zaphrentis.]
CYSTOSTYLUS, Whitfield, 1880, Ann. Rep.
  Geo. Sur. Wis. [Ety. kustis, a bladder; stylos, a stalk.]
  typicus, Whitfield, 1880, Ann. Rep. Geo.
    Sur., Wis., Niagara Gr. [Sig. the
    type.]
DANIA, Edwards & Haime, 1849, Comptes
  Rendus, t. 29, p. 261. [Ety. proper
  name.]
  huronica, Edwards & Haime. 1849,
    Comptes Rendus, t. 29, Up. Silurian.
    [Ety proper name.]
DEKAYIA, Edwards & Haime, 1851, Mon. d.
  Pol. foss. d. Terr. Pal., p. 277. [Ety.
  proper name.]
  aspera, Edwards & Haime, 1851, Mon.
    d. Pol. foss. d. Terr. Pal., Hud. Riv.
    Gr. [Sig. rough.]
  attrita, Nicholson (Chetetes attritus),
    syn. for D. aspera.
DENDROGRAPTUS compactus, Walcott, 1879,
  Utica Slate and related formations,
  Utica Slate. [Sig. compact.]
  gracillimus, Lesquereux, 1877, Proc.
    Am. Phil. Soc. (Psilophyton gracilli-
    mum), Hud. Riv. Gr. [Sig. very
    slender.]

DENDROGRAPTUS novellus, Hall, 1879, Desc.
  New Spec. Foss., Niagara Gr. [Sig.
  young.]
  simplex, Walcott, 1879, Utica Slate and
    related formations, Utica Slate. [Sig.
    simple.]
  tenuiramosus, Walcott, 1879, Utica Slate
    and related formations, Utica Slate.
    [Sig. very branchy.]
DENDROPORA ornata refer to Trachypora
  ornata.
  osculata, Davis (In press), Foss. Corals
    of Ky., in vol. 1, Pal. Ky. St. Geo.
    Sur., Devonian. [Sig. kissed, joined
    by contact.]
DICTYONEMA fenestrata was described in
  1851, in Foster & Whitney's Rep. on
  the Lake Superior Land District.
  pergracilis, Hall and Whitfield, 1872,
    24th Reg. Rep., Niagara Gr. [Sig.
    very slender.]
  tenella, Spencer, 1878, Can. Nat., vol.
    8, Niagara Gr. [Sig. delicate.]
  websteri was described in 1860, Can.
    Nat. & Geo., vol. 5.
DIORRYCHOPORA, Davis (In press), Foss.
  Corals of Ky., in vol. 1, Pal. Ky. St.
  Geo. Sur. [Ety. diorrusso, to dig a
  canal; pora, a tube.]
  tennis, Davis (In press), Foss. Corals of
    Ky. in vol. 1, Pal. Ky. St. Geo. Sur.,
    Niagara Gr. [Sig. slender.]
DIPHYPHYLLUM adnatum, Hall, 1882, Foss.
  Corals, Niagara and Up. Held Gr.,
  Corniferous limestone. [Sig. adnate.]
  apertum, Hall, 1882, Foss. Corals, Ni-
    agara and Up. Held. Gr., Cornifer-
    ous limestone. [Sig. opened.]
  bellis, Davis (In press), Foss. Corals of
    Ky., in vol. 1, Pal. Ky. St. Geo. Sur.,
    Mid. Devonian. [Sig. a daisy.]
  breve, Hall, 1882, Foss. Corals, Niagara
    and Up. Held. Gr., Corniferous
    limestone. [Sig. short.]
  coagulatum, Davis (In press), Foss.
    Corals of Ky., in vol. 1, Pal. Ky. St.
    Geo. Sur., Mid. Devonian. [Sig.
    thickened.]
  coalescens, Davis (In press), Foss.
    Corals of Ky., in vol. 1, Pal. Ky. St.
    Geo. Sur., Mid. Devonian. [Sig.
    growing together.]
  conjunctum, Davis (In press), Foss.
    Corals of Ky., in vol. 1, Pal. Ky. St.
    Geo. Sur., Mid Devonian. [Sig.
    joined intimately.]
  cylindraceum, Hall, 1882, Foss. Corals,
    Niagara and Up. Held. Gr., Cornifer-
    ous limestone. [Sig. like cylinders.]
  dividuum, Davis (In press), Foss. Corals
    of Ky., in vol. 1, Pal. Ky. St. Geo.
    Sur., Niagara Gr. [Sig. separable.]
  fasciculum, Meek, 1877, U. S. Geo. Sur.,
    40th parallel, Devonian. [Sig. a small
    bundle.]
  gracile, McCoy, 1854, Brit. Pal. Foss.,
    Corniferous limestone. [Sig. slender.]
  tumidulum, Hall, 1882, Foss. Corals,

Niagara and Up. Held. Gr., Corniferous limestone. [Sig. tumid.]
DIPLOGRAPTUS hudsonicus, Nicholson, 1875, Pal. Prov. Ont., Hud. Riv. Gr. [Ety. proper name.]
hypniformis, White, 1874 (Graptolithus hypniformis', Rep. Invert. Foss., Trenton Gr. [Sig. like *Hypnum*, from the moss-like aspect of the stipes.]
Diplotrypa, Nicholson, 1879, Tabulate Corals. A subgeneric name, founded upon Monticulipora petropolitana.
DRYMOPORA, Davis (In press), Foss. Corals of Ky., in vol. 1, Pal. Ky. St. Geo. Sur. [Ety. *drumos*, a thicket; *pora*, a tube.]
auloporoidea, Davis (In press), Foss. Corals of Ky., in vol. 1, Pal. Ky. St. Geo. Sur., Up. Devonian. [Sig. resembling *Aulopora*.]
commensalis, Davis (In press), Foss. Corals of Ky., in vol. 1, Pal. Ky. St. Geo. Sur., Mid. Devonian. [Sig. living at the same table.]
frutectosa, Davis (In press), Foss. Corals of Ky., in vol. 1, Pal. Ky. St. Geo. Sur., Up. Devonian. [Sig. shrubby, bushy.]
ELASMOPHYLLUM, Hall, 1882. Foss. Corals, Niagara and Up. Held. Gr. [Ety. *elasma*, lamella; *phyllon*, a leaf.]
attenuatum, Hall, 1882, Foss. Corals, Niagara and Up. Held'Gr., Corniferous limestone. [Sig. attenuated.]
ERIDOPHYLLUM simcoense is from the Corniferous limestone.
FAVISTELLA calicina, Nicholson, 1874, Rep. Brit. Assoc., Hud. Riv. Gr. [Sig. a little cup.]
franklini, Salter, 1852, Sutherland's Jour., vol. 2, Up. Sil. [Ety. proper name.]
reticulata, Salter, 1852. Sutherland's Jour., vol. 2, Up. Sil. [Sig. reticulated.]
FAVOSITES alveolaris may be stricken out as not an American species.
amplissimus, Davis (In press). Foss. Corals of Ky., in vol. 1, Pal. Ky. St. Geo. Sur., Low. and Mid. Devonian. [Sig. very large or ample.]
arbor, Davis (In press), Foss. Corals of Ky., in vol. 1 Pal Ky. St. Geo. Sur., Low. Devonian. [Sig. a tree.]
baculus, Davis (In press), Foss. Corals of Ky., in vol. 1, Pal. Ky. St. Geo. Sur., Low. Devonian. [Sig. a staff, a scepter.]
canadensis, Billings, 1858, Can. Nat. & Geol., vol. 3, instead of Fistulipora canadensis.
cariosus, Davis (In press), Foss. Corals of Ky., in vol. 1, Pal. Ky. St. Geo. Sur., Low. & Mid. Devonian. [Sig. worm eaten.]
cellulata, Castelnau. 1843 (Calamopora cellulata), Syst. Sil. From Point Latour, northeast of Lake Huron, but not recognized by later authors for some reasons not known to the author.
FAVOSITES clelandi, Davis (In press). Foss. Corals of Ky., in vol. 1. Pal. Ky. St. Geo. Sur., Low. Devonian. [Ety. proper name.]
convexus, Davis (In press). Foss. Corals of Ky., in vol. 1. Pal. Ky. St. Geo. Sur., Low. and Mid. Devonian. [Sig. convex.]
cymosus, Davis (In press), Foss. Corals of Ky., in vol. 1, Pal. Ky. St. Geo. Sur., Low. and Mid. Devonian. [Sig. full of sprouts.]
chapmani, Nicholson, 1874, Pal. Prov. Ont., Corniferous limestone. [Ety. proper name.]
divergens, Winchell, 1862, Proc. Acad. Nat. Sci., Low. Carb. [Sig. diverging.]
emmonsi, Rominger, syn. for F. heliolitiformis.
eximius, Davis (In press), Foss. Corals of Ky., in vol. 1, Pal. Ky. St. Geo. Sur., Up. Devonian. [Sig. uncommon, remarkable.]
flabelliformis. Troost, 1843. Not satisfactorily defined.
forbesi, var. occidentalis, Hall, 1876, 28th Reg. Rep., Niagara Gr. [Sig. western.]
forbesi var. *waldronensis*, Nicholson, syn. for F. forbesi, var. occidentalis.
frutex, Davis (In press), Foss. Corals of Ky., in vol. 1, Pal. Ky. St. Geo. Sur., Low. Devonian. [Sig. a shrub, a bush.]
fustiformis, Davis (In press), Foss. Corals of Ky., in vol. 1, Pal. Ky. St. Geo. Sur., Low. Devonian. [Sig. club shaped.]
goldfussi. Castelnau, 1843, Syst. Sil. (Calamopora goldfussi, Up. Sil. [Ety. proper name.]
goodwyni, Davis (In press), Foss. Corals of Ky., in vol. 1, Pal Ky. St. Geo. Sur., Up. Dev. [Ety. proper name.]
hamiltonensis, Rominger, syn. for F. dumosus.
impeditus, Davis (In press), Foss. Corals of Ky., in vol 1, Pal. Ky. St. Geo. Sur., Low. Devonian. [Sig. hindered, of stunted growth.]
louisvillensis, Davis (In press), Foss. Corals of Ky., in vol. 1, Pal. Ky. St. Geo. Sur., Niagara Gr. [Ety. proper name.]
mamillaris, Castelnau, 1843, Syst. Sil. Not recognized.
mancus instead of F. manus.
minuta and *F. minutissima*, Castelnau. Not recognized.
mundus, Davis (In press), Foss. Corals of Ky., in vol. 1, Pal. Ky. St. Geo. Sur., Low, and Mid. Devonian. [Sig. neat, pretty.]
mundus *var.* placentoidens, Davis (In press), Foss. Corals of Ky., in vol. 1, Pal. Ky. St. Geo. Sur., Low. De-

vonian. [Sig. resembling F. placenta.]
FAVOSITES occidens, Whitfield, 1878, Ann. Rep. Geo. Sur. Wis., Niagara Gr. [Sig. western]
  parvo, Davis (In press), Foss. Corals of Ky., in vol. 1, Pal. Ky. St. Geo. Sur., Low. Devonian. [Sig. a peacock, in allusion to the resemblance between the larger round tubes regularly interspersed among the smaller polygonal tubes of the fossil and the dotted circles or "eyes" of a peacock's tail.]
  pirum, Davis (In press), Foss. Corals of Ky., in vol. 1, Pal. Ky. St. Geo. Sur., Mid. Devonian. [Sig. a pea.]
  proximus, Davis (In press), Foss. Corals of Ky., in vol. 1, Pal. Ky. St. Geo. Sur., Low Devonian. [Sig. a near neighbor.]
  quercus, Davis (In press), Foss. Corals of Ky., in vol. 1, Pal. Ky. St. Geo. Sur., Low. Devonian. [Sig. an oak.]
  radians. Castelnau, 1843. Not recognized.
  ramulosus, Davis (In press), Foss. Corals of Ky., in vol. 1, Pal. Ky. St. Geo. Sur., Low. Devonian. [Sig. having small branches.]
  rotundituba. Davis (In press), Foss. Corals of Ky., in vol. 1, Pal. Ky. St. Geo. Sur., Niagara and Devonian. [Sig. round tubed.]
  spiculatus, Davis (In press), Foss. Corals of Ky., in vol. 1, Pal. Ky. St. Geo. Sur., Low. Devonian. [Sig bearing spicules or spikes.]
  spinigerus instead of F. niagarensis, var. spinigerus.
  spongilla, Rominger, syn. for F. spinigerus.
  troosti, Edwards & Haime, 1851, Mon. d. Pol. foss d. Terr. Pal., Devonian. [Ety. proper name.]
  verneuili, Castelnau, 1843, syn. for Monticulipora fibrosa.
Favositopora palæozoica, Kent, 1870, Ann. & Mag. Nat. Hist., 4 ser., vol. 6. Not recognized.
FISTULIPORA canadensis, refer to Favosites canadensis.
  flabellata, Ulrich, 1879, Jour. Cin. Soc. Nat. Hist., vol. 2, Hud. Riv. Gr. [Sig. from flabellum, a fan.]
  lens, Whitfield. 1878, Ann. Rep. Geo. Sur. Wis., Hud. Riv. Gr. [Sig. a lentil.]
  rugosa, Whitfield, 1880, Ann. Rep. Geo. Sur. Wis., Hud. Riv. Gr. [Sig. rugose.]
  solidissima, Whitfield, 1878, Ann. Rep. Geo. Sur. Wis., Hud. Riv. Gr. [Sig. very solid.]
GRAPTOLITHUS annectans, Walcott, 1879, Utica Slate and related formations, Utica Slate Gr. [Sig. connected together.]

GRAPTOLITHUS whitianus, n. sp. Hud. Riv. Gr., from five miles north of Belmont, Nevada. Proposed instead of G. ramulus of White, 1874, Exp. and Sur., W. 100th merid., Prelim. Rep. Invert. Foss. p. 13, and vol. 4, pt. 1, p. 62, pl. iv., figs. 3a, 3b and 3c. The name was preoccupied by Hall, for a distinct species in the same genus.
HADROPHYLLUM glans, instead of Zaphrentis glans.
HALLIA, Edwards & Haime, 1851, Mon. d. Pol. foss. d. Terr. Pal. [Ety. proper name]
  divergens, Hall. 1882, Foss. Corals, Niagara and Up. Held. Gr., Niagara Gr. [Sig. diverging.]
  divisa, Hall, 1882, Foss. Corals, Niagara and Up. Held. Gr., Niagara Gr. [Sig. dividing.]
  insignis, Edwards & Haime, 1851, Mon. d. Pol. foss. d. Terr. Pal., Up. Held. Gr. [Sig. marked.]
  pluma. Hall, 1882, Foss. Corals Niagara and Up. Held. Gr., Niagara Gr. [Sig. a small, soft feather.]
  scitula, Hall, 1882, Foss. Corals, Niagara and Up. Held. Gr., Niagara Gr. [Sig. handsome.]
HALYSITES catenulatus, var. feildeni, Etheridge, 1878, Quar. Jour. Geo. Soc., vol. 34, Upper Silurian. [Ety. proper name.]
  catenulatus, var. harti. Etheridge, 1878, Quar. Jour. Geo. Soc., vol. 34, Up. Sil. [Ety. proper name.]
  labyrinthica, Goldfuss, 1826 (Catenipora labyrinthica), Petref. Germ. Niagara Gr. [Sig. labyrinthine.]
  nexus, Davis (In press), Foss. Corals of Ky., in vol. 1, Pal. Ky. St. Geo. Sur., Niagara Gr. [Sig. linked together.]
  parryi, König, 1824 (Catenipora parryi), Supp. to App. of Capt. Parry's Voyage for the Discovery of a Northwest Passage, Up. Sil. [Ety. proper name.]
HELIOLITES should be credited to Guettard, Mem. 3, p. 454.
HELIOPHYLLUM acuminatum, Hall, 1882. Foss. Corals, Niagara and Up. Held. Gr., Corniferous limestone. [Sig. acuminated.]
  aequale, Hall, 1882, Foss. Corals, Niagara and Up. Held. Gr., Corniferous limestone. [Sig. equal.]
  aequum, Hall, 1882, Foss. Corals, Niagara and Up. Held. Gr., Coruiferous limestone. [Sig. even, level.]
  alternatum, Hall, 1882, Foss. Corals, Niagara and Up. Held. Gr., Corniferous limestone. [Sig. alternated.]
  annulatum, Hall, 1882, Foss. Corals, Niagara and Up. Held. Gr., Corniferous limestone. [Sig. annulated.]
  campaniforme, Hall, 1882, Foss. Corals, Niagara and Up. Held. Gr., Corniferous limestone. [Sig. bell formed.]
  canadense, Billings, 1859, Can. Jour.,

vol. 4, Corniferous limestone. [Ety. proper name.]
HELIOPHYLLUM cancellatum, Hall, 1882, Foss. Corals, Niagara and Up. Held. Gr., Corniferous limestone. [Sig. cancellated.]
cayugaense, Billings, 1859, Can. Jour. vol. 4, Corniferous limestone. [Ety. proper name.]
compactum, Hall, 1882, Foss. Corals, Niagara and Up. Held. Gr., Corniferous limestone. [Sig. compact.]
dentatum, Hall, 1882, Foss. Corals, Niagara and Up. Held. Gr., Corniferous limestone. [Sig. toothed.]
denticulatum, Hall, 1882, Foss. Corals, Niagara and Up. Held. Gr., Corniferous limestone. [Sig. denticulated.]
dentilineatum, Hall, 1882, Foss. Corals, Niagara and Up. Held. Gr., Niagara Gr. [Sig. tooth lined.]
distans, Hall, 1882, Foss. Corals. Niagara and Up. Held. Gr., Corniferous limestone. [Sig. distant.]
fasciculatum, Hall, 1882, Foss. Corals, Niagara and Up. Held. Gr , Corniferous limestone.[Sig. fasciculated.]
fecundum, Hall, 1882, Foss. Corals, Niagara and Up. Held. Gr., Corniferous limestone. [Sig. fruitful.]
fissuratum, Hall, 1882, Foss. Corals, Niagara and Up. Held. Gr., Corniferous limestone. [Sig. fissured.]
gemmatum, Hall, 1882, Foss. Corals, Niagara and Up. Held. Gr., Corniferous limestone. [Sig. budded.]
gemmiferum, Hall, 1882, Foss. Corals, Niagara and Up. Held. Gr., Niagara Gr. [Sig. bud bearing.]
imbricatum, Hall, 1882, Foss. Corals, Niagara and Up. Held. Gr., Corniferous limestone. [Sig. imbricated.]
incrassatum, Hall, 1882, Foss. Corals, Niagara and Up. Held. Gr., Corniferous limestone. [Sig. thickened.]
invaginatum, Hall, 1882, Foss. Corals, Niagara and Up. Held. Gr., Corniferous limestone [Sig. invaginated.]
latericrescens, Hall, 1882, Foss. Corals, Niagara and Up. Held. Gr., Corniferous limestone. [Sig. side growing.]
lincolatum, Hall, 1882, Foss. Corals, Niagara and Up. Held. Gr., Corniferous limestone. [Sig. fine lined.]
mitella, Hall, 1882, Foss. Corals, Niagara and Up. Held. Gr., Niagara Gr. [Sig. a head band.]
nettlerothi, Hall, 1882, Foss. Corals, Niagara and Up. Held. Gr., Corniferous limestone. [Ety. proper name.]
pocillatum, Hall, 1882, Foss. Corals, Niagara and Up. Held. Gr., Corniferous limestone. [Sig. little cupped.]
pravum, Hall, 1882, Foss. Corals, Niagara and Up. Held. Gr., Niagara Gr. [Sig. cooked.]
puteatum, Hall, 1882, Foss. Corals, Niagara and Up. Held. Gr., Niagara Gr. [Sig. having a little well.]
HELIOPHYLLUM scyphulus, Hall, 1882. Foss. Corals, Niagara and Up. Held. Gr., Corniferous limestone. [Sig. a small cup.]
sordidum, Hall, 1882, Foss. Corals Niagara and Up. Held. Gr., Corniferous limestone. [Sig. poor]
tenuimurale, Hall, 1882, Foss. Corals, Niagara and Up. Held. Gr., Corniferous limestone. [Sig. thin walled.]
venatum, Hall, 1882, Foss. Corals, Niagara and Up. Held. Gr., Corniferous limestone. [Sig. veined.]
verticale, Hall, 1882, Foss. Corals, Niagara and Up. Held. Gr., Corniferous limestone. [Sig. vertical.]
Heterotrypa, Nicholson, 1879, Pal. Tab. Cor., p. 293. Proposed as a subgenus of Monticulipora.
INOCAULIS arbuscula, Ulrich, 1879 Jour. Cin. Soc. Nat Hist , vol. 2, Hud. Riv. Gr. [Sig. a little shrub.]
divaricatus, Hall, 1879, Desc. New Species Foss., Niagara Gr. [Sig. spread apart.]
problematicus, Spencer, 1878, Can. Nat., vol. 8, Niagara Gr. [Sig. problematic.]
LEPTOPORA typa should be referred to the Marshall or Kinderhook Gr.
winchelli, White, 1879, Bull. U. S. Sur., vol. 5, No. 2, Carboniferous. [Ety. proper name.]
LIMARIA lunata, Nicholson & Hinde, 1874 (Coenites lunata), Can. Jour., Niagara Gr. [Sig. lunate.]
LINDSTROMIA, Nicholson & Thompson, 1877, Proc. Roy. Soc. Edinb., vol. 9. [Ety. proper name.]
columnaris, Nicholson & Thompson, 1877, Proc. Roy. Soc. Edinb., vol. 9, Devonian. [Sig. columnar.]
LITHOSTROTION harmodites, Edwards & Haime, 1851, Mon. d. Pol. foss. d. Terr. Pal., Carboniferous. [Sig. well-fitting stone.]
junceum, Fleming, 1828 (Caryophyllæa juncea), Brit. Anim., Subcarboniferous. From Feilden Isthmus, the most northern point of land. [Sig. made of rushes.]
microstylum, White, 1880, 12th Rep., U. S. Sur. Terr., Kinderkook Gr. [Sig. having small stalks.]
stokesi, Edwards & Haime, 1851, Mon. d. Pol foss. d. Terr. Pal., Carboniferous. [Ety. proper name.]
whitneyi, Meek, 1875, Wheeler's Sur. W. 100th meridian, vol. 4, Coal Meas. [Ety. proper name.]
LOPHOPHYLLUM *calceola*, see Zaphrentis calceola.
LYELLIA americana is from the Niagara Group.
glabra, Owen, 1840 (Sarcinula glabra), Rep. on Mineral Lands, Devonian.

[Sig. smooth.] This species was too poorly defined for identification.

puella, Davis (In press). Foss. Corals of Ky., in vol. 1, Pal. Ky. St. Geo. Sur., Niagara Gr. [Sig. a young thing.]

MICHELINIA clappi, refer to Chonostegites clappi.

expansa, White, 1880. 12th Rep., U. S. Geo. Sur. Terr., Kinderhook Gr. [Sig. exp nded.]

niagarensis. Davis (In press). Foss. Corals of Ky., in vol. 1, Pal. Ky. St. Geo. Sur., Niagara Gr. [Ety. proper name.]

placenta, White, 1880, 12th Rep., U. S. Geo. Sur. Terr., Kinderhook Gr. [Sig. a cake.]

plana, Davis (In press), Foss. Corals of Ky., in vol. 1, Pal. Ky. St. Geo. Sur., Up. Devonian. [Sig. flat, from the flat tabulæ.]

prima, Davis (In press), Foss. Corals of Ky., in vol. 1, Pal. Ky. St. Geo. Sur., Niagara Gr. [Sig. first.]

trochiscus, Rominger, is a syn. for Pleurodictyum americanum.

MILLERIA, Davis (In press), Foss. Corals of Ky., in vol. 1, Pal. Ky. St. Geo. Sur. [Ety. proper name.]

laminata, Davis (In press), Foss. Corals of Ky., in vol. 1, Pal. Ky. St. Geo. Sur., Niagara Gr. [Sig arranged in thin plates.].

MONOGRAPTUS convolutus var. coppingeri, Etheridge. 1878, Quar. Jour. Geo. Soc., vol. 34, Silurian. [Ety. proper name.]

Monotrypa, Nicholson, 1879, Pal. Tab. Cor., p. 320. Proposed as a subgenus of Monticulipora.

MONTICULIPORA andrewsi, Nicholson, 1881, Struct. and Affin. of Montic., Hud. Riv. Gr. [Ety. proper name.] This is supposed to be the type of M. fibrosa.

barrandei should be referred to the Ham. Gr.

briareus belongs to the Utica Slate Gr., and so far as known, does not pass up into the Hud. Riv. Gr.

calceolus. Miller & Dyer, 1878, Jour. Cin. Soc. Nat. Hist., vol. 1, Hud. Riv. Gr. [Sig. a little shoe.]

dawsoni, Nicholson, 1881, Struct. and Affin. of Montic., Hud. Riv. Gr. [Ety. proper name.]

elegans, Ulrich, 1879 (Chetetes elegans), Jour. Cin. Soc. Nat. Hist., vol. 2, Hud. Riv. Gr. [Sig. elegant.]

filiasa, D'Orbigny, 1850, Prodr. de Pal. Hud. Riv. Gr. [Sig. having offshoots.]

fletcheri is not an American species.

granulifera, Ulrich, 1879 (Chetetes granuliferus). Jour. Cin. Soc. Nat. Hist., vol. 2, Trenton Gr. [Sig. grain bearing.]

implicata, Nicholson, 1881, Struct. and Affin. of Montic., Hud. Riv. Gr. [Sig. implicated.]

MONTICULIPORA irregularis, Ulrich, 1879, (Chetetes irregularis,) Jour. Cin. Soc. Nat. Hist., vol. 2, Hud. Riv. Gr. [Sig. irregular.]

molesta, Nicholson, 1881, Struct. and Affin. of Montic., Hud. Riv. Gr. [Sig. troublesome.] Syn. for M. mammulata (?).

moniliformis, instead of Chetetes moniliformis.

multituberculata, Whitfield, 1878, Ann. Rep. Geo. Sur. Wis., Hud. Riv. Gr. [Sig. having many tubercles.]

petasiformis, Nicholson, 1881, Struct. and Affin. of Montic., Hud. Riv. Gr. [Sig. like a broad-brimmed hat.]

punctata, Whitfield, 1878, Ann. Rep. Geo. Sur. Wis., Hud. Riv. Gr. [Sig. punctured.]

ramosa, D'Orbigny, 1850, Prodr. d. Pal., t. 1, .p. 25, Hud. Riv. Gr. [Sig. ramose.]

rectangularis, Whitfield, 1878, Ann. Rep. Geo. Sur. Wis., Hud. Riv. Gr. [Sig. rectangular.]

selwyni, Nicholson, 1881, Struct. and Affin. of Montic., Trenton Gr. [Ety. proper name.]

selwyni, var. hospitalis, Nicholson, 1881, Struct. and Affin. of Montic., Hud. Riv. Gr. [Sig. relating to a guest.]

subglobosa, Ulrich, 1879 (Chetetes subglobosus), Jour. Cin. Soc. Nat. Hist., vol. 2. Hud. Riv. Gr. [Sig. subglobose.]

trentonensis, Nicholson, 1881, Struct. and Affin. of Montic., Trent. Gr. [Ety. proper name.]

ulrichi, Nicholson, 1881, Structure and Affinities of the genus Monticulipora, Hud. Riv. Gr. [Ety. proper name.] Nicholson subdivided the genus Monticulipora in the above mentioned work into six subgenera, viz.: Heterotrypa, Diplotrypa, Monotrypa, Prasopora, and Peronopora. I endeavored to show, in a notice of the work, published in the Jour. Cin. Soc. Nat. Hist., vol. 5, that this subdivision of the genus is of very little if any value.

undulata, Nicholson, 1875 (Chetetes undulatus) Pal. of Ontario, Hud. Riv. Gr. [Sig. undulated.]

venusta, Ulrich, 1878 (Chetetes venustus), Jour. Cin. Soc. Nat. Hist., vol. 1, Utica Slate Gr. [Sig. beautiful.]

whiteavesi, Nicholson, 1881, Struct. and Affin. of Montic., Trenton Gr. [Ety. proper name.]

NICHOLSONIA, Davis (In press), Foss. Corals of Ky., in vol. 1, Pal. Ky. St. Geo. Sur. [Ety. proper name]

angulata, Davis (In press), Foss. Corals of Ky., in vol 1, Pal. Ky. St. Geo. Sur., Mid. Devonian [Sig. having angles.]

OMPHYMA verrucosa, should be credited to Rafinesque and Clifford, 1820, Monog. d. Turbinolides in Ann. d. phys. d. Brux., t. 5.
PACHYPORA, Lindstrom 1873. Ofversigt af K. Vetensk. Akad. Förhandl. [Ety. *pachys*, thick; *poros*, a pore.] fischeri instead of Alveolites fischeri. frondosa instead of Alveolites frondosus.
PALÆOCYCLUS was defined in Comptes rendus, t. 29.
Peronopora, Nicholson, 1881, Struct. and Affin. of Montic. Proposed as a subgenus of Monticulipora.
PHILLIPSASTREA ingens Davis, (In press), Foss. Corals of Ky., in vol. 1, Pal. Ky. St. Geo. Sur., Niagara Gr. [Sig. very large.] *mamillaris* refer to Strombodes mamillaris.
PLEURODICTYUM americanum. Rœmer, 1876, Lethæa Palæozoica, pl. 33, figs. 2a and 2b, Ham. Gr. [Ety. proper name.]
PLUMALINA is refered by Prof. Hall to the family Plumularidæ. densa, Hall, 1879, 30th Rep., N. Y. State Museum, Ham. Gr. [Sig. dense.]
PRASOPORA, Nicholson & Etheridge, 1877, Ann. and Mag. Nat. Hist., 4th ser., vol. 20, p. 38. A subgenus of Monticulipora including M. selwyni, and M. selwyni *var.* hospitalis.
PROCTERIA, Davis (In press), Foss. Corals of Ky., in vol. 1, Pal. Ky. St. Geo. Sur. [Ety. proper name.]
michelinoiden, Davis (In press), Foss. Corals of Ky., in vol. 1. Pal. Ky. St. Geo. Sur., Mid. Devonian. [Sig. resembling Michelinia.]
papillosa, Davis (In press), Foss. Corals of Ky., in vol. 1, Pal. Ky. St. Geo. Sur., Mid. Devonian. [Sig. having small prominences.]
PTILOGRAPTUS foliaceus, Spencer, 1878, Can. Nat., vol. 8, Niagara Gr. [Sig. leafy.]
PTYCHOPHYLLUM coniferum, Davis (In press), Foss. Corals of Ky., in vol. 1, Pal. Ky. St. Geo. Sur., Low. Devonian. [Sig. cone bearing.]
diaphragma, Davis (In press), Foss. Corals of Ky., in vol. 1, Pal. Ky. St. Geo. Sur., Low. Devonian. [Sig. a diaphragm.]
floriforme, Hall, 1882. Foss. Corals, Niagara and Up. Held. Gr., Niagara Gr. [Sig. flower formed.]
fulcratum, Hall, 1882, Foss. Corals, Niagara and Up. Held. Gr., Niagara Gr. [Sig. stayed with props.]
infundibulum, Meek, 1877, U. S. Geo. Sur., 40th parallel, Devonian. [Sig. a funnel.]
invaginatum, Davis (In press), Foss. Corals of Ky., in vol. 1, Pal. Ky. St. Geo. Sur., Niagara Gr. [Sig. incased, cup in cup.]
PTYCHOPHYLLUM ipomoea, Davis (In press), Foss. Corals of Ky., in vol. 1, Pal. Ky. St. Geo. Sur., Niagara Gr. [Sig. a morning glory.]
striatum, Hall, 1882, Foss. Corals, Niagara and Up. Held. Gr., Corniferous limestone. [Sig. striated.]
tropœum. Davis (In press), Foss. Corals of Ky., in vol. 1, Pal. Ky. St. Geo. Sur., Low. Devonian. [Sig. returning.]
typicum, Davis (In press), Foss. Corals of Ky., in vol. 1, Pal. Ky. St. Geo. Sur., Mid. Devonian. [Sig. typical.]
versiforme, Hall, 1882, Foss. Corals, Niagara and Up. Held. Gr., Corniferous limestone. [Sig. changeable.]
Quenstedtia, Rominger, 1876, being preoccupied by Morris & Lycett, in 1853, Nicholson proposed Romingeria.
RHIZOGRAPTUS, Spencer, 1878, Can. Nat., vol. 8. [Ety. *riza*, a root; *grapho*, to write.]
bulbosus, Spencer, 1878, Can. Nat., vol. 8, Niagara Gr. [Sig. bulbous.]
ROMINGERIA, Nicholson, 1879, Tabulate Corals. Proposed instead of Quenstedtia, Rominger, which was preoccupied.
fasciculata, Davis (In press), Foss. Corals of Ky., in vol. 1, Pal. Ky. St. Geo. Sur., Mid. Devonian. [Sig. arranged in little bundles.]
incrustans, Davis (In press), Foss. Corals of Ky., in vol. 1, Pal. Ky. St. Geo. Sur., Low. and Mid. Devonian. [Sig. incrustating.]
uva, Davis (In press), Foss. Corals of Ky., in vol. 1, Pal. Ky. St. Geo. Sur., Niagara Gr. and Devonian. [Sig. a bunch or cluster of grapes.]
vannula, Davis (In press), Foss. Corals of Ky., in vol. 1, Pal. Ky. St. Geo. Sur., Niagara Gr. [Sig. a little fan.]
SARCINULA *glabra*, see Lyellia glabra.
SMITHIA hennahi, Lonsdale, 1840 (Astræa hennahi), Geo. Trans., vol. 5, Devonian. [Ety. proper name.]
STELLIPORA *limitaris*, Ulrich, syn. for S. antheloiden.
STREPHODES *austini*, see Clisiophyllum austini.
pickthorni, see Cyathophyllum pickthorni.
STREPTELASMA æquidistans, Hall, 1882, Foss. Corals, Niagara and Up. Held. Gr., Corniferous limestone. [Sig. equidistant.]
ampliatum, Hall, 1882, Foss. Corals, Niagara and Up. Held. Gr., Corniferous limestone. [Sig. enlarged.]
coarctatum, Hall, 1882, Foss. Corals, Niagara and Up. Held. Gr., Corniferous limestone. [Sig. compressed.]
conspicuum, Hall, 1882, Foss. Corals, Niagara and Up. Held. Gr., Corniferous limestone. [Sig. conspicuous.]

STREPTELASMA crateriforme, Hall, 1882, Foss Corals, Niagara and Up. Held. Gr., Corniferous limestone. [Sig. crateriform.]
dissimile, Hall, 1882, Foss. Corals, Niagara and Up. Held. Gr., Schoharie Grit. [Sig. dissimilar.]
exstans, Hall, 1882, Foss. Corals, Niagara and Up. Held. Gr., Niagara Gr. [Sig. projecting.]
fossula, Hall, 1882, Foss. Corals, Niagara and Up. Held. Gr., Corniferous limestone. [Sig. a little ditch.]
inflatum, Hall, 1882, Foss. Corals, Niagara and Up. Held. Gr., Corniferous limestone. [Sig. inflated.]
involutum, Hall, 1882, Foss. Corals, Niagara and Up. Held. Gr., Corniferous limestone. [Sig. involuted.]
lamellatum, Hall, 1882, Foss. Corals, Niagara and Up. Held. Gr., Corniferous limestone. [Sig. lamellated.]
laterarium, Hall, 1882, Foss. Corals, Niagara and Up. Held. Gr., Corniferous limestone. [Sig. belonging to the side.]
limitare, Hall. 1882, Foss. Corals, Niagara and Up. Held. Gr., Niagara Gr. [Sig. that is on the border.]
mammiferum, Hall, 1882, Foss. Corals, Niagara and Up. Held. Gr., Corniferous limestone. [Sig. mamma bearing.]
minimum, syn. for Duncanella borealis.
papillatum, Hall, 1882, Foss. Corals, Niagara and Up. Held. Gr., Corniferous limestone. [Sig. papillated.]
simplex, Hall, 1882, Foss. Corals, Niagara and Up. Held. Gr., Corniferous limestone. [Sig. simple.]
tenue, Hall, 1882, Foss. Corals Niagara and Up. Held. Gr., Corniferous limestone. [Sig. slender.]
STRIATOPORA alba, Davis (In press) Foss. Corals of Ky. in vol. 1, Pal. Ky. St. Geo. Sur., Mid. and Up. Devonian. [Sig. White.]
formosa, Billings, 1860, Can. Jour., vol. 5, Corniferous Gr. [Sig. beautiful.]
Stromatocerium is a sponge, and belongs to the Protista.
Stromatopora is a sponge, and belongs to the Protista.
alternata is Stomatopora alternata.
hindei belongs to the Niagara Gr.
STROMBODES incertus, Davis (In press), Foss. Corals of Ky., in vol. 1, Pal. Ky. St. Geo. Sur., Niagara Gr. [Sig. of doubtful affinity.]
intermedius, Davis (In press), Foss. Corals of Ky. in vol. 1, Pal. Ky. St. Geo. Sur., Niagara Gr. [Sig. intermediate, between S. pentagonus and S. mammillatus.]
knotti, Davis (In press), Foss. Corals of Ky. in vol. 1, Pal. Ky. St. Geo. Sur., Low. and Mid. Devonian. [Ety. proper name.]

STROMBODES mammillatus should be written S. mamillaris, Owen, 1840 (Astræa mamillaris), Rep. on Min. Lands, Niagara Gr.
quadrangularis, Davis (In press). Foss. Corals of Ky., in vol. 1, Pal. Ky. St. Geo. Sur. Niagara Gr. [Sig. four angled.]
sinemurus, Davis (In press), Foss Corals of Ky., in vol. 1, Pal. Ky. St. Geo. Sur., Niagara Gr. [Sig. without a wall.]
unicus, Davis (In press), Foss. Corals of Ky., in vol. 1, Pal. Ky. St. Geo. Sur., Niagara Gr. [Sig. single.]
STYLASTREA, Lonsdale, 1845, Geol. and Pal. of Russia and the Ural Mountains. [Ety. *stylos*, a pillar; *aster*, a star.]
anna, Whitfield, 1882, Ann. N. Y. Acad. Sci., vol. 2. Up. Held. Gr. [Ety. proper name.]
SYRINGOLITES, Hinde, 1879, Geo. Mag. vol. 6. [Ety. *syrinx*, a pipe; *lithos*, stone.]
huronensis, Hinde, 1879, Geo. Mag., vol. 6, Niagara Gr. [Ety. proper name.]
SYRINGOPORA aulopora, Salter, 1855, Belcher's last of the Arctic Voyages, vol. 2. Carboniferous. [Sig. a porous pipe.]
harveyi belongs to the Choteau or Kinderhook Gr.
infundibulum, Whitfield, 1878, Ann. Rep. Geo. Sur. Wis., Niagara Gr. [Sig. a funnel.]
intermedia is from the Ham. Gr.
parallela, Etheridge, 1878, Quar. Jour. Geo. Soc., vol. 34, Up. Sil. [Sig. parallel.]
straminea, Davis (In press), Foss Corals of Ky., in vol. 1, Pal. Ky. St. Geo. Sur., Low. Devonian. [Sig. a bundle of straws.]
THAMNOGRAPTUS bartonensis, Spencer, 1878, Can. Nat., vol. 8, Niagara Gr. [Ety. proper name.]
TRACHYPORA ornata instead of Dendropora ornata.
TROCHOPHYLLUM, Edwards & Haime, 1851, Mon. d. Pol. foss. d. Terr. Pal. [Ety. *trochos*, a wheel; *phyllon*, a leaf.]
verneuilianum, Edwards & Haime, 1851, Mon. d. Pol. foss. d. Terr. Pal., Sub-carboniferous. [Ety. proper name.]
ZAPHRENTIS acuta, White & Whitfield, 1862, Proc. Bost. Soc. Nat. Hist., vol. 9, Marshall or Choteau Gr. [Sig. acute.]
annulata, Hall, 1882, Foss. Corals, Niagara and Up. Held. Gr., Corniferous limestone. [Sig. annulated.]
calcariformis, Hall, 1882, Foss. Corals, Niagara and Up. Held. Gr., Corniferous limestone. [Sig. spur formed.]
calceola, White and Whitfield, 1862 (Lophophyllum calceola), Proc. Bost. Soc., Nat. Hist., vol. 9, Marshall or Choteau Gr. [Sig. a little shoe.]

ZAPHRENTIS cannonensis, Winchell, is from the Choteau or Kinderhook Gr.
casssedayi. Milne-Edwards, 1860, Hist. d. Corallaires, t. 3, Warsaw Gr. [Ety. proper name.]
centralis, Edwards & Haime, 1851, Mon. d. Pol. Foss. d. Terr. Pal., Carboniferous. [Sig. central.]
cliffordana, Edwards & Haime, 1851, Mon. d. Pol. Foss. d. Terr. Pal., Carboniferous. [Ety. proper name.]
colletti, Hall, 1882, Foss. Corals. Niagara and Up. Held. Gr., Corniferous limestone. [Ety. proper name.]
complanata, Hall, 1882, Foss. Corals, Niagara and Up. Held. Gr., Corniferous limestone. [Sig. made even.]
compressa, Milne-Edwards, 1860, Hist. d. Corallaires, t. 3, Warsaw Gr. [Sig. compressed.]
concava, Hall, 1882, Foss. Corals, Niagara and Up. Held. Gr., Corniferous limestone. [Sig. concave.]
constricta, Hall, 1882, Foss. Corals, Niagara and Up. Held. Gr., Corniferous limestone. [Sig. constricted.]
contorta, Hall, 1882, Foss. Corals, Niagara and Up. Held. Gr., Corniferous limestone. [Sig. contorted.]
convoluta, Hall, 1882, Foss. Corals, Niagara and Up. Held. Gr., Corniferous limestone. [Sig. convoluted.]
cornalba, Davis (In press), Foss. Corals of Ky., in vol. 1, Pal. Ky. St. Geo. Sur., Up. Devonian. [Sig. white horn.]
cornicula, Lesueur, 1820, Mem. du Museum, vol. 1, instead of corniculum, Edwards & Haime.
corrugata, Hall, 1882, Foss. Corals, Niagara and Up. Held. Gr., Schoharie Grit. [Sig. corrugated.]
cristulata, Hall, 1882, Foss. Corals, Niagara and Up. Held. Gr., Niagara Gr. [Sig. small crested.]
curvata, Hall, 1882, Foss. Corals, Niagara and Up. Held. Gr., Corniferous limestone. [Sig. curved.]
cyathiformis, Hall, 1882, Foss. Corals, Niagara and Up. Held. Gr., Corniferous Gr. [Sig. cup-shaped.]
dalei, Edwards & Haime, 1851, Mon. d. Pol. Foss. d. Terr. Pal., Subcarboniferous. [Ety. proper name.]
denticulata, d'Eichwald, 1857. Probably not American.
desori, Edwards & Haime, 1851, Mon. d. Pol. Foss. de Terr. Pal., Low. Held. Gr. [Ety. proper name.]
duplicata, Hall, 1882, Foss. Corals, Niagara and Up. Held. Gr., Corniferous limestone. [Sig. duplicated.]
elegans, Hall, 1882, Foss. Corals, Niagara and Up. Held. Gr., Corniferous limestone. [Sig. elegant.]
elliptica, Davis (In press), Foss. Corals of Ky., in vol. 1, Pal. Ky. St. Geo. Sur., Low. Devonian. [Sig. elliptical.]
ZAPHRENTIS excentrica, Meek, 1872, Hayden's U. S. Geo. Sur. Terr., Coal Meas. [Sig. excentric.]
exigua, var. elongata. Davis (In press), Foss Corals of Ky., in vol. 1, Pal. Ky., St. Geo. Sur., Low. Devonian. [Sig. lengthened.]
exilis, Davis (In press), Foss. Corals of Ky., in vol. 1, Pal. Ky. St. Geo. Sur., Mid. Devonian. [Sig. thin, fragile.]
explanata, Davis (In press), Foss. Corals of Ky., in vol. 1, Pal. Ky. St. Geo. Sur., Up Devonian. [Sig. spread out.]
fastigata, Hall, 1882, Foss. Corals, Niagara and Up. Held. Gr., Corniferous limestone. [Sig. pointed.]
foliata, Hall, 1882, Foss. Corals, Niagara and Up. Held. Gr., Corniferous limestone. [Sig. leafy.]
frequentata, Hall, 1882, Foss. Corals, Niagara and Up. Held. Gr., Corniferous limestone. [Sig. frequent.]
fusiformis, Hall, 1882, Foss. Corals, Niagara and Up. Held. Gr., Corniferous limestone. [Sig. fusiform.]
gallicalcar, Davis (In press), Foss. Corals of Ky., in vol. 1. Pal. Ky. St. Geo. Sur., Up. Devonian. [Sig. cock's spur.]
genitiva, Billings, 1875, Can. Nat. and Quar. Jour., vol. 7, Corniferous Gr. [Sig. original.]
gigantea, Lesueur, 1820, Mem. du Museum, vol. 6, instead of Rafinesque.
glans refer to Hadrophyllum glans.
gravis, Hall, 1882, Foss. Corals, Niagara and Up. Held. Gr., Corniferous limestone. [Sig. ponderous.]
halli, Edwards & Haime, Mon. d. Pol. foss. d. Terr. Pal., Ham. Gr. [Ety. proper name.]
haysi, Meek, 1865, Am. Jour. Sci. & Arts, 2d ser., vol. 40, Low. Held. Gr. [Ety. proper name.]
herzeri, Hall, 1882, Foss. Corals, Niagara and Up. Held. Gr., Corniferous limestone. [Ety. proper name.]
immanis, Davis (In press), Foss. Corals of Ky., in vol. 1, Pal Ky. St. Geo. Sur., Mid. Devonian. [Sig. immense.]
inaequalis, Hall, 1882, Foss. Corals, Niagara and Up. Held. Gr., Corniferous limestone. [Sig. unequal.]
inclinata, Hall, 1882, Foss. Corals, Niagara and Up. Held. Gr., Corniferous limestone. [Sig. inclined.]
irregularis, Hall, 1882, Foss. Corals, Niagara and Up. Held. Gr., Corniferous limestone. [Sig. irregular.]
knappi, Hall, 1882, Foss. Corals, Niagara and Up. Held. Gr., Corniferous limestone. [Ety. proper name.]
latisinus, Hall, 1882, Foss. Corals, Ni-

agara and Up. Held. Gr., Niagara Gr. [Sig. having a wide sinus.]
ZAPHRENTIS linneyi, Davis (In press), Foss. Corals of Ky., in vol. 1, Pal. Ky. St. Geo Sur., Devonian. [Ety. proper name.]
maconathi, Davis (In press), Foss. Corals of Ky., in vol. 1, Pal. Ky. St. Geo. Sur., Low. and Mid. Devonian. [Ety. proper name.]
marcoui, Edwards & Haime, 1851, Mon. d. Pol. Foss. d. Terr. Pal., Niagara Gr. [Ety. proper name.]
nettlerothi, Davis (In press), Foss. Corals of Ky., in vol. 1, Pal. Ky. St. Geo. Sur., Up. Devonian. [Ety. proper name.]
nitida, Hall, 1882, Foss. Corals, Niagara and Up. Held. Gr., Corniferous limestone. [Sig. neat.]
obliqua, Davis (In press), Foss. Corals of Ky., in vol. 1, Pal. Ky. St. Geo. Sur., Niagara Gr. [Sig. oblique, inclined.]
offleyensis, Etheridge, 1878, Quar. Jour. Geo., vol. 34, Up. Sil. [Ety. proper name.]
ovalis, Hall, 1882, Foss. Corals, Niagara and Up. Held. Gr., Corniferous limestone. [Sig. oval.]
ovibos, Salter, 1855, Belcher's last of the Arctic voyages, vol. 2, Carboniferous. [Sig. the musk ox.]
patella, Davis (In press), Foss. Corals of Ky., in vol. 1, Pal. Ky. St. Geo. Sur., Hud. Riv. and Niagara Gr. [Sig. a dish.]
planima, Hall, 1882, Foss. Corals, Niagara and Up. Held. Gr., Corniferous limestone. [Sig. plane.]
ponderosa, Hall, 1882, Foss. Corals, Niagara and Up. Held. Gr., Corniferous limestone. [Sig. ponderous.]
pressula, Hall, 1882, Foss. Corals, Niagara and Up. Held. Gr., Niagara Gr. [Sig. somewhat compressed.]
profunda, Hall, 1882, Foss. Corals, Niagara and Up. Held. Gr., Corniferous limestone. [Sig. profound.]
prona, Milne-Edwards, 1860, Hist. d. Corollaires, Warsaw Gr. [Sig. bent forward.]
pulmonea, Lesueur, 1820 (Caryophyllia pulmonea, Mem. du Museum, vol. 6, Carboniferous. [Sig. spongy.]
racinensis, Whitfield, 1880, Ann. Rep. Geo. Sur. Wis., Niagara Gr. [Ety. proper name.]
rafinesquei, Edwards & Haime, 1851, Mon. d. Pol. foss. d. Terr. Pal., Up. Held. Gr. [Ety. proper name.]
ZAPHRENTIS rigida, Hall, 1882, Foss. Corals, Niagara and Up. Held. Gr., Niagara Gr. [Sig. rigid.]
roemeri, Edwards & Haime, 1851, Mon. d. Pol. foss. d. Terr. Pal., Delthyris shale. [Ety. proper name.]
scutella, Davis (In press), Foss. Corals of Ky., in vol. 1, Pal. Ky. St. Geo. Sur., Hud. Riv. and Niagara Gr. [Sig. a saucer.]
sentosa, Hall, 1882, Foss. Corals, Niagara and Up. Held. Gr., Corniferous limestone. [Sig. thorny.]
socialis, Davis (In press), Foss. Corals of Ky., in vol. 1, Pal. Ky. St. Geo. Sur. Niagara Gr. [Sig. gregarious, sociable.]
spissa, Hall, 1882, Foss. Corals, Niagara and Up. Held. Gr., Corniferous limestone. [Sig. dense.]
subcompressa, Hall, 1882, Foss. Corals, Niagara and Up. Held. Gr., Corniferous limestone. [Sig. somewhat compressed.]
subvada, Hall, 1882, Foss. Corals, Niagara and Up. Held. Gr., Niagara Gr. [Sig. somewhat shallow.]
subvesicularis, Hall, 1882, Foss. Corals, Niagara and Up. Held. Gr., Niagara, Gr. [Sig. somewhat vesicular.]
tabulata, Hall, 1882, Foss. Corals, Niagara and Up. Held. Gr., Corniferous limestone. [Sig. tabulated.]
torquata, Davis (In press), Foss. Corals of Ky., in vol. 1, Pal. Ky. St. Geo. Sur., Mid. Devonian. [Sig. twisted.]
torta, Hall, 1882, Foss. Corals, Niagara and Up. Held. Gr., Corniferous limestone. [Sig. twisted.]
transversa, Hall, 1882, Foss. Corals, Niagara and Up. Held. Gr., Corniferous limestone. [Sig. transverse.]
trigemma, Davis (In press), Foss. Corals of Ky., in vol. 1, Pal. Ky. St. Geo. Sur., Low. and Mid. Devonian. [Sig. bearing three buds.]
trisutura, Hall, 1882, Foss Corals, Niagara and Up. Held. Gr., Corniferous limestone. [Sig. having three sutures.]
unica, Davis (In press), Foss. Corals of Ky., in vol. 1, Pal. Ky. St. Geo. Sur., Low. Devonian. [Sig. unique, extraordinary.]
venusta, Hall, 1882, Foss. Corals, Niagara and Up. Held. Gr., Corniferous limestone. [Sig. beautiful.]

CLASS ECHINODERMATA.

ORDER CRINOIDEA.

FAMILY ACROCRINIDÆ.—Acrocrinus.
FAMILY ACTINOCRINIDÆ.—Actinocrinus, Agaricocrinus, Alloprosallocrinus, Amphoracrinus, Batocrinus, Cœlocrinus, Dorycrinus, Eretmocrinus, Gennæocrinus, Megistocrinus, Physetocrinus, Saccocrinus, Steganocrinus, Stereocrinus, Strotocrinus, Teleiocrinus.
FAMILY ALLAGECRINIDÆ.—Allagecrinus.
FAMILY ANCYROCRINIDÆ.—Ancyrocrinus.
FAMILY BELEMNOCRINIDÆ.—Belemnocrinus.
FAMILY CALCEOCRINIDÆ.—Calceocrinus, Eucheirocrinus.
FAMILY CALYPTOCRINIDÆ.—Eucalyptocrinus, Hypanthocrinus.
FAMILY CUPRESSOCRINIDÆ.—Aspidocrinus, Edriocrinus, Synbathocrinus.
FAMILY CYATHOCRINIDÆ.—Ampheristocrinus, Arachnocrinus, Barycrinus, Carabocrinus, Cyathocrinus, Nipterocrinus, Pachyocrinus, Palæocrinus, Vasocrinus.
FAMILY DIMEROCRINIDÆ.—Cytocrinus, Dolatocrinus, Macrostylocrinus.
FAMILY GASTEROCOMIDÆ.—Myrtillocrinus.
FAMILY GLYPTOCRINIDÆ.—Archæocrinus, Cupulocrinus, Glyptaster, Glyptocrinus, Lampterocrinus, Retiocrinus, Xenocrinus.
FAMILY HAPLOCRINIDÆ.—Coccocrinus, Haplocrinus.
FAMILY HETEROCRINIDÆ.—Erisocrinus, Graphiocrinus, Heterocrinus, Iocrinus.
FAMILY HYBOCRINIDÆ.—Anomalocrinus, Hybocrinus.
FAMILY ICHTHYOCRINIDÆ.—Cleiocrinus, Homalocrinus, Ichthyocrinus, Lecanocrinus, Mespilocrinus.
FAMILY MELOCRINIDÆ.—Ctenocrinus, Mariacrinus, Melocrinus, Technocrinus.
FAMILY PISOCRINIDÆ.—Catillocrinus, Pisocrinus.
FAMILY PLATYCRINIDÆ.—Cordylocrinus, Cotyledonocrinus, Dichocrinus, Eucladocrinus, Marsupiocrinus, Platycrinus, Pterotocrinus, Talarocrinus.
FAMILY POTERIOCRINIDÆ.—Agassizocrinus, Bursacrinus, Cœliocrinus, Cromyocrinus, Decadocrinus, Dendrocrinus, Eupachycrinus, Homocrinus, Hydreionocrinus, Pachylocrinus, Parisocrinus, Porocrinus, Poteriocrinus, Scaphiocrinus, Scytalocrinus, Zeacrinus.
FAMILY RHODOCRINIDÆ.—Goniasteroidocrinus, Hadrocrinus, Lyriocrinus, Rhodocrinus, Thysanocrinus.
FAMILY STELIDOCRINIDÆ.—Schizocrinus.
FAMILY TAXOCRINIDÆ.—Forbesiocrinus, Onychocrinus, Taxocrinus.
FAMILY AFFINITY UNCERTAIN.—Brachiocrinus, Closterocrinus, Cystocrinus, Dictyocrinus, Syringocrinus.

ORDER CYSTOIDEA.

FAMILY AMYGDALOCYSTIDÆ.—Amygdalocystites, Malocystites, Palæocystites.
FAMILY ANOMALOCYSTIDÆ.—Anomalocystites, Ateleocystites.
FAMILY CARYOCRINIDÆ.—Caryocrinus, Heterocystites.
FAMILY COMAROCYSTIDÆ.—Comarocystites.
FAMILY ECHINOCYSTIDÆ.—Echinocystites.

FAMILY GOMPHOCYSTIDÆ.—Gomphocystites.
FAMILY HOLOCYSTIDÆ.—Crinocystites, Holocystites.
FAMILY HYBOCYSTIDÆ.—Hybocystites.
FAMILY LEPADOCRINIDÆ.—Apiocystites, Callocystites, Echino-encrinites, Glyptocystites, Lepadocrinus, Pleurocystites, Sphærocystites.
FAMILY UNCERTAIN.—Eocystites.

ORDER BLASTOIDEA.

FAMILY NUCLEOCRINIDÆ.—Nucleocrinus.
FAMILY PENTREMITIDÆ.—Blastoidocrinus, Granatocrinus, Pentremites, Troostocrinus.
FAMILY STEPHANOCRINIDÆ.—Codaster, Codonites, Eleutherocrinus, Stephanocrinus.

ORDER PERISHO-ECHINIDÆ.

FAMILY ARCHÆOCIDARIDÆ.—Archæocidaris, Eocidaris, Lepidocidaris, Pholidocidaris.
FAMILY LEPIDECHINIDÆ.—Lepidechinus, Lepidesthes.
FAMILY PALÆCHINIDÆ.—Melonites, Oligoporus, Palæchinus.

ORDER ASTEROIDEA.

FAMILY PALÆASTERIDÆ.—Onychaster, Palæaster, Palæasterina, Petraster, Schænaster, Stenaster.

ORDER OPHIUROIDEA.

Eugaster, Palæocoma, Protaster, Ptilonaster, Tæniaster.

ORDER AGELACRINOIDEA, *n. ord. and n. fam.*

This order is proposed to include, so far as known, only the family Agelacrinidæ, and each may, therefore, be defined as follows:

Body thin, circular and parasitic upon other objects. The lower side consists of a thin, smooth, attaching membrane or plate. The upper side is more or less convex, and composed of thin, squamiform or imbricating plates, usually much smaller at the periphery than toward the center. Ambulacra constituting part of the convex surface furrowed on the interior, and composed of a double series of transverse alternating plates, sometimes having smaller, middle, intercalated ones. Two or more rows of ambulacral pores connect the exterior with the interior of each ambulacrum. The so-called ovarian or anal aperture is situated in one of the inter-ambulacral areas, and is usually surrounded by cuneiform plates forming a depressed circular prominence. The genera belonging to this order and family are Agelacrinus, Edrioaster and Hemicystites.

ORDER LICHENOCRINOIDEA, *n. ord. and n. fam.*

This division of the fossil Echinodermata, and the family Lichenocrinidæ, are established upon the genus Lichenocrinus.

The definition of the order and family will be the same, as both are founded on a single genus.

The body attached during part or all of its life to foreign objects. It is circular, convex upon the upper side, and more or less crateriform surrounding the central stalk-like appendage. The lower side at some period of life possessed a thin attaching plate. The upper side is covered with numerous polygonal plates, without any evidence of the presence of ambulacra, arms, mouth, pectinated rhombs or pores connecting the exterior with the internal cavity. The interior of the visceral cavity contains numerous radiating upright lamellæ that support the polygonal plates of the upper side, and often leave their impression, like the radiations of a star, upon the object to which it was attached. The stalk rises from the central depressed area, and consists, at first, of interlocking plates, but, afterward, of circular ones, like those of a crinoid column, and finally tapers to a point. It was flexible and perforated with a longitudinal channel, though the perforation has not been satisfactorily ascertained at the upper terminating point.

ORDER MYELODACTYLOIDEA, *n. ord.*

This division of the fossil Echinodermata is established as follows:

Body free, discoidal, and possessed of an internal radiating system of pores, which increase, by division, from the center to a tubular channel in the circular margin or surrounding coil. There are two families referred to this order, the Myelodactylidæ and the Cyclocystoididæ. In the former, the radiating and circular systems become complicated, by the connection, between succeeding coils and through the flattened connecting finger-like processes; in the latter, the arrangement is more simple, as the interior radiations connect with a single marginal circular system. The external form and internal structure are so essentially distinct from other well defined orders, that the technical names, used in description, have no ascertained application. That is, we can not intelligently apply the words calyx, ambulacra, arm, etc., to any part of these peculiar organisms. This order has been suggested with hesitation, because there still exists a possibility that Myelodactylus belongs, in some manner, to the vault of a crinoid, but the author thinks there is not much probability of such connection.

FAMILY MYELODACTYLIDÆ, *n. fam.*

This family is founded upon the single genus Myelodáctylus, and defined as follows:

Body free, discoidal, and resembling a coil rolled in the same plane, and covered upon either side by finger-like processes from each succeeding turn overlapping the next inner one. The whorls are composed of a series of plates, having a tubular channel within, and perforated and finger-like processes upon the exterior, directed toward the center, and flattened down upon the next inner whorl to which they are attached, and form a porous connection from the tubular channel of one whorl to the next inner one. The cast of the pores of the inner whorl resemble the radiating spokes of a wheel: they are multiplied in connecting the tubular channels of each succeeding whorl, thus making the internal radiating system doubly complicated. The central aperture, if one exists, has not been discovered, and the structure of the terminal end of the anomalous coil is wholly unknown. The internal radiating system of pores may be compared with that of the family Cyclocystoididæ, and here the analogy in structure, with other families in the class Echinodermata, so far as known, ceases. The terminal end of the coil being unknown has led to the suggestion of the possibility of its having been connected with the vault of a crinoid, but as no genus is known having any such appendage, and some classification seeming desirable, this family has been proposed.

FAMILY CYCLOCYSTOIDIDÆ, *n. fam.*

This family is founded upon the single genus Cyclocystoides, and defined as follows:

Body free, consisting of a circular disk, and having a margin composed of a series of perforated plates. Within this marginal series the disk is covered with an integument of small plates, except, possibly, a small central aperture. The rim or marginal series contains a tubular channel, making the complete circle, which is connected with the interior, by numerous pores, that radiate from the center, and repeatedly bifurcate before reaching it. The inner side of the rim is grooved, for the reception of the internal part of the disk, and the outer side depressed and scarred, either by mammillary elevations or concave depressions, as if for the attachment of ossicular or other processes. The tubular channel is connected with the exterior by minute circular pores which were probably analagous, in their purpose, to the calycine pores in the Cystideæ.

ACROCRINUS wortheni, Wachsmuth, 1882, Bull. No. 1, Ill. St. Mus. Nat. Hist., Coal Meas. [Ety. proper name.]
ACTINOCRINUS andrewsanus, McChesney, 1860, New Pal. Foss., Up. Burlington Gr. [Ety proper name.] Wachsmuth refers it to Batocrinus.
biturbinatus, refer to Batocrinus biturbinatus.
brevis, refer to Agaricocrinus brevis.
calyculus, refer to Batocrinus calyculus.
calyculus var. hardinensis, refer to Batocrinus calyculus var. hardinensis.
caroli, refer to Batocrinus caroli.
clypeatus, refer to Batocrinus clypeatus.
concinnus, refer to Steganocrinus concinnus.
copei, S. A. Miller, 1881, Jour. Cin. Soc. Nat. Hist., vol. 4, Burlington Gr. [Ety. proper name.]
coreyi is from the Keokuk Gr., and was described in 1860.
coronatus, refer to Eretmocrinus coronatus.
dalyanus, S. A. Miller, 1881, Jour. Cin. Soc. Nat. Hist., vol. 4, Burlington Gr. [Ety. proper name.]
ectypus instead of Strotocrinus ectypus.
fiscellus, refer to Agaricocrinus fiscellus.
hageri, refer to Batocrinus hageri.
helice, refer to Agaricocrinus helice.
helice var. eris, refer to Agaricocrinus eris.
indianensis is from the Keokuk Gr. It was described in 1860, and should be referred to Batocrinus indianensis.
leucosia, refer to Eretmocrinus leucosia.
meeki, refer to Macrostylocrinus meeki.
multicornis was described in 1869, from the Up. Held. Gr.
nashvillæ, refer to Batocrinus nashvillæ.
nashvillæ var. subtractus, refer to Batocrinus nashvillæ var. subtractus.
obpyramidalis, refer to Melocrinus obpyramidalis.
pentaspinus was described in 1869, from the Up. Held. Gr.
pyramidatus, refer to Agaricocrinus pyramidatus.
pyriformis var. rudis, being preoccupied the fossil was afterward named Batocrinus pistilliformis.
ramulosus, refer to Eretmocrinus ramulosus.
rotundus, refer to Batocrinus rotundus.
sculptus, refer to Steganocrinus sculptus.
sillimani, is a syn. for A. scitulus.
similis, refer to Batocrinus similis.
steropes, refer to Batocrinus steropes.
tenuisculptus, McChesney, 1867, Chi., Acad. Sci., vol. 1, Low. Burlington Gr. [Sig. finely sculptured.]
ventricosus, refer to Physetocrinus ventricosus.
ventricosus var. cancellatus, refer to Physetocrinus ventricosus var. cancellatus.
ventricosus var. internodus, refer to Physetocrinus ventricosus var. internodus.
ACTINOCRINUS wachsmuthi, White, 1862, syn. for A. scitulus.
wachsmuthi, White, 1880, 12th Rep. U. S. Geo. Sur. Terr., Keokuk Gr. Ety. proper name.]
yandelli, is from the Keokuk Gr.
AGARICOCRINUS americanus is found in the Burlington and Keokuk Groups.
brevis instead of Actinocrinus brevis.
crassus, Wetherby, 1881, Jour. Cin. Soc. Nat. Hist., vol. 4, Keokuk Gr. [Sig. thick.]
elegans, Wetherby, 1881, Jour. Cin. Soc. Nat. Hist., vol. 4, Keokuk Gr. [Sig. elegant]
eris, Hall. 1864, 17th Rep. N. Y. St. Cab. Nat. Hist. (Actinocrinus helice var. eris), Waverly Gr. [Ety. proper name.]
fiscellus instead of Actinocrinus fiscellus.
helice instead of Actinocrinus helice.
pyramidatus instead of Actinocrinus pyramidatus.
springeri, White, 1882, 11th Rep. Geo. & Nat. Hist. Indiana, Keokuk Gr. [Ety. proper name.]
AGASSIZOCRINUS hemisphericus, Worthen, 1882, Bull. No. 1, Ill. St. Mus. Nat. Hist., Kaskaskia Gr. [Sig. hemispherical.]
papillatus, Worthen, 1882, Bull No. 1, Ill. St. Mus. Nat. Hist., Kaskaskia Gr. [Sig. shaped like a bud.]
AGELACRINUS septembrachiatus, Miller & Dyer, 1878, Jour. Cin. Soc. Nat. Hist., vol. 1, Hud. Riv. Gr. [Sig. seven armed.]
ALLAGECRINUS, Etheridge & Carpenter, 1881, Ann. & Mag. Nat. Hist. [Ety. allage, a change; krinon, a lily.]
carpenteri, Wachsmuth, 1882, Bull. No. 1, Ill. St. Mus. Nat. Hist., Kaskaskia Gr. [Ety. proper name.]
ALLOPROSALLOCRINUS euconus, see Batocrinus euconus.
AMPHERISTOCRINUS, Hall, 1879, Desc. New. Spec. Foss. [Ety. ampheristos, doubtful, disputed; krinon, a lily.]
typus, Hall, 1879, Desc. New Spec. Foss., Niagara Gr. [Sig. the type.]
AMPHORACRINUS planobasalis was described in 1860, in Supp. to Geo. Rep. Iowa.
quadrispinus is a syn. for A. divergens.
AMYGDALOCYSTITES huntingtoni, Wetherby, 1881, Jour. Cin. Soc. Nat Hist., vol. 4, Trenton Gr. [Ety. proper name.]
ANOMALOCRINUS caponiformis, Lyon, 1869, Trans. Am. Phil. Soc., vol. 13, Hud. Riv. Gr. [Sig. capon formed.]
incurvus is from the Hud. Riv. Gr.
ANOMALOCYSTITES balanoides is from the Hud. Riv. Gr.

Anomaloides. Ulrich, 1878, Jour. Cin. Soc. Nat. Hist., vol. 1. This is a poorly constructed word for a generic name, beside the fossil is a fragment and not understood.
  reticulatus, Ulrich, 1878, Jour. Cin. Soc. Nat. Hist., vol. 1, Hud. Riv. Gr. [Sig. reticulated.]
ARACHNOCRINUS, Meek & Worthen, 1866, Geo. Sur. Ill., vol. 2. [Ety. *arachne*, a spider; *krinon*, a lily.]
  bulbosus, instead of Cyathocrinus bulbosus.
  extensus, Wachsmuth & Springer, 1879, Revision of the Palæocrinoidea, Ham. Gr. [Sig. extended.]
  knappi, Wachsmuth & Springer, 1879, Rev. of the Palæocrinoidea, Ham. Gr. [Ety. proper name.]
  pisiformis, instead of Poteriocrinus pisiformis.
ARCHÆOCIDARIS dinianl, White, 1880, Proc. U. S. Nat. Mus., vol. 2, Up. Coal Meas. [Ety. proper name.]
  keokuk is from the Keokuk Gr.
  triplex, White, 1882, Rep. Carb. Invert. Foss. New Mex., Coal Meas. [Sig. three-fold.]
ARCHÆOCRINUS, Wachsmuth & Springer, 1881, Proc. Acad. Nat. Sci. [Ety. *archaios*, ancient; *krinon*, a lily.] A generic name proposed for the purpose of including Glyptocrinus lacunosus, G. marginatus, Thysanocrinus microbasilis and T. pyriformis.
BARYCRINUS angulatus instead of Cyathocrinus angulatus.
  bullatus instead of Cyathocrinus bullatus.
  cornutus instead of Caythocrinus cornutus.
  crassibrachiatus instead of Cyathocrinus crassibrachiatus.
  kelloggi instead of Cyathocrinus kelloggi.
  magister instead of Cyathocrinus magister.
  multibrachiatus instead of Cyathocrinus multibrachiatus.
  rhombiferus instead of Poteriocrinus rhombiferus.
  sculptilis instead of Cyathocrinus sculptilis.
  solidus instead of Cyathocrinus solidus.
  spurius instead of Cyathocrinus spurius.
  stellatus instead of Cyathocrinus stellatus.
  tumidus instead of Cyathocrinus tumidus.
  wachsmuthi instead of Cyathocrinus wachsmuthi.
BATOCRINUS was described in 1854, by Casseday, in Deutsche Zeitscher d. Geol. Gesellsch, vol. 6.
  andrewsanus instead of Actinocrinus andrewsanus.
  biturbinatus instead of Actinocrinus biturbinatus.
  calyculus instead of Actinocrinus calyculus.
  calyculus *var.* hardinensis instead of Actinocrinus calyculus *var.* hardinensis.
  caroli instead of Actinocrinus caroli.
  clypeatus instead of Actinocrinus clypeatus.
  euconus instead of Alloprosallocrinus euconus
  hageri instead of Actinocrinus hageri.
  indianensis instead of Actinocrinus indianensis.
  lovei, Wachsmuth & Springer, 1881, Proc. Acad. Nat. Sci., Burlington Gr. [Ety. proper name.]
  nashvillæ instead Actinocrinus nashvillæ.
  nashvillæ var. subtractus, instead of Actinocrinus nashvillæ var. subtractus.
  rotundus instead of Actinocrinus rotundus.
  similis instead of Actinocrinus similis.
  steropes instead of Actinocrinus steropes.
  whitei, Wachsmuth & Springer, 1881, Proc. Acad. Nat. Sci., Keokuk Gr. [Ety. proper name.]
  yandelli instead of Actinocrinus yandelli.
BELEMNOCRINUS florifer, Wachsmuth & Springer, 1877, Am. Jour. Sci. & Arts, 3d ser., vol. 13, Burlington Gr. [Sig. flower bearing.]
  pourtalesi, Wachsmuth & Springer, 1877, Am. Jour. Sci. & Arts, 3d ser., vol. 13, Burlington Gr. [Ety. proper name.
CALCEOCRINUS radiculus. Ringueberg, 1882. Jour. Cin. Soc. Nat. Hist., vol. 5, Niagara Gr. [Sig. a small root.]
Centrocrinus, Wachsmuth & Springer. 1881, Proc. Acad. Nat. Sci. Proposed as a subgenus under Actinocrinus to include A. multicornis and A. pentaspinus, but the name was preoccupied by Austin, in 1843.
CODASTER gratiosus, S. A. Miller, 1880, Jour. Cin. Soc. Nat. Hist., vol. 2, Keokuk Gr. [Sig. agreeable.]
  hindei, Etheridge & Carpenter, 1882, Ann. & Mag. Nat. Hist., Ham. Gr. [Ety. proper name.]
  kentuckiensis is from the Burlington Gr.
  pentalobus, see Stephanocrinus pentalobus.
  pulchellus, see Stephanocrinus pulchellus.
CODONITES was described in 1869, in Proc. Acad. Nat. Sci. It is classed by Zittel, Ludwig. Carpenter, and other European authors, as a syn. for Orophocrinus, described by Von Seebach in 1864, in Nachr. k. Gesellsch. zu Gottingen.
  gracilis should bear the same date as the genus.
CŒLIOCRINUS cariniferus instead of Zeacrinus cariniferus.

CŒLIOCRINUS lyra instead of Zeacrinus lyra.
COMAROCYSTITES shumardi, var. obconicus, Meek & Worthen, 1865, Proc. Acad. Nat. Sci., Trenton Gr. [Sig. obconical.]
CORDYLOCRINUS, Angelin, 1878, Icon. Crin. Suec. [Ety. *kordyle*, a cudgel; *krinon*, a lily.]
parvus instead of Platycrinus parvus.
plumosus instead of Platycrinus plumosus.
ramulosus instead of Platycrinus ramulosus.
Crinosoma antiqua, Castelnau, 1843, Syst. Sil. Probably a fucoid.
CROMYOCRINUS, Trautschold, 1867, Crin. jung. Bergkalkes bei Moskau. [Ety. *kromyon*, an onion; *krinon*, a lily.]
gracilis, Wetherby, 1880, Jour. Cin. Soc. Nat. Hist., vol. 2, Kaskaskia Gr. [Sig. slender.]
CUPULOCRINUS, D'Orbigny, 1850. Prodr. d. Pal. Proposed instead of Scyphocrinus, Hall, that was preoccupied.
heterocostalis instead of Scyphocrinus heterocostalis.
CYATHOCRINUS aemulus, Hall, 1879, Desc. new spec. foss., Niagara Gr. [Sig. emulous.]
angulatus refer to Barycrinus angulatus.
barydactylus, Wachsmuth & Springer, 1878, Proc. Acad. Nat. Sci., Up. Burlington Gr. [Sig. heavy fingered.]
bulbosus refer to Barycrinus bulbosus.
crassibrachiatus refer to Barycrinus crassibrachiatus.
crassus refer to Eupachycrinus crassus.
crawfordsvillensis, S. A. Miller, 1882, Jour. Cin. Soc. Nat. Hist., vol. 5, Keokuk, Gr. [Ety. proper name.]
decadactylus was described in 1860, Am. Jour. Sci., vol. 29, and from the Keokuk Gr.
fasciatus refer to Macrostylocrinus fasciatus.
gilesi, Wachsmuth & Springer, 1878, Proc. Acad. Nat. Sci., Burlington Gr. [Ety. proper name.]
hamiltonensis, Worthen, 1882, Bull. No. 1, Ill. St. Mus. Nat. Hist., Keokuk Gr. [Ety. proper name.]
harrisi, S. A. Miller, 1880, Jour. Cin. Soc. Nat. Hist., vol. 2, Keokuk Gr. [Ety. proper name.]
harrodi, Wachsmuth & Springer, 1879, Proc. Acad. Nat. Sci., Keokuk Gr. [Ety. proper name.]
hexadactylus was described in 1860, from the Keokuk Gr.
inæquidactylus, Whitfield, 1882, Desc. New Sp. Foss. from Ohio, Kaskaskia Gr. [Sig. unequal fingered.]
inflexus, Geinitz, refer to Erisocrinus inflexus.
insperatus is from the Keokuk Gr.
kelloggi, refer to Barycrinus kelloggi.
magister, refer to Barycrinus magister.

CYATHOCRINUS marshallensis, Worthen, 1882, Bull. No. 1, Ill. St. Mus. Nat. Hist., Kinderhook Gr. [Ety. proper name.]
multibrachiatus, refer to Barycrinus multibrachiatus.
nucleus instead of Dendrocrinus nucleus.
pusillus, refer to Lecanocrinus pusillus.
rarus was described in 1869, from the Up. Held. Gr.
sangamonensis, refer to Eupachycrinus sangamonensis.
sculptilis, refer to Barycrinus sculptilis.
solidus, refer to Barycrinus solidus.
somersi, Whitfield, 1882, Desc. New Spec. from Ohio Coal Meas. [Ety. proper name.]
spurius, refer to Barycrinus spurius.
stellatus, refer to Barycrinus stellatus.
stillativus, White, 1880, Proc. U. S. Nat. Mus., vol. 2, Up. Coal Meas. [Sig. dropping.]
tenuibrachiatus, Lyon, 1869, Trans. Am. Phil. Soc., vol. 13, Up. Held. Gr. [Sig. slender armed.]
tumidus, see Barycrinus, tumidus.
vanhornei, S. A. Miller, 1881, Jour. Cin. Soc. Nat. Hist., vol. 4, Niagara Gr. [Ety. proper name.]
wachsmuthi, see Barycrinus wachsmuthi.
waldronensis, Miller & Dyer, 1878, Cont. to Pal. No. 2, Niagara Gr. [Ety. proper name.]
CYCLOCYSTOIDES bellulus, Miller & Dyer, 1878, Jour. Cin. Soc. Nat. Hist., vol. 1, Hud. Riv. Gr. [Sig. beautiful.]
magnus, Miller & Dyer, 1878, Jour. Cin. Soc. Nat. Hist., vol. 1, Hud. Riv. Gr. [Sig. large.]
minus, Miller & Dyer, Jour. Cin. Soc. Nat. Hist., vol. 1, Hud. Riv. Gr. [Sig. small.]
mundulus, Miller & Dyer, 1878, Jour. Cin. Soc. Nat. Hist., vol. 1, Hud. Riv. Gr. [Sig. neat, trim.]
parvus, Miller & Dyer, 1878, Jour. Cin. Soc. Nat. Hist., vol. 1, Hud. Riv. Gr. [Sig. little.]
CYTOCRINUS may be restored, as it is doubtful about its being a syn. for Macrostylocrinus,
DENDROCRINUS ancilla, Hall, 1879, Desc. New Spec. Foss., Niagara Gr. [Sig. a hand-maid.]
curtus, Ulrich, 1879, Jour. Cin. Soc. Nat. Hist., vol. 2, Utica Slate Gr. [Sig. short.]
erraticus, S. A. Miller, 1881, Jour. Cin. Soc. Nat. Hist., vol. 4, Hud. Riv. Gr. [Sig. a wanderer.]
navigiolum, S. A. Miller, 1880, Jour. Cin. Soc. Nat. Hist., vol. 3, Utica Slate Gr. [Sig. a small vessel.]
nucleus, refer to Cyathocrinus nucleus.
Decadocrinus, Wachsmuth & Springer, 1879, Proc. Acad. Nat. Sci. [Ety. *dekas*, number of ten; *krinon*, a lily.] A proposed subgenus of Poteriocrinus.

DICHOCRINUS constrictus is from the Warsaw Group.
coxanus, Worthen, 1882, Bull. No. 1, Ill. St. Mus. Nat. Hist., Keokuk Gr. [Ety. proper name.]
elegans, Casseday & Lyon, was preoccupied by De Koninck & Lehon in 1853, but the species is referred to the genus Talarocrinus, by Wachsmuth & Springer.
hamiltonensis, Worthen, 1882, Bull. No. 1, Ill. St. Mus. Nat. Hist., Keokuk Gr. [Ety. proper name.]
ornatus, Wachsmuth & Springer, 1881, Proc. Acad. Nat. Sci., Keokuk Gr. [Sig. ornate.] This name was proposed instead of D. sculptus, Casseday & Lyon, because the latter name was preoccupied.
sculptus, Casseday & Lyon. The name was preoccupied by De Koninck & Lehon in 1853, see D. ornatus.

DOLATOCRINUS marshi was described in 1869, Trans. Am. Phil. Soc., vol. 13, Up. Held. Gr.
ornatus, Meek, 1871, Proc. Acad. Nat. Sci., Corniferous Gr. [Sig. ornamented.]

DORYCRINUS, Rœmer, 1854, Archiv. f. Naturgesch Jahrg, vol. 19, p. 207.
lineatus, S. A. Miller, 1881, Jour. Cin. Soc. Nat. Hist., vol. 4, Burlington Gr. [Sig. lined.]

ERETMOCRINUS adultus, Wachsmuth & Springer, 1881, Proc. Acad. Nat. Sci., Keokuk Gr. [Sig. adult.]
coronatus instead of Actinocrinus coronatus.
intermedius, Wachsmuth & Springer, 1881, Proc. Acad. Nat. Sci., Keokuk Gr. [Sig. intermediate.]
leucosia instead of Actinocrinus leucosia.
originarius, Wachsmuth & Springer, 1881, Proc. Acad. Nat. Sci., Keokuk Gr. [Sig. original.]
ramulosus instead of Actinocrinus ramulosus.
varsouviensis, Worthen, 1882, Bull. No. 1, Ill. St. Mus Nat. Hist., Warsaw Gr. [Ety. proper name.]

ERISOCRINUS hemisphericus instead of Poteriocrinus hemisphericus.
inflexus, Geinitz, 1866 (Cyathocrinus inflexus), Carb. und Dyas, Coal Meas. [Sig. curving.]
planus, White, 1880, Proc. U. S. Nat. Mus., Coal. Meas. [Sig. even.]

EUCALYPTOCRINUS constrictus, Hall, 1879, syn. for E. tuberculatus.
depressus, S. A. Miller, 1880, Jour. Cin. Soc. Nat. Hist., vol. 3, Niagara Gr. [Sig. depressed.]
egani, S. A. Miller, 1880, Jour. Cin. Soc. Nat. Hist., vol. 3, Niagara Gr. [Ety. proper name.]
ovalis instead of ovatus.
proboscidalis, S. A. Miller, 1882, Jour. Cin. Soc. Nat. Hist., vol. 5, Niagara Gr. [Sig. having a proboscis.]
EUCALYPTOCRINUS rotundus, S. A. Miller, 1882, Jour. Cin. Soc. Nat. Hist., vol. 5, Niagara Gr. [Sig. rotund.]
tuberculatus, Miller & Dyer, 1878, Jour. Cin. Soc. Nat. Hist., vol. 1, Niagara Gr. [Sig. tuberculated.]
turbinatus, S. A. Miller, 1882, Jour. Cin. Soc. Nat. Hist., vol. 5, Niagara Gr. [Sig. turbinate.]

EUCLADOCRINUS, Meek, 1871, U. S. Geo. Sur. Terr. [Ety. eu, very; klados, a branch; krinon, a lily.]
millebrachiatus, Wachsmuth & Springer, 1878, Proc. Acad. Nat. Sci., Burlington and Keokuk Gr. [Sig. many-armed.]
montanensis, Meek, 1871, Hayden's Rep. U. S. Geo. Sur. Terr., Subcarboniferous. [Ety. proper name.]
pleuroviminus instead of Platycrinus pleuroviminus.

EUPACHYCRINUS asperatus, Worthen, 1882, Bull. No. 1, Ill. St. Mus. Nat. Hist., Kaskaskia, Gr. [Sig. roughened.]
crassus instead of Cyathocrinus crassus.
formosus instead of Zeacrinus formosus.
germanus, S. A. Miller, 1879, Jour. Cin. Soc. Nat. Hist., vol. 2, Kaskaskia Gr. [Sig. near of kin.]
monroensis, Worthen, 1882, Bull. No. 1, St. Mus. Nat. Hist., Kaskaskia Gr. [Ety. proper name.]
quatuordecembrachialis instead of Graphiocrinus quatuordecembrachialis, and from Kaskaskia Gr.
sangamonensis instead of Cyathocrinus sangamonensis.
spartarius, S. A. Miller, 1879, Jour. Cin. Soc. Nat. Hist., vol. 2, Kaskaskia Gr. [Sig. of or belonging to a broom.]
subtumidus instead of Zeacrinus subtumidus.

FORBESIOCRINUS asteriformis, refer to Onychocrinus asteriformis.
communis, refer to Taxocrinus communis.
giddingi, refer to Taxocrinus giddingi.
juvenis, refer to Taxocrinus juvenis.
kelloggi, refer to Taxocrinus kelloggi.
lobatus, refer to Taxocrinus lobatus.
lobatus var. tardus. refer to Taxocrinus lobatus var. tardus.
meeki, refer to Taxocrinus meeki, from Keokuk Gr.
multibrachiatus, refer to Taxocrinus multibrachiatus.
parvus, Wetherby, 1879, Jour. Cin. Soc. Nat. Hist., vol. 2, Kaskaskia Gr. [Sig. small.]
ramulosus, refer to Onychocrinus ramulosus.
shumardanus, refer to Taxocrinus shumardanus.
whitfieldi, refer to Taxocrinus whitfieldi.

GENNÆOCRINUS, Wachsmuth & Springer, 1882, Proc. Acad. Nat. Sci. [Ety. gennaios, of noble origin; krinon, a lily.] This

ECHINODERMATA.

was proposed to include certain species of Actinocrinus from the Hamilton Group, to-wit: A. calypso, A. cassedayi, A. cauliculus, A. eucharis, A. kentuckiensis, and A. nyssa.

GLYPTASTER egani, S. A. Miller, 1881, Jour. Cin. Soc. Nat. Hist., vol. 4, Niagara Gr. [Ety. proper name.]
occidentalis, var. crebescens, Hall, 1879, 28th Reg. Rep., Niagara Gr. [Sig. abundant.]

GLYPTOCRINUS angularis, Miller & Dyer, Jour. Cin. Soc. Nat. Hist., vol. 1, Hud. Riv. Gr. [Sig. angular.]
cognatus, S. A. Miller, 1881, Jour. Cin. Soc. Nat. Hist., vol. 4, Hud. Riv. Gr. [Sig. near to, cognate.]
dyeri, var. sublævis, S. A. Miller, 1878, Jour. Cin. Soc. Nat. Hist., vol. 1, Hud. Riv. Gr. [Sig. somewhat smooth.]
gracilis, Wetherby, 1881 (Reteocrinus gracilis), Jour. Cin. Soc. Nat. Hist., vol. 4, Hud. Riv. Gr. [Sig. slender.]
harrisi, S. A. Miller, 1881, Jour. Cin. Soc. Nat. Hist., vol. 1, Hud. Riv. Gr. [Ety. proper name.]
miamiensis, S. A. Miller, 1882, Jour. Cin. Soc. Nat. Hist., vol. 5, Hud. Riv. Gr. [Ety. proper name.]
pattersoni, S. A. Miller, 1882, Jour. Cin. Soc. Nat. Hist., vol. 5, Utica Slate Gr. [Ety. proper name.]
richardsoni, Wetherby, 1880, Jour. Cin. Soc. Nat. Hist., vol. 2, Hud. Riv. Gr. [Ety. proper name.]
sculptus, S. A. Miller, 1882, Jour. Cin. Soc. Nat. Hist., vol. 5, Hud. Riv. Gr. [Sig. carved.]
shafferi var. germanus, S. A. Miller, 1880, Jour. Cin. Soc. Nat. Hist., vol. 3, Hud. Riv. Gr. [Sig. near of kin.] I believe this is a distinct species, and will therefore become G. germanus.

GRANATOCRINUS missouriensis and G. rœmeri, should be referred to the Choteau or Kinderhook Gr.

GRAPHIOCRINUS carbonarius instead of Scaphiocrinus carbonarius.
meadamsi instead of Scaphiocrinus meadamsi.
quatuordecembrachialis, refer to Eupachycrinus quatuordecembrachialis, and to the Kaskaskia Gr.
rudis instead of Scaphiocrinus rudis.
simplex instead of Scaphiocrinus simplex.
spinobrachiatus instead of Scaphiocrinus spinobrachiatus.
striatus instead of Scaphiocrinus striatus.
tortuosus instead of Scaphiocrinus tortuosus.
wachsmuthi instead of Scaphiocrinus wachsmuthi.

HETEROCRINUS crassus, refer to Iocrinus crassus.
geniculatus, Ulrich, 1879, Jour. Cin. Soc. Nat. Hist., vol. 2, Utica Slate Gr. [Sig. geniculated.]

HETEROCRINUS milleri, Wetherby, 1880, Jour. Cin. Soc. Nat. Hist., vol. 3, Trenton Gr. [Ety. proper name.]
œhanus, Ulrich, 1882, Jour. Cin. Soc. Nat. Hist., vol. 5, Hud. Riv. Gr. [Ety. proper name.]
pentagonus, Ulrich, 1882, Jour. Cin. Soc. Nat. Hist., vol. 5, Hud. Riv. Gr. [Sig. pentagonal.]
subcrassus, refer to Iocrinus subcrassus.
taupelli, Wetherby, 1881. I am unable to recognize this species. It appears to be founded upon a fragment of H. simplex.

HOLOCYSTITES baculus, S. A. Miller, 1879, Jour. Cin. Soc. Nat. Hist., vol. 2, Niagara Gr. [Sig. a staff.]
brauni, S. A. Miller, 1878, Jour. Cin. Soc. Nat. Hist., vol. 1, Niagara Gr. [Ety. proper name.]
dyeri, S. A. Miller, 1879, Jour. Cin. Soc. Nat. Hist., vol. 2, Niagara Gr. [Ety. proper name.]
elegans, S. A. Miller, 1878, Jour. Cin. Soc. Nat. Hist., vol. 1, Niagara Gr. [Sig. elegant.]
globosus, S. A. Miller, 1878, Jour. Cin. Soc. Nat. Hist., vol. 1, Niagara Gr. [Sig. globose.]
jolietensis, S. A. Miller, 1882, Jour. Cin. Soc. Nat. Hist., vol. 5, Niagara Gr. [Ety. proper name.]
ornatus, S. A. Miller, 1878, Jour. Cin. Soc. Nat. Hist., vol. 1, Niagara Gr. [Sig. ornate.]
perlongus, S. A. Miller, 1878, Jour. Cin. Soc. Nat. Hist., vol. 1, Niagara Gr. [Sig. very long.]
plenus, S. A. Miller, 1878, Jour. Cin. Soc. Nat. Hist., vol. 1, Niagara Gr. [Sig. full, large.]
pustulosus, S. A. Miller, 1878, Jour. Cin. Soc. Nat. Hist., vol. 1, Niagara Gr. [Sig. full of pustules.]
rotundus, S. A. Miller, 1879, Jour. Cin. Soc. Nat. Hist., vol. 2, Niagara Gr. [Sig. rotund.]
subrotundus, S. A. Miller, 1879, Jour. Cin. Soc. Nat. Hist., vol. 2, Niagara Gr. [Sig. subrotund.]
tumidus, S. A. Miller, 1879, Jour. Cin. Soc. Nat. Hist., vol. 2, Niagara Gr. [Sig. tumid.]
turbinatus, S. A. Miller, 1880, Jour. Cin. Soc. Nat. Hist., vol. 2, Niagara Gr. [Sig. turbinate.]
ventricosus, S. A. Miller, 1879, Jour. Cin. Soc. Nat. Hist., vol. 2, Niagara Gr. [Sig. ventricose.]
wetherbyi, S. A. Miller, 1878, Jour. Cin. Soc. Nat. Hist., vol. 1, Niagara Gr. [Ety. proper name.]

HOMOCRINUS, angustatus instead of angustus, from Hud. Riv. Gr.

HYBOCYSTITES, Wetherby, 1880, Jour. Cin.

Soc. Nat. Hist., vol. 3. [Ety. *hubos*, hump-backed; *kustis*, a bladder.]
HYBOCYSTITES problematicus, Wetherby, 1880, Jour. Cin. Soc. Nat. Hist., vol. 3, Trenton Gr. [Sig. problematical.]
HYDREIONOCRINUS, De Koninck, 1858, Bull. Acad. Royale Belgique, 2me serie, tome 3. Messrs. Wachsmuth & Springer refer to this genus Zeacrinus acanthophorus, Z. armiger, Z. depressus, Z. discus and Z. mucrospinus.
ICHTHYOCRINUS nobilis, Wachsmuth & Springer, 1878, Proc. Acad. Nat. Sci., Upper Burlington and Keokuk Gr. [Sig. noble.]
IOCRINUS, Hall, 1864, Advance sheets, 24th Reg. Rep. N. Y. [Ety. *io*, in triumph; *krinon*, a lily.]
crassus instead of Heterocrinus crassus.
subcrassus instead of Heterocrinus subcrassus.
LAMPTEROCRINUS parvus, Hall, 1879, Desc. New Spec. Foss., Niagara Gr. [Sig. little.]
LECANOCRINUS pusillus, Hall, 1863 (Cyathocrinus pusillus) Trans. Alb. Inst., vol. 4, Niagara Gr. [Sig. very small.]
pusillus, Winchell & Marcy, syn. for L. pusillus, Hall.
Lecythiocrinus, White, 1880, Proc. U. S. Nat. Mus., vol. 2. [Ety. *lekuthion*, a small oil flask; *krinon*, a lily.] This name was preoccupied by Muller in 1858, and by Zittel in 1879.
adamsi, Worthen, 1882, Bull. No. 1, Ill. St. Mus. Nat. Hist., Coal Meas. [Ety. proper name.]
ollicula eformis, White, 1880, Proc. U. S. Nat. Mus., vol. 2, Upper Coal Meas. [Sig. like a little pot.]
LEPIDESTHES colletti, White, 1878, Proc. Acad. Nat. Sci., Keokuk Gr. [Ety. proper name.]
formosus, S. A. Miller, 1879, Jour. Cin. Soc. Nat. Hist., vol. 2, Keokuk Gr. [Sig. beautiful.]
LICHENOCRINUS affinis, S. A. Miller, 1882, Jour., Cin. Soc. Nat. Hist., vol. 5, Hud. Riv. Gr. [Sig. related to, from resemblance to L. crateriformis.]
dubius, S. A. Miller, 1880, Jour. Cin. Soc. Nat. Hist., vol. 3, Utica Slate Gr. [Sig. doubtful.]
pattersoni, S. A. Miller, 1879, Jour. Cin. Soc. Nat. Hist., vol. 2, Utica Slate Gr. [Ety. proper name.]
LYRIOCRINUS melissa instead of Rhodocrinus melissa.
sculptus, S. A. Miller, 1882, Jour. Cin. Soc. Nat. Hist., vol. 5, Niagara or Low. Held. Gr. [Sig. sculptured.]
MACROSTYLOCRINUS fasciatus instead of Cyathocrinus fasciatus.
fusibrachiatus, Ringueberg, 1882. Jour. Cin. Soc. Nat. Hist., vol. 5, Niagara Gr. [Sig. from the fusiform arms.]
meeki instead of Actinocrinus meeki.
striatus var. granulosus, Hall, 1879, 28th Reg. Rep., Niagara Gr. [Sig. granulous.]
MARSUPIOCRINUS tennesseensis instead of Platycrinus tennesseensis.
tentaculatus instead of Platycrinus tentaculatus.
MEGISTOCRINUS *infelix*, refer to Saccocrinus infelix.
marcouanus, refer to Saccocrinus marcouanus.
necis, refer to Saccocrinus necis.
nodosus, Barris, 1879, Proc. Davenport Acad., Nat. Sci., Up. Held. Gr. [Sig. nodose.]
parvirostris is a syn. for M. plenus.
pileatus, S. A. Miller, 1879, Jour. Cin. Soc. Nat. Hist., vol. 2, Up. Held. Gr. [Sig. covered with a cap.]
spinosus should be stricken out.
MELOCRINUS clarkei, Williams, 1882, Proc. Acad. Nat. Sci., Chemung Gr. [Ety. proper name.]
obpyramidalis, Winchell & Marcy, 1865 (Actinocrinus obpyramidalis). Mem. Bost. Soc. Nat. Hist., vol. 1, Niagara Gr. [Sig. obpyramidal.]
MYELODACTYLUS bridgeportensis, S. A. Miller, 1880, Jour. Cin. Soc. Nat. Hist., vol. 3, Niagara Gr. [Ety. proper name.]
OLIGOPORUS coreyi, Meek & Worthen, 1870, Proc. Acad. Nat. Sci., Keokuk Gr. [Ety. proper name.]
Ollacrinus, Cumberland, 1826, Appendix to Reliquiæ Conservata. Figured without description, and subsequently declared by Koninck & Lehon to be a Rhodocrinus. Wachsmuth & Springer claim priority for this name over Goniasteroidocrinus, without good reason, however, and contrary to the laws of nomenclature, as shown by Meek in Ill. Geo. Sur., vol. 2, p. 217.
ONYCHOCRINUS asteriæformis instead of Forbesiocrinus asteriæformis.
distensus, Worthen, 1882, Bull. No. 1, Ill. St. Mus. Nat. Hist., Kaskaskia Gr. [Sig. distended.]
meeki, refer to Taxocrinus meeki, and to Keokuk Gr.
ramulosus instead of Forbesiocrinus ramulosus.
whitfieldi, refer to Taxocrinus whitfieldi.
Pachylocrinus, Wachsmuth & Springer, 1879, Proc. Acad. Nat. Sci. [Ety. *pachylos*. thick; *krinon*, a lily.] A proposed subgenus of Poteriocrinus, founded upon the form of the body.
PALÆASTER clarkanus, S. A. Miller, 1880, Jour. Cin. Soc. Nat. Hist., vol. 3, Hud. Riv. Gr. [Ety. proper name.]
clarki, S. A. Miller, refer to Palæaster clarkanus.
crawfordsvillensis, S. A. Miller, 1880, Jour. Cin. Soc. Nat. Hist., vol. 2, Keokuk Gr. [Ety. proper name.]
dubius, Miller & Dyer, 1878, Cont. to

ECHINODERMATA. 285

Pal., No. 2, Utica Slate Gr. [Sig. doubtful.]
PALÆASTER exculptus, S. A. Miller, 1881, Jour. Cin. Soc. Nat. Hist., vol. 4, Hud. Riv. Gr. [Sig. chiseled out.]
finei, Ulrich, 1879, Jour. Cin. Soc. Nat. Hist., vol. 2, Utica Slate Gr. [Ety. proper name.]
harrisi, S. A. Miller, 1879, Jour. Cin. Soc. Nat. Hist., vol. 2, Hud. Riv. Gr. [Ety. proper name.]
longibrachiatus, S. A. Miller, 1878, Jour. Cin. Soc. Nat. Hist., vol. 1, Hud. Riv. Gr. [Sig. long armed.]
miamiensis, S. A. Miller, 1880, Jour. Cin. Soc. Nat. Hist., vol. 3, Hud. Riv. Gr. [Ety. proper name.]
simplex, S. A. Miller, 1878, Jour. Cin. Soc. Nat. Hist., vol. 1, Hud. Riv. Gr. [Sig. simple.]
spinulosus, Miller & Dyer, 1878, Jour. Cin. Soc. Nat. Hist., vol. 1, Hud. Riv. Gr. [Sig. full of spines.]
PALÆASTERINA approximata, Miller & Dyer, 1878, Jour. Cin. Soc. Nat. Hist., vol. 1, Hud. Riv. Gr. [Sig. near to.]
speciosa, Miller & Dyer, 1878, Jour. Cin. Soc. Nat. Hist., vol. 1, Hud. Riv. Gr. [Sig. beautiful.]
Parisocrinus, Wachsmuth & Springer, 1879, Proc. Acad. Nat. Sci. [Ety. parisos, resembling; krinon, a lily.] A proposed subgenus under Poteriocrinus.
PENTREMITES angularis, and P. robustus, are from the Kaskaskia Group.
bipyramidalis, P. clavatus, P. grosvenori, P. lineatus, P. reinwardti, P. subcylindricus, P. subtruncatus, and P. wortheni, are referred to the subgenus Troostocrinus.
abbreviatus, Hambach, 1880, Trans. St. Louis Acad. Sci., vol. 4, Kaskaskia Gr. [Sig. abbreviated.]
basilaris, Hambach, 1880, Trans. St. Louis Acad. Sci., vol. 4, Kaskaskia Gr. [Sig. relating to the base.]
broadheadi, Hambach, 1880, Trans. St. Louis Acad. Sci., vol. 4, Kaskaskia Gr. [Ety. proper name.]
chesterensis, Hambach, 1880, Trans. St. Louis Acad. Sci., vol. 4, Kaskaskia Gr. [Ety. proper name]
clavatus, Hambach, 1880, Trans. St. Louis Acad. Sci , vol. 4, Kaskaskia Gr. [Sig. clavated.]
hemisphericus, Hambach, 1880, Trans. St. Louis Acad. Sci., vol. 4, Kaskaskia Gr. [Sig. hemispherical.]
nodosus, Hambach, 1880, Trans. St. Louis Acad. Sci., vol. 4, Kaskaskia Gr. [Sig. nodose.]
potteri, Hambach, 1880, Trans. St. Louis Acad. Sci., vol. 4, Burlington Gr. [Ety. proper name.]
spinosus, Hambach, 1880, Trans. St. Louis Acad. Sci., vol. 4, Kaskaskia Gr. [Sig. spinous.]
sulcatus, is said by Hambach to be a

syn. for P. laterniformis. The latter was founded upon a cast.
Pereichocrinus, Austin, 1843, Ann. & Mag. Nat. Hist., vol. 11. Not defined so as to be recognized, though some authors, disregarding the rules of nomenclature, have used the name in an attempt to supplant, with it, Saccocrinus.
PHYSETOCRINUS ornatus instead of Actinocrinus ornatus.
ventricosus instead of Actinocrinus ventricosus.
ventricosus var. cancellatus instead of Actinocrinus ventricosus var. cancellatus.
ventricosus var. internodus instead of Actinocrinus ventricosus var. internodus.
reticulatus instead of Actinocrinus reticulatus.
PISOCRINUS, DeKoninck, 1858, Bull. Acad. Roy. Belgique, 2me ser., tome 3. [Ety. pisos, a pea; krinon, a lily.]
gemmiformis, S. A. Miller, 1879, Jour. Cin. Soc. Nat. Hist., vol. 2, Niagara Gr. [Sig. bud shaped.]
PLATYCRINUS bloomfieldensis, S. A. Miller, 1880, Jour. Cin. Soc. Nat. Hist., vol. 2, Keokuk Gr. [Ety. proper name.]
bonoensis, White, 1878, Proc. Acad. Nat. Sci., Keokuk Gr. [Ety. proper name.]
leai was described in 1860, Trans. Am. Phil. Soc., vol. 13, from Up. Held. Gr.
monroensis, Worthen, 1882, Bull. No. 1, Ill. St. Mus. Nat. Hist., St. Louis Gr. [Ety. proper name.]
montanaensis, refer to Eucladocrinus montanensis.
multibrachiatus, is from the Burlington Group.
ornigranulus, McChesney, 1860, Desc. New Pal. Foss., Burlington Gr. [Sig. having granules.]
parvus, refer to Cordylocrinus parvus.
pleurovimineus, refer to Eucladocrinus pleurovimineus.
plumosus, refer to Cordylocrinus plumosus.
poculum, S. A. Miller, 1881, Jour. Cin. Soc. Nat. Hist., vol. 4, Burlington Gr. [Sig. a cup.]
praenuntius, Wachsmuth & Springer, 1878, Proc. Acad. Nat. Sci., Burlington Gr. [Sig. a harbinger.]
pratteni, Worthen, 1860, Trans. St. Louis Acad. Sci., Burlington Gr. [Ety. proper name.]
ramulosus, refer to Cordylocrinus ramulosus.
siluricus, Hall, 1879, Desc. New Spec. Foss., Niagara Gr. [Ety. proper name.]
tennesseensis, refer to Marsupiocrinus tennesseensis.
tentaculatus, refer to Marsupiocrinus tentaculatus.
vexabilis, white, 1875, Wheeler's U. S.

Sur. W. 100th. Meridian, Subcarboniferous. [Sig. disturbed.]
POROCRINUS crassus, is from the Hud. Riv. Gr., and P. pentagonus, from the Trenton Gr.
smithi, Grant, 1881, Trans. Ottawa Field Naturalists' Club, No. 2, Trenton Gr. [Ety. proper name.]
POTERIOCRINUS anomalos, Wetherby, 1880, Jour. Cin. Soc. Nat. Hist., vol. 3, Kaskaskia Group. [Sig. anomalous.]
arachnæformis, Worthen, 1882, Bull. No. 1, Ill. St. Mus. Nat. Hist., Keokuk Gr. [Sig. spider-like.]
asper, Worthen, 1882, Bull. No. 1, Ill. St. Mus. Nat. Hist., Keokuk Gr. [Sig. rough.]
asperatus, Worthen, 1882, Bull. No. 1, Ill. St. Mus. Nat. Hist., Keokuk Gr. [Sig. made rough, uneven.]
briareus, Worthen, 1882, Bull. No. 1, Ill., St. Mus. Nat. Hist., Keokuk Gr. [Ety. mythological name.]
burketi, Worthen, 1882, Bull. No. 1, Ill., St. Mus. Nat. Hist., Keokuk Gr. [Ety. proper name.]
calyx, Hall, 1879, Desc. New Spec. Foss., Niagara Gr. [Sig. a cup.]
clarkei, Williams, 1882, Proc. Acad. Nat. Sci., Chemung Gr. [Ety. proper name.]
claytonensis, Worthen, 1882, Bull No. 1, Ill. St. Mus. Nat. Hist., Warsaw Gr. [Ety. proper name.]
clytis, Worthen, 1882, Bull. No. 1, Ill. St. Mus. Nat. Hist., St. Louis Gr. [Ety. proper name.]
columbiensis, Worthen, 1882, Bull. No. 1, Ill. St. Mus. Nat. Hist., Kaskaskia Gr. [Ety. proper name.]
cornellanus, Williams, 1882, Proc. Acad. Nat. Sci., Chemung Gr. [Ety. proper name.]
coxanus, Worthen, 1882, Bull. No. 1, Ill. St. Mus. Nat. Hist., Keokuk Gr. [Ety. proper name.]
cylindricus was described in 1869 from the Up. Held. Gr.
davisanus, S. A. Miller, 1882, Jour. Cin. Soc. Nat. Hist., vol. 5, Up. Held. Gr. [Ety. proper name.]
fountainensis, Worthen, 1882, Bull. No. 1, Ill. St. Mus. Nat. Hist., St. Louis Gr. [Ety. proper name.]
gregarius, Williams, 1882, Proc. Acad. Nat. Sci., Chemung Gr. [Sig. gregarious.]
hamiltonensis, Worthen, 1882, Bull. No. 1, Ill. St. Mus. Nat. Hist., Keokuk Gr. [Ety. proper name.]
hemisphericus, refer to Erisocrinus hemisphericus.
illinoisensis, Worthen, 1882, Bull. No. 1, Ill. St. Mus. Nat. Hist., Warsaw Gr. [Ety. proper name.]
iowensis, Worthen, 1882, Bull. No. 1, Ill. St. Mus. Nat. Hist., Keokuk Gr. [Ety. proper name.]

POTERIOCRINUS jesupi, Whitfield, 1881, Bull. No. 1, Ann. Mus. Nat. Hist., Burlington Gr. [Ety. proper name.]
kaskaskensis, Worthen, 1882, Bull. No. 1, Ill. St. Mus. Nat. Hist., Kaskaskia Gr. [Ety. proper name.]
latidactylus, Worthen, 1882, Bull. No. 1, Ill. St. Mus. Nat. Hist., Keokuk Gr. [Sig. wide armed.]
milleri, Wetherby, 1880, Jour. Cin. Soc. Nat Hist., vol. 3, Kaskaskia Gr. [Ety. proper name.]
nauvooensis, Worthen, 1882, Bull. No. 1, Ill. St. Mus. Nat. Hist., Keokuk Gr. [Ety. proper name.]
occidentalis, Worthen, 1882, Bull. No. 1, Ill. St. Mus. Nat. Hist., Keokuk Gr. [Sig. western.]
okawensis, Worthen, 1882, Bull. No. 1, Ill. St. Mus. Nat. Hist., Kaskaskia Gr. [Ety. proper name.]
orestes, Worthen, 1882, Bull. No. 1, Ill. St. Mus. Nat. Hist., Keokuk Gr. [Ety. proper name.]
otterensis, Worthen, 1882, Bull. No. 1, Ill. St. Mus. Nat. Hist., Keokuk Gr. [Ety. proper name.]
peculiaris, Worthen, 1882, Bull. No. 1, Ill. St. Mus. Nat. Hist., Kaskaskia Gr. [Sig. peculiar.]
penicilliformis, Worthen, 1882, Bull. No. 1, Ill. St. Mus. Nat. Hist., Keokuk Gr. [Sig. like a painter's brush.]
pisiformis, refer to Arachnocrinus pisiformis.
popensis, Worthen, 1882, Bull. No. 1, Ill. St. Mus. Nat. Hist., Kaskaskia Gr. [Ety. proper name.]
posticus, refer to Dendrocrinus posticus.
propinquus, Worthen, 1882, Bull. No. 1, Ill. St. Mus. Nat. Hist., Kaskaskia Gr. [Sig. near to.]
rhombiferus, refer to Barycrinus rhombiferus.
richfieldensis, Worthen, 1882, Bull. No. 1, Ill. St. Mus. Nat. Hist., Kinderhook Gr. [Ety. proper name.]
salteri, Worthen, 1882, Bull. No. 1, Ill. St. Mus. Nat. Hist., Kaskaskia Gr. [Ety. proper name.]
sculptus, Worthen, 1882, Bull. No. 1, Ill. St. Mus. Nat. Hist., Kaskaskia Gr. [Sig. sculptured.]
similis, Worthen, 1882, Bull. No. 1, Ill. St. Mus. Nat. Hist., Keokuk Gr. [Sig. similar.]
simplex was described in 1869 from the Up. Held. Gr.
spinobrachiatus, Worthen, 1882, Bull. No. 1, Ill. St. Mus. Nat. Hist., Kaskaskia Gr. [Sig. spine armed.]
subramulosus, Worthen, 1882, Bull. No. 1, Ill. St. Mus. Nat. Hist., Keokuk Gr. [Sig. somewhat full of branches.]
talboti, Worthen, 1882, Bull. No. 1, Ill. St. Mus. Nat. Hist., St. Louis Gr. [Ety. proper name.]
tentaculatus, Worthen, 1882, Bull. No. 1,

Ill. St. Mus. Nat. Hist., Keokuk Gr. [Sig. tentacled.]
POTERIOCRINUS tenuidactylus, Worthen, 1882, Bull. No. 1, Ill. St. Mus. Nat. Hist., Keokuk Gr. [Sig. slender-fingered.]
validus, Worthen. 1882, Bull. No. 1, Ill. St. Mus. Nat. Hist., Warsaw Gr. [Sig strong.]
varsouviensis, Worthen, 1882, Bull. No. 1, Ill. St. Mus. Nat. Hist., Warsaw Gr. [Ety. proper name.]
venustus, Worthen, 1882, Bull No. 1, Ill. St. Mus. Nat. Hist., Kaskaskia Gr. [Sig. beautiful.]
wachsmuthi refer to Graphiocrinus wachsmuthi.
wetherbyi, S. A. Miller, 1879, Jour. Cin. Soc. Nat. Hist., vol. 2, Kaskaskia Gr. [Ety. proper name.]
zethus, Williams, 1882, Proc. Acad. Nat. Sci., Chemung Gr. [Ety. mythological name.]
PROTASTER flexuosus, Miller & Dyer, 1878, Jour. Cin. Soc. Nat. Hist., vol. 1, Utica Slate and Hud. River Gr. [Sig. full of turnings.]
miamiensis, S. A. Miller, 1882, Jour. Cin. Soc. Nat. Hist., vol. 5, Hud. Riv. Gr. [Ety. proper name.]
Protasterina, Ulrich, syn. for Protaster.
fimbriata, Ulrich, syn. for Protaster flexuosus.
PTEROTOCRINUS acutus, Wetherby, 1879, Jour. Cin. Soc. Nat. Hist., vol. 2, Kaskaskia Gr. [Sig. acute.]
bifurcatus, Wetherby, 1879, Jour. Cin. Soc. Nat. Hist., vol. 2, Kaskaskia Gr. [Sig. bifurcated.]
cornigerus, refer to Talarocrinus cornigerus.
sexlobatus, refer to Talarocrinus sexlobatus.
spatulatus, Wetherby, 1879, Jour. Cin. Soc. Nat. Hist., vol. 2, Kaskaskia Gr. [Sig. spatulate.]
RETIOCRINUS *gracilis*, Wetherby, see Glyptocrinus gracilis.
RHODOCRINUS coxanus, Worthen, 1882, Bull., No. 1, Ill. St. Mus. Nat. Hist., Keokuk Gr. [Ety. proper name.]
melissa, refer to Lyriocrinus melissa.
vesperalis, White, 1880, Proc. U. S., Nat. Mus., vol. 2, Upper Coal Meas. (?) [Sig. belonging to evening.]
SACCOCRINUS egani, S. A. Miller, 1881, Jour. Cin. Soc. Nat. Hist., vol. 4, Niagara Gr. [Ety. proper name.]
infelix, Winchell & Marcy, 1865 (Megistocrinus infelix), Mem. Bost. Soc. Nat. Hist., Niagara Gr. [Sig. unhappy.]
marcouanus, Winchell & Marcy, 1865 (Megistocrinus marcouanus), Mem. Bost. Soc. Nat. Hist., Niagara Gr. [Ety. proper name.]
necis, Winchell & Marcy, 1865 (Megistocrinus necis), Mem. Bost. Soc. Nat. Hist., Niagara Gr. [Sig. death.]
SACCOCRINUS pyriformis, S. A. Miller, 1882, Jour. Cin. Soc. Nat. Hist., vol. 5, Niagara Gr. [Sig. pear shaped.]
urniformis, S. A. Miller, 1881, Jour. Cin. Soc. Nat. Hist., vol. 4, Niagara Gr. [Sig. urn shaped.]
SCAPHIOCRINUS *carbonarius*, refer to Graphiocrinus carbonarius.
gibsoni, White, 1878, Proc. Acad. Nat. Sci., Keokuk Gr. [Ety. proper name.]
gurleyi, White, 1878, Proc. Acad. Nat. Sci., Keokuk Gr. [Ety. proper name.]
macadamsi, refer to Graphiocrinus macadamsi.
rudis, refer to Graphiocrinus rudis.
simplex, refer to Graphiocrinus simplex.
spinifer, Wetherby, 1880, Jour. Cin. Soc. Nat. Hist., vol. 3, Kaskaskia Gr. [Sig. spine bearing.]
spinobrachiatus, refer to Graphiocrinus spinobrachiatus
striatus, refer to Graphiocrinus striatus.
tortuosus, refer to Graphiocrinus tortuosus.
wachsmuthi, refer to Graphiocrinus wachsmuthi.
Scyphocrinus, Hall. This name was preoccupied by Zenker in 1833, and D'Orbigny proposed instead of it Cupulocrinus.
heterocostalis refer to Cupulocrinus heterocostalis.
Scytalocrinus Wachsmuth & Springer, 1879, Proc. Acad. Nat. Sci. [Ety. *skutale*, a staff or club; *krinon*, a lily.] A subgenus of Poteriocrinus of doubtful utility.
wachsmuthi, Wetherby, 1880, Jour. Cin. Soc. Nat. Hist., vol. 3, Kaskaskia Gr. [Ety. proper name.]
STEGANOCRINUS concinnus instead of Actinocrinus concinnus
sculptus, Hall, instead of Actinocrinus sculptus.
STEPHANOCRINUS osgoodensis, S. A. Miller, 1879, Jour. Cin. Soc. Nat. Hist., vol. 2, Niagara Gr. [Ety. proper name.]
pentalobus, Hall, 1879 (Codaster pentalobus), Desc. New Spec. Foss., Niagara Gr. [Sig. five lobed.]
pulchellus, Miller & Dyer, 1878 (Codaster pulchellus), Jour. Cin. Soc. Nat. Hist., vol. 1, Niagara Gr. [Sig. beautiful.]
STEREOCRINUS, Barris, 1879, Proc. Davenport Acad. Sci., vol. 2. [Ety. *stereos*, firm; *krinon*, a lily.]
triangulatus, Barris, 1879, Proc. Davenport Acad. Sci., vol. 2, Upper Helderberg Gr. [Sig. triangular.]
triangulatus var. liberatus, Barris, 1879, Proc. Davenport Acad. Sci., vol. 2. Upper Helderberg Gr. [Sig. furrowed.]

STROTOCRINUS bloomfieldensis, S. A. Miller, 1882, Jour. Cin. Soc. Nat. Hist., vol. 2, Keokuk Gr. [Ety. proper name.] *ectypus*, refer to Actinocrinus ectypus.

SYNBATHOCRINUS granuliferus, Wetherby, 1880, Jour. Cin. Soc. Nat. Hist., vol. 2, Kinderhook Gr. [Sig. granule bearing.]
oweni is from the Kinderhook Gr.
robustus is from the Keokuk Gr.

TÆNIASTER elegans, S. A. Miller, 1882, Jour. Cin. Soc. Nat. Hist., vol. 5, Hud. Riv. Gr. [Sig. elegant.]

TALAROCRINUS, Wachsmuth & Springer, 1881, Proc. Acad. Nat. Sci. [Ety. *talaros*, a basket; *krinon*, a lily.]
cornigerus instead of Dichocrinus cornigerus.
elegans instead of Dichocrinus elegans.
ovatus, Worthen, 1882, Bull. No. 1, Ill. St. Mus. Nat. Hist., Kaskaskia Gr. [Sig. ovate.]
sexlobatus instead of Dichocrinus sexlobatus.
symmetricus instead of Dichocrinus symmetricus.

TAXOCRINUS communis instead of Forbesiocrinus communis.
curtus, Williams, 1882, Proc. Acad. Nat. Sci, Chemung Gr. [Sig. short.]
fletcheri, Worthen, 1882, Bull. No. 1, Ill. St. Mus. Nat. Hist., Keokuk Gr. [Ety. proper name.]
giddingi instead of Forbesiocrinus giddingi.
ithacensis, Williams, 1882, Proc. Acad. Nat. Sci., Chemung Gr. [Ety. proper name.]
juvenis instead of Forbesiocrinus juvenis.
kelloggi instead of Forbesiocrinus kelloggi.
lobatus instead of Forbesiocrinus lobatus.
lobatus var. tardus instead of Forbesiocrinus lobatus var. tardus.
meeki instead of Forbesiocrinus meeki.
multibrachiatus instead of Forbesiocrinus multibrachiatus.

TAXOCRINUS multibrachiatus var. colletti, White, 1881, 2d Ann. Rep. Bureau of Statistics of Indiana, Keokuk Gr. [Ety. proper name.]
shumardanus instead of Forbesiocrinus shumardanus.
whitfieldi instead of Forbesiocrinus whitfieldi.

Teleiocrinus, Wachsmuth & Springer, 1881, Proc. Acad. Nat. Sci. [Ety. *teleios*, perfect; *krinon*, a lily.] A subgenus of Strotocrinus, of doubtful utility.

TROOSTOCRINUS, Shumard, 1865, Trans. St. Louis Acad. Sci., vol 2. [Ety. proper name.] A proposed subgenus of Pentremites.

VASOCRINUS lyoni, instead of Cyathocrinus lyoni.
macropleurus instead of Cyathocrinus macropleurus.

XENOCRINUS, S A. Miller, 1881, Jour. Cin. Soc. Nat. Hist., vol. 4. [Ety. *xenos*, strange, new; *krinon*, a lily.]
penicillus, S. A. Miller, 1881, Jour. Cin. Soc. Nat. Hist., vol. 4, Hud. Riv. Gr. [Sig. a painter's brush.]

ZEACRINUS cariniferus, refer to Cœliocrinus cariniferus.
compactilis is from the Kaskaskia Gr.
coxanus, Worthen, 1882, Bull. No. 1, Ill. St. Mus. Nat. Hist., Keokuk Gr. [Ety. proper name.]
crassus, refer to Eupachycrinus crassus.
formosus, refer to Eupachycrinus formosus.
keokuk, Worthen, 1882, Bull. No. 1, Ill. St. Mus. Nat. Hist., Keokuk Gr. [Ety. proper name.]
lyra, refer to Cœliocrinus lyra.
moorei, Whitfield, 1882, Desc. New Spec. Foss. from Ohio, Coal Meas. [Ety. proper name.]
pikensis, Worthen, 1882, Bull. No. 1, Ill. St. Mus. Nat. Hist., Burlington Gr. [Ety. proper name.]
subtumidus, refer to Eupachycrinus subtumidus.

SUBKINGDOM MOLLUSCA.

CLASS BRYOZOA.

ACANTHOCLADIA, King, 1849, Ann. & Mag. Nat. Hist., 2d ser., vol. 3, p. 389. [Ety. *akantha*, a spine; *klados*, a branch.]
americana, Swallow, 1858, Trans. St. Louis Acad. Sci., vol. 1, Permian. [Ety. proper name.]
ALECTO was preoccupied by Leach in the class Echinodermata, when Lamoureux used it, and hence Stomatopora is used in its place.
canadensis, see Aulopora canadensis.
AMPLEXOPORA, Ulrich, 1882, Jour. Cin. Soc. Nat. Hist., vol. 5. [Ety. *amplexus*, an encircling; *porus*, a pore.]
cingulata, Ulrich, 1882, Jour. Cin. Soc. Nat Hist., vol. 5, Hud. Riv. Gr. [Sig. encircled.]
septosa, Ulrich, 1879, Jour. Cin. Soc. Nat. Hist., vol. 2 (Atactopora septosa), Hud. Riv. Gr. [Sig. partitioned.] Mr. Ulrich also refers to this genus Monticulipora discoidea.
ARTHROCLEMA spiniforme, Ulrich, 1882, Jour. Cin. Soc. Nat. Hist., vol. 5, Trenton Gr. [Sig. spiniform.]
ARTHRONEMA, Ulrich, 1882, Jour. Cin. Soc. Nat. Hist., vol. 5. [Ety. *arthron*, a joint; *nema*, a thread.]
curtum, Ulrich. 1882, Jour. Cin. Soc. Nat. Hist., vol. 5, Hud. Riv. Gr. [Sig. short.]
tenue, James, as figured by Ulrich, 1882 (Helopora tenuis), Jour. Cin. Soc. Nat. Hist., vol. 5, Trenton and Utica Slate Gr. [Sig. thin.]
ASCODICTYON, Nicholson, 1877, Ann. and Mag. Nat. Hist., 4th ser., vol. 19. [Ety. *askos*, a leather bottle; *dictyon*, a net.]
fusiforme, Nicholson, 1877, Ann. and Mag. Nat. Hist., 4th ser., vol. 19, Ham. Gr. [Sig. fusiform.]
stellatum, Nicholson, 1877, Ann. and Mag. Nat. Hist., 4th ser., vol. 19, Ham. Gr. [Sig. starred.]
ATACTOPORA, Ulrich, 1879, Jour. Cin. Soc. Nat. Hist., vol. 2. [Ety. *atactos*, without regularity; *porus*, a pore.]
hirsuta, Ulrich, 1879, Jour. Cin. Soc. Nat. Hist., vol. 2, Hud. Riv. Gr. [Sig. rough.]
maculata, Ulrich, 1879, Jour. Cin. Soc. Nat. Hist., vol. 2, Hud. Riv. Gr. [Sig. spotted.]
multigranosa, Ulrich, 1879, Jour. Cin. Soc. Nat. Hist., vol. 2, Hud. Riv. Gr. [Sig. many grained.]
mundula, Ulrich, 1879, Jour. Cin. Soc Nat. Hist., vol. 2, Hud. Riv. Gr. [Sig. neat.]
ATACTOPORA *septosa*, see Amplexopora septosa.
subramosa, Ulrich, 1879, Jour. Cin. Soc. Nat. Hist., vol. 2, Hud. Riv. Gr. [Sig. somewhat branchy.]
tenella, Ulrich, 1879. Jour. Cin. Soc. Nat. Hist., vol. 2, Hud. Riv. Gr. [Sig. neat.]
BATOSTOMA, Ulrich, 1882, Jour. Cin. Soc. Nat. Hist., vol. 5, founded upon Monticulipora jamesi, and M. implicata.
BATOSTOMELLA, Ulrich, 1882, Jour. Cin. Soc. Nat. Hist., vol. 5, founded upon Monticulipora gracilis, M. granulifera, and Trematopora annulifera.
BERENICEA, Lamoureux, 1821, Exp. Meth. des. genres d. pol., p. 80. [Ety. mythological name.]
primitiva, Ulrich, 1882, Jour. Cin. Soc. Nat. Hist., vol. 5, Hud. Riv. Gr. [Sig. primitive.]
vesiculosa, Ulrich, 1882, Jour. Cin. Soc. Nat. Hist., vol. 5, Utica Slate or Lower part Hud. Riv. Gr. [Sig. vesiculous.]
BYTHOPORA, Miller & Dyer, 1878, Cont. to Pal., No. 2. [Ety. *buthos*, depths of the sea; *poros*, a pore.]
arctipora instead of Ptilodictya arctipora.
fruticosa, Miller & Dyer, 1878, Cont. to Pal., No. 2, Hud. Riv. Gr. [Sig. shrubby.]
nashvillensis, S. A. Miller, 1880, Jour. Cin. Soc. Nat. Hist., vol. 3, Trenton Gr. [Ety. proper name.]
CALLOPORA aculeolata, Hall, 1881, Bryozoans of the Up. Held. Gr. [Sig. thorny.]
cervicornis, Hall, 1879, Desc. New Spec. Foss., Niagara Gr. [Sig stag horned.]
cincinnatensis, Ulrich, 1878, Jour. Cin. Soc. Nat. Hist., vol. 1, Hud. Riv. Gr. [Ety. proper name.]
diversa, Hall, 1879. Desc. New Spec. Foss., Niagara Gr. [Sig. diverse.]
exsul instead of Alveolites exsul.
irregularis, Hall, 1881, Bryozoans of the Up. Held. Gr. [Sig irregular.]
minutissima, Nicholson. 1875. Pal. Prov. Ont., Ham Gr. [Sig. very minute.]
multiseriata, Hall, 1881, Bryozoans of the Up. Held. Gr. [Sig. having many series.]
subplana, Ulrich, 1882, Jour. Cin. Soc. Nat. Hist., vol. 5, Hud. Riv. Gr. [Sig. somewhat level.] Mr. Ulrich refers to this genus Monticulipora ramosa, M. dalei, M. sigillarioides, M. nodulosa, and M. andrewsi.

CERAMOPORA explanata, Hall, 1879, Desc. New Spec. Foss., Niagara Gr. [Sig. spread out.]
nothus, Hall, 1879, Desc. New Spec. Foss., Niagara Gr. [Sig. spurious.]
rariporn, Hall, 1879, Desc. New Spec. Foss., Niagara Gr. [Sig. having few pores.]
CLATHROPORA intermedia, Nicholson & Hinde, 1874, Can. Jour., Niagara Gr. [Sig. intermediate.]
CLONOPORA, Hall, 1881, Bryozoans of the Up. Held. Gr. [Ety. *klonos*, confusion; *poros*, pore.]
incurva, Hall, 1881, Bryozoans of the Up. Held. Gr. [Sig. incurved]
semireducta, Hall, 1881, Bryozoans of the Up. Held. Gr. [Sig. half drawn back.]
Crateripora, Ulrich, 1879. *C. erecta*, *C. lineata*, and *C. lineata* var. *expansa*, represent basal fragments of Bryozoa, and are not entitled to rank as species.
CRISINA scrobiculata, Hall, 1881, Bryozoans of the Up. Held. Gr. [Sig. having small furrows.]
CYSTODICTYA, Ulrich, 1882, Jour. Cin. Soc. Nat. Hist., vol. 5. [Ety. *kustis*, a bladder; *dictyon*, a net.]
ocellata, Ulrich, 1882, Jour. Cin. Soc. Nat. Hist., vol. 5, Keokuk Gr. [Sig. having little eyes.]
CYSTOPORA, Hall, 1881, Bryozoans of the Up. Held. Gr. [Ety. *kustis*, a bladder; *poros*, a pore.]
geniculata, Hall, 1881, Bryozoans of the Up. Held. Gr. [Sig. geniculated.]
DICRANOPORA, Ulrich, 1882, Jour. Cin. Soc. Nat. Hist., vol. 5. [Ety. *dikranos*, two pointed; *poros*, a pore.]
lata, Ulrich, 1882, Jour. Cin. Soc. Nat. Hist., vol. 5, Hud. Riv. Gr. [Sig. wide.]
trentonensis, Ulrich, 1882, Jour. Cin. Soc. Nat. Hist., vol. 5, Trenton Gr. [Ety. proper name.]
ERIDOPORA, Ulrich, 1882, Jour. Cin. Soc. Nat. Hist., vol. 5. [Ety. *eridos*, in dispute; *poros*, a pore.]
macrostoma, Ulrich, 1882, Jour. Cin. Soc. Nat. Hist., vol. 5, Kaskaskia Gr. [Sig. long mouth.]
punctifera, Ulrich, 1882, Jour. Cin. Soc. Nat. Hist., vol. 5, Kaskaskia Gr. [Sig. bearing punctures]
ESCHAROPORA angusta, Hall, 1879, Desc. New Spec. Foss., Niagara Gr. [Sig. narrow.]
DIPLOTRYPA milleri, Ulrich, 1882, Jour. Cin. Soc. Nat. Hist., vol. 5, Niagara Gr. [Ety. proper name.]
DISCOTRYPA, Ulrich, 1882, Jour. Cin. Soc. Nat. Hist., vol. 5, founded upon Chetetes elegans.
FENESTELLA acaulis, Hall, 1881, Bryozoans of the Up. Held. Gr. [Sig. without a stem.]
aculeata, Hall, 1881, Bryozoans of the Up. Held. Gr. [Sig. thorny.]

FENESTELLA adnata, Hall, 1881, Bryozoans of the Up. Held. Gr. [Sig. close to.]
æqualis, Hall, 1881, Bryozoans of the Up. Held. Gr. [Sig. equal]
ambigua, Hall, 1879, 28th Reg. Rep., Niagara Gr. [Sig. doubtful.]
angulata, Hall, 1881, Bryozoans of the Up. Held. Gr. [Sig. angular.]
anonyma, Hall, 1881, Bryozoans of the Up. Held. Gr. [Sig. nameless.]
arctica, Salter, 1855, Belcher's last Arctic Voyage, vol. 2, Carboniferous. [Sig. arctic.]
bellistriata, Hall, 1879, Desc. New Spec. Foss., Niagara Gr. [Sig. beautifully striated.]
bi-imbricata, Hall, 1881, Bryozoans of the Up Held. Gr. [Sig. double imbricated.]
biseriata, Hall, 1881, Bryozoans of the Up. Held. Gr. [Sig. having a double series.]
biserrulata, Hall, 1881, Bryozoans of the Up. Held. Gr. [Sig. double serrulated.]
brevisulcata, Hall, 1881, Bryozoans of the Up. Held. Gr. [Sig. having a short furrow.]
celsipora, Hall, 1881, Bryozoans of the Up. Held. Gr. [Sig. having high pores.]
celsipora var. minima, Hall, 1881, Bryozoans of the Up. Held. Gr. [Sig. the least.]
celsipora var. minor, Hall, 1881, Bryozoans of the Up. Held. Gr. [Sig. less.]
conferta, Hall, 1879, Desc. New Spec. Foss., Niagara Gr. [Sig. pressed together.]
conjunctiva, Hall, 1881, Bryozoans of the Up. Held. Gr. [Sig. connecting.]
corticata. F. intermedia, F. subretiformis, F. trituberculata and F. variabilis, are from the Coal Mens.
cribrosa, Hall, 1881, Bryozoans of the Up. Held. Gr. [Sig. sieve like.]
cultellata, Hall, 1881, Bryozoans of the Up. Held. Gr. [Sig. being like a little knife.]
cultrata, Hall, 1881, Bryozoans of the Up. Held. Gr. [Sig. knife-formed.]
curvijunctura, Hall, 1881, Bryozoans of the Up. Held. Gr. [Sig. curve juncture.]
cylindracea, Hall, 1881, Bryozoans of the Up. Held. Gr. [Sig. like a cylinder.]
davidsoni instead of F. dawsoni.
depressa, Hall, 1881, Bryozoans of the Up. Held. Gr. [Sig. depressed.]
distans, Hall, 1881, Bryozoans of the Up. Held. Gr. [Sig. distant.]
elegantissima, Hall, 1881, Bryozoans of the Up. Held. Gr. [Sig. most elegant.]
erectipora, Hall, 1881, Bryozoans of the Up. Held. Gr. [Sig. having erect pores.]
fastigata Hall, 1881, Bryozoans of the Up. Held. Gr. [Sig. pointed.]

FENESTELLA favosa, Hall, 1881. Bryozoans of the Up. Held. Gr. [Sig. honeycomb like.]
flabelliformis, Hall, 1881, Bryozoans of the Up. Held. Gr. [Sig. fan like.]
granifera, Hall, 1881, Bryozoans of the Up. Held. Gr. [Sig. grain bearing.]
granulinea, Hall, 1881, Bryozoans of the Up. Held. Gr. [Sig. grain lined.]
granulosa, Whitfield, 1878, Ann. Rep. Geo. Sur. Wis., Hud. Riv. Gr. [Sig. full of grains.]
hexagonalis, Hall, 1881, Bryozoans of the Up. Held. Gr. [Sig. hexagonal.]
hexagonalis var. foraminulosa, Hall, 1881, Bryozoans of the Up. Held. Gr. [Sig. full of openings.]
interrupta, Hall, 1881, Bryozoans of the Up. Held. Gr. [Sig. interrupted.]
largissima, Hall, 1881, Bryozoans of the Up. Held. Gr. [Sig. largest.]
lata, Hall, 1881, Bryozoans of the Up. Held. Gr. [Sig. wide.]
latijunctura, Hall, 1881, Bryozoans of the Up. Held. Gr. [Sig. wide jointed.]
levinodata, Hall, 1881, Bryozoans of the Up. Held. Gr. [Sig. having smooth knots.]
lineanoda, Hall, 1881, Bryozoans of the Up. Held. Gr. [Sig. having lined knots.]
lunulata, Hall, 1881, Bryozoans of the Up. Held. Gr. [Sig. resembling a small crescent.]
mutabilis, Hall, 1881, Bryozoans of the Up. Held. Gr. [Sig. changeable.]
nervia. [Sig. having strong ribs, instead of proper name.]
nexa Hall, 1881, Bryozoans of the Up. Held. Gr. [Sig. interlaced.]
oxfordensis, Ulrich, 1882, Jour. Cin. Soc. Nat. Hist., vol. 5, Hud. Riv. Gr. [Ety. proper name.]
parallella, Hall, 1881, Bryozoans of the Up. Held. Gr. [Sig. parallel.]
perangulata, Hall, 1881, Bryozoans of the Up. Held. Gr. [Sig. very angular.]
permarginata, Hall, 1881, Bryozoans of the Up. Held. Gr. [Sig. having a large border.]
pernodosa, Hall, 1881, Bryozoans of the Up. Held. Gr. [Sig. very nodose.]
perplexa, Hall, 1881, Bryozoans of the Up. Held. Gr. [Sig. intricate.]
pertenuis, Hall, 1879, Desc. New Spec. Foss., Niagara Gr. [Sig. very thin.]
pertenuis, Hall, 1881. The name was preoccupied. See F. proutana.
perundata, Hall, 1881, Bryozoans of the Up. Held. Gr. [Sig. very wavy.]
porosa, Hall, 1881, Bryozoans of the Up. Held. Gr. [Sig full of pores.]
prolixa, Hall, 1879. Desc. New Spec. Foss., Niag. Gr. [Sig. Stretched out.]
propria, Hall, 1881, Bryozoans of the Up. Held. Gr. [Sig. proper.]
proutana, n. sp., Up. Held. Gr. Named in respect for the work of Dr. Hiram A. Prout. Proposed instead of F. pertennis, Hall, 1881, Bryozoans of the Up. Held. Gr., page 29, which was preoccupied.
FENESTELLA quadrangularis, Hall, 1881, Bryozoans of the Up. Held. Gr. [Sig. quadrangular.]
rhombifera, Hall, 1881, Bryozoans of the Up. Held. Gr. [Sig. rhomb bearing.]
rigida, Hall, 1881, Bryozoans of the Up. Held. Gr. [Sig. rigid.]
robusta, Hall, 1881, Bryozoans of the Up. Held. Gr. [Sig. robust.]
semirotunda, Hall, 1881, Bryozoans of the Up. Held. Gr. [Sig. half rotund.]
serrata, Hall, 1881, Bryozoans of the Up. Held. Gr. [Sig. serrated.]
singularitas, Hall, 1881, Bryozoans of the Up. Held. Gr. [Sig. single.]
stellata, Hall, 1881, Bryozans of the Up. Held. Gr. [Sig. starred.]
stipata, Hall, 1881, Bryozoans of the Up. Held. Gr. [Sig. pressed together.]
striatopora, Hall, 1881, Bryozoans of the Up. Held. Gr. [Sig. having striæ and pores.]
submutans, Hall, 1881, Bryozoans of the Up. Held. Gr. [Sig. somewhat changing.]
substriata, Hall, 1881, Bryozoans of the Up. Held. Gr. [Sig. somewhat striated]
tantulus, Hall, 1879, Desc. New Spec. Foss., Niagara Gr. [Sig. so little.]
tegulata, Hall, 1881, Bryozoans of the Up. Held. Gr. [Sig. tiled.]
torta, Hall, 1881, Bryozoans of the Up. Held. Gr. [Sig. twisted.]
variopora, Hall, 1881, Bryozoans of the Up. Held. Gr. [Sig. having different pores.]
GLAUCONOME nodata, Hall, 1881, Bryozoans of the Up. Held. Gr. [Sig. knotty.]
sinuosa, Hall, 1881, Bryozoans of the Up. Held. Gr. [Sig. sinuous.]
tenuistriata, Hall, 1881, Bryozoans of the Up. Held. Gr. [Sig. fine lined.]
GORGONIA anticorum, Castelnau, 1843, Syst. Sil. Not recognized.
retiformis, Hall, 1843, Geo. 4th Dist. N. Y., Niagara Group. Supposed to be a Graptolite.
siluriana, Castelnau, 1843, Syst. Sil. Not recognized.
GRAPTODICTYA, Ulrich, 1882, Jour. Cin. Soc. Nat. Hist., vol. 5. [Ety. grapho, to write; dictyon, a net.] I am unable to distinguish the generic difference between this and Stictopora.
nitida, Ulrich, 1882, Jour. Cin. Soc. Nat. Hist., vol. 5, Hud. Riv. Gr. [Sig. neat.]
HEMITRYPA dubia, Hall, 1876, is a Fenestella, and the name dubia being preoccupied, the species is called F. ambigua.

HOMOTRYPA, Ulrich, 1882, Jour. Cin. Soc. Nat. Hist., vol. 5. [Ety. *homos*, similar; *trypa*, a perforation.]
curvata, Ulrich, 1882, Jour. Cin. Soc. Nat. Hist., vol. 5, Hud. Riv. Gr. [Sig. curved.]
obliqua, Ulrich, 1882, Jour. Cin. Soc. Nat. Hist., vol. 5, Hud. Riv. Gr. [Sig. oblique.]
INTRAPORA. Hall, 1881. Bryozoans of the Up. Held. Gr. [Ety. *intra*, within; *porus*, a pore.]
puteolata, Hall, 1881, Bryozoans of the Up. Held. Gr. [Sig. pitted.]
INTRICARIA clathrata, Miller & Dyer, 1878, Cont. to Pal., No. 2, Hud. Riv. Gr. [Sig. latticed.]
LEIOCLEMA, Ulrich, 1882 Jour. Cin. Soc. Nat. Hist., vol. 5 [Ety. *leios*, smooth; *klema*, a twig.] Proposed to include Callopora punctata, Hall.
LICHENALIA alternata, Hall, 1881, Bryozoans of the Up. Held. Gr. [Sig. alternated.]
alveata, Hall, 1881, Bryozoans of the Up. Held. Gr. [Sig. hollowed out.]
bistriata, Hall, 1881, Bryozoans of the Up. Held. Gr. [Sig. double striated.]
carinata, Hall, 1881, Bryozoans of the Up. Held. Gr. [Sig. carinated.]
circincta, Hall, 1881, Bryozoans of the Up. Held. Gr. [Sig. encompassed.]
clivulata, Hall, 1881, Bryozoans of the Up. Held. Gr. [Sig. having little Hills.]
complexata, Hall, 1881, Bryozoans of the Up. Held. Gr. [Sig. encircled.]
concentrica var. maculata, Hall, 1879, 28th Rep. N. Y. St. Mus., Niagara Gr. [Sig. spotted.]
conulata, Hall, 1881, Bryozoans of the Up. Held. Gr. [Sig. having little cones.]
crustacea, Hall, 1881, Bryozoans of the Up. Held. Gr. [Sig. having a crust]
denticulata, Hall, 1881, Bryozoans of the Up. Held. Gr. [Sig. denticulated.]
granifera, Hall, 1881, Bryozoans of the Up. Held. Gr. [Sig grain bearing.]
longispina, Hall, 1881, Bryozoans of the Up. Held. Gr. [Sig. long thorned.]
lunata, Hall, 1881, Bryozoans of the Up Held. Gr. [Sig. crescent shaped.]
palliformis, Hall, 1881, Bryozoans of the Up. Held. Gr. [Sig. stake formed.]
permarginata, Hall, 1881, Bryozoans of the Up. Held. Gr. [Sig. having a large margin]
pyriformis, Hall, 1881, Bryozoans of the Up. Held Gr. [Sig. pear shaped]
radiata, Hall, 1881, Bryozoans of the Up. Held. Gr. [Sig. radiated].
subcava, Hall, 1881, Bryozoans of the Up. Held. Gr. [Sig. somewhat excavated]
substellata Hall, 1881, Bryozoans of the Up. Held. Gr. [Sig. somewhat starred.]

MITOCLEMA, Ulrich 1882, Jour. Cin. Soc. Nat. Hist., vol 5. [Ety. *mitos*, a thread; *klema*, a twig.]
cinctosa, Ulrich, 1882. Jour. Cin. Soc. Nat. Hist., vol. 5, Trenton Gr. [Sig. girded.]
MONOTRYPELLA, Ulrich, 1882, Jour. Cin. Soc., Nat. Hist., vol. 5. [Ety. diminutive of *Monotrypa*]
æqualis, Ulrich, 1882, Jour. Cin. Soc. Nat. Hist., vol. 5, Hud. Riv. Gr. [Sig. equal]
subquadrata, Ulrich. 1882, Jour. Cin. Soc. Nat. Hist., vol. 5, Hud. Riv. Gr. [Sig. subquadrate.] Mr. Ulrich also refers to this genus Monticulipora quadrata.
MONTICULIPORA consimilis. Ulrich, 1882, Jour. Cin. Soc. Nat. Hist., vol. 5, Hud. Riv. Gr. [Sig. entirely similar.] Mr. Ulrich refers this genus to the Bryozoa.
lævis, Ulrich, 1882, Jour. Cin. Soc. Nat. Hist., vol. 5, Hud. Riv. Gr. [Sig. smooth.]
parasitica. Ulrich, 1882, Jour. Cin. Soc. Nat. Hist., vol. 5, Hud. Riv. Gr. [Sig. parasitic.]
wetherbyi. Ulrich, 1882, Jour. Cin. Soc. Nat. Hist., vol. 5, Hud. Riv. Gr. [Ety. proper name.]
PACHYDICTYA, Ulrich, 1882. Jour. Cin. Soc. Nat Hist., vol. 5. [Ety. *pachys*, thick; *dictyon*, a net.]
robusta, Ulrich, 1882. Jour. Cin. Soc. Nat. Hist., vol. 5, Trenton Gr. [Sig. robust.]
PALESCHARA incrassata. Hall.1879,28th Reg. Rep., Niagara Gr. [Sig. thickened.]
PERONOPORA uniformis, Ulrich, 1882, Jour. Cin. Soc. Nat. Hist., vol. 5, Hud. Riv. Gr. [Sig. one formed.] To this genus Mr. Ulrich refers Monticulipora compressa.
PETIGOPORA, Ulrich, 1882, Jour. Cin. Soc. Nat. Hist., vol. 5. Founded upon Monticulipora petechialis.
PHYLLODICTYA, Ulrich. 1882, Jour. Cin. Soc. Nat. Hist., vol. 5. [Ety. *phyllon*, a leaf; *dictyon*, a net]
frondosa, Ulrich, 1882. Jour. Cin. Soc. Nat. Hist., vol. 5, Trenton Gr. [Sig. frondose.]
PHRACTOPORA, Hall, 1881, Bryozoans of the Up. Held Gr. [Ety. *phraktos*, enclosed; *porus*, a pore.]
cristata. Hall, 1881, Bryozoans of the Up. Held. Gr. [Sig. crested.]
PHYLLOPORA, King, 1849, Ann. & Mag. Nat. Hist., 2d ser. vol. 3 p. 389 [Ety. *phyllon*, a leaf; *poros*, a perforation.]
variolata, Ulrich, 1882, Jour. Cin. Soc. Nat Hist., vol. 5, Hud. Riv. Gr. [Sig. variolated]
POLYPORA arkonensis, n. sp., Ham. Gr. [Ety. proper name.] Proposed instead of P. tuberculata, Nicholson, in Geo. Mag. for Apr., 1874, and Rep.

Pal. Prov. Ont., p. 100, figs. 37, *a*, *b*, *c*. Found at Arkona, township of Bosanquet, Canada.
POLYPORA grandis, Toula, 1875, N. Jahrbuch, Carboniferous. [Sig. large.]
megastoma, DeKoninck, 1863, Quar. Jour. Geo. Soc., vol. 19, Carboniferous. [Sig. large mouth.]
tuberculata, Prout, is from the Kaskaskia Gr
tuberculata, Nicholson, see P. arkonensis.
PRASOPORA nodosa, Ulrich. 1882, Jour. Cin. Soc. Nat. Hist., vol. 5, Hud. Riv. Gr. [Sig. nodose.]
PRISMOPORA, Hall, 1881, Bryozoans of the Up. Held. Gr. [Ety. *prismos*, the hole made by a cylindrical saw; *poros*, a pore.]
pauciranna, Hall, 1881, Bryozoans of the Up. Held. Gr. [Sig. having few branches.]
triquetra, Hall, 1881, Bryozoans of the Up. Held. Gr. [Sig. three cornered.]
PTILODICTYA *acuminata*, is Stictopora acuminata.
arctipora is Bythopora arctipora, from the Hud. Riv. Gr.
briareus, Ulrich, 1882, Jour. Cin. Soc. Nat. Hist., vol. 5, Trenton Gr. [Ety. mythological name.]
hilli, James, 1882 (as figured by Ulrich), Jour. Cin. Soc. Nat. Hist , vol. 5, Trenton Gr. [Ety. proper name.]
internodia, is Stictopora internodia.
maculata, Ulrich, 1882, Jour. Cin. Soc. Nat. Hist., vol. 5, Hud. Riv. Gr. [Sig. maculated.]
magnifica, S. A. Miller, 1878, Jour. Cin. Soc. Nat. Hist., vol. 1, Hud. Riv. Gr. [Sig. magnificent.]
nodosa, James, 1882 (as figured by Ulrich), Jour. Cin. Soc. Nat. Hist., vol. 5, Hud. Riv. Gr. [Sig nodose.]
perelegans, see Stictopora perelegans.
plumaria, James, 1882 (as figured by Ulrich), Jour. Cin. Soc. Nat. Hist., vol. 5, Hud. Riv. Gr. [Sig. belonging to a feather.]
punctata, Nicholson & Hinde, 1878, Can. Jour., Clinton Gr. [Sig. having punctures.]
ramosa, Ulrich, 1882, Jour. Cin. Soc. Nat. Hist., vol. 5, Trenton Gr. [Sig. ramose.]
triangulata, White, 1878, Proc. Acad. Nat. Sci., Coal Meas. [Sig. triangulated.]
RAMIPORA, Toula, 1875, N. Jahrbuch. [Ety. *ramus*, branch; *porus*, pore.]
hochstetteri, Toula, 1875, N. Jahrbuch, Carboniferous. [Ety. proper name.]
RETEPORA fenestrata is from the Trenton Gr.
RHINIDICTYA, Ulrich, 1882, Jour. Cin. Soc. Nat. Hist., vol. 5. [Ety. *rhine*, a file; *dictyon*, a net.]
nicholsoni, Ulrich, 1882, Jour. Cin. Soc.

Nat. Hist., vol. 5, Trenton Gr. [Ety. proper name.]
Ropalonaria, Ulrich, 1879, Jour. Cin. Soc. Nat. Hist., vol. 2. [Ety. *ropalon*, a club.]
venosa, Ulrich, 1879, Jour. Cin. Soc. Nat. Hist., vol. 2, Hud. Riv. Gr. [Sig. veiny.]
SAGENELLA ambigua, Walcott, 1879, Utica Slate and related formations, Utica Slate. [Sig. ambiguous.]
SCALARIPORA, Hall, 1881, Bryozoans of the Up. Held. Gr. [Ety. *scalare*, a staircase, ladder; *porus*, a pore.]
scalariformis, Hall, 1881, Bryozoans of the Up. Held. Gr. [Sig. ladder formed.]
subconcava, Hall, 1881, Bryozoans of the Up. Held. Gr. [Sig. somewhat concave.]
SCENELLOPORA, Ulrich, 1882, Jour. Cin. Soc. Nat. Hist., vol. 5. [Ety. *scene*, a tent; *ellus*, diminutive; *poros*, a pore.]
radiata, Ulrich, 1882, Jour Cin. Soc. Nat. Hist., vol. 5, Trenton Gr. [Sig. radiated]
SPATIOPORA, Ulrich, 1882, Jour. Cin. Soc. Nat. Hist., vol. 5. Founded upon Monticulipora tuberculata.
STICTOPORA basalis, Ulrich, 1882, Jour. Cin. Soc. Nat. Hist., vol. 5, Trenton Gr. [Sig. basal.]
fruticosa, Hall, 1881, Bryozoans of the Up. Held. Gr. [Sig. bushy.]
internodia, Miller and Dyer, 1878 (Ptilodictya internodia), Cont. to Pal., No. 2, Hud. Riv. Gr. [Sig. between nodes.]
invertis, Hall, 1881, Bryozoans of the Up. Held Gr. [Sig. inverted.]
linearis, Hall, 1881, Bryozoans of the Up. Held. Gr. [Sig. linear.]
orbipora, Hall, 1870, Desc. New Spec. Foss., Niagara Gr. [Sig. having round pores]
ovatipora, Hall, 1881, Bryozoans of the Up. Held Gr. [Sig. having ovate pores.]
perarcta, Hall, 1881, Bryozoans of the Up. Held. Gr. [Sig. very close.]
perelegans, Ulrich, 1878 (Ptilodictya perelegans). Jour. Cin. Soc. Nat. Hist , vol. 1, Hud. Riv. Gr. [Sig. very elegant.]
rhomboidea, Hall, 1881, Bryozoans of the Up. Held. Gr. [Sig rhomb like]
rigida, Hall 1881, Bryozoans of the Up. Held. Gr. [Sig. rigid.]
semistriata, Hall, 1881, Bryozoans of the Up. Held. Gr. [Sig. half striated]
Stictoporella, Ulrich, 1882, Jour. Cin. Soc. Nat. Hist., vol. 5. [Ety. diminutive of Stictopora.] I am unable to distinguish this from Stictopora.
interstincta, Ulrich, 1882, Jour. Cin. Soc. Nat. Hist., vol. 5, Utica Slate Gr. [Sig. divided.]

STOMATOPORA, Bronn, 1825, System d. urwetl. Pflanzenthiere. [Ety. *stoma*, mouth; *poros*, a perforation.) This name is preferred to Alecto, because the latter name had been used prior to its application among the Bryozoa.
alternata instead of Stromatopora alternata.
proutana, S. A. Miller, 1882, Jour. Cin. Soc. Nat. Hist., vol. 5, Hud. Riv. Gr. [Ety. proper name.]
SYNOCLADIA rectistyla, Whitfield, 1882, Desc. New Spec. Foss. from Ohio, Kaskaskia Gr. [Sig. having straight stiles.]
THALLISTIGMA, Hall, 1881, Bryozoans of the Up. Held. Gr. (Ety. *thallos*, a young branch or twig; *stigma*, a spot.]
intercellatum, Hall, 1881, Bryozoans of the Up. Held. Gr. [Sig. being intercellular.]
lamellatum, Hall, 1881, Bryozoans of the Up. Held. Gr. [Sig lamellated.]
sparsipora, Hall, 1881, Bryozoans of the Up Held. Gr. [Sig. having few pores.]
THAMNISCUS multiramus, Hall, 1881, Bryozoans of the Up. Held. Gr. [Sig. many branched]
nanus, Hall, 1881, Bryozoans of the Up. Held. Gr. [Sig. a dwarf.]
THAMNOPORA, Hall, 1881, Bryozoans of the Up. Held. Gr. [Ety. *thamnos*, a bush; *poros*, a pore.]
divaricata, Hall, 1881, Bryozoans of the Up. Held. Gr. [Sig. divaricated.]

TREMATOPORA alternata, Hall, 1881, Bryozoans of the Up. Held. Gr. Sig. alternated.]
americana, S. A. Miller, 1881, Jour. Cin. Soc. Nat. Hist., vol. 4, Burlington Gr. [Ety. proper name.]
annulifer, Whitfield, 1878, Ann. Rep. Geo. Sur. Wis., Hud. Riv. Gr. [Sig. ring bearing.]
annulata, Hall, 1881, Bryozoans of the Up. Held. Gr. [Sig. annulated.]
annulata var. pronaspina, Hall, 1881, Bryozoans of the Up. Held. Gr. [Sig. having the prickles bent forward.]
arborea, Hall, 1881, Bryozoans of the Up. Held. Gr. [Sig. tree like.]
crebipora, Hall, 1879, Desc. New Spec. Foss., Niagara Gr. [Sig. having close pores.]
fragilis, refer to Kinderhook Gr.
granulata, Whitfield. 1878, Ann. Rep. Geo. Sur. Wis., Hud. Riv. Gr. [Sig. granulated.]
macropora, Hall, 1879, Desc. New Spec. Foss., Niagara Gr. [Sig. having long pores.]
rectilinea, Hall, 1881, Bryozoans of the Up. Held. Gr. [Sig. straight lined.]
scutulata, Hall, 1881, Bryozoans of the Up. Held. Gr. [Sig. checkered.]
subimbricata, Hall, 1879, Desc. New Spec. Foss., Niagara Gr. [Sig. somewhat imbricated]
vesiculosa is from the Burlington Gr.

CLASS BRACHIOPODA.

The Family SPIRIFERIDÆ in the 1st edition of this work should be subdivided as follows:
Family ATHYRIDÆ—Athyris, Merista, Meristella, Meristina. Family ATRYPIDÆ—Atrypa, Cœlospira, Zygospira. Family NUCLEOSPIRIDÆ—Nucleospira, Retzia, Trematospira. Family SPIRIFERIDÆ—Spirifera, Spiriferina, etc.

Aegilops, Hall. This name was preoccupied for a genus in Botany, beside the species is supposed to be founded upon the cast of a Lamellibranch.
ANASTROPHIA internascens. Hall, 1879, 28th Reg. Rep., Niagara Gr. [Sig. growing between.]
ATHYRIS americana. A. euzona and A. formosa, are from the Kaskaskia Gr.
chloe is from the Ham. Gr.
claytoni, Hall & Whitfield, 1877, U. S. Geo. Expl., 40th parallel, Waverly Gr. [Ety. proper name.]
clusia, Billings, 1860, Can. Jour., vol. 5, Corniferous limestone. [Sig. pertaining to *Clusium*, the name of a town in Etruria.]
corpulenta, A. hannibalensis, and A. prouti, are from the Kinderhook Gr.
obmaxima, McChesney, 1860, Desc. New Pal. Foss., p. 80, Low. Carb. [Sig. large in front.]
ATHYRIS ohioensis is from the Waverly Gr.
papilloniformis, McChesney, 1867, Trans. Chi. Acad. Sci., vol. 1, Kaskaskia Gr. [Sig. butterfly formed]
pectinifera is from the Keokuk Gr.
persinuata, Meek, 1877, U. S. Geo. Sur., 40th parallel, Carboniferous. [Sig. very sinuate.]
reflexa is from the Warsaw Gr.
trinuclea instead of Terebratula trinuclea.
ATRYPA *inflata* was not defined.
mansoni, Salter, 1852 (Rhynchonella mansoni), Sutherland's Jour., vol. 2, Devonian. [Ety. proper name.]
nustella, Castelnau, 1843, Syst. Sil. Not recognized.
phoca, Salter, 1852 (Rhynchonella phoca), Sutherland's Jour., vol. 2, Devonian. [Sig. a sea dog.]

Calceola. This genus belongs to the Polypi and the family Cyathophyllinæ.

CAMARELLA ortoni is from the Niagara Gr.
primordialis, Whitfield, 1878 (Triplesia primordialis), Ann. Rep. Geo. Sur. Wis., Potsdam Gr. [Sig primordial.]
waldronensis, Miller & Dyer, 1878 (Spirifera (?) waldronensis). Jour. Cin. Soc. Nat. Hist., vol. 1, Niagara Gr. [Ety. proper name.]

CAMAROPHORIA giffordi, Worthen, 1882, Bull. No. 1, Ill. St. Mus. Nat. Hist., Middle Coal Meas. [Ety. proper name.]
occidentalis, S. A. Miller, 1881, Jour. Cin. Soc. Nat. Hist , vol. 4, Burlington Gr. [Sig. western.]

CENTRONELLA allei is from the Kinderhook Gr.
crassicardinalis, Whitfield, 1882, Bull. Ann. Mus. Nat. Hist., No. 3, Warsaw Gr. [Sig. from the thick cardinal edges.]
flora, Winchell, 1879, Notices and Desc. Foss., from the Marshall Gr. [Ety. mythological name.]

CHONETES loganensis, Hall & Whitfield, 1877, U. S. Geo. Expl. 40th parallel, Waverly Gr. [Ety. proper name.]
michiganensis, C. multicosta, C. ornatus, belong to the Marshall or Choteau Group.
minima, Hall, 1876, being preoccupied, is now C. undulatus.
reversus, Whitfield, 1882, Desc. New Spec. Foss., from Ohio, Marcellus shale [Sig. reversed]
striatellus, Dalman, 1827 (Orthis striatella), Kongl. Svenska Ak. Handl., Up. Sil. [Sig. finely channeled.]
undulatus, Hall, 1879, 28th Reg. Rep., Niagara Gr. [Sig. undulated.]
vernulanus var. utahensis, Meek, 1876, Simpson's Rep. on Gt. Basin of Utah, Carboniferous. [Ety. proper name.]

CRANIA carbonaria, Whitfield, 1882, Desc. New Spec. Foss., from Ohio, Coal Meas. [Sig. from the Carboniferous.]
• granulosa, Winchell. 1880. 8th Rep. Geo. Sur. Minn., Trenton Gr. [Sig. full of granules.]
parallella, Ulrich, 1878. Jour. Cin. Soc. Nat. Hist., vol. 1, Hud. Riv. Gr. A doubtful species.
percarinata, Ulrich, 1878, Jour. Cin. Soc. Nat. Hist , vol. 1, Hud. Riv. Gr. A doubtful species.
socialis, Ulrich, 1878. Jour Cin. Soc. Nat Hist., vol. 1, Hud. Riv. Gr. [Sig. social.] Found upon crinoid columns.
spinigera, Hall, 1879, Desc. New Spec. Foss., Niagara Gr. [Sig. spine bearing.]

CRYPTONELLA lens instead of Terebratula lens.
lincklæni instead of Terebratula lincklæni.

CYRTINA acutirostris is from the Choteau Gr.
euphemia, Billings, 1863, Can. Nat. and Geol., vol. 8, Corniferous Gr. [Sig. of good omen.]

DISCINA manhattensis, Meek & Hayden, 1859, Proc. Acad. Nat. Sci., Coal Meas. [Ety. proper name.]
marginalis, Whitfield, 1880, Ann. Rep. Geo. Sur. Wis., Ham. Gr. [Sig. margined.]
meekana, Whitfield, 1882, Desc. New Spec. Foss. from Ohio, Coal Meas. [Ety. proper name.]
microscopica, Shumard, 1861, Am. Jour. Sci. and Arts., vol. 32, Potsdam Gr. [Sig. microscopic.]
newberryi and D. patellaris are from the Waverly or Marshall Gr.
saffordi, Winchell, is from the Marshall Gr.
sublamellosa, Ulrich, 1878, Jour. Cin. Soc. Nat. Hist. Probably the cast of a Trematis.
tenuilineata, Meek & Hayden, 1859, Proc. Acad. Nat. Sci., Coal Meas. [Sig. fine lined.]
tenuistriata, Ulrich, 1878, Jour. Cin. Soc. Nat. Hist. Probably the cast of a Trematis.

GYPIDULA munda, Calvin, 1878, Bull. U. S. Geo. Sur., vol. 4, No. 3, Low. Devonian. [Sig. elegant.]

Hemipronites americanus, refer to Streptorhynchus americanum.

KONINCKIA americana is from the Kaskaskia Gr.

KUTORGINA minutissima, Hall & Whitfield, 1877, U. S. Geo. Expl. 40th parallel, Potsdam Gr. [Sig. very small.]

LEIORHYNCHUS laura, Billings, May, 1860 (Rhynchonella laura), Can. Jour., Ham. Gr. [Ety. proper name.]
multicosta was regarded by Billings as a syn. for L. laura.

LEPTÆNA melita, Hall & Whitfield, 1877, U. S. Geo. Expl. 40th parallel, Potsdam Gr. [Ety. mythological name.]
plicatella, Ulrich, 1879, Jour. Cin. Soc. Nat. Hist., vol. 2, Utica Slate Gr. [Sig. having small folds.]
vicina, Castelnau, 1843, Syst. Sil. Not recognized.]

LINGULA acutangula, Rœmer, 1852, Kreid. von Texas, Silurian. [Sig. acute angled.]
billingsana, Whiteaves, 1878, Am. Jour. Sci. & Arts, 3d ser., vol. 16, St. John's Gr. [Ety. proper name.]
densa was described in 1863, in 16th Reg. Rep.
elderi, Whitfield, 1880, Am. Jour. Sci. & Arts, 3d ser., vol. 19 Trenton Gr. [Ety. proper name.]
gibbosa, Hall, 1879, Desc. New Spec. Foss., Niagara Gr. [Sig. gibbous.]
hurlbuti, Winchell, 1880, Geo Sur.

Minn., 8th Rep., Galena Gr. [Ety. proper name.]
LINGULA lowensis, Owen, 1840, Rep. Min. Lands, Iowa, Wis., and Ill., Galena Gr. [Ety. proper name.]
melie was described in 1863, in 16th Reg. Rep.
membranacea, is from the Marshall Gr.
rectilatera. Restore it.
striata, Emmons, 1856, Am. Geol., Quebec Gr. [Sig. striated.]
LINGULEPIS cuneolus, Whitfield, 1877, Prelim. Rep. Pal., Black Hills, Potsdam Gr. [Sig. a little wedge.]
ella, Hall & Whitfield, 1877, U. S. Expl. 40th parallel, Quebec Gr. [Sig. small.]
mæra, Hall & Whitfield, 1877, U. S. Expl. 40th parallel, Potsdam Gr. [Ety. mythological name.]
minuta, Hall & Whitfield, 1877, U. S. Expl. 40th parallel, Potsdam Gr. [Sig. minute.]
perattenuata, Whitfield, 1877, Prelim. Rep. Pal., Black Hills, Potsdam Gr. [Sig. very attenuated.]
MERISTELLA rectirostra, Hall, 1879, Desc. New Spec. Foss., Niagara Gr. [Sig. straight beak.]
NUCLEOSPIRA rotundata, Whitfield, 1882, Desc. New Spec. Foss. from Ohio, Low. Held. Gr. [Sig. rounded.]
OBOLELLA desquamata, is founded according to Ford on the dorsal valve of O. crassa, and should be referred to the Lower Potsdam Gr.
discoidea, Hall & Whitfield, 1877, U. S. Geo. Expl. 40th parallel, Potsdam Gr. [Sig. discoid.]
ORBICULOIDEA conica, Dwight, 1880, Am. Jour. Sci. and Arts, 3d Ser., vol. 19, Trenton Gr. [Sig. conical.]
ORTHIS *alternans*, Castelnau, 1843. Not recognized.
amœna, Winchell, 1880, Geo. Sur. Minn. 8th Rep., Hud. Riv. Gr. [Sig. pleasant.]
charlottæ, Winchell, 1880, Geo. Sur. Minn. 8th Rep., Hud. Riv. Gr. [Ety. proper name.]
cincinnatensis, n sp Hud. Riv. Gr., Cincinnati, Ohio. Proposed instead of *Orthis costata*, Hall, 1845, Am. Jour. Sci. and Arts. This is a very small species, found associated with other minute fossils on Vine street hill.
circularis, Winchell, 1880, Geo. Sur. Min. 8th Rep., Hud. Riv. Gr. [Sig. circular.]
clytie is from the Trenton Gr.
conradi, Castelnau, 1843. Not recognized.
conradi, Winchell, 1880, Geo. Sur. Minn. 8th Rep., Hud. Riv. Gr. [Ety. proper name.]
cooperensis is from the Warsaw Gr.
costata, Hall, being preoccupied, see Orthis, cincinnatensis.

ORTHIS dalyana, S. A. Miller, 1881, Jour. Cin. Soc. Nat. Hist., vol. 4, Burlington. Gr. [Ety. proper name.]
ella, O. emacerata, and O. jamesi, are from the Hud. Riv. Gr.
eryna was described in 1863 in the 16th Reg. Rep.
flabellum, Sowerby, 1839, in Murch. Sil. Syst., Niagara Gr. [Sig. a small fan.]
flava, and O. occasus are from the Kinderhook Gr.
huronensis, Castelnau, 1843, Syst. Sil. Not recognized.
inferа, Calvin, 1878, Bull. U. S. Geo. Sur. Terr., vol. 4, No. 3, Low. Devonian. [Sig. underground.]
kassubæ, Winchell, 1880, Geo Sur. Minn. 8th Rep., Hud. Riv. Gr. [Ety. proper name.]
media, Winchell, 1880. Geo. Sur. Minn. 8th Rep., Hud. Riv. Gr. [Sig. in the middle.]
minneapolis, Winchell, 1880, Geo. Sur. Minn., 8th Rep., Hud. Riv. Gr. [Ety. proper name.]
pogonipensis, Hall & Whitfield, 1877, U. S. Geo. Expl. 40th parallel, Chazy Gr. [Ety. proper name.]
schohariensis, Castelnau, 1843, Syst. Sil. Not recognized.
scovillei, S. A. Miller, 1882, Jour. Cin. Soc. Nat. Hist., vol. 5, Hud. Riv. Gr. [Ety. proper name.]
sectostriata, Ulrich, syn. for O. ella.
striatella, see Chonetes striatellus.
subelliptica, White & Whitfield, 1862, Proc. Bost. Soc. Nat. Hist., vol. 8, Kinderhook Gr. [Sig. subelliptical.]
subnodosa, Hall, 1879, Desc. New Spec. Foss., Niagara Gr. [Sig. subnodose.]
PENTAMERUS *beaumonti*, Castelnau, 1843, Syst. Sil. Not recognized.
coppingeri, Etheridge, 1878, Quar. Jour. Geo. Soc., vol. 34, Up. Silurian. [Ety. proper name.]
deshayesi, Castelnau, 1843, Syst. Sil. Not recognized.
lenticularis, White & Whitfield, 1862, Proc. Bost. Soc. Nat. Hist., vol. 8, Kinderhook Gr. [Sig. lenticular.]
pesovis. Whitfield, 1882, Desc. New Spec. Foss., from Ohio, Low. Held. Gr. [Sig. sheep foot.]
trisinuatus, McChesney, 1861, Desc. New Pal. Foss., p. 86, Niagara Gr. [Sig. three furrowed.]
PORAMBONITES obscurus, Hall & Whitfield, 1877, U. S. Geo. Expl. 40th parallel, Quebec Gr. [Sig. obscure.]
PRODUCTUS callawayensis, is from the Ham. Gr.
cooperensis, P. curtirostratus, P. dolorosus, and P. parvulus, are from the Choteau or Kinderhook Gr.
coriformis, is from the Kaskaskia Gr.
depressus, and P. fentonensis, are from the Keokuk Gr.

PRODUCTUS *hepar*, Morton, 1836, Am. Jour.
   Sci. & Arts., vol. 29, Coal Meas. Not
   recognized.
   nanus, Meek & Worthen. 1860, Proc.
   Acad. Nat. Sci., Coal Meas. [Sig. a
   dwarf.]
   nevadensis, Meek, 1877, U. S., Geo. Sur.,
   40th parallel, Carboniferous. [Ety.
   proper name.]
   pocillum, Morton, 1836, Am. Jour. Sci.
   & Arts, vol. 29, Coal Meas. Not
   recognized.
   pyxidiformis, DeKoninck, 1847. Monographie du genre Productus, Subcarboniferous. [Sig. box like.]
   subhorridus, Meek, 1877, U. S. Geo. Sur.,
   40th parallel, Carboniferous. [Sig.
   somewhat like *P. horridus.*]
   sulcatus, Castelnau, 1843, Syst. Sil. Not
   recognized.
RENSSELÆRIA condoni instead of R. conradi.
RETZIA evax instead of Rhynchospira evax.
   marcyi, is from the Kaskaskia Gr.
   osagensis, and R. popana, are from the
   Choteau Gr.
   sexplicata, White & Whitfield, 1862,
   Proc. Bost. Soc. Nat. Hist., vol. 8,
   Kinderhook Gr. [Sig. having six
   folds.]
   woosteri, White, 1879, Bull. U. S. Sur.,
   vol. 5, No. 2, Coal Meas. [Ety.
   proper name.]
RHYNCHONELLA ambigua, Calvin, 1878,
   Bull. U. S. Geo. Sur., vol. 4, No. 3,
   Low. Devonian. [Sig. ambiguous.]
   arctirostrata and R. perrostellata are
   from the Kaskaskia Gr.
   boonensis and R. ringens are from the
   Burlington Gr.
   cooperensis. R heteropsis, R. micropleura, R. missouriensis, R. obsolescens, R. persinuata and R. unica
   are from the Choteau or Kinderhook
   Gr.
   cuneata, refer to Rhynchotreta cuneata
   var. americana.
   eatoniformis is a Syn for R. rockymontana.
   emmonsi, Hall & Whitfield, 1877, U. S.
   Geo. Expl. 40th parallel, Devonian.
   [Ety. proper name.]
   hydraulica, Whitfield, 1882, Desc. New
   Spec. Foss. from Ohio, Low. Held.
   Gr. [Sig. hydraulic, from the
   hydraulic limestone]
   internodia, Barris, 1879, Proc. Davenport Acad. Sci., Corniferous Gr.
   [Sig. between nodes.]
   laura, see Lelorhynchus laura.
   mansoni, see Atrypa mansoni.
   medea, Billings, 1860, Can. Jour., vol. 5,
   Corniferous limestone. [Ety. mythological name.]
   neenah, Whitfield, 1880, Ann. Rep. Geo.
   Sur. Wis., Trenton Gr. Ety. proper
   name.]
   neglecta var scobina is from the Niagara
   Gr.

RHYNCHONELLA nucula, Sowerby, 1839
   (Terebratula nucula), Murch. Sil.
   Syst., Up. Sil. [Sig. a small nut.]
   opposita, White & Whitfield, 1862, Proc.
   Bost. Soc. Nat. Hist., vol. 8, Kinderhook Gr. [Sig. opposite.]
   perlamellosa, Whitfield, 1878, Ann. Rep.
   Geo. Sur. Wis., Hud. Riv. Gr. [Sig.
   very lamellose]
   phoca, see Atrypa phoca.
   raricosta, Whitfield, 1882, Desc. New
   Spec. Foss., from Ohio, Up. Held.
   Gr. [Sig. having few costæ.]
   rockymontana, Marcou, 1858 (Terebratula rockymontana), Geo. North
   America, Coal Meas. [Ety. proper
   name.]
   sinuata may be erased.
   stricklandi, Sowerby, 1839 (Terebratula
   sticklandi), Murch. Sil. Syst., Niagara Gr. [Ety. proper name.]
   tuta, S. A. Miller, 1881, Jour. Cin. Soc.
   Nat. Hist., vol. 4, Burlington Gr.
   [Sig. examined.]
   warrenensis is from the Ham. Gr.
   whitii, Hall, 1863, being preoccupied, see
   Rhynchonella whitiana.
   whitiana, n. sp., Niagara Gr., from
   Waldron, Indiana. Proposed instead
   of R. whitii. Hall, 1863, Trans. Alb.
   Inst., vol. 4, p. 216, and also in 28th
   Rep. N.Y. St. Mus. Nat. Hist., p. 164,
   pl. 26, figs. 23-33, and again in the
   11th Ann. Rep. Geol. & Nat. Hist. of
   Indiana, p. 307, pl. 26, figs. 23-33.
Rhynchospira is a syn. for Retzia.
   evax, refer to Retzia evax.
   sinuata was described by Hall, in 1860,
   Can. Nat. & Geo., vol. 5.
RHYNCHOTRETA, Hall, 1879, 28th Reg. Rep.
   [Ety. *rhynchos*, beak; *tretos*, with a
   hole in it.]
   cuneata var. americana. Hall. 1879, 28th
   Reg. Rep., Niagara Gr. [Ety. proper
   name.]
   quadriplicata, S. A. Miller, instead of
   Trematospira quadriplicata.
SPIRIFERA alata, Castelnau, 1843, Syst. Sil.
   Not recognized.
   albapinensis, Hall & Whitfield, 1877, U.
   S. Geo. Expl. 40th parallel, Waverly
   Gr. [Ety. proper name.]
   aldrichi, Etheridge, 1878, Quar. Jour.
   Geo. Soc., vol. 34, Devonian. [Ety.
   proper name.]
   amara. S. cooperensis, S. hannibalensis,
   S. marionensis, S. missouriensis, S.
   semiplicata, and S. vernonensis, are
   from the Choteau or Kinderhook Gr.
   arctica, Haughton, 1857, Jour. Roy. Soc.
   Dub , vol. 1, Devonian. [Sig. arctic.]
   argentaria, Meek, 1877, U. S. Geo. Sur.,
   40th parallel, Devonian. [Sig. pertaining to silver.]
   atwaterana, S. A. Miller, 1878, Proc.
   Davenport Acad. Sci., Ham. Gr. [Ety.
   proper name.] Proposed instead of
   S. pennata, which was preoccupied.

SPIRIFERA bicostata var. petila, Hall, 1879, Desc. New Spec. Foss., Niagara Gr. [Sig. thin.] clara and S. translata are from the Kaskaskia Gr.
conradana, n. sp, Oriskany, Up. Held. and Ham. Gr. Proposed instead of S. fimbriata of Conrad in Jour. Acad. Nat Sci., vol. 8, p. 263, which was preoccupied.
costalis, Castelnau, 1843, Syst. Sil. Not recognized.
crispa var. simplex, Hall, 1879, Desc. New Spec. Foss., Niagara Gr. [Sig. simple.]
fastigata, Morton, 1836. Am Jour. Sci. and Arts, vol. 29, Coal Meas. [Sig. sloping up to a point.]
fastigata, Meek & Worthen. The name was preoccupied by Morton. See S. mortonana.
fimbriata, Morton, 1836, Am. Jour. Sci. and Arts, vol. 29, Coal Meas. [Sig. fringed.]
fimbriata, Conrad. The name was preoccupied. See S. conradana.
fischeri, Castelnau, 1843, Syst. Sil. Not recognized.
hirtus, White & Whitfield, 1862, Proc. Bost. Soc. Nat. Hist., vol. 8, Kinderhook Gr. [Sig. hairy, rough]
huronensis, Castelnau, 1843, Syst. Sil. Not recognized.
inæquivalvis Castelnau, 1843, Syst. Sil. Not recognized.
ligus, Owen, 1852, Rep. Geo. Sur. Wis., Iowa and Minn., Ham. Gr. [Ety. proper name.]
macropleura, Castelnau, 1843. The name was preoccupied.
meeki, is from the Burlington Gr.
mortonana, n. sp., Keokuk Gr. Proposed instead of S. fastigata, of Meek and Worthen, 1870, in Proc. Acad. Nat. Sci., p. 36, and afterward in Geo. Sur. Ill., vol. 6, p. 521, pl. 30, fig. 3, from Crawfordsville, Indiana.
mucronata, Conrad, syn. for S. pennata.
multicostata, Castelnau, 1843, Syst. Sil. Not recognized.
murchisoni, Castelnau, 1843, Syst. Sil. Not recognized.
novamexicana, S. A. Miller, 1881, Jour. Cin. Soc. Nat. Hist., vol. 4, Burlington Gr. [Ety. proper name.]
pennata, Atwater, 1820 (Terebratula pennata) Am. Jour. Sci. & Arts, vol. 2, Ham. Gr. [Sig. winged.]
pennata, Owen. The name was preoccupied, see S. atwaterana.
rostrata, Morton, 1836, Am. Jour. Sci. & Arts, vol. 29, Coal Meas. [Sig. beaked.]
semiplicata, is a syn. for S. cooperensis.
sheppardi, Castelnau, 1843, Syst. Sil. Not recognized.
sowerbyi, Castelnau, 1843, Syst. Sil. Not recognized.

SPIRIFERA strigosa, Meek, 1860, Proc. Acad. Nat. Sci., Devonian. Proposed instead of S. macra, Meek, which was preoccupied. [Sig. thin.]
subvaricosa, is from the Chemung Gr.
taneyensis, Swallow, 1860, Trans. St. Louis Acad. Sci., vol. 1, Kinderhook Gr. [Ety. proper name.]
temeraria, S. A. Miller, 1881, Jour. Cin. Soc. Nat. Hist., vol. 4, Burlington Gr. [Sig. casual.]
troosti, Castelnau, 1843, Syst. Sil. Not recognized.
utahensis, Meek, 1860, syn. for S. norwoodi.
waldronensis, Miller & Dyer, refer to Camarella waldronensis.
waverlyensis, Winchell, 1870, Notices and Desc. Foss., from Marshall Gr. [Ety. proper name.]

SPIRIFERINA spinosa is from the Kaskaskia Gr.

STREPTORHYNCHUS americanum, Whitfield, 1878 (Hemipronites americanus), Ann. Rep. Geo. Sur. Wis., Hud. Riv. Gr. [Ety. proper name.]
cardinale, Whitfield, 1880, Ann. Rep. Geo. Sur. Wis., Hud. Riv. Gr. [Sig. cardinal.]
flabellum, Whitfield, 1882, Desc. New Spec. Foss., from Ohio, Up. Held. Gr. [Sig. a small fan.]
hydraulicum, Whitfield, 1882, Desc. New Spec. Foss., from Ohio, Low. Held. Gr. [Sig. hydraulic, from the hydraulic limestone.]
inflatum, White & Whitfield, 1862, Proc. Bost. Soc. Nat. Hist., vol. 8, Kinderhook Gr. [Sig. inflated.]
occidentalis, Newberry, syn. for Meekella striato-costata.
pyramidalis, Newberry, syn. for Meekella striato-costata.

STRICKLANDINIA multilirata, Whitfield, 1878, Ann. Rep. Geo. Sur. Wis., Niagara Gr. [Sig. many furrowed.]

STROPHALOSIA numularis, is from the Marshall Gr.

STROPHODONTA altidorsata, S. boonensis, S. cymbiformis, S. inflexa, S. kemperi, S. navalis and S. subcymbiformis are from the Ham. Gr.
calvini, n. sp. Upper Helderberg Gr. Proposed instead of S. quadrata, Calvin, 1878, in Bull. U. S. Geo. Sur. Terr., vol. 4, No. 3, which was preoccupied.
feildeni, Etheridge, 1878, Quar. Jour. Geo. Soc., vol 34, Up. Sil. [Ety. proper name]
quadrata, Swallow, 1860, Trans. St. Louis Acad. Sci., vol. 1, Ham. Gr. [Sig. quadrate.]
quadrata, Calvin, 1878, Bull. U. S. Geo. Sur. Terr., vol. 4, No. 3. The name was preoccupied, see S. calvini.
semifasciata, refer to Strophonella semifasciata.

STROPHODONTA striata instead of Strophomena striata.
 vari(bilis, Calvin, 1878, Bull. U. S. Geo. Sur., vol. 4, No. 3, Up. Held. Gr. [Sig. variable.]
STROPHOMENA donneti. Salter, 1852, 'Sutherland's Jour, vol. 2, Devonian. [Ety. proper name.]
 kingi, Whitfield, 1878, Ann. Rep. Geo. Sur. Wis., Hud. Riv. Gr. [Ety. proper name.]
 nemea, Hall & Whitfield, 1877, U. S. Geo. Sur. 40th parallel, Quebec Gr. [Ety. proper name.]
 rhomboidalis, should be credited to Wilckens, 1769 (Conchites rhomboidalis, Nachricht von Seltener Verst. Its range is from the Lower Silurian to the Keokuk Gr.
 striata refer to Strophodonta striata
 wisconsinensis, Whitfield, 1880, Ann. Rep. Geo. Sur. Wis., Hud. Riv. Gr. [Ety. proper name.]
STROPHONELLA, Hall, 1879. 28th Reg. Rep. [Ety. diminutive of Strophos twisted.] Hall refers to this genus Strophodonta ampla, S. cœlata, S. cavumbona, S. leavenworthana, S. punctulifera, S. reversa and S. semifascinta.
SYRINGOTHYRIS halli, is from the Kinderhook Gr., and S. typus is found in the Kinderhook and Keokuk Groups.
TEREBRATULA acuminatissima, Castelnau. Not recognized.
 arcuata, Swallow, see T. shumardana.
 borealis, Castelnau. Not recognized.
 brevilobata, is from the Kaskaskia Gr.
 gracilis, Swallow, see T. swallovana.
 lapillus, Morton, 1836, Am. Jour. Sci. & Arts, vol. 29, Coal Meas. [Sig. a little stone.]
TEREBRATULA lens, refer to Cryptonella lens.
 lincklæni, refer to Cryptonella lincklæni.
 nuciformis, Morton, 1836, Am. Jour. Sci. & Arts, vol. 29, Coal Meas. [Sig. nut formed.]
 nucula, see Rhynchonella nucula.
 pennata, see Spirifera pennata.
 rockymontana, see Rhynchonella rockymontana.
 shumardana, n. sp., Kaskaskia Gr. Proposed instead of T. arcuata, Swallow, 1863, Trans. St. Louis Acad. Sci., vol. 2, p. 83. which was preoccupied.
 stricklandi, see Rhynchonella stricklandi.
 swallovana, n. sp., Kaskaskia Gr. Proposed instead of T. gracilis, Swallow, 1863, Trans. St. Louis Acad. Sci., vol. 2, p. 83, which was preoccupied.
 trinuclea, refer to Athyris trinuclea.
 utah, Hall & Whitfield, 1877, U. S. Geo. Expl. 40th parallel, Waverly Gr. [Ety. proper name.]
 valenciennei, Castelnau, 1843, Syst. Sil. Not recognized.
TREMATIS quincuncialis, Miller & Dyer, 1878, Cont. to Pal. No. 2, Hud. Riv.Gr. [Sig. from the quincunx punctures.]
 Trematospira quadriplicata, S. A. Miller, refer to Rhynchotreta quadriplicata.
 Triplesia primordialis, see Camarella primordialis.
 putillus, Hall, 1879, Desc. New Spec. Foss., syn. for Camarella waldronensis.
ZYGOSPIRA concentrica, Ulrich, 1879, Jour. Cin. Soc. Nat. Hist., vol. 2, Niagara Gr. [Sig. concentric.]
 minima, Hall, 1879, Desc. New Spec. Foss., Niagara Gr. [Sig. the least.]

CLASS PTEROPODA.

CLATHROCŒLIA. Hall, 1879, Pal. N. Y., vol. 5, pt. 2. [Ety. clathro, latticed; koilia, the belly.]
 eborica, Hall, 1879, Pal. N. Y., vol. 5, pt. 2. Ham. Gr. [Ety. proper name.]
COLEOLUS, Hall, 1879, Pal. N. Y., vol. 5, pt. 2. [Ety. koleos. a sheath.]
 acicula instead of Orthoceras acicula.
 aciculatus instead of dentalium aciculatum.
 crenatocinctus, Hall, 1879, Pal. N. Y., vol. 5, pt. 2, Up. Held. Gr. [Sig notched around.]
 gracilis, Hall, 1879, Pal. N. Y., vol. 5, pt. 2, Chemung Gr. [Sig. slender.]
 mohri, Hall, 1879, Pal. N. Y., vol. 5, pt. 2, Up. Held. Gr. [Ety. proper name.]
 spinulus, Hall, 1879, Desc. New Spec. Foss., Niagara Gr. [Sig. a little thorn.]
 tenuicinctus instead of Coleoprion tenuicinctum.
COLEOPRION, tenue, Hall, 1879, Pal. of N. Y., vol. 5, pt. 2, Ham. Gr. [Sig. thin.]
CONULARIA continens var. rudis, Hall, 1879, Pal. N. Y., vol. 5, pt. 2, Ham Gr. [Sig. rude]
 crustula, White, 1880, 12th Rep. U. S. Geo Sur. Terr., Coal Meas. [Sig. a little crust.]
 formosa, Miller & Dyer, 1878, Jour. Cin. Soc. Nat. Hist., vol. 1, Hud. Riv. Gr. [Sig. beautiful.]
 indentata, Conrad, 1854, Proc. Acad. Nat. Sci., vol. 7, Trenton Gr. [Sig. indented.]
 infrequens. Hall, 1879, Desc. New Spec. Foss., Niagara Gr. [Sig. infrequent.]
 multicostata, Meek & Worthen, 1865. Proc. Acad. Nat. Sci., Waverly Gr. [Sig. many ribbed.]
 osagensis, is from the Kaskaskia Gr.
 subcarbonaria, was first described in 1865, in Proc. Acad. Nat. Sci.

CONULARIA vernouliana, Emmons. 1846, Am. Quar. Jour. Agr. & Sci., vol. 4, Low. Carb. [Ety. proper name.]
whitei, Meek & Worthen, 1865, Proc. Acad. Nat. Sci., Waverly Gr. [Ety. proper name.]
HYOLITHES, Eichwald, 1840, Sil. Schicht. Syst. in Ehstl.
aculeatus instead of Theca aculeata, and it is from the Kinderhook Gr.
baconi, Whitfield, 1878, Ann. Rep. Geo. Sur. Wis., Trenton Gr. [Ety. proper name.]
gregarius instead of Theca gregaria.
parviusculus instead of Theca parviuscula.
singulus, Hall, 1879, Pal. N.Y., vol. 5, pt. 2, Ham. Gr. [Sig. single.]
triliratus, Hall, 1879, Pal. N.Y., vol. 5, pt. 2, Ham. Gr. [Sig. three furrowed.]
STYLIOLA, Lesueur, 1826. [Ety. from *stylos*, a pillar.]
STYLIOLA fissurella, Hall, 1843, instead of Tentaculites fissurellus.
fissurella var. intermittens, Hall, 1879, Pal. N.Y., vol. 5, pt. 2, Gennessee Slate. [Sig. intermitting.]
fissurella var. obsolescens, Hall, 1879, Pal. N.Y., vol. 5, pt. 2, Ham. Gr. [Sig. obsolete.]
fissurella var. strigata, Hall, 1879, Pal. N. Y., vol. 5, pt. 2, Marcellus shale. [Sig. fluted.]
obtusa, Hall. 1879, Pal. N. Y., vol. 5, pt. 2, Ham. Gr. [Sig. obtuse.]
TENTACULITES gracilistriatus, Hall, 1879. Pal N. Y., vol. 5 pt. 2, Marcellus shale. [Sig. slender furrowed.]
neglectus, Nicholson & Hinde, 1874, Can. Jour., Clinton Gr. [Sig. neglected.]
ornatus is not an American species.
THECA. The species under this genus may be referred to Hyolithes.

CLASS GASTEROPODA.

Aclis, Loven, 1846. Index Mollusc. litora Scandin. occid. habit. Refer A. minuta, A. robusta, and A. swallovana to the genus Aclisina.
stevensoni, see Aclisina stevensoni.
ACLISINA DeKoninck, 1881, Faune du Calcaire Carbonifere de la Belgique Ann. d. Mus. Roy. d'Hist. Nat., tome 6. [Ety. from the genus Aclis.]
stevensoni, White, 1882 (Aclis stevensoni). Rep. Invert. Foss. New Mex., Coal Meas. [Ety. proper name.]
ANTHRACOPUPA, Whitfield, 1881, Am. Jour. Sci. & Arts, 3d ser., vol. 21. [Ety. *anthrax*, coal; *Pupa*, a genus.]
ohioensis, Whitfield, 1881, Am. Jour. Sci. & Arts, 3d ser., vol 21, Coal Meas. [Ety. proper name.]
BELLEROPHON alternodosus, Whitfield, 1882, Desc. New Spec. Foss., from Ohio. Kaskaskia Gr. [Sig. having alternate nodes.]
antiquatus, Whitfield, 1878, Ann. Rep. Geo. Sur. Wis., Potsdam Gr. [Sig. antiquated.]
bilabiatus, White & Whitfield, 1862, Proc. Bost. Soc. Nat. Hist., vol. 8, Kinderhook Gr. [Sig. two lipped.]
explanatus, Hall, 1879, Pal. N. Y., vol. 5, pt. 2, Chemung Gr. [Sig. spread out.]
gibsoni, White, 1882. 11th Rep. Geol. & Nat. Hist., Indiana, St. Louis Gr. [Ety. proper name.]
helena. Hall, 1879, Pal. N. Y., vol. 5, pt. 2, Ham. Gr. [Ety. mythological name.]
hyalina, Hall, 1879, Pal. N. Y., vol. 5, pt. 2, Up. Held. Gr. [Sig. of glass.]
inspeciosus, White, 1882, Rep. Invert. Foss., New Mex., Coal Meas. [Sig. not handsome.]
BELLEROPHON lincolatus is from the Kinderhook Gr.
morrowensis, Miller & Dyer, 1878, Cont. to Pal. No. 2, Hud. Riv. Gr. [Ety. proper name.]
nactus, Hall, 1879, Pal. N. Y., vol. 5, pt. 2, Chemung Gr. [Sig. stumbled on.]
natator instead of Phragmostoma natator.
pelops var. exponens, Hall, 1879. Pal. N. Y., vol. 5, pt. 2, Up. Held. Gr. [Sig. exposed.]
perelegans, White & Whitfield, 1862, Proc. Bost. Soc. Nat. Hist., vol. 8, Kinderhook Gr. [Sig. very elegant.]
perforatus, Winchell & Marcy, 1866, syn. for Bucania chicagoensis.
punctifrons, refer to Bucania punctifrons.
repertus, Hall, 1879, Pal. N.Y., vol. 5, pt. 2, Ham. Gr. [Sig discovered.]
rotalinea, Hall, 1879, Pal. N.Y., vol. 5, pt. 2, Ham Gr. [Sig. round lined.]
rugosus, Emmons, 1856, Am Geol., Utica Slate Gr. [Sig. rugose.]
subpapillosus, White, 1879, Bull. U. S. Geo. Sur. Ter., vol. 5, Carboniferous. [Sig. somewhat papillated.]
vinculatus, White and Whitfield, 1862, Proc. Bost. Soc. Nat. Hist., vol. 8, Kinderhook Gr. [Sig. bound.]
wisconsinensis, Whitfield, 1878, Ann. Rep. Geo. Sur. Wis., Trenton Gr. [Ety. proper name.]
BUCANIA buelli, Whitfield, 1878, Ann Rep. Geo. Sur. Wis., Trenton Gr. [Ety. proper name.]
devonica is from the Upper Held. Gr.

BUCANIA punctifrons should be referred to Emmons, 1842 (Bellerophon punctifrons), Geo. Rep. N. Y., Black River and Trenton Gr.

BULIMORPHA, Whitfield, 1882, Bull. Am. Mus. Nat. Hist., No. 3.
bullmiformis instead of Polyphemopsis bulimiformis.
canaliculata instead of Polyphemopsis canaliculata.
elongata instead of Polyphemopsis elongata.

CALLONEMA, Hall, 1879, Pal. N.Y., vol. 5, pt. 2. [Ety. *kallos*, beautiful ; *nema*, thread.]
bellatulum instead of Loxonema bellatulum, and from the Up. Held. Gr. [Sig. pretty.]
imitator instead of Pleurotomaria imitator. and from Ham. Gr.
lichas instead of Platystoma lichas.

CHEMNITZIA tenuilineata, is from the Choteau Gr.

CHITON occidentalis. Hildreth. 1837. Am. Jour. Sci. & Arts, vol. 31, Coal Meas. [Sig. western.]

CLISOSPIRA occidentalis, Whitfield, 1878, Ann. Rep. Geo. Sur. Wis., Trenton Gr. [Sig. western.]

CYCLONEMA cincinnatense, S. A. Miller, 1882. Jour. Cin. Soc. Nat. Hist., vol. 5, Utica Slate Gr. [Ety. proper name.]
doris instead of Pleurotomaria doris.
leavenworthanum instead of Pleurotomaria leavenworthana.
subangulatum instead of Pleurotomaria subangulata.

CYCLORA depressa, Ulrich, 1879, Jour. Cin. Soc. Nat. Hist., vol. 2, Hud. Riv. Gr. [Sig. depressed.]
pulcella, S. A. Miller, 1882, Jour. Cin. Soc. Nat. Hist., vol. 5, Hud. Riv. Gr. [Sig. beautiful little.]

CYRTOLITES magnus, S. A. Miller, 1878, Jour. Cin. Soc. Nat. Hist., vol. 1, Hud. Riv. Gr. [Sig. large.]
nitidulus, Ulrich, 1878, Jour. Cin. Soc. Nat. Hist., vol. 2, Utica Slate Gr. [Sig. neat.]
sinuatus, Hall & Whitfield, 1877, U. S. Geo. Expl. 40th parallel, Quebec Gr [Sig. sinuated.]

CYRTONELLA, Hall, 1879, Pal. N. Y., vol. 5, pt. 2. A subgenus under Cyrtolites to include C. mitella and C. pileolus.

DENTALIUM *aciculatum*, refer to Coleolus · aciculatus.
grandaevum is from the Marshall Gr.
martini, Whitfield, 1882, Desc. New Spec. Foss., from Ohio, Up. Held. Gr. [Ety. proper name.]

Discolites, Emmons, syn. for Cyclora.
minutus, see Cyclora minuta.

EUNEMA priscum, is from the Calciferous Gr.

EUOMPHALUS ammon, White & Whitfield, 1862, Proc. Bost. Soc. Nat. Hist., vol. 8, Kinderhook Gr. [Ety. mythological name.]
boonensis, is from the Burlington Gr.
decewi, refer to Pleuronotus decewi.
gyroceras Rœmer, 1852, Kreid. von Tex., Silurian. [Sig. a circular horn.]
becale var. corpulens, Hall, 1879, Pal. N. Y., vol. 5, pl. 27, Chemung Gr. [Sig. corpulent.]
luxus, White, 1875, Expl. W. 100th meridian, Low. Carb. [Sig. dislocated.]
macrolineatus, Whitfield, 1878, Ann. Rep. Geo. Sur. Wis., Niagara Gr. [Sig. long lined.]
minutissimus, Castelnau, 1843, Syst. Sil. Not recognized.
ophirensis, Hall & Whitfield, 1877, U. S. Geo. Expl. 40th parallel, Waverly Gr. [Ety. proper name.]
polygyratus, Rœmer, 1852, Kreid. von Texas, Silurian. [Sig. many coiled.]
sanctisabæ, Rœmer, Kreid. von Texas, Silurian. [Ety. proper name.]
spirorbis, is from the Kinderhook Gr.
strongi, Whitfield, 1878, Ann. Rep. Geo. Sur. Wis., Magnesian Gr. [Ety. proper name.]
trochiscus, refer to Raphistoma trochiscus.
utahensis, refer to Straparollus utahensis.
verneuili, Castelnau, 1843, Syst. Sil. Not recognized.

EOTROCHUS, Whitfield, 1882, Bull. Am. Mus. Nat. Hist., No. 3. [Ety. *eos*, dawn ; *Trochus*, a genus of shells.]
concavus, Hall, instead of Pleurotomaria concava.

FUSISPIRA compacta, Hall & Whitfield, 1877, U. S. Expl. 40th parallel, Quebec Gr. [Sig. compact.]

HELICOTOMA naresi, Etheridge, 1878, Quar. Jour. Geo. Soc., vol. 34, Up. Sil. [Ety. proper name.]
serotina, is from the Corniferous limestone.

HOLOPEA conica, is from the Marshall Gr., and H. eriensis from the Corniferous limestone.
magniventra, Whitfield, 1878, Ann. Rep. Geo. Sur. Wis., Niagara Gr. [Sig. large bellied]
newtonensis, Whitfield, 1882, Desc. New Spec. Foss., from Ohio, Kaskaskia Gr. [Ety. proper name.]
reversa, is from the Up. Silurian.
sweeti, Whitfield, 1880, Ann. Rep. Geo. Sur. Wis , Potsdam Gr. [Ety. proper name.]

HOLOPELLA mira, is from the Marshall Gr.
Inachus percetus, see Euomphalus pervetus.

ISONEMA *bellatulum*, refer to Callonema bellatulum.
lichas, refer to Callonema lichas.

LEPETOPSIS, Whitfield, 1882, Bull. Am. Mus. Nat. Hist., No. 3.

levettei, White, 1882 (Patella levettei), 11th Rep. Geo. of Indiana, Warsaw Gr. [Ety. proper name.]
LOXONEMA attenuatum var. semicostatum, Meek, 1871, Proc. Acad. Nat. Sci., Coal Meas. [Sig. half ribbed.]
breviculum, Hall, 1879, Pal. N. Y., vol. 5, pt. 2, Ham. Gr. [Sig. somewhat short.]
cotterana, Billings, 1861, Can. Jour., vol 6, Corniferous limestone. [Ety. proper name.]
kanei, Meek, 1865, Am. Jour. Sci. & Arts, 2d ser., vol. 40, Low. Held. Gr. [Ety. proper name.]
læviusculum, Hall, 1879, Pal. N. Y., vol. 5, pt. 2, Ham. Gr. [Sig. somewhat smooth.]
laxum, Hall, 1879, Pal. N. Y., vol. 5, pt. 2, Chemung Gr. [Sig. wide.]
magnum, Whitfield, 1878, Ann. Rep. Geo. Sur. Wis., Niagara Gr. [Sig. large.]
macellntochi, Haughton, 1857, Jour. Roy. Dub. Soc., vol. 1, Devonian. [Ety. proper name.]
parvulum, Whitfield, 1882, Desc. New Spec. Foss., from Ohio, Up. Held. Gr. [Sig. small.]
plicatum, Whitfield, 1882, Desc. New Spec. Foss., from Ohio, Coal Meas. [Sig. plicated.]
postrenum, Hall, 1879, Pal. N. Y., vol. 5, pt. 2, Chemung Gr. [Sig. the last.]
rectistriatum, Hall, 1879, Pal. N. Y., vol 5, pt. 2, Ham. Gr. [Sig. having straight furrows.]
rossi, Haughton, 1857, Jour. Roy. Soc. Dub., vol. 1, Devonian. [Ety. proper name.]
sicula, Hall 1879, Pal. N. Y., vol. 5, pt. 2, Up. Held. Gr. [Sig. a little dagger.]
turriforme is from the Kinderhook Gr.
vincta, refer to Murchisonia vincta.
MACLUREA cuneata, Whitfield, 1878, Ann. Rep. Geo. Sur. Wis., Trenton Gr. [Sig. wedge formed.]
minima, Hall & Whitfield, 1877, U. S. Geo. Expl., 40th parallel, Chazy Gr. [Sig. the least.]
subrotunda, Whitfield, 1878, Ann. Rep. Geo. Sur. Wis., Trenton Gr. [Sig. somewhat rotund.]
MACROCHEILUS. This generic name was preoccupied by Hope, in 1838, for a genus of Coleopterous insects. Bayle has proposed Macrochilina, to which all the species may be referred.
cooperense, is from the Kaskaskia Gr.
pinquis, is from the Marshall or Kinderhook Gr.
priscum, Whitfield, 1882, Desc. New Spec. Foss. from Ohio, Up. Held. Gr. [Sig. ancient.]
subcorpulentum, Whitfield, 1882, Desc. New Spec. Foss. from Ohio, Kaskaskia Gr. [Sig. somewhat corpulent.]
MACROCHILINA, Bayle, 1880, Journal de Conchyliologie, 3me. ser., t. 19. Proposed instead of Macrocheilus of Phillips, which was preoccupied by Hope. [Ety. diminutive of Macrocheilus.]
METOPTOMA baraduensis, Whitfield, 1878, Ann. Rep. Geo. Sur. Wis., Low Magnesian Gr [Ety. proper name.]
cornutiformis, Walcott, 1879, 33d Reg. Rep., Calciferous Gr. [Sig. horn shaped.]
perovalis, Whitfield, 1878, Ann. Rep. Geo. Sur. Wis., Trenton Gr. [Sig. oval.]
recurvus, Whitfield, 1878, Ann. Rep. Geo. Sur. Wis., Low. Mag. Gr. [Sig. recurved.]
retrorsus, Whitfield, 1880, Ann. Rep. Geo. Sur Wis., Potsdam Gr. Sig. turned back.]
similis, Whitfield, 1878, Ann. Rep. Geo. Sur. Wis., Low Mag. Gr. [Sig. similar]
MICROCERAS minutissimum, Ulrich, 1879, Jour. Cin. Soc. Nat. Hist., vol. 2, Hud. Riv. Gr. [Sig. very small.] I think this is identical with M. inornatum.
MURCHISONIA aciculata, and M. arisaigensis, are from the Upper Silurian.
chamberlini, Whitfield, 1878, Ann. Rep. Geo. Sur. Wis., Niagara Gr. [Ety. proper name.
copei, White, 1882, Rep. Invert. Foss. New Mex., Coal Meas. [Ety. proper name.]
elegantula, refer to Pleurotomaria elegantula.
intercedens, Hall, 1879, Pal. N.Y., vol. 5, pt. 2. Up. Held. Gr. [Sig. intervening.]
latifasciata, Etheridge, 1878, Quar. Jour. Geo. Soc., vol. 34, Up. Sil. [Sig. wide banded]
limitaris is from the Kinderhook Gr.
marcouana, Geinitz, 1866, Carb. und Dyas in Neb., Coal Meas. [Ety. proper name.]
multigruma, S. A. Miller, 1878, Jour. Cin. Soc. Nat. Hist., vol. 1, Hud. Riv. Gr. [Sig. much heaped up]
neglecta, M. quadricincta and M. Shumardana, are from the Marshall or Kinderhook Gr.
obsoleta, Meek, 1871, Proc. Acad. Nat. Sci., Coal Meas. [Sig. obsolete.]
prolixa, White & Whitfield, 1862, Proc. Bost Soc. Nat Hist , vol. 8, Kinderhook Gr. [Sig. prolix.]
terebra, White, 1879, Bull. U. S. Geo. Sur. Terr., vol. 5, No. 2, Carboniferous. [Sig. an auger.]
vincta instead of Loxonema vincta.
worthenana, S. A. Miller, 1882, Jour.

GASTEROPODA.

Cin. Soc. Nat. Hist., vol. 5, Niagara Gr. [Ety. proper name.]
NATICOPSIS aequistriata was first published in 1871, in Proc. Acad. Nat. Sci.
compacta, Hall, 1879, Pal. N. Y., vol. 5, pt. 2, pl. 29, Up. Held. Gr. [Sig. ascertained.]
depressa is from the Marshall or Kinderhook Gr.
littonana var. genevievensis, Meek & Worthen, 1866, Proc. Acad. Nat. Sci., Kaskaskia Gr. [Ety. proper name.]
monilifera, White, 1880, 12th Rep. U. S. Geo. Sur. Terr., Up. Coal Meas. [Sig. bead bearing.]
ortoni, Whitfield, 1882, Desc. New Spec. Foss. from Ohio, Coal Meas. [Ety. proper name.]
ziczac, Whitfield, 1882, Desc. New Spec. Foss. from Ohio, Kaskaskia Gr. [Sig. having short, sharp turns.]
OPHILETA complanata var. nana, Meek, 1870, Hayden's U. S. Geo. Sur. Terr., Calciferous Gr. [Sig. a dwarf.]
ORMATHICHNUS, S. A. Miller, 1880, Jour. Cin. Soc. Nat. Hist., vol. 2. [Ety. *ormathos*, a string of beads: *ichnos*, a track.] Supposed to be the trail of a Gasteropod.
moniliformis, S. A. Miller. 1880, Jour. Cin. Soc. Nat. Hist., vol 2, Utica Slate Gr. [Sig. like a necklace.]
PALÆACHLEA irvingi, Whitfield, 1878, Ann. Rep. Geo. Sur. Wis., Potsdam Gr. [Ety. proper name.]
PALÆOTROCHUS, Hall, 1879, Pal. N. Y., vol. 5, pt. 2. [Ety. *palaios*, ancient: *Trochus*, a genus.]
kearneyi instead of Pleurotomaria kearneyi.
PATELLA Linnaeus, 1738, Syst. Nat. 10th Ed. [Ety. *patella*, a dish.]
leveillei, White, 1882, 11th Rep. Geol. & Nat. Hist. Indiana, Warsaw Gr., refer to Lepetopsis leveillei.
PHANEROTINUS paradoxus is from the Marshall or Kinderhook Gr.
PHRAGMOSTOMA cumulus (a little cradle) instead of P. cumulus
Pileopsis consobrius, P. naticoides, P. rotundatus, and *P. spiralis*. Castelnau, 1843. Syst. Sil. Not recognized.
PLATYCERAS bivolve, White & Whitfield, 1862, Proc. Bost. Soc. Nat. Hist., vol. 8, Kinderhook Gr. [Sig. with two rolls.]
chesterense, Meek & Worthen, 1866, Proc. Acad. Nat. Sci., Kaskaskia Gr. [Ety. proper name.]
cornuforme and P. vomerium are from the Marshall or Kinderhook Gr.
berneri, Winchell, 1870, Notices and Desc. Foss. from Marshall Gr. [Ety. proper name.]
lævigatum, Meek & Worthen, 1866, Proc. Acad. Nat. Sci., Kaskaskia Gr. [Sig. smoothed.]
minutissimum, Walcott, 1879, 33d Reg.

Rep., Calciferous Gr. [Sig. the smallest.]
PLATYCERAS naticoides, Etheridge, 1878, Quar. Jour. Geo. Soc., vol. 34, Up. Sil. [Sig. like a *Natica*.]
paralium, White & Whitfield, 1862, Proc. Bost. Soc. Nat. Hist., vol. 8, Kinderhook Gr. [Sig. that grows by the seaside.]
squalodens, Whitfield, 1882, Desc. New Spec. Foss. from Ohio, Up. Held. Gr. [Sig. a kind of fish tooth.]
subsinuosum, Worthen, 1882, Bull. No. 1, Ill. St. Mus. Nat. Hist., Low. Held. Gr. [Sig. somewhat sinuous.] Proposed instead of P. subundatum, M. & W.
tribulosum, White, 1880, 12th Rep. U. S. Geo. Sur. Terr., Burlington Gr. [Sig. thorny.]
PLATYSTOMA grayvillense, Worthen, 1882, Bull. No. 1, Ill. St. Mus. Nat. Hist., Coal Meas. Proposed instead of P. tumidum, M. & W., which was preoccupied.
PLEURONOTUS, Hall, 1879, Pal. N. Y., vol. 5, pt. 2. [Ety. *pleura*, side; *notos*, back.]
decewi, instead of Euomphalus decewi.
PLEUROTOMARIA adjutor, Hall, 1879, Pal. N. Y., vol. 5, pt. 2, Up. Held. Gr. [Sig. a helper.]
arata var. clausa, Hall, 1879, Pal. N. Y., vol. 5, pt. 2, Up. Held. Gr. [Sig. an inclosed place.]
broadheadi, White, 1880, 12th Rep. U. S. Geo. Sur. Terr., Coal Meas. [Ety. proper name.]
coniformis, Worthen, 1882, Bull. No. 1, Ill. St. Mus. Nat. Hist., Coal Meas. [Sig. cone shaped.] Proposed instead of P. conoides, M. & W.
doris, refer to Cyclonema doris.
concava, of Hall, refer to Eotrochus concavus.
elegantula, instead of Murchisonia elegantula.
gurleyi, Meek, 1871, Proc. Acad. Nat. Sci., Coal. Meas. [Ety. proper name.]
imitator, refer to Callonema imitator.
isys var. tenuispira, Hall, 1879, Pal. N. Y., vol. 5, pt. 2. Ham. Gr. [Sig. having a slender spire.]
kearneyi, in the 14th Reg. Rep., refer to Palæotrochus kearneyi.
laphami, Whitfield, 1878, Ann. Rep. Geo. Sur. Wis., Niagara Gr. [Ety. proper name.]
leavenworthana, refer to Cyclonema leavenworthanum.
moeta, Meek and Worthen, 1865, Proc. Acad. Nat. Sci., Keokuk Gr. [Sig. pyramidal.]
mississippiensis, White and Whitfield, 1862 Proc. Bost. Soc. Nat. Hist., vol. 8, Kinderhook Gr. [Ety. proper name.]
mitigata, P. quinquesulcata, P. rota, P.

tectoria, and P. vadosa, are from the Marshall or Kinderhook Gr.
PLEUROTOMARIA muralis is from the Trenton Gr.
  newportensis, White, 1880, 12th Rep. U. S. Geo. Sur. Terr., Coal Meas. [Ety. proper name.]
  nitela, Hall, 1879, Pal. N. Y., vol. 5, pt. 2. Up. Held. Gr. [Sig. brightness]
  perizomata, White, 1882, Rep. Invert. Foss , New Mex., Coal Meas. [Sig. girdled.]
  quadricarinata instead of P. quadricostata.
  quadrilix, Hall, 1879, Pal. N.Y., vol. 5, pt. 2, Up. Held. Gr. [Sig. having four whorls (?).]
  racinensis, Whitfield, 1878, Ann. Rep. Geo. Sur. Wis., Niagara Gr. [Ety. proper name]
  riddelli instead of P. riddlei.
  rugulata, Hall, 1860, 13th Reg. Rep., Ham. Gr [Sig. having wide furrows.]
  subangulata, refer to Cyclonema subangulatum.
  taggarti, Meek, 1874, 7th Rep. Hayden's U. S. Geo. Sur. Terr., Coal Meas. [Ety. proper name.]
  tenuimarginata, refer to Eotrochus concavus.
  tumida, refer to Platystoma grayvillense.
POLYPHEMOPSIS buliniformis, refer to Bulimorpha buliniformis.
  canaliculata, refer to Bulimorpha canaliculata.
  elongata, refer to Bulimorpha elongata.
  melanoides, Whitfield, 1882, Desc. New Spec. Foss., from Ohio, Kaskaskia Gr. [Sig. like a Melania.]
PORCELLIA crassinoda, White & Whitfield, 1862 Proc. Bost. Soc. Nat. Hist., vol. 8, Kinderhook Gr. [Sig. thick noded.]
  rectinoda is from the Marshall or Kinderhook Gr.
  rotatoria, refer to Goniatites plebeiformis.
PUPA bigsbyi, Dawson, 1880, Am. Jour. Sci. & Arts, 3d ser., vol. 20, Coal Meas. [Ety. proper name.]
RAPHISTOMA acutum, Hall & Whitfield, 1877, U. S. Geo. Expl., 40th parallel, Chazy Gr. [Sig. acute.]
  niagarense, Whitfield, 1878, Ann. Rep. Geo. Sur. Wis., Niagara Gr. [Ety. proper name.]
  trochiscus, Meek, 1870 (Euomphalus trochiscus), Proc. Acad. Nat. Sci. Calciferous or Trenton Gr. [Sig. a small round ball.]
ROTELLA verruculifera, White, 1882, Rep. Invert. Foss., New Mex., Coal Meas. [Sig. bearing little eminences.]

SCÆVOGYRA, Whitfield, 1878, Ann. Rep. Geo. Sur. Wis. [Ety. scœrus toward the left; gyrus, a circle.]
  elevata, Whitfield, 1878, Ann. Rep. Geo. Sur. Wis., Low. Mag. Gr. [Sig. elevated.]
  obliqua, Whitfield, 1878, Ann. Rep. Geo. Sur. Wis., Low. Mag. Gr. [Sig. oblique.]
  swezeyi, Whitfield, 1878, Ann. Rep. Geo. Sur. Wis., Low. Mag. Gr. [Ety. proper name]
SOLENISCUS brevis, White, 1882, Rep. Invert. Foss., New Mex., Coal Meas. [Sig. short.]
  planus, White, 1882, Rep. Invert. Foss., New Mex., Coal Meas. [Sig. flat.]
STRAPAROLLUS barrisi, and S. macromphalus from the Marshall or Kinderhook Gr.
  planispira, S. quadrivolvis, S. spergenensis, and S. spergenensis var. planorbiformis, refer to Euomphalus.
  utahensis, Hall & Whitfield, 1877. U. S. Geo. Expl., 40th parallel, Waverly Gr. [Ety. proper name.]
STROPHITES, Dawson, 1880, Am. Jour. Sci. and Arts. [Ety. from the genus Strophia.]
  grandævus, Dawson, 1880, Am. Jour. Sci. and Arts, 3d ser., vol. 20, Devonian. [Sig. old aged.]
STROPHOSTYLUS cycloastomus var. disjunctus, Hall, 1879. 28th Reg. Rep., Niagara Gr. [Sig. disjoined.]
  obliquus, S. ovatus, and S subglobosus, are from the Corniferous limestone.
SUBULITES gracilis, S. A. Miller, 1882, Jour. Cin. Soc. Nat. Hist., vol. 5, Niagara Gr. [Sig. slender.]
  Tremanotus is a syn. for Bucania. The supposed openings on the cast represent the spines upon the back of the anterior part of the last whorl of the shell, and the fossil is a true Bucania.
  alpheus, syn. for Bucania chicagoensis.
TROCHITA carbonaria from the Kaskaskia Gr.
TROCHONEMA beloitense, Whitfield, 1878, Ann. Rep. Geo. Sur. Wis., Trenton Gr. [Ety. proper name.]
  beachi, Whitfield, 1878, Ann. Rep. Geo. Sur. Wis., Trenton Gr. [Ety. proper name.]
TURBO huronensis, Castelnau, 1843. Not recognized.
  shumardi, de Verneuil, 1846, Bulletin de la Soc. Geol. de France, Up. Held. Gr. [Ety. proper name.]
TURRITELLA schohariensis, Castelnau, 1843. Not recognized.

CLASS CEPHALOPODA.

ACTINOCERAS BEAUDANTI, Cast'enau, 1843, Systeme Silurien, Niagara Gr. [Ety. proper name.]
  beaumonti, Castelnau, 1843, Systeme Silurien, Niagara Gr. [Ety. proper name.]
  blainvillei, Castelnau, 1843, Systeme Silurien, Hud. Riv. Gr.(?) [Ety. proper name.]
  cordieri, Castelnau, 1843, Systeme Silurien, Hud. Riv. Gr.(?) [Ety. proper name.]
  deshayesi, Castelnau, 1843, Systeme Silurien, Hud. Riv. Gr.(?) [Ety. proper name.]
  dufresnoyi, Castelnau, 1843, Systeme Silurien, Niagara Gr. [Ety. proper name.
Ammonites bellicosus. Morton. 1836, Am. Jour. Sci. and Arts, vol. 29, Coal Meas. Not recognized.
  colubrellus, see Goniatites colubrellus.
  hildrethi, see Goniatites hildrethi.
BACTRITES, Sandberger, 1841, Leonh. u. Bronn's Jahrb. [Ety. *baktron*, a staff.]
  clavus, Hall, 1879, Pal. N.Y., vol. 5, Ham. Gr. [Sig. a spike.]
Beatricea should be referred to the sponges.
COLPOCERAS clarkei, Wetherby, 1881, Jour. Cin Soc. Nat. Hist., vol 4, Trenton Gr. [Ety. proper name.]
CYRTOCERAS *absens*, refer to Gomphoceras absens.
  æmulum, Hall, 1879, Pal. N.Y., vol. 5, Up. Held. Gr [Sig. emulous.]
  alternatum, Hall, 1879. Pal. N.Y., vol. 5, Marcellus Shale. [Sig. changed by turns.] Proposed instead of C. undulatum of Hall.
  ammon, Billings, 1861, Can. Jour., vol. 6, Corniferous limestone. [Ety. mythological name.]
  amœnum, S. A. Miller, 1878. Jour. Cin. Soc. Nat. Hist., vol. 1, Hud. Riv. Gr. [Sig. welcome.]
  annulatum, Hall, see C. subannulatum.
  arcuatum, Hall, see C. subarcuatum.
  aristides is a proper name instead of mythological.
  bannisteri, Winchell & Marcy, see Trochoceras bannisteri.
  belua, Billings, 1861, Can. Jour, vol. 6, Corniferous Gr. [Ety. proper name.]
  beta is Gomphocer s, beta.
  cancellatum, Hall, see C. subcancellatum.
  cessator. Hall & Whitfield, 1877, U. S. Expl. Exped. 40th parallel, Coal Meas. [Sig a loiterer.]
  citum, Hall, 1879, Pal. N.Y., vol. 5, Up. Held. Gr. [Sig. speedy.]
  clavatum, refer to Gomphoceras clavatum.
  conoidale, Wetherby, 1881, Jour. Cin. Soc. Nat Hist., vol. 4, Hud. Riv. Gr. [Sig. conoidal.]

CYRTOCERAS *conradi*, Hall, is Gomphoceras conr di.
  corniculum, Hall, see C. tenuistriatum.
  cretaceum, Whitfield, 1882, Desc. New Spec. Foss. from Ohio, Up. Held. Gr. [Sig. chalky.]
  densum, Hall, 1879, Pal. N.Y., vol. 5, Ham. Gr. [Sig. thick.]
  formosum, Hall, 1879, Pal. N.Y., vol. 5, Ham. Gr. [Sig. beautiful.]
  gibbosum, Hall, is a syn. for Gomphoceras oviforme.
  hallanum instead of C. lamellosum, Hall.
  hector, Hall, 1879, Pal. N.Y., vol. 5, Up. Chemung Gr. [Ety. proper name.]
  infundibulum, Whitfield, 1880, Ann. Rep. Geo. Sur. Wis., Niagara Gr. [Sig. a funnel.]
  irregulare, Wetherby, 1881, Jour. Cin. Soc. Nat. Hist., vol. 4, Hud. Riv. Gr. [Sig. irregular.]
  jason, refer to Gyroceras jason.
  markœi, Castelnau. 1843, Systeme Silurien, Trenton Gr. (?) [Ety. proper name.]
  olenus, Hall, in the addenda to the 1st edition of this work, may be stricken out as the species is now referred to Trochoceras.
  orion, Hall, refer to Trochoceras orion.
  planidorsatum, Whitfield, 1880, Ann. Rep. Geo. Sur. Wis., Trenton Gr. [Sig. having a level back.]
  rectum, Whitfield, 1880, Ann. Rep. Geo. Sur. Wis., Niagara Gr. [Sig. straight.]
  sacculus, Meek and Worthen, refer to Gomphoceras sacculus.
  unicorne is from the Marshall or Kinderhook Gr.
DISCITES ammonis, Hall, 1879, Pal. N.Y., vol. 5, Up. Held. Gr. [Sig of Ammon.]
  inopinatus, Hall, 1879, Pal. N.Y., vol. 5, Up. Held. Gr. [Sig. unexpected.]
  marcellensis instead of Nautilus marcellensis.
ENDOCERAS bristolense. S. A. Miller, 1882, Jour. Cin. Soc. Nat. Hist., vol. 5, Hud. Riv. Gr. [Ety. proper name.]
  egani, S. A. Miller. 1882, Jour. Cin. Soc. Nat. Hist., vol. 5, Hud. Riv. Gr. [Ety. proper name.]
  inæquabile, S. A. Miller, 1882, Jour. Cin. Soc. Nat. Hist., vol. 5, Hud. Riv. Gr. [Sig. unequal.]
  lamarcki, Billings, instead of Orthoceras lamarcki.
  montrealense, Billings, instead of Orthoceras montrealense.
  subannulatum, Whitfield, 1880, Ann. Rep. Geo. Sur. Wis., Trenton Gr. [Sig. somewhat annulated.]
GOMPHOCERAS abruptum, Hall, 1879, Pal. N.Y., vol. 5, Ham. Gr. [Sig abrupt.]

GOMPHOCERAS absens instead of Cyrtoceras absens.
ajax, Hall, 1879, Pal. N. Y., vol. 5, Portage Gr. [Ety. mythological name.]
amphora, Whitfield, 1882, Desc. New Spec. Foss. from Ohio, Up. Held. Gr. [Sig. a bottle.]
beta instead of Cyrtoceras beta.
cammarus, Hall, 1879, Pal. N. Y., vol. 5, Up. Held. Gr. [Sig. a lobster.]
clavatum instead of Cyrtoceras clavatum.
conradi instead of Cyrtoceras conradi.
cruciferum, Hall, 1879, Pal. N. Y., vol. 5, Schoharie Grit. [Sig. cross bearing.]
fax, Hall, 1879, Pal. N. Y., vol. 5, Schoharie Grit. [Sig. a torch.]
gomphus, Hall, 1879, Pal. N. Y., vol. 5, Up. Held. Gr. [Sig. a club.]
hyatti, Whitfield, 1882, Desc. New Spec. Foss. from Ohio, Up. Held. Gr. [Ety. proper name.]
illænus, Hall, 1879, Pal. N. Y., vol. 5, Schoharie Grit. [Ety. supposed to be from the genus *Illænus*, but quere ?.]
impar, Hall, 1879, Pal. N. Y., vol. 5, Up. Held. Gr. [Sig. unequal.]
lunatum, Hall, 1879, Pal. N. Y., vol. 5, Ham. Gr. [Sig. lunate.]
manes, Hall, 1879, Pal. N. Y., vol. 5, Genessee Slate. [Ety. mythological.]
mitra, Hall, 1879, Pal. N. Y., vol. 5, Up. Held. Gr. [Sig. a head band.]
pingue, Hall, 1879, Pal N. Y., vol. 5, Ham. Gr. [Sig. fat.]
planum, Hall, 1879, Pal. N. Y., vol. 5, Ham. Gr. [Sig. even.]
poculum, Hall, 1879, Pal. N. Y., vol. 5, Ham. Gr. [Sig. a cup.]
potens, Hall, 1879, Pal. N. Y., vol. 5, Waverly Gr. [Sig. powerful.]
raphanus, Hall, 1879, Pal. N. Y., vol. 5, Ham. Gr. [Sig. a radish.]
rude, Hall, 1879, Pal. N. Y., vol. 5, Ham. Gr. [Sig. rough.]
sacculus instead of Cyrtoceras sacculus.
sciotoense, Whitfield, 1882, Desc New Spec. Foss., from Ohio, Up. Held. Gr. [Ety. proper name.]
solidum, Hall, 1879, Pal. N. Y., vol. 5, Marcellus Shale. [Sig. solid.]
tumidum, Hall, 1879, Pal. N. Y., vol. 5, Chemung Gr. [Sig. tumid.]

GONIATITES andrewsi, Winchell, 1870, Notices and Desc., from Marshall Gr., etc., Marshall Gr. [Ety. proper name.]
canadensis, Castelnau, 1843, Syst. Sil. Probably a syn. for Bellerophon bilobatus.
colubrellus, Morton, 1836 (Ammonites colubrellus), Am. Jour. Sci. & Arts, vol. 29, Waverly Gr. [Sig. a little snake.]
expansus, Vanuxem, being preoccupied by von Buch, in 1838. G. vanuxemi has been proposed instead of it.
GONIATITES goniolobus, Meek, 1877, U. S. Geo. Sur., 40th parallel, Carboniferous. [Sig having angular lobes.]
hildrethi, Morton, 1836 (Ammonites hildrethi), Am. Jour. Sci. & Arts, vol. 29, Waverly Gr. [Ety. proper name.]
holmesi, G. ixion, G. morganensis, G. osagensis. G. oweni, and G. oweni var. parallelus, are from the Choteau or Kinderhook Gr.
hyas is a syn. for G. lyoni.
kingi, Hall & Whitfield, 1877, U. S. Geo. Expl. Exped., 40th parallel, Coal Meas. [Ety. proper name.]
nundaia is a syn. for G. sinuosus.
ohioensis, Winchell, 1870, Notices and Desc. Foss. from Marshall Gr. [Ety. proper name.]
opimus, White & Whitfield, 1862, Proc. Bost. Soc. Nat. Hist., vol. 8, Kinderhook Gr. [Sig. plump.]
plebeiformis, Hall, 1879, Pal. N. Y., vol. 5, Marcellus Shale. [Sig. from resemblance to *G. plebeius*.] Proposed instead of Porcellia rotatoria, Hall, which was preoccupied in this genus.
vanuxemi, Hall, 1879, Pal. N. Y., vol. 5, Marcellus Shale. [Ety. proper name.] Proposed instead of *G. expansus*, of Vanuxem, which was preoccupied by Von Buch.

GYROCERAS abruptum, Hall, 1879, Desc. New Spec. Foss., Niagara Gr. [Sig. abrupt.]
burlingtonense, may be referred to Nautilus burlingtonensis.
columbiense, Whitfield, 1882, Desc. New Spec. Foss. from Ohio, Up. Held. Gr. [Ety. proper name.]
duplicostatum, Whitfield, 1878, Ann. Rep. Geo Sur. Wis., Trenton Gr. [Sig. double ribbed.]
elrodi, White, 1882, 11th Ann. Rep. Geol. and Nat. Hist. Indiana, Niagara Gr. [Ety. proper name.]
gracile is from the Kinderhook Gr.
jason, was described by Hall in 1862, in 15th Reg Rep., under the name of Cyrtoceras jason.
laciniosum, Hall, 1879. Pal. N. Y., vol. 5, Up. Held. Gr. [Sig. full of points.]
liratum, refer to Nautilus liratus.
matheri is from the Up. Held. Gr.
pratti, Barris, 1879, Proc. Dav. Acad. Sci., vol. 2, Up. Held. Gr. [Ety. proper name.]
seminodosum, Whitfield, 1882, Desc. New Spec. Foss. from Ohio, Up. Held. Gr. [Sig. half nodose.]
subliratum, refer to Nautilus subliratus.
undulatum is from the Up. Held. Gr., and dated 1842.
vagrans instead of vagans.
Hortholus americanus, refer to Lituites americanus.

Huronia stokesi, Castelnau, 1843, Syst. Sil. Schoharie Grit. Not recognized.
LITUITES americanus, Emmons, 1856 (Hortholus americanus) Am. Geol., Black Riv. Gr. [Ety. proper name.]
convolvans, as identified by Hall, in Pal. N. Y., vol. 1, is described as L. americanus.
multicostatus, Whitfield, 1880. Ann. Rep. Geo. Sur. Wis., Niagara Gr. [Sig. many ribbed.]
murchisoni, Troost. Not defined so as to be recognized.
Melia cancellatus, Emmons, 1856, Am. Geol. Not defined so as to be recognized.
NAUTILUS acræus, Hall, 1879, Pal. N.Y., vol. 5, Ham. Gr. [Sig. occupying a height.]
barrandi, Hall, being preoccupied by Von Hauer in 1850, N. magister has been proposed in its place.
burlingtonensis instead of Gyroceras burlingtonense.
cavus, Hall, 1879, Pal. N.Y., vol. 5, Ham. Gr. [Sig. concave.]
danvillensis, White, 1878, Proc. Acad. Nat. Sci., Coal Meas. [Ety. proper name.]
digonus, refer to Trematodiscus digonus.
liratus instead of Gyroceras liratum.
liratus var juvenis, Hall, 1879, Pal. N. Y., vol. 5, Ham. Gr. [Sig. young.]
magister, Hall, 1879, Pal. N. Y., vol 5. Ham. Gr. [Sig. the chief.] Proposed instead of N. barrandi, which was preoccupied.
marcellensis, refer to Discites marcellensis.
oceanus, Hall, 1879, Desc. New Spec. Foss., Niagara Gr. [Sig. the ocean.]
ortoni, Whitfield. 1882. Desc. New Spec. Foss. from Ohio, Coal Meas. [Ety. proper name.]
pauper, Whitfield, 1882, Desc. New Spec. Foss. from Ohio, Kaskaskia Gr. [Sig. poor.]
subliratus instead of Gyroceras subliratum.
Nelimenia incognita, Castelnau, 1843. Syst. Sil. Probably a fragment of Phragmoceras or Oncoceras.
ONCOCERAS brevicurvatum, Whitfield, 1880, Ann. Rep. Geo. Sur. Wis., Trenton Gr. [Sig. short curved.]
mummiforme, Whitfield, 1880, Ann. Rep. Geo. Sur. Wis. Trenton Gr. Sig. mummiform.]
ORTHOCERAS *acicula*, refer to Coleolus acicula.
amycus, Hall, 1879, Desc. New Spec. Foss., Niagara Gr. [Ety. mythological name.]
anguis, Hall, 1879, Pal. N. Y., vol. 5, Chemung Gr. [Sig. a serpent.]
annulato-costatum, Meek & Worthen, is O. randolphense.
arcuatellum, Sandberger, identified by Winchell, in Am. Jour. Sci. & Arts, in 1862, is not an American species.
ORTHOCERAS atreus, Hall, 1879, Pal. N. Y., vol. 5, Portage Gr. [Ety. mythological name.]
aulax, Hall, 1879, Pal. N. Y., vol. 5, Ham. Gr. [Sig. a furrow.]
bebryx var. cayuga, Hall, 1877, Pal. N. Y., vol. 5, Chemung Gr. [Ety. proper name.]
beloitense, Whitfield. 1878, Ann. Rep. Geo. Sur. Wis., Trenton Gr. [Ety. proper name.]
bipartitum, Hall, 1879, Pal. N. Y., vol. 5, Up. Chemung Gr. [Sig. two parted.]
cælamen, Hall, 1879, Pal. N. Y., vol. 5, Ham. Gr. [Sig. a bass relief.]
cancellatum, Hall, refer to O. subcancellatum.
carltonense, Whitfield, 1878, Ann. Rep. Geo. Sur. Wis., Niagara Gr. [Ety. proper name.]
carnosum, Hall, 1879, Pal. N. Y., vol. 5, Schoharie Grit. [Sig. fleshy.]
chemungense and O. indianense are from the Kinderhook Gr.
cingulum, Hall, 1879, Pal. N. Y., vol. 5, Schoharie Grit. [Sig. a zone.]
clavatum, Hall, refer to O. desideratum.
cochleatum, Hall, 1879, Pal. N. Y., vol. 5, Chemung Gr. [Sig. screw formed.]
collatum, Hall, 1879, Pal. N. Y., vol. 5, Schoharie Grit. [Sig. collected.]
columnare, Hall, refer to O. orus.
conicum, Castelnau, 1843, Syst. Sil., Niagara Gr. [Sig. conical.]
constrictum, Vanuxem, 1842, Geo. Rep. 3d Dist. N. Y., Ham. Gr. [Sig. constricted.]
constrictum, Conrad. Not defined so as to be recognized, and it may be stricken from the list.
creon, Hall, 1879, Pal. N. Y., vol. 5, Schoharie Grit. [Ety. proper name.]
dawsonanum, n.sp., Carboniferous. Proposed instead of O. perstrictum, Dawson, in Acadian Geology, p. 312, fig. 120, as the name was preoccupied by Barrande.
demus, Hall, 1879, Pal. N. Y., vol. 5, Chemung Gr. [Sig. at last, solely.]
elegantulum is in vol. 5, Can. Nat., and from the Up. Sil.
filiforme, Castelnau, 1843, Syst. Sil., Niagara Gr. [Sig. filiform.]
fluctum, Hall, 1879, Pal. N. Y., vol. 5, Schoharie Grit. [Sig. waved.]
fulgidum, Hall, 1879, Pal. N. Y., vol. 5, Chemung Gr. [Sig. shining.]
fustis, Hall, 1879, Pal. N. Y., vol. 5, Marcellus shale. [Sig. a club.]
griffithi, Haughton, 1857. Jour. Roy. Dub. Soc., vol. 1, Devonian. ? [Ety. proper name.]
harttanum, n. sp., Carboniferous. Proposed instead of O. laqueatum, Hartt,

CEPHALOPODA.

in Acadian Geol., p. 312, fig. 128, which was preoccupied.
ORTHOCERAS herculaneum. Verneuil, 1846, Bull. de la Soc. Geol. de France. vol. 4, Low. Sil. [Sig. large of its kind.]
hercules, Castelnau, 1843, Syst. Sil. Up. Sil. [Ety. mythological name.]
hyas is a syn. for·O. thoas.
idmon, Hall, 1879, Pal. N. Y., vol. 5, p. 302, Ham. Gr. [Ety. mythological name.]
inoptatum, Hall, 1879, Pal. N. Y., vol. 5, Up. Held. Gr. [Sig. undesired]
isogramma, Meek, 1871, Proc. Acad. Nat. Sci., Coal Meas. [Sig. equal weight.]
jaculum, Hall, 1879, Pal N. Y., vol. 5, Up. Held. Gr. [Sig. a dart.]
kingi, Meek, 1877, U. S. Geo. Sur., 40th parallel, Devonian. [Ety. proper name.]
læve, refer to O. sublævæ.
laqueatum. Hartt, refer to O. harttanum.
leander, Hall, 1879, Pal. N. Y., vol. 5, Chemung Gr. [Ety. proper name.]
lima, Hall, 1879, Pal. N. Y., vol. 5, Ham. Gr. [Sig. a file.]
linteum, Hall, 1879, Pal. N. Y., vol. 5, Ham. Gr. [Sig. a girdle.]
masculum, Hall, 1879, Pal. N. Y., vol. 5, Schoharie Grit. [Sig.·masculine.]
medium, Hall, 1879, Pal. N. Y., vol. 5, Schoharie Grit. [Sig. middle.]
michiganense, n. sp., Marshall Gr., in the Southern part of Michigan. Proposed instead of O. multicinctum, Winchell. Proc. Acad. Nat. Sci. Phil., Sept.. 1862. p. 421.
moniliforme, Swallow, refer to O. swallovanum.
multicinctum. Winchell, being preoccupied by Hall, I have proposed O. michiganense.
nobile, Meek & Worthen, 1865, Proc. Acad. Nat Sci., Kaskaskia Gr. [Sig. famous, noted.]
occidentale, Winchell, being preoccupied by Swallow, I have proposed O. vinchellanum.
œdipus, Hall, 1879, Pal. N.Y.. vol. 5, Ham. Gr. [Ety. mythological name.]
ohioense instead of O. pelops var. ohioense.
omnanceyi, Salter, 1852, in Sutherland's Jour., vol. 2, Devonian. [Ety. proper name.]
oneidaense, Walcott, 1879, Trans. Alb. Inst., vol. x, Utica Slate Gr. [Ety. proper name.]
oppletum, Hall, 1879, Pal. N.Y., vol. 5, Schoharie Grit. [Sig. filled up.]
pacator, Hall, 1879, Pal. N.Y.. vol. 5, Portage Gr. [Sig. a peacemaker.]
palmatum, Hall, 1879, Pal. N.Y., vol. 5, Chemung Gr. [Sig. marked with the palm of a hand.]
pelops var. *ohioense*, refer to O. ohioense.

ORTHOCERAS *perstrictum*, Dawson, being preoccupied by Barrande I have proposed O. dawsonanum.
pertextum, Hall, 1879, Pal. N. Y., vol. 5, Chemung Gr. [Sig. woven throughout.]
pervicax, Hall, 1879, Pal. N. Y., vol. 5, Schoharie Grit. [Sig. firm.]
pravum, Hall, 1879, Pal. N. Y., vol. 5, Schoharie Grit. [Sig. crooked.]
punctostriatum is from the Up. Sil.
randolphense, Worthen, 1882, Bull. No. 1. Ill. St. Mus. Nat Hist., Kaskaskia Gr. [Ety. proper name.] Proposed instead of O. annulato-costatum, Meek & Worthen, which was preoccupied.
scintilla, Hall, 1879, Pal N. Y., vol. 5, Ham. Gr. [Sig. a spark.]
sicinus, Hall, 1879, Pal. N. Y., vol. 5, Marcellus Shales. [Ety.proper name.]
sirpus, Hall, 1879, Pal. N. Y., vol. 5, Up. Held. Gr. [Sig. a bull rush.]
spissum, Hall, 1879, Pal. N. Y., vol. 5, Ham. Gr. [Sig. compact.]
swallovanum, n. sp., Coal Measures in the Valley of Verdigris in Kansas. [Ety. proper name.] Proposed instead of O. moniliforme, Swallow, in Trans. St. Louis Acad. Sci., vol. 1. p. 200, which was preoccupied by Hall.
tantalus, Hall, 1879, Pal. N. Y., vol. 5, Schoharie Grit. [Ety. mythological name.]
telamon, Hall, 1879, Pal. N. Y., vol. 5, Ham. Gr. [Ety. mythological name.]
tenere, Hall, 1879, Pal. N. Y., vol. 5, Ham. Gr. [Sig. delicate.]
tersum, Hall, 1879, Pal. N. Y., vol. 5, Ham. Gr. [Sig. neat.]
textum, Hall, 1879, Pal. N. Y., vol. 5, Ham. Gr. [Sig. that which is braided]
thestor, Hall, 1879, Pal. N. Y., vol. 5, Marcellus Shales. [Ety mythological name.]
thyestes. Hall, 1879, Pal. N. Y., vol. 5, Portage Gr. [Ety. mythological name.]
varum, Hall, 1879, Pal. N. Y., vol. 5, Schoharie Grit. [Sig bent.]
vastator, Hall, 1879, Pal. N. Y., vol. 5, Schoharie Grit. [Sig. a destroyer.] Correct in the Index, but printed O. *obliquum*, on page 243.
viator, Hall, 1879, Pal. N. Y., vol. 5, Up. Held. Gr. [Sig. a traveller.]
vinchellanum, n. sp., Marshall Gr. in Southern Michigan. [Ety. proper name.] Proposed instead of O. *occidentale*, Winchell, 1862, Am. Jour. Sci. & Arts, 2d ser., vol. 33, p. 356, which was preoccupied by Swallow.
wauwatosense, Whitfield, 1880, Ann. Rep. Geo. Sur. Wis., Niagara Gr. [Ety. proper name.]
woodworthi was proposed by McChesney, in 1865, instead of O. irreg-

ulare of McChesney, which was preoccupied.
PETALICHNUS, S. A. Miller, 1880, Jour. Cin. Soc. Nat. Hist., vol. 2. [Ety. *petalos*, spread out; *ichnos*. track.]
multipartitus, S. A. Miller, 1880, Jour. Cin. Soc. Nat. Hist., vol. 2. Utica Slate Gr. [Sig. many parted.]
PHRAGMOCERAS hoyi, Whitfield, 1878, Ann. Rep. Geo. Sur. Wis., Niagara Gr. [Ety. proper name.]
hoyi var. compressum. Whitfield. 1878, Ann. Rep. Geo. Sur. Wis., Niagara Gr. [Sig. compressed.]
labiatum, Whitfield, 1878. Ann. Rep. Geo. Sur. Wis., Niagara Gr. [Sig. lipped.]
PILOCERAS amplum, Dawson, 1881, Can. Nat., vol 10, Calciferous Gr. [Sig. of large extent.]
Sidemina infundibuliforme, Castelnau, 1843, Syst. Sil. Probably the fragment of an Endoceras.
TERATICHNUS, S. A. Miller, 1880 Jour. Cin. Soc. Nat. Hist., vol. 2. [Ety. *teras*, a wonder; *ichnos*, track.]
confertus, S. A. Miller, 1880, Jour. Cin. Soc. Nat. Hist, vol. 2, Utica Slate Gr. [Sig. pressed together.]
TRACHOMATICHNUS, S. A. Miller, 1880, Jour. Cin. Soc. Nat. Hist, vol. 2. [Ety. *trachoma*, that which is made rough; *ichnos*, track.]
cincinnatensis, S. A. Miller, 1880, Jour. Cin. Soc. Nat. Hist., vol. 2, Utica Slate Gr. [Ety. proper name.]
numerosus, S. A. Miller, 1880, Jour. Cin. Soc. Nat. Hist, vol 2, Utica Slate Gr. [Sig. numerous.]

TRACHOMATICHNUS permultus, S. A. Miller, 1880, Jour. Cin. Soc. Nat. Hist., vol. 2, Utica Slate Gr. [Sig. very many.]
TREMATOCERAS, Whitfield, 1882, Desc. New Spec. Foss., from Ohio. [Ety. *trema*, a hole; *keras*, a horn.]
ohioense, Whitfield, 1882, Desc. New Spec. Foss., from Ohio, Up. Held. Gr. [Ety. proper name.]
TREMATODISCUS digonus instead of Nautilus digonus.
konincki, Wetherby, 1881, Jour. Cin. Soc. Nat. Hist., vol. 4, Waverly Gr. [Ety. proper name]
rockymontanus, S. A. Miller, 1881, Jour. Cin. Soc. Nat. Hist., vol. 4, Burlington Gr. [Ety. proper name.]
TROCHOCERAS æneus, Hall, 1870, Rev. Ed. 20th Reg. Rep. Expl., pl. 25, Niagara Gr. [Ety. mythological name.]
barrandei, Hall, 1879, Pal. N. Y., vol. 5, Schoharie Grit. [Ety. proper name.]
biton, Hall, 1879, Pal. N. Y., vol. 5, Schoharie Grit. [Ety. mythological name.]
expansum, Hall, 1879, Pal. N. Y., vol. 5, Schoharie Grit. [Sig. expanded.]
orion instead of Cyrtoceras orion.
pandum, Hall, 1879. Pal. N. Y., vol. 5, Schoharie Grit. [Sig. crooked.]
TROCHOLITES circularis, Miller & Dyer, 1878, Cont. to Pal. No. 2, Hud. Riv. Gr. [Sig. circular.]
minusculus, Miller and Dyer. 1878, Cont. to Pal. No. 2, Utica Slate Gr. [Sig. rather small.]

CLASS LAMELLIBRANCHIATA.

ACTINODESMA subrectum, Whitfield, 1882, Desc. New Spec. Foss. from Ohio, Ham. Gr. [Sig. somewhat erect.]
ALLORISMA andrewsi, Whitfield, 1882, Desc. New Spec. Foss. from Ohio, Kaskaskia Gr. [Ety. proper name.]
antiquum is from the Kaskaskia Gr.
elongatum, Morton, 1836 (Pholadomya elongata), Am. Jour. Sci. & Arts. vol. 29, Coal Meas. [Sig. elongated.]
gilberti, White, 1879, Bull. U. S. Geo. Sur., vol. 5, No. 2, Carboniferous. [Ety. proper name.]
maxvillense, Whitfield. 1882, Desc New Spec. Foss from Ohio, Kaskaskia Gr. [Ety. proper name.]
AMBONYCHIA retrorsa, S. A. Miller, 1878, Jour. Cin. Soc. Nat. Hist., vol. 1, Hud. Riv. Gr. [Sig. turned back.]
robusta, S. A. Miller, 1880, Jour. Cin. Soc. Nat. Hist., vol. 3, Hud. Riv. Gr. [Sig. robust.]
ANATINA leda is from the Kinderhook Gr.
ANGELLUM, S. A. Miller, 1878, Jour. Cin. Soc. Nat. Hist., vol. 1. [Ety. *aggos*, a pail; *ellus*, diminutive.]
cuneatum, S. A. Miller, 1878, Jour. Cin. Soc. Nat. Hist., vol. 1, Hud. Riv. Gr. [Sig. wedge formed.]
ANTHRACOPTERA polita, White, 1880, 12th Rep. U. S Geo. Sur. Terr., Coal Meas. [Ety. polished.]
Arca modesta is from the Kinderhook Gr.
ASTARTELLA gurleyi, White, 1878, Proc. Acad. Nat. Sci., Coal Meas. [Ety. proper name.]
AVICULA æsopus, A. angustirostra and A. trilobata, from the Ham. Gr., A. magrm is from the Kaskaskia Gr., A. subquadrans is Devonian A. textilis var.

arenaria is from the Oriskany sandstone, and A. whitei is from the Kinderhook Gr.
AVICULA cancellata, Barris, see Pterinea cancellata.
pecteniformis, Hall, 1843, Geol. of N.Y., Chemung Gr. The species is an Aviculopecten, and the name was preoccupied by Conrad.
pinnæformis, Geinitz, 1848 (Solen pinnæformis), Versteinerungen d. deutsch Zechsteingebirg, Coal Meas. [Sig. wing-formed.]
AVICULOPECTEN caroli, and A. tenuicostus, are from the Kinderhook Gr., and A. oblongus is from the Warsaw Gr.
catactus, Meek, 1877, U. S. Geo. Expl. 40th par., Carboniferous. [Sig. frail.]
curtocardinalis, Hall & Whitfield, 1877, U. S. Geo. Expl. 40th parallel, Coal Meas. [Sig. short cardinal.]
iowensis, n. sp., Marshall or Kinderhook Gr., at Burlington, Iowa. Proposed instead of A. occidentalis, of Winchell, in 1863, in Proc. Acad. Nat. Sci., Phil., p. 9, which was preoccupied by Shumard.
newarkensis, Winchell, 1870, Notices & Desc. Foss. from Marshall Gr., Marshall Gr. [Ety. proper name.]
nodocostatus, White & Whitfield, 1862, Proc. Bost. Soc. Nat. Hist., vol. 8, Kinderhook Gr. [Sig. having nodes and ribs.]
occidaneus, Meek, 1877, U. S. Geo. Expl. 40th parallel, Carboniferous. [Sig. western]
parvulus, Hall & Whitfield. 1877, U. S. Geo. Expl. 40th parallel, Coal Meas. [Sig. little.]
spinuliferus, Meek & Worthen, 1870, Proc. Acad. Nat. Sci. Keokuk Gr. [Sig. spine bearing]
weberensis, Hall & Whitfield, 1877, U. S., Geo. Sur., 40th parallel, Coal Meas. [Ety. proper name]
CARDINIA æquimarginalis, refer to Edmondia æquimarginalis.
occidentalis is from the Kinderhook or Choteau Gr.
CARDIOLA salteri, Haughton, 1857, Jour. Roy. Soc. Dub., vol. 1, Devonian. [Ety. proper name.]
CARDIOMORPHA triangulata and C. trigonalis are from the Choteau or Marshall Gr.
CARDIOPSIS crenistriata, refer to Pterinea crenistriata.
CARDIUM nautiloides, Castelnau, 1843, Syst. Sil., Seneca Lake, N. Y. [Sig. like a Nautilus.]
CHÆNOMYA maria, Worthen, 1882, Bull. No. 1, Ill. St. Mus. Nat. Hist., Up. Coal Meas. [Ety. proper name.]
CLEIDOPHORUS chicagoensis, S. A. Miller, 1880, Jour. Cin. Soc. Nat. Hist., vol 3, Niagara Gr. [Ety. proper name.]
ellipticus, Ulrich, 1879, Jour. Cin. Soc. Nat. Hist., vol. 2, Hud. Riv. Gr. [Sig. elliptical.]
CLEIDOPHORUS major, Ulrich, 1879, Jour. Cin. Soc. Nat. Hist., vol. 2, Hud. Riv. Gr. [Sig. larger.]
semiradiatus and C. subovatus are from the Arisaig series of the Upper Silurian.
CONOCARDIUM antiquum, Owen, 1852, Geo. Wis., Iowa & Minn., Silurian. [Sig. ancient.]
pulchellum, White & Whitfield, 1862, Proc. Bost. Soc. Nat. Hist., vol. 8, Kinderhook Gr. [Sig. beautiful.]
CUNEAMYA curta, Whitfield, 1878, Jour. Cin. Soc. Nat. Hist., vol. 1, Hud. Riv. Gr. [Sig. short.]
elliptica. S. A. Miller, 1881, Jour. Cin. Soc. Nat. Hist., vol. 4, Hud. Riv. Gr. [Sig. elliptical.]
neglecta instead of Sedgwickia neglecta.
parva, S. A. Miller, 1880, Jour. Cin. Soc. Nat. Hist., vol. 3, Hud. Riv. Gr. [Sig. small.]
CYPRICARDELLA plicata, refer to Goniophora plicata.
quadrata, White & Whitfield, 1862, Proc. Bost. Soc. Nat. Hist., vol. 8, Kinderhook Gr. [Sig. quadrate.]
CYPRICARDIA chotauensis and C. ventricosa, are from the Choteau or Kinderhook Gr., and C. shumardana, from the St. Genevieve limestone or St. Louis Group.
indianensis, refer to Cypricardinia (?) indianensis.
subplana, refer to Edmondia subplana.
swallovana, n. sp., Coal Measures of Harrison county, Missouri. Proposed instead of C. occidentalis, Swallow, 1863, in Trans. St. Louis Acad. Sci., p. 96.
CYPRICARDINIA indianensis instead of Cypricardia indianensis.
subovata. Miller & Dyer. 1878, Cont. to Pal. No. 2, Niagara Gr. [Sig. subovate.]
CYPRICARDITES biyfieldi, C. inconstans and C. montrealensis may be restored to the genus Vanuxemia, as it is probably distinct from this genus.
chemungensis, refer to Goniophora chemungensis.
megambonus, Whitfield, 1878, Ann. Rep. Geo. Sur. Wis., Trenton Gr. [Sig. having a large umbo.]
quadrangularis, Whitfield, 1878, Jour. Cin. Soc. Nat. Hist., vol. 1, Hud. Riv. Gr. [Sig. quadrangular.]
sigmoideus is from the Hud. Riv. Gr.
DEXIOBIA halli, and D. whitei are from the Marshall Gr.
EDMONDIA æquimarginalis, Winchell, 1862 (Cardinia æquimarginalis) Proc. Acad. Nat. Sci., Marshall Gr. [Sig. equal margined.]
bicarinata. E. elliptica. E. marionensis, E. nitida, E. nuptialis, and E. stri-

gillata, are from the Marshall or Choteau Gr.
EDMONDIA inlesi is a typographical error for E. nilesi.
  pinonensis, Meek, 1877, U. S. Geo. Expl. 40th parallel, vol. 4, Devoulan. [Ety. proper name.]
  subplana instead of Cypricardia subplana.
GONIOPHORA chemungensis, Vanuxem, 1842 (Cypricardites chemungensis), Geo. Rep. N. Y., Chemung Gr. [Ety. proper name.]
  plicata instead of Cypricardella plicata.
  speciosa, Hall, 1879, Desc. New Spec. Foss., Niagara Gr. [Sig. beautiful.]
Isocardia, jennæ is from the Marshall Gr.
LEPTODOMUS undulatus, Whitfield, 1878, Ann. Rep. Geo. Sur. Wis., Niagara Gr. [Sig. undulated.]
LITHOPHAGA illinoisensis, Worthen, 1882, Bull. No. 1, Ill. St. Mus. Nat. Hist. [Ety. proper name.] Proposed instead of the form identified as L. lingualis of Phillips.
  lingualis is not an American species.
LUCINA billingsana, n. sp. Devonian. This name is proposed instead of L. occidentalis, Billings. 1859, Assiniboine & Saskatchewan, Ex. Exped., p. 187, figs. b and c. It is from Snake Island, Lake Winnipegosis. The specific name occidentalis was preoccupied by Morton for a Cretaceous species.
LUNULICARDIUM fragosum, Meek, 1877 (Posidonomya fragosa), U. S. Geo. Expl., 40th parallel, Carboniferous. [Sig. rough.]
MACRODON cochlearis is from the Marshall Gr.
MEGALOMUS compressus, Nicholson & Hinde, 1874, Can. Jour., vol. 14, Niagara Gr. [Sig. compressed.]
MODIOLOPSIS cancellata, Walcott, 1879, Trans. Alb. Inst., vol. 10, Utica Slate Gr. [Sig. cancellated.]
  carrollensis, Worthen, 1882, Bull. No. 1, Ill. St. Mus. Nat. Hist., Galena Gr. [Ety. proper name.] Proposed instead of M. subunasuta of Meek & Worthen, 1870, Proc. Acad. Nat. Sci., p. 41, which was preoccupied.
  rectiformis, Worthen, 1882, Bull. No. 1, Ill. St. Mus. Nat. Hist., Trenton Gr. [Sig. straight formed.] proposed instead of M. orthonota, Meek & Worthen, 1868, Geo. Sur. Ill., vol. 3, which was preoccupied.
MODIOMORPHA concentrica, should be Conrad, 1838 (Pterinea concentrica) Ann. Rep. Geo. Sur., N. Y., etc.
MONOTIS septentrionalis, Haughton, 1857, Jour. Roy. Dub. Soc., vol. 1. [Sig. northern.]
MYALINA apachesi is from the Subcarboniferous; M. concentrica from the Warsaw Gr., M. iowensis from the Marshall Gr., and M. perattenuata was described in 1858.
MYALINA iowensis, Winchell, 1865, Proc. Acad. Nat. Sci., Burlington Gr. [Ety. proper name.]
MYTILARCA percarinata, Whitfield, 1882, Desc. New Spec. Foss. from Ohio, Up. Held. Gr. [Sig. very carinate.]
NUCULA was described in 1801; N. hians, N. houghtoni and N. microdonta are from the Marshall or Kinderhook Gr.
  iowensis was described in the Proc. Bost. Soc. Nat. Hist., vol. 8, and is a Tellinomya. N. stella is also a Tellinomya.
  nasuta, refer to Nuculana nasuta.
  perumbonata, White, 1879, Bull. U. S. Geo. Sur., vol. 5, No. 2, Carboniferous. [Sig. having a very convex umbo.]
NUCULANA dens-mamillata, N. nuculiformis, N. paudoriformis, and N. saccata are from the Marshall Gr.
  nasuta instead of Nucula nasuta.
  obesa, White, 1879, Bull. U. S. Geo. Sur., vol. 5, No. 2, Carboniferous. [Sig. plump.]
NUCULITES mactroides is from the Marshall Gr.
  sulcatinus is from the Marshall Gr., and is a Tellinomya.
  triangularis, Hall & Whitfield, 1877, U. S. Geo. Expl, 40th parallel, Devonian. [Sig. triangular.]
  yoldiiformis, Ulrich, 1879, Jour. Cin. Soc. Nat. Hist., vol. 2, Hud. Riv. Gr. [Sig. shaped like a Yoldia.] It is not a Nuculites.
ORTHODESMA byrnesi, S. A. Miller, 1881, Jour. Cin. Soc. Nat. Hist., vol. 4, Hud. Riv. Gr. [Ety. proper name.]
  cuneiforme, S. A. Miller, 1880, Jour. Cin. Soc. Nat. Hist., vol. 3, Hud. Riv. Gr. [Sig. wedge formed.]
  mickelborough, Whitfield. 1878, Jour. Cin. Soc. Nat. Hist., vol. 1, Hud. Riv. Gr. [Ety. proper name.]
  occidentale, S. A. Miller, 1880, Jour. Cin. Soc. Nat. Hist., vol. 3, Hud. Riv. Gr. [Sig. western.]
  subovale, Ulrich, 1879, Jour. Cin. Soc. Nat. Hist., vol. 2, Hud. Riv. Gr. [Sig. suboval.]
ORTHONOTA phasclia is from the Marshall Gr.
  ventricosa, White & Whitfield, 1862, Proc. Bost. Soc. Nat. Hist., vol. 8, Kinderhook Gr. [Sig. ventricose.]
ORTHONOTELLA, S. A. Miller, 1882, Jour. Cin. Soc. Nat. Hist., vol. 5. [Ety. orthos, straight; notos, back; ellus, diminutive.]
  faberi, S. A. Miller, 1882, Jour. Cin. Soc. Nat. Hist., vol. 5, Hud. Riv. Gr. [Ety. proper name.]
Ostrea patercula, Winchell, is from the Marshall Gr.
PALÆONEILO similis, Whitfield, 1882, Desc. New Spec. Foss. from Ohio, Erie Shale, Portage (?) Gr. [Sig. similar.]

PARACYCLAS peroccidens, Hall & Whitfield, 1877, U. S. Geo. Expl., 40th parallel. Devonian. [Sig. far western.] sabini is from the Chemung Gr.
PERNOPECTEN limatus is from the Marshall Gr.
Pholadomya elongata is Allorisma elongatum.
PINNA ludlovi, Whitfield, 1876, in Ludlow's Carroll to Yellowstone Park, Coal Meas. [Ety. proper name.]
maxvillensis, Whitfield, 1882, Desc. New Spec. Foss. from Ohio, Kaskaskia Gr. [Ety. proper name.]
missouriensis is from the St. Genevieve limestone, or St. Louis Gr.
Pleurorhynchus antiqua is Conocardium antiquum.
POSIDONIA clathrata, P. distans, and P. perstriata were described in 1853.
POSIDONOMYA ambigua is from the Marshall Gr.
fragosa, Meek, 1877, U. S. Geo. Expl., 40th parallel, refer to Lunulicardium fragosum.
PRISCONAIA, Conrad, 1867, Am. Jour. Conch., vol. 3. [Ety. proper name.]
ventricosa, Conrad, 1867, Am. Jour. Conch., vol. 3, Coal Meas. [Sig. ventricose.]
Pseudomonotis curta is jurassic.
PTERINEA brisa is probably a syn. for P. striæcosta.
cancellata, Barris, 1879 (Avicula cancellata) Proc. Dav. Acad. Sci. Corniferous limestone. [Sig. cancellated.]
concentrica is Modiomorpha concentrica.
crenistriata, Winchell, 1862 (Cardiopsis crenistriata), Proc. Acad. Nat. Sci., Marshall Gr. [Sig. having wrinkled striæ.]
mucronata, Ulrich, 1879, Jour. Cin. Soc. Nat. Hist., vol. 2, Hud. Riv. Gr. [Sig. pointed.]
neglecta, McChesney, 1861, New Palæozoic Fossils, Niagara Gr. [Sig. overlooked.]
newarkensis, Meek, 1871, Proc. Acad. Nat. Sci., Waverly Gr. [Ety. proper name.]
similis, Whitfield, 1882, Desc. New Spec. Foss. from Ohio, Marcellus Shale. [Sig. similar.]
spinalata, Winchell, 1863, Proc. Acad. Nat. Sci., Burlington (?) Gr. [Sig. spine winged.]
PTERONITES spergenensis, Whitfield, 1882, Bull. Am. Mus. Nat. Hist., No. 3, Warsaw Gr. [Ety. proper name.]

PYANOMYA, S. A. Miller, 1881, Jour. Cin. Soc. Nat. Hist., vol 4. .[Ety. *pyanos*, a bean; *Mya*, a genus.]
gibbosa, S. A. Miller, 1881, Jour. Cin. Soc. Nat. Hist., vol. 4., Hud. Riv. Gr. [Sig. gibbous.]
PTYCHODESMA knappanum is from the Ham. Gr.
SANGUINOLARIA leptogaster is from the Marshall or Kinderhook Gr.
SANGUINOLITES chemungensis, refer to Goniophora chemungensis.
amygdalinus, S. iowensis, S. jejunus, S. missouriensis, S. nasutus, S. strigatus and S. sulciferus are from the Marshall or Kinderhook Gr.
naiadiformis, Winchell, 1870, Notices & Desc. Foss, from the Marshall Gr. [Sig. like a water nymph.]
securis, Winchell, 1870, Notices & Desc. Foss. from Marshall Gr. [Sig. broad-edged axe.]
sulciferus is from Proc. Acad. Nat. Sci., 1863.
SCHIZODUS subtrigonalis, Meek, 1871, Proc. Acad. Nat. Sci., Waverly Gr., [Sig. subtrigonal.]
SEDGWICKIA lunulata, Whitfield, 1878, Jour. Cin. Soc. Nat Hist., vol. 1, Hud. Riv. Gr. [Sig. resembling a little crescent.]
neglecta, is Cuneamya neglecta.
TELLINOMYA angustata, and T. attenuata are from the Upper Silurian.
cingulata, Ulrich, 1879, Jour. Cin. Soc. Nat. Hist., vol. 2, Hud. Riv. Gr. [Sig. girded.]
iowensis instead of Nucula iowensis.
stella instead of Nucula stella.
sulcatina instead of Nuculites sulcatinus.
Unio orthonotus is Modiolopsis orthonota.
primigenius is Modiolopsis primigenia.
VANUXEMIA, Billings, 1858, Can. Nat. & Geol., vol. 3. [Ety. proper name.] This genus may be restored as it is probably distinct from Cypricardites though related to it. The species are V. bayfieldi, V. inconstans, V. dixonensis, V. montrealensis and V. tomkinsi.
tomkinsi, Billings, 1860, Can. Jour , vol. 6, Corniferous limestone. [Ety. proper name.]
YOLDIA rushensis, McChesney, instead of Y. gibbosa, McChesney, as the latter was preoccupied. And Y. knoxensis instead of Y. polita for like reason.

SUBKINGDOM ARTICULATA.*

CLASS ANNELIDA.

ARABELLITES, Hinde, 1879, Quar. Jour. Geo. Soc. Lond., vol. 35.
ascialis, Hinde, 1879, Quar. Jour. Geo. Soc. Lond., vol. 35, Hud. Riv. Gr. [Sig. axe shaped.]
cervicornis, Hinde, 1879, Quar. Jour. Geo. Soc. Lond., vol. 35, Hud. Riv. Gr. [Sig. deer horned.]
cornutus, Hinde, 1879, Quar. Jour. Geo. Soc. Lond., vol. 35, Hud. Riv. Gr. [Sig. horned.]
crenulatus, Hinde, 1879, Quar. Jour. Geo. Soc. Lond., vol. 35, Hud. Riv. Gr. [Sig. crenulated.]
cristatus, Hinde, 1879, Quar. Jour. Geo. Soc. Lond., vol. 35, Hud. Riv. Gr. [Sig. tufted.]
cuspidatus, Hinde, 1879, Quar. Jour. Geo. Soc. Lond., vol. 35, Hud. Riv. Gr. [Sig. pointed.]
elegans, Hinde, 1879, Quar. Jour. Geo. Soc. Lond., vol. 35, Clinton Gr. [Sig. elegant.]
gibbosus, Hinde, 1879, Quar. Jour. Geo. Soc. Lond., vol. 35, Hud. Riv. Gr. [Sig. gibbous.]
hamatus, Hinde, 1879, Quar. Jour. Geo. Soc. Lond., vol. 35, Hud. Riv. Gr. [Sig. hooked]
lunatus, Hinde, 1879, Quar. Jour. Geo. Soc. Lond., vol. 35, Hud Riv. Gr. [Sig. lunate.]
obliquus, Hinde, 1879, Quar. Jour. Geo Soc. Lond., vol. 35, Hud. Riv. Gr. [Sig. oblique.]
ovalis, Hinde, 1879, Quar. Jour. Geo. Soc. Lond., vol. 35, Hud. Riv. Gr. [Sig. oval.]
pectinatus, Hinde, 1879, Quar. Jour. Geo. Soc. Lond., vol. 35, Hud. Riv. Gr. [Sig. pectinated.]
quadratus, Hinde, 1879, Quar. Jour. Geo. Soc. Lond., vol. 35, Silurian [Sig. quadrate.]
rectus, Hinde, 1879, Quar. Jour. Geo. Soc. Lond., vol. 35, Hud. Riv. Gr. [Sig. straight.]
scutellatus, Hinde, 1879, .Quar. Jour. Geo. Soc. Lond., vol. 35, Hud. Riv. Gr. [Sig. scutellated.]
similis, Hinde, 1879. Quar. Jour. Geo. Soc. Lond., vol. 35, Niagara Gr. [Sig. similar.]
CONCHICOLITES is regarded by Prof. Hall as a syn. for Cornulites.
CORNULITES clintoni, Hall, 1879, 28th Reg. Rep., Clinton Group. [Ety proper name.] This name was proposed instead of C. flexuosus which is preoccupied, when Conchicolites is regarded as synonymous with Cornulites.
DISTACODUS, Hinde, 1879, Quar. Jour. Geo. Soc. Lond., vol. 35. [Ety. distazo to doubt; odous a tooth.]
incurvus, Pander, 1856, (Machairodus incurvus,) Monogr. d. foss. Fische. d. Silur. syst., Hud. Riv. Gr. [Sig incurved.]
DREPANODUS, Pander, 1856, Monogr. d. foss. Fische. d. Silur. syst. [Ety. drepane, a sickle; odous tooth.]
arcuatus, Pander, 1856, Monogr. d. foss. Fische. d. Silur. syst., Hud. Riv. Gr. [Sig. arcuate.]
Eotrophonia, Ulrich, 1878, Jour. Cin. Soc. Nat. Hist., vol. 1. Not satisfactorily defined.
setigera, Ulrich, 1878, Jour. Cin. Soc. Nat. Hist., vol. 1. Not satisfactorily defined, and specimen too poor for definition.
EUNICITES, Ehlers, 1868, Palaeontographica vol. 17. [Ety. from the genus Eunice, a Nereid; and lithos, stone.]
alveolatus, Hinde, 1879, Quar. Jour. Geo. Soc. Lond., vol. 35, Ham. Gr. [Sig. hollowed out like a tray.]
chiromorphus, Hinde, 1879, Quar. Jour. Geo. Soc. Lond. vol. 35, Clinton Gr. [Sig. hand formed.]
clintonensis, Hinde, 1879, Quar. Jour. Geo. Soc. Lond., vol. 35, Clinton Gr. [Ety. proper name.]
compactus, Hinde, 1879, Quar. Jour. Geo. Soc. Lond., vol. 35, Ham. Gr. [Sig. compact.]
contortus, Hinde, 1879, Quar. Jour. Geo. Soc. Lond., vol., 35, Hud. Riv. Gr. [Sig. contorted.]
coronatus, Hinde, 1879, Quar. Jour. Geo. Soc. Lond., vol. 35, Clinton Gr. [Sig. coronated.]
digitatus, Hinde, 1879, Quar. Jour. Geo. Soc. Lond., vol. 35, Hud. Riv. Gr. [Sig. digitated.]
gracilis, Hinde, 1879, Quar. Jour. Geo. Soc. Lond., vol. 35, Hud. Riv. Gr. [Sig. slender.]
major, Hinde, 1879, Quar. Jour. Geo. Soc. Lond., vol. 35, Hud. River. Gr. [Sig. larger.]
nanus, Hinde, 1879, Quar. Jour. Geo. Soc. Lond., vol. 35, Ham. Gr. [Sig. a dwarf.]

---

* NOTE.—I have included here the Conodonts, because there is no good reason why they should be placed in the Class Pisces.

EUNICITES palmatus, Hinde, 1879, Quar. Jour. Geo. Soc. Lond., vol. 35, Ham. Gr. [Sig. palmate.]
perdentatus, Hinde, 1879, Quar. Jour. Geo. Soc. Lond., vol. 35, Hud. Riv. Gr. [Sig. many toothed.]
politus, Hinde, 1879, Quar. Jour. Geo. Soc. Lond., vol. 35, Ham. Gr. [Sig. polished.]
similis, var. arcuatus, Hinde, 1879, Quar. Jour. Geo. Soc. Lond., vol. 35, Ham. Gr. [Sig. arcuate.]
simplex, Hinde, 1879, Quar. Jour. Geo. Soc. Lond., vol. 35, Hud. Riv. Gr. [Sig. simple.]
tumidus, Hinde, 1879, Quar. Jour. Geo. Soc. Lond., vol. 35, Ham. Gr. [Sig. tumid.]
GLYCERITES, Hinde, 1879, Quar. Jour. Geo. Soc. Lond., vol. 35. [Ety. from the genus *Glyceris*; and *lithos*, stone.]
calceolus, Hinde, 1879, Quar. Jour. Geo. Soc. Lond., vol. 35, Niagara Gr. [Sig. a little shoe.]
sulcatus, Hinde, 1879, Quar. Jour. Geo. Soc. Lond., vol. 35, Hud. Riv. Gr. [Sig. furrowed.]
sulcatus, var. excavatus, Hinde, 1879, Quar. Jour. Geo. Soc. Lond., vol. 35, Hud. Riv. Gr. [Sig. excavated.]
LUMBRICONEREITES, Ehlers, 1868, Palæontographica, vol. 17. [Ety. from the genera *Lumbricus* and *Nereis*; and *lithos*, stone.]
armatus, Hinde, 1879, Quar. Jour. Geo. Soc. Lond., vol. 35, Clinton Gr. [Sig. armed.]
basalis, Hinde, 1879, Quar. Jour. Geo. Soc. Lond., vol. 35, Clinton Gr. [Sig. pertaining to the base.]
dactylodus, Hinde, 1879, Quar. Jour. Geo. Soc. Lond., vol. 35, Hud. Riv. Gr. [Sig. finger-toothed.]
triangularis, Hinde, 1879, Quar. Jour. Geo. Soc. Lond., vol. 35, Clinton Gr. [Sig. triangular.]
Machairodus, Pander, 1856. This name was preoccupied. See Distacodus.
incurvus, see Distacodus incurvus.
MONOCRATERION, Torell, 1869, Acta universitatis lundensis. [Ety *monos*, one; *kraterion*, a small basin.]
lesleyi, Prime, 1878, Geo. Sur. Pa. DD, Calciferous (?). Gr. [Ety. proper name.]
NEREIDAVUS, Grinnell, 1877, Am. Jour. Sci. and Arts, 3d ser., vol. 14. [Ety. *Nereis*, a genus; *avus*, grandfather.]
solitarius, Hinde, 1879, Quar. Jour. Geo. Soc. Lond., vol. 35, Ham. Gr. [Sig. solitary.]
varians, Grinnell, 1877, Am. Jour. Sci. and Arts, 3d ser., vol 14, Hud. Riv. Gr. [Sig. variable.]
OENITES, Hinde, 1879, Quar. Jour. Geo. Soc. Lond., vol. 35. [Ety. *Oenas*, a genus; *lithos*, stone.]
amplus, Hinde, 1879, Quar. Jour. Geo. Soc. Lond., vol. 35, Clinton Gr. [Sig. ample.]
OENITES carinatus, Hinde, 1879, Quar. Jour. Geo. Soc. Lond., vol. 35, Hud. Riv. Gr. [Sig. carinated.]
cuneatus, Hinde, 1879, Quar. Jour. Geo. Soc. Lond., vol. 35, Hud. Riv. Gr. [Sig. wedged.]
curvidens, Hinde, 1879, Quar. Jour. Geo. Soc. Lond., vol. 35, Hud. Riv. Gr. [Sig. bent toothed.]
fragilis, Hinde, 1879, Quar. Jour. Geo. Soc. Lond., vol. 35, Clinton Gr. [Sig. fragile.]
inæqualis, Hinde, 1879, Quar. Jour. Geo. Soc. Lond., vol. 35, Hud. Riv. Gr. [Sig. unequal.]
infrequens, Hinde, 1879, Quar. Jour. Geo. Soc. Lond., vol. 35, Clinton Gr. [Sig. infrequent.]
rostratus, Hinde, 1879, Quar. Jour. Geo. Soc. Lond., vol. 35, Hud. Riv. Gr. [Sig. beaked.]
serratus, Hinde, 1879, Quar. Jour. Geo. Soc. Lond., vol. 35, Hud. Riv. Gr. [Sig. serrated.]
Planolites, Nicholson, 1873, Proc. Roy. Soc. [Ety. *planos*, a wanderer; *lithos*, stone.] Syn. for Palæophycus.
vulgaris, Nicholson, a Palæophycus.
POLYGNATHUS, Hinde, 1870, Quar. Jour. Geo. Soc. Lond., vol. 35. [Ety. *polys*, many; *gnathos*, a jaw.]
coronatus, Hinde, 1879, Quar. Jour. Geo. Soc. Lond., vol. 35, Ham. Gr. [Sig. coronated.]
crassus, Hinde, 1879, Quar. Jour. Geo. Soc. Lond., vol. 35, Ham. Gr. [Sig. thick.]
cristatus, Hinde, 1879, Quar. Jour. Geo. Soc. Lond., vol. 35, Ham. Gr. [Sig. tufted.]
curvatus, Hinde, 1879, Quar. Jour. Geo. Soc. Lond., vol. 35, Ham. Gr. [Sig. curved.]
dubius, Hinde, 1879, Quar. Jour. Geo. Soc. Lond., vol. 35, Ham. Gr. [Sig. doubtful.]
duplicatus, Hinde, 1879, Quar. Jour. Geo. Soc. Lond., vol. 35, Ham. Gr. [Sig. duplicated.]
eriensis, Hinde, 1879, Quar. Jour. Geo. Soc. Lond., vol. 35, Ham. Gr. [Ety. proper name.]
immersus, Hinde, 1879, Quar. Jour. Geo. Soc. Lond., vol. 35, Ham. Gr. [Sig immersed.]
linguiformis, Hinde, 1879, Quar. Jour. Geo. Soc. Lond., vol. 35, Ham. Gr. [Sig. tongue-shaped.]
nasutus, Hinde, 1879, Quar. Jour. Geo. Soc. Lond., vol. 35, Ham. Gr. [Sig. nasute.]
palmatus, Hinde, 1879, Quar. Jour. Geo. Soc. Lond., vol. 35, Ham. Gr. [Sig. palmate.]
pennatus, Hinde, 1879, Quar. Jour. Geo.

ANNELIDA. 315

Soc. Lond., vol. 35, Ham. Gr. [Sig. winged.]
POLYGNATHUS princeps, Hinde, 1879, Quar. Jour. Geo. Soc. Lond., vol. 35, Ham. Gr. [Sig. the chief.]
punctatus, Hinde, 1879, Quar. Jour. Geo. Soc. Lond., vol. 35, Ham. Gr. [Sig. punctated.]
radiatus, Hinde, 1879, Quar. Jour. Geo. Soc. Lond., vol. 35, Ham. Gr. [Sig. radiated.]
serratus, Hinde, 1879, Quar. Jour. Geo. Soc. Lond., vol. 35, Ham. Gr. [Sig. serrated.]
simplex, Hinde, 1879, Quar. Jour. Geo. Soc. Lond., vol. 35, Ham. Gr. [Sig. simple.]
solidus, Hinde, 1879, Quar. Jour. Geo. Soc. Lond., vol. 35, Ham. Gr. [Sig. solid.]
truncatus, Hinde, 1879, Quar. Jour. Geo. Soc. Lond., vol. 35, Ham. Gr. [Sig. truncated.]
tuberculatus, Hinde, 1879, Quar. Jour. Geo. Soc. London. vol. 35, Ham. Gr. [Sig. tuberculated.]
PRIONIODUS, Pander, 1856, Monogr. d. foss. Fische d. Silur. Syst. [Ety. *prionion*, a small saw; *odous*, a tooth.]
abbreviatus, Hinde, 1879, Quar. Jour. Geo. Soc. Lond., vol. 35, Ham. Gr. [Sig. abbreviated.]
acicularis, Hinde, 1879, Quar. Jour. Geo. Soc. Lond., vol. 35, Ham. Gr. [Sig. acicular.]
alatus, Hinde, 1879, Quar. Jour. Geo. Soc. Lond., vol. 35, Ham. Gr. [Sig. winged.]
angulatus, Hinde, 1879, Quar. Jour. Geo. Soc. Lond., vol. 35, Ham. Gr. [Sig. angulated.]
armatus, Hinde, 1879, Quar. Jour. Geo. Soc. Lond., vol. 35, Ham. Gr. [Sig. armed.]
elegans, Hinde, 1879, Quar. Jour. Geo. Soc. Lond., vol. 35, Hud. Riv. Gr. [Sig. elegant.]
erraticus, Hinde, 1879, Quar. Jour. Geo. Soc. Lond., vol. 35, Ham. Gr. [Sig. erratic.]
furcatus, Hinde, 1879, Quar. Jour. Geo. Soc. Lond., vol. 35, Hud. Riv. Gr. [Sig. forked.]
panderi, Hinde, 1879, Quar. Jour. Geo. Soc. Lond., vol. 35, Ham. Gr. [Ety. proper name.]
politus, Hinde, 1879, Quar. Jour. Geo. Soc. Lond., vol. 35, Hud. Riv. Gr. [Sig. polished.]

PRIONIODUS radicans, Hinde, 1879, Quar. Jour. Geo. Soc. Lond., vol. 35, Hud. Riv. Gr. [Sig. rooting.]
spicatus, Hinde, 1879, Quar. Jour. Geo. Soc. Lond., vol. 35, Ham. Gr. [Sig. spiked.]
PROTOSCOLEX, Ulrich, 1878, Jour. Cin Soc. Nat Hist., vol. 1. [Ety. *protos*, first; *skolex*, a worm.]
covingtonensis, Ulrich, 1878, Jour. Cin. Soc. Nat. Hist., vol. 1, Utica Slate Gr. [Ety. proper name.]
ornatus, Ulrich, 1878, Jour. Cin. Soc. Nat. Hist., vol. 1, Utica Slate Gr. [Sig. ornate.]
simplex, Ulrich, 1878, Jour. Cin. Soc. Nat. Hist., vol. 1, Utica Slate Gr. [Sig. simple.]
tenuis, Ulrich, 1878, Jour. Cin. Soc. Nat. Hist., vol. 1, Utica Slate Gr. [Sig. slender.]
SCOLITHUS is doubtless the work of some kind of a borer.
tuberosus, Miller & Dyer, 1878, Cont. to Pal. No. 2, Hud. Riv. Gr. [Sig. full of humps.]
woodi, Whitfield, 1880, Ann. Rep. Geo. Sur. Wis., Potsdam Gr. [Ety. proper name.]
SERPULA insita, White, 1878, Proc. Acad. Nat. Sci., Coal Meas. [Sig. inserted.]
SPIRORBIS anthracosia, Whitfield, 1881, Am. Jour. Sci. and Arts, 3d ser., vol. 21, Coal Meas.[Sig. pertaining to coal.]
cincinnatensis, Miller & Dyer, 1878, Jour. Cin. Soc. Nat. Hist., vol. 1, Hud. Riv. Gr. [Ety. proper name.]
omphalodes, Goldfuss, 1826, Germ. Petref., Up. Held. and Ham. Gr. [Sig. like a navel or boss.]
spinuliferus, Nicholson, 1875, Pal. Prov. Ont., Ham. Gr. [Sig. spine bearing.]
STAUROCEPHALITES, Hinde, 1879, Quar. Jour. Geo. Soc. Lond., vol. 35. [Ety. *stauros*, a cross; *kephale*, head; *lithos*, stone.]
niagarensis, Hinde, 1879, Quar. Jour. Geo. Soc. Lond., vol. 35, Niagara Gr. [Ety. proper name.]
WALCOTTIA, Miller & Dyer, 1878, Jour. Cin. Soc. Nat. Hist., vol. 1. [Ety. proper name.]
cookaua, Miller & Dyer, 1878, Cont. to Pal. No. 2, Hud. Riv. Gr. [Ety. proper name.]
rugosa, Miller & Dyer, 1878, Jour. Cin. Soc. Nat. Hist., vol. 1, Hud. Riv. Gr. [Sig. rugose.]

CLASS CRUSTACEA.

ACIDASPIS fimbriata, Hall, 1879, Desc. New Spec. Foss. from Niagara Gr. [Sig. fimbriated.]
parvula, Walcott, 31st Reg. Rep.
AGLASPIS eatoni, Whitfield, 1880, Ann. Rep. Geo. Sur. Wis., Potsdam Gr. [Ety. proper name.]
AGNOSTUS communis, Hall & Whitfield, 1877, U. S. Geo. Expl. 40th parallel, Potsdam Gr. [Sig. common.]
neon, Hall & Whitfield, 1877, U. S. Geo. Expl. 40th parallel, Potsdam Gr. [Ety. proper name.]
prolongus, Hall & Whitfield, 1877, U. S. Geo. Expl. 40th parallel, Potsdam Gr. [Sig. prolonged.]
tumidosus, Hall & Whitfield, 1877, U. S. Geo. Expl. 40th parallel, Potsdam Gr. [Sig. high swelling.]
AGRAULOS woosteri, Whitfield, 1878, Ann. Rep. Geo. Sur. Wis., Potsdam Gr. [Ety. proper name.]
AMPHIPELTIS is from *Amphi*, on both sides, instead of doubtful.
Arctinurus, Castelnau, syn. for Lichas.
ARGES, signifies bright or shining, instead of a mythological name.
ARIONELLUS convexus, Whitfield, 1878, Ann. Rep. Geo. Sur. Wis., Potsdam Gr. [Sig. convex.]
pustulatus, Walcott, 31st Reg. Rep.
ASAPHOIDICHNUS, S. A. Miller, 1880, Jour. Cin. Soc. Nat. Hist., vol. 2. [Ety. *Asaphus*, a genus; *eidos*, form; *ichnos*, track.]
dyeri, S. A. Miller, 1880, Jour. Cin. Soc. Nat. Hist., vol. 2, Utica Slate Gr. [Ety. proper name.]
trifidus, S. A. Miller, 1880, Jour. Cin. Soc. Nat. Hist., vol. 2, Utica Slate Gr. [Sig. trifid.]
ASAPHUS *caudatus*, Green, syn. for Dalmanites limulurus, (?)
cordieri, Castelnau, syn. for Dalmanites limulurus.
crypturus was described by Green in 1834, Trans. Geo. Soc. Penn., vol. 1, pt. 1.
ditmarsiæ, Honeyman, 1879, Proc. Nova Scotia Inst., vol. 5, Lower Silurian. [Ety. proper name.]
edwardsi, Castelnau, syn. for Dalmanites limulurus.
homalonotoides is in 31st Reg. Rep. N. Y.
murchisoni, Castelnau, syn. for A. gigas.
nodostriatus, Hall, 1847, Pal. N. Y., vol. 1. Not defined so as to establish a species.
triangulatus, Whitfield, 1880, Ann. Rep. Geo. Sur. Wis , Trenton Gr. [Sig. triangular.]

ATOPS, Emmons, 1844, Taconic System. This genus should probably be restored because it is distinct from Triarthrus. There is only one species defined. Atops trilineatus. It is related to Conocephalites, and as a generic name has priority.
BATHYURUS pogonipensis, Hall and Whitfield, 1877, U. S. Geo. Expl. 40th parallel, Quebec Gr. [Ety. proper name.]
BEYRICHIA lithofactor should be defined as maker of stone, and so also B. petrifactor.
persulcata, Ulrich, 1879, Jour. Cin. Soc. Nat. Hist. vol. 2, Hud. Riv. Gr. [Sig. very much furrowed.]
regularis was named from the regular bars instead of "formed in bars."
BRONTEUS was defined by Goldfuss in 1839, in Nova Act. Phys. Med. Cæsareæ Leop-Carol. Nat. Curios. xlx., pt. 1, p. 360.
canadensis, Logan, 1846, Rep. Geo. Sur. Canada, Low. Held. Gr. [Ety. proper name.]
flabellifer, Goldfuss, Nova. Acta Acad. Caes. Leop. Nat. Cur. vol., 19 Up. Silurian. [Sig a fan bearer.]
laphami, Whitfield, 1878, Ann. Rep. Geo. Sur. Wis., Niagara Gr. [Ety. proper name.]
CALYMENE christyi is from the Hud. Riv. Gr.
conradi, Emmons, 1856, Am. Geol., Lorraine Shales or Hud. Riv. Gr. [Ety. proper name.]
nasuta, Ulrich, 1879, Jour. Cin. Soc. Nat. Hist vol. 2, Niagara Gr. [Sig. nasute.]
rostrata, Vogdes. 1880, Proc. Acad. Nat. Sci., Clinton Gr. [Sig. hooked.]
CERATIOCARIS grandis, Pohlman, 1881, Bull. Buf. Soc. Nat. Hist., vol. 4, Waterlime Gr. [Sig. grand.]
CERAURUS rarus is in 31st Reg. Rep.
CHARIOCEPHALUS tumifrons, Hall & Whitfield, 1877, U. S. Geo. Expl., 40th parallel, Potsdam Gr. [Sig. having a tumid front.]
CONOCEPHALITES binodosus instead of C. binodus.
calciferus, Walcott, 1879, 32 Reg. Rep., Calciferous Gr. [Sig. calciferous.]
calymenoides, Whitfield, 1878. Ann. Rep. Geo. Sur. Wis., Potsdam Gr. [Sig. like a Calymene.]
explanatus, Whitfield, 1880, Ann. Rep. Geo. Sur. Wis., Potsdam Gr. [Sig. spread out.]
harti, Walcott, 1879, 32d Reg. Rep, Calciferous Gr. [Ety. proper name.]
laticeps, Hall & Whitfield, 1877, U. S.

Geo. Expl., 40th parallel, Potsdam Gr. [Sig. having a wide head.]
CONOCEPHALITES quadratus, Whitfield, 1880, Ann. Rep. Geo. Sur. Wis., Potsdam Gr. [Sig. quadrate.]
  subcoronatus, Hall & Whitfield, 1877, U. S. Geo. Expl., 40th parallel, Quebec Gr. [Sig. somewhat coronated.]
CREPICEPHALUS angulatus, Hall & Whitfield, 1877, U. S. Geo. Expl., 40th parallel, Potsdam Gr. [Sig. angulated.]
  anytus, Hall & Whitfield, 1877, U. S. Geo. Expl., 40th parallel, Potsdam Gr. [Ety. proper name.]
  centralis, Whitfield, 1877, Rep. on Pal. of the Black Hills, Potsdam Gr. [Sig. central.]
  gibbesi, Whitfield. 1880, Ann. Rep. Geo. Sur. Wis., Potsdam Gr. [Ety. proper name.]
  granulosus, Hall & Whitfield, 1877, U. S. Geo. Expl., 40th parallel, Potsdam Gr. [Sig. granulous.]
  haguei, Hall & Whitfield, 1877, U. S. Geo. Expl., 40th parallel, Potsdam Gr. [Ety. proper name.]
  maculosus, Hall & Whitfield, 1877, U. S. Geo. Expl., 40th parallel, Potsdam Gr. [Sig. spotted.]
  nitidus, Hall & Whitfield, 1877, U. S. Geo. Expl., 40th parallel, Potsdam Gr. [Sig. neat.]
  onustus, Whitfield, 1878, Ann. Rep. Geo. Sur. Wis., Potsdam Gr. [Sig. full.]
  planus, Whitfield, 1877, Rep. on Pal. of Black Hills, Potsdam Gr. [Sig. plane.]
* quadrans, Hall & Whitfield, 1877, U. S. Geo. Expl., 40th parallel, Quebec Gr. [Sig. a quarter.]
  simulator, Hall & Whitfield, 1877, U. S. Geo. Expl., 40 parallel, Potsdam Gr. [Sig. an imitator.]
  unisulcatus, Hall & Whitfield, 1877, U. S. Geo. Expl., 40th parallel, Potsdam Gr. [Sig. one furrowed.]
CYTHERE carbonaria, refer to Leperditia carbonaria.
  irregularis, S. A. Miller, 1878, Jour. Cin. Soc. Nat. Hist., vol. 1, Hud. Riv. Gr. [Sig. irregular.] The species does not belong to this genus.
CYTHERELLINA, Jones & Hall.
  glandella, Whitfield, 1882, Bull. No. 3, Am. Mus Nat. Hist., Warsaw Gr. [Sig a small kernel]
CYTHEROPSIS rugosa, is Primitia rugosa.
DALMANITES calliteles, signifies a beautiful tail.
  intermedius is in 31st Reg Rep.
DICELLOCEPHALUS, see Pterocephalia, which has priority.
  baraboensis, Whitfield, 1878, Ann. Rep. Geo. Sur. Wis., Low. Magnesian Gr. [Ety. proper name.]
  bilobatus, Hall & Whitfield, 1877, U. S. Geo. Expl., 40th parallel, Potsdam Gr. [Sig. two lobed.]
  eatoni, Whitfield, 1878, Ann. Rep. Geo. Sur. Wis., Low. Magnesian Gr. [Ety. proper name.]
DICELLOCEPHOLUS flabellifer, Hall & Whitfield, 1877, U. S. Geo. Expl., 40th parallel, Potsdam Gr. [Sig. a fan bearer.]
  gothicus, Hall & Whitfield, 1877, U. S. Geo. Expl., 40th parallel, Potsdam Gr. [Sig. gothic.]
  lodensis, Whitfield, 1880, Ann. Rep. Geo. Sur. Wis., Potsdam Gr. [Ety. proper name.]
  multicinctus, Hall & Whitfield, 1877, U. S. Geo. Expl., 40th parallel, Potsdam Gr. [Sig. many girded.]
  quadriceps, Hall & Whitfield, 1877, U. S. Geo. Expl., 40th parallel, Quebec Gr. [Sig. square headed.]
  sancti-sabæ, Rœmer, 1849, Texas Mit. naturwissench Anhang. (Pterocephalia sancti-sabæ), and 1852, Kreid von Texas, Potsdam Gr. [Ety. proper name.]
  wahsatchensis, Hall & Whitfield, 1877, U. S. Geo. Expl., 40th parallel, Quebec Gr. [Ety. proper name.]
DOLICHOPTERUS mansfieldi, Hall, 1877, Trans. Am. Phil. Soc., Lower Coal Meas. [Ety. proper name.]
ECHINOCARIS, Whitfield, 1880, Am. Jour. Sci. and Arts, 3d ser., vol. 19. [Ety. echinos, the sea urchin; karis, a shrimp.]
  multinodosa, Whitfield, 1880, Am. Jour. Sci. and Arts, 3d ser., vol. 19, Erie Shales. [Sig. many noded.]
  pustulosa, Whitfield, 1880, Am. Jour. Sci. and Arts, 3d ser., vol. 19, Erie Shales. [sig. pustulous.]
  sublævis, Whitfield. 1880, Am. Jour. Sci. and Arts, 3d ser., vol. 19, Erie Shales. [Sig. somewhat smooth.]
ECHINOGNATHUS, Walcott, 1882, Am. Jour. Sci. and Arts, 3d ser., vol. 23. [Ety. echinos, sea urchin; gnathos, the jaw.]
  clevelandi, Walcott, 1882, Am. Jour. Sci. and Arts, 3d ser., vol. 23, Utica Slate Gr. [Ety. proper name.]
Elliptocephala curta is Olenellus curtus, and E. asapholdes is O. asapholdes. Olenellus is a syn. for Elliptocephala of Emmons, but the latter name being preoccupied, Olenellus must be used.]
ENCRINURUS egani, S. A. Miller, 1880, Jour. Cin. Soc. Nat., Hist., vol. 2, Niagara Gr. [Ety. proper name.]
  mirus. This is not an Encrinurus, but I am not able to refer it satisfactorily to any genus.
  trentonensis, and E. varicostatus are in the 31st Reg. Rep.
Enoploura, Wetherby, 1878, Jour. Cin. Soc. Nat. Hist., vol. 1. Proposed instead of Anomalocystites upon the ground that it is a Crustacean instead of a Cystidean.
ESTHERIA, Ruppell and Straus-Durckheim, 1837, Mus. Senckenberg, vol. 2, p. 119. [Ety. proper name.]

ESTHERIA pulex, Clarke, 1882, Am. Jour. Sci. and Arts, 3d ser., vol. 23, Ham. Gr. [Sig. a flea.]
EURYPTERUS eriensis, Whitfield, 1882, Desc. New Spec. Foss. from Ohio, Low. Held. Gr. [Ety. proper name.]
  pennsylvanicus, Hall, 1877, Trans. Am. Phil. Soc., Devonian. [Ety. proper name.]
  pulicaris, signifies like a flea.
EUSARCUS, Grote and Pitt, 1877, Bull. Buf. Soc. Nat. Hist., vol. 4. [Ety. *eu*, well off; *sarkos*, flesh.]
  grandis, Grote and Pitt, 1877, Bull. Buf. Soc. Nat. Hist., vol. 4, Waterlime Gr. [Sig. grand.]
  scorpionis, Grote and Pitt, 1877, Bull. Buf. Soc. Nat. Hist., vol. 4, Waterlime Gr. [Sig. scorpion.]
Helminthoidichnites marinus, is Gordia marina, Emmons, 1844, Taconic Syst.
HOMALONOTUS atlas, *H. giganteus* and *H. herculaneus* of Castelnau are synonyms for H. delphinocephalus, or they are not recognized for want of proper definition.
ILLÆNURUS convexus, Whitfield, 1878, Ann. Rep. Geo. Sur. Wis., Low. Mag. Gr. [Sig. convex.]
ILLÆNUS indeterminatus is in 31st. Reg. Rep. N. Y.
  niagarensis, Whitfield, 1880, Ann. Rep. Geo. Sur. Wis., Niagara Gr. [Ety. proper name.]
  pterocephalus, Whitfield, 1878, Ann. Rep. Geo. Sur. Wis., Niagara Gr. [Sig. winged head.]
ISOCHILINA jonesi, Wetherby, 1881, Jour. Cin. Soc. Nat. Hist., vol. 4, Trenton Gr. [Ety. proper name.]
ISOTELUS canalis is from the Chazy Gr.
  megistos was described in the Trans. Assoc. Am. Geol. and Naturalists.
LEPERDITIA angulifera, Whitfield, 1882, Desc. New. Spec. Foss. from Ohio, Low. Held. Gr. [Sig. bearing angles.]
  billingsi, Jones, 1881, Ann. & Mag. Nat. Hist., 5th ser., vol. 18, Trenton Gr. [Ety. proper name.]
  bivertex, Ulrich, 1879, Jour. Cin. Soc. Nat. Hist., vol. 2, Utica Slate Gr. [Sig. two headed.]
  cæcigena, S. A. Miller, 1881, Jour. Cin. Soc. Nat. Hist., vol. 4, Hud. Riv. Gr. [Sig. born blind.]
  carbonaria, instead of Cythere carbonaria.
  crepiformis, Ulrich, 1879, Jour. Cin. Soc. Nat. Hist., vol. 2, Hud. Riv. Gr. [Sig. boot form.]
  radiata, Ulrich, 1879, Jour. Cin. Soc. Nat. Hist., vol. 2, Utica Slate Gr. [Sig. radiated.]
  unicornis, Ulrich, 1879, Jour. Cin. Soc. Nat. Hist., vol. 2, Utica Slate Gr. [Sig. one horned.]
LICHAS emarginatus, Hall, 1879, 28th Reg. Rep., Niagara Gr. [Sig. emarginated.]
  harrisi, S. A. Miller, 1878, Jour. Cin. Soc. Nat. Hist., vol. 1, Hud. Riv. Gr. [Ety. proper name.]
Lisgocaris, Clarke, 1882, Am. Jour. Sci. and Arts, 3d. ser., vol.23. [Syn. for Spathiocaris.]
  lutheri, Clarke, 1882, Am. Jour. Sci. and Arts, 3d ser., vol. 23, Ham. Gr., refer to Spathiocaris lutheri.
MICRODISCUS lobatus is from the Taconic or Lower Potsdam, and M. speciosus was described in 1873.
OGYGIA parabola, Hall & Whitfield, 1877, U. S. Geo. Expl., 40th parallel, Quebec Gr. [Sig. a parabola.]
  producta, Hall & Whitfield, 1877, U. S. Geo. Expl., 40th parallel, Quebec Gr. [Sig. extended.]
OLENELLUS asaphoides was described in the Taconic System.
  curtus, Whitfield, 1878 (Elliptocephalus curtus), Ann. Rep. Geo. Sur. Wis, Potsdam Gr. [Sig. short.]
PALÆOPALÆMON, Whitfield, 1880, Am. Jour. Sci. and Arts, 3d ser., vol. 19. [Ety. *palaios*, ancient; *Palæmon*, a genus.]
  newberryi, Whitfield, 1880, Am. Jour. Sci and Arts., 3d ser., vol. 19, Erie Shales. [Ety. proper name.]
PHILLIPSIA tennesseensis is from the Kinderhook Gr.
  tuberculata, Meek & Worthen, 1870, Proc. Acad. Nat. Sci., Burlington Gr. [Sig. tuberculated.]
PLUMULITES was described by Barrande, 1872, Syst. Sil. Boh. Instead of Plumulites, English authors use Turrilepis, proposed by Woodward in 1865, but not defined so as to be understood.
  devonicus, Clarke, 1882, Am. Jour. Sci. and Arts, 3d ser., vol. 23, Ham. Gr. [Sig. devonian.]
  newberryi, Whitfield, 1882, Desc. New Spec. Foss. from Ohio, Portage Gr. [Ety. proper name.]
PRIMITIA rugosa instead of Cytheropsis rugosa.
PROETUS auriculatus, P. doris and P. swallovi are from the Waverly, Choteau or Kinderhook Gr.
  davenportensis, Barris, 1879, Proc. Dav. Acad. Sci., Corniferous limestone. [Ety. proper name.]
  denticulatus, Meek, 1877, U. S. Geo. Expl., 40th parallel, Devonian. [Sig. denticulated.]
  granulatus, Wetherby, 1881, Jour. Cin. Soc. Nat. Hist., vol. 4, Kaskaskia Gr. [Sig. granulated.]
  loganensis, Hall & Whitfield, 1877, U.S. Geo. Expl., 40th parallel, Waverly Gr. [Ety. proper name.]
  parviusculus was described in the 13th Reg. Rep. 1800, from the Hud. Riv. Gr.
  peroccidens, Hall & Whitfield, 1877, U. S. Geo. Expl., 40th parallel, Waverly Gr. [Sig. from the far west.]

PROTICHNITES alternans, P. latus, P. lineatus, P. multinotatus, P. octo-notatus and P. septem-notatus are from the Potsdam Gr.

PTEROCEPHALIA Rœmer, 1849, Texas, Mit naturwissench. Anhang. Bonn., and afterward in 1852. Kreid von Texas. [Ety. *pteron* wing; *kephale* head.] It is identical with Dicellocephalus, and has priority of definition, and was illustrated the same year.

sancti-sabæ, Rœmer, 1849, Texas. Mit naturwissench. Anhang., and in 1852, Kreid von Texas., Potsdam Gr. [Ety. proper name.]

PTERYGOTUS buffaloensis, Pohlman, 1881, Bull. Buf. Soc. Nat. Hist., vol. 4, Waterlime Gr. [Ety. proper name.]

cummingsi, Grote & Pitt, 1877, Bull Buf. Soc. Nat. Hist., vol. 4, Waterlime Gr. [Ety. proper name.]

PTYCHASPIS minuta, Whitfield, 1878, Ann. Rep. Geo. Sur. Wis., Potsdam Gr. [Sig. minute.]

pustulosa, Hall & Whitfield, 1877, U. S. Geo. Expl., 40th Parallel, Potsdam Gr. [Sig. pustulous.]

speciosa, Walcott, 1879, 32d Reg. Rep. [Sig beautiful.]

striata, Whitfield, 1878, Ann. Rep. Geo. Sur. Wis., Potsdam, Gr. [Sig. striated.]

RHACHURA, Scudder, 1878, Proc. Bost. Soc. Nat. Hist. [Ety. *rachis*, a ridge; *oura*, tail.]

RHACHURA venosa, Scudder, 1878, Proc. Bost. Soc. Nat. Hist., Coal Meas. [Sig. full of veins.]

RUSICHNITES carbonarius, Dawson, 1868, Acadian Geology,Carboniferous. [Sig. pertaining to Carbon.]

grenvillensis, Dawson, Chazy Gr. [Ety. proper name.]

SOLENOPLEURA nana, Ford, 1878, Am. Jour. Sci. and Arts, 3d ser., vol. 15, Potsdam Gr. [Sig. a dwarf.]

SPATHIOCARIS, Clarke, 1882, Am. Jour. Sci. and Arts, 3d ser., vol. 23. [Ety. *spathe*, a spathe; *karis*, a shrimp.]

emersoni, Clarke, 1882, Am. Jour. Sci. and Arts, 3d ser., vol. 23, Portage Gr. [Ety. proper name.]

lutheri instead of Lisgocaris lutheri.

TRIARTHRUS becki is from the Utica Slate Gr.

fischeri, Billings, 1865, Pal. Foss., Quebec Gr. [Ety. proper name.]

glaber, Billings, 1859, Can. Nat. and Geol., vol. 4, Utica Slate Gr. [Sig. smooth.]

spinosus is from the Utica Slate Gr.

trilineatus should be referred to Atops trilineatus.

TRINUCLEUS *bellulus*, Ulrich, 1878, Jour. Cin. Soc. Nat. Hist., vol. 1, is from the Utica Slate, and seems to be the young of T. concentricus.

CLASS ARACHNIDA.

EOSCORPIUS was published in vol. 46, Am. Jour. Sci. and Arts.

CLASS MYRIAPODA.

EUPHOBERIA was published in vol. 46, Am. Jour. Sci. and Arts.

CLASS INSECTA.

ARCHIMYLACRIS mantis is a typographical or accidental mistake, and the name should be stricken out.

EPHEMERITES primordialis, Scudder, 1878, Proc. Bost. Soc. Nat. Hist., Coal Meas. [Sig. primordial.]

TERMES, Linnaeus, 1748, Systema Naturae, p. 610, and older authors. [Ety.*termes*, a worm that eats wood.]

contusus, Scudder, 1878, Proc. Bost. Soc. Nat. Hist., Coal Meas. [Sig. broken or bruised.]

CLASS PISCES.

ACONDYLACANTHUS (?) mudgianus, St. John & Worthen (In press), Geo Sur. Ill., vol. 7. Up. Coal Meas. [Ety. proper name.]

nuperus, St. John & Worthen (In press), Geo. Sur Ill., vol. 7, Up. Coal Meas. [Sig. new.]

ACONDYLACANTHUS rectus, St. John & Worthen (In press), Geo. Sur. Ill., vol. 7, Up. Coal Meas. [Sig. straight.]

xiphias, St. John & Worthen (In press), Geo. Sur. Ill., vol. 7, Keokuk Gr. [Sig. a sword fish.]

AMACANTHUS can not be derived from *ana-*

PISCES.

kanthos, and its etymology is not apparent unless it is from *ama*, together; and *kanthos*, the felly of a wheel.
ASTEROPTYCHIUS tenellus, St. John & Worthen (In press), Geo. Sur. Ill., vol. 7, Up. Coal Meas. [Sig. young, delicate.]
BATACANTHUS (?) necis, St. John & Worthen (In press), Geo. Sur. Ill., vol. 7, Keokuk Gr. [Sig. death.]
Catopterus macrurus is from the Triassic, and the specific name signifies long tailed.
CHITONODUS, St. John & Worthen (In press), Geo. Sur. Ill., vol. 7. [Ety. *chiton*, a smock or coat; *odous*, a tooth.]
  antiquus, St. John & Worthen (In press). Geo. Sur Ill., vol. 7, Low. Burlington Gr. [Sig. ancient.]
  latus, Leidy. 1856 (Cochliodus latus), Trans. Am. Phil. Soc., vol 11, p. 87. Pl. 5, fig. 17, Keokuk Gr. [Sig. wide.]
  liratus, St. John & Worthen (In press), Geo. Sur. Ill., vol. 7, St. Louis Gr. [Sig. furrowed]
  rugosus instead of Poecilodus rugosus, P. ornatus and P. convolutus.
  springeri, St. John & Worthen (In press), Geo. Sur. Ill., vol. 7, Up. Burlington Gr. [Ety. proper name.]
  tribulis, St. John & Worthen (In press), Geo. Sur. Ill., vol. 7, Keokuk Gr. [Sig. one of the same tribe.]
CLADODUS occidentalis, Leidy, 1859, Proc. Acad Nat Sci., Up. Coal Meas. [Sig. western.]
COCHLIODUS is derived from *kochlias*, anything spiral; *odous*, a tooth.
  crassus, Newberry & Worthen, is a syn. for Sandalodus lævissimus.
  latus. refer to Chitonodus latus.
  nitidus, Leidy. 1856, refer to Deltoptychius nitidus.
  nobilis, Newberry & Worthen, syn. for Chitonodus latus.
  obliquus, St. John & Worthen (In press), Geo. Sur. Ill., vol. 7, St. Louis Gr. [Sig. oblique]
  occidentalis, Leidy, 1856 refer to Deltodus occidentalis.
  validus, St. John & Worthen (In press), Geo. Sur. Ill., vol. 7, Kaskaskia Gr. [Sig. strong, stout.]
  vanhornii, St. John & Worthen (In press), Geo Sur. Ill., vol. 7, St. Louis Gr. [Ety. proper name.]
COPODUS pusillus, St. John & Worthen (In press), Geo. Sur. Ill., vol. 7, Kaskaskia Gr. [Sig. very small.]
  vanhornii, St. John & Worthen (In press), Geo. Sur. Ill., vol. 7, St. Louis Gr. [Ety proper name.]
CTENACANTHUS buttersi, St. John & Worthen (In press), Geo. Sur. Ill., vol. 7. [Ety. proper n me.]
  cannaliratus, St. John & Worthen (In press), Geo. Sur. Ill.. vol. 7, Kaskaskia Gr. [Sig. reed furrowed.]
CTENACANTHUS costatus, refer to Eunemacanthus costatus.
  coxanus, St. John & Worthen (In press), Geo. Sur. Ill., vol. 7, Keokuk Gr. [Ety. proper name.]
  deflexus, St. John & Worthen (In press), Geo. Sur. Ill., vol. 7, St. Louis Gr. [Sig. deflexed]
  harrisoni, St. John & Worthen (In press), Geo. Sur. Ill., vol. 7, St. Louis Gr. [Ety. proper name.]
  pellensis, St. John & Worthen (In press), Geo. Sur. Ill., vol. 7, St. Louis Gr. [Ety. proper name.]
CTENODUS dialophus, Cope, 1878. Pal. Bull. No. 29, Permian. [Sig through the neck.]
  fossatus, Cope, 1877, Proc. Am. Phil. Soc., Permian. [Sig. dug out.]
  gurleianus, Cope, 1877. Proc. Am. Phil. Soc., Permian. [Ety. proper name.]
  periprion, Cope. 1878, Pal Bull. No. 29, Permian. [Sig. a round saw.]
  porrectus, Cope, 1878. Pal. Bull. No. 29, Permian. [Sig. prolonged.]
  pusillus, Cope, 1877, Pal. Bull. No. 26, Permian [Sig. very small.]
CTENOPTYCHIUS digitatus. Leidy, 1856, Trans. Am. Phil. Soc., vol. 11, St. Louis Gr. [Sig digitated.]
  semicircularis, refer to Peripristis semicircularis.
DELTODOPSIS, St. John & Worthen (In press), Geo. Sur. Ill , vol. 7. [Ety. from the resemblance to the genus *Deltodus*.]
  affinis, St. John & Worthen (In press), Geo. Sur. Ill., vol. 7, Warsaw Gr. [Sig. near to.]
  angusta, Newberry & Worthen, instead of Deltodus angustus.
  bialveata, St. John & Worthen (In press), Geo. Sur. Ill., vol. 7. Up. Burlington Gr [Sig. twice hollowed out.]
  convexa, St. John & Worthen (In press), Geo. Sur. Ill., vol. 7, Up. Burlington and Keokuk Gr. [Sig. convex.]
  convoluta, St. John & Worthen (In press), Geo. Sur. Ill., vol. 7. Up. Burlington Gr. [Sig. convoluted.]
  exornata, St. John & Worthen (In press), Geo Sur. Ill., vol. 7, Warsaw Gr. [Sig. adorned.]
  keokuk, St. John & Worthen (In press), Geo. Sur Ill., vol. 7, Keokuk Gr. [Ety. proper name]
  stludovici, St. John & Worthen (In press), Geo. Sur. Ill., vol. 7, St. Louis Gr. [Ety. proper name]
DELTODUS alutus, Newberry & Worthen, syn. for Chitonodus latus.
  angularis, Newberry & Worthen, syn. for Orthopleurodus carbonarius.
  angustus, refer to Deltodopsis angusta.

DELTODUS cinctulus, St. John & Worthen (In press), Geo. Sur. Ill., vol. 7, Warsaw Gr. [Sig. small girt.]
fasciatus, refer to Tæniodus fasciatus.
intermedius, St. John & Worthen (In press), Geo. Sur. Ill., vol. 7, St. Louis Gr. [Sig. intermediate.]
latior, St. John & Worthen (In press), Geo. Sur. Ill., vol. 7, Keokuk Gr. [Sig. to be concealed.]
occidentalis. Leidy, 1856 (Cochliodus occidentalis), Trans. Am. Phil. Soc. vol. 11, Warsaw and Keokuk Gr. [Sig. western.]
parvus, St. John & Worthen (In press), Geo. Sur. Ill., vol. 7, St. Louis Gr. [Sig. small.]
powelli, St. John & Worthen (In press), Geo. Sur. Ill., vol. 7, Up. Carb. [Ety. proper name]
propinquus, St. John & Worthen (In press), Geo. Sur. Ill., vol. 7, Coal Meas [Sig. near.]
rhomboideus, Newberry & Worthen, is a syn. for Sandalodus spatulatus.
stellatus, Newberry & Worthen, syn. for D. occidentalis.
trilobus, St. John & Worthen (In press), Geo. Sur. Ill., vol. 7, Warsaw Gr. [Sig. three lobed.]
DELTOPTYCHIUS expansus, St. John & Worthen (In press), Geo. Sur. Ill., vol. 7, St. Louis Gr. [Sig. expanded]
nitidus, Leidy, 1856 (Cochliodus nitidus), Trans. Am. Phil. Soc., vol. 11, Kaskaski ι Gr. [Sig. neat.]
primus, St. John & Worthen (In press), Geo. Sur. Ill., vol. 7, Up. Burlington Gr. [Sig. first.]
varsoviensis, St. John & Worthen (In press), Geo. Sur. Ill., vol. 7, Warsaw Gr. [Ety. proper name.]
wachsmuthi, St. John & Worthen (In press), Geo. Sur. Ill., vol. 7, Keokuk Gr. [Ety. proper name.]
DIPLODUS penetrans is from the Can. Nat. & Geol., vol. 5.
ECTOSTEORHACHIS, Cope, 1880. Pal. Bull. No. 32. [Ety. ektos, without; rachis, a ridge.]
nitidus, Cope, 1880. Pal. Bull. No. 32, Permian. [Sig neat.]
EUNEMACANTHUS, St. John & Worthen (In press). Geo Sur. Ill., vol. 7. [Ety. eu, well; nema, a line; akantha, a spine.]
costatus, Newberry & Worthen, instead of Ctenacanthus costatus.
GLYMMATACANTHUS petrodoides, St. John & Worthen (In press). Geo. Sur. Ill., vol. 7, Kaskaskia Gr. [Sig. like Petrodus.]
rudis, St. John & Worthen (In press), Geo. Sur. Ill , vol. 7, Keokuk Gr. [Sig. a slender stick.]
GYRACANTHUS cordatus, St. John & Worthen (In press), Geo. Sur. Ill., vol. 7, Keokuk Gr. [Sig. Judicious.]

HELODUS consolidatus, Newberry & Worthen, syn. for Chitonodus la us.
gibbus, Leidy, 1856, Trans. Am. Phil. Soc., vol. 11, Keokuk Gr. [Sig. gibbous.]
placenta, refer to Psephodus placenta.
JANASSA. Münster, 1830, Beiträge zur Petrefakten-kunde, vol. 1. and in Ag. Polss. Foss., vol.3, p. 375. [Ety. mythological name.]
gurleiana, Cope. 1877 (Strigilina gurleiana). Proc. Am. Phil. Soc., Permian. [Ety. proper name.]
LEPTOPHRACTUS was accidentally printed in this class, but it also appeared correctly among the Reptilia.
ORACANTHUS consimilis, St. John & Worthen, syn. for O. vetustus.
rectus, St. John & Worthen (In press), Geo. Sur. Ill., vol. 7, Kaskaskia Gr. [sig. straight.]
vetustus, Leidy. 1856, Jour. Acad. Nat. Sci., Coal Meas. [Sig. ancient.]
ORTHACANTHUS quadriseriatus, Cope, 1877, Pal. Bull., No. 26, Permian. [Sig. having four series.]
ORTHOPLEURODUS, St. John & Worthen (In press), Geo Sur. Ill., vol. 7. [Ety. orthos, straight; pleura, side; odous, a tooth.]
carbonarius, Newberry & Worthen, instead of Sandalodus carbonarius.
convexus, St John & Worthen (In press), Geo. Sur. Ill., vol. 7, Coal Meas. [Sig. convex.]
novomexicanus, St. John & Worthen (In press), Geo. Sur. Ill., vol. 7, Low. Carb. [Ety. proper name.]
PALÆOBATIS, Leidy, 1856, Trans. Am. Phil. Soc., vol 11. [Ety. palaios, ancient; batis, a prickly kind of roach or ray.]
insignis, Leidy, 1856 Trans. Am. Phil. Soc., vol. 11, Keokuk Gr. [Sig. marked]
PALÆONISCUS alberti, refer to Rhadinichthys alberti.
cairnsi, refer to Rhadinichthys cairnsi.
jacksoni, Dawson, 1877. Can. Nat. Quar. Jour. Sci., vol. 8, Carboniferous. [Ety. proper name]
modulus, see Rhadinichthys modulus.
PERIPRISTIS Agassiz, 1870, Proc. Am. Phil. Soc , vol. 11. [Ety peri, around; pristis, a saw.]
semicircularis, Newberry & Worthen, 1866 (Ctenoptichius semicircularis), Geo. of Ill., vol. 2, Coal Meas. [Sig. semicircular.]
PETALODUS destructor is a syn. for P. alleghaniensis.
PETALORHYNCHUS being neuter the specific names should be made to correspond.
PHYSONEMUS falcatus, St. John & Worthen (In press), Geo. Sur., Ill., vol. 7, St. Louis Gr [Sig. scythe shaped.]
PNIGEACANTHUS trigona is, St. John & Worthen (In press), Geo. Sur. Ill., vol. 7, St. Louis Gr. [Sig. trigonal.]

PŒCILODUS carbonarius, St. John & Worthen (In press), Geo. Sur. Ill., vol. 7, Coal Meas. [Sig. pertaining to coal.]
cestriensis, St. John & Worthen (In press), Geo. Sur. Ill., vol. 7, Kaskaskia Gr. [Ety. proper name.]
convolutus, Newberry & Worthen, is a syn. for Chitonodus rugosus.
ornatus, Newberry & Worthen, syn. for Chitonodus rugosus.
springeri, St. John & Worthen (In press), Geo. Sur. Ill., vol. 7, Low. Carb. [Ety. proper name.]
studovici, St John & Worthen (In press), Geo. Sur. Ill., vol. 7, St. Louis Gr. [Ety. proper name.]
varsoviensis, St. John & Worthen (In press), Geo. Sur. Ill., vol. 7, Warsaw Gr. [Ety. proper name.]
wortheni, St. John (In press), Geo Sur. Ill., vol. 7, Kaskaskia Gr. [Ety. proper name.]
PSAMMODUS cœlatus, St. John & Worthen (In press), Geo. Sur. Ill., vol. 7. St. Louis Gr. [Sig. sculptured.]
crassidens, St. John & Worthen (In press), Geo. Sur. Ill., vol. 7, St. Louis Gr. [Sig. thick tooth.]
glyptus, St. John & Worthen (In press), Geo. Sur. Ill., vol. 7, Up. Burlington Gr. [Sig. sculptured.]
grandis, St. John & Worthen (In press) Geo. Sur. Ill., vol. 7, Keokuk Gr. [Sig. large.]
lovianus, St. John & Worthen (In press), Geo. Sur. Ill., vol. 7, Burlington Gr. [Ety. proper name.]
plenus, St. John & Worthen (In press), Geo. Sur. Ill., vol. 7, St. Louis Gr. [Sig. large.]
rhomboideus, Newberry & Worthen, syn. for Sandalodus lævissimus.
semicylindricus, Newberry & Worthen, syn. for Sandalodus lævissimus.
springeri, St. John & Worthen (In press), Geo. Sur. Ill., vol. 7, Up Burlington Gr. [Ety. proper name.]
tumidus, St. John & Worthen (In press), Geo. Sur Ill, vol 7, Up. Burlington Gr. [Sig. tumid.]
turgidus, St. John & Worthen (In press), Geo. Sur. Ill., vol. 7, Keokuk Gr. [Sig. inflated.]
PSEPHODUS latus, St. John & Worthen (In press) Geo. Sur. Ill., vol. 7, St. Louis Gr. [Sig. wide.]
lunulatus, St. John & Worthen (In press), Geo. Sur. Ill., vol. 7. Kaskaskia Gr. [Sig. resembling a small crescent.]
obliquus. St. John & Worthen (In press), Geo. Sur. Ill., vol 7, Kinderhook Gr. [Sig. oblique.]
placenta. Newberry & Worthen, instead of Helodus placenta.
symmetricus. St. John & Worthen (In press), Geo. Sur. Ill., vol. 7, Kinderhook Gr. [Sig. symmetrical.]
PTERICHTHYS canadensis, Whiteaves, 1890, Am. Jour. Sci. & Arts, 3d ser., vol. 20, Devonian. [Ety. proper name.]
PTERICHTHYS norwoodensis is a syn. for Macropetalichthys rapheidolabis.
PTYONODUS, Cope, 1877, Proc. Am. Phil. Soc. [Ety. ptyon, a fan; odous, a tooth.]
paucicristatus, Cope, 1877, Proc. Am. Phil. Soc., Permian. [Sig. few crested.]
vinslovi. Cope, 1876. Proc. Acad. Nat. Sci. Phil., Permian. [Ety. proper name.]
RHADINICHTHYS, Traquair, 1877, Quar. Jour. Geo. Soc. Lond., vol. 33, p. 548. [Ety. rhadinos, slender; ichthys, a fish.]
alberti instead of Palaeoniscus alberti.
cairnsi instead of Palaeoniscus cairnsi.
modulus, Dawson, 1877 (Palaeoniscus modulus), Can. Nat. and Quar. Jour. Sci., vol. 8, Carboniferous. [Sig. a small measure]
RHYNCHODUS excavatus, Newberry, 1877, Geo. of Wis., Ham. Gr. [Sig. excavated.]
SANDALODUS complanatus. Newberry & Worthen, instead of Deltodus complanatus.
crassus, Newberry & Worthen, is a syn. for S. spatulatus.
grandis, Newberry & Worthen, is a syn. for S. laevissimus.
Sicarius extinctus, Leidy, 1855, Proc. Acad. Nat. Sci., vol. 7, not satisfactorily defined.
STENACANTHUS nitidus was defined in 1856, in the Jour. Acad. Nat. Sci., vol. 3, and is from the Catskill Gr.
STENOPTERODUS, St. John & Worthen (In press), Geo. Sur. Ill., vol. 7. [Ety. stenos, narrow; pteron, a wing; odous, tooth]
e'ongatus, St. John & Worthen (In press). Geo. Sur. Ill., vol. 7, Warsaw Gr. [Sig elongated.]
parvulus instead of Sandalodus parvulus.
planus, St. John & Worthen (In press), Geo. Sur. Ill., vol. 7, Up. Burlington Gr. [Sig. flat.]
Strigilina gurleiana, refer to Janassa gurleiana.
TÆNIODUS, DeKoninck, MSS., and in press in Geo. Sur. Ill., vol. 7. [Ety. tænia, a ribbon; odous, a tooth.]
fasciatus instead of Deltodus fasciatus.
obliquus, St. John & Worthen (In press), Geo. Sur. Ill., vol. 7, Kaskaskia Gr. [Sig. oblique.]
regularis, St. John & Worthen (In press), Geo. Sur. Ill., vol. 7, Warsaw Gr. [Sig. regular.]
TOMODUS limitaris, St. John & Worthen (In press), Geo. Sur. Ill., vol. 7, Up. Burlington Gr. [Sig. that is on the border.]
TRIGONODUS major. Newberry & Worthen, is a syn. for Sandalodus complanatus.

VATICINODUS, St. John & Worthen (In press), Geo. Sur Ill., vol. 7. [Ety. *vaticinus*, prophetical; *odous*, a tooth.]
carbonarius, St. John & Worthen (In press), Geo. Sur. Ill., vol. 7, Coal Meas. [Sig. pertaining to coal.]
discrepans, St. John & Worthen (In press), Geo. Sur. Ill., vol. 7, Up. Burlington Gr. [Sig. varying.]
lepis, St. John & Worthen (In press), Geo. Sur. Ill., vol. 7, Up. Coal Meas. [Sig. a scale.]
similis, St. John & Worthen (In press), Geo. Sur. Ill., vol. 7, Kaskaskia Gr. [Sig. similar.]
simplex, St. John & Worthen (In press), Geo. Sur. Ill., vol. 7, St. Louis Gr. [Sig. simple.]
vetustus, St. John & Worthen (In press), Geo. Sur. Ill., vol. 7, Kinderhook Gr. [Sig. ancient.]
XYSTRODUS and X. occidentalis were defined in 1870, in Proc. Am. Phil. Soc., vol. 11.
bellulus, St. John & Worthen (In press), Geo. Sur. Ill., vol. 7, Coal Meas. [Sig. elegant.]
imitatus, St. John & Worthen (In press), Geo. Sur. Ill., vol. 7, St. Louis Gr. [Sig. imitating.]
inconditus, St. John & Worthen (In press), Geo. Sur. Ill., vol. 7, Keokuk Gr. [Sig. irregular.]
simplex, St. John & Worthen (In press), Geo. Sur. Ill., vol. 7, Up. Burlington Gr. [Sig. simple.]
verus, St. John & Worthen (In press), Geo. Sur. Ill., vol. 7, Kaskaskia Gr. [Sig. genuine.]

CLASS REPTILIA.

ACHELOMA, Cope, 1882, Pal. Bull. No. 35. [Ety. *achos*, trouble; *loma*, the border.]
cumminsi, Cope, 1882, Pal. Bull. No. 35. Permian. [Ety. proper name.]
ANISODEXIS, Cope, 1882, Pal. Bull. No. 35. [Ety. *anisos*, unequal; *dexios*, on the right.]
imbricarius, Cope, 1882, Pal. Bull. No. 35, Permian. [Sig. having imbrications.]
ARCHÆOBELUS, Cope, 1877, Proc. Am. Phil. Soc. [Ety. *archaios*, ancient; *belos*, the house itself.]
vellicatus, Cope, 1877, Proc. Am. Phil. Soc., Permiar. [Sig. vellicated.]
BAPHETES and B. planiceps were defined in Jour. Geo. Soc. Lond., vol. 10.
BOLOSAURUS, Cope, 1878, Pal. Bull. No. 29. [Ety. *bolos*, the casting of teeth; *sauros*, a sea fish.]
rapidens, Cope, 1878, Pal. Bull. No. 29, Permian. [Sig. rapid.]
striatus, Cope, 1878, Pal. Bull. No. 29, Permian. [Sig. striated.]
CLEPSYDROPS, Cope, 1876, Proc. Acad. Nat. Sci. [Ety. *klepsydra*, an hour glass; *ops*, view.]
colletti, Cope, 1876, Proc. Acad. Nat. Sci., Permian. [Ety. proper name]
gigas refer to Dimetrodon gigas.
limbatus, Cope, 1877, Proc. Am. Phil. Soc., Permian. [Sig. bordered.]
natalis, Cope, 1878, Pal. Bull. No. 29, Permian. [Sig. natal.]
pedunculatus, Cope, 1876, Proc. Acad. Nat. Sci., Permian. [Sig. pedunculated.]
vinslovi, Cope, 1877, Proc. Am. Phil. Soc., Permian. [Ety. proper name.]
CLEPSYSAURUS pennsylvanicus is from the Triassic.
CRICOTUS, Cope, 1876, Proc. Acad. Nat. Sci. [Ety. *krikotos*, ringed.]
discophorus, Cope, 1877, Pal. Bull. No 26, Permian. [Sig. a dish bearer.]
gibsoni, Cope, 1877, Pal. Bull. No. 26, Permian. [Ety. proper name.]
heteroclitus, Cope, 1876, Proc. Acad. Nat. Sci., Permian. [Sig. anomalous.]
DENDRERPETON obtusum, Cope, 1868, Proc. Acad. Nat. Sci., Coal Meas. [Sig. obtuse.]
DIADECTES, Cope, 1878, Pal. Bull. No. 29. [Ety. *dia*, through; *dektes*, a biter.]
latibuccatus, see Empedocles latibuccatus.
molaris, see Empedocles molaris.
phaseolinus, Cope, 1880, Pal. Bull. No. 32, Permian. [Sig. resembling a bean.]
sideropelicus, Cope, 1878, Pal. Bull. No. 29, Permian. [Sig. having great strength.]
DIMETRODON, Cope, 1878, Pal. Bull. No. 29. [Ety. *dimetros*, of two measures; *odous*, tooth.]
cruciger, Cope, 1878, Am. Nat., Permian. [Sig cross bearer.]
gigas. Cope, 1878 (Clepsydrops gigas), Am. Nat., Permian. [Sig. large.]
incisivus, Cope, 1878, Pal. Bull. No. 29, Permian. [Sig. having the quality of cutting or biting.]
rectiformis, Cope, 1878, Pal. Bull. No. 29, Permian. [Sig. straight formed.]
semiradicatus, Cope, 1881, Bull. U. S. Geo.Sur. Terr., vol. 6, No. 1, Permian. [Sig. half radicated.]
DIPLOCAULUS, Cope, 1877, Pal. Bull. No. 26. [Ety. *diploos*, double; *kaulos*, shaft.]

DIPLOCAULUS magnicornis, Cope, 1882, Pal. Bull. No. 35, Permian. [Sig. large horned.]
salamandroides, Cope, 1877, Pall. Bull. No. 26, Permian. [Sig. like a salamander.]
ECTOCYNODON, Cope, 1878, Pal. Bull. No 29. [Ety. *ektos*, far from; *kunos*, dog; *odous*, tooth.]
aguti, Cope, 1882, Pal. Bull. No. 35, Permian. [Ety. proper name.]
ordinatus, Cope, 1878, Pal. Bull. No. 29, Permian. [Sig. ordinated.]
EDAPHOSAURUS, Cope, 1882, Pal. Bull. No. 35. [Ety. *edaphos*, a foundation; *sauros*, a sea fish.]
pogonias, Cope, 1882, Pal. Bull. No. 35, Permian. [Sig. bearded.]
EMBOLOPHORUS, Cope, 1878, Pal. Bull. No. 29. [Ety. *embolos*, anything running to a point; *phoros*, bearing.]
fritillus, Cope, 1878, Pal. Bull. No. 29, Permian [Sig. a dice box.]
EMPEDOCLES, Cope, 1878, Pal. Bull. No. 29. [Ety. proper name.]
alatus, Cope, 1878, Pal. Bull. No. 29. Permian. [Sig. winged.]
molaris, Cope, 1880, Pal. Bull., No. 32, Permian. [Sig. a grinder.]
latibuccatus, Cope. 1878 (Diadectes latibuccatus), Pal. Bull. No. 29, Permian. [Sig. side cheeked.]
EPICORDYLUS, Cope, 1878, Pal Bull. No. 29. [Ety. *epi*, upon; *kordylos*, a water lizard.]
erythroliticus, Cope. 1878, Pal. Bull. No. 29, Permian. [Sig. red stone.]
ERYOPS, Cope, 1877, Proc. Am. Phil. Soc. [Ety. *eryos*, a shoot; *ope*, view.]
megacephalus, Cope, 1877, Proc. Am. Phil. Soc., Permian. [Sig. large headed.]
reticulatus, Cope, 1881, Am. Naturalist, p. 1020, Permian. [Sig. reticulated.]
HELODECTES, Cope, 1880, Pal. Bull. No. 32. [Ety. *helos*, a nail; *dectes*, a biter]
isaaci, Cope, 1880, Pal. Bull. No. 32, Permian. [Ety. proper name.]
paridens. Cope, 1880, Pal. Bull. No. 32, Permian [Sig. equal toothed.]
HYLONOMUS wymani instead of H. hymani.
ICHTHYCANTHUS, Cope, 1877, Pal. Bull No. 24, and Proc. Am. Phil. Soc. [Ety. *ichthys*, a fish; *kanthos*, the corner of the eye.]
ohioensis, Cope, 1877. Proc. Am. Phil Soc., Coal Meas. [Ety. proper name.]
platypus, Cope. 1877. Proc. Am. Phil. Soc., Coal Meas. [Ety. broad footed.]
LEPTOPHRACTUS lineolatus, Cope, 1877, Proc. Am. Phil. Soc., Coal Meas. [Sig. fine lined.]
LYSOROPHUS, Cope, 1877, Pal. Bull. No. 26. [Ety. *lysis*, setting free; *rophos*, supped up.]
tricarinatus, Cope, 1877, Pal. Bull. No. 26, Permian. [Sig. three carinated.]
METARMOSAURUS. Cope. 1878, Pal. Bull No. 29, p. 516.
fossatus, Cope, 1878, Pal. Bull. No. 29, Permian. [Sig. dug. out.]
OPHIACODON, Marsh, 1878, Am. Jour. Sci. and Arts, 3d. ser., vol. 15. [Ety. *ophiakos*, belonging to serpents; *odous*, tooth.]
grandis, Marsh. 1878. Am. Jour. Sci. and Arts, 3d. ser., vol. 15, Permian. [Sig. grand.]
mirus, Marsh, 1878. Am. Jour. Sci. and Arts, 3d ser., vol. 15, Permian. [Sig. wonderful]
PANTYLUS, Cope, 1881, Bull. U. S. Geo. Sur. Terr. vol. 6, No. 1. [Ety. *pan*, all; *tylos* a knob.]
cordatus, Cope, 1881, Bull. U. S. Geo. Sur. Terr. vol. 6, No. 1, Permian. [Sig. cordated.]
PARIOTICHUS, Cope, 1878, Pal. Bull. No. 29. [Ety. *parios*, parian; *tychos*, a hammer.]
brachyops, Cope, 1878, Pal. Bull. No. 29, Permian. [Sig. short sighted.]
PARIOXYS, Cope. 1878, Pal. Bull. No. 29. [Ety. *para*, beside; *oxys*, sharp.]
ferricolus, Cope, 1878, Pal Bull. No. 29, Permian. [Sig. iron distaff.]
RHACHITOMUS, Cope, 1878, Pal. Bull. No. 29. [Ety. *rachis*, a ridge; *tomos*, sharp.]
valens, Cope, 1878, Pal. Bull. No. 29, Permian. [Sig. vigorous.]
SAUROPLEURA longipes, Cope, 1874, Trans. Am. Phil. Soc., vol. 13, Coal Meas. [Sig. long footed.]
SPHENACODON, Marsh, 1878, Am. Jour. Sci. and Arts, 3d. ser., vol. 15. [Ety. *sphen*, a wedge; *akis*, a barb; *odous*, tooth.]
ferox, Marsh, 1878, Am. Jour. Sci. and Arts, 3d. ser., vol. 15, Permian. [Sig. fierce.]
THEROPLEURA, Cope, 1878, Pal. Bull. No. 29. [Ety. *theros*, summer; *pleura*, a rib.]
obtusidens, Cope, 1880, Pal. Bull. No. 32, Permian. [Sig. having obtuse teeth.]
retroversa, Cope, 1878, Pal Bull. No. 29, Permian. [Sig. turned back.]
triangulata, Cope, 1878, Pal. Bull. No. 29, Permian. [Sig. triangular.]
uniformis. Cope, 1878, Pal. Bull. No. 29, Permian. [Sig. uniform.]
TRIMERORHACHIS, Cope, 1878, Pal Bull., No. 29, Permian. [Ety. *trimeres*, tripartite; *rachis*, a ridge.]
insignis, Cope. 1878, Pal. Bull. No. 29, Permian. [Sig. marked.]
TUDITANUS *mordax* is a syn. for Ceraterpeton punctolineatum.
tabulatus. Cope, 1877, Proc. Am. Phil. Soc., Coal Meas [Sig. tabulated.]
ZATRACHYS, Cope, 1878, Pal. Bull. No. 29. [Ety. *za*, an intensive; *trachys*, rough.]
apicalis. Cope. 1881, Am. Naturalist, p. 1020, Permian. [Sig. apical.]
serratus, Cope, 1878, Pal. Bull., No. 29, Permian. [Sig. serrated.]

OMISSION.

POTERIOCRINUS NETTLEROTHANUS, S. A. Miller, 1882, Jour. Cin. Soc. Nat. Hist., vol. 5, Up. Held. Gr. [Ety. proper name.]

—NOTE.—

The author has been informed that Mr. J. M. Clarke has described a new genus and three new species of Phillocaridæ, which will probably appear in the Am. Jour. Sci. & Arts, for February, 1833, viz: Dipterocaris pescervæ, D. pennidædali, and D. procne—all from the Chemung.

Prof. E. W. Claypole says there is little or no classic ground for printing *æ* in the beginning of the second part of a compound word, such as *rugæstriatus*. It should be *rugistriatus*. An exception occurs when the first part of the word is merely a prefix, as in *subæqualis*. That *nebraskensis* should be written instead of *nebrascensis*, as there is no ground for using the *c* instead of *k*. And that the generic name *Chariocephalus* should be written *Charitocephalus*.

Mr. Henry Nettleroth has described, under the title of "Fossil Mollusca of Kentucky," for vol. 1 of that State, which is now in press, Aviculopecten cancellatus, A. concentricus, Paracyclas octerlonyi, Tellinomya striata, Modiomorpha charlestownensis, Millerella (n. gen.) jonesi, Spirifera euruteines, var. erecta, S. euruteines, var. elongata, S. byrnesi, S. hobbsi, S. davisi, S. mcconathyi, Rhynchonella gainesi, Euomphalus protteri, Cyclonema clarki, Cyrtoceras hydraulicum, Murchisonia tuberculata, Platyceras nodulosum, P. elongatum, P. compressum, and Orthis goodwini, from the Devonian formation; and Spirifera knotti, Pentamerella schwartzi, Rhynchonella compressa, Pentamerus foggi, P. trigonalis, P. louisvillensis, and Stricklandia louisvillensis, from the Niagara Group; and Cypricardites halli, from the Hudson River Group.

INDEX OF GENERA.

In addition to indexing the Genera in the whole work, the gender of each genus is designated—*m*, for masculine; *f*, feminine; *n*, neuter.

Genus	PAGE.	Genus	PAGE.	Genus	PAGE.
Acambona, *f*	103	Amplexus, *m*	46, 243, 262	Archimylacris, *f*	225, 319
Acanthaspis, *f*	227	*Ampullaria*	143	Architarbus, *n*	224
Acanthocladia, *f*	289	Ampyx, *m*	209	Archiulus, *m*	224
Acanthograptus, *m*	262	Amygdalocystites, *m*. 70, 279		*Arctinurus*	316
Acantholepis, *f*	227	Anaclitacanthus, *m*	228	Arenicolites, *m*	206
Acantholoma	208	Anarthrocanna, *f*	22	Arges, *m*	210, 316
Acanthophyton, *n*	21	Anastrophia, *f*	104, 294	Arionellus, *m*	210, 316
Acanthotelson, *n*	208	Anatina, *f*	181, 309	*Aristophycus*	248
Acervularia, *f*	46, 262	Ancyrocrinus, *m*	70	Arthraria, *f*	22
Acheloma, *n*	323	Ancimites, *m*	22	Arthrocleina, *n*	95, 289
Acidaspis, *f*	208, 316	Angellum, *n*	309	Arthrolycosa, *f*	224
Aclis	143, 300	Anisodexis, *m*	323	Arthronema, *n*	289
Aclisina, *f*	300	Anisophyllum, *n*	262	Arthrophycus, *n*	23
Acondylacanthus, *m* 227, 319		Annularia, *f*	22, 248	Arthrostigma, *n*	23
Acrocrinus, *m*	66–279	Anodontopsis, *f*	182	Artisia, *f*	23
Acroculia	143	Anomalocrinus, *m*	70, 279	*Asaphiscus*	210
Acrolepis, *f*	227	Anomalocystites, *m*	70, 279	Asaphoidichnus, *m*	316
Acrophyllum, *n*	46, 262	Anomalodonta, *f*	182	Asaphus, *m*	210, 316
Acrotreta, *f*	103	*Anomaloides*	280	Ascoceras, *n*	165
Actinoceras, *n*	165, 305	*Anomia*	104	Ascodictyon, *n*	280
Actinocrinus, *m*	66, 279	*Anomites*	104	*Asolanus*	248
Actinodesma, *n*	180, 309	Anomphalus, *m*	143	Aspidella, *f*	141
Aegilops	103, 294	Anopolenus, *m*	209	Aspidichthys, *m*	228
Agaricia	46	Antholithes, *m*	22, 248	Aspidocrinus, *m*	71
Agaricocrinus, *m*	69, 279	Anthophyllum, *n*	46	Aspidodus, *m*	228
Agassichthys	227	Anthracerpes, *m*	224	Asplenites, *m*	23
Agassizocrinus, *m*	69, 279	Anthracomya, *f*	182	Astarte, *f*	182
Agassizodus, *m*	228	Anthraconectes, *m*	209	Astartella, *f*	182, 309
Agelacrinus, *m*	69, 279	Anthracopalæmon, *m*	210	Asteracanthus, *m*	228
Aglaspis, *f*	209, 316	Anthracoptera, *f*	182, 309	*Asterias*	71
Agnostus, *m*	209, 316	Anthracopupa, *f*	300	Asterocarpus, *m*	23
Agraulos, *m*	209, 316	Anthracosia, *f*	182	*Asterocrinus*	71
Alecto	95, 289	Antliodus, *m*	228	Asterophycus, *n*	23, 248
Alethopteris, *f*	21, 248	Apedodus, *m*	228	Asterophyllites, *f*	23, 248
Allagecrinus, *m*	279	Aphlebia, *f*	248	Asteropteris, *f*	248
Alloprosallocrinus, *m* 70, 279		Apiocystites, *m*	70	Asteroptychius, *m*	228, 320
Allorisma, *n*	180, 309	Arabellites, *m*	313	Asterosteus, *m*	228
Alveolites, *m*	46, 262	Arachnocrinus, *m*	280	Astræa	46, 262
Amacanthus, *m*	228, 319	Arachnophyllum, *n*	262	Astræophyllum, *n*	262
Amblypterus, *m*	228	Araucarites, *m*	22	Astræospongia, *f*	42
Ambocoelia, *f*	103	*Arca*	182, 309	*Astrios*	71
Ambonychia, *f*	181, 309	Archaeobelus, *m*	323	Astrocerium, *n*	47
Ammonites	305	Archaeocaris, *f*	210	Astroconia, *n*	260
Ampheristocrinus, *m*	279	Archaeocidaris, *f*	70, 280	*Astrocrinites*	71
Amphibamus, *m*	240	Archæocrinus, *m*	280	Astylospongia, *f*	42, 260
Amphicoelia, *f*	181	Archaeopteris, *f*	22, 248	Atactopora, *f*	289
Amphigenia, *f*	104	Archegogryllus, *m*	225	*Ataxocrinus*	71
Amphion, *m*	209	Archaeocyathellus, *m*	260	Ateleocystites, *m*	71
Amphipeltis, *f*	209, 316	Archaeocyathus, *m*	42	Athyris, *f*	104, 294
Amphoracrinus, *m*	70, 279	Archimedes, *m*	95	Atops, *f*	211, 316
Amplexopora, *f*	289	*Archimedipora*	95	Atrypa, *f*	105, 294

INDEX OF GENERA.

Aulacophyllum, *n*......262
Aulocopina, *f*............42
Aulophyllum, *n*.........47
Aulopora, *f*.....47, 64, 263
Aulosteges, *m* 106
Avicula, *f* 182, 309
Aviculopecten, *m*..184, 310
Aviculopinna, *f*........185
Azinura 47
Azinus 185
Axophyllum, *n* 47, 64
Bactrites, *m* 305
Baiera, *f*..............248
Bakevellia, *f*..........185
Balanocrinus 71
Baphetes, *m*.....240, 323
Barrandia............211
Barycrinus, *m* 71, 280
Baryphyllum, *n* 47, 263
Batacanthus, *m*...228, 320
Bathycheilodus, *m*.....229
Bathynotus, *m*.........211
Bathyurellus, *m*........211
Bathyurus, *m*......211, 316
Batocrinus, *m*......71, 280
Batostoma, *n*..........280
Batostomella, *f*........280
Beatricea, *f*...165, 260, 305
Bechera, *f*............ 23
Beinertia, *f*........... 23
Belemnocrinus, *m* 72, 94, 280
Belinurus, ...212
Bellerophon, *m*.....143, 300
Berenicea, *f* 289
Bergeria..............248
Beyrichia, *f*......212, 316
Blastoidocrinus, *m* 72
Blastophycus, *n*........240
Blattina, *f* 225
Blothrophyllum, *n*..47, 263
Blumenbachium........42
Bolboporites, *m*........ 47
Bolosaurus, *m*........ 323
Bornia, *f*..............249
Botryllopora, *f* 96
Brachiocrinus, *m* 72
Brachiospongia, *f*..42, 64
Brachydectes, *m*.......240
Brachymerus.........106
Brachyphyllum, *n*...23, 249
Brachyprion..........107
Brongniartia 212
Bronteus, *m*......212, 316
Bruckmannia.........249
Bucanella, *f*...........145
Bucania, *f*......145, 300
Bulimella 145
Bullmorpha, *f* 301
Bumastus 212
Bursacrinus, *m*.........72
Buthograptus, *m*........47
Buthotrephis, *f* ..23, 249
Bythiacanthus, *m*......229
Bythopora, *f*..........280
Cacabocrinus 72
Calamites, *m*......24, 249
Calamocladus, *m*... 24
Calamodendron, *n*...24, 249

Calamophycus, *n*......249
Calamopora........47, 263
Calamostachys, *m*......249
Calapœcia, *f*............47
Calathium, *n*...........43
Calathocrinus...........72
Calcarina, *f*...........260
Calceocrinus, *m* ..72, 280
Calceola. *f*....107, 263, 295
Calcisphæra, *f*.......260
Callipteridium, *n*.......249
Callipteris, *f*......24, 249
Callocystites, *m*........73
Callograptus, *m*.....47, 263
Callonema, *n*..........301
Callopora, *f*........96, 289
Calophyllum, *n*........263
Calopodus, *m*..........229
Calymene, *f*......213, 316
Calyptograptus, *m*.....264
Camarella.......107, 295
Camarium, *n*..........107
Camaroceras, *n*........165
Camarophoria, *f*...107, 295
Campanulites...........73
Campophyllum, *n* ..48, 64
Caninia..................48
Cannapora, *f*......48, 264
Capulus, *m*............146
Caraboerinus, *m*........73
Carbonarca, *f*..........185
Carcharopsis, *f*.......229
Cardinia, *f*......185, 310
Cardiocarpon, *n*...24, 249
Cardiola, *f*.......186, 310
Cardiomorpha, *f* .186, 310
Cardiopsis, *f*......186, 310
Cardiopteris, *f*........249
Cardium, *n*......186, 310
Carinaropsis, *f*........146
Carinopora, *f*..........96
Carpolithes, *m*....25, 250
Caryocrinus, *m* 73
Caryocystites........... 73
Caryophyllia 264
Catenipora 48, 264
Casuarinites 250
Catillocrinus, *m*........73
Catopterus,.....229, 320
Caulerpites, *m*......25, 250
Caulopteris. *f*......25, 250
Caunopora, *f*..........48
Celluloxylon, *n*........250
Centrocrinus..........280
Centronella, *f*......107, 295
Cephalaspis, *f*........229
Ceramopora. *f*...96, 290
Ceraterpeton, *n*.......240
Ceratocaris, *f*.....213, 316
Ceratocephala........213
Ceraurus, *m*......213, 316
Ceriopora, *f*............96
Chænocardia, *f*........186
Chænomya, *f*.....186, 310
Chætetes, *m*.....48, 264
Chariocephalus, *m*.214, 316
Charionella............108
Cheirocrinus..............73

Cheirodus, *m*..........229
Cheirotherium, *n*......240
Cheirurus..............214
Chemnitzia, *f* ...146, 301
Chiton, *m*........141, 301
Chitonodus, *m* 320
Chloephycus..........250
Cholodus, *m*..........229
Choinatodus, *m*........229
Chondrites, *m*.....25, 250
Chonetes, *m*.....108, 295
Chonograptus, *m*...48, 64
Chonophyllum, *n* 48, 64, 264
Chonostegites......48, 264
Chrestotes, *f*..........225
Cimitaria, *f*..........187
Cladodus, *m*.....229, 320
Cladograptus, *m*........ 49
Cladopora, *f*...49, 64, 264
Clathrocœlia, *f*......209
Clathropora, *f*.....96, 290
Cleidophorus, *m* ..187, 310
Cleiocrinus, *m*.........73
Clepsydrops, *f*.......323
Clepsysaurus, *m*...240, 323
Climactichnites, *m*.....214
Climacograptus, *m*.....49
Climacodus, *m*........230
Clinopistha, *f*..........187
Clioderma 141
Clisiophyllum, *n*...49, 265
Clisospira, *f*......146, 301
Clonopora, *f* 290
Closterocrinus, *m*.......73
Clymenia..............165
Cnemidium, *n*.......43, 64
Coccocrinus, *m* 73
Coccosteus, *m*.........230
Cochliodus, *m*.....230, 320
Cocytinus, *m*,........ 24
Codaster, *m*......73, 280
Codonites. *m*.......74, 280
Cœlacanthus, *m*.......230
Cœliocrinus, *m* ... 74, 280
Cœlocrinus, *m*.........74
Cœlospira, *f*......... 109
Coenites 265
Coleolus, *m*............299
Coleoprion, *m* 141, 299
Collettosaurus, *m*......240
Colosteus, *m*..........240
Colpocaris 214
Colpoceras, *n*....165, 305
Columnaria, *f* ...49, 265
Columnopora, *f*.....49, 265
Comarocystites, *m* ..74, 281
Combophyllum, *m*.....49
Compsacanthus, *n* 231
Conchicolites, *m*...206, 313
Conchiopsis.............231
Conchodus, *m* 231
Conchopeltis, *f*.......146
Conilites..............165
Conocardium, *n* .187, 310
Conocephalites, *m*..214, 316
Conocephalus.........214
Conoceras, *n*..........165
Conocoryphe, *f*........214

INDEX OF GENERA.

Conocrinus 74	Cypricardia, *f*.....188, 310	Dinichthys, *m* 232
Conophyllum 50	Cypricardinia, *f*...188, 310	Dinoholus, *m* 110
Conopterium, *n* 43	Cypricardites, *m*...188, 310	Dionide, *f* 217
Conostichus, *m*....25, 250	Cyrtacanthus, *m*.......231	Diorrychopora, *f*......267
Conotubularia 165	Cyrtia, *f*..............110	Diphyphyllum, *n*...52, 267
Constellaria 50	Cyrtina, *f*......110, 295	*Diplazites* 27
Conularia, *f*....141, 299	Cyrtoceras, *n*.....166, 305	*Dipleura*.............217
Conulites 165	Cyrtoceriua, *f*........167	Diplichnites, *m* 217
Copodus, *m*..........320	*Cyrtodonta* 190	Diplocaulus, *m*.......323
Cordaianthus. *m*.....250	Cyrtolites, *m*......147, 301	Diploceras, *n* 168
Cordaicarpus, *m*......250	Cyrtonella, *f*.........301	Diplodus, *m*......232, 321
Cordaistrobus, *m* 250	Cystiphyllum, *n*...50, 266	Diplograptus, *m*....52, 268
Cordaites, *m*. ...25, 250	Cystocrinus, *m*........75	*Diplophyllum*..........52
Cordylocrinus, *m*....281	Cystodictya. *f*........290	Diplostegium, *n*.......27
Cornulites, *m*....206, 313	Cystopora, *f*.........290	Diplostylus, *m*.......217
Coronocrinus, *m*......74	Cystostylus, *m* 267	Diplotrypa, *f*....268, 290
Coscinium, *n*..........96	Cythere, *f*......215, 317	Dipterocaris, *f*........325
Coscinopora...........43	Cytherellina, *f*.......317	Dipterus, *m* 232
Cotyledonocrinus, *m*..74	Cytherina, *f*.........215	Discina, *f*........111, 295
Crania, *f*........109, 295	Cytherodon, *m* 190	Discites, *m* 168, 305
Craspedophyllum, *n*..265	Cytheropsis, *f*....215, 317	*Discolites* 301
Crateripora..........290	Cytocrinus, *m*.....76, 281	Discophycus, *n*.......251
Cremutopteris, *f*.....25	Dactylodus, *m*........232	Discophyllum, *n* 52
Crepicephalus, *m*..215, 317	Dactylophycus, *n*.....251	Discosorus, *m* 168
Crepidophyllum 265	Dadoxylon, *n*..........26	Discotrypa, *f* 290
Cricotus, *m*..........323	*Dæmonocrinites*.......76	Distacodus, *m*........313
Crinocystites, *m*......74	*Dalmania* 215	Dithyrocaris, *f* 217
Crinosoma 281	Dalmanites, *m*...215, 317	Dolabra, *f*...........190
Crisina, *f* 290	Danæites, *m*.....26, 251	Dolatocrinus, *m*...77, 282
Cromyocrinus, *m*.....281	Dania, *f*.............267	Dolichometopus, *m*....217
Crumenæcrinites 74	Dawsonella, *f*........147	Dolichopterus, *m* ..217, 317
Cruziana, *f*...........26	Dawsonia, *f*...........51	*Donacicrinites* 77
Cryphæus.............215	*Decadactylocrinites* ..76	Dorycrinus, *m*.....77, 282
Cryptoceras 166	*Decadocrinus*.........281	Drepanacanthus, *m*...232
Cryptolithus.........215	Decheni, *f*...........251	Drepanodus, *m*.......313
Cryptonella, *f*...109, 295	Dekayia, *f*...........267	Drymopora, *f*........268
Cryptopora, *f*.........97	Deltodopsis, *f*.......320	Duncanella, *f*.........52
Ctenacanthus, *m*..231, 320	Deltodus, *m*.....232, 320	Dyscritus, *m*.........225
Ctenocrinus, *m* 74	*Delthyris* 110	*Dystactella*, *f*......190
Ctenodonta..........187	Deltoptychius, *m*.....321	Dystactophycus, *n*....251
Ctenodus, *m*.....231, 320	Dendrerpeton, *n*..240, 323	Dystactospongia, *f*...260
Ctenopetalus, *m*.....231	Dendrocrinus, *m*..76, 281	Eatonia, *f* 111
Ctenoptychius, *m*..231, 320	Dendrographus, *m*..51, 267	Ecculiomphalus, *m* ..147
Cucullæa............188	Dendropora, *f*.....51, 267	Echinocaris, *f*........317
Cuneamya, *f*.....188, 310	Dentalina, *f*..........43	Echinocystites, *m*.....77
Cupellæcrinus........74	Dentalium, *m*.....147, 301	Echinoeucrinites, *m* ..77
Cupulocrinus, *m* 281	Desmiodus, *m* 232	Echinognathus, *m*....317
Cyathaxonia, *f*....50, 265	Desmiophyllum, *n*....251	*Echinus* 77
Cyathocrinus, *m*..74, 281	Dexiobia, *f*.....190, 310	Ectocynodon, *m*......324
Cyathophycus, *n*......266	Diadectes, *m*........323	Ectosteorachis, *m* ...321
Cyathophyllum, *n*50, 64, 265	*Diamesopora* 97	Edaphosaurus, *m*....324
Cyathopora..........50	Dicellocephalus, *m*.216, 317	Edestes, *m*..........233
Cyathospongia, *f*...260	Dichocrinus, *m*...76, 282	Edmondia, *f* 190-310
Cybele 215	*Dichograptus* 51	Edrioaster, *m* 77
Cyclaster 75	*Dicraniscus* 110	Edriocrinus, *m*........77
Cycloconcha, *f*......188	Dicranographus, *m* ...51	Eichwaldia, *f* 111
Cyclocystoides, *m* ..75, 281	Dicranophyllum, *n*...251	*Elæacrinus*............78
Cyclolites, *m* 50	Dicranopora, *f* 290	Elasmophyllum, *n* ...268
Cyclonema, *n* ...146, 301	*Dicranurus* 217	Eleutherocrinus, *m*....78
Cyclopora, *f*..........97	Dictyocrinus, *m*.......77	Elliptocephala.....217, 317
Cyclopteris, *f*....26, 250	*Dictyolites* 26	*Elonichthys* 233
Cyclora, *f*......147, 301	Dictyonema, *n*...51, 64, 267	Embolophorus, *m* ...324
Cyclostigma, *n*...26, 251	Dictyophyllum, *n*....251	Emmonsia, *f*..........52
Cyclostoma 147	Dictyophyton, *n* 26, 251, 260	Empedocles, *m*. 324
Cymatodus, *m* 232	Dictyopteris, *f*.......26	Encrinurus, *m*....217, 317
Cymoglossa, *f* 251	Dictyostroma, *n*....52, 64	Endoceras, *n* ...168, 305
Cyphaspis, *f* 215	Didymograptus, *m* ...52	*Endolobus* 168
Cypricardella, *f*....188, 310	Didymophyllum, *n*..27, 251	Endothyra, *n* 261
	Dimetrodon, *m* 323	

INDEX OF GENERA.

	PAGE.		PAGE.		PAGE.
Endymion	217	Gampsacanthus, m	233	*Hipparionyx*	112
Endymionia, f.	217	Geisacanthus, m	233	Hippodophycus, n	27
Enoploura, f.	317	*Gennæocrinus*	282	*Hippothoa*	99
Entolium, n	191	Gerephemera, f.	225	Holocystites, m	81, 283
Eocidaris, f.	78	Gervillia, f	191	Holometopus, m	218
Eocystites, m	78	Glauconome, f	98, 291	Holopea. f.	149, 301
Eodon, m	244	Glossoceras, n	168	Holopella, f.	149, 301
Eophyton, n	27	*Glossogryptus*	54	Holoptychius, m	234
Eopteria, f.	191	Glycerites, m	314	*Homæcanthus*	234
Eosaurus, m	241	Glymmatacanthus, m	233, 321	Homalonotus, m	219, 318
Eoscorpius, m	224, 319	Glyptaster, m	79, 283	Homocrinus, m	81, 283
Eospongia, f.	43	Glyptocrinus, m	79, 283	Homothetus, m	225
Eotrochus, m	301	Glyptocystites, m	80	Homotrypa, f.	292
Eotrophonia	313	Glyptodendron. n	251	Hornera, f.	99
Eozoon, n	43	Gomphoceras, n	168, 305	*Hortholus*	306
Ephemerites, m	225, 319	Gomphocystites, m	80	*Houghtonia*	56
Epicordylus, m	324	Goniasteroidocrinus, m	80	Huronia, f.	170, 307
Equisetites, m	27, 251	Goniatites, m	169, 306	Hybocladodus, m	234
Equisetum	27	Gonioceras, n	170	Hybocrinus, m	82
Eremopteris, f	27, 251	*Goniocælia*	112	Hybocystites, m	283
Eretmocrinus, m	78, 282	Goniophora, f	192, 311	*Hydnoceras*	170
Eridophyllum, n	53, 268	*Goniopteris*	251	Hydrelonocrinus, m	82, 284
Eridopora, f.	290	Gordia	27	Hylerpeton, n	241
Erismacanthus, m	233	Gorgonia. f	98, 291	Hylonomus, m	241, 324
Erisocrinus, m	78, 282	Grammysia, f.	192	Hymenophyllites, m	27, 252
Eryops. f.	324	Granatocrinus, m	80, 283	Hyphasma, n	241
Eschara, f.	97	Graphiocrinus, m	81, 283	Hyolithellus, m	141
Escharopora, f	97, 290	Graptodictya, f.	291	Hyolithes. m	141, 306
Estheria, f.	317	Graptolithus, m	54, 244, 269	*Hypanthocrinites*	82
Ethmophyllum, n	53	*Gryphorhynchus*	192	Ichnophycus, n	28
Eucalyptocrinus, m	78, 282	Guillelmites	257	Ichthyacanthus, m	324
Euchasma, n	191	Gypidula, f.	112, 295	Ichthyocrinus, m	82, 284
Euchondria, f.	191	Gyracanthus, m	233, 321	Ichthyorachis, f.	99
Eucladocrinus, m	94, 282	Gyroceras, n	170, 244, 306	*Icosidactylocrinites*	82
Eugaster, m	78	Hadrocrinus, m	81	Idiophyllum, n	252
Eulima	147	Hadrophyllum, n	56, 269	*Ilionia. f*	192
Eumicrotis	191	Haimeophyllum, n	56	Illænurus, m	218, 318
Eunema, n	148, 301	Hallia, f	269	Illænus, m	218, 318
Eunemacanthus, m	321	Halonia, f.	27, 251	*Inachus*	149, 301
Eunicites, m	313	Halysites, m	56, 64, 244, 269	Inocaulis, m	56, 270
Euomphalus, m	148, 244, 301	Haplocrinus, m	81	*Inoceramus*	192
Eupachycrinus, m	79, 282	Haplophlebium, n	225	Intrapora, f.	292
Euphoberia, f	224, 319	Harlania, f.	27	Intricaria. f	99, 292
Euproops, f.	217	Harpacodus, m	233	Iocrinus, m	284
Eurylepis, f.	233	Harpes. m	218	Iphidea, f	112
Eurypterus, m	217, 318	Harpides, m	218	*Ischadites*	43, 261
Eurythorax, m	241	Helicotoma, f.	149, 301	Ischyrinia, f.	192
Eusarcus, m	244, 318	Heliodus, m	233	*Isocardia*	192, 311
Evactinopora, f.	97	Heliolites. m	56, 269	Isochilina, f	219, 318
Exochorhynchus	191	Heliophycus, n	262	Isonema, m	149, 301
Faviphyllum, n	53	Heliophyllum, n	54, 244, 269	Isotelus, n	219, 318
Favistella, f.	53, 268	Helminthoidichnites, m	218	Janassa, f	321
Favosites, m	53, 64, 244, 268		318	Knorria, n	28
Favositopora	269	Helolectes, m	324	Koninckia, f.	112, 295
Fenestella, f	97, 290	Helodus, m	233, 321	Kutorgina, f.	112, 295
Fenestralia, f	98	Helopora, f	98	Lamblodus, m	234
Ficoidites	251	*Heterotrypa, f.*	270	Lamellipora, f.	56
Filicites	27, 54, 251	Hemeristia, f	225	Lamperocrinus, m	82, 284
Fissodus, m	233	Hemicosmites. m	81	Leaia, f	219
Fistulipora, f.	54, 269	*Hemicrypturus*	218	Lecanocrinus, m	82, 284
Flabellaria	27	Hemicystites, m	81	Lecracanthus, m	234
Flustra, f.	98	*Hemipronites*	112, 295	*Lecythiocrinus*	284
Forbesiocrinus, m	79, 282	Hemitrypa, f.	98, 291	*Leda*	193
Fucoides	27	Heterocrinus, m	81, 283	Leioclema, n	292
Fusispira, f.	148, 301	Heterocystites, m	81	Leiodus. m	243
Fusilina, f.	43, 261	*Heterodictya*	99	Leiopteris, f.	193
Fusus	149	Heterophrentis, f.	56	Leiorhynchus, n	112, 295
Galium	27	Hindeia, f	261	Lepadocrinus, m	82

INDEX OF GENERA. 331

	PAGE.		PAGE.		PAGE.
Leperditia, *f.*	219, 318	Lyropora, *f.*	90	Myelodactylus, m	84, 244
Lepetopsis, *f.*	301	Lysorophus, m	324	Mylacris, *f.*	226
Lepidechinus, m	82	Machæracanthus, m	235	Myrianites, m	31, 253
Lepidesthes, *f.*	82, 284	*Machairodus*	314	Myrtillocrinus, m	84
Lepidocidaris	82	Maclurea, *f.*	150, 302	Mytilarca, *f.*	197, 311
Lepidodendron, n	28, 252	Macrocheilus, n	151, 244, 302	Mytilus, m	197
Lepidodiscus	82	Macrochilina, *f.*	302	*Naiadites,*	197
Lepidocystis, m	252	Macrodon, n	194, 311	*Natica*	154
Lepidolites, m	261	Macropetalichthys, m	235	Naticopsis, *f.*	154, 303
Lepidophloios, m	29, 252	Macrostachya, *f.*	253	Nautilus, m	171, 307
Lepidophyllum, n	30, 252	Macrostylocrinus, m	83, 284	Nebulipora, *f.*	58
Lepidostrobus, m	30, 253	*Madrepora*	57	*Nelimenia*	307
Lepidoxylon, n	253	Malocystites, m	83	Nemagraptus, m	58
Lepocrinites	82	Mariacrinus, m	83	Nemapodia, *f.*	31, 253
Leptacanthus	234	Marracanthus, m	235	Nematophycus, n	31
Leptæna, *f.*	112, 295	Marsuplocrinus, m	83, 284	Nematophyllum, n	253
Leptobolus, m	113	Martinia *f.*	115	Nematoxylon, n.	31
Leptocœlia, *f.*	113	Matheria, *f.*	194	*Nephropteris*	31
Leptodomus, *f.*	193, 311	Mazonia, *f.*	224	Nereidavus, m	314
Leptophlœum, n	29	Mecolepis, *f.*	235	Nereites, m	31, 253
Leptophractus, m	234, 241, 321, 324	Meekella, *f.*	115	Nereograptus, m	58
		Megalaspis, *f.*	220	Neriopteris, *f.*	31
Leptopora, *f.*	56, 64, 270	Megalograptus, m	57	Neuropteris, *f.*	31, 253
Lescuropteris, *f.*	30, 253	Meg.lomus, m	194, 311	Nicholsonia, *f.*	271
Lesleya, *f.*	253	Megalopteris, *f.*	31, 253	Nileus, m	220
Libellula, *f.*	225	Megambonia, *f.*	194	Nipterocrinus, m	84
Lichas, m	220, 318	*Meganteris.*	115	Nodosinella, *f.*	261
Lichenalia, *f.*	99, 202	Megaphyton, n	31, 253	Nœggerathia, *f.*	32, 254
Lichenocrinus, m	82, 284	*Megaptera*	194	Nucleocrinus, m	84
Licrophycus, n	30, 253	Megathentomum, n	235	Nucleospira, *f.*	116, 206
Lima, *f.*	193	Megistocrinus, m	83, 284	Nucula, *f.*	197, 311
Limaria, *f.*	56, 270	*Melia*	307	Nuculana, *f.*	198, 311
Limoptera, *f.*	193	Melocrinus, m	83, 284	Nuculites, m	198, 311
Lindstromia, *f.*	270	Melonites, m	84	Nullipora, *f.*	43
Lingula, *f.*	113, 244, 295	Menocephalus, m	220	*Nuttainia*	220
Lingulella, *f.*	115	Merista, *f.*	115	Nyassa, *f.*	198
Lingulepis, *f.*	115, 296	Meristella, *f.*	115, 296	Obolella, *f.*	116, 296
Lingulops, *f.*	115	Meristina, *f.*	116	*Obolellina*	116
Linipora	57	Mesodmodus, m	235	Obolus, m	116
Liognathus, m	234	Mespilocrinus, m	84	*Odontocephalus*	221
Liagocaris,	318	Metarmosaurus, m	324	*Odontochile*	221
Liagodus, m	234	Metoptoma, *f.*	151, 302	Odontopteris, *f.*	33, 254
Listracanthus, m	235	Miamia, *f.*	225	Oenites, m	314
Lithentomum, n	225	Michelinia, *f.*	57, 244, 271	Oestocephalus, m	241
Lithophaga, *f.*	193, 311	Microceras, n	152, 302	Ogygia, *f.*	221, 318
Lithostrotion, n	57, 270	Microcyclus, m	57	Oldhamia, *f.*	58
Litnites, m	170, 307	Microdiscus, m	220, 318	Olenellus, m	221, 318
Littorina,	150	Microdoma, *f.*	152	Olenus, m	221
Loftusia, *f.*	261	*Microdon,*	194	Oligocarpia, *f.*	254
Loganellus	220	Microspongia, *f.*	261	Oligoporus, m	84, 284
Lonchocephalus	220	*Millepora*	57	*Olivanites*	84
Lonchopteris, *f.*	30	Millerella, *f.*	325	*Ollacrinus.*	284
Lonsdaleia, *f.*	57	Milleria, *f.*	271	*Omphalotrochus*	154
Lophodus	235	Mitoclema, n	202	Omphyma, *f.*	58, 272
Lophophyllum, n	57, 270	Modiola, *f.*	194	*Onchus*	235
Loxonema, n	150, 244, 302	Modiolopsis, *f.*	195, 311	Oncoceras, n	172, 307
Lucina, *f.*	193, 311	Modiomorpha, *f.*	195, 311	Onycha-ter, m	84
Lumbriconereites, m	314	Molgophis, m	241	Onychocrinus, m	84, 284
Lunatipora, *f.*	57	Monocraterion, n	314	Onychodus, m	235
Lunulicardium, n	193, 311	Monograptus, m	57, 271	Ophiacodon, m	324
Lunulites	43	Monomerella, *f.*	116	Ophileta, *f.*	154, 303
Lycopodiolithes	253	Monopteria, *f.*	196	*Opisthoptera.*	199
Lycopodites, m	31, 253	Monotis, *f.*	196, 311	Oracanthus, m	235, 321
Lyellia, *f.*	57, 270	*Monotrypa,*	271	Orbicula	116
Lyonsia	193	Monotrypella, *f.*	202	Orbiculoidea, *f.*	296
Lyriocrinus, m	83, 284	Monticulipora, *f.*	57, 271, 202	*Orbitulites*	43
Lyriopecten, m	193	Murchisonia, *f.*	152, 244, 302	Ormathichnus, m	303
Lyrodesma, n	193	Myalina, *f.*	196, 311	Ormoceras, n	173

INDEX OF GENERA.

Name	Page
Ormoxylon, n	33
Ornithichnites, m	241
Orodus, m	235
Orthacanthus, m	236, 321
Orthis, f	116, 296
Orthisina, f	120
Orthoceras, n	173, 244, 307
Orthogoniopteris, f	33
Orthodesma, n	199, 311
Orthonema, n	155
Orthonota, f	199, 311
Orthonotella, f	311
Orthonychia	155
Orthopleurodus, m	321
Orthostoma, n	155
Ortonia	206
Ostrea	199, 311
Pachycrinus, m	84
Pachydictya, f	292
Pachylocrinus	284
Pachyphyllum	33
Pachyphyllum, n	59
Pachypora, f	272
Pachypteris, f	33
Palæacis, f	43, 261
Palæacmæa, f	155, 303
Palæanatima, f	199
Palæarca	199
Palæaster, m	84, 284
Palæasterina, f	85, 285
Palæchinus, m	85
Palæobatis, m	321
Palæocampa, f	226
Palæocardia, f	199
Palæocaris, f	221
Palæochorda, f	33, 254
Palæocoma, f	85
Palæocrinus, m	85
Palæocyclus, m	59, 272
Palæocystites, m	85
Palæomanon, f	43, 261
Palæonello, f	199, 311
Palæoniscus, m	236, 321
Palæopalæmon, m	318
Palæophycus, n	33, 254
Palæophyllum, n	59
Palæopteris	33
Palæotrochis	59
Palaeotrochus, m	303
Palæoxyris, f	33, 254
Paleschara, f	99, 292
Palmacites	254
Panopæa	199
Pantylus, m	324
Paolia, f	226
Paracyclas, f	199, 311
Paradoxides, m	221
Pariotichus, m	324
Parioxys, m	324
Parisocrinus	285
Psceolus, m	43
Patella, f	303
Pattersonia, f	261
Pecopteris, f	34, 254
Pecten	200
Pelion, m	241
Peltodus, m	236
Peltura	221
Pemphigaspis, f	221
Pentacrinites	85
Pentagonia, f	120
Pentagonites	85
Pentamerella, f	120
Pentamerus, m	120, 296
Pentremites, m	85, 285
Peplorhina, f	236
Pereichocrinus	285
Periplectrodus, m	236
Peripristis, m	321
Pernopecten, m	200, 312
Peronopora, f	272, 292
Petalichnus, m	309
Petalodus, m	236, 321
Petalorhynchus, n	237, 321
Petigopora, f	292
Petraia, f	59
Petraster, m	86
Petrodus, m	237
Phacops, f	221
Phænopora, f	99
Phanerotinus, m	155, 303
Phœbodus, m	237
Phillipsia, f	221, 245, 318
Phillipsastrea, f	59, 272
Philocrinus	86
Phlegethontia, f	241
Pholadella, f	200
Pholadomya	312
Pholidocidaris, m	86
Pholidops, f	121
Phractopora, f	292
Phragmoceras, n	178, 309
Phragmolites	155
Phragmostoma, n	155, 303
Phthonia, f	200
Phyllodictya, f	292
Phyllograptus, m	59
Phyllopora, f	99, 292
Phyllopteris, f	35
Physetocrinus, m	86, 285
Physonemus, m	237, 321
Physophycus, n	35, 255
Phytolithus	35, 255
Phytopsis, f	35
Pileopsis	155, 303
Piliolites	222
Piloceras, n	178, 309
Pinna, f	200, 312
Pinnopsis	200
Pinnularia, f	35
Pisocrinus, m	285
Placunopsis, f	200
Planolites	314
Planorbis	155
Plasmopora, f	59
Platephemera, f	226
Platyceras, n	155, 245, 303
Platycrinus, m	86, 285
Platyodus, m	237
Platynotus	222
Platyschisma, f	157
Platysomus, m	237
Platystoma, n	157, 245, 303
Platystrophia	121
Plasmopora, f	59
Plectambonites	121
Plectostylus	157
Pleuracanthus, m	237
Pleurocystites, m	88
Pleurodictyum, n	59, 272
Pleuronotus, m	303
Pleurophorus, m	200
Pleuroptyx, f	241
Pleurorhynchus	201, 312
Pleurotomaria, f	157, 245, 303
Plicatula	121
Plumulina, f	59, 272
Plumulites, m	222, 318
Pnigeacanthus, m	237, 321
Poeciloduns, m	237, 322
Polycronites	178
Polydilasma	59
Polygnathus, m	314
Polyphemopsis, f	161, 245, 304
Polypora, f	99, 292
Polyporites, m	35, 255
Polyrhizodus, m	237
Polyspora	255
Porambonites, m	121, 296
Porcellia, f	161, 304
Porites	59
Porocrinus, m	88, 286
Posidonia, f	201, 312
Posidonomya, f	201, 312
Poteriocrinus, m	88, 286, 325
Prasopora, f	272, 293
Primitia, f	222, 318
Prioniodus, m	315
Prionotus	59
Prisconaia, f	312
Primopora, f	293
Pristicladodus, m	238
Pristodus, m	238
Procteria, f	272
Productella, f	121
Productus, m	122, 245, 296
Proetus, m	222, 318
Promacrus, m	201
Protaræa, f	60
Protaster, m	89, 287
Protasterina	287
Prothyris, f	201
Protichnites, m	222, 319
Protoblechnum, n	255
Protocyathus, m	261
Protoscolex, m	315
Protostigma, f	255
Prototaxites, m	35
Psammodus, m	238, 322
Psaronius, m	35
Psephodus, m	238, 322
Pseudocrania, f	124
Pseudomonotis, f	201, 312
Pseudopecopteris, f	255
Psilophyton, n	35, 256
Pterichthys, m..?	238, 322
Pterinea, f	201, 312
Pterocephalia, f	319
Pteronautilus, m	178
Pteronitella, f	202
Pteronites, m	202, 312
Pterotheca, f	142
Pterotocrinus, m	89, 287
Pterygotus, m	223, 319

INDEX OF GENERA.

Genus	PAGE
Ptilocarpus, m	35, 256
Ptilodictya, f	100, 293
Ptilograptus, m	60, 272
Ptilonaster	89
Ptilophyton	256
Ptilopora, f	100
Ptychaspis, f	223, 319
Ptychodesma, n	202, 312
Ptychophyllum, a	60, 272
Ptyctodus, m	238
Ptyonius, m	241
Ptyonodus, m	322
Pugiunculus	142
Pupa, f	162, 304
Pyanomya, f	312
Pygopterus, m	238, 241
Pygorhynchus, n	89
Pyrenomoeus, m	202
Quenstedtia	60, 245, 272
Ramipora, f	293
Raniceps	241
Raphistoma, n	162, 304
Rastrites, m	60
Receptaculites, m	43, 261
Remopleurides, m	223
Rensselæria, f	124, 297
Retepora, f	100, 293
Retiocrinus, m	90, 287
Retiograptus, m	60
Retiolites, m	60
Retzia, f	124, 297
Rhabdaria, f	44
Rhabdichnites	223
Rhabdocarpus, m	36, 256
Rhachiopteris, f	36, 256
Rhachitomus, m	324
Rhachura, f	319
Rhacophyllum, a	36, 256
Rhadinichthys, m	322
Rhinidictya, f	293
Rhinipora, f	100
Rhizodus, m	238
Rhizograptus, m	272
Rhizolithes	36
Rhizomopteris, f	36
Rhizomorpha, f	257
Rhodocrinus, m	90, 287
Rhombopora, f	60
Rhynchodus, m	238, 322
Rhynchonella, f	125, 297
Rhynchospira	128, 297
Rhynchotreta, f	297
Rhynobolus	128
Ribeiria, f	44
Rinodus	238
Romingeria, f	272
Ropalonaria	293
Rotalia, f	44, 261
Rotella, f	304
Rotularia	36
Rusichnites, m	223, 319
Rusophycus, n	36, 257
Saccammina, f	261
Saccocrinus, m	90, 287
Særichnites, m	178
Sagenaria	36
Sagenella, f	100, 293
Salterella, f	206
Sandalodus, m	238, 322
Sanguinolaria, f	202, 312
Sanguinolites, m	202, 312
Saportæa, f	257
Sarcinula, f	60, 272
Sauripteris	238
Sauropleura, f	241, 324
Sauropus, m	241
Scævogyra, f	304
Scalaripora, f	293
Scalites, m	162
Scaphiocrinus, m	91, 287
Scenella, f	142
Scenellopora, f	293
Schænaster, m	91
Schizocrania, f	128
Schizocrinus, m	91
Schizodus, m	203, 312
Schizopteris	36
Schutzia, f	36, 257
Scolithus, m	36, 257, 315
Scolopendrites, m	36, 257
Scyphia	44
Scyphocrinus, m	91, 287
Scytalocrinus	287
Sedgwickia, f	203, 312
Selaginites, m	36, 257
Selenoides	44
Semicoscinium, n	101
Serpula, f	206, 315
Serpulites, m	206
Septopora, f	101
Shumardia, f	223
Sicarius	322
Sidemina	309
Sigillaria, f	37, 257
Sigillarioides, m	38, 258
Siphonia	44
Skenidium, n	128
Smithia, f	60, 272
Solenoula, f	258
Solarium	162
Solemya	204
Solen, m	204
Soleniscus, m	162, 304
Solenocaris	223
Solenochellus, n	178
Solenomya, f	204
Solenopleura, f	223, 319
Solenopsis, f	204
Solenoula, f	258
Sorocladus, m	258
Spathiocaris, f	319
Spatiopora, f	293
Sphærexochus, m	223
Sphærocoryphe, f	223
Sphærocrinus	91
Sphærocystites, m	91
Sphærolites, m	60
Sphenacodon, m	324
Sphenophyllum, n	38, 258
Sphenopteris, f	38, 258
Sphenopterium	44
Sphenothallus, m	39
Spheropezium, n	242
Spirangium, n	259
Spirifera, f	128, 297
Spiriferina, f	133, 298
Spirigera	133
Spirophyton, n	39, 259
Spirorbis, m	207, 315
Spirula	178
Spongia	44
Sporangites, m	39, 259
Sporocystis, m	259
Staphylopteris, f	39, 259
Staurocephallites, m	315
Staurograptus, m	60
Steganocrinus, m	91, 287
Stellipora, f	60, 272
Stemmatodus, m	239
Stemmatopteris, f	259
Stenacanthus, m	239, 322
Stenaster, m	91
Stenopora, f	60
Stenopterodus, m	322
Stenoschisma, n	133
Stenotheca, f	142
Stephanocrinus, m	92, 287
Stereocrinus, m	287
Sternbergia, f	39
Stictopora, f	101, 293
Stictoporella	293
Stigmaria, f	40, 259
Stigmarioides, m	40, 259
Stomatopora, f	294
Straparollina, f	162
Straparollus, m	162, 304
Streblopteria, f	204
Strephodes	272
Streptaxis, m	163
Streptelasma, n	61, 245, 272
Streptoceras, n	178
Streptorhynchus, m	134, 298
Striatopora, f	61, 245, 273
Stricklandia	134
Stricklandinia, f	134, 298
Strigilina	322
Strobilocystites, m	92
Strobilus	259
Stromatocerium n 61, 261, 273	
Stromatopora, f	61, 261, 273
Strombodes, m	62, 273
Strophalosia, f	135, 298
Strophites, m	304
Strophodonta, f	135, 298
Strophomena, f	136, 299
Strophonella, f	299
Strophostylus, m	163, 304
Strotocrinus, m	92, 288
Styleastrea, f	273
Stylifer	163
Styliola, f	300
Subulites, m	163, 304
Synbathocrinus, m	92, 288
Synocladia, f	101, 294
Syntrielasma, n	138
Syringocrinus, m	92
Syringodendron, n	40, 259
Syringolites, m	273
Syringopora, f	62, 273
Syringostroma, n	62
Syringothyris, f	138, 299
Syringoxylon, n	40
Tæniaster, m	92, 288
Tæniodus, m	322

INDEX OF GENERA.

	PAGE.		PAGE.		PAGE.
Taeniophyllum, n.	259	Thysanocrinus, m	93	Tubipora, f.	63
Tæniopora, f.	101	Tomodus, m	322	Tuditanus, m.	242, 324
Tæniopteris, f.	40, 259	Trachomatichnus. m.	309	Turbo, m.	164, 304
Talarocrinus, m	288	Trachydomia, n.	163	*Turbonilla*	164
Tanaodus, m	239	Trachypora, f.	63, 273	Turritella, f.	164, 304
Taonurus, m	259	Trachyum, n.	44	Ulodendron, n	41, 260
Taxocrinus, m.	92, 288	Tremanotus.	164, 304	*Ungulina*	205
Technocrinus, m	92	Trematis, f.	139, 299	*Unio*	312
Teleiocrinus	288	Trematoceras, n.	309	Uphantænia, f.	41, 260, 261
Telephus, m	223	*Trematocrinus*	93	Valvulina, f.	261
Tellina	204	Trematodiscus, m.	179, 309	Vanuxemia, f.	205, 312
Tellinomya, f.	204, 312	Trematopora, f.	101, 294	Vasocrinus, m.	93, 288
Tellinopsis, f.	205	Trematospira, f.	139, 299	Vaticinodus, m	323
Temnocheilus, n.	178	Triarthrella, f.	223	Venustodus, m	239
Tentaculites, m	142, 300	Triarthrus, m.	223, 319	Vermipora, f.	63
Terataspis, f.	223	Trichomanites, m.	40	Vesicularia. f.	63
Teratichnus, m.	309	Trichophycus, m.	260	Vitulina, f.	140
Terebratula, f.	138, 245, 299	Trichospongia, f.	44	Walchia, f.	41, 260
Terebratulites	139	Trigonocarpum, n.	40, 260	Walcottia, f.	315
Termes, m	319	Trigonodus, m.	239, 322	Waldheimia, f.	140
Tetradium, n.	62	Trigonotreta, f.	139	Whittleseya, f.	41, 260
Tetragraptus, m	62	Trimerella, f	139	Xenocrinus, m.	288
Textilaria.	261	Trimerorhachis, f.	324	Xenoneura, f.	226
Thalcops, f.	223	*Trimerus*	223	Xenophora, f.	164
Thallistigma, n.	294	Trinucleus. m.	223, 319	Xylobius, m	224
Thamniscus, m.	101, 294	Triphyllopteris, f.	260	Xystracanthus, m.	239
Thamnograptus, m.	63, 273	*Triplesia*.	140, 299	Xystrodus, m.	239, 323
Thamnopora, f.	294	Trochita, f.	164, 304	Yoldia, f.	205, 312
Theca, f.	142, 300	Trochoceras, n.	179, 309	Zamites, m	41, 260
Thecia, f.	63	Trochonema, n.	164, 304	Zaphrentis, f.	63, 245, 273
Thecostegites, m.	63	Trocbolites, m.	179, 309	Zatrachys, n.	324
Thenaropus, m	242	Trochophyllum, n	41, 260 273	Zeacrinus, m	93, 288
Theropleura, f.	324	*Trochus*	164	Zonites, m.	162
Thrinacodus, m.	239	Troostocrinus, m	288	Zygospira, f.	140, 499
Thyrsidium, n	242	Tropidoleptus, m	140		

www.ingramcontent.com/pod-product-compliance
Lightning Source LLC
Chambersburg PA
CBHW030300240426
43673CB00040B/1012